CRC HANDBOOK SERIES IN NUTRITION AND FOOD

Miloslav Rechcigl, Jr.

Editor-in-Chief

SECTION OUTLINE

SECTION A: Science of Nutrition and Food

Nomenclature, Nutrition Literature, Organization of Research, Training and Extension in Nutrition and Food, Nutrition Societies, and Historical Milestones in Nutrition.

SECTION B. Part 1: Ecological Aspects of Nutrition

The Living Organisms, Their Distribution and Chemical Constitution, Habitats and Their Ecology, Biotic Associations and Interactions, Cycling of Water in the Biosphere, Ecosystem Productivity, and Factors Affecting Productivity of Animals and Plants.

SECTION B. Part 2: Physiological Aspects of Nutrition

Feeding and Digestive Systems in Various Organisms, Food Assimilation Processes in Microorganisms and Plants, Selection of Food in Various Organisms, Factors Affecting Food and Water Intake, Gastrointestinal Tract-Anatomical, Morphological and Functional Aspects, Passage of Ingesta in Various Organisms, Digestive Enzymes, Nutrient Absorption, Transport, and Excretion, Cellular Digestion and Metabolism.

SECTION C: The Nutrients and Their Metabolism

Chemistry and Physiology of Nutrients and Growth Regulators, Antinutrients and Antimetabolites, Nutrient Metabolism in Different Organisms, Regulatory Aspects of Nutrient Metabolism, Nutritional Adaptation, Biogenesis of Specific Nutrients, and Nutrient Interrelationships.

SECTION D: Nutritional Requirements

Comparative Nutrient Requirements, Qualitative Requirements of Specific Organisms, Tissues, and Cells, Quantitative Requirements (Nutritional Standards) of Selected Organisms, Nutritional Requirements for Specific Processes and Functions, and Requirements and Utilization of Specific Nutrients.

SECTION E: Nutritional Disorders

Nutritional Disorders in Specific Organisms, Nutritional Disorders in Specific Tissues, Effect of Specific Nutrient Deficiencies and Toxicities (3 Vol.), and Nutritional Aspects of Disease.

SECTION F: Food Composition, Digestibility, and Biological Value

Nutrient Content and Energy Value of Foods and Feeds, Factors Affecting Nutrient Composition of Plants, Factors Affecting Nutrient Composition of Animals, Effect of Processing on Nutrient Content of Foods and Feeds, and Utilization and Biological Value of Food.

SECTION G: Diets, Culture Media, and Food Supplements

Diets for Animals, Diets for Invertebrates, Culture Media for Microorganisms and Plants, Culture Media for Cells, Organisms and Embryos, and Nutritional Supplements.

SECTION H: The State of World Food and Nutrition

World Population, Natural and Food Resources, Food Production, Food Losses, Food Usage and Consumption, Socioeconomic, Cultural, and Psychological Factors Affecting Nutrition, Geographical Distribution of Nutritional Diseases, Nutrient Needs - Current and Projected, Agricultural Inputs - Current and Projected, Food Aid, and Food Marketing and Distribution.

SECTION I: Food Safety, Food Spoilage, Food Wastes, Food Preservation, and Food Regulation

Naturally Occurring Food Toxicants, Food Contaminants, Food Additives, Food Spoilage and Deterioration, Disposal of Food Wastes, Food-Borne Diseases, Detoxication of Foreign Chemicals, Food Sanitation and Preservation, Food Laws, and Nutrition Labeling.

SECTION J: Production, Utilization, and Nutritive Value of Foods

Plant and Animal Sources.

SECTION K: Nutrition and Food Methodology

Assessment of Nutritional Status of Organisms, and Measuring Nutritive Value of Food.

CRC Handbook Series

in

Nutrition and Food

Miloslav Rechcigl, Jr., Editor-in-Chief

Nutrition Advisor and Director
Interregional Research Staff
Agency for International Development
U.S. Department of State

Section E: Nutritional Disorders

Volume I

Effect of Nutrient Excesses and Toxicities
in Animals and Man

CRC Press, Inc.
2255 Palm Beach Lakes Boulevard · West Palm Beach, Florida 33409

Library of Congress Cataloging in Publication Data

Main entry under title:

Effect of nutrient excesses and toxicities in animals and man.

(Nutritional disorders; v. 1)
Bibliography: p.
Includes index.
1. Nutritionally induced diseases. 2. Vitamins
— Toxicology. 3. Trace elements — Toxicology.
4. Minerals in nutrition — Toxicology.
I. Rechcigl, Miloslav. II. Series.
RC620.A1N87 vol. 1 [RC622] 616.3'9'008s
ISBN 0-8493-2726-1 [616.3'99] 77-19146

© 1978 by CRC Press, Inc.

International Standard Book Number 0-8493-2700-8 (Complete Set)
International Standard Book Number 0-8493-2796-2 (Volume I)

Library of Congress Card Number 77-19146
Printed in the United States

PUBLISHER'S PREFACE

In 1913, when the First Edition of the *Handbook of Chemistry and Physics* appeared, scientific progress, particularly in chemistry and physics, had produced an extensive literature but its utility was seriously handicapped because it was fragmented and unorganized. The simple but invaluable contribution of the *Handbook of Chemistry and Physics* was to provide a systematic compilation of the most useful and reliable scientific data within the covers of a single volume. Referred to as the "bible," the Handbook soon became a universal and essential reference source for the scientific community. The latest edition represents more than 65 years of continuous service to millions of professional scientists and students throughout the world.

In the years following World War II, scientific information expanded at an explosive rate due to the tremendous growth of research facilities and sophisticated analytical instrumentation. The single-volume Handbook concept, although providing a high level of convenience, was not adequate for the reference requirements of many of the newer scientific disciplines. Due to the sheer quantity of useful and reliable data being generated, it was no longer feasible or desirable to select only that information which could be contained in a single volume and arbitrarily to reject the remainder. **Comprehensiveness** had become as essential as **convenience.**

By the late 1960's, it was apparent that the solution to the problem was the development of the multi-volume Handbook. This answer arose out of necessity during the editorial processing of the *Handbook of Environmental Control*. A hybrid discipline or, to be more precise, an interdisciplinary field such as Environmental Science could be logically structured into major subject areas. This permitted individual volumes to be developed for each major subject. The individual volumes, published either simultaneously or by some predetermined sequence, collectively became a multi-volume Handbook series.

The logic of this new approach was irrefutable and the concept was promptly accepted by both the scientist and science librarian. It became the format of a growing number of CRC Handbook Series in fields such as Materials Science, Laboratory Animal Science, and Marine Science.

Within a few years, however, it was clear that even the multi-volume Handbook concept was not sufficient. It was necessary to create an information structure more compatible with the dynamic character of scientific information, and flexible enough to accommodate continuous but unpredictable growth, regardless of quantity or direction. This became the objective of a "third generation" Handbook concept.

This latest concept utilizes each major subject within an information field as a "Section" rather than the equivalent of a single volume. Each Section, therefore, may include as many volumes as the quantity and quality of available information will justify. The structure achieves permanent flexibility because it can, in effect, expand "vertically" and "horizontally." Any section can continue to grow (vertically) in number of volumes, and new sections can be added (horizontally) as and when required by the information field itself. A key innovation which makes this massive and complex information base almost as convenient to use as a single-volume Handbook is the utilization of computer technology to produce up-dated, cumulative index volumes.

The *Handbook Series in Nutrition and Food* is a notable example of the "sectionalized, multi-volume Handbook series." Currently underway are additional information programs based on the same organizational design. These include information fields such as Energy and Agricultural Science which are of critical importance not only to scientific progress but to the advancement of the total quality of life.

We are confident that the "third generation" CRC Handbook comprises a worthy contribution to both information science and the scientific community. We are equally certain that it does not represent the ultimate reference source. We predict that the most dramatic progress in the management of scientific information remains to be achieved.

B. J. Starkoff
President
CRC Press, Inc.

PREFACE
CRC HANDBOOK SERIES IN NUTRITION AND FOOD

Nutrition means different things to different people, and no other field of endeavor crosses the boundaries of so many different disciplines and abounds with such diverse dimensions. The growth of the field of nutrition, particularly in the last two decades, has been phenomenal, the nutritional data being scattered literally in thousands and thousands of not always accessible periodicals and monographs, many of which, furthermore, are not normally identified with nutrition.

To remedy this situation, we have undertaken an ambitious and monumental task of assembling in one publication all the critical data relevant in the field of nutrition.

The *CRC Handbook Series in Nutrition and Food* is intended to serve as a ready reference source of current information on experimental and applied human, animal, microbial, and plant nutrition presented in concise tabular, graphical, or narrative form and indexed for ease of use. It is hoped that this projected open-ended multivolume set will become for the nutritionist what the *CRC Handbook of Chemistry and Physics* has become for the chemist and physicist.

Apart from supplying specific data, the comprehensive, interdisciplinary, and comparative nature of the *CRC Handbook Series in Nutrition and Food* will provide the user with an easy overview of the state of the art, pinpointing the gaps in nutritional knowledge and providing a basis for further research. In addition, the *Handbook* will enable the researcher to analyze the data in various living systems for commonality or basic differences. On the other hand, an applied scientist or technician will be afforded the opportunity of evaluating a given problem and its solutions from the broadest possible point of view, including the aspects of agronomy, crop science, animal husbandry, aquaculture and fisheries, veterinary medicine, clinical medicine, pathology, parasitology, toxicology, pharmacology, therapeutics, dietetics, food science and technology, physiology, zoology, botany, biochemistry, developmental and cell biology, microbiology, sanitation, pest control, economics, marketing, sociology, anthropology, natural resources, ecology, environmental science, population, law, politics, nutritional and food methodology, and others.

To make more facile use of the *Handbook,* the publication has been divided into sections of one or more volumes each. In this manner the particular sections of the *Handbook* can be continuously updated by publishing additional volumes of new data as they become available.

The Editor wishes to thank the numerous contributors, many of whom have undertaken their assignment in pioneering spirit, and the Advisory Board members for their continuous counsel and cooperation. Last but not least, he wishes to express his sincere appreciation to the members of the CRC editorial and production staffs, particularly President Bernard J. Starkoff, Mr. Robert Datz, Mr. Paul R. Gottehrer, and Ms. Marsha Baker, for their encouragement and support.

We invite comments and criticism regarding format and selection of subject matter, as well as specific suggestions for new data (and additional contributors) which might be included in subsequent editions. We should also appreciate it if the readers would bring to the attention of the Editor any errors or omissions that might appear in the publication.

Miloslav Rechcigl, Jr.
Editor-in-Chief
August 1978

PREFACE
SECTION E: NUTRITIONAL DISORDERS

This section provides systematic and detailed information on all relevant aspects of the relationship between nutrition and disease. One subsection is devoted to natural and foodborne diseases in various taxa of organisms, ranging from single-cellular type to higher organisms. Another subsection concerns itself with the effect of specific nutrient deficiencies and excess, described in terms of gross, morphological, and biochemical alterations. The changes in specific tissues and organs, due to malnutrition, and the nutritional aspects of disease form the theme of the remaining subsections.

Miloslav Rechcigl, Jr.,
Editor
May 1978

MILOSLAV RECHCIGL, JR., EDITOR

Miloslav Rechcigl, Jr. is Nutrition Advisor and Director of the Interregional Research Staff in the Agency for International Development, U.S. Department of State.

He has a B.S. in Biochemistry (1954), a Master of Nutritional Science degree (1955), and a Ph.D. in nutrition, biochemistry, and physiology (1958), all from Cornell University. He was formerly a Research Biochemist in the National Cancer Institute, National Institutes of Health and subsequently served as Special Assistant for Nutrition and Health in the Health Services and Mental Health Administration, U.S. Department of Health, Education, and Welfare.

Dr. Rechcigl is a member of some 30 scientific and professional societies, including being a Fellow of the American Association for the Advancement of Science, Fellow of the Washington Academy of Sciences, Fellow of the American Institute of Chemists, and Fellow of the International College of Applied Nutrition. He holds membership in the Cosmos Club, the Honorary Society of Phi Kappa Pi, and the Society of Sigma Xi, and is recipient of numerous honors, including an honorary membership certificate from the International Social Science Honor Society Delta Tau Kappa. In 1969, he was a delegate to the White House Conference on Food, Nutrition, and Health and in the last two years served as President of the District of Columbia Institute of Chemists and a Councilor of the American Institute of Chemists.

His bibliography extends over 100 publications, including contributions to books, articles in periodicals, and monographs in the fields of nutrition, biochemistry, physiology, pathology, enzymology, and molecular biology. Most recently he authored and edited *World Food Problem: A Selective Bibliography of Reviews* (CRC Press, 1975), *Man, Food, and Nutrition: Strategies and Technological Measures for Alleviating the World Food Problem* (CRC Press, 1973), *Food, Nutrition and Health: A Multidisciplinary Treatise Addressed to the Major Nutrition Problems from a World Wide Perspective* (Karger, 1973), following his earlier pioneering treatise on *Enzyme Synthesis and Degradation in Mammalian Systems* (Karger, 1971), and that on *Microbodies and Related Particles. Morphology, Biochemistry and Physiology* (Academic Press, 1969). Dr. Rechcigl also has initiated and edits a new series on Comparative Animal Nutrition and is Associated Editor of *Nutrition Reports International.*

CONTRIBUTORS
SECTION E: NUTRITIONAL DISORDERS
VOLUME I

Nancy J. Auda
  Regional Hospital
  Yuma, Arizona

Paul G. Barash
  Department of Anesthesiology
  School of Medicine
  Yale University
  New Haven, Connecticut

Harold D. Battarbee
  Department of Physiology and Biophysics
  Louisiana State University Medical Center
  School of Medicine in Shreveport
  Shreveport, Louisiana

R. Raines Bell
  Department of Nutrition
  College of Biological Science
  University of Guelph
  Guelph, Ontario, Canada

Donald R. Bennett
  Department of Drugs
  American Medical Association
  Chicago, Illinois

P. R. Bird
  Pastoral Research Station
  Department of Agriculture, Victoria
  Hamilton, Victoria, Australia

D. W. Briggs
  Winton Hill Technical Center
  The Procter and Gamble Company
  Cincinnati, Ohio

Michael H. Briggs
  Deakin University
  Geelong, Victoria, Australia

Robert E. Burch
  Department of Medicine
  Marshall University School of Medicine
  Veterans Administration Hospital
  Huntington, West Virginia

Martha Byington
  Department of Laboratory Medicine
  University of Washington
  Seattle, Washington

William H. Crosby
  Division of Hematology/Oncology
  L. C. Jacobson Blood Center
  Scripps Clinic and Research Foundation
  La Jolla, California

Joseph C. Dougherty
  Moses Taylor Kidney and Hypertension
   Institute
  Scranton, Pennsylvania

H. H. Draper
  Department of Nutrition
  College of Biological Science
  University of Guelph
  Guelph, Ontario, Canada

Philip O. Ettinger
  Division of Cardiovascular Diseases
  College of Medicine and Dentistry of New
   Jersey
  New Jersey Medical School
  Newark, New Jersey

J. J. Franxman
  Miami Valley Laboratories
  The Procter & Gamble Company
  Cincinnati, Ohio

Shiro Goto
  Department of Nutrition
  Tokyo University of Agriculture
  Setagaya, Tokyo, Japan

Harold E. Harrison
The Children's Medical and Surgical Center
The Johns Hopkins Hospital
Baltimore, Maryland

Betty E. Haskell
Department of Food Science
College of Agriculture
University of Illinois at Urbana-Champaign
Urbana, Illinois

R. A. Hiles
Miami Valley Laboratories
The Procter & Gamble Company
Cincinnati, Ohio

Yoshinori Itokawa
Department of Hygiene
Faculty of Medicine
Kyoto University
Kyoto, Japan

M. Young Jenkins
Division of Nutrition
Food and Drug Administration
Department of Health, Education, and
  Welfare
Washington, D.C.

Hans Kaunitz
Department of Pathology
College of Physicians and Surgeons of
  Columbia University
New York, New York

H.-J. Lantzsch
Institut fur Tierernahrung
Universitat Stuttgart-Hohenheim
Federal Republic of Germany

Steven Leeson
Department of Animal and Poultry Science
Ontario Agricultural College
University of Guelph
Guelph, Ontario, Canada

Philip J. Lipsitz
Long Island Jewish-Hillside Medical Center
New Hyde Park, New York

Robert W. Longton
Naval Regional Dental Center
Miramar Branch Dental Clinic
Naval Air Station Miramar
San Diego, California

John L. Martin
Department of Foods and Nutrition
Texas Technological University
Lubbock, Texas

G. R. Meneely
Department of Physiology and Biophysics
Louisiana State University Medical Center
School of Medicine in Shreveport
Shreveport, Louisiana

A. E. Moffitt, Jr.
Environmental Quality Control
Bethlehem Steel Corporation
Bethlehem, Pennsylvania

Keitaro Nishiyama
Department of Hygiene
School of Medicine
Tokushima University
Tokushima, Japan

Edwin L. Overholt
Gundersen Clinic, Limited
LaCrosse, Wisconsin

P. R. Paul
Department of Zoology
University of Delhi
Delhi, India

Harry J. Preuss
Nephrology Division
Georgetown University Medical Center
Washington, D.C.

Sheldon Reiser
Carbohydrate Nutrition Laboratory
Nutrition Institute
Agricultural Research Service
U.S. Department of Agriculture
Beltsville, Maryland

Richard S. Rivlin
Institute of Human Nutrition
College of Physicians and Surgeons of
  Columbia University
New York, New York

G. L. Romoser
Animal Nutritional Products
Monsanto Industrials Chemicals Company
St. Louis, Missouri

H. Schenkel
Institut fur Tiernernahrung
Universitat Stuttgart-Hehenheim
Federal Republic of Germany

Steven E. Silvis
Special Diagnostic and Treatment Unit
Veterans Administration Hospital
Minneapolis, Minnesota

James F. Sullivan
Creighton University School of Medicine
Veterans Administration Hospital
Omaha, Nebraska

Takako Tomita
Shizuoka College of Pharmaceutical
  Sciences
Shizuoka, Japan

G. I. Vidor
Launceston General Hospital
Launceston, Tasmania, Australia

R. A. Waterman
Animal Physiology and Genetics Institute
Agricultural Research Service
U.S. Department of Agriculture
Beltsville, Maryland

Charles W. Weber
Department of Nutrition and Food Science
College of Agriculture
University of Arizona
Tucson, Arizona

C. O. Weiskittel
Miami Valley Laboratories
The Procter & Gamble Company
Cincinnati, Ohio

To my inspiring teachers at Cornell: Harold H. Williams, John K. Loosli, Richard H. Barnes, the late Clive M. McCay, and the late Leonard A. Maynard.
And to my supportive and beloved family: Eva, Jack, and Karen.

TABLE OF CONTENTS
SECTION E: NUTRITIONAL DISORDERS
VOLUME I

Water-soluble Vitamins

EFFECT OF NUTRIENT TOXICITIES IN ANIMALS AND MAN: THIAMINE

Y. Itokawa

PHYSIOLOGICAL ACTION OF THIAMINE

When attempting to discuss the physiological actions of a substance, the toxicity-related problems are always worthy of a great deal of attention. There is now ample documented evidence for two independent roles of thiamine. One is the well-established role of thiamine pyrophosphate as a cofactor in intermediary carbohydrate metabolism, and the other is a function in the nerve excitation process.

For the latter function of thiamine, Binet and Minz liberated a particular substance from the vagus and other nerve trunks, both in vivo and in vitro, by repetitive electrical stimulation.[1] This substance, which potentiates various effects of acetylcholine, was later identified by Minz as thiamine.[2] Recently, Itokawa et al. clarified that this releasable thiamine is localized in the purified membrane fraction of the nerve.[3] As tetrodotoxin promotes the release of thiamine from the nerve preparation, the current view is that thiamine occupies a fixed site in the sodium channel of nerve membrane or at least in a proximate area.[3-10] Various other experiments, procedures, and equipment have been used to demonstrate the role of thiamine in nerve excitation.[11-23] Although the pharmacological and toxicological actions of thiamine are for the greater part related to the nervous function, the physiological action of the vitamin appears to be independent.

PHARMACOLOGICAL ACTION OF THIAMINE IN EXPERIMENTAL ANIMALS

The range between the therapeutic and lethal doses of thiamine is wide in comparison with other nutrients or remedies. In normal animals, pharmacodynamic effects are produced only by doses more than 1000 times larger than those needed for a therapeutic effect.

Regarding acute toxicity of thiamine, the lethal effect of thiamine has been determined by several workers.[24-28] LD_{50} level of thiamine in mice, rats, and rabbits is much the same, and the values are 2000 to 3000 mg/kg orally, 80 to 120 mg/kg intravenously, and 300 to 500 mg/kg intraperitoneally.[24-28] Haley reported that the symptoms of toxicity of thiamine in rabbits were restlessness, labored respiration, vasodilation, cyanosis, muscular twitching, clonic convulsions, and death by respiratory paralysis. The respiratory paralysis was of central origin because electrical stimulation of both the muscle of the diaphragm and the phrenic nerve showed that the muscle was still capable of contraction.[25] Visual sign of anoxia was a gradually deepening bluish coloration of the ears and other areas. In animals, cardiac arrhythmias were evident at the time of thoracotomy. Death could be induced by a direct paralyzing action on the respiratory center, after which cardiac failure ensued.[25] Hayashi,[29] Minesita and Ueda,[30] and Minesita[31] observed a similar phenomenon but assumed that death was due to an excessive administration of thiamine originating from lesions in neuromuscular transmission rather than central in origin.

Thiamine, by virtue of its thiazolium ring structure, is a quaternary ammonium compound and is thus related to compounds which act as neuromuscular and ganglionic blocking agents. Smith et al. reported the curare-like action of thiamine.[32] Regarding respiration, blood pressure, and muscle response recorded from the toe during sciatic nerve stimulation in dogs injected with thiamine or D-tubocurarine, the latter in doses of

0.15 mg/kg produced effects closely resembling those of thiamine (150 mg/kg).[32] Di Palma and Hitchcook injected thiamine hydrochloride intravenously in doses of 20 mg/kg into cats and 80 mg/kg into dogs and found that these dosages caused a complete neuromuscular paralysis.[33] At this dose level, a 50 to 95% ganglionic block also existed, as confirmed by a fall in blood pressure and a decrease in nictitating membrane response. After a single dose of thiamine (50 mg/kg into cats), neuromuscular block lasted on the average from 15 to 20 min and the ganglionic block 5 to 10 min.[33] Ngai et al. offered an explanation for the mechanism of this action by thiamine.[34] They found that thiamine, pyrithiamine, and pyrithiamine analogues in which there is a hydroxy group on the pyridinium ring (see Table 1) caused a decrease in the twitch response of the tibialis muscle to peroneal nerve stimulation in cats without initial potentiation, depressed respiration, and lowered arterial pressure. These compounds also antagonized both D-tubocurarine and decamethonium. On the other hand, methyl thiazolium iodide, methylpyridinium iodide, and the analogues of thiamine and pyrithiamine in which there is no hydroxy group on the ring bearing the quarternary nitrogen produced an increase in the twitch with muscular fasciculation, and, with larger doses, a decrease in the twitch. Changes in respiratory movements generally paralleled those of the twitch. Arterial pressure was usually elevated. These compounds antagonized D-tubocurarine and potentiated decamethonium. The thiazolium fragment, at the dose level studied, increased the twitch only slightly, antagonized D-tubocurarine, and potentiated decamethonium. The pyrimidine fragment apparently had no effect on the twitch. From these findings Ngai et al. concluded that the activity of thiamine and pyrithiamine on the neuromuscular junction is related to the quarternary nitrogen by coupling with pyrimidine and the presence of a hydroxy group on the quaternary nitrogen bearing ring. The antagonistic action of these compounds against D-tubocurarine and decamethonium also depends on the same structural components.[34] Hayashi and colleagues also examined the blocking action of various thiamine derivatives on the neuromuscular transmission of bullfrogs.[35] They applied two components of thiamine, 2-methyl-4-amino-5-hydroxy-methylpyrimidine (OMPm) and thiazole, on sciatic nerve-sartorius muscle preparations and found that these components never blocked neuromuscular transmission. They concluded that thiamine acts upon neuromuscular junction as a whole structure, not a segmented one. This view is consistent with Ngai's hypothesis, but the following hypothesis did not parallel that of Ngai. Hayashi et al. considered that the quaternary nitrogen of the thiazole ring in ammonium form is not always necessary for the blocking action of analogues of thiamine on neuromuscular transmission.[35] These workers found that several derivatives of thiamine (CET, CBT, TPD, and TTFD) which had no quaternary nitrogen had a blocking action similar to that of thiamine. In addition, they observed a similar blocking action on the neuromuscular junction with the application of DCET, a thiamine derivative which has a carbethoxy group instead of a hydroxyethyl group in the thiazole ring. Thus, they were dubious as to the function of the hydroxyethyl group in the thiazole ring which would potentiate the blocking action of quaternary nitrogen.[35] They observed that the critical concentration and blocking time of free thiamine and thiamine pyrophosphate (TPP) on the neuromuscular junction were the same and claimed that the neuromuscular blocking action of thiamine is independent of the enzymatic action as cocarboxylase. In contrast, when a thiamine derivative with a disulfide bond (TPD and TTFD) was applied to a nerve-muscle preparation in vitro, the blocking action was never recovered even after the preparation had been washed with Ringer's solution. Recovery of the blocking action occurred with thiamine. It was thus assumed that thiamine derivatives having the disulfide bond would have a stronger affinity to tissue as compared to other types of thiamine.[35] Other studies have been done on the effect of thiamine on neuromuscular blocking, and several authors have reported their findings.[36-42] Although there is room for further clarification, it is most likely that

Table 1

STRUCTURAL FORMULA OF THIAMINE ANALOGUES CITED IN THE TEXT

Structural formula	Name	Abbreviation
	Thiamine HCl, 3-(4-amino-2-methyl pyrimidyl-5-methyl)-4-methyl-5-β-hydroxyethyl thiazolium chloride hydrochloride (mol wt = 337)	V.B$_1$
	Pyrithiamine HBr, 1-(4-amino-2-methyl pyrimidyl-5-methyl)-2-methyl-3-β-hydroxyethyl pyridinium bromide hydrobromide (mol wt = 420)	
	Oxythiamine Cl, 3-(4-hydroxy-2-methyl pyrimidyl-5-methyl)-4-methyl-5-β-hydroxyethyl thiazolium chloride (mol wt = 302)	
	Methyl thiazolium iodide (mol wt = 227)	
	3:4-Dimethyl-5-β-hydroxyethyl thiazolium iodide (mol wt = 285)	

Table 1 (continued)
STRUCTURAL FORMULA OF THIAMINE ANALOGUES CITED IN THE TEXT

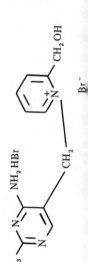

Structural formula	Name	Abbreviation
	Methyl pyridinium iodide (mol wt = 221)	
	3-(4-Amino-2-methyl pyrimidyl-5-methyl)-thiazolium bromide hydrobromide (mol wt = 368)	
	1-(4-Amino-2-methyl pyrimidyl-5-methyl)-pyridinium bromide hydrobromide (mol wt = 362)	
	1-(4-Amino-2-methyl pyrimidyl-5-methyl)-2-methyl pyridinium bromide hydrobromide (mol wt = 376)	
	1-(4-Amino-2-methyl pyrimidyl-5-methyl)-2-hydroxymethyl pyridinium bromide hydrobromide (mol wt = 392)	

Table 1 (continued)
STRUCTURAL FORMULA OF THIAMINE ANALOGUES CITED IN THE TEXT

Structural formula	Name	Abbreviation
	1-(4-Amino-2-methyl pyrimidyl-5-methyl)-3-hydroxymethyl pyridinium bromide hydrobromide (mol wt = 392)	
	1-(4-Amino-2-methyl pyrimidyl-5-methyl)-4-hydroxymethyl pyridinium bromide hydrobromide (mol wt = 392)	
	1-(4-Amino-2-methyl pyrimidyl-5-methyl)-3-hydroxymethyl pyridinium bromide hydrobromide (mol wt = 378)	
	Cocarboxylase; thiamine diphosphate thiamine pyrophosphate (mol wt = 478)	TPP

Table 1 (continued)
STRUCTURAL FORMULA OF THIAMINE ANALOGUES CITED IN THE TEXT

Structural formula	Name	Abbreviation
	2-Methyl-4-amino 5-hydroxymethylpyrimidine (mol wt = 139)	OMPm
	4-Methyl-5-β-hydroxyethyl thiazole (mol wt = 85)	Th
	Thiamine propyl disulfide (mol wt = 392)	TPD
	Thiamine tetrahydrofurfuryl disulfide (mol wt = 398)	TTFD
	Thiamine-8-methyl-6-acetyl dihydrothioctate) disulfide (mol wt = 545)	TATD

Table 1 (continued)
STRUCTURAL FORMULA OF THIAMINE ANALOGUES CITED IN THE TEXT

Structural formula	Name	Abbreviation
(chemical structure)	S-Carboethoxythiamine hydrochloride (mol wt = 391)	CET
(chemical structure)	O,S-Dicarboethoxythiamine hydrochloride (mol wt = 481)	DCET
(chemical structure)	S-Carbobutoxythiamine hydrochloride (mol wt = 419)	CBT
(chemical structure)	S-Benzoylthiamine O-monophosphate (mol wt = 467)	BTMP

the immediate cause of death after administration of an excessive amount of thiamine to laboratory animals is respiratory paralysis due to blocking of neuromuscular transmission. There are adequate data indicating that spontaneous respiration is resumed and animals recover when artificial respiration is maintained in animals given fatal doses of thiamine.[43-46]

When the curare-like action of thiamine is eliminated by artificial respiration, various other effects of thiamine appear. Smith et al. reported that an intravenous injection of large doses of thiamine (125 mg/kg) caused a temporary fall in blood pressure in dogs when artificial respiration was used.[43] When artificial respiration was not used, there was an inevitable fall to zero. As there was a simultaneous increase in the voltage of the T-wave, they assumed these phenomena were due to anoxia. Large doses of thiamine decreased the voltage of the QRS complex on the electrocardiogram.[43] Various studies have also been reported on the effect of thiamine on the vascular system. Jaros et al. injected thiamine (25 mg/kg), acetylcholine, and histamine intravenously into anesthetized dogs and noted a uniform fall in blood pressure, which could be reversed by epinephrine. The fall in blood pressure produced by acetylcholine and histamine was blocked by atropine and an antihistaminic agent, respectively, but these drugs did not antagonize the hypotension after thiamine. The effect of thiamine on blood pressure thus appeared to be independent of the effects of acetylcholine and histamine.[47] Haley et al. also recognized a fall in blood pressure in rabbits.[25] Bradycardia was reported by Smith et al.[44] and Boissier et al.[46] Haley and Flesher found that an excessive amount of thiamine injected intravenously into rabbits produced symptoms of vasodilation.[48] Smith et al. stated that the vasodilation is probably due to a direct action on blood vessels, but may be central in origin.[44] Itokawa et al. reported that rats fed a thiamine-excess and magnesium-deficient diet manifested symptoms of peripheral vasodilation characterized by erythema in the ear and nose. It has been clarified that serotonin levels increased in the blood, stomach, and intestine of thiamine-excess, magnesium-deficient rats, while both thiamine- and magnesium-deficient rats revealed no elevation of blood serotonin when the peripheral vasodilation symptom was not present. It is possible that excess thiamine promotes magnesium deficiency and liberates serotonin into the bloodstream from the mast cells, since magnesium is required for the maintenance of mast cells.[49] The possibility that these pharmacologically active substances play the same role in toxicity in animals given excess amounts of thiamine should not be overlooked.

Among the various pharmacologically active substances, the relationship between acetylcholine and thiamine has long been noted. Agid and Balkanyi found an acetylcholine-like effect of thiamine on the isolated frog heart in which a solution containing more than 3×10^{-5} M of thiamine diminished, inhibited, or reversed the action of acetylcholine.[50] Byer and Harpuder reported that thiamine contracted the eserinized dorsal muscle of the leech and postulated that thiamine and acetylcholine have different modes of action since the sensitizing effect on acetylcholine was observed only when thiamine brought about an independent contraction.[51] Further, the combined application of acetylcholine and thiamine in subthreshold doses had no effect on the eserinized leech, while thiamine had a sensitizing action for acetylcholine on rat intestine. Thiamine was inactive on the heart and rectus muscles of frogs, intestine of guinea pigs, and denervated ears of cats. These experiments, however, contain no evidence of an anticholinesterase action of thiamine.[51] In relation to the effect of thiamine on choline esterase, Glick and Antopol studied the effect of thiamine on this enzyme in horse and rat serum in vitro and observed that inhibition was obtained only with high concentrations of thiamine (i.e., 3.6×10^{-4} M inhibits 20%, 2.2×10^{-3} M inhibits 50%).[52] The affinity of thiamine for choline esterase was calculated to be 26 times that of acetylcholine for the enzyme.[52] Torda and Wolff investigated the synthesis of acetylcholine in vitro in the presence of thiamine hydrochloride and thiamine pyrophosphate.[53,54] Both thiamine hydrochloride and thiamine pyrophosphate in concentrations of 3 × 10^{-6} M and 2 × 10^{-6} M, respectively, slightly increased the synthesis of acetylcholine, while in higher concentrations thiamine compounds depressed the synthesis of acetylcholine. These authors assumed that the slight increase of synthesis was due to the direct effect of cocarboxylase on carbohydrate metabolism, which in turn resulted in an increased supply of the acetyl radical, and that decrease of synthesis was due to a direct reaction of large amounts of thiamine with the enzyme involved in the synthesis of acetylcholine.[53,54] Other reports also describe the sensitizing effect of thiamine on acetylcholine.[55-57] Thus, it is probable that thiamine works on acetylcholine synthesis through coenzyme A metabolism which is influenced by the energy supply derived from carbohydrate metabolism.

As for the relationship between thiamine and histamine, Post and Smith, and Smith and Sohn injected thiamine (100 mg/kg) intravenously into dogs and observed that a pronounced hypotension and bronchoconstriction occurred in all animals.[58,59] This was attributed to the liberation of histamine, since in other dogs with low blood pressure which were given benadryl (an antihistamine agent), thiamine caused no further fall in blood pressure and no bronchoconstriction. Thus, it was assumed that bronchoconstriction produced by thiamine can be due to the block of the sympathetic system as well as the liberation of histamine.[58,59] This view does not parallel that of Jaros et al.[47]

Several studies have been done on the antagonistic effect of thiamine on nicotine. Haimovici and Pick investigated the effect of thiamine on nicotine-induced vasoconstriction using frog vessels.[60] The vitamin, in a low concentration (0.3 mM), inhibited the vasoconstrictor action of nicotine; the inhibiting action was linked with the thiazole moiety of the molecule, as the pyrimidine group was ineffective. These authors assumed that the site of action of thiamine was the myoneural junction in the striated muscles and at the postganglionic nerve endings or in the muscle elements of the vessel walls in the smooth muscles.[60] Pick and Unna demonstrated in intact cats that thiamine also inhibited the stimulation of vasomotor synapses and the rise in blood pressure caused by nicotine.[61] In agreement with the findings of Haimovici, they showed that this effect was caused solely by the thiazole component of thiamine, whereas the pyrimidine component remained ineffective. They were inclined to think that there was a competitive action between the pyridine group of nicotine and the thiazole group of thiamine. This competitive action between nicotine and thiamine is much the same as the displacement of thiamine by pyrithiamine. It was assumed that the blocking effect of thiamine concerns specifically the sympathetic ganglia and not the peripheral nerve ends, as drugs which acted on sympathetic or parasympathetic nerve endings retained their effectiveness despite complete inhibition of the nicotinic action.[61] Yamamoto and co-workers also studied the antagonistic effect of thiamine on nicotine in various organs in vitro and provided data indicating that antagonism of thiamine to nicotine was stronger than thiamine to acetylcholine or histamine and that this antagonistic effect was by no means referable to the action of the cocarboxylase-dependent enzyme system.[62-67]

To determine the action of thiamine on the central nervous system, Dias applied thiamine directly to the cerebral "motor" cortex of the dog.[68] When thiamine hydrochloride (2 to 10%) was applied, after 1 to 2 min a motor reaction occurred which consisted of rhythmic contraction (clonus) of the muscular group corresponding to the cortical "motor" point submitted to the action of the substance. Initially, the rhythmic muscular reactions were weak, but within 1 to 3 min the intensity increased. Afferent, repetitive, mechanical stimulation of the cutaneous region connected with the muscle in rhythmic action increased the intensity or frequency of the motor reaction, after which motor phenomena presented a localized convulsive reaction. With further stimulations, it was possible to obtain progressive generalization of the motor convulsive reactions to other muscular groups, and when all skeletal musculature was involved, a generalized epileptiform convulsion took place. Epileptiform convulsions occurred in 34 out of 45 dogs. Solutions of 2 to 5% thiamine pyrophosphate produced identical results; however, the two separate thiamine moieties, pyrimidine and thiazole, gave negative results.[68] A series of studies were done by Hayashi in an attempt at clarification.[69-72] When 500 to 800 mg of thiamine was injected intravenously or intra-arterially into cats, the animals did not convulse, yet died within 2 min. When 20 to 25 mg of thiamine was administered into the cerebrospinal fluid of cats, convulsions occurred and the animals died. The mode of convulsion was similar to that caused by administration of D-tubocurarine; however, administration of thiamine resulted in convulsions when introduced into the motor-regulating system in the brain cortex. This did not occur with D-tubocurarine. Thiamine pyrophosphate effects were similar to those of thiamine. However, oxythiamine and pyrithiamine, antithiamine agents, caused a convulsion when administered intrathecally in a high concentration (about two to three times the dose of thiamine), but simultaneous administration of a lower concentration of these compounds, together with thiamine, suppressed the convulsion. Thiamine derivatives, such as carbethoxythiamine, which convert to thiamine in the brain, caused convulsion after a latent period of 10 to 20 min. Dicarbethoxythiamine, a thiamine derivative which does not readily convert to thiamine in the brain, did not produce a convulsion. The mechanism of thiamine-induced convulsion, therefore, is still open to question.

Information on the adverse effects of excessive administration of thiamine is also available from studies on the interrelationships among the vitamin B group. Unna and Clark administered large amounts of thiamine to rats in diets either entirely or partly deficient in one or more factors of the other B complex vitamins, riboflavin, pyridoxine, nicotinamide, and calcium pantothenate.[73] They found no evidence of adverse effects following excessive administration of thiamine in the presence of deficiencies of other vitamins of the B complex. Morrison and Sarett also reported the lack of influence of excess levels of thiamine in rats fed diets deficient in pyridoxine, pantothenate, or riboflavin on weight gain and food efficiency.[74] Studies by Morrison and Scott on reproductive performance revealed no untoward effects of excess thiamine.[75] They fed a diet of 7.5 mg of thiamine per 100 g of food to female weanling rats of McCollum-Wisconsin strain for 12 weeks, after which these animals were mated to provide males for the study of reproductive performance. The results showed that excess thiamine had no effect on weight gain, reproductive performance, or levels of riboflavin, pyridoxine, pantothenate, and vitamin B_{12} in the livers of the animals after parturition and lactation; however, liver thiamine levels in these animals and their young at weanling were markedly increased by excess thiamine.[75] Miyagawa et al. observed similar results regarding growth, riboflavin content in liver, and xanthine oxidase and D-amino acid oxidase activities of liver after rats ingested large amounts of thiamine for 2 months.[76] Contrary to these negative results, Richards found that administration of excess thiamine to female rats fed a white flour-casein diet, simulating a low-quality human diet, had no effect on growth, but adversely affected reproduction, as evidenced by high mortality and poor growth rates among the young.[77] Convulsions similar to those found in pyridoxine deficiency were observed in some of the offspring, which could be prevented by giving the dams additional pyridoxine during lactation.[77] Fukutomi injected subcutaneously 0.5 mg of thiamine to pyridoxine-deficient rats every day for 50 days and observed a greater depression in weight gain as compared to rats deficient in pyridoxine without excess thiamine.[78] Shinagawa administered 97.5 mg of thiamine subcutaneously into rabbits for five consecutive days.[79] Riboflavin content in urine increased to 208% on the first day, 185% on the second day, and after 3 days became normal. Niacin in the urine showed a pattern similar to that of riboflavin.[79] However, subcutaneous injection of thiamine pyrophosphate did not produce such an increase in riboflavin and niacin excretion.[80] With respect to the mechanism of these phenomena, Chevillard and Thoai, using yeast phosphokinase, showed that esterification of pyridoxine was prevented by the presence of 10 to 20 times the amount of thiamine in the reaction mixture.[81] The explanation for the etiology of the phenomenon mentioned above may be the competition in the phosphorylation mechanism between thiamine and other B vitamins. However, Ando proposed another explanation for this phenomenon.[82-84] He administered a large amount of thiamine derivatives to rats and found a significant decrease in hepatic coenzyme A and total pantothenic acid, while the vitamin B_6 value remained unchanged. Administration of pantothenic acid to these rats reverted the levels of pantothenic acid and coenzyme A precursors to the original levels, while administration of pyridoxine had no effect. As there was no inhibitory action of thiamine on the phosphorylation of pantothenic acid in vitro and a significant decrease was found in vivo in activities of the pantothenate 4'-phosphotransferase, which plays a role in phosphorylation of pantothenic acid, he proposed that a massive amount of ingested thiamine directly affects the phosphorylation enzyme.

Grasse and Wade investigated the effect of a large amount of thiamine on drug-metabolizing pathways.[85] On a high thiamine diet that supplied an average daily intake of 2.0 mg, rats showed significant reductions in aniline, zoxazolamine, and aminopyrine metabolic rates in vitro. Since the V_{max} but not the K_m for aniline was depressed, they assumed that although the quantity of enzyme responsible for aniline hydroxylase was decreased, the characteristics of the enzyme were qualitatively unaltered in rats fed a diet high in thiamine. The hexobarbital metabolic rate was not significantly altered. Both microsomal cytochrome b_5 and microsomal cytochrome p450 levels were significantly decreased in the rats on a high thiamine diet. Likewise, NADPH-cytochrome C reductase, a hepatic microsomal enzyme necessary for electron transport to cytochrome p450, exhibited a decreased activity in rats fed a diet high in thiamine. Thus, dietary thiamine-induced alterations in hepatic microsomal aniline hydroxylase activity may be partially or wholly due to altered levels of cytochrome b_5, cytochrome p450, and the activity of NADPH-cytochrome C reductase. However, it does not appear that either of these components is rate limiting in the case of hexobarbital oxidase. Grasse and Wade hypothesized that excess thiamine in diets has an effect on the synthesis or maintenance of the specific enzymes involved in various metabolic pathways or on the electron transport system, which is rate limiting only for specific pathways. However, the possibility that thiamine works indirectly through some other mechanism to produce these alterations must be considered.

Regarding other pharmacological actions of thiamine, Hecht and Weese reported that subcutaneous injections of thiamine (500 mg/kg) in rabbits increased blood sugar slightly, whereas a 200-mg/kg injection of thiamine had no effect.[24] They also found that subcutaneous injection of 100 mg/kg of thiamine in rabbits promotes urination (about twice the amount of controls).[24] Oligouria and hyperkaliemia were reported by Boissier et al. in rabbits given an excessive amount of thiamine intravenously.[46] The insect repellent action of thiamine was also reported but could not be proved under controlled laboratory conditions.[86,87]

According to the data of Fukumoto, when 500 μg of thiamine was administered subcutaneously to rats for 20 days, about 59% of the thiamine was excreted into the urine, 2.6% into the feces, 2.6% remained in the body, and 35.8% was metabolized in the body.[88] Therefore, the toxicity of metabolic products of thiamine must be given attention. Abderhalden found that 2-methyl-4-amino-hydroxymethylpyrimidine (OMPm) was a product of the hydrolysis of thiamine, and injection of this compound (12 mg/100 g) into rats and mice caused convulsions.[89] Makino and colleagues, and Shintani, found that a very small amount of pyridoxine (1/100 of OMPm) completely suppressed the toxicity of OMPm.[90-92] Thus, toxicity of OMPm was increased when animals were deficient in pyridoxine.[93,94] OMPm and its metabolites were rapidly excreted in the urine after intravenous injection of OMPm.[95] In a similar manner, pyramine (2,5-dimethyl-4-aminopyrimidine) excreted in the urine after thiamine administration,[96,97] had a convulsive effect which could be prevented by the administration of vitamin B_6.[98,99] In spite of the considerable toxicity of intermediary products, animals given a massive amount of thiamine did not manifest these severe symptoms. The following explanations may be considered:

1. The amount of OMPm and pyramine produced from thiamine in the body is too little to manifest toxic effects.
2. Vitamin B_6 and other substances in the body and diet counteract the toxicity of these intermediary compounds.
3. These products are metabolized quickly and retention time in the body is too short to manifest a toxicity.

Although the toxicity of OMPm and pyramine is negligible after administration of thiamine, there is still the possibility that an excessive amount of thiamine increases the toxicity of these substances. For clarification, Fukutomi studied the effect of thiamine administration in rats poisoned with OMPm, and he found that an excessive amount of thiamine promotes to some extent the toxicity of OMPm.[78]

Habituation to excessive doses of thiamine was tested by Murakami using love birds (*Uroloncha domestica*) and rats. He administered large amounts of thiamine, 100 to 1000 times as much as the requirement, for a long period to love birds and rats. After the administration, they were fed a thiamine-deficient diet, and the body weight as well as days of survival were observed. The results showed that no special difference was observed between the dosed group and the control, which was not accustomed to the profuse administration of thiamine.[100,101] These findings suggest that long-term administration of large doses of thiamine did not cause habituation.

As described in the following section, rare cases of anaphylactic shock have been reported following parenteral administration of thiamine to man. Molitor investigated the sensitizing effect of thiamine in dogs.[102] Dogs were divided into two groups, one of which was given 50 mg/kg of thiamine hydrochloride intravenously for 10 consecutive days. After ten daily injections, an interval of 10 to 60 days was allowed before the final injection was performed. The second group of dogs was given no pretreatment. There was no significant difference between the two groups; however, upon autopsy it appeared that in some of the dogs given ten doses of thiamine, the right heart was dilated and the abdominal organs showed a greater congestion, while such evidence was nil in the second group.[102] Haley and Flesher stated that there was no evidence of anaphylaxis of thiamine toxicity in rabbits, but injection of a sensitizing dose apparently increased the resistance of the animal to toxic injection of thiamine.[48]

TOXICITY OF THIAMINE IN MAN

It has been speculated that thiamine may influence nervous diseases and various forms of pain, since this vitamin has a specific role in nervous function. Jolliffe et al. considered the relationship of diminished thiamine intake to neuritic symptoms to be definite.[103,104] In their study of many alcoholics, every alcoholic who developed polyneuritis had an inadequate thiamine intake, while those who did not develop polyneuritis had an adequate thiamine intake. They assumed that alcohol had no direct toxic action on nervous tissue but that thiamine deficiency causes polyneuritis.[103,104] Goodhart and Jolliffe observed an additional group of 17 alcoholics who revealed an uncomplicated, mild polyneuritis and were given diets containing four times the estimated thiamine requirement.[105] Alternate subjects were given thiamine intravenously, 10 mg daily for 10 days. By every method of comparison, the response of those given the thiamine supplement was better than that of the control group.[105] Metildi reported six cases of severe lightning-like pains in patients with tabes dorsalis who found relief after intravenous administration of massive doses of thiamine.[106] Recently, Pincus et al. showed that massive doses of thiamine, or of the Japanese drug thiamine propyl disulfide (a lipotropic thiamine derivative), produced good results in treating children with Leigh's disease (subacute necrotizing encephalomyelopathy), a rare and fatal neurological disease.[107-110] In the majority of cases, the symptomatology of the disease is noted before the age of 2 years and is characterized by a large variability in the clinical picture. Feeding problems, weakness, visual disturbances, ataxia, convulsions, swallowing difficulties, and peripheral neuropathy have been reported in different combinations. Thiamine doses were started at 100 mg/day orally and stepped up to 1 to 1.5 g orally. Thiamine propyl disulfide was also started at 100 mg/day orally and later increased to an oral dose of 300 mg/day. The propyl disulfide derivative of thiamine appears to be much more effective than thiamine in maintaining high thiamine levels in the cerebrospinal fluid and improving the clinical status of the patients. With long-term therapy, several patients became refractory to treatment, which correlated with their inability to maintain high cerebrospinal fluid levels of thiamine. Of the 22 patients treated with thiamine derivatives, 11 are still living, 3 of whom had been on therapy for more than 3 years at the time of the presentation (4th International Meeting of the International Society of Neurochemistry, 1973).[109] Oguro stated that large doses of thiamine by intramuscular or intravenous injection are necessary for treatment of patients with a neurological disease; however, by intraspinal injection, a smaller amount of thiamine is effective as thiamine slowly penetrates the blood brain barrier.[111] Stern reported 120 intraspinal subarachnoid injections of thiamine given to 28 patients in doses of 1 to 100 mg. The cases included inoperable cancer, von Recklinghausen's disease, multiple sclerosis, degeneration of the pyramidal system of unknown etiology, thromboangiitis obliterans with pregangrenous condition of the feet, duodenal ulcer with pulmonary tuberculosis, alcoholic neuritis of the supraorbital nerve, neuritis of the sciatic nerve, cardiac decompensation with uremia and anuria, tabes dorsalis, hypertrophic spondylitis, osteoporosis of the spine, Paget's disease, intractable pruritus ani and vulvae, beriberi, and acute poliomyelitis. From his observations in these cases, Stern concluded that many chronic or incurable conditions of the central and peripheral nervous system respond favorably to the intraspinal subarachnoid injection of synthetic thiamine. This vitamin appears to be particularly effective in cases of multiple sclerosis, encephalitis, syphilis and poliomyelitis.[112] However, Stern also stated that when thiamine was injected into the lumbar subarachnoid space of human adults at the dosage of 30 mg, undesirable symptoms such as weakness, conjunctivitis, loss of appetite, buzzing in the ears, muscle stiffness, and slight fever occurred for a short period of time.[112] Robie treated the following: multiple sclerosis, amyotrophic lateral sclerosis, postencephalitic syndrome, spastic quadriplegia, poliomyelitis, progressive muscular dystrophy, spina bifida, and alcoholic polyneuritis.[113] Complete recovery was obtained only in the case of alcoholic polyneuritis, and the results indicate that if the nerve degeneration has reached the stage of irreversibility, motor or sensory function will not return even though massive doses of thiamine are injected intraspinally over a long period of time. Moreover, although a local anesthetic was used, the intraspinal injections were painful, and reactions such as headache or vomiting frequently developed, lasting for as long as 2 days.[113] Odom and McEachern observed signs of meningial irritation in four patients to whom thiamine hydrochloride had been administered into the subarachnoid space in the lumbar region.[114] When 6.0 to 8.0 mg of thiamine was injected, the patients complained of some back stiffness and leg pains; at the same time, protein and cell counts in cerebrospinal fluid were elevated. However, serious or persisting results of subarachnoid injection of thiamine were absent.[114]

In contrast to these neurological diseases, another indication of thiamine therapy is acidosis due to the disturbances of cocarboxylase-dependent metabolism. Increased serum pyruvate and lactic acid levels have been found in many pathologic conditions, such as diabetic acidosis, liver diseases, cardiac decompensation, various forms of toxicosis, myocardial infarct, and radiation therapy.[115-118]

There is considerable literature on the use of thiamine in treatment. Zbinden surveyed the literature and found 696 published papers from 1936 to 1960 reporting thiamine therapy in more than 230 different diseases.[27] In Japan, 183 reports have described the use of massive doses of thiamine given for treatment from 1884 to 1960.[119]

Since doses in excess of 10 mg daily of thiamine to a normal person are not metabolized and are excreted unchanged in the urine, it is generally believed that thiamine may be administered without risk of harmful effect. However, Mills reported the first death in man due to an intramuscular injection of thiamine hydrochloride; at autopsy, multiple ecchymoses were found beneath the pia over both cerebral hemispheres, with areas of encephalomalacia and perivascular hemorrhages.[120] Reingold's patient had received three previous injections of 100 mg of thiamine hydrochloride at frequent intervals and the necessary latency period for anaphylaxis had passed.[121] Following the fourth injection, the patient complained of generalized burning, perspired profusely, became dyspneic and cyanotic, and died. Her death occurred less than 10 min after the injection of thiamine. It is unlikely that pyrogens were present in the vitamin preparation, since two similar injections from the same bottle had been given to two other patients on the same day without untoward reactions. The death is considered an example of anaphylactic shock, with thiamine hydrochloride as the anaphylactogen. Pathologically, the anaphylaxis probably manifested itself in constriction of the smooth muscles of the pulmonary arteries and bronchioles, resulting in pulmonary engorgement and right heart dilation and failure, and in mild pulmonary emphysema. In addition, small foci of polymorphonuclear leukocytes and round cells were seen in thickened septa of pulmonary alveoli. Foci of pigmented macrophages probably indicated that previous hemorrhages had occurred following, and anaphylactizing, the injection.[121]

Laws reported on a woman given a parenteral injection of thiamine daily at first, and then weekly.[122] Following many of these injections, she would sneeze violently several times, and the site of injection would itch for several hours. Ten days elapsed before her final injection, after which she developed generalized urticaria, facial edema, dyspnea, cyanosis, and wheezing. Intracutaneous and passive transfer tests indicated that the patient was sensitive to thiamine. Epinephrine was given within a few minutes and at the end of 5 or 6 hr the entire reaction had subsided.[122] Stiles observed two patients in whom prolonged administration of thiamine hydrochloride produced hypersensitivity.[123] One patient complained of nervousness, associated with flushing and tachycardia, and another complained of a sensation of heat and profuse perspiration. Oral administration of thiamine had produced similar, but milder symptoms.[123] Schiff's patient became nauseated, vomited, and collapsed; respiration ceased, and there was no pulse.[124,125] A total of 59 thiamine injections (25 mg) had been given intramuscularly. After instituting artificial respiration, epinephrine hydrochloride was given intravenously and the patient eventually recovered.[124,125] Stein and Morgenstern reported that a young man given large parenteral doses of thiamine, at first several days apart and later at approximately monthly intervals, always noticed severe itching and the appearance of a large reddened wheal at the site of injection, lasting several hours. After the eighth injection, a wave of intense itching began at the site of injection and passed over the entire body. He felt very weak, became dyspneic, and lost consciousness. His pulse was very rapid and hardly perceptible; wheezing was heard throughout both lung fields. He recovered the next morning, after he had been given epinephrine, caffeine sodium benzoate, and codeine sulfate. After 48 hr, an intradermal test with a saline solution of thiamine hydrochloride was positive.[126] A patient of Shapero and Gwinner who was given 50 mg of thiamine also developed severe itching and redness of the skin and eyes, dryness of the throat, and severe shortness of breath with wheezing respiration, and then collapsed.[127] In regard to the value of a positive intradermal test with thiamine, Shapero and Gwinner carried out a comparative study of skin reaction in thiamine-administered patients and control groups and suggested that solutions of no greater concentration than 5 mg/ml should be used for testing.[127] Many other workers reported patients who reacted in a similar fashion.[128-134] In all the above reported allergic cases, the hypersensitivity developed gradually. Thus, it was demonstrated that thiamine given parenterally is capable of sensitizing human beings and that a subsequent injection of thiamine which exceeds the latent period can produce anaphylactic shock.

Conversely, Jaros et al. reviewed the literature and summarized 203 reported cases of reactions to thiamine.[47] Based on their experimental work, they concluded that thiamine overdosage causes a chemical toxicity due to excessive amounts of acetylcholine and, possibly, histamine. Furthermore, they felt that patients with impaired excretion of thiamine and low physiologic tolerance to acetylcholine experience cumulative effects and, consequently, manifest toxic reactions. They summarized their findings by stating that it was hardly feasible that thiamine could act as an allergen.[47] However, Mitrani[129] and Kawasaki et al.[133] succeeded in desensitizing patients to thiamine. They injected increasing doses of a solution of thiamine daily for 10 and 178 consecutive days to patients who recovered from thiamine-induced anaphylactic shock; finally, a considerable dosage of thiamine was tolerated. The presence of Prausnitz-Küstner antibodies was established by a definite reaction to 0.02 cc of thiamine hydrochloride in a nonallergic individual.[130] From these findings it is conceivable that thiamine may, occasionally, not only act as an allergen when given parenterally, but also may cause an anaphylaxis in a previously sensitized individual. The response of the patient to parenteral administration of thiamine followed the familiar pattern of sensitization to proteins, particularly horse serum; therefore, thiamine binds to the protein group.[126] The intervals between the injections of thiamine were far enough apart to create a latent period capable of producing sensitization and suitable for the production of anaphylactic shock. When sensitization had been acquired, the last injection was the trigger mechanism that produced the violent and shocking reaction.

Leitner assumed that untoward effects of thiamine therapy may be either purely allergic or similar to a thyrotoxic state.[128] The hyperthyroid-like irritability indicated by continued restlessness, insomnia, and fatigue was another feature of the case. The observed patients seldom showed both manifestations, but in the allergic cases the anaphylactic shock may have been so prominent that the observers omitted lesser clinical manifestations.[128] Mills[135] and Shapero et al.[127] also observed hyperthyroid-like symptoms after administration of thiamine.

Another problem which should be considered as a toxic effect of excessive thiamine is whether the long-term administration of large amounts of thiamine to human beings would eventually adduce clinical evidence of other B vitamin deficiencies.

Klopp et al. investigated the excretion of riboflavin after administration of thiamine over various periods of time to human subjects and observed that a relationship exists between the thiamine intake and riboflavin excretion in certain persons; however, the magnitude of this loss of riboflavin was not great enough to produce any clinical evidence of riboflavin deficiency.[136] Fujiwara administered 5 to 100 mg of thiamine subcutaneously to both patients with various diseases and normal subjects and determined the riboflavin content in blood and urine.[137] In normal subjects, administration of thiamine produced no change in riboflavin levels, but decrease in blood riboflavin and increase in urinary riboflavin were observed in patients with liver disease, tuberculosis, or anemia.[137] Regarding thiamine derivatives, Murata et al. administered orally massive amounts (200 to

400 mg) of TPD, BTDS, and TTFD to healthy subjects, but urinary excretion of riboflavin was not increased.[138,139] Nakayama tested the same effect on infants.[140,141] In the case of oral administration via a catheter of 0.4 mg/kg of TPD for 3 to 4 weeks to healthy infants being artificially fed, urinary excretion of riboflavin increased in the first week but tended to decrease in the second week, and a transient decrease in riboflavin was evident in the blood. In the case of the administration of 2.0 mg/kg of body weight of TPD to healthy bottle-fed infants, an increase of urinary free riboflavin was observed, especially 6 hr after the administration.[140,141] Thus, clinical symptoms of riboflavin deficiency were not demonstrated. Inoue et al. gave massive doses of TPD parenterally and consecutively for 1 to 4 weeks to patients with neurological disorders.[142] They found that these patients showed symptoms of riboflavin deficiency, such as cheilosis and a beefy red tongue, as well as a decrease in levels of blood riboflavin. Injections of riboflavin produced a return to normal.[142]

In regard to the effect of thiamine on the nervous system, Hamamoto et al. injected a large amount of thiamine to nine children with various types of mental deficiencies and determined the electroencephalographic effect.[143] No difference in the EEG was noted.

On the other hand, regarding the cocarboxylase effect of thiamine, Maeda found that the blood glucose level decreased after an intravenous injection of thiamine pyrophosphate (100 mg) to healthy persons.[144] He assumed that this effect was due to the acceleration of the keto-acid oxidation system resulting from the stimulating effect of excessive thiamine.[144]

From these data, oral administration is clearly the safest route for massive doses of thiamine. When a massive amount of thiamine must be given parenterally, skin tests for patients may be a wise precautionary method to prevent anaphylactic shock, although these experiences are rare. Also, physicians should pay attention to the nutritional status of the other members of the vitamin B group in patients.

ACKNOWLEDGMENT

Thanks are due to M. Ohara, Kyoto University, for assistance in preparation of the manuscript.

REFERENCES

1. **Binet, L. and Minz, B.,** Sur les réactions biochimiques du nerf au repos et au cours d'une excitation électrique, *Arch. Int. Physiol.,* 42, 281–300, 1936.
2. **Minz, B.,** Sur la libération de la vitamine B_1 par le tronc isolé du nerf pneumogastrique soumis à l'excitation électrique, *C. R. Soc. Biol.,* 127, 1251–1253, 1938.
3. **Itokawa, Y., Schulz, R. A., and Cooper, J. R.,** Thiamine in nerve membranes, *Biochim. Biophys. Acta,* 266, 293–299, 1972.
4. **Itokawa, Y. and Cooper, J. R.,** Thiamine release from nerve membrane by tetrodotoxin, *Science,* 166, 759–761, 1969.
5. **Itokawa, Y. and Cooper, J. R.,** On a relationship between ion transport and thiamine in nervous tissue, *Biochem. Pharmacol.,* 18, 545–547, 1969.
6. **Itokawa, Y. and Cooper, J. R.,** Ion movements and thiamine in nervous tissue. I. Intact nerve preparations, *Biochem. Pharmacol.,* 19, 985–992, 1970.
7. **Itokawa, Y. and Cooper, J. R.,** Ion movement and thiamine. II. The release of the vitamin from membrane fragments, *Biochim. Biophys. Acta,* 196, 274–284, 1970.
8. **Itokawa, Y.,** Is calcium deficiency related to thiamine-dependent neuropathy in pigeon?, *Brain Res.,* 94, 475–484, 1975.
9. **Itokawa, Y.,** Role of thiamine in excitable membrane of nerve, *Vitamins,* 49, 415–427, 1975.
10. **Itokawa, Y.,** Thiamine and nerve membrane, *J. Nutr. Sci. Vitaminol.,* Suppl. 22, 17–19, 1976.
11. **von Kunz, H. A.,** Über die Wirkung von Antimetaboliten des Aneurins auf die einzelne markhaltige Nervenfaser, *Helv. Physiol. Pharmacol. Acta,* 14, 411–423, 1956.
12. **von Muralt, A.,** The role of thiamine (vitamin B_1) in nerve excitation, *Exp. Cell Res. Suppl.,* 5, 72–79, 1958.

13. Petropulos, S. F., The action of an antimetabolite of thiamine on single myelinated nerve fibers, *J. Cell. Comp. Physiol.,* 56, 7–13, 1960.
14. Lofland, H. B., Jr., Goodman, H. O., Clarkson, T. B., and Prichard, R. W., Enzyme studies in thiamine-deficient pigeons, *J. Nutr.,* 79, 188–194, 1963.
15. Armett, R. W. and Cooper, J. R., The role of thiamine in nervous tissue: effect of antimetabolites of the vitamin on conduction in mammalian nonmyelinated nerve fibers, *J. Pharmacol. Exp. Ther.,* 148, 137–143, 1965.
16. Cooper, J. R., Itokawa, Y., and Pincus, J. H., Thiamine triphosphate deficiency in subacute necrotizing encephalomyelopathy, *Science,* 164, 368–373, 1969.
17. Eichenbaum, J. W. and Cooper, J. R., Restoration by thiamine of the action potential in ultraviolet irradiated nerves, *Brain Res.,* 32, 258–260, 1971.
18. Murphy, J. V., Craig, L. J., and Glew, R. H., Leigh disease. Biochemical characteristics of the inhibitor, *Arch. Neurol.,* 31, 220–227, 1974.
19. Goldberg, D. J., Begenisich, T. B., and Cooper, J. R., Effects of thiamine antagonists on nerve conduction. II. Voltage clamp experiments with antimetabolites, *J. Neurobiol.,* 6, 453–462, 1975.
20. Fox, J. M. and Duppel, W., The action of thiamine and its di- and triphosphates on the slow exponential decline of the ionic currents in the node of ranvier, *Brain Res.,* 89, 287–302, 1975.
21. Berman, K. and Fishman, R. A., Thiamine phosphate metabolism and possible coenzyme-independent functions of thiamine in brain, *J. Neurochem.,* 24, 457–465, 1975.
22. Dreyfus, P. M., Thiamine deficiency encephalopathy: thoughts on its pathogenesis, in *Thiamine,* Gubler, C. J., Fujiwara, M., and Dreyfus, P. M., Eds., John Wiley & Sons, New York, 1976, 229–243.
23. Sasa, M., Takemoto, I., Nishino, K., and Itokawa, Y., The role of thiamine on excitable membrane of crayfish giant axon, *J. Nur. Sci. Vitaminol.,* Suppl. 22, 21–24, 1977.
24. Hecht, G. and Weese, H., Pharmakologisches über Vitamin B$_1$, *Klin. Wochenschr.,* 16, 414–415, 1937.
25. Haley, T. J., A comparison of the acute toxicity of two forms of thiamine, *Proc. Soc. Exp. Biol. Med.,* 68, 153–157, 1948.
26. Minesita, K., Ueda, M., and Matsumura, S., Pharmacological effect of carbalkoxythiamine, *Vitamins,* 25, 522–524, 1962.
27. Zbinden, G., Therapeutic use of vitamin B$_1$ in diseases other than beriberi, *Ann. N.Y. Acad. Sci.,* 98, 550–561, 1962.
28. Inoue, K. and Itokawa, Y., Metabolism of O-benzoylthiamine in animal body, *Biochem. Med.,* 8, 450–463, 1973.
29. Hayashi, T., Mechanism of death by massive dose of thiamine, *Vitamins,* 26, 76–77, 1962.
30. Minesita, K. and Ueda, M., On the acute toxicity of large doses of thiamine, *Vitamins,* 26, 77, 1962.
31. Minesita, K., Pharmacological studies on carbalkoxythiamine, *Vitamins,* 27, 319–320, 1963.
32. Smith, J. A., Foa, P. P., and Weinstein, H. R., The curare-like action of thiamine, *Science,* 108, 412, 1948.
33. Di Palma, J. R. and Hitchcook, P., Neuromuscular and ganglionic blocking action of thiamine and its derivatives, *Anesthesiology,* 19, 762–769, 1958.
34. Ngai, S. H., Ginsburg, S., and Katz, R. L., Action of thiamine and its analog on neuromuscular transmission in the cat, *Biochem. Pharmacol.,* 7, 256–264, 1961.
35. Hayashi, T., Kurahashi, Y., and Takeuchi, H., Blocking action of thiamine and its derivatives upon neuromuscular transmission of cold blooded animals, *J. Vitaminol.,* 11, 30–36, 1965.
36. Wolfson, S. K., Jr. and Ellis, S., Thiamine toxicity and ganglionic blockage, *Fed. Proc. Fed. Am. Soc. Exp. Biol.,* 13, 418, 1954.
37. Nastuk, W. L. and Kahn, N., Blocking action of thiamine on neuromuscular transmission and membrane conduction, *Fed. Proc. Fed. Am. Soc. Exp. Biol.,* 19, 260, 1960.
38. Hayashi, T., Blocking action of thiamine on neuromuscular transmission, *Vitamins,* 26, 76, 1962.
39. Hayashi, T., Comparison of action of thiamine and curare on neuromuscular transmission, *Vitamins,* 27, 227–228, 1963.
40. Hayashi, T., Action of thiamine and oxythiamine on neuromuscular transmission, *Vitamins,* 27, 476, 1963.
41. Hayashi, T., Effect of pyrithiamine on neuromuscular transmission, *Vitamins,* 28, 119, 1963.
42. Hayashi, T., Effect of thiamine derivatives on neuromuscular transmission, *Vitamins,* 28, 195, 1963.
43. Smith, J. A., Foa, P. P., and Weinstein, H. R., Some toxic effects of thiamine, *Fed. Proc. Am. Soc. Exp. Biol.,* 6, 204, 1947.

44. **Smith, J. A., Foa, P. P., Weinstein, H. R., Ludwig, A. S., and Wertheim, J. M.,** Some aspects of thiamine toxicity, *J. Pharmacol. Exp. Ther.,* 93, 294–304, 1948.
45. **Smith, J. A., Foa, P. P., and Weinstein, H. R.,** Observations on the curare-like action of thiamine, *Am. J. Physiol.,* 155, 469–470, 1948.
46. **Boissier, J. R., Tillement, J. P., Viars, P., Merlin, L., and Simon, P.,** Quelques aspects de la toxicité et du métabolism de la thiamine administrée à fortes doses, *Anesth. Analog. Reanim.,* 24, 515–527, 1967.
47. **Jaros, S. H., Wnuck, A. L., and DeBeer, E. J.,** Thiamine intolerance, *Ann. Allergy,* 10, 291–307, 1952.
48. **Haley, T. J. and Flesher, A. M.,** A toxicity study of thiamine hydrochloride, *Science,* 104, 567–568, 1946.
49. **Itokawa, Y., Tanaka, C., and Kimura, M.,** Effect of thiamine on serotonin levels in magnesium-deficient animals, *Metabolism,* 21, 375–379, 1972.
50. **Agid, R. and Balkanyi, J.,** Action de la vitamin B_1 sur le coeur isolé de grenouille, *C. R. Soc. Biol.,* 127, 680–683, 1938.
51. **Byer, J. and Harpuder, K.,** The sensitizing effect of thiamine for acetylcholine, *J. Pharmacol. Exp. Ther.,* 70, 328–333, 1940.
52. **Glick, D. and Antopol, W.,** The inhibition of choline esterase by thiamine (vitamin B_1), *J. Pharmacol. Exp. Ther.,* 65, 389–394, 1939.
53. **Torda, C. and Wolff, H. G.,** Effect of vitamin B_1 and cocarboxylase on synthesis of acetylcholine, *Proc. Soc. Exp. Biol. Med.,* 56, 88–89, 1944.
54. **Torda, C. and Wolff, H. G.,** Effect of vitamins on acetylcholine synthesis. The apparently specific action of vitamin E, *Proc. Soc. Exp. Biol. Med.,* 58, 163–165, 1945.
55. **Abderhalden, E. and Abderhalden, R.,** Vitamin B_1 and Acetylcholine, *Klin. Wochenschr.,* 17, 1480, 1938.
56. **Minz, B.,** La role de la vitamin B_1 dans la régulation humorale du systeme nerveux, *Presse Med.,* 76, 1406–1407, 1938.
57. **Minz, B.,** L'influence de l'aneurine sur l'activité des éléments-cholinergiques, *C. R. Soc. Biol.,* 131, 1156–1159, 1939.
58. **Post, M. and Smith, J. A.,** Bronchoconstrictor effect of thiamine, *Am. J. Physiol.,* 163, 742–743, 1950.
59. **Smith, J. A. and Sohn, H.,** Dual role of thiamine in producing bronchoconstriction, *Fed. Proc. Fed. Am. Soc. Exp. Biol.,* 10, 128–129, 1951.
60. **Haimovici, H. and Pick, E. P.,** Inhibitory effect of thiamine on vasoconstrictor action of nicotine tested in the Laewen-Trendelenburg preparation, *Proc. Soc. Exp. Biol. Med.,* 62, 234–237, 1946.
61. **Pick, E. P. and Unna, K.,** Blockade of the nicotine action on the blood pressure by thiazole-compounds (sulfathiazole and thiamine), *J. Pharmacol. Exp. Ther.,* 87, 138–148, 1946.
62. **Yamamoto, I., Iwata, H., Tamori, Y., and Hirayama, M.,** On the nicotine antagonistic effect of thiamine. I. *Folia Pharmacol. Jap.,* 52, 429–435, 1956.
63. **Yamamoto, I., Iwata, H., Tamori, Y., and Hirayama, M.,** On the nicotine antagonistic effect of thiamine. II. *Folia Pharmacol. Jap.,* 53, 307–310, 1957.
64. **Tamori, Y.,** Studies on the anti-nicotine effect of thiamine, *Folia Pharmacol. Jap.,* 54, 571–584, 1958.
65. **Yamamoto, I., Inoki, R., and Mizoguchi, K.,** Studies on nicotine. Relationship between nicotine and thiamine on pyruvate oxidation, *Folia Pharmacol. Jap.,* 58, 120–126, 1962.
66. **Yamamoto, I.,** Thiamine as a nicotine antagonist, *Jpn. J. Pharmacol.,* 13, 240–252, 1963.
67. **Iwata, H. and Inoue, A.,** Antagonistic effect of thiamine and its derivatives to nicotine on atria of guinea pig, *Folia Pharmacol. Jap.,* 64, 46–51, 1968.
68. **Dias, M. V.,** Action of thiamine applied directly to the cerebral cortex, *Science,* 105, 211–213, 1947.
69. **Hayashi, T.,** Effect of carbalkoxythiamine administration into cerebrospinal fluid, *Vitamins,* 25, 526–528, 1962.
70. **Hayashi, T.,** Comparison of convulsive effect of thiamine and curare, *Vitamins,* 27, 228, 1963.
71. **Hayashi, T.,** Convulsion due to administration of oxythiamine and pyrithiamine, *Vitamins,* 32, 183, 1965.
72. **Hayashi, T.,** Convulsive effect of thiamine and its derivatives after administration into cerebrospinal fluid, *Vitamins,* 31, 426–427, 1965.
73. **Unna, K. and Clark, J. D.,** Effect of large amounts of single vitamins of the B group upon rats deficient in other vitamins, *Am. J. Med. Sci.,* 204, 364–371, 1942.
74. **Morrison, A. B. and Sarett, H. P.,** Studies on B vitamin interrelationships in growing rats, *J. Nutr.,* 68, 473–484, 1959.

75. **Morrison, A. B. and Sarett, H. P.,** Effect of excess thiamine and pyridoxine on growth and reproduction in rats, *J. Nutr.,* 69, 111–116, 1959.

76. **Miyagawa, K., Ikehata, H., and Murata, K.,** Studies on the administration of large amount of thiamine propyldisulfide, *Vitamins,* 23, 103–108, 1961.

77. **Richards, M. B.,** Imbalance of vitamin B factors. Pyridoxine deficiency caused by additions of aneurin and chalk, *Br. Med. J.,* 1, 4395–4398, 1945.

78. **Fukutomi, H.,** Clinical and experimental studies on the excessive administration of thiamine. V. Excessive administration of thiamine for pyridoxine deficient rats, *Vitamins,* 33, 156–159, 1966.

79. **Shinagawa, T.,** Influence of the over-dose of thiamine and thiamine allyldisulfide on the urinary excretion of riboflavin and niacin, *Vitamins,* 11, 175–178, 1956.

80. **Shinagawa, T.,** Effect of the over-dose of diphosphothiamine on the urinary excretion of B-vitamins, and the effect of over-dose of thiamine on the urinary excretion of niacin and methylnicotinamide, *Vitamins,* 12, 243–245, 1957.

81. **Chevillard, L. and van Thoai, N.,** Synthèse enzymatique de la cocarboxylase et de la tyrosine codécarboxylase par les phosphokinase de levures, *Bull. Soc. Chim. Biol.,* 33, 1147–1151, 1951.

82. **Ando, H.,** Effect of excessive administration of thiamine derivatives on the coenzyme A metabolism in rat. I. Effect of excessive administration of thiamine derivatives on hepatic coenzyme A concentration in young rat, *Vitamins,* 38, 158–162, 1968.

83. **Ando, H.,** Effect of excessive administration of thiamine derivatives on the coenzyme A metabolism in rat. II. Effect of excessive administration of thiamine derivatives on the hepatic coA, pantothenic acid and vitamin B_6 concentration in young rat, *Vitamins,* 38, 163–168, 1968.

84. **Ando, H.,** Effect of excessive administration of thiamine derivatives on the coenzyme A metabolism in rat. III. Effect of thiamine on the phosphorylation of pantothenic acid and effect of excessive administration of thiamine derivatives on the pantothenate-4-phosphotransferase activity in rat, *Vitamins,* 38, 169–174, 1968.

85. **Grasse, W., III and Wade, A. E.,** The effect of thiamine consumption on liver microsomal drug-metabolizing pathways, *J. Pharmacol. Exp. Ther.,* 176, 758–765, 1971.

86. **Müting, D.,** Über die Verhütung von Müchenstichen durch Einnahme von Vitamin B_1, *Med. Klin.* (Munich), 53, 1023, 1958.

87. **Rahm, U.,** Besitzt Vitamin B_1 insektenabhaltende Eigenschaften? *Schweiz. Med. Wochenschr.,* 26, 634–635, 1958.

88. **Fukumoto, H.,** Clinical and experimental studies on the excessive administration of thiamine. IV. On the absorption, excretion and accumulation of thiamine after the excessive administration, *Vitamins,* 33, 151–155, 1966.

89. **Abderhalden, E.,** Beitrag zum Problem der für die B - (Aneurin bzw. Thiamin-) Avitaminose charakterischer Erscheinungen. *Pfluegers Arch. Gesamte Physiol. Menschen. Tiere.,* 240, 647–652, 1938.

90. **Makino, K., Kinoshita, T., Aramaki, Y., and Shinya, S.,** On the atoxopyrimidine action of B_6 group vitamins, *Vitamins,* 7, 670, 1954.

91. **Shintani, S.,** On the relationship between hydroxymethylpyrimidine and pyridoxine. I. Inhibitory effect of pyridoxine on the convulsion by hydroxymethylpyrimidine in mice, *J. Pharm. Soc. Jpn.,* 76, 13–15, 1956.

92. **Makino, K., Kinoshita, T., Sasaki, T., and Shioi, T.,** Atoxopyrimidine, *Nature,* 173, 34–35, 1954.

93. **Shintani, S.,** On the relationship between hydroxymethylpyrimidine and pyridoxine. II. On the agumentation of convulsant effect of hydroxymethylpyrimidine by pyridoxine-deficient synthetic diet in mice, *J. Pharm. Soc. Jpn.,* 76, 15–17, 1956.

94. **Shintani, S.,** On the relationship between hydroxymethylpyrimidine and pyridoxine. III. The antipyridoxine activity of hydroxymethylpyrimidine in rats, *J. Pharm. Soc. Jpn.,* 76, 18–20, 1956.

95. **Shintani, S.,** Studies on the change of hydroxymethylpyrimidine in vivo. I. Determination of hydroxymethylpyrimidine in urine and its excretion in rabbit, *J. Pharm. Soc. Jpn.,* 77, 781–784, 1957.

96. **Pollack, H., Ellenberg, M., and Dolger, H.,** Excretion of thiamine and its degradation products in man, *Proc. Soc. Exp. Biol. Med.,* 47, 414–417, 1941.

97. **Mickelson, O., Caster, W. O., and Keys, A.,** A statistical evaluation of the thiamine and pyramin excretions of normal young men on controlled intakes of thiamine, *J. Biol. Chem.,* 168, 415–431, 1947.

98. **Shintani, S.,** On the relationships between the analogs of hydroxymethylpyrimidine and vitamin B_6, II. Accelerating effect of the analogs of hydroxymethylpyrimidine upon the development of vitamin B_6-deficiency syndrome in rat, *J. Pharm. Soc. Jpn.,* 77, 993–996, 1957.

99. **Nishizawa, Y.,** Studies on substances of the toxopyrimidine group, *J. Vitaminol.,* 4, 63–64, 1958.

100. **Murakami, M.,** Studies on the effects of excessive administration of thiamine upon the living body. I. Study on habituation of Uroloncha Striata var. Domestica in the metabolism of thiamine as a result of successive profuse administration of thiamine, *Vitamins,* 11, 211–216, 1956.

101. **Murakami, M.,** Studies on the effects of excessive administration of thiamine upon the living body. II. Study on habituation of albino rats in the metabolism of thiamine as a result of successive profuse administration of the vitamin, *Vitamins,* 11, 216–219, 1956.

102. **Molitor, H.,** Vitamins as pharmacological agents, *Fed. Proc. Fed. Am. Soc. Exp. Biol.,* 1, 309–315, 1942.

103. **Jolliffe, N. and Colbert, C. N.,** The etiology of polyneuritis in the alcohol addict, *JAMA,* 107, 642–647, 1936.

104. **Jolliffe, N., Colbert, C. N., and Joffe, P. M.,** Observations on the etiologic relationship of vitamin B (B_1) to polyneuritis in the alcohol addict, *Am. J. Med. Sci.,* 191, 515–524, 1936.

105. **Goodhart, R. and Jolliffe, N.,** Effect of vitamin B (B_1) therapy on the polyneuritis of alcohol addicts, *JAMA,* 110, 414–419, 1938.

106. **Metildi, P. F.,** The treatment of tabetic lightning pain with thiamine chloride, *Am. J. Syph. Gonorrhea Vener. Dis.,* 23, 1–6, 1939.

107. **Pincus, J. H., Itokawa, Y., and Cooper, J. R.,** Enzyme-inhibiting factor in subacute necrotizing encephalomyelopathy, *Neurology,* 19, 841–845, 1969.

108. **Pincus, J. H., Cooper, J. R., Itokawa, Y., and Gumbinas, M.,** Subacute necrotizing encephalomyelopathy, *Arch. Neurol.,* 24, 511–517, 1971.

109. **Cooper, J. R. and Pincus, J. H.,** Treatment of Leigh's disease (subacute necrotizing encephalomyelopathy) with thiamine derivatives, in *Abstr. 4th Int. Meet. Int. Soc. Neurochemistry,* The Local Organizing Committee of I.S.N., Tokyo, 1973, 382.

110. **Pincus, J. H., Cooper, J. R., Murphy, J. V., Rabe, E. F., Lonsdale, D., and Dunn, H. G.,** Thiamine derivatives in subacute necrotizing encephalomyelopathy, *Pediatrics,* 51, 716–721, 1973.

111. **Oguro, Y.,** Mechanism on the effect of intraspinal injection of thiamine in treatment of nervous disease, *Vitamins,* 16, 196–201, 1959.

112. **Stern, E. L.,** The intraspinal (subarachnoid) injection of vitamin B_1 for the relief of intractable pain, and for inflammatory and degenerative diseases of the central nervous system, *Am. J. Surg.,* 45, 495–511, 1938.

113. **Robie, T. R.,** The administration of vitamin B_1 by intraspinal injection, *Am. J. Surg.,* 48, 398–409, 1940.

114. **Odom, G. and McEachern, D.,** Subarachnoid injection of thiamine in cats; unmasking of brain lesions by induced thiamine deficiency, *Proc. Soc. Exp. Biol. Med.,* 50, 28–31, 1942.

115. **Murata, T.,** Influence of thiamine on the amount of lactic acid in urine, *J. Vitaminol.,* 4, 109–113, 1958.

116. **Serge, A.,** Cocarboxylase. I. Mitteilung, *Arzneim. Forsch.,* 9, 1–9, 1959.

117. **Serge, A.,** Cocarboxylase. II. Mitteilung, *Arzneim. Forsch.,* 9, 102–113, 1959.

118. **Wendel, O. W.,** A study of urinary lactic acid levels in humans. I. Influence of thiamine and pyrithiamine, *J. Vitaminol.,* 6, 16–23, 1960.

119. **Takada, R. and Katsura, E.,** *Japanese Literature on Vitamins,* The Vitamin Society of Japan, Kyoto, 1962, 1385–1449.

120. **Mills, C. A.,** Discussion on vitamin therapy, *JAMA,* 117, 1500–1501, 1941.

121. **Reingold, I. M. and Webb, F. R.,** Sudden death following intervenous injection of thiamine hydrochloride, *JAMA,* 130, 491–492, 1946.

122. **Laws, C. L.,** Sensitization to thiamine hydrochloride, *JAMA,* 117, 176, 1941.

123. **Stiles, M. H.,** Hypersensitivity to thiamine chloride, with a note on sensitivity to pyridoxine hydrochloride, *J. Allergy,* 12, 507–509, 1941.

124. **Schiff, L.,** Collapse following parenteral administration of solution of thiamine hydrochloride, *JAMA,* 117, 609, 1941.

125. **Schiff, L.,** Discussion on vitamin therapy, *JAMA,* 117, 1501, 1941.

126. **Stein, W. and Morgenstern, M.,** Sensitization to thiamine hydrochloride: report of another case, *Ann. Intern. Med.,* 20, 826–828, 1944.

127. **Shapero, W. and Gwinner, M. W.,** Sensitivity to thiamine hydrochloride, *Ann. Allergy,* 5, 349–352, 1947.

128. **Leitner, Z. A.,** Untoward effects of vitamin B_1, *Lancet*, 2, 474, 1943.
129. **Mitrani, M. M.,** Vitamin B_1 hypersensitivity with desensitization, *J. Allergy*, 15, 150–153, 1944.
130. **Engelhardt, H. T. and Baird, V. C.,** Sensitivity to thiamine hydrochloride. A potential hazard in a common office procedure, *Ann. Allergy*, 4, 291–292, 1946.
131. **Miura, O.,** Allergy by vitamins, *Sogo Rinsho*, 6, 985–992, 1956.
132. **Tetrault, A. F. and Beck, I. A.,** Anaphylactic shock following intramuscular thiamine chloride, *Ann. Intern. Med.*, 45, 134–138, 1956.
133. **Kawasaki, T., Asano, T., and Makita, S.,** A case of anaphylactic shock due to intravenous thamine hydrochloride, *J. Jpn. Soc. Int. Med.*, 51, 246–252, 1962.
134. **Shioda, K., Yamada, K., Ohgai, Y., Matsumoto, Y., Yamamoto, Y., and Maeda, Y.,** Anaphylactic shock following injection of thiamine, *Nippon Rinsho*, 18, 543–546, 1960.
135. **Mills, C. A.,** Thiamine overdosage and toxicity, *JAMA*, 116, 2101, 1941.
136. **Klopp, C. T., Abels, J. C., and Rhods, C. P.,** The relationship between riboflavin intake and thiamine excretion in man, *Am. J. Med. Sci.*, 205, 852–857, 1943.
137. **Fujiwara, M.,** Influence of thiamine on riboflavin metabolism, *Vitamins*, 7, 206–208, 1954.
138. **Murata, K., Suzuki, S., Irimajiri, S., Miyagawa, K., Miyamoto, T., and Ikehata, H.,** Studies on the oral administration of large amounts of thiamine to men, *Vitamins*, 27, 282–287, 1963.
139. **Murata, K., Suzuki, S., Miyatake, K., Miyamoto, T., and Ikehata, H.,** Studies on the oral administration of large amounts of thiamine in men, *Vitamins*, 30, 33–37, 1964.
140. **Nakayama, Y.,** Studies on the relationship between thiamine propyl disulfide and riboflavin. I. Effect of oral administration of TPD on riboflavin content in the urine of normal children and in the urine as well as blood of healthy infant under artificial feeding, *Vitamins*, 16, 294–300, 1959.
141. **Nakayama, Y.,** Studies on the relationship between thiamine propyl disulfide and riboflavin. II. Effect of parenteral administration of TPD and esters of riboflavin on the urinary riboflavin content and its forms in healthy infants under artificial feeding, *Vitamins*, 16, 300–309, 1959.
142. **Inoue, K., Katsura, E., and Kariyone, S.,** Secondary riboflavin deficiency, *Vitamins*, 10, 69, 1956.
143. **Hamamoto, E., Inaba, M., Oka, E., Ohno, M., Ban, T., and Ohtahara, S.,** Studies on the effects of large amounts of thiamine and thiamine propyldisulfide on the EEG of mentally retarded children, *Vitamins*, 39, 145–154, 1969.
144. **Maeda, T.,** Clinical and experimental studies on the effect of the administration of large amount of B-vitamins on the blood sugar and insulin content of blood, *Vitamins*, 37, 439–455, 1968.

EFFECT OF NUTRIENT TOXICITIES (EXCESS) IN ANIMALS AND MAN: RIBOFLAVIN

R. S. Rivlin

A large body of evidence has accumulated that treatment with riboflavin in excess of nutritional requirements has very little toxicity either for experimental animals or for man. Early studies consisted of administering lactoflavin (riboflavin) in large doses by oral or parenteral routes to experimental animals and then examining both the survival rates of these animals and any evident metabolic or pathological abnormalities. Because of the very low solubility of riboflavin in water, namely 10 to 13 mg/100 ml at room temperature and less than 19 mg/100 ml at body temperature,[1] the vitamin was usually given either in aqueous suspension or prepared in various other solvents. Kuhn[2,3] attributed the previously reported high toxicity of riboflavin to the toxic solvents used rather than to the vitamin itself, because he observed that mice treated with doses of riboflavin as high as 340 mg/kg body weight exhibited no ill effects. Similarly, Demole[4] independently investigated the effects of riboflavin administration in large doses upon a variety of animal species, including fish, frogs, mice, rats, rabbits, cats and dogs, without observing any ill effects. Others[5] also showed that administration of 10 g/kg body weight riboflavin in aqueous suspension orally or 5 g/kg subcutaneously to rats had no demonstrable ill effects, nor did 2 g/kg of riboflavin when fed to dogs.

Only when doses of riboflavin (600 mg/kg body weight) were administered intraperitoneally to rats were ill effects evident.[5] Anuria developed and at autopsy crystals were apparent in the collecting tubules and renal pelvis. The significance of these findings is doubtful because the solubility of riboflavin was probably exceeded under these circumstances. These workers also observed that oral administration of riboflavin in aqueous suspension produced intensely yellow-colored feces, which suggested that only a small proportion of the riboflavin was being absorbed through the intestine. The important observation that the intestine has a limited capacity to absorb riboflavin was subsequently documented by tracer studies.[6,7] The intestinal absorption of riboflavin appears to involve a specialized transport process which is easily saturated, thereby preventing excessive blood levels of riboflavin from developing after dietary intake of large quantities of the vitamin. A further mechanism preventing excessive blood levels of riboflavin under ordinary circumstances is the rapid elimination of the vitamin through biliary and urinary excretion.[6,7] It is likely that similar mechanisms are operative both in animals and in man.

Even when large amounts of riboflavin are administered to normal animals, tissue concentrations are not appreciably increased. Studies have shown that the tissue levels of riboflavin as well as those of the coenzymes derived from riboflavin, flavin mono-nucleotide (FMN) and flavin adenine dinucleotide (FAD), have an apparent, upper limit in normal animals which cannot be exceeded by oral or parenteral administration of riboflavin.[8,9] The exact mechanisms accounting for these apparent limits have not been determined.

Riboflavin appears to be safe when administered in large doses to patients therapeutically. It has been suggested that riboflavin deficiency should be treated with a daily oral dose of 10 to 15 mg.[10] Clinical toxicity has not been observed at these dosage levels. Therapy with riboflavin by the parenteral route also appears to be relatively free of hazard. In one Russian dermatological clinic apparent allergy to parenteral riboflavin was observed,[11] but this finding has not been reported elsewhere.

It is of interest that a patient with multiple myeloma was recently reported in whom a specific monoclonal immunoglobulin was detected which bound riboflavin with high affinity and high specificity.[12] This patient had given a history of prior ingestion of multivitamin tablets which included riboflavin, and her skin and hair were an obviously yellow color. A delay in the excretion and turnover of dietary riboflavin was demonstrated in this unique patient.[13] The role of riboflavin in the causation of this disease or in any of its pathological features is not known.

Riboflavin administration should be expected to provide a certain degree of protection against toxicity due to a variety of drugs if the body stores of riboflavin are not adequate, because flavin coenzymes are components of drug-metabolizing enzymes. Riboflavin deficiency reduces the activity of TPNH-cytochrome C reductase,[14] an FAD-requiring enzyme which is involved in the degradative metabolism of many drugs and carcinogens. Riboflavin has marked effects on the activity of the mixed-function oxidase system.[15] Attention has been focused on the multiplicity of nutritional factors which regulate drug metabolism and on the likelihood that deficiency of riboflavin and other nutrients might potentiate toxicity due to certain drugs.[16]

Riboflavin administration exerts a specific protective effect against the toxicity of boric acid. Mice, rats, and chicks fed boric acid in the diet undergo significant retardation of growth which can be ameliorated by treatment with riboflavin.[17] The mechanism of the effect appears to be that of binding between boric acid and riboflavin with the formation of a stable riboflavin-boric acid complex which is excreted in urine.[17,18] Boric acid feeding results in a substantial increase in urinary riboflavin excretion, and those animals which receive a low or marginally adequate diet of riboflavin appear to be more susceptible to the toxic effects of boric acid than those with dietary riboflavin supplementation. Massive riboflavinuria occurs after accidental boric acid ingestion in man,[19] which raises the possibility that riboflavin may be therapeutically useful in such circumstances.

In summary, riboflavin appears to be safe and effective when administered to either animals or humans in doses greater than the daily requirements. When massive amounts of riboflavin are administered orally, only a small fraction of the dose is absorbed, the remainder being excreted in the feces. Excessive amounts of riboflavin, when given parenterally, are rapidly excreted in the urine and bile. The capacity of the tissues to store riboflavin and its coenzyme derivatives appears to be limited when excessive amounts are administered.

REFERENCES

1. **Jauregg-Wagner, J.,** Riboflavin, in *The Vitamins — Chemistry, Physiology, Pathology, Methods,* Vol. 5, 2nd ed., Sebrell, W. H., Jr. and Harris, R. S., Eds., Academic Press, New York, 1972, 3–43.
2. **Kuhn, R. and Boulanger, P.,** Uber die Giftigkeit der Flavine, *Hoppe-Seyler's Z. Physiol. Chem.,* 241, 233–238, 1936.
3. **Kuhn, R.,** Ist Lactoflavin giftig? *Klin. Wochenschr.,* 17, 222–223, 1938.
4. **Demole, V. V.,** Vertraglichkeit des Lactoflavins, *Z. Vitam. Horm. Fermentforsch.,* 7, 138–142, 1938.
5. **Unna, K. and Greslin, J. G.,** Studies on the toxicity and pharmacology of riboflavin, *J. Pharmacol. Exp. Ther.,* 76, 75–80, 1942.
6. **Jusko, W. J. and Levy, G.,** Absorption, protein binding, and elimination of riboflavin, in *Riboflavin,* Rivlin, R. S., Ed., Plenum Press, New York, 1975, 99–152.
7. **Christensen, S.,** The biological fate of riboflavin in mammals, *Acta Pharmacol. Toxicol.,* 32 (Suppl. 11), 1–68, 1973.
8. **Ellis, L. N., Zmachinsky, A., and Sherman, H. C.,** Experiments upon the significance of liberal levels of intake of riboflavin, *J. Nutr.,* 25, 153–160, 1943.
9. **Burch, H. B., Lowry, O. H., Padilla, A. M., and Combs, A. M.,** Effects of riboflavin deficiency and realimentation on flavin enzymes of tissues, *J. Biol. Chem.,* 223, 29–45, 1956.
10. **Goldsmith, G.,** Riboflavin deficiency, in *Riboflavin,* Rivlin, R. S., Ed., Plenum Press, New York, 1975, 221–244.

11. **Soloshenko, E. N. and Brailovski, A.,** Allergic skin reactions caused by group B vitamins, *Sov. Med.,* 10, 141, 1975.
12. **Farhangi, M. and Osserman, E. F.,** Myeloma with xanthoderma due to an IgG$_\lambda$ monoclonal anti-flavin antibody, *N. Engl. J. Med.,* 294, 177–183, 1976.
13. **Pinto, J., Huang, Y. P., Chaudhuri, R., and Rivlin, R. S.,** Riboflavin excretion and turnover in an unusual case of multiple myeloma, *Clin. Res.,* 23, 426A, 1975.
14. **Rivlin, R. S., Menendez, C. E., and Langdon, R. G.,** Biochemical similarities between hypothyroidism and riboflavin deficiency, *Endocrinology,* 83, 461–469, 1968.
15. **Campbell, T. C. and Hayes, J. R.,** Role of nutrition in the drug-metabolizing enzyme system, *Pharmacol. Rev.,* 26, 171–197, 1974.
16. **Basu, T. K. and Dickerson, J. W. T.,** Inter-relationships of nutrition and the metabolism of drugs, *Chem.-Biol. Interactions,* 8, 193–206, 1974.
17. **Roe, D. A., McCormick, D. B., and Lin, R. T.,** Effects of riboflavin on boric acid toxicity, *J. Pharm. Sci.,* 61, 1081–1085, 1972.
18. **Roe, D. A.,** *Drug-Induced Nutritional Deficiencies,* AVI Press, Westport, Conn., 1976.
19. **Rivlin, R. S., Huang, Y. P., Pinto, J., and McConnell, R. J.,** Increased excretion of riboflavin due to boric acid intoxication, *Clin. Res.,* 24, 424A, 1976.

NUTRIENT TOXICITIES IN ANIMALS AND MAN: NIACIN

R. A. Waterman

INTRODUCTION

The use of niacin as a drug began shortly after its discovery as a vitamin. Because it could markedly reduce the diarrhea and mental disturbances of pellagra, niacin was administered in increasing doses for the treatment of sprue and schizophrenia. Based in part on its easily observable vasomotor effect, niacin has been used to treat various circulatory deficiencies such as hemicrania, claudicato intermittens, cerebral thrombosis, emboli, hypertension, and bronchial asthma. Niacin has also been administered for various anemias, porphyria, arthritis, and skin diseases. The ability of niacin to reduce hyperlipemia, associated with atherosclerotic problems, has led to much of the present-day investigation into the pharmacologic effects of nicotinic acid.

METABOLISM

Nicotinic acid is readily absorbed in pharmacologic doses when orally administered, as shown by the prompt rise in blood and urinary metabolites.[1] Blood transport of nicotinic acid is associated mainly with the red blood cells. Lee et al.[2] i.v. administered labeled nicotinic acid to a 71 kg man and noted only 337 dpm/ml plasma at 1.5 min. Approximately 50% (4960 dpm/ml) of the injected dose was present in the red cells at this time. Nicotinic acid rapidly leaves the blood stream and enters kidney, liver, and adipose tissues. Carlson and Nye[3] administered a single s.c. dose of nicotinic acid (250 mg/kg) to rats and noted a peak of free nicotinic acid at 15 min in plasma and liver, which declined to control levels by 6 hr. Adipose tissue concentrations peaked at 1 hr but had also declined to normal by 6 hr.

Basal excretory forms of nicotinic acid include nicotinamide, nicotinamide-N-oxide, N-methylnicotinamide, 2- and 4-pyridones, 6-hydroxynicotinamide, and 6-hydroxynicotinic acid with qualitative and quantitative species variations.[1] Herbivorous animals generally do not excrete methylated compounds.[4] For example, sheep urine does not contain N-methyl-2-pyridone-5-carboxyamide, prevalent in pig urine.[5] However, the rabbit can excrete appreciable amounts of the 2-pyridone, a transformation product of N-methylnicotinamide.[1] Metabolites normally present in trace amounts under basal conditions increase greatly with increasing nicotinic acid doses, assuming the significance of detoxification products. Pharmacologic doses of nicotinic acid rapidly increase the excretion of nicotinuric acid in most species,[5] as well as N-methylnicotinamide and the 2- and 4-pyridones in omnivorous animals.[1] Large dosages of nicotinic acid also lead to increased excretion of the unaltered nicotinic acid.[1,6]

DERIVATIVES

A number of derivatives of nicotinic acid have been prepared to reduce its undesirable side effects without influencing its pharmacologic properties. Subjection of polyalcohols to multiple esterification with nicotinic acid results in compounds with presumably longer duration of action while reducing adverse circulatory and gastrointestinal effects. Inositol hexanicotinate, for example, shows pharmacologic actions on blood vessels and lipids comparable to nicotinic acid,[7] without its undesirable effects.[8] A hexanicotinate prepared with myoinositol reduced serum cholesterol and the α/β lipoprotein ratio,[9] while increasing the walking ability of patients with arteriosclerosis obliterans.[10]

Other compounds displaying pharmacologic activities of nicotinic acid are derived from sorbitol, xylotol, D-glucose, and erythritol. Benzenehexol, inulin, riboflavin, pyridoxine, tocopherol, and unsaturated fatty acid derivatives have also been prepared.[11]

Substitutions in the pyridine nucleus have also been studied. 5-chloro-, 2-hydroxy- and 2-methylnicotinic acid inhibit fatty acid and cholesterol biosynthesis in rats to a greater extent than does pure nicotinic acid.[12] The potency of the nicotinic alcohol β-pyridylcarbinol is four times that of nicotinic acid,[13] although thin layer chromatographic analyses have shown that 80 to 90% of the resulting labeled circulatory material is nicotinic acid.[14] 3-Pyridine acetate shares the pharmacologic properties of β-carbinol but is not metabolized in the body. Rather, it is excreted unaltered with a urinary peak at 2 to 4 hr after ingestion.[15]

Although nicotinamide has no vasomotor or hypolipemic effects in man,[18,25] it can decrease serum cholesterol, triglycerides, and free fatty acids in rats, dogs, and rabbits.[16,17] Presumably, these species can effect conversion of nicotinamide to nicotinic acid in sufficient amounts to exert pharmacologic effects. Ricci and Pallini[19] found an increase in free nicotinic acid in rat liver after nicotinamide injection.

TOXICITY

Many early studies with nicotinic acid were designed to include a measure of its toxic effects in animals. Table 1 illustrates that the harmful effects of nicotinic acid occur at levels far in excess of vitamin doses. For example, short-term i.v. administration of niacin at a dose of 2.5 g/kg was needed before 50% of the test mice died.[20] Respective values for s.c. and oral routes of administration were 2.8 and 4.5 g/kg in mice. Nicotinamide is two to three times more toxic than the free acid.[20,22]

A relative tolerance of nicotinic acid in pharmacologic doses has also been reported for longer periods of administration. Nath et al.[23] fed nicotinic acid to rats for 3 weeks at 0.5, 1, and 2.5% of the diet (about 0.043, 0.09, and 0.21 g/day) and noticed no toxic reactions except for a slight inhibition of weight gain in the third group. They further noted that no effect was observed in liver lipids, contrasting to Handler and Dann,[24] who noted an increase in fatty acids from 4.2 to 11.8% in rats receiving 2% dietary nicotinic acid. Altschul[25] reported only a slight increase in total liver fat following nicotinic acid treatment in rats and rabbits and further noted that nicotinic acid inhibited to varying degrees the fatty liver development associated with cholesterol feeding in rabbits. Unna[22] fed ten 6-week-old rats 1 g/kg sodium nicotinate for 40 days and obtained normal growth rates. Gross and microscopic examinations of heart, lung, liver, spleen, kidney, intestinal tract, and genital organs revealed no pathological changes.

Three young dogs fed 1 g/kg sodium nicotinate for 63 days grew as well as three littermates and revealed no toxic reactions.[22] In contrast, Chen et al.[26] fed two dogs weighing 13.8 and 15 kg, respectively, nicotinic acid at 2 g/day. The smaller dog became sick on day 11 and died on day 19 after loosing 4.2 kg. The larger dog developed convulsions and bloody feces (symptoms noted in the first dog) and was sacrificed on day 20 after loosing 3.6 kg. Both dogs displayed a fatty metamorphosis of the liver, and the larger one displayed erosions in the stomach and petechiae in the mucosa of the colon. Norcia et al.[27] found no histological changes in liver biopsies from dogs fed 1 to 4 g/day nicotinic acid for 7 months.

Studies of the effects of nicotinic acid in high doses on conception and gestation are rare. Altschul[25] reported that rabbits treated with 0.3 g/day from before mating through lactation had no deformed living or dead offspring, and serum cholesterol in the surviving offspring was not altered at 90 days of age. Several cogeners of nicotinic acid are, however, used as experimental teratogens, including 6-aminonicotinamide and N-ethyl-nicotinamide.

Table 1
ANIMAL TOXICITY STUDIES

Species	Treatment	Notes	Ref.
Short-term, LD_{50} (g/kg/day)			
Young rats	5 s.c. sodium salt	Nicotinamide 3 times more toxic	21
Mice and rats	4–5 s.c. 5–7 oral	Nicotinamide 2 times more toxic	22
Mice	2.5 i.v. 2.8 s.c. 4.5 oral		20
Rabbits	2.5 i.v.		
Short-term, high doses (g/6 hr, oral)			
Goats	Up to 20	Random deaths in fed or fasted states	73
Cows	Up to 200	No adverse effects during lactation	6, 62
Long-term, high doses			
Rats	2% of diet	Fatty liver development	24
	2.5% of diet for 3 weeks 0.21 g/day	Slight growth inhibition, no fatty liver	23
Young rats	Sodium salt, 1 g/day for 40 days	Normal growth, no pathological changes	22
Young dogs	Sodium salt, 1 g/kg/day for 63 days	Normal growth	
Adult dogs	Sodium salt, 2 g/kg/day for 35 days	No adverse reactions	
Dogs	1–4 g/day for 3 months	No toxic reactions, liver biopsy normal	27
Dogs	2 g/day	Weight, 13.8 kg, died on day 19; 15 kg, sacrificied on day 20 in poor condition	26
Mice	500 mg/kg/day i.v. 5 times/week for 4 weeks	Weight gains	
Rabbits	0.3 g/day from before mating through lactation	No abnormal offspring	25

Note: s.c., subcutaneous and i.v., intravenous administration. LD_{50}, dosage at which 50% of experimental animals died.

SIDE EFFECTS IN MAN

Skin

An initial skin flush (Table 2) with elevated surface temperatures is usually noted in the upper half of the body upon initiation of nicotinic acid therapy. In most cases, the skin flush will decrease in intensity after a few days of treatment although chronic flushing may persist in up to 30% of the patients. In seven experiments,[28] an increase of the plasma nicotinic acid concentration to 0.1 to 0.2 μg/ml was associated with a marked flushing of the face and neck and an increased blood flow in the hand (skin) and forearm

Table 2
SIDE EFFECTS IN MAN

Subjects	Treatment (g/day)	Skin	Liver	Gastrointestinal	Carbohydrate	Other	Ref.
61 patients (7–66 years old)	3–6 for 1 year or more	Initial flush, all; prolonged flushing or pruritis, 8/61	At least one abnormal BSP, SGOT, AP, 23/52	Prolonged anorexia nausea, or vomit, 11/61	GTT abnormal, 14/27; FBS abnormal, 10/30	Uric acid abnormal, 16/26	38
48 patients (35–70 years old)	1–4 for 1–10 months	Initial flush, all; pruritis, 12/48		Minor upsets, 19/48; ulcer, 2/48		Nervous or panic reaction, 8/48	29
28 patients (30–72 years old)	3.3 for 1 year or more	Initial flush, all; pruritis 1/28	Abnormal BSP, 45%; abnormal enzymes, 2/15; jaundice, 2/28	Minor upsets, 1/28	GTT abnormal, 6/7; abnormal FBS, 5/10	Abnormal uric acid, 4/8	31
5 patients	4.5		Group average for enzymes abnormal		Group average for GTT abnormal		74
90 patients	3–6 for 6 months or more	Initial flush	Abnormal BSP, 32/90; abnormal SGOT, 11/17; abnormal AP, 9/68				37
36 patients	3–7.5 for 1 year or more	Initial flush	Abnormal BSP, 8/36; abnormal SGOT, 3/36; abnormal AP, 2/36			Abnormal urate, prior, 7/14 during, 9/14	46

Note: BSP, bromsulfalein retention; SGOT, serum glutamic oxylacetic transaminase; AP, alkaline phosphatase; GTT, glucose tolerance test; FBS, fasting blood sugar.

(muscle). The flushing was present only with increasing plasma nicotinic acid levels and disappeared when a constant level was attained. Cardiac output was also temporarily increased in association with the vasodilatory effect, presumably resulting in no decline of blood flow to other areas of the body. Itching and pruritis are not uncommon in patients receiving nicotinic acid therapy. Belle and Halpern[29] reported that among 48 patients, 35% discontinued treatment due to flushing and 25% suffered from pruritis. However, Rivin[30] stressed that none of 31 cases discontinued treatment for those reasons. Cases of hyperpigmentation in varying degrees[30,32] and acanthosis nigricans,[32,34] which disappear upon withdrawal of nicotinic acid, have been reported. Altschul[25] has noted that rabbits, rats, ground squirrels, and sheep do not display the skin reactions observed in man.

Gastrointestinal Tract

More serious are the effects on the gastrointestinal system noted in 20 to 40% of the patients. Transitory or chronic nausea, diarrhea, and vomiting are common, especially during establishment of the treatment regime. As was the case with skin problems, these symptoms spontaneously disappear upon withdrawal of nicotinic acid and, except in dogs, have not been observed in animal studies. Parsons[35] has reported nine ulcers in a series of 68 patients, while Belle and Halpern[29] have observed two cases out of 48 patients. In a detailed study of seven cases receiving 3 to 7.5 g/day nicotinic acid, duodenal ulcers occurred after 96 to 130 weeks of treatment.[36] All seven cases were prone to ulcer attacks: two had a previous ulcer history; one had displayed previous symptoms; two had other gastrointestinal symptoms; and two had suffered severe emotional trauma. Although all seven cases improved when the patients were switched from pure nicotinic acid to a buffered compound, it should be noted that most sustained release preparations appear to increase the incidence of upper gastrointestinal irritations.

Liver

Hepatic function may be disturbed by nicotinic acid therapy. However, evidence indicates that the alterations in hepatic function are readily reversible and do not represent permanent dysfunction or morphologic damage in nearly all cases. The most common abnormalities reported include increased sulfobromophthalein (BSP) retention and transaminase (SGOT, SGPT) activities. Parsons[37] has indicated that one or more of these clinical tests may become abnormal during nicotinic acid therapy and further believes that preexisting abnormal hepatic function tests do not contraindicate nicotinic acid treatment. Most common is an abnormal BSP retention, usually seen at least once in about 50% of the cases. Abnormal enzyme activities usually noted in decreasing incidence include SGOT, SGPT, lactate dehydrogenase, and alkaline phosphatase. Berge et al.[38] have reported that of 52 patients on nicotinic acid therapy for at least 1 year only 29 had no abnormal hepatic function tests.

Liver biopsies of patients on nicotinic acid therapy are usually normal based on histologic examination. Baggenstoss et al.[39] reported on eight patients who had received nicotinic acid for 22 months to 9 years and found normal or very minor alterations in both treated and untreated patients. However, in the treated patients, electron microscopy studies revealed dilation of the endoplasmic reticulum with formation of numerous vesicles and sacs and a diminuation of the parallel rays of the rough endoplasmic reticulum. The mitochondria varied greatly in size and shape but were present in normal numbers, and the Golgi apparatus appeared normal. The alterations were not interpreted as being severe enough to interfere seriously with cellular function or produce hepatic dysfunction.

Three cases of jaundice and one case of severe hepatic dysfunction during nicotinic acid therapy have been reported. Christensen et al.[31] reported that two cases of jaundice appeared during therapy with slow release preparations, and the question has been raised as to whether the jaundice was caused by nicotinic acid or by severe side reactions.[25] The third case of jaundice, as reported by Rivin,[40] occurred in a 23-year-old patient with familial hypercholesterolemia. After 14 months of treatment, the jaundice was noted and nicotinic acid therapy was discontinued. During the ensuing 20 days, bilirubin, alkaline phosphatase, and transaminase levels and urinalysis returned to normal without supplemental treatment. Pardue[41] reported one case of severe hepatic dysfunction in a 58-year-old man receiving 3 g nicotinic acid for 6 months. The patient manifested general edema, a reversed albumin/globulin ratio, abnormal SGOT and alkaline phosphatase activities, and abnormal prothrombin time. These symptoms cleared within 1 month after nicotinic acid withdrawal.

Carbohydrate Metabolism

Hyperglycemia and abnormal glucose tolerance tests have been reported in a high percentage of nondiabetic patients receiving nicotinic acid therapy. Berge and Molnar[42] observed in a study with six diabetic patients receiving 3 g/day for 5 to 6 months that all had glycemia, glycosuria, elevated free fatty acids, and an increased insulin requirement. The diabetogenic effect was readily reversible upon cessation of nicotinic acid treatment. Nikkila and Miettinen[43] noted that five hyperlipemic patients given 1.5 g/day of a slow release nicotinic acid preparation had an enhanced hyperglycemia from oral glycerol loading that was not apparent prior to nicotinic acid treatment. Gurian and Aldersburg[44] and Parsons and Flinn[33] have reported that their diabetic patients, in contrast to Berge and Molnar,[42] displayed no increased need for insulin.

Uric Acid

Berge[45] and Parsons and Flinn[33] report an increase in serum uric acid in their patients treated with nicotinic acid. Cases of hyperuricemia, independent of nicotinic acid treatment, are common in hypercholesterolemia, angina pectoris, and in elderly people. However, Parsons[46] noted a further increase in uric acid during nicotinic acid therapy. Fourteen patients had pretreatment uric acid levels of 6.3 mg/100 ml and 6.9 mg/100 ml during therapy. During nicotinic acid therapy, 25 patients averaged 7.6 mg/100 ml, and 4 weeks after cessation of nicotinic acid treatment they averaged 6.3 mg/100 ml. Uric acid excretion was within normal limits in 32 patients with hyperuricemia. Berge et al.[38] reported that 16 of 26 patients had hyperuricemia on at least one occasion during nicotinic acid therapy and that two persons experienced single episodes of acute gouty arthritis. This is in contrast to Parsons[37] who reported that two patients with previous cases of gout did not display any adverse effects after 1 and 3 years of nicotinic acid therapy.

LIPID METABOLISM

Cholesterol

Experimental evidence indicates that hypercholesterolemia is at least one important factor in the development of arteriosclerosis, and many regimes of diet and medication have been designed to reduce elevated levels of blood cholesterol. Altschul observed that ultraviolet radiation[47] and oxygen inhalation[48] had a hypocholesterolemic effect in man. Based on the hypothesis that increased levels of a respiratory coenzyme (NAD) might influence cholesterol metabolism by intensified oxidation to "oxycholesterol", Altschul et al.[18] administered nicotinic acid to man. In the first experiment, 11 normal men received four 1-g doses, and an 8.4% drop in serum cholesterol was observed during the first 24 hr. Nicotinamide had no effect on serum cholesterol.

Nicotinic acid effects of cholesterol metabolism vary with dosage, duration, and species. In rats receiving 1% dietary nicotinic acid, Hardy et al.[49] observed a 47% reduction in serum cholesterol on day 2, cholesterol levels 120% of controls on day 14, and no difference between control and treated rats on day 28. Rats receiving 0.1% dietary nicotinic acid displayed a similar pattern with comparably smaller responses. In eight hypercholesterolemic patients maintained on a low cholesterol diet, cholesterol synthesis may have been enhanced by chronic nicotinic acid therapy.[50] After a single i.v. injection of 150 mg nicotinic acid, however, cholesterol synthesis remained unchanged or reduced in three patients. Guinea pigs are similar to rats with respect to the influence of nicotinic acid on cholesterol. Chickens and dogs also show diversified results after nicotinic acid treatment depending on experimental conditions and diets. Rabbits are more uniform in that nicotinic acid retards development of atheromata and lowers serum cholesterol levels, especially under experimentally induced hypercholesterolemic conditions.

Reductions of serum cholesterol (Table 3) by nicotinic acid may be accomplished by an inhibition of cholesterol synthesis, alteration of cholesterol absorption and excretion, or by a redistribution between tissues and plasma. In a cholesterol-free diet, 1% dietary nicotinic acid had no effect on serum cholesterol in rats fed for 3 weeks, although hepatic cholesterolgenesis was reduced by more than 50%.[51] Parsons[52] and Nunn et al.[53] have reported that nicotinic acid reduced conversion of acetate to serum cholesterol in a high percentage of cases. The block appears to be at an early stage of synthesis, probably between β-hydroxymethylglutarate and mevalonate.[54] Duncan and Best[55] found that the absorption of labeled cholesterol was not significantly different between control and nicotinic acid-fed rats. They also noted that the radioactivity remaining in serum and liver after 2 days was not significantly different.

Most studies indicate that nicotinic acid leads to alterations in cholesterol synthesis and especially excretion, but not in bile acid metabolism. Wollenweber et al.[56] found no significant alteration in the combined metabolism of cholic and chenodeoxycholic acids but did note a drop in the glycholic/taurocholic ratio from 4.6 to 3.9. Miettinen[57] has demonstrated that the nicotinic acid induced hypocholesterolemia is initially associated with a significant increase in fecal excretion of neutral steroids of endogenous origin. Increased cholesterol mobilization is also suggested by Berge et al.[38] and Parsons,[37] who observed reductions of xanthomas during nicotinic acid therapy. Carlson and Walldius[58] observed a 28% reduction in skeletal muscle triglycerides but no change in phospholipids after 2 to 4 months of nicotinic acid therapy in 15 patients.

Free Fatty Acids

The case for nicotinic acid causing general alterations in lipid metabolism received renewed support from Carlson and co-workers in the early 1960s. At that time, the importance of free fatty acids as precursors for plasma triglyceride synthesis had just been established. These workers attempted to treat hyperlipemia by lowering free fatty acid influx into the liver and used nicotinic acid as one test compound. Three subjects ingested 200 mg nicotinic acid, and a prompt decline in arterial free fatty acids was observed.[59] This effect lasted nearly 60 min and was followed by a pronounced overshoot or rebound in free fatty acid levels. The depression of plasma free fatty acids was subsequently interpreted as due to inhibition of adipose tissue triglyceride lipolysis since norepinephrine stimulated free fatty acid mobilization could be blocked by nicotinic acid.[50]

The rebound effect in dogs appeared to be dose dependent, as between 1 and 3.2 mg/kg i.v. nicotinic acid increasingly intensified the depression. Levels between 3.2 and 32 mg/kg increasingly prolonged the depression and intensified the rebound effect.[61] Pereira[61] further noted that adrenalectomized rats receiving 10 mg/kg i.v. nicotinic acid had a slightly delayed rebound time; hypophysectomized rats had a markedly delayed and less severe rebound; and adrenohypophysectomized rats had a very prolonged

Table 3
LIPID METABOLISM

Subjects	Treatment (g/day)	Cholesterol (mg/100 ml)[a]		Triglycerides (mmol/l)[a]		Others (mg%)		Ref.
		Control	Final	Control	Final	Control	Final	
8 patients (32–76 years old)	3–6, for 2–9 years; hyperlipemic	345	245			Total fatty acids		39
86 patients	3 for 1 month hyperlipemia:					580	424	58
	17 type IIA	377	299	1.63	1.20			
	26 type IIB	336	270	2.93	1.76			
	4 type III	574	286	5.26	3.62			
	39 type IV	258	232	2.71	1.92			
5 patients (31–56 years old)	3 for 10 days, low cholesterol diet, hypercholesterolemia	440	316	182 mg%	82 mg%	Sterol balance (mg/day) −599	−119	75
51 patients	3.9 for 9 months, hypercholesterolemia	327	250			Phospholipid 290	249	38
10 patients	4.5 g/day for 1–2 months, hyperlipoproteinemia	308	215	8.29	2.69	Total fatty acids 523 i.v. Fat tolerance 2.6%/min	421 4.2%/min	72
5 patients (9–52 years old)	1–3 for 3–12 months, hyperlipemia	358	213	615 mg%	431 mg%	Phospholipid 410	279	76
31 patients	3 for 3 weeks, hypercholesterolemia	330	24% Decrease					77
14 normal patients		?	9% Decrease					

[a] Unless otherwise noted.

depression with no apparent rebound. Thyroidectomized rats had a normal depression and rebound curve when compared to control rats.

The turnover of plasma free fatty acids is usually slowed during the depression period in man, sheep, and goats (Table 4). During chronic nicotinic acid therapy, basal free fatty acid levels in plasma return to normal,[59a] although the ability to inhibit the actions of norepinephrine remains active.[60]

Ketosis is supported by an accelerated mobilization of adipose tissue fatty acids and should thus be susceptible to nicotinic acid treatment. Schultz and co-workers[62,63] have experimentally treated bovine ketosis with four 40 g/2 hr oral doses of nicotinic acid. Their confirming study[64] reported a prompt decline in plasma free fatty acids and ketone bodies, followed by a marked overshoot at 36 to 60 hr. By days 14 to 21, ketone bodies and free fatty acids had returned to normal in both clinically and subclinically ketotic cows. Carlson et al.[65] measured the splanchnic β-hydroyxbutyrate production in seven patients and found it to be suppressed in a parallel fashion to free fatty acid depression after nicotinic acid administration. Thus, the antiketogenic effect of nicotinic acid appears to be primarily indirect via inhibited fatty acid availability, but also may be direct. Rats fed olive oil and then injected with heparin displayed a 40% reduction in plasma ketone bodies but only a 15% reduction in plasma free fatty acids.[66] Normally, β-hydroxybutyrate is the major circulating ketone body but the proportions of acetoacetate are usually elevated during nicotinic acid therapy.[66] An increase in hepatic NAD+ could result in increased tricarboxylic acid cycle activity with a concomitant decrease in hepatic ketogenesis.

The rate of adipose tissue-triglyceride lipolysis is controlled by intracellular levels of cyclic AMP under a variety of experimental conditions.[67] Since insulin and prostaglandins decrease adipose tissue-cyclic AMP levels and lipolytic rates in a parallel fashion, Butcher et al.[68] reasoned that the action of nicotinic acid also could involve depressed cyclic AMP levels. They reported that nicotinic acid at an in vitro concentration of about 10^{-7}M could lower cyclic AMP levels in isolated fat cells previously treated with caffeine and epinephrine, ACTH, or glucagon. Chmelar and Chmelarova[69] noted that, although nicotinic acid concentrations higher than 10^{-5}M inhibited adipose tissue hormone-sensitive lipase activity, nicotinic acid at 10^{-7} or $10^{-6}M$ significantly activated this enzyme. This is not to be confused with the overshoot phenomena which remains very difficult to demonstrate in vitro. Carlson and Micheli[70] have reported that 1 μg/ml PGE$_1$ markedly stimulated in vitro lipolysis in adipose tissues from rats pretreated with nicotinic acid, while tissues from rats pretreated with saline failed to show this response. Whether nicotinic acid acts on adenylcyclase or phosphodiesterase activities remains unclear.

Hormone-sensitive lipase activity in rat liver is enhanced by all in vitro concentrations of nicotinic acid and by 10 mg/kg in vivo in rats.[69] This is in agreement with the observation that, in rats receiving 0.5 mg/min i.v. nicotinic acid, the fraction of free fatty acids recirculating as triglycerides was decreased both in vitro using isolated perfused liver and in vivo.[71] In man, postheparin lipoprotein lipase activity was not affected by nicotinic acid treatment,[72] although α/β lipoprotein ratios are significantly altered,[25] and improved i.v. fat tolerances are often observed during chronic nicotinic acid therapy.[46]

Three actions of nicotinic acid administration are generally agreed on: the reductions in plasma fatty acids and cholesterol and the transitory vasomotor effects in man. These observations and their possible implications in the prevention of arteriosclerosis have stimulated much of the study in this area. Species differences, tissue differences, acute vs. chronic actions, and a clearly defined mechanism of action remain to be defined.

Table 4
FREE FATTY ACID METABOLISM

Subjects/treatment		Plasma FFA	FFA turnover	Plasma ketones	Plasma nicotinate	Ref.
6 normal men, 250 µg/kg/min i.v. for 2 hr	Ca	0.33 meq/l	4.07 µeq/kg/min			78
6 control men		0.71 meq/l	7.31 µeq/kg/min			
6 sheep, 500 mg i.v. sodium salt	C	0.62 meq/l	804 µmol/min			79
	D	0.08 meq/l	199 µmol/min		8.4 mg/l	
	R	1.41 meq/l	1510 µmol/min		0.7 mg/l	
5 humans, 500 mg oral	C	0.64 meq/l	549 µmol/min			60
	D	0.19 meq/l	151 µmol/min		7.6 mg/l	
	R	0.24 meq/l	763 µmol/min		1.1 mg/l	
9 humans, 1 g oral	D	0.26 mmol/l			25 mg/l	
	R	1.70 mmol/l			<1.0 mg/l	
2 goats, 4 × 1 g/2 hr oral, 48 hr fast, lactating	D	17 mg/l	0.74 mg/kg/min	12 mg/l		80
	R	683 mg/l	10.14 mg/kg/min	166 mg/l		
6 rats, 10 mg/kg i.v., fasted	C	0.5 meq/l				61
	D	0.15 meq/l				
	R	0.85 meq/l				
4 dogs, 10 mg/kg i.v., fasted	C	1.0 meq/l				
	D	0.25 meq/l				
	R	1.7 meq/l				
9 rats, 48 hr fasted, 1 ml olive oil at 48 hr, 50 mg/kg i.p. nicotinate at 52 hr, sampled at 53 hr	C	1.05 meq/l		2.52 mmol/l		66
	T	0.88 meq/l		1.32 mmol/l		
7 humans, 4 × 200 mg/20 min i.v., overnight fast			splanchnic FFA uptake	β-hydroxybutyrate production		65
	C		0.20 mmol/min	0.20 mmol/min		
	T		0.04 mmol/min	0.01 mmol/min		

a C, control group; D, depression or period of minimum plasma FFA; R, rebound or period of maximum plasma FFA; T, treated group.

REFERENCES

1. **Fumagalli, R.**, Pharmacokinetics of nicotinic acid and some of its derivatives, in *Metabolic Effects of Nicotinic Acid and Its Derivatives,* Gey, K. F. and Carlson, L. A., Eds., Hans Huber, Bern, Switzerland, 1971, 33–49.

2. **Lee, K. W., Abelson, D. M., and Kwon, Y. O.,** Nicotinic acid-6-^{14}C metabolism in man, *Am. J. Clin. Nutr.,* 21, 223–225, 1968.

3. **Carlson, L. A. and Nye, E. R.,** Acute effects of nicotinic acid in the rat. I. Plasma and liver lipids and blood glucose, *Acta Med. Scand.,* 179, 453–461, 1966.

4. **Perlzweig, W. A., Rosen, F., and Pearson, P. B.,** Comparative studies in niacin metabolism: The fate of niacin in man, rat, dog, pig, rabbit, guinea pig, goat, sheep, and calf, *J. Nutr.,* 40, 453–469, 1950.

5. **Chang, W. M. L. and Johnson, B. C.,** Metabolism of C[14]-nicotinic acid in pigs and sheep, *J. Nutr.,* 76, 512–514, 1962.

6. **Waterman, R. and Schultz, L. H.,** Nicotinic acid loading of normal cows: Effects on blood metabolites and excretory forms, *J. Dairy Sci.,* 55, 1511–1513, 1972.

7. **Miller, O. N. and Hamilton, J. G.,** Nicotinic acid and derivatives, in *Lipid Pharmacology,* Paoletti, R., Ed., Academic Press, New York, 1964, 275–298.

8. **Sommer, H.,** Nicotinic acid levels in the blood and fibrinolysis under the influence of the hexa nicotinic acid ester of meso-inositol, *Arzneim. Forsch.,* 15, 1337, 1964.

9. **Guarneri, B.,** Esperienze terapeutiche e risultati di laboratorio in psoriasici trattati con prodotti ad azione normolipemica ed eudermica, *Minerva Med.,* 62, 691–694, 1971.

10. **Lumish, S. H., Blyn, C., and Nodine, J. H.,** Inositol nicotinate as a peripheral vasodilator, *Cur. Ther. Res. Clin. Exp.,* 4, 243–248, 1962.

11. **Witte, E. C.,** Antihyperlipidaemic agents, in *Progress in Medicinal Chemistry,* Vol. 11, Ellis, G. P. and West, G. B., Eds., North-Holland, Amsterdam, 1975, 119–191.

12. **Miller, O. N., Gutierrez, M., Sullivan, A., and Hamilton, J. G.,** The effect of nicotinic acid and ring-substituted analogues on the *in vivo* biosynthesis of cholesterol and fatty acids in rat liver, in *Metabolic Effects of Nicotinic Acid and Its Derivatives,* Gey, K. F. and Carlson, L. A., Eds., Hans Huber, Bern, Switzerland, 1971, 609–620.

13. **Zeller, W.,** A long term study of the effect of nicotinic acid and Ronical retard on serum lipids and fatty liver, in *Progress in Biochemical Pharmacology,* Vol. 2, Kritchevsky, D., Paoletti, R., and Steinberg, D., Eds., S. Karger, Basel, 1967, 401–475.

14. **Raaflaub, J.,** Zur pharmakokinetic von Rocinal retard, *Med. Pharmacol. Exp.,* 16, 393–398, 1967.

15. **Ginoulhiac, E., Tenconi, L. T., and Chiancone, F. M.,** 3-Pyridine acetic acid and nicotinic acid: Blood levels, urinary elimination and excretion of nicotinic acid derivatives in man, *Nature,* 193, 948–949, 1962.

16. **Dalton, C.,** Antilipaemic effect of nicotinamide, *Nature,* 216, 825, 1967.

17. **Fontenot, R., Redetski, H., Deupree, R.,** Effects of nicotinic acid and nicotinamide on serum cholesteral and erythrocyte nicotinamide adenine dinucleotide levels of rabbits, *Proc. Soc. Exp. Biol. Med.,* 119, 1053–1055, 1965.

18. **Altschul, R., Hoffer, A., and Stephen, J. D.,** Influence of nicotinic acid on serum cholesterol in man, *Arch. Biochem.,* 54, 558–559, 1955.

19. **Ricci, C. and Pallini, V.,** Occurrence of free nicotinic acid in the liver of nicotinamide-injected rats, *Biochem. Biophys. Res. Commun.,* 17, 34–38, 1964.

20. **Fromherz, K. and Spielgelberg, H.,** Pharmakologische wirkungen des β-pyridylcarbinols unds verwandter β-pyridylverbindungen, *Helv. Physiol. Pharmacol. Acta,* 6, 42–54, 1948.

21. **Brazda, F. G. and Coulson, R. A.,** Toxicity of nicotinic acid and some of its derivatives, *Proc. Soc. Exp. Biol. Med.,* 62, 19–20, 1946.

22. **Unna, K.,** Studies on the toxicity and pharmacology of nicotinic acid, *J. Pharmacol. Exp. Ther.,* 65, 95–103, 1939.

23. **Nath, N., Harper, A. E., and Elvehjem, C. A.,** Diet and cholesterolemia. Four effects of carbohydrate and nicotinic acid, *Proc. Soc. Exp. Biol. Med.,* 102, 571–574, 1959.

24. **Handler, P. and Dann, W. J.,** The inhibition of rat growth by nicotinamide, *J. Biol. Chem.,* 146, 357–368, 1942.

25. **Altschul, R.,** Influence of nicotinic acid (niacin) on hypercholesterolemia and hyperlipemia and on the course of atherosclerosis, in *Niacin in Vascular Disorders and Hyperlipemia,* Altschul, R., Ed., Charles C Thomas, Springfield, Ill, 1964, 3–135.

26. **Chen, K. K., Rose, C. L., and Robbins, E. B.,** Toxicity of nicotinic acid, *Proc. Soc. Exp. Biol. Med.,* 38, 241–245, 1938.

27.   Norcia, L. N., Brown, H. J., and Furman, R. H., Non-hypocholesterolemic action of nicotinic acid in dogs, *Lancet*, 1, 1255, 1959.
28.   Svedmyr, N., Harthon, L., and Lundholm, L., Dose response relationship between concentration of free nicotinic acid concentration of plasma and some metabolic and circulatory effects after administration of nicotinic acid and pentoerythritol tetranicotinate in man, in *Metabolic Effects of Nicotinic Acid and Its Derivatives*, Gey, K. F. and Carlson, L. A., Eds., Hans Huber, Bern, Switzerland, 1971, 1085–1098.
29.   Belle, M. and Halpern, M. M., Oral nicotinic acid for hyperlipemia, *Am. J. Cardiol.*, 2, 449–452, 1958.
30.   Rivin, A. U., Hypercholesterolemia – use of niacin and niacin combinations in therapy, *Calif. Med.*, 96, 267–269, 1962.
31.   Christensen, N. A., Achor, R. W. P., Berge, K. G., and Mason, H. L., Nicotinic acid treatment of hypercholesterolemia, *JAMA*, 77, 546–550, 1961.
32.   Wittenborne, J. R., Weber, E. S. P., and Brown, M., Niacin in the long-term treatment of schizophrenia, *Arch. Gen. Psychiatry*, 28, 308–315, 1973.
33.   Parsons, W. B. and Flinn, J. H., Reduction of serum cholesterol levels and beta-lipoprotein cholesterol levels by nicotinic acid, *Arch. Intern. Med.*, 103, 783–790, 1959.
34.   Tromovitch, T. A., Jacobs, P. H., and Kern, S., Acanthosis nigricans-like lesions from nicotinic acid, *Arch. Dermatol.*, 89, 222–223, 1964.
35.   Parsons, W. B., Use of nicotinic acid to reduce serum lipid levels, *J. Am. Geriatr. Soc.*, 10, 850–864, 1962.
36.   Parsons, W. B., Activation of peptic ulcer by nicotinic acid, *JAMA*, 173, 1466–1470, 1960.
37.   Parsons, W. B., The effect of nicotinic acid on the liver: evidence favoring functional alteration of enzymatic reactions without hepatocellular damage, in *Niacin in Vascular Disorders and Hyperlipemia*, Altschul, R., Ed., Charles C Thomas, Springfield, Ill., 1964, 263–286.
38.   Berge, K. G., Achor, R. W. P., Christensen, N. A., Mason, H. L., and Barker, N. W., Hypercholesterolemia and nicotinic acid: A long-term study, *Am. J. Med.*, 31, 24–36, 1961.
39.   Baggenstoss, A. H., Christensen, N. A., Berge, K. B., Baldus, W. P., Spiekerman, R. E., and Ellefson, R. D., Fine structural changes in the liver in hypercholesterolemic patients receiving long-term nicotinic acid therapy, *Mayo Clin. Proc.*, 42, 385–399, 1967.
40.   Rivin, A. U., Jaundice occurring during nicotinic acid therapy for hypercholesterolemia, *JAMA*, 170, 2088–2089, 1959.
41.   Pardue, W. P., Severe liver dysfunction during nicotinic acid therapy, *JAMA*, 175, 137–138, 1961.
42.   Berge, K. G. and Molnar, G. D., Effects of nicotinic acid on clinical aspects of carbohydrate metabolism, in *Niacin in Vascular Disorders and Hyperlipemia*, Altschul, R., Ed., Charles C Thomas, Springfield, Ill., 1964, 136–155.
43.   Nikkila, E. A. and Miettinen, T. A., Nicotinic acid and gluconeogenesis in man, in *Metabolic Effects of Nicotinic Acid and Its Derivatives*, Gey, K. F. and Carlson, L. A., Eds., Hans Huber, Bern, Switzerland, 1971, 753–756.
44.   Gurian, H. and Aldersberg, D., The effect of large doses of nicotinic acid on circulating lipids and carbohydrate tolerance, *Am. J. Med. Sci.*, 237, 12–22, 1959.
45.   Berge, K. G., Side effects of nicotinic acid in the treatment of hypercholesterolemia, *Geriatrics*, 16, 416–422, 1961.
46.   Parsons, W. B., Studies on nicotinic acid use in hypercholesterolemia, *Arch. Intern. Med.*, 107, 653–667, 1961.
47.   Altschul, R., Lowering of serum cholesterol by ultraviolet irradiation, *Geriatrics*, 10, 208–214, 1955.
48.   Altschul, R. and Herman, I. H., Influence of oxygen inhalation on cholesterol metabolism, *Arch. Biochem. Biophys.*, 51, 308–309, 1954.
49.   Hardy, R. W. F., Gaylor, J. L., and Bauman, C. A., Biosynthesis of sterols and fatty acids as affected by nicotinic acid and related compounds, *J. Nutr.*, 71, 159–167, 1960.
50.   Miettinen, T. A., Influence of nicotinic acid on cholesterol synthesis in man, in *Metabolic Effects of Nicotinic Acid and Its Derivatives*, Gey, K. F. and Carlson, L. A., Eds., Hans Huber, Bern, Switzerland, 1971, 649–658.
51.   Kritchevsky, D. and Tepper, S. A., Influence of nicotinic acid, picolinic and pyridine-3-sulfonic acids on cholesterol metabolism in the rat, *J. Nutr.*, 82, 157–161, 1964.
52.   Parsons, W. B., Reduction in heptic synthesis of cholesterol from C^{14}-acetate in hypercholesterolemia patients by nicotinic acid, *Circulation*, 24, 1099, 1961.
53.   Nunn, S. L., Tauxe, W. N., and Juergens, J. L., Effect of nicotinic acid on human cholesterol biosynthesis, *Circulation*, 24, 1099, 1961.

54. **Gamble, W. and Wright, L. D.,** Effect of nicotinic acid and related compounds on incorporation of mevalonic acid into cholesterol, *Proc. Soc. Exp. Biol. Med.,* 107, 160–162, 1961.

55. **Duncan, C. H. and Best, M. M.,** Lack of nicotinic acid on pool size and turnover of taurocholic acid in normal and hypothyroid dogs, *J. Lipid Res.,* 1, 159–166, 1960.

56. **Wollenweber, J., Kottke, B. A., and Owen, C. A.,** The effect of nicotinic acid on pool sized and turnover of taurocholic acid normal and hypothyroid dogs, *Proc. Soc. Exp. Biol. Med.,* 122, 1070–1075, 1966.

57. **Miettinen, T. A.,** Effect of nicotinic acid on the fecal excretion of neutral sterols and bile acids, in *Metabolic Effects of Nicotinic Acid and Its Derivatives,* Gey, K. F. and Carlson, L. A., Eds., Hans Huber Publishers, Bern, Switzerland, 1971, 677–686.

58. **Carlson, L. A. and Walldius, G.,** Serum and tissue lipid metabolism and effect of nicotinic acid in different types of hyperlipidemia, *Adv. Exp. Med. Biol.,* 26, 165–178, 1972.

59. **Carlson, L. A. and Oro, L.,** The effect of nicotinic acid on the plasma free fatty acids. Demonstration of a metabolic type of sympathicolysis, *Acta Med. Scand.,* 172, 641–645, 1962.

59a. **Froberg, S. O., Boberg, J., Carlson, L. A., and Eriksson, M.,** Effect of nicotinic acid on the diurnal variation of plasma levels of glucose, free fatty acids, triglycerides, and cholesterol and of urinary excretion of catecholamines, in *Metabolic Effects of Nicotinic Acid and Its Derivatives,* Gey, K. F. and Carlson, L. A., Eds., Hans Huber, Bern, Switzerland, 1971, 167–181.

60. **Carlson, L. A.,** Nicotinic acid: Its metabolism and its effects on plasma free fatty acids, in *Metabolic Effects of Nicotinic Acid and Its Derivatives,* Gey, K. F. and Carlson, L. A., Eds., Hans Huber, Bern, Switzerland, 1971, 157–166.

61. **Pereira, J. N.,** The plasma free fatty acid rebound induced by nicotinic acid, *J. Lipid Res.,* 8, 238–244, 1967.

62. **Waterman, R., Schwalm, J. W., and Schultz, L. H.,** Nicotinic acid treatment of bovine ketosis. I. Effects on circulatory metabolites and interrelationships, *J. Dairy Sci.,* 55, 1447–1453, 1972.

63. **Waterman, R. and Schultz, L. H.,** Nicotinic acid treatment of bovine ketosis. II. Effects on long-chain fatty acid compositions of plasma lipid fractions, *J. Dairy Sci.,* 55, 1454–1460, 1972.

64. **Nurmio, P., Roine, K., Juokslahti, T., and Loman, A.,** A study of the effects of nicotinic acid in cattle, with special reference to ketosis therapy, *Nord. Veterinaermed.,* 26, 370–381, 1974.

65. **Carlson, L. A., Freyschuss, U., Kjellberg, J., and Ostman, J.,** Suppression of splanchinic ketone body production in man by nicotinic acid, *Diabetologia,* 3, 494–499, 1967.

66. **Williamson, D. H., Mayor, F., Veloso, D., and Page, M. A.,** Effects of nicotinic acid and related compounds on ketone body metabolism, in *Metabolic Effects of Nicotinic Acid and Its Derivatives,* Gey, K. F. and Carlson, L. A., Eds., Hans Huber, Bern, Switzerland, 1971, 227–236.

67. **Robison, G. A., Butcher, R. W., and Sutherland, E. W.,** Cyclic AMP, *Annu. Rev. Biochem.,* 37, 149–174, 1968.

68. **Butcher, R. W.,** Effects of nicotinic acid or cyclic AMP Levels in rat adipose tissue, in *Metabolic Effects of Nicotinic Acid and Its Derivatives,* Gey, K. F. and Carlson, L. A., Eds., Hans Huber, Bern, Switzerland, 1971, 347–355.

69. **Chmelar, M. and Chmelarova, M.,** Interactions of nicotinic acid with hormone-sensitive lipases of different mammalian tissue, in *Metabolic Effects of Nicotinic Acid and Its Derivatives,* Gey, K. F. and Carlson, L. A., Eds., Hans Huber, Bern, Switzerland, 1971, 236–282.

70. **Carlson, L. A. and Micheli, H.,** Stimulatory effect of prostaglandin E$_1$ on fat mobilizing lipolysis in adipose tissue of rats treated with nicotinic acid, in *Metabolic Effects of Nicotinic Acid and Its Derivatives,* Gey, K. F. and Carlson, L. A., Eds., Hans Huber, Bern, Switzerland, 1971, 995–1001.

71. **Nikkila, E. A.,** Effect of nicotinic acid on hepatic lipogenesis and triglyceride synthesis and release, in *Metabolic Effects of Nicotinic Acid and Its Derivatives,* Gey, K. F. and Carlson, L. A., Eds., Hans Huber, Bern, Switzerland, 1971, 471–477.

72. **Boberg, J., Carlson, L. A., Forberg, S., Olsson, A., Oro, L., and Rossner, S.,** Effects of chronic treatment with nicotinic acid on intravenous fat tolerance and post-heparin lipoprotein lipase activity in man, in *Metabolic Effects of Nicotinic Acid and Its Derivatives,* Gey, K. F. and Carlson, L. A., Eds., Hans Huber, Bern, Switzerland, 1971, 465–470.

73. **Gill, R. A.,** Studies on Lipid Metabolism in Ruminants, Ph.D. thesis, University of Wisconsin, Madison, 1970.

74. **Gant, Z. N., Solomon, H. M., and Miller, O. N.,** The influence of antilipemic doses of nicotinic acid on carbohydrate tolerance and plasma insulin levels in man, in *Metabolic Effects of Nicotinic Acid and Its Derivatives,* Gey, K. F. and Carlson, L. A., Eds., Hans Huber, Bern, Switzerland, 1971, 923–927.

75. **Miettinen, T. A.,** Effect of nicotinic acid on catabolism and synthesis of cholesterol in man, *Clin. Chim. Acta,* 20, 43–51, 1968.
76. **Feldman, E. B.,** Nicotinic acid in the management of frank disorders of lipid metabolism, in *Niacin in Vascular Disorders and Hyperlipemia,* Altschul, R., Ed., Charles C Thomas, Springfield, Ill., 1964, 208–224.
77. **Iliescu, C. C., Iliescu, M., Roman, L., Jacobini, P., Constantinescu, S., and Nutu, S.,** Effect of nicotinic acid on blood lipids in atherosclerosis, *Med. Interna,* (Bucur.), 15, 39–43, 1963.
78. **Sailer, S. and Bolzano, K.,** The action of nicotinic acid on the esterification rate of plasma free fatty acids to plasma triglycerides, in *Metabolic Effects of Nicotinic Acid and Its Derivatives,* Gey, K. F. and Carlson, L. A., Eds., Hans Huber, Bern, Switzerland, 1971, 479–480.
79. **Nye, E. R. and Buchanan, H.,** Short term effect of nicotinic acid on plasma level and turnover of free fatty acids in sheep and man, *J. Lipid Res.,* 10, 193–196, 1969.
80. **Waterman, R. and Schultz, L. H.,** Carbon [14]-labeled palmitic acid metabolism in fasted, lactating goats following nicotinic acid administration, *J. Dairy Sci.,* 56, 1569–1574, 1973.

TOXICITY OF VITAMIN B$_6$

B. E. Haskell

The toxicity of vitamin B$_6$ is low. Humans are reported to tolerate daily doses of 20 to 1000 mg per day of pyridoxine·HCl for prolonged periods without deleterious effects.[1-8] These doses are 10 to 500 times greater than the recommended daily dietary allowance of vitamin B$_6$ for adults.[9] Side effects are usually absent; however, one case of anaphylactic shock attributed to intramuscular injection of vitamin B$_6$ has been described.[10]

The possibility that excessively high doses of vitamin B$_6$ during pregnancy might have adverse effects on the fetus has been investigated. No teratogenic effects were associated with the use during human pregnancy of an antinauseant drug supplying 10 mg pyridoxine·HCl per tablet (daily dosage not specified).[11] When pregnant rats were fed 0, 20, 40, 60, or 80 mg pyridoxine·HCl per kilogram body weight per day on days 6 through 15 of gestation, the incidence of fetal anomalies was no greater than in controls.[12]

Feeding rats excess vitamin B$_6$ (6.25 or 7.5 mg pyridoxine·HCl per 100 g diet) during pregnancy did not increase the vitamin B$_6$ requirement of the pups.[13,14] Nor did it adversely affect birth weight, litter size, or survival to weaning.[13,14]

Acute toxicity of vitamin B$_6$ has been investigated in the rat, mouse, rabbit, dog, cat, and pigeon. For LD$_{50}$ values, see Table 1. Daily intraperitoneal injection of 1 g per kilogram body weight of pyridoxine·HCl impairs heat perception in rats after 8 to 18 days.[15] Incoordination and convulsions occur in rats after a massive dose of pyridoxine·HCl (4 to 6 g per kilogram body weight).[16] Degenerative changes in the posterior columns of the spinal cord are observed at autopsy in dogs and rats treated with 2.5 to 3 g pyridoxine·HCl per kilogram body weight.[17] In the cat, intracisternal injection of pyridoxine·HCl (10 mg) is reported to have a sedative effect; 3 mg pyridoxal·HCl or 0.8 to 5.0 mg pyridoxal-5'-phosphate·H$_2$O administered intracisternally produces convulsive seizures.[18]

Prolonged feeding of pyridoxine·HCl to rats (2.5 mg per day), puppies (20 mg per kilogram body weight per day), and monkeys (10 mg per kilogram body weight per day) produced no toxic symptoms.[19] In mice, repeated daily intravenous injections of 100 mg pyridoxine·HCl per kilogram body weight produced no pathological changes detectable at autopsy.[20]

Table 1
TOXICITY OF VITAMIN B_6 AND ITS METABOLITES[a]

| | LD$_{50}$ (mg/kg body weight) | | | | | |
	Mouse	Rat	Cat	Rabbit	Dog	Pig
Pyridoxine·HCl						
Oral	5500	4000–6000[19]	<1000	–	>500	
Subcutaneous	2450	3000–3700[19]	–	–	–	
Intravenous	545[20]–600	530–675[20]	560	464	–	1.
Intramuscular	–	–	500	–	–	
Pyridoxal·HCl						
Oral	1800	2150	>500	–	>500	
Subcutaneous	530	530	–	–	–	
Intravenous	390	320	160	465	–	2(
Intramuscular	–	–	152	–	–	
Pyridoxamine·2HCl						
Oral	5100	7500	>1000	–	–	
Subcutaneous	2950	4720	–	–	–	
Intravenous	590	800	540	700	–	7!
Intramuscular	–	–	–	–	–	
4-Pyridoxic acid						
Oral	2900	7500	–	–	–	
Sucutaneous	2500	5400	–	–	–	
Intravenous	500	2050	–	–	–	
Intramuscular	–	–	–	–	–	
Pyridoxal-5'-phosphate·H$_2$O						
Oral	4640	5900	–	–	–	
Subcutaneous	870	850	–	–	–	
Intravenous	–	–	–	–	–	
Intramuscular	–	–	250	–	–	
Pyridoxamine-5'-phosphate·HCl						
Oral	5800	>7500	–	–	–	
Subcutaneous	3700	6300	–	–	–	
Intravenous	500	610	215	–	–	69
Intramuscular	–	–	–	–	–	

[a] Unless indicated otherwise, data are from Reference 18.

REFERENCES

1. Committee on Nutrition, American Academy of Pediatrics, Vitamin B_6 requirements in man, *Pediatrics,* 38, 1068–1076, 1966.
2. **Jolliffe, N.,** Treatment of neuropsychiatric disorders with vitamins, *JAMA,* 117, 1496–1500, 1941.
3. **Canales, E. S., Soria, J., Zarate, A., Mason, M., and Molina, M.,** The influence of pyridoxine on prolactin secretion and milk production in woman, *Br. J. Obstet. Gynaecol.,* 83, 387–388, 1976.
4. **Frimpter, G. W., Andelman, R. J., and George, W. F.,** Vitamin B_6 dependency syndromes, *Am. J. Clin. Nutr.,* 22, 794–805, 1969.
5. **Wynn, V.,** Vitamins and oral contraceptive use, *Lancet,* 1, 561–564, 1975.
6. **Vilter, R. W.,** The vitamin B_6-hydrazine relationship *Vitam. Horm.* (N.Y.), 22, 797–805, 1964.
7. **Harris, J. W. and Horrigan, D. L.,** Pyridoxine-responsive anemia — prototype and variations on the theme, *Vitam. Horm.* (N.Y.), 22, 721–753, 1964.
8. **Barber, G. W. and Spaeth, G. I.,** The successful treatment of homocystinuria with pyridoxine, *J. Pediatr.,* 75, 463–478, 1969.
9. National Research Council, National Academy of Sciences, *Recommended Dietary Allowances,* 8th ed., (rev.), National Academy of Sciences, Washington, D.C., 1974, 74–76.
10. **Danilov, L. N.,** Anaphylactic shock during vitamin B_6 treatment, *Klin. Med.* (Moscow), 51, 139, 1973.

11. Miklovich, L. and van den Berg, B. J., An evaluation of the teratogenicity of certain antinauseant drugs, *Am. J. Obstet. Gynecol.*, 125, 244–248, 1976.
12. Khera, K. S., Teratogenicity study in rats given high doses of pyridoxine (vitamin B_6) during organogenesis, *Experientia*, 31, 469–470, 1975.
13. Morrison, A. B. and Sarett, H. P., Effects of excess thiamine and pyridoxine on growth and reproduction in rats, *J. Nutr.*, 69, 111–120, 1959.
14. Schumacher, M. F., Williams, M. A., and Lyman, R. L., Effect of high intakes of thiamine, riboflavin and pyridoxine on reproduction in rats and vitamin requirements of offspring, *J. Nutr.*, 86, 343–349, 1965.
15. Alder, S. and Zbinden, G., Use of pharmacological screening tests in subacute neurotoxicity studies of isoniazid, pyridoxine·HCl and hexachlorophene, *Agents Actions*, 3, 233–243, 1973.
16. Unna, K., Studies on the toxicity and pharmacology of vitamin B_6 (2-methyl-3-hydroxy-4,5-bis (hydroxymethyl)-pyridine), *J. Pharmacol. Exp. Ther.*, 70, 400–407, 1940.
17. Antopol, W. and Tarlov, I. M., Experimental study of the effects produced by large doses of vitamin B_6, *J. Neuropathol. Exp. Neurol.*, 1, 330–336, 1942.
18. Kraft, H. G., Fiebig, L., and Hotovy, R., Zur Pharmakologie des Vitamin B_6 und seiner Derivate, *Arzneim. Forsch.*, 11, 922–929, 1961.
19. Unna, K. and Antopol, W., Toxicity of vitamin B_6, *Proc. Soc. Exp. Biol. Med.*, 43, 116–118, 1940.
20. Weigand, C. G., Eckler, C. R., and Chen, K. K., Action and toxicity of vitamin B_6 hydrochloride, *Proc. Soc. Exp. Biol. Med.*, 44, 147–151, 1940.

EFFECT OF NUTRIENT TOXICITIES IN ANIMALS AND MAN: BIOTIN

P. K. Paul

INTRODUCTION

The adverse effects of biotin deficiency on various physiological functions, including reproduction in mammals, are well documented.[1,2] Insects, in general, are also known to require adequate amounts of biotin in their diet for normal growth and development.[3] The dietary requirement of biotin for animals and man has not been established; however, diets containing from 150 to 300 μg of biotin daily are considered adequate for man.[4] Brewer and Edwards[5] have reported that the requirement of biotin is 100 μg/kg of food for optimal hatching in hens. The requirement of biotin for weight increase in turkey chicks is estimated to be 120 to 170 μg/kg of food.[6] Terroine[1] has observed that a minimum daily level 3 μg per rat of biotin is needed to support pregnancy during the last days, which represents approximately 100 μg/kg of food. Therefore, it seems that the requirement of biotin varies depending upon the physiological state of the animal. However, in any case, the dietary requirement of biotin for normal physiological functions possibly does not exceed microgram levels per animal per day. Administration of biotin at dose levels of milligrams can thus be considered as excess.

The B-group vitamins are not known to cause hypervitaminosis or any toxic effects; however, a few recent reports show biotin toxicity mainly on reproduction of some species of insects and rats of the female sex.

BIOTIN TOXICITY ON REPRODUCTION

Insects

An overdose of dietary biotin* causes a decline in fertility to complete sterility, particularly in the female, in a number of insect species, such as the Mexican fruit fly,[7] house fly,[8] hide beetle,[9] yellow fever mosquito,[10] and *Trogoderma*.[11] The mortality rate of all the biotin-fed female insects of different species has been very high. Biotin-induced sterility in insects seems to be due to an adverse effect of biotin on the development and maturation of the follicle and egg hatching (Figures 1 and 2). There have been insignificant effects or no effects at all of biotin excess found in the male of these species. The mechanism of action in excess of biotin-induced fertility in the female hide beetle appears to be partial inactivation of egg proteases during the last stages of embryogenesis. Experiments with dietary biotin-carbonyl ^{14}C suggest that an excess of biotin forms a complex with insoluble yolk proteins, which interferes with the hydrolysis and subsequent use of egg protein by the developing embryo[12,13] (Table 1).

Rat

An acute dose of biotin** has been reported to cause irregularities of the estrus cycle with heavy infiltration of leukocytes in the vagina of the rat up to 14 days after treatment.[14] The ovarian histological studies in these biotin-treated rats indicate enhanced luteinization associated with atrophy of the corpora lutea and stroma. An increase in leukocyte infiltration to the vaginal lumen normally occurs during the diestrous stage when functional luteal tissues appear in the ovary. Therefore, these results suggest that an acute dose of biotin may cause sterilizing effects on reproductive functions in the female rat.

---

* 1 to 2% in food.
** 5 mg/100 g body weight of D-(+)-biotin (E. Merck) dissolved in 0.1 *N* NaOH and injected sub-cutaneously.

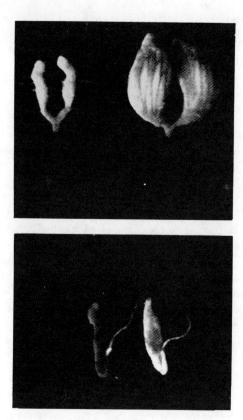

FIGURE 1.   Ovaries (top) and testes (bottom)
of Mexican fruit fly, abnormal in biotin fed on
left and normal on right. Greatly enlarged.
(From Benschoter, C. A. and Paniague, R. G.,
Ann. Entomol. Soc. Am., 59, 289–300, 1966.
With permission.)

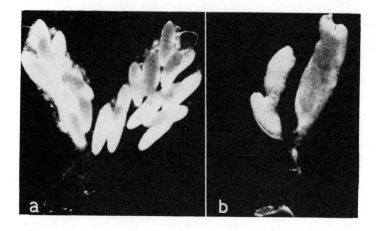

FIGURE 2.   Ovaries of *A. aegypti;* (a) normal; (b) biotin fed. (Magnification x 25.) (From Pillai, M. K. K. and Madhukar, B. V. R.,
Naturwissenschaften, 4, 218–219, 1969. With permission.)

Table 1

**DISTRIBUTION OF RADIOACTIVITY AMONG EGG FRACTIONS
AFTER ADMINISTRATION OF BIOTIN-^{14}C TO FEMALE *D. MACULATUS***

Egg fraction	Biotin-^{14}C injected		Biotin-^{14}C fed	
	Dis/min[a]	Incorporation (%)	Dis/min[a]	Incorporation (%)
Water insoluble protein[b]	1300	17.6	1900	79.2
Soluble protein[c]	40	0.5	0	0
Soluble non-protein	4500	61.0	0	0
Lipid[d]	0	0	0	0
Total	5800	79.1	1900	79.2
Eggs	7400	100	2400	100

Note: Biotin-carbonyl-^{14}C was injected at a rate of 1700 dis/min per female beetle or fed to females at a level of 37,000 dis/min per 100 mg of diet.

[a] Each sample represents 100 eggs.
[b] Including egg shell residues.
[c] Precipitated by TCA.
[d] Et$_2$O extract.

From Levinson, H. Z. and Cohen, E., *J. Insect Physiol.*, 19, 551–558, 1973. With permission.

Studies on maintenance of pregnancy following an excess of biotin treatment (5 or 10 mg/100 g body weight), before or after mating in rats, demonstrate its antifertility effect by showing resorption of fetuses and placentae in most of the animals at the end of 21 days[15,16] (Tables 2 and 3). Estrogen* or progesterone** therapy under such biotin-treated conditions restores maintenance of pregnancy in terms of number of fetuses and placentae and their weights. An excess of biotin, even at a postimplantation stage (day 14 and 15), exhibits its growth-inhibitory effect on fetuses and placentae[17] (Table 4). The administration of estrogen in these biotin-treated rats, although continued for a short period (day 15 to 21), has resisted the biotin-induced inhibition of fetal and placental growth, while progesterone remained ineffective. Therefore, these observations suggest that an excess of biotin probably causes deficiency of estrogen, which is an important anabolic hormone during pregnancy. Estrogen may have a role in inducing progesterone secretion during pregnancy in the rat, as it has been observed in the rabbit.[18] However, Mittelholzer[19] has not observed any abnormal effects of high doses of biotin on the reproductive performance of the female rat when repeating the experiments of Paul et. al.[15] It is difficult to explain the discrepancy between these results. The source and purity of biotin and the nutritional standard of the animals could be factors which contribute to the discrepancy.

BIOTIN TOXICITY ON TISSUE GLYCOGEN, RNA, AND PROTEIN

The uterine glycogen (Table 5) and the hepatic and uterine RNA and protein concentrations (Table 6) decrease following an acute dose of biotin in the nonpregnant and pregnant rat.[16,17] Estrogen administration under such biotin-treated conditions restores glycogen, RNA, and protein in these organs at levels comparable to those of controls. Progesterone, however, does not rectify biotin-induced depletion of the liver and uterine RNA and protein and uterine glycogen. During normal pregnancy in rodents,

* 0.1 μg/rat/day of Estradiol-17β.
** 4 mg/rat/day.

Table 2
EFFECTS OF BIOTIN ON MATING AND MAINTENANCE OF PREGNANCY IN THE RAT

Treatment	No. rats mated	No. rats with fetuses	No. fetuses/ rat[a][c]	No. implantation sites/rat[b]	Fetal weight[a] (g)	Placental[a] weight (g)
Day 14 of Pregnancy[c]						
Mating allowed						
7 days after biotin	6	6	8.0 ±2.8	11.0 ±1.0	0.126 ±0.062	0.109 ±0.024
14 days after biotin	6	4	5.6 ±1.4	6.5 ±3.2	0.101 ±0.027	0.191 ±0.074
21 days after biotin	6	3	8.0 ±0.0	8.6 ±1.4	0.163 ±0.002	0.205 ±0.001
Untreated controls	6	6	10.2 ±2.0	10.3 ±1.7	0.721 ±0.132	0.251 ±0.026
Day 21 of Pregnancy[c]						
Mating allowed						
7 days after biotin	6	2	11.0 ±0.0	11.0 ±0.0	3.41	0.501
14 days after biotin	6	3	11.0 ±1.1	11.6 ±1.2	4.26 ±0.21	0.536 ±0.034
21 days after biotin	6	1	9.0	9.0 ±1.0	5.56	0.614
7 days after biotin + E_2[d]	6	6	9.8 ±1.8	10.8 ±1.8	5.33 ±0.83	0.621 ±0.051
Untreated controls	6	6	9.2 ±3.3	10.12 ±1.3	5.86 ±0.54	0.602 ±0.153

Mating allowed 7 days after biotin[e]   None of the 6 rats mated as observed for 2 months

a Average of the number of rats with fetuses.
b Average of the number of rats mated.
c 5 mg biotin per 100 g body weight.
d Estradiol – 17β (1 μg/rat/day starting from day 6 of pregnancy).
e 10 g mg biotin per 100 g body weight.

From Paul, P. K., Duttagupta, P. N., and Agarwal, H. C., *Curr. Sci.*, 42, 613–615, 1973. With permission.

Table 3
EFFECTS OF AN ACUTE DOSE OF BIOTIN ON PREGNANCY MAINTENANCE
AND ITS RELATION WITH ESTROGEN AND PROGESTERONE IN THE RAT

Treatment	No. implanta-tion sites	No. fetuses	Fetal weight (g)	Placental weight (g)
	With fetuses (9)-none resorbed			
Untreated control run for 21 days	9.19 ± 1.98	7.55 ± 1.45	6.02 ± 0.95	0.602 ± 0.153
	Without fetuses (8)-1 with fetuses			
Biotin[a]	9.30 ± 1.00	Resorbed	—	Resorbed
	With fetuses (6)-1 resorbed			
Biotin + 0.1 μg estradiol from day 6–21	9.11 ± 1.10	9.11 ± 1.10	4.72 ± 0.50	0.482 ± 0.075
	With fetuses (7)-none resorbed			
Biotin + 4 mg progesterone from day 6–21	9.29 ± 1.71	7.45 ± 1.55	3.54 ± 0.90	0.479 ± 0.051

Note: The figures in the parentheses indicate number of rats. The animals were sacrificed on day 22 of pregnancy in the morning.

[a] 10 mg/100 g of body weight in 2 injections on day 1 and 2 of pregnancy.

From Paul, P. K. and Duttagupta, P. N., *J. Nutr. Sci. Vitaminol.,* 21, 89–101, 1975. With permission.

there are increases in the hepatic, uterine, and placental RNA, DNA and protein,[20-22] as well as the hepatic and uterine glycogen.[23,24] It is known that estrogen and pregnancy induce glycogen accumulation in the uterus and liver, but progesterone exerts a glycogenolytic effect in these organs and also causes hyperglycemia,[25-27] yet essential for the maintenance of pregnancy. These observations suggest that the elevated level of estrogen, which normally occurs during gestation, plays a special role in storage of carbohydrate and protein in tissues during pregnancy. Biotin-induced reduction of the uterine glycogen and the hepatic and uterine RNA and protein, therefore, indicates that the vitamin overdose may at least partially inhibit estrogen synthesis or release.

Table 4
EFFECTS OF AN ACUTE DOSE OF BIOTIN ON PREGNANCY MAINTENANCE AND ITS RELATION WITH ESTROGEN AND PROGESTERONE

Treatment	No. implantation sites	No. fetuses	Fetal weight (g)	Placental weight (g)
Untreated control run for 21 days	(9)[a] 9.2 ±1.2	7.55 ±1.45	6.02 ±0.25	0.602 ±0.153
Biotin[b]	(9)[c] 9.7 ±1.3	7.5 ±0.75	2.79 ±0.13	0.409 ±0.205
Biotin + 0.1 μg estradiol from day 15 to 21	(7) 9.5 ±2.04	9.5 ±2.4	5.55 ±0.14	0.593 ±0.036
Biotin + 4 mg progesterone from day 15 to 21	(7) 11.5 ±1.2	10.4 ±3.4	3.00 ±0.12	0.406 ±0.042

[a] The figures in the parentheses indicate number of rats. The animals were sacrificed on day 22 of pregnancy in the morning.
[b] 10 mg/100 g B.W. in 2 injections on day 14 and 15 of pregnancy.
[c] 2 out of 11 rats resorbed their fetuses and placentae. Data obtained from these 2 rats are not included.

From Paul, P. K. and Duttagupta, P. N., *J. Nutr. Sci. Vitaminol.*, 22, 181–186, 1976. With permission.

Table 5
EFFECTS OF AN ACUTE DOSE OF BIOTIN ON TISSUE GLYCOGEN LEVEL IN THE PREGNANT AND NONPREGNANT RAT AND ITS RELATION WITH ESTROGEN AND PROGESTERONE

Pregnant

	Glycogen (mg/100 mg tissue)			
		Uterus		
Treatment	Liver	With implantation site	Without implantation site	Placenta
With fetuses (9)				
Untreated control run for 21 days	1.34 ± 0.168	0.565 ± 0.084	0.269 ± 0.010	0.387 ± 0.063
Without fetuses (8)				
Biotin[a]	1.14 ± 0.058	0.248 ± 0.010	—	—
With fetuses (6)				
Biotin + 0.1 μg estradiol for 16 days from day 6–21	1.21 ± 0.073	0.395 ± 0.013	0.195 ± 0.027	0.340 ± 0.072

Table 5 (continued)
EFFECTS OF AN ACUTE DOSE OF BIOTIN ON TISSUE GLYCOGEN LEVEL IN THE PREGNANT AND NONPREGNANT RAT AND ITS RELATION WITH ESTROGEN AND PROGESTERONE

Pregnant (continued)

| | Glycogen (mg/100 mg tissue) | | | |
| | | Uterus | | |
Treatment	Liver	With implantation site	Without implantation site	Placenta
		With fetuses (7)		
Biotin + 4 mg progesterone for 16 days from day 6—21	1.06 ± 0.012	0.238 ± 0.078	0.172 ± 0.032	0.208 ± 0.004

Nonpregnant

| | Glycogen (mg/100 mg tissue) | | | |
| | | Uterus | | |
Treatment	Liver	With implantation site	Without implantation site	Placenta
Untreated control				
Proestrus	0.715 ± 0.103 (10)	–	0.153 ± 0.014	–
Estrus	1.36 ± 0.105 (10)	–	0.302 ± 0.026	–
Metaestrus	0.899 ± 0.049 (10)	–	0.146 ± 0.017	–
Diestrus	1.000 ± 0.110 (10)	–	0.108 ± 0.010	–
Biotin[a]	1.13 ± 0.093 (6)	–	0.116 ± 0.005	–
0.1 μg estradiol for 16 days	0.350 ± 0.050 (6)	–	0.250 ± 0.030	–
4 mg progesterone for 16 days	0.648 ± 0.041 (6)	–	0.066 ± 0.009	–

Note: The figures in parentheses indicate number of rats. The animals were sacrificed on the 22nd day of the experiment in the morning.

[a] 10 mg/100 g body weight in two injections on day 1 and 2 of the experimental period. Estradiol or progesterone was given during the last 16 days of the experiment.

From Paul, P. K. and Duttagupta, P. N., *J. Nutr. Sci. Vitaminol.*, 21, 89–101, 1975. With permission.

Table 6
EFFECTS OF AN ACUTE DOSE OF BIOTIN ON TISSUE LEVELS OF RNA, DNA AND PROTEIN IN THE PREGNANT AND NONPREGNANT RAT AND ITS RELATION WITH ESTROGEN AND PROGESTERONE

Pregnant

Treatment	Liver (mg/g tissue)			Uterus (mg/g tissue)			Placenta (mg/g tissue)		
	RNA	DNA	Protein	RNA	DNA	Protein	RNA	DNA	Protei
With fetuses (9)									
Untreated control run for 21 days	13.4 ±1.2	1.98 ±0.09	191.4 ±8.07	4.67 ±0.05	1.62 ±0.07	90.4 ±6.7	4.07 ±1.00	1.63 ±0.05	99.4 ±7.
Without fetuses (8)									
Biotin[a]	9.9 ±0.9	2.86 ±0.12	72.3 ±3.2	5.33 ±1.00	4.87 ±0.72	66.2 ±5.2	—	—	—
With fetuses (6)									
Biotin + 0.1 μg estradiol for 16 days from day 6 to 21	15.4 ±1.3	1.57 ±0.01	154.5 ±12.8	8.33 ±0.95	1.81 ±0.02	88.0 ±7.7	8.09 ±1.15	2.09 ±0.05	97.0 ±6.
With fetuses (7)									
Biotin + 4 mg progesterone for 16 days from day 6 to 21	11.2 ±1.1	2.85 ±0.07	140.9 ±10.0	4.75 ±0.98	1.16 ±0.05	60.4 ±8.1	4.82 ±0.87	1.47 ±0.04	68.7 ±7.

Nonpregnant

Treatment	Liver (mg/g tissue)			Uterus (mg/g tissue)			Placenta (mg/g tissue)		
	RNA	DNA	Protein	RNA	DNA	Protein	RNA	DNA	Protei
Untreated control									
Proestrus	4.23(10) ±0.59	2.46 ±0.25	18.0 ±0.9	1.96 ±0.12	1.02 ±0.13	4.0 ±0.2	—	—	—
Estrus	8.66(10) ±0.74	3.90 ±0.29	24.1 ±1.2	3.77 ±0.38	1.52 ±0.16	10.1 ±1.0	—	—	—
Metaestrus	4.43(10) ±0.60	2.57 ±0.21	17.0 ±1.7	2.57 ±0.39	1.70 ±0.18	3.0 ±1.1	—	—	—
Diestrus	4.44(10) ±0.52	2.43 ±0.25	15.1 ±1.6	0.83 ±0.15	0.70 ±0.13	1.1 ±0.2	—	—	—
Biotin[a]	4.52(6) ±0.43	2.72 ±0.16	21.6 ±1.2	1.00 ±0.04	0.71 ±0.01	1.7 ±0.2	—	—	—
0.1 μg estradiol for 16 days	6.46(6) ±0.24	2.49 ±0.54	64.8 ±6.4	3.07 ±0.52	2.27 ±0.20	30.2 ±3.8	—	—	—
4 mg progesterone for 16 days	2.70(6) ±0.18	2.52 ±0.01	22.8 ±1.3	0.85 ±0.05	0.88 ±0.19	0.80 ±0.1	—	—	—

Note: The figures in parentheses indicate number of rats. The animals were sacrificed on the 22nd day of the experimen in the morning.

[a] 10 mg/100 g body weight in 2 injections on day 1 and 2 of the experimental period. Estradiol or progesterone was give during last 16 days of the experiment.

From Paul, P. K. and Duttagupta, P. N., *J. Nutr. Sci. Vitaminol.*, 21, 89–101, 1975. With permission.

BIOTIN TOXICITY ON TISSUE GLUCOSE-6-PHOSPHATE DEHYDROGENASE ACTIVITY

Recently it has been reported that an acute dose of biotin (10 mg/100 g body weight), when administered at the pre- (day 1 and 2) (Table 7) or post- (day 14 and 15) implantation stage of pregnancy in the rat, causes reduction of glucose-6-phosphate dehydrogenase (G-6-PD) activity in the ovary, uterus, and liver.[16,17] A reduction in the activity of the enzyme is observed in the adrenal of the pregnant rat treated with biotin at the preimplantation stage. The vitamin treatment does not affect the adrenal enzyme activity when it is executed at the postimplantation stage. Mistry and Dakshinamurti,[2] however, have observed an increased activity of G-6-PD in the liver of biotin-deficient rats which is considered to be due to reduced availability of glucose-6-phosphate for metabolism. McKerns[28] has observed that the ovarian G-6-PD activity of the rat is about twice as high during estrus as during diestrous. The enzyme shows its highest activity in the rat ovary during cycle, at parturition and after weaning, but during pregnancy its activity is low.[29] Glock and McLean[30] first noted a markedly higher activity of G-6-PD in the liver of the female rat as compared with that of the male. Subsequently, it has been observed that estrogen increases the enzyme activity in the liver of the ovariectomized and adrenalectomized female and the castrated male rat.[31] On the other hand, adrenal G-6-PD activity is enhanced by ovariectomy and depressed by estrogen replacement.[32] Paul and Duttagupta[16,17] have observed that although both estrogen and progesterone therapy of the acutely biotin-treated pregnant rat increases G-6-PD activity in the ovary,

Table 7

EFFECTS OF AN ACUTE DOSE OF BIOTIN ON TISSUE GLUCOSE-6-PHOSPHATE DEHYDROGENASE ACTIVITY AND ITS RELATION WITH ESTROGEN AND PROGESTERONE

Pregnant

Treatment	Glucose-6-phosphate dehydrogenase activity mM TPNH/100 mg tissue/min at room temperature				
	Ovary	Adrenal	Liver	Uterus	Placenta
	With fetuses (6)				
untreated control run for 21 days	138.0 ± 6.0	195.0 ± 8.0	156.0 ± 5.0	73.0 ± 4.0	68.0 ± 5.2
	Without fetuses (6)				
biotin[a]	50.4 ± 2.1	90.0 ± 4.1	83.1 ± 5.0	20.5 ± 1.5	—
	With fetuses (6)				
biotin + 0.1 µg estradiol for 16 days from day 6–21	122.0 ± 8.0	129.0 ± 11.0	138.0 ± 10.0	82.5 ± 4.3	72.0 ± 6.0
	With fetuses (6)				
biotin + 4 mg progesterons for 16 days from day 6–21	101.7 ± 15.3	138.0 ± 14.0	120.0 ± 8.0	62.0 ± 6.0	54.0 ± 9.0

10 mg/100 g body weight in 2 injections on day 1 and 2 of the experimental period. Estradiol or progesterone was given during the last 16 days of the experiment.

Table 7 (continued)
EFFECTS OF AN ACUTE DOSE OF BIOTIN ON TISSUE GLUCOSE-6-PHOSPHATE DEHYDROGEN ACTIVITY AND ITS RELATION WITH ESTROGEN AND PROGESTERONE

Nonpregnant

Treatment	Glucose-6-phosphate dehydrogenase activity mM TPNH/100 mg tissue/min at room temperature				
	Ovary		Adrenal	Liver	Uterus	Place
Untreated control						
Proestrus	112.2 ± 3.2	(8)	129.0 ± 3.0	80.8 ± 3.2	55.6 ± 2.5	—
Estrus	174.5 ± 6.2	(8)	145.0 ± 2.0	89.1 ± 5.9	82.3 ± 2.7	—
Metaestrus	89.5 ± 2.5	(8)	170.9 ± 5.7	101.3 ± 8.2	81.0 ± 4.0	—
Diestrus	77.4 ± 1.6	(8)	257.8 ± 8.2	102.3 ± 6.5	64.5 ± 1.5	—
Biotin[a]	77.8 ± 3.1	(6)	154.1 ± 8.3	134.6 ± 6.7	24.5 ± 2.8	—
0.1 μg estradiol for 16 days	89.6 ± 3.2	(6)	140.9 ± 3.1	103.2 ± 2.8	82.6 ± 4.6	—
4 mg progesterone for 16 days	80.8 ± 1.5	(6)	193.2 ± 4.5	106.1 ± 5.5	51.9 ± 1.2	—

Note: The figures in parentheses indicate number of rats. The animals were sacrificed on the 22nd day of the experir in the morning.

From Paul, P. K. and Duttagupta, P. N., *J. Nutr. Sci. Vitaminol.*, 21, 89—101, 1975. With permission.

uterus, liver, and adrenal, estrogen is more potent than progesterone in this regard except for the adrenal. Progesterone therapy of the rats treated with a high dose of biotin at the postimplantation stage (day 14 and 15) causes reduction of the adrenal enzyme activity.[17] In the nonpregnant rat, either estrogen treatment or estrous condition enhances activity of this enzyme in the ovary and uterus.[15] G-6-PD activity in the adrenal, unlike other organs, appears to be more responsive to the progestrone and diestrous condition, while estrogen administration and the proestrus stage exhibit the lowest activity of the enzyme. It is well known that the pentose phosphate cycle, which is initiated by G-6-PD, provides the cell with essential NADPH for synthetic and reductive purposes and pentose sugar for nucleotide synthesis. Therefore, the reduction of G-6-PD activity following an acute dose of biotin treatment probably indicates that the cellular synthetic mechanisms regulated by the pentose phosphate pathway are adversely affected by the vitamin overdose.

CONCLUSION

There cannot be any disagreement regarding the adequate need of biotin for growth and development of animals and man. Although toxicities of biotin in animals and man have not been established, there are some reports suggesting that an overdose of the vitamin may adversely affect metabolic processes concerned with growth and develop-ment of some species of insects and of rats. To date, most of the studies in this area have been concerned with reproduction in some insects and, among mammals, in the female rat. There are a few reports on the relationship between excesses of biotin, hyperchol-esterolemia, and atherosclerosis in man, rabbit, and rat. However, these studies cannot be regarded as toxicity studies, since the aim and objectives were concerned with therapeutic values of biotin in hypercholesterolemia and atherosclerosis. There is no indication of any toxic effect of biotin therapy in these pathological conditions either. On the contrary, disagreement exists on the issue whether biotin has a favorable effect or no effect at all in hypercholesterolemia and atherosclerosis.

In insects, the sterility induced by biotin overdose appears to be due to inactivation of egg proteases during the last stage of embryogenesis. On the other hand, biotin overdose in the rat seems to cause deficiency of estrogen and progesterone, which are essential hormones for the regulation of the reproductive cycle, growth, and development of the embryo. Studies on glucose-6-phosphate dehydrogenase activity in the ovary, adrenal, uterus, and liver indicate that excess of biotin may, at least partially, inactivate the pentose phosphate pathway, which is involved in steroid synthesis and tissue growth.

REFERENCES

1. Terroine, T., Physiology and biochemistry of biotin, in *Vitamins and Hormones,* Vol. 18, Harris, R. S. and Ingle, D. J., Eds., Academic Press, New York, 1960, 1–37.
2. Mistry, S. P. and Dakshinamurti, K., Biochemistry of biotin, in *Vitamins and Hormones,* Vol. 22, Harris, R. S., Wool, I. G., and Loraine, J. A., Eds., Academic Press, New York, 1964, 1–49.
3. House, H. L., Nutrition, in *The Physiology of Insecta,* Vol. 2, 2nd ed., Rockstein, M., Ed., Academic Press, New York, 1974, 1–62.
4. Wagner, A. F. and Folkers, K., *Vitamins and Coenzymes* Interscience, New York, 1964, 138–159.
5. Brewer, L. E. and Edwards, H. M., Jr., Studies on the biotin requirement of broiler breeders, *Poult. Sci.,* 51, 619–624, 1972.
6. Ahrends, L. G., Kienholz, E. W., Shutze, J. V., and Taylor, D. D., Effect of supplemental biotin on reproductive performance of turkey breeder hens and its effect on the subsequent progeny's performance, *Poult. Sci.,* 50, 208–214, 1971.
7. Benschoter, C. A. and Paniagua, R. G., Reproduction and longevity of Mexican fruit flies, *Anastrepha ludens* (Diptera: Tephritidae) fed biotin in the diet, *Ann. Entomol. Soc. Am.,* 59, 289–300, 1966.
8. Benschoter, C. A., Effect of dietary biotin on reproduction of the house fly, *J. Econ. Entomol.,* 60, 1326–1328, 1967.
9. Cohen, E. and Levinson, H. Z., Disrupted fertility of the hide beetle, *Dermester maculatus* (Deg) due to dietary overdose of biotin, *Experientia,* 24, 367–368, 1968.
10. Pillai, M. K. K. and Madhukar, B. V. R., Effects of biotin on the fertility of the yellow fever mosquito, *Aedes aegypti, Naturwissenschaften,* 4, 218–219, 1969.
11. Sehgal, S. S., Agarwal, H. C., and Pillai, M. K. K., Sterilizing effect of a dietary surplus of biotin in *Trogoderma granarium everts, Curr. Sci.,* 39, 551–552, 1970.
12. Cohen, E. and Levinson, H. Z., Studies on the chemosterilizing effects of biotin on the hide beetle, *Dermestes maculatus* (Dermestidae; Coleoptera), *Comp. Biochim. Physiol. B,* 43, 143–149, 1972.
13. Levinson, H. Z. and Cohen, E., The action of overdose biotin on reproduction of the hide beetle, *Dermestes maculatus, J. Insect Physiol.,* 19, 551–558, 1973.
14. Paul, P. K., Duttagupta, P. N., and Agarwal, H. C., Effect of an acute dose of biotin on reproductive organs of the female rat, *Curr. Sci.,* 42, 206–208, 1973.
15. Paul, P. K., Duttagupta, P. N., and Agarwal, H. C., Antifertility effect of biotin and its amelioration by estrogen in the female rat, *Curr. Sci.,* 42, 613–615, 1973.
16. Paul, P. K. and Duttagupta, P. N., The effect of an acute dose of biotin at the pre-implantation stage and its relation with female sex steroids in the rat, *J. Nutr. Sci. Vitaminol.,* 21, 89–101, 1975.
17. Paul, P. K. and Duttagupta, P. N., The effect of an acute dose of biotin at the post-implantation stage and its relation with the female sex steroids in the rat, *J. Nutr. Sci. Vitaminol.,* 22, 181–186, 1976.
18. Hilliard, J., Spies, H. G., and Sawyer, C. H., Hormonal factors regulating ovarian cholesterol metabolism and progestin secretion in intact and hypophysectomized rabbit, in *The Gonads, Part 1,* McKems, K. W., Ed., Appleton-Century-Crofts, New York, 1969, 55–92.
19. Mittelholzer, E., Absence of influence of high dose of biotin on reproductive performance in female rats, *Int. J. Vitam. Nutr. Res.,* 46, 33–39, 1976.
20. Drasher, M. L., Uterine and placental nucleic acids and protein during pregnancy in the mouse, *J. Exp. Zool.,* 122, 385–408, 1953.
21. Poo, L. J., Lew, W., and Addis, T., Protein anabolism of organs and tissues during pregnancy and lactation, *J. Biol. Chem.,* 128, 69–77, 1939.
22. Campbell, R. M., Innes, I. R., and Costerlitz, H. W., Species differences in the desoxyribonucleic acid and ribonucleic acid contents of liver on non-pregnant and pregnant mice, guinea pigs and rats, *J. Endocrinol.,* 9, 45–51, 1953.

23. **Paul, P. K.,** Influence of estrogen and pregnancy on uterus: a comparative study, *Endokrinologie,* 57, 348–352, 1971.

24. **Paul, P. K.,** Dynamics of hepatic glycogen: estrogen and pregnancy, *Acta Endocrinol.* (Copenhagen), 71, 385–392, 1972.

25. **Fotherby, K.,** The biochemistry of progesterone, in *Vitamins and Hormones,* Vol. 22, Harris, R. S., Wool, I. G., and Loraine, J. A., Eds., Academic Press, New York, 1964, 153–197.

26. **Yang, M. M. P.,** The effect of single dose of progesterone on blood glucose of rat, *Endocrinology,* 86, 924–927, 1970.

27. **Paul, P. K. and Duttagupta, P. N.,** Inhibition of estrogen-induced increase in hepatic and uterine glycogen by progesterone in the rat, *Acta Endocrinol.* (Copenhagen), 72, 762–770, 1973.

28. **McKerns, K. W.,** Gonadotropin regulation of the activities of dehydrogenase enzymes of the ovary, *Biochim. Biophys. Acta,* 97, 542–550, 1965.

29. **Lunaas, T., Baldwin, R. L., and Cupps, P. T.,** Ovarian activities of Pyridine nucleotide dependent dehydrogenases in the rat during pregnancy and lactation, *Acta Endocrinol.* (Copenhagen), 58, 521–531, 1968.

30. **Glock, G. E. and McLean, P.,** Further studies on properties and assay of glucose-6-phosphate dehydrogenase of rat liver, *Biochim. J.,* 55, 400–408, 1953.

31. **Huggins, C. and Yao, F.,** Influence of hormone on liver. I. Effect of steroids and thyroxine on pyridine nucleotide linked dehydrogenases, *J. Exp. Med.,* 110, 899–919, 1959.

32. **Kitay, J. I.,** Effects of estrogen and androgen on the adrenal cortex of the rat, in *Functions of Adrenal Cortex,* Vol. 2, McKerns, K. W., Ed., Appleton-Century-Crofts, New York, 1968, 775–811.

EFFECT OF NUTRIENT TOXICITIES IN ANIMALS AND MAN: CHOLINE

Martha Byington

The acute toxicity of orally ingested choline is relatively low (5 g/Kg for rats) in comparison with that of some of its esters and many other quaternary ammonium compounds.[1] Choline qualitatively has the same pharmacological actions as acetylcholine, but is much less active.[2] Experimental sows were fed 3.36 g choline per day with no adverse side effect.[3] An apparent pyridoxine deficiency was produced in broiler chicks fed 490 mg per pound or more of choline for 7 weeks. The animals showed depressed growth, hyperexcitability, spasticity of the legs and wings, and torticollis. These abnormal clinical signs were overcome by withdrawal of the choline or by provision of extra pyridoxine.[4]

The LD_{50} for man is estimated to be of the order of 200 to 400 g. Single oral doses of 10 g produce no obvious pharmacodynamic response in man.[2] Symptoms of cholinergic stimulation, sweating and salivation, reportedly occurred when 16 g of choline were administered per day for 8 days in a patient with tardive dyskinesia. The choline at that level did, however, seem to decrease the abnormal movements associated with that disease.[5]

Experimental work with intravenous choline administration has been conducted to a limited extent. Cats were injected with 40 mg/Kg choline in an attempt to relieve arterial spasm. In this study the drug produced a depolarizing blockage of skeletal muscle with respiratory paralysis. Copious secretions of tracheobronchial mucous and saliva and transient hypotension occurred following the injection.[6] The injection into humans of 2 g of choline in two hours was shown to be without serious side effects.[7]

REFERENCES

1. Best, C. H., Lucas, C. C., and Ridout, J. H., Vitamins and the protection of the liver, *Br. Med. Bull.,* 12, 9–14, 1956.
2. Goodman, L. S. and Gilman, A., *The Pharmacological Basis of Therapeutics,* 4th ed., MacMillan, New York, 1970, 1660–1661.
3. Dobson, K. J., Failure of choline to prevent splayleg in piglets, *Aust. Vet. J.,* 47, 587–590, 1971.
4. Saville, D. G., Solvyns, A., and Humphries, C., Choline induced pyridoxine deficiency in broiler chickens, *Aust. Vet. J.,* 43, 346–348, 1967.
5. Davis, K. L., Berger, P. A., and Hollister, L. E., Choline for tardive dyskinesia, *N. Engl. J. Med.,* 293, 152, 1975.
6. Kapp, J., Mahaley, M. S., and Odom, G. L., Experimental evaluation of potential spasmolytic drugs, *J. Neurosurg.,* 32, 468–472, 1970.
7. Steigmann, F., Firestein, R., and De La Huerga, J., Intravenous choline therapy, *Pharmacol. Exp. Ther.,* 11, 393, 1952.

EFFECT OF NUTRIENT TOXICITIES – EXCESS
IN ANIMALS AND MAN: FOLIC ACID

H. G. Preuss

Folic acid is used predominately in man to treat hematopoietic disorders and to replace decreased folate concentrations produced by anticonvulsant medications. Aside from an occasional allergic reaction, its ability to precipitate neurologic disorders in humans having vitamin B_{12} deficiency and its association with epileptic recurrences in patients on anticonvulsants, folic acid generally has been regarded as a nontoxic vitamin. However, 13 of 14 normal subjects receiving 15 mg of folic acid every day developed adverse reactions.[1] Within a month, 8 of 14 developed gastrointestinal problems (anorexia, nausea, abdominal distension, discomfort and flatulence), 9 of 14 sleep disturbances, 8 of 14 vivid dreaming, 8 of 14 malaise and irritability, and 5 of 14 hyperexcitability. Some also complained of an unpleasant bitter taste. Two studies attempted to reproduce these conditions and found no adverse effects in subjects receiving folic acid 15 mg every day for 3 months.[2,3]

Animal studies have noted three organs affected by parenteral folate – brain, heart, and kidneys. Sodium folate (45 to 120 mg) administered i.v. to 180- to 200-g female Wistar rats induces epilepsy.[4,5] Lower doses of 1 to 15 mg can produce epilepsy in rats having an intracortical electrode or a brain lesion produced by heat. The lower doses may be effective because of damage to the blood brain barrier. Metabolism of the folate in the liver may be necessary to produce the convulsant effect, since ligature of the portal vein or partial hepatectomy reduces the epileptogenic effect of i.v. folic acid.[6]

Folic acid (30 mg/kg i.v.) increases systolic blood pressure and the force of cardiac contraction in Wistar rats weighing 200 to 260 g.[7] In addition, this agent causes a diminution of diastolic relaxation. Many of these reactions were mimicked by perfusing lower doses of folic acid through an isolated heart preparation.[7]

A single bolus of folic acid given i.v. or i.p. augments renal DNA and RNA synthesis in kidney tissue of rats, mice, guinea pigs, and rabbits.[8] Kidney weight increases within 1 day and reaches a maximum at 4 days.[9] Although the predominant increase is in wet weight, a true increase in dry weight, bulk RNA, and bulk DNA has been noted. Originally, this was called a "chemically induced hypertrophy." Since folic acid stimulates DNA synthesis in kidneys, not gut,[10] and produces growth only when evidence of kidney damage is present,[11] the compensatory growth is attributed to renal damage from folate – most likely tubular damage.[11] Folate induced acute renal failure is reproducible: the animals survive this maneuver; and the tubular damage is mild.[12,13]

REFERENCES

1. **Hunter, R., Barnes, J., Oakeley, H. F., and Matthews, D. M.,** Toxicity of folic acid given in pharmacological doses to healthy volunteers, *Lancet,* 1, 61–63, 1970.

2. **Ralston, A. J., Snaith, R. P., and Hinley, J. B.,** Effects of folic acid on fit-frequency and behavior in epileptics on anticonvulsants, *Lancet,* 1, 867–868, 1970.

3. **Hellstrom, L.,** Lack of toxicity of folic acid given in pharmacological doses to healthy volunteers, *Lancet,* 1, 59–61, 1971.

4. **Hommes, O. R., Obbens, E. A. M. T., and Wijffels, C. C. B.,** Epileptogenic activity of sodium-folate and the blood-brain barrier in the rat, *J. Neurol. Sci.,* 19, 63–71, 1973.

5. **Obbens, E. A. M. T. and Hommes, O. R.,** The epileptogenic effects of folate derivatives in the rat, *J. Neurol. Sci.,* 20, 223–229, 1973.

6. **Hommes, O. R. and Obbens, E. A. M. T.,** Liver function and folate epilepsy in the rat, *J. Neurol. Sci.,* 20, 269–272, 1973.
7. **Jenkins, D. and Spector, R. G.,** The action of folate and phenytoin on the rat heart in vivo and in vitro, *Biochem. Pharmacol.,* 22, 1813–1816, 1973.
8. **Haddow, A.,** Chemistry and Biology of Pteridines, Churchill, Livingstone, London, 1954.
8. **Haddow, A.,** *Chemistry and Biology of Pteridines,* Churchill, Livingstone, London, 1954.
9. **Threlfall, G., Taylor, D. M., and Buck, A. T.,** The effect of folic acid on growth and deoxyribonucleic acid synthesis in the rat kidney, *Lab. Invest.,* 15, 1477–1485, 1966.
10. **Tilson, M. D.,** A dissimilar effect of folic acid upon growth of the rat kidney and small bowel, *Proc. Soc. Exp. Biol. Med.,* 134, 95–97. 1970.
11. **Preuss, H. G., Weiss, F. R., Janicki, R. H., and Goldin, H.,** Studies on the mechanism of folate induced growth in rat kidneys, *J. Pharmacol. Exp. Ther.,* 180, 754–758, 1972.
12. **Schubert, G. E.,** Folic acid-induced acute renal failure in the rat: morphological studies, *Kidney Int.,* 10, S46–S50, 1976.
13. **Schmidt, U. and Duback, U. C.,** Acute renal failure in the folate-treated rat: early metabolic changes in various structures of the nephron, *Kidney Int.,* 10, S39–S45, 1976.

EFFECT OF NUTRIENT TOXICITIES IN ANIMALS AND MAN: INOSITOL

T. Tomita

TOXICITY AND EXCESS

The toxicity of inositol is extremely low. Diarrhea is one acute symptom observed in animals following ingestion of an excessive dose of inositol.[1-4] Regarding chronic toxicity, there is some evidence[5-6] suggesting that growth inhibition in young animals may be caused by large amounts of inositol. It is interesting to note that inositol aggravates choline-deficient symptoms, such as renal necrosis[7] and perosis,[5,8] in developing animals fed a diet deficient in choline, while inositol and choline act synergistically on fatty liver.[9-11]

Diarrhea

Martin et al.[1] radiographically studied the effect of inositol on the activity of the stomach and small intestine of dogs, and they found a considerable increase in the peristalsis of these organs. This stimulating action inositol has on the G.I. tract is supported by other observations. Anderson[3] and Bly et al.[2] noted diarrhea in dogs after administration of inositol. For 1 month, Maeda[4] fed weaning rats (Wistar strain) a synthetic diet containing 0, 0.1, 0.5, or 2.5% inositol and found that the diet containing 2.5% caused diarrhea and loss of appetite, resulting in slight inhibition of growth.

Growth Inhibition

A large dosage of dietary inositol seemed to inhibit the growth of young rats tested about 20 days after birth.[6] It was found that the body weight increase of young animals fed a large excess of inositol (1g/Kg/day) was significantly lower than that of those fed the standard diet (10 mg/Kg/day) and even lower than that of those fed the inositol-deficient diet. However, such growth inhibition was not present when older rats (3 to 5 months of age) were fed large amounts of myo-inositol (5 g/Kg/day). Agranoff and Fox[5] also observed growth inhibition caused by inositol in newborn chicks which were fed a choline-deficient diet. Since the large dosage of dietary inositol increased the P/N ratio of hepatic lipids,[6] perhaps the inhibitory effect of massive inositol may be related to interference with metabolism of the specific phosphatides required for formation of cellular and subcellular constituents, particularly in young animals.

CHOLINE ANTAGONISM

Growth suppression of yeast that requires inositol, caused by choline, has been reported by Taylor and McKibbin.[12] In studies of renal necrosis in developing rats fed a choline-deficient diet, the addition of inositol greatly aggravated the necrosis, while reducing the amount of liver fat.[7] In a study of perosis of 1-day-old chicks kept on a choline-deficient diet for 4 weeks, the subsequent addition of inositol to the diet caused a marked aggravation of the disease, both in rate of appearance and in degree. The degree of perosis intensified when the amount of dietary inositol was increased.[5,8]

These results suggest that choline and inositol are not interchangeable in the process of formation of specific phosphatides required during rapid growth. However, in the lipotropic function, they can be reversed.

REFERENCES

1. Martin, G. J., Thompson, M. R., and de Carvajal-Forero, J., Influence of inositol and other B-complex factors on the motility of the gastrointestinal tract, *Am. J. Dig. Dis.,* 8, 290–295, 1941.
2. Bly, C. G., Heggeness, E. W., and Nasset, E. S., The effects of pantothenic acid and inositol added to whole-wheat bread on evacuation time, digestion and absorption in the upper gastrointestinal tract of dogs, *J. Nutr.,* 26, 161–173, 1943.
3. Anderson, R. J., The utilization of inosite in the dog, *J. Biol. Chem.,* 25, 391–397, 1916.
4. Maeda, T., The physiological functions of *Myo*-inositol: The Alteration of Lipid Metabolism Due to Myo-inositol Deficiency in Rats, Ph.D. thesis, Shizuoka College of Pharmaceutical Sciences, Shizuoka, Japan, 1976.
5. Agranoff, B. W. and Fox, M. R. S., Antagonism of choline and inositol, *Nature,* 183, 1259–1260, 1959.
6. Natume, K., Studies on myo-inositol. III. Effect of the excess dosage of myo-inositol on the pattern of lipids in the liver of young rat, *Vitamins,* 32, 363–368, 1965.
7. Handler, P., Factors affecting the occurrence of hemorrhagic kidneys due to choline deficiency, *J. Nutr.,* 31, 621-633, 1946.
8. Jukes, T. M., Perosis in turkeys, I. Experiments related to choline, *Poult. Sci.,* 20, 251–254, 1941.
9. Gavin, G. and McHenry, E. W., Inositol: a lipotropic factor, *J. Biol. Chem.,* 139, 485, 1941.
10. Best, C. H., Lucas, C. C., Patterson, J. M., and Ridout, J. H., The rate of lipotropic action of choline and inositol under special dietary conditions, *Biochem. J.,* 48, 452–458, 1951.
11. Kotaki, A., Sakurai, T., Kobayashi, M., and Yagi, K., Studies on myo-inositol. IV. Effect of myo-inositol on the cholesterol metabolism of rats suffering from experimental fatty liver, *Vitamins,* 14, 87–94, 1968.
12. Taylor, W. E. and McKibbin, J. M., Effect of choline on yeast bioassay of inositol, *Proc. Soc. Exp. Biol. Med.,* 79, 95–96, 1952.

EFFECT OF SPECIFIC NUTRIENT TOXICITIES IN
ANIMALS AND MAN: VITAMIN C

M. H. Briggs

INTRODUCTION

The major form of vitamin C in animal tissues is L-ascorbic acid, though dehydro-L-ascorbic acid also shows some vitamin C activity in animal bioassays. The latter compound is a metabolite of ascorbic acid and its activity may depend on conversion in vivo to the parent compound.

Vitamin C is unusual in being required in the diet by only a small minority of animal species; for the great majority it is formed in the liver from glucose. This fact presents considerable problems to the investigation of ascorbic acid toxicity, for the only convenient laboratory animal which shows an absolute requirement for ascorbic acid is the guinea pig. This is a species not commonly used in such toxicological investigations as carcinogenicity, teratogenicity, etc. The cost of primates rules them out of any large-scale, prolonged studies on vitamin C toxicity. Investigations into the effects of high dose ascorbic acid have been conducted in rodents, but these animals show high endogenous synthesis of this compound, while the effects of exogenous vitamin on this synthesis do not appear to be known. Extrapolation of results from these animals to species showing a dietary requirement for vitamin C are therefore difficult.

Most reported studies are of L-ascorbic acid, though there are a few results with exogenous dehydro-L-ascorbic acid. Pharmaceutical products for human use sometimes contain L-ascorbic acid, but more recently sodium ascorbate and calcium ascorbate have been introduced. These appear to be better tolerated in high dose than the free acid.

ANIMAL STUDIES

There can be no doubt that L-ascorbic acid and its salts are compounds of very low systemic toxicity. There appear to be some circumstances, however, when toxic manifestations of high doses can be demonstrated. Dehydro-L-ascorbic acid is more toxic than L-ascorbic acid, though no systematic investigation appears to have been published.

There seems general agreement that oral ascorbic acid may be administered to most laboratory animals at doses of several grams per kilogram of body weight without the appearance of any very obvious general effect on health.[1] More sensitive studies of organ function, however, often reveal signs of significant changes, some of which could lead to toxicity in time.

In the rat, for example, the administration of ascorbic acid at 2.5 g/kg daily leads to damaged liver lysosomes.[2] In both the rat[3] and guinea pig,[4] there is depresssion of thyroid activity in animals receiving 0.4 g/kg ascorbic acid daily and fed either normal or high protein diets.

No change in cardiac morphology or function appears to have been reported following high dose vitamin C, but guinea pigs, though not rats, show a marked reduction in heart noradrenaline concentration following 1.0 g/kg ascorbic acid.[5] Brain noradrenaline concentration is not significantly changed by vitamin C in either species.[5]

Bone is a further tissue for which there is considerable evidence of significant changes in several species following high doses of ascorbic acid.[6] For example, in chicks ascorbic acid at 0.22 g/kg diet increases release of calcium and phosphorus from bone, combined with increased glucose uptake. These are early signs of bone resorption.[7] Similarly, in the young pig, oral administration of ascorbic acid at 1.0 g/kg daily for 32 days increased bone collagen turnover, and this was accompanied by increased urinary output of hydroxyproline.[6]

Table 1
TOXIC EFFECTS OF VITAMIN C

Species	Compound[a]	Effects of high doses	Reference
Rat	AA	Damaged lysosomes in liver	2
Guinea pig	AA	Inhibition of thyroid	3
Rat	AA	Inhibition of thyroid	4
Guinea pig	AA	Decreased heart noradrenaline	5
Pig	AA	Increased bone collagen turnover	6
Chick	AA	Increased bone resorption	7
Rat	AA	Decreased utilization of β-carotene	8
Guinea pig	AA	Enhances scorbutic effect of deficient diet	9
	AA	Reduced conceptions, dead fetus	10
	AA	Shortened gestation: increased stillbirths	11
	AA	Increased abortions and stillbirths	12
Rat	DHA	Alloxan-type diabetes	14
	DHA	Diabetic-type cataracts	15

[a] AA = L-ascorbic acid; DHA = dehydro-L-ascorbic acid.

Large amounts of ascorbic acid may also influence vitamin status. There is evidence for impaired utilization of β-carotene in rats given excess ascorbic acid,[8] while guinea pigs pretreated with vitamin C became scorbutic sooner when transferred to a deficient diet than untreated controls.[9] It has been suggested several times that animals adapted to high vitamin C intake are at a disadvantage when transferred to a low vitamin C diet, though the point is controversial.

A rather more urgent problem arises from reports[10,11] that high doses of ascorbic acid given to the pregnant guinea pig shorten the gestation period and increase the incidence of stillbirths. One investigation used ascorbic acid at 0.25 g/kg,[11] but the other was at the much lower dose of 25 mg/kg. As similar findings of increased stillbirths and abortions have been published for a study[12] of ascorbic acid given at doses from 50 mg to 0.5 g daily to pregnant guinea pigs, it appears that toxicity during pregnancy occurs at a surprisingly low level in these animals.

In a more recent study,[13] doses up to 1.0 g/kg ascorbic acid were shown to be without toxic effect on the fetus, and were shown not to induce malformations when given to pregnant rats or mice. There may be, therefore, a marked species difference in the effects of vitamin C during embryonic development.

For dehydro-L-ascorbic acid, there is good evidence of diabetogenic actions in rats. Relatively low doses induce pancreatic damage similar to alloxan,[14] and similar structural features between dehydroascorbic acid and alloxan have been pointed out. Continued treatment of rats with dehydroascorbic acid leads to the development of diabetic-type cataracts.[15]

In summary, it can be said that L-ascorbic acid has very low toxicity for most species and that, while gross tissue changes have not been reported at high doses, there is evidence of undesirable toxic effects on the liver, thyroid, and bone. There are also changes in heart biochemistry. No carcinogenicity has been reported, but at high doses there is the induction in pregnant guinea pigs, though not in rats or mice, of fetal damage characterized by shortened gestation, abortions, and increased stillbirths. Dehydro-ascorbic acid appears to be a diabetogenic agent similar to alloxan.

A summary of the reported changes is shown in Table 1.

CLINICAL STUDIES

Very large doses of vitamin C have been given to normal volunteers as well as to patients with a wide variety of diseases. There is also much self-prescribed vitamin C taken in high doses for the common cold. Very few toxic effects have been reported though there are occasional problems, some of them serious.

A physician who has prescribed vitamin C in daily doses of 3 to 30 g for long periods has reported that the commonest side effect is diarrhea.[16] At slightly lower doses (0.25 to 4 g daily) two careful double-blind investigations involving 1171 adults in the first study[17] and 407 in the second[18] reported a very low incidence of diarrhea and other abdominal symptoms that was similar in treated and placebo groups. However, two cases of skin rash associated with vitamin C were seen and could be a rare individual allergic response.

The most serious reported side effect appears to be the development of intestinal ascorbate lithiasis, with blockage of the ileocecal valve, which required surgical correction.[19] The patient had a personal and family history of urinary lithiasis, but why high dose vitamin C should have produced intestinal stones containing ascorbate is unclear, unless this patient was unable to absorb the particular vitamin C formulation she took.

There has been much speculation concerning the possible role of vitamin C in oxalate lithiasis, for it is well known[20] that ascorbic acid is metabolized to oxalate and causes a dose related increase in urinary oxalate excretion, though the effect is small in most people. However, a few individuals show a massive rise in 24-hr urinary oxalate (up to 8 mmol) after a large dose of ascorbic acid.[21] There is at least one report[22] of urinary stone formation in an otherwise healthy man following a short course of vitamin C (2 g/day).

Following animal studies, reported above, which indicated that vitamin C supplemented guinea pigs more readily developed scurvy when switched to a scorbutic diet, similar results have been reported for humans. During the Siege of Leningrad there was apparently a higher incidence of scurvy among people who had previously received vitamin C supplements.[23] Two cases of frank scurvy have also been reported[24] in otherwise healthy men who abandoned vitamin C supplements and returned to a normal dietary intake. The same condition developed in a volunteer who deliberately made this change. There is also evidence of conditioning to high-dose vitamin C in 18 adults who showed gradual lowering of blood and red cell ascorbate, combined with increasing urinary loss, during treatment.[25]

Other side effects of vitamin C are most difficult to evaluate. On the suggestion that ascorbic acid secreted into the mucus of the uterine cervix may disrupt disulfide bonding of glycoprotein fibrils, and so impede sperm penetration, a request was published for details on conceptions occurring during high dose vitamin C in young women. Details on four women who failed to conceive while receiving vitamin C were reported;[26] several of these conceived shortly after withdrawal of the supplement. While vitamin C is undoubtedly widely used during human pregnancy, there are no convincing data that it is entirely safe. There is one short report[12] of 16 pregnant women who aborted after receiving ascorbic acid at 6 g daily for 3 days, but full clinical details are lacking.

Thrombosis is a further suspected adverse association of high-dose vitamin C. A deep vein thrombosis was reported by a young, athletic physiologist who took 3 g ascorbic acid at night and woke with a thrombosed leg,[27] though no cause and effect relationship could be proved. A double blind study of 200 mg vitamin C daily versus a placebo in 538 geriatric patients revealed that thrombotic episodes, coronary or cerebrovascular accidents, occurred in 11.8% of the treated group compared to 9.4% of those receiving placebos.[28] This possible association deserved more detailed study, for there is evidence

Table 2
INTERFERENCE BY VITAMIN C WITH CLINICAL CHEMISTRY

Clinical chemistry test	Effect of excess ascorbic acid	Reference
Serum bilirubin	Result suppressed	31, 32
Aminotransferases	Result suppressed	32, 33
Lactate dehydrogenase	Result suppressed	32
Serum glucose	Result suppressed	34
Serum uric acid	Result enhanced	35
Urine glucose	Result suppressed or enhanced depending on method	36
17-Hydroxycorticosteroids	Result suppressed	37
Urobilinogen	Result suppressed	38
Occult stool blood	Result suppressed	43

Table 3
DRUG INTERACTIONS WITH EXCESS ASCORBIC ACID

Drug	Effect with ascorbic acid	Reference
Amino salicyclic acid	May enhance risk of drug crystalluria	39
Amphetamines	Renal tubular reabsorption of amphetamine is decreased	38
Warfarin	Two reports of impaired response	40, 41
Tricyclic anti-depressants	Renal tubular reabsorption is decreased	38
Salicylates	Renal tubular reabsorption is increased	38
Vitamin B_{12}	Substantial amounts destroyed	42

from in vitro work with human platelets[29] of an adverse effect of ascorbic acid on their structure and function.

There are a few further possible indirect indications of possible long-term hazards from excessive vitamin C. A study of 1 g daily ascorbic acid in healthy young men for 2 weeks reported increased urinary excretion of calcium, iron, and manganese, but reductions in copper and zinc.[30] There were wide individual variations and it is not clear whether these changes reflect any real effect of the vitamin.

Two final aspects of vitamin C actions in the human will be mentioned, though neither are direct toxic actions. The first is the interference by high concentrations of ascorbic acid with clinical chemistry measurements so that a potentially toxic change may be hidden, or a false positive finding be obtained. Table 2 lists some interferences of this type.

Finally, there is the possibility that high doses of vitamin C given with another drug may initiate or enhance toxicity by synergism. Table 3 lists several reported interactions.

In summary, large amounts of vitamin C have little toxic effect in humans, though diarrhea may be a problem at very high dose. Occasional rare individuals may develop an allergic rash. There is one reported case of intestinal lithiasis and another of urinary lithiasis, though a few individuals show high capacity to metabolize ascorbic acid to oxalic acid. Sudden stopping vitamin C supplements and changing to a normal intake may precipitate scurvy. Daily doses in excess of 2 g may reduce fertility in young women, while there is a report of abortions induced by 6 g daily. A possible relationship to inappropriate coagulation requires further study. Large amounts of vitamin C may also give false clinical chemistry results in some tests, while interaction with several drugs has been reported. A summary is provided in Table 4.

Table 4
TOXIC EFFECTS OF EXCESS VITAMIN C
IN HUMANS

Subjects	Effects of high doses	References
Psychiatric patients	Occasional diarrhea	16
Normal adults	Rare allergic skin rash (?)	18
Adult woman	Intestinal lithiasis with blockage of ileocecal valve	19
Normal adults	3 of 67 showed massive oxaluria following vitamin C	21
Normal man	Urinary lithiasis	22
Normal adults	Scurvy following cessation of vitamin C supplements	23, 25, 34
Normal women	Reduced fertility (?)	26
Pregnant women	Abortion	12
Normal man	Thrombosis	27
Geriatric patients	Increased thrombosis (?)	28
Normal men	Altered mineral balance (?)	30
Human fibroblasts (in vitro)	Chromosome aberrations	46
Normal adults	Increased uric acid clearance	45
Normal adults	Increased red cell hemolysis in vitro	46
Man with red cell G6PD deficiency	Death from in vivo red cell hemolysis and disseminated intravascular coagulation	47

REFERENCES

1. **Korner, W. F. and Weber, F.,** Zur Toleranz hoher Ascorbinsäuredosen, *Int. Z. Vit. Eru. Forsch.,* 42, 528, 1972.
2. **Lippi, U., Pulido, E., and Guidi, G.,** Lesioni lisosomiali da dosi elevate di acido ascorbico, *Acta Vitaminol.,* 5, 177, 1966.
3. **Mallick, N. and Deb, C.,** Effect of different doses of ascorbic acid on thyroid activity in rats at different levels of protein intake, *Endocrinologie,* 65, 333, 1975.
4. **Deb, C. and Mallick, N.,** Effect of chronic graded doses of ascorbic acid on thyroid activity, protein bound iodine of blood and deiodinase enzyme of peripheral tissues of guinea pigs, *Endocrinologie,* 63, 231, 1974.
5. **Dashman, T., Horst, D., Bautz, G., and Kamm, J. J.,** Ascorbic acid: Effect of high doses on brain and heart catecholamine levels, *Experientia,* 29, 832, 1973.
6. **Brown, R. G.,** Possible problems of large intakes of ascorbic acid, *J.A.M.A.,* 224, 1529, 1973.
7. **Thornton, P. A. and Omdahl, J. L.,** Further evidence of skeletal response to exogenous ascorbic acid, *Proc. Soc. Exp. Biol. Med.,* 132, 618, 1969.
8. **Mayfield, H. L. and Roehm, R. R.,** Influence of ascorbic acid and the source of B vitamins on the utilization of carotene, *J. Nutr.,* 58, 203, 1956.
9. **Gordonoff, T.,** Can water-soluble vitamins be overdosed? Research on vitamin C, *Schweiz Med. Wochenschr.,* 90, 726, 1960.
10. **Neuweiler, W.,** Hypervitaminosis and its relation to pregnancy, *Int. Z. Vitaminforsch.,* 22, 392, 1951.
11. **Mouriquand, G. and Edel, V.,** Hypervitaminosis C, *C. R. Soc. Biol.* (Paris), 147, 1432, 1953.
12. **Samborskaia, E. P.,** Effect of large doses of ascorbic acid on course of pregnancy in the guinea pig, *Bull. Exp. Biol. Med.* (Moskva), 57, 105, 1966.
13. **von Frohberg, H., Gleich, J., and Kieser, H.,** Reproduktionstoxikologische Studien mit Ascorbinsäure, au Mausen und Ratten, *Arzneim-Forsch.,* 23, 1081, 1973.
14. **Patterson, J. A.,** Diabetogenic effect of dehydroascrobic acid and dehydroisoascorbic acid, *J. Biol. Chem.,* 183, 81, 1950.
15. **Patterson, J. A.,** Course of diabetes and development of cataracts after injecting dehydroascorbic acid and related substances, *Am. J. Physiol.,* 165, 61, 1951.
16. **Hoffer, A.,** Ascorbic acid and toxicity, *N. Engl. J. Med.,* 285, 635, 1972.

17.   **Anderson, T. W., Surany, G., and Beaton, G. H.,** The effect on winter illness of large doses of vitamin C, *Can. Med. Assoc. J.,* 111, 31, 1974.
18.   **Anderson, T. W., Reid, D. B. W., and Beaton, G. H.,** Vitamin C and the common cold: a double-blind trial, *Can. Med. Assoc. J.,* 107, 503, 1972.
19.   **Vickery, R. E.,** Unusual complication of excessive ingestion of vitamin C tablets, *Int. Surg.,* 58, 422, 1973.
20.   **Lamden, M. P. and Chrystowski, G. A.,** Urinary oxalate excretion by man following ascorbic acid ingestion, *Proc. Soc. Exp. Biol. Med.,* 85, 190, 1954.
21.   **Briggs, M. H.,** Vitamin C induced hyperoxaluria, *Lancet,* i, 154, 1976.
22.   **Briggs, M. H.,** Side-effects of vitamin C, *Lancet,* ii, 1439, 1973.
23.   **Jakovlieu, N.,** Scurvy following nutritional stress, *Ernaehrungsforschung,* 3, 446, 1958.
24.   **Rhead, W. J. and Schrauzer, G. N.,** Risks of long term ascorbic acid overdosage, *Nutr. Rev.,* 11, 262, 1971.
25.   **Schrauzer, G. N. and Rhead, W. J.,** Ascorbic acid abuse: Effects of long-term ingestion of excessive amounts on blood levels and urinary excretion, *Int. J. Vit. Nutr. Res.,* 43, 201, 1973.
26.   **Briggs, M. H.,** Fertility and high-dose vitamin C, *Lancet,* ii, 1083, 1973.
27.   **Horrobin, D. G.,** D.V.T. after vitamin C? *Lancet,* ii, 317, 1973.
28.   **Andrews, C. T. and Wilson, T. S.,** Vitamin C and thrombotic episodes, *Lancet,* ii, 39, 1973.
29.   **Cowan, D. H., Graham, R. C., Shook, P., and Griffiths, R.,** Influence of ascorbic acid on platelet structure and function, *Thromb. Diath. Haemorrh.,* 34, 50, 1975.
30.   **Hanck, A. B.,** Der Einfluss von 1000 mg Vitamin C pro Tag auf das renale Ausscheidungsverhalten einiger Elektrolyte im Harn des gesunden Meuschen, *Int. Z. Vit. Ern. Forsch.,* 43, 34, 1972.
31.   **Briggs, M. H., Garcia-Webb, P., and Johnson, J.,** Dangers of excess vitamin C, *Med. J. Aust.,* ii, 48, 1973.
32.   **Spiegel, H. E. and Pinili, E.,** Effects of vitamin C on SGOT, SQPT, LDH and bilirubin, *Med. J. Aust.,* ii, 117, 1974.
33.   **Singh, H. P., Herbert, M. A., and Gault, M. H.,** Effect of some drugs on clinical laboratory values as determined by the Technicon SMA-12/60, *Clin. Chem.,* 18, 137, 1972.
34.   **Rodriguez, J. A., Robinson, C. A., Smith, M. S., and Frye, J. H.** Evaluation of an automated glucose-oxidase procedure, *Clin. Chem.,* 21, 1513, 1975.
35.   **Carroll, J.,** A simplified alkaline phosphotungstate assay for uric acid in serum, *Clin. Chem.,* 17, 158, 1971.
36.   **Mayson, J. J.,** False negative tests for urinary glucose in the presence of ascorbic acid, *Am. J. Clin. Pathol.,* 58, 297, 1972.
37.   **Hansten, P. D.,** *Drug Interactions,* 2nd ed., Lea & Febiger, Philadelphia, 1972.
38.   **Milne, M. D.,** Influence of acid-base balance on efficacy and toxicity of drugs, *Proc. R. Soc. Med.,* 58, 961, 1965.
39.   **Meyers, F. H.,** *Review of Medical Pharmacology,* 3rd ed., Lange, Los Altos, Calif., 1972, 518.
40.   **Rosenthal, G.,** Interaction of ascorbic acid and warfarin, *J.A.M.A.,* 215, 1671, 1971.
41.   **Smith, E. C.,** Interaction of ascorbic acid and warfarin, *J.A.M.A.,* 221, 1166, 1972.
42.   **Herbert, V. and Jacob, E.,** Destruction of vitamin B_{12} by ascorbic acid, *J.A.M.A.,* 230, 241, 1974.
43.   **Jaffe, R. M., Kasten, B., Young, D. S., and MacLowry, J. D.,** Fake negative stool occult blood tests caused by ingestion of ascorbic acid (vitamin C), *Ann. Intern. Med.,* 83, 824, 1975.
44.   **Stich, H. F., Karim, J., Koropatnick, J., and Lo, L.,** Mutagenic action of ascorbic acid, *Nature,* 260, 722, 1976.
45.   **Stein, H. B., Hassan, A., and Fox, I. H.,** Ascorbic acid-induced uricosuria: A consequence of megavitamin therapy, *Ann. Intern. Med.,* 84, 385, 1976.
46.   **Mengel, C. E. and Green, H. L., Jr.,** Ascorbic acid effects on erythrocytes, *Ann. Intern. Med.,* 84, 490, 1976.
47.   **Campbell, G. D., Jr., Steinberg, M. H., and Bower, J. D.,** Ascorbic acid-induced hemolysis in G6PD deficiency, *Ann. Intern. Med.,* 82, 810, 1975.

Fat-soluble Vitamins

EFFECT OF NUTRIENT TOXICITIES (EXCESS)
IN ANIMALS AND MAN: VITAMIN A

M. Y. Jenkins

Adverse effects due to excessive vitamin A intake have been observed in animals and humans. Although the likelihood of toxicity from ingesting naturally occurring foods is very small, Arctic explorers and others have described the occurrence of symptoms of acute vitamin A intoxication following ingestion of liver rich in vitamin A (e.g., liver from polar bear, bearded seal, halibut, whale, tuna, cod, Husky dog, shark, and Greenland fox).[1-5] Toxic reactions have also been known to occur following administration of large doses of vitamin A preparations.

Preparations containing large amounts of vitamin A have been used for the management of a wide variety of skin disorders, for some gynecological problems, as a public health measure in some countries, by food faddists, as a prophylactic, and to improve vision.[5-8] Vitamin A toxicity appears as either the acute or chronic form; acute vitamin A intoxication is caused by short-term ingestion of excessive amounts and chronic hypervitaminosis A results from long-term intake of smaller excessive doses. The most frequently observed symptoms of acute vitamin A intoxication are increased cerebrospinal fluid pressure (bulging of the fontanels in infants and children; headache predominantly in the occipital region in adolescents and adults), anorexia, nausea, vomiting, and desquamation of the skin and mucous membranes.[5] In chronic vitamin A intoxication, the most prominent features are intracranial hypertension, skin and hair deviations, pain in the musculoskeletal system, and fatigue.[9,10]

The dose of vitamin A required to produce toxicity in man is variable. One patient had toxic symptoms within 3 days with a total intake of 1,300,000 IU,[11] while another patient tolerated up to 1,250,000 IU daily for 5 years.[12] A recent review of the evaluation of 517 cases of vitamin A toxicity indicated a significant positive correlation between the dose administered and the duration of treatment.[5] Emulsified or equivalent preparations, because of their better absorbability, caused the appearance of the symptoms of hypervitaminosis A significantly earlier (by a factor of 6) than did comparable doses of oily emulsions.

Many cases of vitamin A toxicity in children and adults are described in the literature.[1-3,5,6,8-35] Typical reports of effects in humans and animals are given in Table 1. For additional data on vitamin A toxicity, the reader is referred to review articles.[5,9,10,36-51]

TOXICITY IN ANIMALS

One of the first reports of vitamin A toxicity in animals was described by Japanese investigators over 50 years ago as a result of testing crude vitamin A concentrates in rats and mice. Large oral doses produced loss of hair, emaciation, and paralysis, with death following within days or weeks.[52] Since then, numerous animal experiments have demonstrated the toxic effects of large doses of vitamin A.[38,39,43,44,47,49,52-93]

Ability to induce toxicity varies widely with species. The calf and pig are in a range of susceptibility comparable to man (approximately 1000 to 3000 μg/kg body weight/day), whereas the rat and carnivores such as the dog, cat, and probably polar bear and seal can tolerate extremely large (20,000 to 60,000 μg/kg) daily doses before being clinically affected.[39] Some of the characteristic effects of excess vitamin A in experimental animals are depressed growth, anemia, loss of hair, skin lesions, hemorrhages related to secondary vitamin K deficiency, degenerative atrophy of various organs, and congenital malformations.[44] In addition, skeletal lesions occur which may include reduction in oppositional bone formation due to suppression of osteoblastic activity and reduction in longitudinal growth of the long bones due to generative changes in cartilaginous epiphyseal plates.[38,44,49]

In recent years, much emphasis has been placed on the teratological effects of excess vitamin A. Many congenital abnormalities have been produced by treatment with excess vitamin A.[53-65] The main malformations noted in the early studies include resorption of the fetus, anencephaly, cleft palates, and cataracts.[53,56] The types and incidence of malfunctions depend on the stage of pregnancy, dose, and to a lesser extent on species and strain.[53-65] Reported effects on structures which are analogous to some malformations observed in humans include the following: brain (anencephaly), spinal cord (spina bifida), face (cleft lip, cleft palate, micrognathia), eye (microphthalmia), all parts of the ear, teeth, salivary glands, aortic arch (malformations of several types), heart (ventricular septal defect), lungs, gastrointestinal tract (imperforate anus, omphalocele), liver and gallbladder, urinary system (renal "agenesis," hydronephrosis), genitalia, pituitary, thyroid, thymus, skull, vertebrae, ribs, extremities (phocomelia, digit malformations), muscles, and situs inversus.[65]

There are indications of interaction between vitamin A and several other substances. Studies in rats show that vitamins A and D diminish the toxic effects of each other.[84] The combination of large amounts of vitamin A with vitamin D partially prevents the pathological changes in the skeleton that occur in hypervitaminosis D. It is suggested that the beneficial effect of vitamin A results from increased mucopolysaccharide and collagen turnover.

Depending on the dosage, vitamin E partially counteracts the toxic effect of vitamin A in rats and chicks.[85-87] It has also been suggested that excess vitamin A causes deficiencies of vitamins C, E, and K.[88-90] Pretreatment of rats with pregnenolone-16α-carbonitrile (a microsomal enzyme inducer) prevents bone lesions and decreases the number, size, and lipid content of perisinusoidal cells caused by excess vitamin A.[91]

Studies in rats, monkeys, and humans suggest that vitamin A toxicity occurs when excessive amounts of vitamin A are presented to cell membranes in association with plasma lipoproteins, rather than specifically bound to retinol-binding protein. Retinol-binding protein may not only regulate the supply of retinol to tissues but also protect tissues from the surface-active properties of the vitamin.[39,92-94]

The information presented shows the nutritional and metabolic disturbances caused by excess vitamin A. The complexity of nutrient interaction emphasizes the need for caution in consumption of excesses of any nutrient.

RESPONSES TO VITAMIN A TAKEN ORALLY IN EXCESS OF REQUIREMENTS

Species	Age	Oral vitamin A intake	Conditions and effects	Reference
Human	2 months	Pregnant mother had taken 25,000 IU/day for 3 months and 50,000 IU/day during last 6 months.	Congenital renal anomalies resulting in obstructive salt-losing nephropathy.	15
Human	4½ months	17,000 IU/day for 3½ months	Slightly bulged fontanels, plasma vitamin A 470 IU/100 ml, wrist and ankle metaphyses widened and sharply demarcated. At 9 months of age, conditions in the wrist and ankles were normal.	27
Human	3–7½ months	350,000 IU	Three persons, within 12 hr after ingestion, hydrocephalus with intensive bulging of the fontanel, frequent vomiting, and agitation or insomnia. Disturbances subsided within 24–48 hr either spontaneously or after lumbar puncture.	23
Human	12–32 months	75,000–500,000 IU/day for 6–15 months	Seven persons. Pruritis, hyperirritability, swelling, and pain of the extremities accompanied by hyperostoses.	16
Human	Infant	30,000–600,000 IU/day for approximately 3–33 months	Thirty-six persons. In general all age groups shared common symptoms and signs. All infants and toddlers, but only one of nine in the older group, had changes in long bones. Central nervous system symptoms and signs due to increased intracranial pressure occurred in 20% of the total group.	45
	Young	200,000–463,040 IU/day for 6 weeks–5½ years		
	Adult	50,000–600,000 IU/day for 2 months–8½ years		
Human	15 months	15,000 IU/day for 7 months	Enlarged head, shiny skin, cheilitis, severe alopecia and bone tenderness. At 13 years of age there was permanent deformity of long bones and scoliosis.	30

Table 1 (continued)

RESPONSES TO VITAMIN A TAKEN ORALLY IN EXCESS OF REQUIREMENTS

Species	Age	Oral vitamin A intake	Conditions and effects	Reference
Human	30 months	57,000 IU/day for a year	Obviously large head, stiff neck, slightly enlarged liver and spleen, patches of alopecia, inability to walk because of pain in both shins; serum vitamin A, 520 μg/100 ml. Increased intracranial pressure and low cerebrospinal fluid protein concentrations (6 mg/100 ml; normal 11.5–23.9). Slight cupping of metaphyses in feet. Clinical condition improved dramatically when all vitamin supplements were stopped. Two months later the cupped metaphyses in feet were unchanged.	32
Human	6 years	625,000 IU/week for over 4 years	Anemia, dry skin, sparse scalp and eyebrow hair, anorexia, hepatosplenomegaly, hepatic dysfunction, and hypersplenism. Liver biopsy specimen displayed large lipid-laden Kupffer's cells which were shown to contain vitamin A by fluorescence microscopy. Good growth of hair, mild anemia, and hypersplenism were observed 1 year later.	29
Human	15 years	Approximately 200,000–300,000 IU/day for about a year	Diplopia. Pseudotumor cerebri indicated by severe headaches, sixth-nerve palsy and papilledema. After 4 weeks patient was significantly improved.	22
Human	16 years	5,000 IU/day for 3 years; 50,000–75,000 IU/week for 8 months; 25,000 IU for unknown period; 50,000–200,000 IU/day during preceding 4 months	Emotional lability, weight loss, serum vitamin A 525 IU/100 ml, hypercalcemia and vague skeletal pains. Multiple sites of periosteal calcification and pronounced demineralization of sella turcica. Two months after all vitamin supplements were discontinued, periosteal calcification had disappeared and marked remineralization of the sella turcica had occurred.	19
Human	18 years	300,000 IU/day for one month; 100,000–200,000 IU/day for approximately 4 months	Painful muscular stiffness and fatigue, nosebleeds, anorexia, swollen feet, generalized pruritus. Severe headache. A bright red discoloration of the gingival margin. Severe hepatic fibrosis and	9

Table 1 (continued)

RESPONSES TO VITAMIN A TAKEN ORALLY IN EXCESS OF REQUIREMENTS

Species	Age	Oral vitamin A intake	Conditions and effects	Reference
Human	18 years	100,000—150,000 IU/day for 20 months	Frontal headaches, blurred vision, sleep disturbances, irregular menses and ringing in the ears. Anorexia, weight loss, painful sensitiveness to sounds and difficulty with balance. Development of a severe toxic psychotic reaction followed almost immediately by pseudo tumor with resolution all symptoms upon cessation of vitamin A.	28
Human	19 years	300,000 IU/day for 2 years	Somnolence and tiredness, a holo cranial murmur, bilateral papilloedema with suffusions, enlargement of blind spot, sixth-nerve paresis, dry and chapped lips, dry and squamous skin. Completely recovered 4 weeks after discontinuing medication.	10
Human	20 years	50,000 IU/day increased to 400,000 IU/day during 2-year period.	Anorexia, increasingly severe headaches and diffuse scaling, erythematous dermatitis, marked bilateral hemorrhagic papilledema and hepatomegaly. Asymptomatic 8 months after being placed on a diet low in vitamin A and carotene.	94
Human	Adult	High vitamin A halibut liver (ca. 20—300 g/meal). 2—30 million IU	Eleven people, 5 hr after eating, had dull heavy headaches, dizziness, and nausea; 4 men vomited. Skin was red and desquamated 24 hr later. By 3 weeks desquamation of the skin had entirely disappeared.	3
Human	28 years	120,000 IU/day	Lesions in the peripheral nervous system.	34
Human	28 years	1,300,000 IU vitamin A acetate within 3 days	Nausea, vomiting, intense headache, blurring of vision associated with papilledema. Subsidence of the acute symptoms followed by exfoliation of the superficial skin.	11

Table 1 (continued)

RESPONSES TO VITAMIN A TAKEN ORALLY IN EXCESS OF REQUIREMENTS

Species	Age	Oral vitamin A intake	Conditions and effects	Reference
Human	36 years	30,000 IU/day for 8 months; 125,000 IU/day for 4 months	Irritability, depression. Excessive fatiguability, pain in legs, rough dry and scaly skin, dark brown pigmentation around mouth, and hair loss. Condition improved following cessation of vitamin A therapy.	8
Human	41 years	100,000 IU/day for 6 months	Bilateral papilledema with suffusions and nystagmus when looking aside. Completely cured 8 months after discontinuance of vitamin A intake.	10
Human	48 years	Up to 1,500,000 µg retinol/day for 3 months. Large doses of carrot juice for approximately 6 years	Autopsy revealed severe jaundice with liver enlargement. A high liver concentration of retinol remained even after fixation.	95
	54 years	Up to 1,250,000 IU/day for 5 years	Dry skin, fatigue, hair loss, hepatosplenomegaly and ascites. Liver vitamin A, 1,700 µg/g liver (controls 100—130 µg/g). Perisinusoidal fibrosis and massive accumulation of lipid storing cells (Ito cells) in liver. Six months later, although hepatosplenomegaly remained, liver function showed slight improvement.	12, 31
Goat	Adult	17,640 µg retinol/kg live weight/day for 16 weeks	Epidermal and bone lesions	67
Dog	4 to 6 weeks	100,000—300,000 IU/kg bw/day for 88—98 days	Loss of body weight, dullness, emaciation, roughened coat, evident pain in limb joints and retarded growth. Decreases in overall length and thickness of long bones, development of osteophytes, periosteal reactions, and premature closure of epiphyses. Degenerative epiphyseal plate, hemorrhage and exostotic proliferation of periosteum, fatty liver, and microcalculi in kidney.	66
Calf	1 day	8,800—17,600 µg retinol/kg/day for 12 weeks	Arachnoid granulations, small and covered by thin and small fibrous caps. Structural changes and	70

Table 1 (continued)
RESPONSES TO VITAMIN A TAKEN ORALLY IN EXCESS OF REQUIREMENTS

Species	Age	Oral vitamin A intake	Conditions and effects	Reference
Calf	21 weeks	8,800—17,600 μg retinol/kg/day for 12 weeks	Enlarged adrenals and kidneys. Supraspinatus muscles contained less dry matter; greater ash per unit of dry matter; and greater sodium per unit of fresh tissue, dry matter, and ash than controls.	68
Calf	2½ months	603—16,000 μg retinol/lb bw/day for 12 weeks	Decreased growth and food consumption. Elevation of heart rate. Decrease in cerebrospinal fluid pressure. Increase in plasma and liver vitamin A. Decrease in plasma ascorbic acid. Enlarged liver, heart, and kidney. Change in composition of metacarpal bones.	83
Chick	7 days	20 mg retinyl acetate/100 g bw/day for 14—28 days	Inflamed skin around external nares and commisures of beak and eyes. Shorter, narrower, and lighter bones than normal. Inhibition of osteoblasts. Increased amount of epiphyseal cartilage.	71
Duck	3—22 days	100,000—250,000 IU/day for 1—19 days	Decreased growth and food intake. Combined width of epiphyseal cartilage and metaphyseal zones decidedly less than normal, accelerated endochondral ossification.	72
Cat		High vitamin A in commercial food, mostly from fish liver.	Massive hemorrhages and death.	69
Cat	Young	High vitamin A in raw liver from sheep and cattle, 17—35 μg/g bw/day	Marked lipid infiltration of reticulohistiocytic cells of liver, lungs, spleen, and hepatic lymph node and of the tubular epithelium of the renal cortex. High vitamin A levels in plasma, liver, and kidney. Deforming cervical spondylosis within 10 months.	74

Table 1 (continued)
RESPONSES TO VITAMIN A TAKEN ORALLY IN EXCESS OF REQUIREMENTS

Species	Age	Oral vitamin A intake	Conditions and effects	Reference
Cat	Young	High vitamin A in raw liver from sheep and cattle, 35 μg/g bw/day; synthetic vitamin A 30—150 μg/g bw/day	Loss of appetite, lethargy, torpor, marked irritability, exophthalmos, scurfiness and dullness of the coat. Extensive exostoses of the cervical spine with palpable rigidity of the cervical spine, hyperesthesia, lameness of one or both forelimbs, and abnormalities of gait. Lesions of teeth and skin. Testicular degeneration and temporary loss of fertility in males. Substantial transfer of vitamin A to offspring in milk but less via placenta.	75
Mouse	Pregnant adult and fetuses	80 mg retinoic acid/kg on 10th—16th day of gestation.	Phocomelia and micromelia in fetuses when administered to pregnant mice on the 12th or 13th day of gestation. Cleft palate in 68% of fetuses from dams treated on 10th day and cleft palate in 100% of fetuses from dams treated on 12th day.	60
Rat	Pregnant adult and fetuses	50,000 IU vitamin A palmitate on 7th—10th day of gestation	Abnormal flexures of the neural tube and malformation of the skull base.	59
Rat	Young and adult	15 mg/day; 30 mg/day; 7.5 mg/day; 15 mg/day	No effect of acute dose in young rats. No effect of acute dose in adult rats. Both young and adult rats treated chronically for 6—8 days had a 93% decrease in cerebrospinal fluid.	73
Rat		500 IU vitamin A palmitate/g bw/day for 3—4 weeks	Changes in cardiac intracellular action potential: decrease in amplitude, in depolarization rate, in fast and slow repolarization rate, and increase in partial and total duration. Increase in spontaneous heart rate and systole duration and decrease in diastole duration.	80

RESPONSES TO VITAMIN A TAKEN ORALLY IN EXCESS OF REQUIREMENTS

Species	Age	Oral vitamin A intake	Conditions and effects	Reference
Rat		250,000 IU retinyl palmitate	Significant reductions in erythrocyte, hemoglobin, hematocrit, and thrombocyte values and in the concentrations of coagulation factors II, V, VII and X with a consequent prolongation in the coagulation-, Quick-, and partial-thromboplastin times.	79
Rat	Young	30,000 IU retinol for 2 days	Marked increase in hepatic glycogen, cholesterol, and glycerides, elevated plasma free fatty acids, increased adrenal weight, and decreased cholesterol in the adrenals.	78
Rat	Pregnant adult and fetuses	100,000—200,000 IU vitamin A on 12th—16th day of gestation.	Cleft palate formation, disorganization of mandibular and maxillary growth, and relative macroglossia.	64
Rat	Pregnant adult and fetuses	10,000—40,000 IU/kg/day on 8th—10th day of gestation.	High prenatal and postnatal mortality rates and marked growth retardation and abnormal gait in survivors. Offspring showed hypoactivity compared with saline- and nontreated controls. Decreased avoidance and discrimination acquistion when tested in a Y-maze avoidance task.	63
Rat	Adult	Vitamin A palmitate or vitamin A alcohol (50,000 IU/day and 75,000 IU/day for 17 days)	Mortality, bone lesions, and calcification of lungs, heart, kidneys, and liver.	82
Rat	Adult and young	10,000 IU/100 g bw for 30 days	Decreased mucopolysaccharide and collagen content of bone.	84
Guinea pig	Adult	2.5—3 million IU vitamin A palmitate	Death within 10—17 days; loss of weight, appetite, and hair, a spastic walk and considerable rigidity	77

Table 1 (continued)
RESPONSES TO VITAMIN A TAKEN ORALLY IN EXCESS OF REQUIREMENTS

Species	Age	Oral vitamin A intake	Conditions and effects	Reference
Guinea pig	Young	330 mg retinol/kg diet for 40 days	with marked curvature of the spine. Severe damage in the cerebellum, thalamus, and mesencephalon. Reduced growth, metastatic calcinosis of the kidney and liver, and severe atrophy of the liver.	81
Guinea pig Hamster	Pregnant adult and fetuses	75,000—400,000 IU vitamin A palmitate/kg bw	Extremely retarded limb and tail development, imperforate ani, hydronephrosis, and anephrosis.	58
Hamster	Pregnant adult and fetuses	20,000 IU vitamin A on 8th day of gestation.	Somite necrosis and mesodermal alterations within 12 hr.	62
Rabbit	Pregnant adult and fetuses	80,000 IU/kg bw 150,000 IU/kg bw	50% abortions and some malformations. Total abortions.	57
Pig	7 weeks	19,842 µg/kg bw/day for 5 weeks	Anorexia, cutaneous erythema, and hemorrhage, and posterior lameness. Tubular and flat bones decreased in overall length and thickness. At necropsy, the bones were more fragile, broke and could be cut readily.	76

REFERENCES

1. Cleland, J. B. and Southcott, R. V., Illnesses following the eating of seal liver in Australian waters, *Med. J. Aust.*, 1, 760–763, 1969.
2. Lugg, D. J., Antarctic medicine 1775–1975. II, *Med. J. Aust.*, 2, 335–337, 1975.
3. Nater, J. P. and Doeglas, H. M. G., Halibut liver poisoning in 11 fishermen, *Acta Derm. Venereol.*, 50, 109–113, 1970.
4. Rodahl, K. and Moore, T., The vitamin A content and toxicity of bear and seal liver, *Biochem. J.*, 37, 166–168, 1943.
5. Körner, W. F. and Vollm, J., New aspects of the tolerance of retinol in humans, *Int. J. Vitam. Nutr. Res.*, 45, 363–372, 1975.
6. Olson, J. A., The prevention of childhood blindness by the administration of massive doses of vitamin A, *Isr. J. Med. Sci.*, 8, 1199–1206, 1972.
7. Kusin, J. A., Reddy, V., and Sivakumar, B., Vitamin E supplements and the absorption of a massive dose of vitamin A, *Am. J. Clin. Nutr.*, 27, 774–776, 1974.
8. Gupta, M. C. and Kumar, S., Chronic hypervitaminosis A in an adult, *J. Assoc. Physicians India*, 22, 865–868, 1974.
9. Muenter, M. D., Perry, H. O., and Ludwig, J., Chronic vitamin A intoxication in adults, *Am. J. Med.*, 50, 129–136, 1971.
10. Lombaert, A. and Carton, H., Benign intracranial hypertension due to A-hypervitaminosis in adults and adolescents, *Eur. Neurol.*, 14, 340–350, 1976.
11. Furman, K. I., Acute hypervitaminosis A in an adult, *Am. J. Clin. Nutr.*, 26, 575–577, 1973.
12. Hruban, Z., Russell, R. M., Boyer, J. L., Glagov, S., and Bagheri, S. A., Ultrastructural changes in livers of two patients with hypervitaminosis, *Am. J. Pathol.*, 76, 451–461, 1974.
13. Gal, I., Vitamin A, pregnancy, and oral contraceptives, *Br. Med. J.*, 2, 560–561, 1974.
14. Greer, M., Management of benign intracranial hypertension (pseudotumor cerebri), *Clin. Neurosurg.*, 15, 161–174, 1968.
15. Bernhardt, I. B. and Dorsey, D. J., Hypervitaminosis A and congenital renal anomalies in a human infant, *Obstet. Gynecol.*, 43, 750–755, 1974.
16. Caffey, J., Chronic poisoning due to excess of vitamin A, *Pediatrics*, 5, 672–688, 1950.
17. Di Benedetto, R. J., Chronic hypervitaminosis A in an adult, *J.A.M.A.*, 201, 700–702, 1967.
18. Fisher, G. and Skillern, P. G., Hypercalcemia due to hypervitaminosis A, *J.A.M.A.*, 227, 1413–1414, 1974.
19. Frame, B., Jackson, C. E., Reynolds, W. A., and Umphey, J. E., Hypercalcemia and skeletal effects in chronic hypervitaminosis A, *Ann. Intern. Med.*, 80, 44–48, 1974.
20. Josephs, H. W., Hypervitaminosis A and carotenemia, *Am. J. Dis. Child.*, 67, 33–43, 1944.
21. Katz, C. M. and Tzagournis, M., Chronic adult hypervitaminosis A with hypercalcemia, *Metabolism*, 21, 1171–1176, 1972.
22. Lascari, A. D. and Bell, W. E., Pseudotumor cerebri due to hypervitaminosis A, *Clin. Pediatr. (Philadelphia)*, 9, 627–628, 1970.
23. Marie, J. and See, G., Acute hypervitaminosis of the infant. Its clinical manifestation with benign acute hydrocephalus and pronounced bulge of fontanel, *Am. J. Dis. Child.*, 87, 731–736, 1954.
24. Muenter, M. D., Hypervitaminosis A, *Ann. Intern. Med.*, 80, 105–106, 1974.
25. Morrice, G., Jr., Havener, W. H., and Kapetansky, F., Vitamin A intoxication as a cause of pseudotumor cerebri, *J.A.M.A.*, 173, 1802–1805, 1960.
26. Naha, P. N., Vitamin A, *J. Assoc. Physicians India*, 22, 860–863, 1974.
27. Persson, B., Tunell, R., and Ekengren, K., Chronic vitamin A intoxication during the first half year of life, *Acta Paediatr. Scand.*, 54, 49–60, 1965.
28. Restak, R. M., Pseudotumor cerebri psychosis, and hypervitaminosis A, *J. Nerv. Ment. Dis.*, 155, 72–75, 1972.
29. Rubin, E., Florman, A. L., Degnan, T., and Diaz, J., Hepatic injury in chronic hypervitaminosis A, *Am. J. Dis. Child.*, 119, 132–138, 1970.
30. Ruby, L. K. and Mital, M. A., Skeletal deformities following chronic hypervitaminosis A, *J. Bone Jt. Surg. Am. Vol.*, 56-A, 1283–1287, 1974.
31. Russell, R. M., Boyer, J. L., Bagheri, S. A., and Hruban, Z., Hepatic injury from chronic hypervitaminosis A resulting in portal hypertension and ascites, *N. Engl. J. Med.*, 291, 435–440, 1974.
32. Siegel, N. J. and Spackman, T. J., Chronic hypervitaminosis A with intracranial hypertension and low cerebrospinal fluid concentration of protein, *Clin. Pediatr. (Philadelphia)*, 11, 580–584, 1972.
33. Sulzberger, M. B. and Lazar, M. P., Hypervitaminosis A, *J.A.M.A.*, 146, 788–793, 1951.
34. Warren, A. G., Letter: Hypervitaminosis A, *Leprosy Rev.*, 44, 220–222, 1973.

35. **Woodard, W. K., Miller, L. J., and Legant, O.,** Acute and chronic hypervitaminosis in a 4-month-old infant, *J. Pediatr.,* 59, 260–264, 1961.

36. **Bartolozzi, G., Bernini, G., Marianelli, L., and Corvaglia, E.,** Ipervitaminosi A cronica nel lattante e nel bambino. Descrizione – di due casi e rassegna critica della letteratura, *Riv. Clin. Pediatr.,* 80, 231–290, 1967.

37. **Bauernfeind, J. C., Newmark, H., and Brin, M.,** Vitamins A and E nutrition via intramuscular or oral route, *Am. J. Clin. Nutr.,* 27, 234–253, 1974.

38. **Clark. L.,** Hypervitaminosis A: A review, *Aust. Vet. J.,* 47, 568–571, 1971.

39. **Hayes, K. C. and Hegsted, D. M.,** Toxicity of the vitamins, in *Toxicants Occurring Naturally in Foods,* 2nd ed., National Academy of Sciences, Washington, D.C., 1973, 236–239.

40. **Jaffee, S. J. and Filer, L. J.,** The use and abuse of vitamin A, *Nutr. Rev.,* 32, *Suppl.* 1, 41–43, 1974.

41. **Jeghers, H. and Marraro, H.,** Hypervitaminosis A: Its broadening spectrum, *Am. J. Clin. Nutr.,* 6, 335–339, 1958.

42. **Knudson, A. G. and Rothman, P. E.,** Hypervitaminosis A. A review with a discussion of vitamin A, *Am. J. Dis. Child.,* 85, 316–334, 1953.

43. **Moore, T.,** *Vitamin A,* Elsevier, London, 1957, 340–351, 444–455.

44. **Nieman, H. J. and Klein, O.,** The biochemistry and pathology of hypervitaminosis A, *Vitam. Horm.,* 12, 69–99, 1954.

45. **Oliver, T. K.,** Chronic vitamin A intoxication. Report of a case in an older child and review of the literature, *Am. J. Dis. Child.,* 95, 57–68, 1958.

46. **Anon.,** Hypervitaminosis A, *Nutr. Rev.,* 9, 183–184, 1951.

47. **Rodahl, K.,** Hypervitaminosis A in the rat, *J. Nutr.,* 41, 399–421, 1950.

48. **Rodriguez, M. S. and Irwin, M. I.,** Hypervitaminosis A, *J. Nutr.,* 102, 919–920, 1972.

49. **Stewart, R. J. C.,** Bone pathology in experimental malnutrition, *World Rev. Nutr. Diet.,* 21, 39–40, 1975.

50. **Stimson, W. H.,** Vitamin A intoxication in adults. Report of a case with a summary of the literature, *N. Engl. J. Med.,* 265, 369–373, 1961.

51. **Young, C. M.,** Overnutrition, *World Rev. Nutr. Diet.,* 16, 187–202, 1973.

52. **Takahashi, K., Nakamiya, Z., Kawakimi, K., and Kitasato, T.,** Physical and chemical properties of biosterin and its physiological significance, *Sci. Pap. Inst. Phys. Chem. Res. (Jpn.),* 3, 81–148, 1925.

53. **Cohlan, S. Q.,** Excessive intake of vitamin A as a cause of congenital anomalies in the rat, *Science,* 117, 535–536, 1953.

54. **Giroud, A. and Martinet, M.,** Fentes du palais chez l'embryon de rat par hypervitaminose A, *C. R. Soc. Biol.,* 148, 1742, 1954.

55. **Giroud, A. and Martinet, M.,** Diverse malformations of the rat fetus in relation to the time of administration of vitamin A in excess, *C. R. Soc. Biol.,* 149, 1088–1090, 1955.

56. **Giroud, A. and Martinet, M.,** Vitamin-A hypervitaminose A und anomalien beim foetus der ratte, *Int. Z. Vitaminforsch.,* 26, 10–18, 1955.

57. **Giroud, A. and Martinet, M.,** Vitamin A effects of hypervitaminosis A on rabbit embryos, *C. R. Soc. Biol.,* 152, 931–932, 1958.

58. **Robens, J. F.,** Teratogenic effects of hypervitaminosis A in the hamster and the guinea pig, *Toxicol. Appl. Pharmacol.,* 16, 88–99, 1970.

59. **Geelen, J. A. G.,** Skullbase malformations in rat fetuses with hypervitaminosis A-induced exencephaly, *Teratology,* 7, 49–56, 1973.

60. **Kochhar, D. M.,** Limb development in mouse embryos. I. Analysis of teratogenic effects of retinoic acid, *Teratology,* 7, 289–298, 1973.

61. **Nolen, G. A.,** Variations in teratogenic response in hypervitaminosis A in three strains of the albino rat, *Food Cosmet. Toxicol.,* 7, 209–214, 1969.

62. **Marin-Padilla, M. and Ferm, V. H.,** Somite necrosis and developmental malformations induced by vitamin A in the golden hamster, *J. Embryol. Exp. Morphol.,* 13, 1–8, 1965.

63. **Vorhees, C. V.,** Some behavioral effects of maternal hypervitaminosis A in rats, *Teratology,* 10, 269–274, 1974.

64. **Yarington, C. T., Jr. and Stivers, F. E.,** Lathyrogenic effects of vitamin A in the rat embryo, *Laryngoscope,* 84, 1310–1315, 1974.

65. **Shenefelt, R. E.,** Animal model: Treatment of various species with a large dose of vitamin A at known stages in pregnancy, *Am. J. Pathol.,* 66, 589–592, 1972.

66. **Cho, D. Y., Frey, R. A., Guffy, M. M., and Leipold, H. W.,** Hypervitaminosis A in the dog, *Am. J. Vet. Res.,* 36, 1597–1603, 1975.

67. **Frier, H. I., Gorgacz, E. J., Hall, R. C., Jr., Gallina, A. M., Rousseau, J. E., Eaton, H. D., and Nielsen, S. W.,** Formation and absorption of cerebrospinal fluid in adult goats with hypo- and hypervitaminosis A, *Am. J. Vet. Res.,* 35, 45–55, 1974.

68.   Hall, R. C., Jr., Rousseau, J. E., Jr., Gorgacz, E. J., and Eaton, H. D., Sodium and potassium in supraspinatus muscle from hypervitaminotic A Holstein calves, *J. Dairy Sci.*, 56, 252–254, 1973.

69.   Gershoff, S. N., Nutritional problems of household cats, *J. Am. Vet. Med. Assoc.*, 166, 455–458, 1975.

70.   Gorgacz, E. J., Nielsen, S. W., Frier, H. I., Eaton, H. D., and Rousseau, J. E., Jr., Morphologic alterations associated with decreased cerebrospinal fluid pressure in chronic bovine hypervitaminosis A, *Am. J. Vet. Res.*, 36, 171–180, 1975.

71.   Baker, J. R., Howell, J. McC., and Thompson, J. N., Hypervitaminosis A in the chick, *Br. J. Exp. Pathol.*, 48, 507–512, 1967.

72.   Wolbach, S. B. and Hegsted, D. M., Hypervitaminosis A in young ducks, *Arch. Pathol.*, 55, 47–54, 1953.

73.   Maddux, G. W., Foltz, F. M., and Nelson, S. R., Effect of vitamin A intoxication on intracranial pressure and brain water in rats, *J. Nutr.*, 104, 478–482, 1974.

74.   Seawright, A. A., English, P. B., and Gartner, R. J. W., Hypervitaminosis A and deforming cervical spondylosis of the cat, *J. Comp. Pathol.*, 77, 29–39, 1967.

75.   Seawright, A. A., English, P. B., and Gartner, R. J. W., Hypervitaminosis A of the cat, *Adv. Vet. Sci.*, 14, 1–24, 1970.

76.   Wolke, R. E., Nielsen, S. W., and Rousseau, J. E., Jr., Bone lesions of hypervitaminosis A in the pig, *Am. J. Vet. Res.*, 29, 1009–1024, 1968.

77.   Brusa, A. and Testa, F., Lesions of the central nervous system of guinea pigs with hypervitaminosis A, *Int. Z. Vitaminforsch.*, 25, 55–62, 1953.

78.   Singh, V. N., Singh, M., and Venkitasubramanian, T. A., Early effects of feeding excess vitamin A: Mechanism of fatty liver production in rats, *J. Lipid Res.*, 10, 395–401, 1969.

79.   Soliman, M. K., Vitamin A overdosing. II. Cytological and biochemical changes in blood of rats treated with high doses of vitamin A and alpha-tocopherol, *Int. J. Vitam. Nutr. Res.*, 42, 576–582, 1972.

80.   Ventura, U., Ceriani, I., Zelaschi, F., and Rindi, G., Action potential modifications in rat myocardial cells induced by hypervitaminosis A, *Q. J. Exp. Physiol.*, 56, 147–155, 1971.

81.   Gil, A., Briggs, G. M., Typpo, J., and Mackinney, G., Vitamin A requirement of the guinea pig, *J. Nutr.*, 96, 359–362, 1968.

82.   Leelaprute, V., Boonpucknavig, V., Bhamarapravati, N., and Weerapradist, W., Hypervitaminosis A in rats: Varying responses due to different forms, doses and routes of administration, *Arch. Pathol.*, 96, 5–9, 1973.

83.   Hazzard, D. G., Woelfel, C. G., Calhoun, M. C., Rousseau, J. E., Jr., Eaton, H. D., Nielsen, S. W., Grey, R. M., and Lucas, J. J., Chronic hypervitaminosis A in Holstein male calves, *J. Dairy Sci.*, 47, 391–401, 1964.

84.   Clark, I. and Smith, M. R., Effects of hypervitaminosis A and D on skeletal metabolism, *J. Biol. Chem.*, 239, 1266–1271, 1964.

85.   Cox, R. P., Deuel, H. J., Jr., and Ershoff, B. H., Potentiating effects of DPPD, bile salts and sulfasuxidine on hypervitaminosis A in the rat, *Exp. Med. Surg.*, 15, 328–334, 1957.

86.   McCuaig, L. W. and Motzok, I., Excessive dietary vitamin E: Its alleviation of hypervitaminosis A and lack of toxicity, *Poult. Sci.*, 49, 1050–1052, 1970.

87.   Jenkins, M. Y. and Mitchell, G. V., Influence of excess vitamin E on vitamin A toxicity in rats, *J. Nutr.*, 105, 1600–1606, 1975.

88.   Primbs, E. R., Sinnhuber, R. O., and Warren, C. E., Hypervitaminosis A in the rainbow trout: Counteraction by vitamin C, *Int. J. Vitam. Nutr. Res.*, 41, 331–338, 1971.

89.   Krjukova, L. V., Ulasevie, I. I., and Medvedskaja, V. S., Interrelation of vitamins A and E in the live animal, *Byull. Eksp. Biol. Med.*, 67, 59–60, 1969.

90.   Matschiner, J. T., Amelotti, J. M., and Daisy, E. A., Jr., Mechanism of the effect of retinoic acid and squalene on vitamin K deficiency in the rat, *J. Nutr.*, 91, 303–306, 1967.

91.   Tuchweber, B., Garg, B. D., and Salas, M., Microsomal enzyme inducers and hypervitaminosis A in rats, *Arch. Pathol. Lab. Med.*, 100, 100–105, 1976.

92.   Anon., The transport of vitamin A in hypervitaminosis A, *Nutr. Rev.*, 34, 119–120, 1976.

93.   Mallia, A. K., Smith, J. E., and Goodman, D. S., Metabolism of retinol-binding protein and vitamin A during hypervitaminosis A in the rat, *J. Lipid Res.*, 16, 180–188, 1975.

94.   Smith, F. R. and Goodman, D. S., Vitamin A transport in human vitamin A toxicity, *N. Engl. J. Med.*, 294, 805–808, 1976.

95.   Leitner, Z. A., Moore, T., and Sharman, I. M., Fatal self-medication with retinol and carrot juice, *Proc. Nutr. Soc.*, 34, 44A–45A, 1975.

EFFECT OF NUTRIENT TOXICITIES IN ANIMALS AND MAN: VITAMIN D

H. E. Harrison

Vitamin D is produced by conversion of skin 7-dehydrocholesterol to cholecalciferol by absorption of short-wave ultraviolet (UV) energy, but toxic concentrations of vitamin D in man have not resulted from unlimited exposure to sunshine. It has been hypothesized that one of the protective mechanisms is the concentration and state of aggregation of the pigment melanin in the skin epithelium, which absorbs the short-wave UV energy and thus prevents conversion of the 7-dehydrocholesterol in the basal cell layers to cholecalciferol. Loomis[1] has suggested that the deep melanin pigmentation of peoples indigenous to the tropics is an evolutionary adaptation, as toxic concentrations of vitamin D might otherwise be produced. Similarly, the marked lack of skin pigmentation and thin skin-keratin layer of peoples living in the subarctic zones (such as the Scandinavians) could be an adaptation to the need for maximal vitamin D formation during the relatively brief time of possible exposure to sunshine. The tanning of the skin resulting from exposure to sunshine is also due to formation of melanin aggregates which shield the skin 7-dehydrocholesterol from activation by UV energy.

Although certain fish liver and body oils are good sources of vitamin D, the concentrations are not so high that toxic amounts of vitamin D would likely be ingested. The liver oil of fishes of the Percomorph order contains a higher concentration of vitamin D than that of other fishes including the cod. Percomorph liver oil was once an important source of vitamin D supplementation in infant's and children's diets, and toxic amounts could possibly have been ingested. The vitamin D concentrates now available for diet supplementation or treatment of patients are solutions of ergocalciferol or cholecalciferol. Highly concentrated preparations of vitamin D are thus available, and important vitamin D toxicity has been seen since such concentrates have been in use. Vitamin D toxicity is cumulative, i.e., the metabolic inactivation of vitamin D is relatively slow, so that doses which are not toxic produce toxicity when ingested daily over weeks or months. With continued intake of vitamin D in high dosage, serum vitamin D concentrations rise to extremely high values. When measured by bioassay, the normal serum vitamin D concentration is approximately 1 to 2 units/ml.[2] Of course, this bioassay measures not only vitamin D, but also its active metabolites 25-OH vitamin D and 1,25-diOH vitamin D. In one study[3] the concentration of 25-OH vitamin D in normal subjects with usual exposure to sunshine was found to be 20.5 ± 3.0 ng/ml. In terms of units of activity this would roughly correspond to 1 unit/ml, so that it is likely that most of the circulating vitamin D under physiologic conditions is 25-OH vitamin D. When large doses of vitamin D are given over a period of time, the serum 25-OH vitamin D concentrations are increased to 250 to 600 ng/ml or 10 to 30 times the normal value.[4] This suggests that, in vivo, the hepatic vitamin D 25-hydroxylase is not under negative feedback control.

The vitamin D content of human serum following 50,000 units/day for 1 month was found by bioassay to be 16.5 units/ml, equivalent to 330 ng 25-OH vitamin D per milliliter. In other subjects on much larger doses of vitamin D (200,000 to 400,000 units/day for several months), the serum vitamin D was as high as 130 units/ml.[5] It is probable that some of this was circulating as vitamin D as well as 25-OH vitamin D. In experimental studies on animals given excessive doses of vitamin D, high concentrations of vitamin D were found in adipose tissue as well as plasma.[6] Unlike fishes, the liver is not a major site of storage of vitamin D in mammals.

The toxicity of these high concentrations of vitamin D is presumably due to the pharmacologic effects of 25-OH vitamin D in high concentration and, possibly, of vitamin D itself. Although hypercalcemia can be produced by administration of 1,25-diOH vitamin D in doses of 1 to 2 μg/day, it is thought that the negative feedback control of the activity of the 25-OH vitamin D 1-hydroxylase in the kidney prevents excessive accumulation of this compound; indeed, it has been found that the circulating concentration of 1,25-diOH vitamin D, unlike that of 25-OH vitamin D, is not greatly increased in vitamin D-intoxicated patients.[7]

Once high concentrations of vitamin D in plasma and adipose tissue are produced, many months may elapse before the excess of vitamin D is metabolized and inactivated or excreted. One mode of elimination is glycuronidation in the liver and excretion of the glycuronide in the bile. Drugs such as phenobarbital and phenytoin, which induce liver microsomal enzymes, reduce the concentration of 25-OH vitamin D in serum, possibly by inducing liver microsomal enzymes which hydroxylate and glycuronidate steroids.[8]

The toxicity of hypervitaminosis D is primarily due to hypercalcemia resulting from intestinal hyperabsorption of calcium. 25-OH Vitamin D also has a parathyroid hormone-like action on bone, causing increased bone-mineral solubilization, probably by stimulation of bone-cell metabolism and proliferation. A major effect of hypercalcemia and associated hypercalciuria is deposition of calcium phosphate in the kidney, both in interstitial tissue and within the tubule lumen. This nephrocalcinosis leads to progressive reduction of kidney function, scarring, and eventual renal insufficiency. Metastatic calcification also occurs in other tissues such as subcutaneous tissue, especially around joints, media of blood vessels, gastric mucosa, lungs, and heart. Arterial spasm with severe hypertension and hypertensive encephalopathy may also be present. Nonspecific symptoms such as anorexia, weakness, and constipation occur. Increased calcium concentration inhibits the action of vasopressin on the distal tubules, so that vasopressin-resistant polyuria can be seen. The usual cause of death is renal insufficiency.

The toxic dose of vitamin D is quite variable. An important factor is duration of intake, as this is a cumulative toxicity. However, there is marked variation among individuals in tolerance to excessive doses of vitamin D, the mechanism of which is unknown. In general, when doses are compared in terms of amounts per unit of body weight, young, growing animals show less evidence of toxicity than adult animals. In infants and children, doses of 2000 to 5000 units of vitamin D per kilogram body weight given daily for several months will produce severe toxicity. However, a single oral dose of 600,000 units or more than 100,000 units/kg vitamin D has been given to normal infants for prevention of rickets without toxicity. Perhaps much of this is unabsorbed or inactivated by the liver, with excretion of the inactive products in the bile. Much smaller amounts of vitamin D than the usual toxic doses have resulted in hypercalcemia and renal injury in infants with the entity of "idiopathic hypercalcemia."[9] These infants presumably have an unusual sensitivity to vitamin D and daily doses of only a few thousand units (only 500 to 1000 units/kg) may produce severe manifestations. The basis for this hyperreactivity is unknown, and children who have recovered from an episode of hypercalcemia do not show a recurrence with ordinary intake of vitamin D.

For various reasons, adults have received doses of 100,000 to 300,000 units of vitamin D per day. In one report,[14] every adult patient with hypoparathyroidism given 100,000 units of vitamin D per day over an extended period of time became hypercalcemic at one time or another. Serious toxicity has also been seen in subjects receiving 200,000 to 300,000 units/day for 1 month or longer. Again, there appears to be considerable variation among individuals in the tolerance to these large amounts of vitamin D. Hyperreactivity to vitamin D is seen in patients with the disease sarcoidosis,[10] the cause of which is as yet unknown. These patients show hypercalcemia, which has been found to be associated with vitamin D intake in the physiologic range. In fact, significantly higher serum calcium concentrations are found in these patients following continued exposure to sunshine in the summer months. The toxic manifestations of hypercalcemia in these patients are the same as those of subjects fed excessively large quantities of vitamin D, with renal insufficiency as the major lesion. The hypercalcemia of sarcoidosis, like that of vitamin D overdose, is reduced by adrenocortical steroid treatment. The effective steroids are those with predominantly glucocorticoid activity (such as cortisol, cortisone, prednisone, and dexamethasone) in so-called suppressive doses, i.e., doses which are of the order of ten times the physiologic replacement amount and which suppress inflammatory reactions among other effects. These large doses of steroids with glucocorticoid activity are known to inhibit calcium transport by the intestine,[11] and probably also have an action opposite to that of vitamin D on bone cells.[12]

Some vitamin D toxicity results from mistaken use of excessive amounts of vitamin D concentrates because of the idea that large amounts of vitamins promote growth and body vigor. At one time, excessive amounts of vitamin D were given to patients with rheumatoid arthritis, but the major source of toxicity in the United States at present is the necessary pharmacologic use of vitamin D in the treatment of hypoparathyroidism, pseudohypoparathyroidism, chronic renal disease, and primary hypophosphatemic vitamin D-resistant rickets. Pharmacologic amounts of vitamin D are required in the first three categories because of deficiency in the renal 1-hydroxylation of 25-hydroxyvitamin D. Pharmacologic doses of vitamin D are needed in the fourth category in order to maintain intestinal calcium absorption in the presence of large phosphate loads. In these disorders, doses of 50,000 to 100,000 units per day or more may be given to older children and adults, and hypercalcemia and renal injury are important complications which must be guarded against by careful monitoring.

Vitamin D toxicity has been studied experimentally in animals. The manifestations in rat, dog, and rabbit are similar to those in man, with hypercalcemia and extensive extraskeletal calcification, especially in the kidney and walls of blood vessels. Natural vitamin D toxicity has been observed[13] in Argentina in cattle feeding on the plant *Solanum malacoxylon*. This plant contains a glycoside of 1,25-diOH vitamin D, and the toxicity is due to this vitamin D metabolite.

REFERENCES

1. **Loomis, W. F.,** Rickets, *Sci. Am.,* 223, 76–91, 1970.
2. **Warkany, J. and Mahon, H. E.,** Estimation of vitamin D in blood serum, *Am. J. Dis. Child.,* 60, 606–614, 1940.
3. **Avioli, L. V. and Haddad, J. G.,** Vitamin D: current concepts, *Metab. Clin. Exp.,* 22, 507, 1973.
4. **Haddad, J. G. and Stamp, T. C. B.,** Circulating 25-hydroxy-vitamin D in man, *Am. J. Med.,* 57, 57–62, 1974.
5. **Warkany, J., Guest, G. M., and Grahill, F. J.,** Vitamin D in human serum during and after periods of ingestion of large amounts of vitamin D, *J. Lab. Clin. Med.,* 27, 557–565, 1941–1942.
6. **Rosenstreich, S. J., Rich, C., and Volwider, W.,** Deposition and release of vitamin D_3 from body fat: evidence for a storage site in the rat, *J. Clin. Invest.,* 50, 679–687, 1971.
7. **Hughes, M. R., Baylink, D. J., Jones, P. G., and Haussler, M. R.,** Radioligand receptor assay for 25-hydroxyvitamin D_2/D_3 and 1α,25-dihydroxyvitamin D_2/D_3. Application to hypervitaminosis D, *J. Clin. Invest.,* 58, 61–70, 1976.
8. **Conney, A. H.,** Pharmacological implications of microsomal enzyme induction, *Pharmacol. Rev.,* 319, 317–366, 1967.
9. American Academy of Pediatrics Committee on Nutrition, The relation between infantile hypercalcemia and vitamin D – public health implications in North America, *Pediatrics,* 40, 1050–1061, 1967.
10. **Henneman, P. H., Dempsey, E. F., Carroll, E. L., and Albright, F.,** The cause of hypercalcemia in sarcoid and its treatment with cortisone and sodium phytate, *J. Clin. Invest.,* 35, 1229–1242, 1956.

11. **Harrison, H. E. and Harrison, H. C.,** Transfer of Ca^{45} across the intestinal wall *in vitro* in relation to action of vitamin D and cortisol, *Am. J. Physiol.,* 199, 265–271, 1960.
12. **Raisz, L. G., Trummel, C. L., Wener, J. A., and Simmons, H.,** Effect of glucocorticoids on bone resorption in tissue culture, *Endocrinology,* 90, 961–967, 1972.
13. **Haussler, M. R., Wasserman, R. H., McCain, T. A., Peterlik, M., Bursac, K. M., and Hughes, M. R.,** 1,25-Dihydroxyvitamin D_3/glycoside: identification of a calcinogenic principle of solanum malacoxylon, *Life Sci.,* 18, 1049–1056, 1976.
14. **Ireland, A. W., Clubb, J. S., Neale, F. C., Posen, S., and Reeve, T. S.,** Calciferol requirements of patients with surgical hypoparathyroidism, *Ann. Intern. Med.,* 69, 81–89, 1968.

EFFECT OF SPECIFIC NUTRIENT TOXICITIES
IN ANIMALS AND MAN: TOCOPHEROLS

M. H. Briggs

INTRODUCTION

Vitamin E activity is shown by a variety of naturally occurring tocopherols (α-, β-, γ-, δ-, and others), all of which are chemical antioxidants.[1] In animal tissues the major compound is RRR-α-tocopherol (previously named D-α-tocopherol), which shows the greatest vitamin E activity in biological assays such as the prevention of infertility in deficient rats, or of nutritional muscular dystrophy in deficient rabbits.[1]

All tocopherols of plant and animal tissues belong to the RRR-series, but chemical synthesis yields racemates, i.e., RRR, SSS-compounds (previously named DL-tocopherols). Some pharmaceutical preparations contain the naturally occurring RRR-α-tocopherol, but others make use of the racemate. The biological activity of the racemate is about 70% that of the naturally occurring isomer, but as the activity is greater than one half, this suggests that the unnatural SSS-α-tocopherol has some vitamin E properties.

To complicate the situation further, most pharmaceutical products containing vitamin E use tocopherol esters, particularly the acetate or succinate, which are more stable and better absorbed than the parent tocopherol alcohols. There appears to have been no published systematic study on the toxicology of the many different tocopherols, natural and synthetic. Most investigators have not distinguished between the many different forms, but as all tocopherols have very low systemic toxicity, this is probably of little practical importance.

In recent years, a number of tocopherolamines have been synthesized[2] and shown to possess considerable vitamin E activity.[3] As they have so far been used only in experimental studies, they will be excluded from the following discussion.

ANIMAL STUDIES

Unless otherwise noted, the studies discussed below are all concerned with α-tocopherol, sometimes given as the racemate, sometimes as the natural stereoisomer, and sometimes as an ester. Much of the earlier work used vitamin E concentrates from plant oils, such as wheat-germ oil, so that tocopherols other than the α-compound were undoubtedly present.

The earliest systematic investigator of tocopherol toxicology appears to be Demole,[4] who gave oral doses for 2 months to rats (up to 4 g/kg) and mice (up to 50 g/kg). No adverse effects were noted in these experiments where the amount administered was approximately 120 and 200 times the usual human dose level. Early reports that high doses of vitamin E in the form of wheat-germ oil extracts induced sarcomas in rats proved unrepeatable.[5,6] More recently, toxic effects of α-tocopherol in rats were looked for at 50 times the recommended daily allowance,[7] again without the detection of any significant effect.

While tocopherols appear to lack any inherent carcinogenicity, there is some evidence[8] that they may act as procarcinogens under certain circumstances. Mice treated with the carcinogen, dibenzanthracene, showed a higher incidence of lung tumors when given 2 mg α-tocopherol every 2 days than when fed a vitamin E-deficient diet (100% incidence versus 71%). Subcutaneous tumors were also more common in the animals receiving α-tocopherol (47%) than in the vitamin E-deficient group (33%).

Surprisingly, only one study of potential teratogenic effects of α-tocopherol appears to have been published.[9] When 591 i.u. RRR-α-tocopherol was given by stomach tube on days 7 to 11 to pregnant ICR strain mice, one malformed animal was seen in a total of 91 offspring from seven litters. This animal showed exencephaly, open eye, and micrognathia, which are malformations never seen in untreated control mice of this strain, though they can be induced by known teratogens. Further studies are urgently needed, for this evidence suggests that α-tocopherol is indeed a mild teratogen.

Several further studies of tocopherols in rodents have demonstrated the induction of undesirable changes in tissue lipids. An early study in rats[10] found fatty infiltration of the liver, with deposition of cholesterol, in animals receiving about 50 mg of a vitamin E concentrate by mouth every 7 days. Intimal sclerosis of the aorta was also seen, with over-development of collagenous tissue at the base of the aortic valve and in the medial coat of the aorta. These changes were not seen in rats fed large amounts of cholesterol or bile salts. These findings have been partially confirmed in a more recent study[11] where rats fed high doses of α-tocopherol were found to have high liver cholesterol and altered tissue fatty acids.

A further study[12] has shown that a diet containing 500 ppm RRR, SSS-α-tocopherol acetate is strongly synergistic to the hepatotoxic actions of ethanol. Rats were fed a basal diet and given 20% ethanol in water as drinking fluid. Animals receiving the tocopherol supplemented diet showed an average of 23 mg/g of liver triglyceride, compared to a mean value of 10 mg/g in rats on the basal diet with normal vitamin E content.

A quite different type of toxicity is suggested by investigations of rats fed irradiated or nonirradiated beef, sometimes supplemented with large amounts of vitamin E.[13] All the supplemented animals showed increased mortality and coagulopathy, with depression of plasma prothrombin.

While rodents have been the animals most investigated for toxic effects of tocopherols, there is also a study in the chick.[14] In this species quite clear-cut evidence of hypervitaminosis E was found. Chicks receiving high doses of α-tocopherol showed depressed growth and, like the rats mentioned above, had increased clotting times. Thyroid function was depressed, while the excess vitamin E significantly increased requirements for both vitamins D and K.

Viewed as a whole, the animal evidence indicates that α-tocopherol is toxic at very high doses. It does not appear to be a carcinogen, but may be a procarcinogen, and also a mild teratogen. Vitamin E tends to induce lipid deposition in the liver and aorta and may act synergistically with other hepatotoxins such as ethanol. Evidence from rats and chicks suggests an interference with blood coagulation, possibly mediated by an interaction with vitamin K.[15]

The data discussed above are summarized in Table 1.

CLINICAL STUDIES

Vitamin E preparations have been available in many countries for over 40 years, often without requiring a prescription. They have been very widely used, often in unnecessary high dose. No general toxicity has been reported, but there is sufficient published evidence to suggest that tocopherols are not entirely free of undesirable side effects in humans under certain circumstances. A minority of these side effects have been reported in healthy individuals receiving high doses of vitamin E; the majority came from patients with a variety of diseases, some of them serious. In the latter cases, unequivocal relationship to vitamin E therapy is usually lacking.

Despite the widespread use of vitamin E supplements, it was not until 1973 that reports[16,17] appeared describing severe weakness and fatigue as side effects induced in healthy adults by daily doses of around 800 i.u. α-tocopherol. The author claimed to have

Table 1
TOXICOLOGY OF α-TOCOPHEROL (ANIMAL STUDIES)

Species	Effects of high doses	Reference
Mouse Rat	No adverse effects	4
Rat	No adverse effects	7
Rat	No tumors induced	5
Rat	No tumors induced	6
Mouse	Increased incidence of lung and subcutaneous tumors in animals pretreated with dibenzanthracene	8
Mouse	One malformed offspring out of 91 born to 7 mothers treated during day 7 to 11 of pregnancy	9
Rat	Fatty infiltration of liver and aortic intimal sclerosis	10
Rat	Cholesterol deposits in liver	11
Rat	Enhanced development of fatty liver by ethenol	12
Rat	Increased mortality and coagulopathy	13
Chick	Depressed growth, increased clotting time, depressed thyroid function	14

experienced the effect personally, as well as seeing many such patients in a busy general practice. It was stated that fatigue induced by excess vitamin E was more common than this condition induced by either anemia or hypothyroidism. While this side effect was not encountered by others[18] using 1600 i.u. α-tocopherol daily in elderly men with ischemic heart disease, it was encountered in two healthy young men receiving 800 i.u. α-tocopherol daily in a double-blind study.[19,20] In the latter case the symptoms were associated with an increase in serum creatine kinase activity and greatly increased 24-hr urinary excretion of creatine. There is an earlier report of increased creatine excretion in a young man taking high doses of vitamin E, though in this case there were apparently no clinical symptoms.[21]

It seems likely that there is a marked individual difference in the induction of muscular weakness and fatigue by excess vitamin E. It certainly occurs in only a minority of people and only at daily doses of 800 i.u. or more. It appears to have a biochemical basis suggestive of some pathological change in skeletal muscles.

A further individual toxic response to α-tocopherol is the rare, but serious, allergic reaction shown by some individuals to topical creams or sprays containing vitamin E. Several reports[22-24] have now appeared of allergic contact dermatitis which followed use of such preparations and was shown by patch testing to be due to α-tocopherol. As skeletal muscles possess relatively large amounts of α-tocopherol, it is possible that some individuals develop a systemic allergic response to high doses of synthetic vitamin E which involves some type of immunological muscle dysfunction. It should be noted that the development of skin rashes, associated with gastrointestinal tract irritation, were reported in early studies[25] of oral vitamin E supplementation, though as the preparation used was wheat-germ oil, the role of components other than tocopherols cannot be excluded.

Turning now to patients with established diseases, in whom vitamin E was being used therapeutically, there are many reports of individuals whose condition deteriorated while the vitamin was being given. For example, a study[26] of 22 patients with various cardiac diseases who received 200 to 400 mg vitamin E daily for several weeks found that four were definitely worse during treatment, but this cannot be certainly linked to tocopherol excess. Similarly, fairly nonspecific symptoms, such as headache, giddiness, and intestinal pain, which have been reported by cardiac patients receiving high-dose vitamin E[27,28] cannot be linked to tocopherols, for they are reported with approximately equal frequency by patients on a placebo.[29]

Table 2

TOXICOLOGY OF α-TOCOPHEROL (CLINICAL STUDIES)

Subjects	Effects of high doses	References
Mainly healthy adults	Severe weakness and fatigue	16, 17
Young healthy men	Weakness and fatigue, accompanied by increased serum creatine kinase and urinary creatine	19, 20
Healthy man	Increased urinary creatine	21
Healthy adults	Allergic contact dermatitis	22—24
Healthy women	Skin rash with gastrointestinal irritation	25
Cardiac patients	Nonspecific symptoms (e.g., headache, giddiness), but same with inert placebo	27—29
Children with iron-deficient anemia	Poor response to iron therapy	30
Male cardiac patient	Ecchymoses and increased prothrombin time	31
Geriatric patients	Increased serum cholesterol	32
Patients with porphyria cutanea tarda	Increased urinary androgen excretion	33
Healthy adults	Raised serum lipids and significantly increased carotenoids	34

Rather rarer, but more clear-cut, adverse associations of α-tocopherol therapy have been reported for small groups of rather special patients. There seems little doubt that the hematologic response to iron therapy by children with iron-deficiency anemia is significantly impaired if vitamin E is given simultaneously.[30] A further hematologic side effect has been reported in a 55-year-old man, who developed ecchymoses and a prolonged prothrombin time while receiving high dose α-tocopherol (up to 1200 i.u. daily) in combination with warfarin and clofibrate.[31] The side effects disappeared when the warfarin and vitamin E were discontinued, and did not redevelop when warfarin was reinstituted. Challenge with vitamin E again induced the changes.

Finally, there are a number of reported biochemical effects induced by high-dose vitamin E which could lead to adverse tissue changes. A group of 52 patients (average age 72 years) showed a mean increase in serum cholesterol of 74 mg/dl while receiving 300 mg daily of α-tocopherol.[32] However, no increase in serum cholesterol was seen in a much younger, small group of healthy men taking 800 i.u. (588 mg) daily.[19,20] Patients with porphyria cutanea tarda were given 1.0 g α-tocopherol daily for 3 months and 24-hr urinary excretion of various steroids was measured.[33] During treatment there was a marked increase in urinary androgens (androsterone and etiocholanolone plus dehydro-epiandrosterone) from a mean value of 3.5 to 4.6 mg at the end of 3 months. In contrast, mean 24-hr pregnanediol excretion fell from 2.2 to 0.5 mg. The significance of these endocrine changes is uncertain, but could be important for patients with endocrine sensitive tumors.

The effects reported above are summarized in Table 2.

CONCLUSIONS

Animal and human studies indicate that tocopherols have very low toxicity but are not entirely devoid of undesirable effects. An early claim that prolonged high dose tocopherols induce sarcoma in rats has not been confirmed, but a procarcinogenic action in mice treated with dibenzanthracene has been reported. No carcinogenic, or procarcinogenic, effect in humans has been found, but experimental data are few. Similarly, α-tocopherol is a suspected weak teratogen in the mouse, but no human study has been reported. High doses induce fatty infiltration of the liver in rats and synergize with the hepatotoxicity of ethanol. In humans, no hepatic effects of excess vitamin E have been reported, though geriatric patients may show increased serum cholesterol. Both chick and rat show interference with normal coagulation mechanisms when large doses of tocopherols are given. There appears to be only a single reported human case of such an interference.

Major adverse effects in humans are not predicted by the animal findings. Some individuals develop severe contact dermatitis to α-tocopherol, while others show muscular weakness and fatigue, accompanied by biochemical signs of muscle damage, when receiving high dose vitamin E. There is also an interaction with iron therapy in anemic children to reduce the hematologic response.

REFERENCES

1. Sebrell, W. H. and Harris, R. S., Eds., "Tocopherols," in *The Vitamins: Chemistry, Physiology, Pathology,* Vol. 5, Academic Press, New York, 1972, chap. 16.
2. Schwieter, U., Tamm, R., Weiser, H., and Wiss, O., Synthesis and vitamin E activity of tocopheramines and their N-alkyl derivatives, *Helv. Chim. Acta,* 49, 2297, 1966.
3. Bieri, J. G., Biological activity and metabolism of N-substituted tocopheramines: implications on vitamin E function, in *The Fat-Soluble Vitamins,* DeLuca, H. F. and Suttie, J. W., Eds., University of Wisconsin Press, Madison, 1969, 307.
4. De mole, V., Pharmacology of vitamin E, *Int. Z. Vitaminforsch.,* 8, 338, 1939.
5. Dingemanse, E. and Van Eck, W. S., Wheat-germ oil and tumor formation, *Proc. Soc. Ex. Biol. Med.,* 41, 622, 1939.
6. Evans, H. M. and Emerson, G. A., Failure to produce abdominal neoplasms in rats receiving wheat-germ oil extracted in various ways, *Proc. Soc. Exp. Biol. Med.,* 41, 318, 1939.
7. Dysmsza, H. A. and Park, J., Excess dietary vitamin E in rats, *Fed. Proc.,* 34, 912, 1975.
8. Telford, I. R., The effects of hypo- and hyper-vitaminosis E on lung tumor growth in mice, *Ann. N.Y. Acad. Sci.,* 52, 132, 1949.
9. Hook, E. B., Healy, K. M., Niles, A. M., and Skalko, R. C., Vitamin E: teratogen or anti-teratogen?, *Lancet,* 1, 809, 1974.
10. Marxs, W., Marks, L., Meserve, E. R., Shimoda, F., and Deuel, H. J., Effects of the administration of a vitamin E concentrate and of cholesterol and bile salts on the aorta of the rat, *Arch. Pathol.,* 47, 440, 1947.
11. Alfin-Slater, R. B., Aftergood, L., and Kishineff, S., Investigations on hypervitaminosis E in rats, *Abst. Int. Congr. Nutr.,* 1X, 191, 1972.
12. Levander, O. A., Morris, V. C., Higgs, D. J., and Varma, R. N., Nutritional interrelationships among vitamin E, selenium, antioxidants and ethyl alcohol in the rat, *J. Nutr.,* 103, 536, 1973.
13. Mellette, S. J. and Leone, L. A., Influence of age, sex, strain of rat, and fat soluble vitamins on hemorrhagic syndromes in rats fed irradiated beef, *Fed. Proc.,* 19, 1045, 1960.
14. March, B. E., Wong, E., Seier, L., Sim, J., and Biely, J., Hypervitaminosis E in the chick, *J. Nutr.,* 103, 371, 1973.
15. Anon., Hypervitaminosis E and coagulation, *Nutr. Rev.,* 33, 269, 1975.
16. Cohen, H. M., Fatigue caused by vitamin E?, *Calif. Med.,* 199, 72, 1973.
17. Cohen, H. M., Effects of vitamin E: good and bad, *N. Engl. J. Med.,* 289, 980, 1973.
18. Toone, W. M., Effects of vitamin E: good and bad, *N. Engl. J. Med.,* 289, 979, 1973.
19. Briggs, M. H., Vitamin E supplements and fatigue, *N. Engl. J. Med.,* 290, 579, 1974.
20. Briggs, M. H. and Briggs, M., Are vitamin E supplements beneficial?, *Med. J. Aust.,* 1, 434, 1974.
21. Hillman, R. W., Tocopherol excess in man. Creatinuria associated with prolonged ingestion, *Am. J. Clin. Nutr.,* 5, 597, 1957.
22. Brodkin, R. H. and Bleiberg, J., Sensitivity to topically applied vitamin E, *Arch. Dermatol.,* 92, 76, 1965.
23. Minkin, W., Cohen, H. J., and Frank, S. B., Contact dermatitis from deodorants, *Arch. Dermatol.,* 107, 774, 1973.

24.   **Aeling, J. L., Panagotacos, P. J., and Andreozzi, R. J.,** Allergic contact dermatitis to vitamin E aerosol deodorant, *Arch. Dermatol.,* 108, 579, 1973.
25.   **Shute, E.,** Wheat-germ oil therapy. I. Dosage idiosyncrasy, *Am. J. Obstet. Gynecol.,* 35, 249, 1938.
26.   **Baer, S., Heine, W. I., and Gelfond, D. B.,** The use of vitamin E in heart disease, *Am. J. Med. Sci.,* 215, 542, 1948.
27.   **Levy, H. and Boas, E. P.,** Vitamin E in heart disease, *Ann. Intern. Med.,* 28, 1117, 1948.
28.   **Vogelsang, A. B., Shute, E. V., and Shute, W. E.,** Vitamin E in heart disease, *Med. Rec.,* 160, 279, 1947.
29.   **Anderson, T. W.,** Vitamin E in angina pectoris, *Can. Med. Assoc. J.,* 110, 401, 1974.
30.   **Melhorn, D. K. and Gross, S.,** Relationships between iron-dextran and vitamin E in iron-deficiency anemia in children, *J. Lab. Clin. Med.,* 74, 789, 1969.
31.   **Corrigan, J. J. and Marcus, F. I.,** Coagulopathy associated with vitamin E ingestion, *J.A.M.A.,* 230, 1300, 1974.
32.   **Dahl, S.,** Vitamin E in clinical medicine, *Lancet,* 1, 465, 1974.
33.   **Pinelli, A., Pozzo, G., Formento, M. L., Favalli, L., and Coglio, G.,** Effect of vitamin E on urine porphyrin and steroid profiles in porphyria cutanea tarda; report of four cases, *Eur. J. Pharmacol.,* 5, 100, 1972.
34.   **Farrell, P. M. and Bieri, J. G.,** Megavitamin E supplementation in man, *Am. J. Clin. Nutr.,* 28, 1381, 1975.

NUTRIENT TOXICITIES OF VITAMIN K

P. G. Barash

The toxic effects of the Vitamin K family are manifested mainly as hematologic and circulatory derangements. Not only is species variation encountered, but profound differences are observed in the ability of the various Vitamin K compounds to evoke a toxic response. In animal studies, oral ingestion of large amounts of Vitamin K_1 (phytonadione) (25 g/kg) produced no fatalities, whereas Vitamin K_3 (menadione) had an LD_{50} (in mice) equal to 500 mg/kg.[1] Acute hematologic studies in cats, dogs, and monkeys demonstrate dose-dependent abnormalities following menadione treatment.[2] Small doses (less than 5 mg/kg) resulted in a mild anemia with no reported pathologic changes noted at post mortem examination. However, anemia, hemoglobinuria, urobilinuria, and urobilinoguria were observed with oral doses of 25 to 50 mg/kg. These doses are approximately 125 times the recommended clinical dose (0.05 mg/kg) in man. The anemia seen following Vitamin K_3, appears to be reversible following withdrawal of the drug.[3] Mice treated with a lethal dose of Vitamin K_3 (Hykinone), developed lacrimation, exophthalmus, and convulsions leading to respiratory paralysis and death. Chronic administration of large, but not lethal doses of menadione to dogs, resulted in cyanosis and methemoglobinemia.[3] In addition, hyperprothrombinemia, hepatic, and renal damage were noted. There appears to be a biphasic prothrombin time response. Treatment with Hykinone on a short-term basis resulted in a decreased prothrombin time. This became elevated following prolonged therapy.[3] The hepato-renal dysfunction in rats resulted in hepatomegaly (fatty infiltration) and hemosiderosis which also resulted in extensive renal tubular damage.[4]

In man, the major subsystems of Vitamin K toxicity are hematologic (newborn infants) and circulatory (adults). In 1953, hematologic derangements with Heinz body formation and hemolysis were noted following Vitamin K treatment in premature infants.[5] (Heinz body formation is associated with hemolytic anemia and impending cellular disintegration following exposure to specific drugs including Vitamin K.) However, it was not until Allison observed kernicterus (bilirubin encephalopathy) and hemolytic anemia following large-dose menadione therapy for hemorrhagic disease of the newborn that an association was made.[6] In confirming these findings, Laurence noted a 38% incidence of elevated serum bilirubin values (greater than 18 mg/100 ml) in a group of newborn infants who received menadione (10 mg three times a day for three days) whereas only a 4% incidence was observed in a similar group of infants who received only 1 mg menadione shortly after delivery.[7] As a result of the elevated bilirubin, two infants died of kernicterus and three had kernicteric manifestations. In the low-dose group, no deaths or toxic effects were observed. A similar reaction was observed in the newborn whose mothers were treated with large doses of menadione during labor. In comparing vitamin K_1 and vitamin K_3, Asteriadou-Samartzis demonstrated that phytonadione administration resulted in serum bilirubin levels equal to the control group, whereas menadione had a 40% increase above control values.[9] In a report issued by the American Academy of Pediatrics, immaturity of the infant is one of the most important predisposing factors to the development of vitamin K toxicity.[10] Full-term infants appear to have a greater tolerance to vitamin K than premature infants. They concluded that all Vitamin K analogues are safe when administered in the proper dosage; however, the formulation with the greatest margin of safety is vitamin K_1.

Several theories have been advanced to explain the toxic effects seen in the newborn. In Vitamin E-deficient infants, autooxidation of menadione analogues may result in hemolysis.[11] Vitamin E-deficient rats treated with menadione developed severe hemolysis, hemogloinuria, and a marked decrease in serum hemoglobin. Phytonadione, however, appeared to be harmless in these Vitamin E-deficient rats. A second hypothesis centers around decreased blood levels of reduced glutathione (GSH).[12] GSH instability has been observed in newborns in association with low blood glucose which may make these patients susceptible to Vitamin K- induced hemolysis. Other theories of toxicity in infants include functional overloading of the liver's capacity to metabolize bilirubin,[13] inhibition of bilirubin conjugation by large doses of Vitamin K,[14] and interference with rbc redox systems which results in a decreased intracellular osmotic pressure leading to hemolysis.[15]

The major toxicity in adult man is apparently in the cardiopulmonary system. In an early report, Beamish notes an injection of intravenous Mephyton (Vitamin K_1) was followed by hypotension, rigor, anemia, jaundice, and hemoglobinuria.[16] In adults, rapid intravenous administration of phytonadione has been associated with flushing, dyspnea, chest pain, and cardiovascular collapse.[17-18] Most case reports are poorly documented. However, recently a patient with severe liver disease, who was extensively monitored during surgery for cancer of the vocal cords, sustained peripheral vascular collapse following a slow intravenous injection of phytonadione (Figure 1).[19] At the conclusion of the operation, phytonadione (10 mg) was administered (1 mg/min). There was a

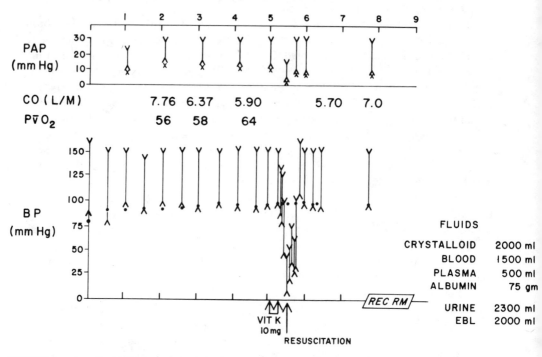

FIGURE 1.   A patient with Laennec's cirrhosis undergoing laryngectomy demonstrates peripheral vascular collapse following intravenous phytonadione. (ROS = review of systems, OP = operation, ANES = anesthesia, PAP = pulmonary artery pressure, CO = cardiac output, $P\bar{v}O_2$ = oxygen tension (mmHg) in mixed venous blood, BP = blood pressure, dot = pulse, Rec Room = recovery room, EBL = estimated blood loss. (Reproduced with permission by Barash et al., *Anesth. Analg.*, Cleveland, 55, 304–306, 1976.)

precipitous decrease in blood pressure 5 minutes later associated with cardiac data indicating peripheral vascular collapse. Resuscitation was accomplished with fluid and vasopressor administration. An anaphylactic reaction is doubtful since the patient had two subsequent phytonadione injections without any untoward effects. The authors hypothesized that either the colloid emulsion (polyethylated fatty acid derivative) or the active drug has a profound peripheral vasodilating effect.

In addition to the circulatory effects, abnormalities in clotting have been observed following Vitamin K.[20] The prolongation of the prothrombin time is the basis of a Vitamin K test for detection of subclinical hepatic dysfunction.[3] In severe liver disease, the abnormal prothrombin time is not fully correctible by Vitamin K administration. However, in individuals with preexisting subclinical hepatic dysfunction and a normal prothrombin time, administration of vitamin K paradoxically results in a prolongation. Richards also reports that, in normal man, doses as high as 10 times the recommended therapeutic dose induce no significant changes in hematologic coagulation function.

In summary, Vitamin K administration has been asssociated with hematologic and circulatory effects. Following large doses of Vitamin K and its analogues, hemolytic anemia, methemoglobinuria, polycythemia, and organ dysfunction (brain, liver, spleen, and kidney) have been observed. The injury is not directly to the hemopoietic system; it is, rather, to the circulating red blood cells. Vitamin K_1 appears to have minimal toxic effects when compared to Vitamin K_3. In infants, hematologic manifestation of toxicity predominate, while in adults circulatory toxicity is encountered. The quinone radical is thought to be an offending agent.[3] The variation in toxicity of the various Vitamin K compounds is explained on the modification of molecular structure.[2] As molecular size increases, the relative toxicity decreases more rapidly than relative potency. N-alkyl esters are more potent than similar isoalkyl esters of menadione. Methylation of naphthquinone results in a 75% decrease in toxicity and a 2000-fold increase in prothrombogenic activity. The doses reported in animal studies and early investigation in human infants were quite large, and it is doubtful that the smaller doses currently used are associated with the degree of toxicity previously reported.

REFERENCES

1. **Molitor, H. and Robinson, H.,** Oral and parenteral toxicity of vitamin K_1, phthiocol and 2-methyl-1,4-naphthoquinone, *Proc. Soc. Exp. Biol. Med.,* 43, 125–128, 1940.
2. **Ansbacher, S., Corwin, W., and Thomas, B.,** Toxicity of menadione, menadiol and esters, *J. Pharmacol. Exp. Ther.,* 75, 111–124, 1942.
3. **Richards, R. K. and Shapiro, S.,** Experimental and clinical studies on the action of high doses of Hykinone and other menadione derivatives, *J. Pharmacol. Exp. Ther.,* 84, 93–104, 1945.
4. **Finkel, M.,** Vitamin K_1 and Vitamin K analogues, *Clin. Pharmacol. Ther.,* 2, 794–814, 1961.
5. **Gasser, C.,** Heinz body anemia and related phenomena, *J. Pediatr.,* 54, 673–690, 1959.
6. **Allison, A. C.,** Danger of vitamin K to newborn, *Lancet,* 1, 669, 1955.
7. **Laurance, B.,** Danger of vitamin K analogues to newborn, *Lancet,* 1, 819, 1955.
8. **Lucey, J. and Dolan, R.,** Hyperbilirubinemia of newborn infants associated with parenteral administration of a vitamin K analogue to the mothers, *Pediatrics,* 23, 553–560, 1959.
9. **Asteriadou-Samartzis, E. and Leikin, S.,** The relations of vitamin K to hypervilirubinemia, *Pediatrics,* 21, 397–402, 1958.
10. **Blumberg, R., Forbes, G., Fraser, D., Hansen, A., May, C., Smith, C., Smith, N., Sweeney, M., and Fomon, S.,** Vitamin K compounds and the water-soluble analogues, *Pediatrics,* 28, 501–507, 1961.

11.   **Allison, A., Moore, T., and Sharman, I.,** Haemolysis and haemoglobinuria in vitamin E deficient rats after injections of vitamin K substitutes, *Br. J. Haematol.,* 2, 197–204, 1956.

12.   **Zinkham, W.,** The mechanism and clinical significance of an abnormality in glutathione metabolism of erythrocytes from normal newborns, *J. Dis. Child.,* 96, 621–624, 1958.

13.   **Gottsegen, G.,** Use of vitamin K in the newborn, *Lancet,* 1, 1010, 1956.

14.   **Waters, J., Dunham, R., and Bowen, W.,** Inhibition of bilirubin conjugation in vitro, *Proc. Soc. Exp. Biol.,* 99, 175–177, 1958.

15.   **Vest, M.,** Vitamin K in medical practice: pediatrics., *Vitam. Horm.,* N.Y., 24, 649–663, 1966.

16.   **Beamish, R. and Storrie, V.,** Severe hemolytic reaction following intravenous administration of emulsified vitamin K (mephyton), *Can. Med. Assoc. J.,* 74, 149–152, 1956.

17.   **Cohn, V. A.,** Fat soluble vitamins III, vitamin K and vitamin E, in *The Phamacologic Basis of Therapeutics,* Goodman, L. S., Gilman, A., Eds., Macmillan, New York, 1970, 1690–1694.

18.   **Douglas, A. S. and Brown, A.,** Effect of vitamin K preparations on hypoprothrombinemia induced by Dicoumarol and Tromexan., *Br. Med. J.,* 1, 412–415, 1952.

19.   **Barash, P., Kitahata, L., and Mandel, S.,** Acute cardiovascular collapse after intravenous phytonadione, *Anesth. Analg.,* Cleveland, 55, 304–306, 1976.

20.   **Smith, A. and Custer, R.,** Toxicity of vitamin K, *JAMA,* 173, 108–110, 1960.

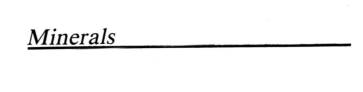

Minerals

EFFECT OF NUTRIENT TOXICITIES IN ANIMALS: CALCIUM

S. Goto

INTRODUCTION

There appears to be abundant concern with the influence of increasing intakes of dietary calcium on the utilization of calcium. Most investigations have shown that as the dietary calcium increases, percent absorption decreases, but total absorption and retention may increase. On the other hand, a consideration of the influence of high intakes of calcium on the absorption of other ingredients leads to the conclusion that this is a broad field which requires definitive study.

H. A. Keener,[1] reporting to the Cornell Nutrition Conference in 1953, gave the results of two experiments with a limited number of animals in which he observed a consistent depression of protein digestion when the diets had approximately 1% of calcium carbonate added to them; moreover, he reported that "No information on the effect of excess calcium in the diet on the digestion of common feed ingredients was found in the literature." The picture today has not changed greatly, although there has been some additional work indicating that the effect of calcium additions may differ with different diets.

EFFECT ON GROWTH

There is somewhat more evidence that levels of calcium exceeding 1% in the diet may have an adverse effect on growth in animals. Richards and Greig[2] noted that addition of calcium carbonate to the diet of mice at the level of 1% sharply depressed the growth and reproduction of this species. In 1956 Swenson and co-workers[3] found that increasing the calcium-phosphorus ratio of the diet by addition of calcium carbonate caused a decrease in weight gain in beef cattle. Fleischman et al.[4] by use of calcium carbonate as the calcium source, reported that fecal excretion increased progressively with increasing dietary calcium and depressed the growth rate. In 1973 Goto and Sawamura[5] indicated through the addition of calcium carbonate to the diet that average daily food intake was lower in the high-calcium and phosphorus diet-fed group than in the other groups for both young and growing rats and the average daily weight gain of this group was very low as compared with the other groups. Fecal excretion increased with high dietary calcium, and high dietary calcium-phosphorus ratios showed an increase in the quantity of feces.

Inorganic calcium salts, especially calcium carbonate or calcium phosphate are generally used as a calcium source in the experimental diet. However, no mention has been made in the earlier literature of the effect of organic calcium salts. The addition of calcium lactate at a level of 2% to the diet of rats sharply depressed the food intake and growth and the high-calcium-fed group showed increased fecal weight as compared with the control group of young rats.[6] High calcium intake had no effect on weight gain of the adult rats, but it had an effect on the fecal excretion.[7] (See Table 1.)

EFFECT ON VARIOUS NUTRIENTS

Haldi et al.[8] reported that increasing the calcium or the calcium and phosphorus levels of the diet caused a reduction in the deposition of fat in body. Recent evidence[5-7] suggested that calcium carbonate or calcium lactate at level of approximately 2% in the diet exhibited a decrease with age of the percent of absorption and retention of calcium and phosphorus, even if the animals were in a stage of active growth. For young rats, net

Table 1

EFFECT OF DIETARY CALCIUM ON WEIGHT GAIN AND FECES WEIGHT IN THE RAT

Animal	Experimental period (days)	Ca level (%)	Dietary Ca source	P level (%)	Initial body weight (g)	Mean weight gain (g/day)	Feces weight, dry (g/day)	Ref.
Male albino rats, Holtzman strain	21	0.08 0.2 1.2	Carbonate	0.35 0.35 0.35	500 500 500	1.29 1.19 1.05	3.13 3.57 4.30	4
Male albino rats, Wistar strain	35	0.08 0.2 1.2 2.0	Carbonate	0.29 0.29 0.29	500 500 500 500	4.06 4.88 1.60 2.12	3.3 4.4 5.2 5.4	15
	10	0.52 1.99 1.99	Carbonate	0.52 0.52 1.97	55 55 55	4.9 4.2 0.03	0.25 0.69 0.32	5
	10	0.52 1.99 1.99	Carbonate	0.52 0.52 1.97	165 165 165	5.0 4.6 −1.8	0.42 1.08 0.47	5
	30	0.54 2.0 2.0	Lactate	0.5 0.5 2.0	60 60 60	4.5 2.0 0.4[a]	0.26 0.57 0.38[a]	6
	30	0.54 2.0 2.0	Lactate	0.5 0.5 2.0	320 320 320	1.3 0.8 0.9	0.45 1.20 1.34	7

[a] This value is that of the 10-day experimental period, as rats fed this diet died between the 12th and 15th days of the experiment.

absorption and retention of phosphorus was higher in the high-calcium and -phosphorus diet-fed group than in high-calcium, normal-phosphorus diet-fed group regardless of the calcium source. However, in growing rats, phosphorus retention yielded a negative value and calcium lactate-fed adult rats exhibited the same positive value as the young rats. However, a different metabolic phenomenon was observed. This may be due to aging.

The effect of excess calcium-phosphorus balance may be an indication that excess calcium will result in the formation of insoluble tricalcium phosphates and, especially with marginal intakes of phosphorus, may result in a phosphorus deficiency condition. Where higher levels of phosphorus accompany the higher levels of calcium, precipitation of a phosphorus deficiency is unlikely, although the efficiency of phosphorus utilization may be markedly reduced by the high levels of calcium intake.

Schryver et al.[9] reported in 1971 that young ponies retained similar amounts of phosphorus when fed a low-calcium (0.15%) or a high-calcium (1.5%) diet. However, according to Goto and Sawamura,[6] the young rats fed with the high level of calcium accompany the high level of phosphorus died between the 12th to 15th days of the experiment. This may be due to the low level of magnesium in the diet. Several investigators have shown that high levels of dietary calcium increase the severity of magnesium deficiency symptoms in the rat[10-12] and guinea pig.[13] An excess of dietary phosphorus has also been shown to increase the magnesium requirement of the guinea pig.[13] Bunce et al.[14] observed that an increase in the dietary phosphorus level (1.0%) caused severe symptoms of magnesium deficiency in the weanling dog.

In regard to the nitrogen balance, Keener[1] reported in 1953, that there was a decrease of nitrogen digestion in excess (approximately 1%) calcium-fed animals. Goto and Sawamura[6] and Goto and Sugai[7] observed that both young and adult rats fed an excess-calcium lactate diet exhibited the decreasing absorption and retention of nitrogen. They also suggested that both young and growing rats fed on excess-calcium carbonate diet exhibited the decrease in absorption of nitrogen.

In regard to the fat balance, excess calcium in the diet depresses absorption of fat in rats. Fleischman et al.[4,15] reported that fecal lipid significantly increased at the 1.2% calcium level, but not at the 0.2% level, and fecal fatty acids became progressively more saturated with increased calcium intake through the use of beef tallow and cocoa butter.

In 1969 Tadayyon and Lutwak[16] demonstrated that when either calcium or phosphorus intake was high (Ca, 0.08, 0.50, and 2.06%; P, 0.15, 0.58, and 1.67%) while the other was low, absorption of the long-chain saturated triglycerides was decreased, but there was no effect on unsaturated triglycerides.

The excess levels of calcium and phosphorus in the diet enhanced the excretion of fecal neutral fat in adult rats and the high calcium-phosphorus ratio in the diet through the use of calcium lactate enhanced the excretion of combined fatty acid in the feces of both young and adult rats; furthermore, a dietary high calcium-phosphorus ratio induced negative phosphorus balance in adult rats.[6,7] The ingested fat may be hydrolyzed free fatty acids, forming insoluble calcium soaps which are excreted, thus interfering with the utilization of both calcium and fat. When excess phosphorus is present in the alimentary tract, a calcium-fatty acid-phosphate complex may be formed, as suggested by Swell et al.[17]

There is abundant information on the effect of high levels of calcium intake on the utilization of calcium. These results may be summarized as indicating that high levels of calcium increase the net retention of calcium by the animal, but there is a marked reduction in efficiency.

OTHERS

According to recent reports,[6,7] the kidneys of the excess-calcium and -phosphorus

Table 2
pH OF STOMACH AND DUODENUM

Dietary level of Ca (%)[a]

Dietary level of P (%)	Young rat			Growing rat		
	0.52	1.99	1.99	0.52	1.99	1.99
0.52						
Stomach	1.8–2.8 (2.2)	1.6–3.0 (2.2)		1.6–2.4 (1.8)	2.6–3.8 (3.3)	
Duodenum	6.4–6.8 (6.6)	6.4–6.6 (6.6)		5.4–6.8 (6.4)	6.4–7.0 (6.8)	
1.97						
Stomach			2.4–4.6 (4.0)			3.6–3.8 (3.6)
Duodenum			6.0–6.8 (6.4)			6.4–6.8 (6.6)

[a] Parentheses indicate average values.

From Goto, S. and Sawamura, T., *Nutr. Rep. Int.,* 7, 103–110, 1973. With permission.

diet-fed group showed higher calcium contents in adult rats; in young rats, the excess-calcium and -phosphorus diet caused the functional obstruction of kidney, i.e., phosphorus and calcium contents were extremely high and showed a high dry matter weight of kidney. Increasing calcium in the above diet by 1% had a sharp depressing effect on the utilization of other nutrients in the diet, including protein, fats, phosphorus, and magnesium. The high-calcium, high-phosphorus diet also decreased the utilization of other nutrients as compared with the normal-calcium, normal-phosphorus diet-fed group. In those instances in which the intake of these nutrients is limited, increasing the calcium or phosphorus content of the diet may have a markedly adverse effect and may produce a clinical deficiency of other nutrients.

There is as yet no clear picture as to how the excess calcium or phosphorus may exert its influence. It is possible that a high level of calcium in the form of calcium carbonate causes an increase in the pH of the intestinal tract, thereby changing solubilities (see Table 2), but it is not clear that the organic calcium salts affect the pH of the intestinal tract.

REFERENCES

1. Keener, H. A., Cornell Nutrition Conference, 1953; as cited in Davis, G. K., Effects of high calcium intakes on the absorption of other nutrients, *Fed. Proc. Fed. Am. Soc. Exp. Biol.,* 18, 1119–1122, 1957.
2. Richards, M. B. and Greig, W. A., The effects of additions of calcium carbonate to the diet of breeding mice, *Br. J. Nutr.,* 6, 265–280, 1952.
3. Swenson, J. J., Underbjerg, G. K. L., and Goetsch, D. D., Effects of dietary trace minerals, excess calcium and various roughages on weight gains of Hereford heifers, *Am. J. Vet. Res.,* 17, 207–212, 1956.
4. Fleishchman, A. I., Yacowitz, H., Hayton, T., and Bierenbaum, M. L., Effects of dietary calcium upon lipid metabolism in mature male rats fed beef tallow, *J. Nutr.,* 88, 255–260, 1966.
5. Goto, S. and Sawamura, T., Effect of excess calcium intakes on absorption of nitrogen, phosphorus and calcium, *Nutr. Rep. Int.,* 7, 103–110, 1973.
6. Goto, S. and Sawamura, T., Effect of excess calcium intake on absorption of nitrogen, fat, phosphorus, and calcium in young rats. The use of organic calcium salt, *J. Nutr. Sci. Vitaminol.,* 19, 355–360, 1973.

7. Goto, S. and Sugai, T., Effect of excess calcium intake on absorption of nitrogen, fat, phosphorus and calcium in adult rats, *Nutr. Rep. Int.,* 11, 49–54, 1975.

8. Haldi, J., Bachmann, G., Wynn, W., and Ensor, C., The effects produced by an increase in the calcium and phosphorus content of the diet on the calcium and phosphorus balance and on various bodily constituents of the rat, *J. Nutr.,* 18, 399–409, 1939.

9. Schryver, H. F., Hintz, H. F., and Craig, P. H., Phosphorus metabolism in ponies fed varying levels of phosphorus, *J. Nutr.,* 101, 1257–1264, 1971.

10. Tufts, E. V. and Greenberg, D. M., The biochemistry of magnesium deficiency. II. The minimal magnesium requirement for growth, gestation and lactation and the effect of the dietary calcium level thereon, *J. Biol. Chem.,* 122, 715–726, 1938.

11. Colby, R. W. and Frye, C. M., Effect of feeding high levels of protein and calcium in rat rations on magnesium deficiency syndrome, *Am. J. Physiol.,* 166, 408–412, 1951.

12. Hegsted, D. M., Vitale, J. J., and McGrath, H., The effect of low temperature and dietary calcium upon magnesium requirement, *J. Nutr.,* 58, 175–188, 1956.

13. O'Dell, B. L., Morris, E. R., and Regan, W. O., Magnesium requirement of guinea pigs and rats. Effect of calcium and phosphorus and symptoms of magnesium deficiency, *J. Nutr.,* 70, 103–111, 1960.

14. Bunce, G. E., Chiemchaisri, Y., and Phillips, P. H., The mineral requirements of the dog. IV. Effect of certain dietary and physiologic factors upon the magnesium deficiency syndrome, *J. Nutr.,* 76, 23–29, 1962.

15. Fleischman, A. I., Yacowitz, H., Hayton, T., and Bierenbaum, M. L., Long-term studies on the hypolipemic effect of dietary calcium in mature male rats fed cocoa butter, *J. Nutr.,* 91, 151–158, 1967.

16. Tadayyon, B. and Lutwak, L., Interrelationship of triglycerides with calcium, magnesium and phosphorus in the rat, *J. Nutr.,* 97, 246–254, 1969.

17. Swell, L., Trout, C. E., Jr., Field, H., Jr., and Treadwell, C. R., Effect of dietary fat and fatty acid on fecal excretion of a calcium oleate phosphate complex, *Proc. Soc. Exp. Biol. Med.,* 92, 613–615, 1956.

NUTRIENT TOXICITIES IN ANIMALS AND MAN: PHOSPHORUS

H. H. Draper and R. Raines Bell

Phosphorus is not only one of the most ubiquitous elements in nature but is also unique among mineral elements in the extent of its involvement in major metabolic processes in living organisms. In animals and man it is essential in the synthesis of nucleic acids, proteins, phospholipids, polysaccharides, and adenosine triphosphate, as well as a great variety of compounds involved in enzymatic phosphate transfer reactions. The fact that these processes seem not to be markedly affected by dietary phosphorus deficiency or excess implies that there are efficient control mechanisms regulating the concentration of phosphate at specific reaction sites.

This resistance of phosphate-dependent enzymatic processes to fluctuations in phosphorus intake is explainable in part by the presence of about 80% of body phosphorus in the skeleton, which serves as a reservoir for maintenance of phosphate homeostasis in the blood serum and extracellular fluids. In addition, there are regulatory mechanisms, still of an obscure nature, which facilitate the entry of phosphate into the cell and its metabolic compartments, or which reduce phosphate loss, under conditions of hypophosphatemia. While the soft tissues are, by virtue of these mechanisms, adaptable to fluctuations in the concentration of serum phosphorus, chronic hyperphosphatemia leads to entry of excess calcium into the cells and to pathological calcification.

Elemental phosphorus does not occur free in nature but can be prepared in two forms, red and yellow (white), the former being essentially nontoxic and the latter being a deadly poison. The history of yellow phosphorus toxicity has been reviewed by Warnet et al.,[1] who point out that prior to discovery of the toxic properties of arsenic in the mid-nineteenth century, yellow phosphorus was the prime murder and suicide poison. Because of its toxic properties, yellow phosphorus was banned as an ingredient of most products (including matches) which could serve as a source of accidental poisoning, but its use in rodenticides has continued. Indeed, development of resistance to warfarin-based rodenticides has led to increased use of yellow phosphorus rodenticides and there have been several recent reports of lethal poisoning of children and household pets following consumption of bread or crackers coated with commercial yellow phosphorus pastes. Such pastes generally contain 2 to 5% yellow phosphorus.

The symptoms of yellow phosphorus poisoning in children, as described by Simon and Pickering,[2] are characterized by three phases. The first may include cutaneous and oral burns from surface contact with the poison, thirst, vomiting, diarrhea, and abdominal pain, leading in severe cases to cardiovascular collapse caused by damage to the myocardium. The stools are black and exude a white vapor ("smoking stool syndrome"). Survivors of this phase generally experience a symptom-free period of several weeks, then suffer from latent systemic effects on the liver, heart, kidneys, and central nervous system. There is no specific treatment. The acute lethal dose has been estimated at about 1 mg per kilogram body weight.

A rising incidence of accidental poisoning of children with yellow phosphorus rodenticides has led to recommendations that such products be banned. Yellow phosphorus also has been found in sea and fresh water where it has been shown to be extremely toxic to some species of invertebrates and fish.[3]

The acute toxicity of orthophosphates and polyphosphates for animals has been extensively reviewed by the FAO/WHO Expert Committee on Food Additives[4] and more recently by Ellinger.[5] When administered as the sodium salts, the LD_{50} of inorganic phosphates decreases with increasing chain length and is much higher by the oral than by parenteral routes. In general, the acute oral toxicity of inorganic phosphates is not substantially different from that of several other common inorganic salts including sodium chloride.

An inhibitory effect of excess dietary phosphorus on bone development in domestic animals has long been recognized. Chronic ingestion of high phosphate diets causes secondary hyperparathyroidism, bone resorption, and, in some cases, calcification of kidney and heart tissue.[5-7] The level of dietary phosphate required to produce these effects is dependent upon the concentrations of calcium, magnesium, and other ingredients. Excess phosphate exacerbates the effects of magnesium deficiency[8] and increases the intake of calcium required to maintain normocalcemia. Within limits, the adverse effects of increasing phosphate intake can be offset by increasing dietary calcium, but the lower efficiency with which calcium is absorbed at high intakes eventually leads to a relative excess of phosphorus in the blood when the consumption of both elements is increased proportionately. Excess serum phosphorus leads to hypocalcemia and parathyroid stimulation.[7] At the nominally adequate concentration of 0.6% calcium in the diet of the adult rat, kidney calcification has been observed by some investigators from feeding 1.8% phosphorus;[9] others have found renal and aortic calcification from feeding 1.2% phosphorus.[10] A common assertion that excess dietary phosphate causes a reduction in calcium absorption by formation of insoluble salts in the lumen of the intestine, however, appears to have little actual experimental support.

Oral or parenteral administration of phosphates to adult human subjects as the sodium or potassium salts causes a depression of serum calcium (and in some cases magnesium) with a consequent stimulus to parathyroid function (secondary hyperparathyroidism).[11] Parathormone-induced bone resorption ensues as part of the normal adaptational mechanism for restoration of normocalcemia. The demonstration that adult animals fed high phosphorus diets, even in the presence of normally adequate concentrations of calcium, undergo an enhanced rate of bone resorption and a net loss of bone mass has raised the possibility that diets high in phosphorus, particularly those that are concomitantly low in calcium, may contribute to "aging bone loss" or osteoporosis in man. The fact that some human foods contain detrimental excesses of phosphate is illustrated by the occurrence of hypocalcemic tetany in infants fed cow's milk, a condition which has been attributed to the high phosphorus content of the milk of this species.[12] An unusually rapid rate of aging bone loss has been recorded in Eskimos which may be related to the high phosphorus, low calcium content of their native diet.[13]

In recent years the widespread use of phosphate food additives (mainly sodium salts of orthophosphates, pyrophosphates, and "polyphosphates") as acidifiers, emulsifiers, chelators, leavening agents, and water-binders has significantly increased the phosphorus content of the food supply of some countries. Such additives currently appear to contribute an additional 0.5 to 1.0 g phosphorus per day to the diet of U.S. adults, over and above an estimated 1.0 to 2.0 g phosphorus per day from natural sources. A balanced diet formulated to include a high proportion of foods containing phosphate additives recently has been observed to produce evidence of parathyroid stimulation and increased bone resorption in adults.[14]

The public health implications of the increase in phosphorus intake associated with the use of phosphate food additives are currently unclear. Chapman and Rugsley[15] have calculated that if phosphate additives were used at maximum permissible levels in all foods to which their addition is authorized, the phosphorus intake by adults in some countries would approach the maximum conditional acceptable daily intake prescribed by the FAO/WHO Expert Committee on Food Additives (4.2 g for a 60 kg man).[16] This figure, which makes no distinction between natural organic phosphates and more efficiently absorbed phosphate additives, is double the amount of phosphorus (as inorganic phosphate) found to lower serum calcium in experimental subjects.[14] These authors cite ". . . a great need for up-to-date information on the amounts of phosphates being consumed by individuals originating from natural sources as well as resulting from the use of phosphates in foods."[15]

Certain natural organophosphate compounds present in cereals and fibrous foods (notably phytic acid, a hexaphosphoric acid ester of hexahydroxy cyclohexane) reduce the absorption of divalent cations through the formation of insoluble salts in the intestine. Cereal diets high in phytin have been shown to reduce the absorption of calcium, magnesium, and zinc in animals and man, in some cases with the development of symptoms of a deficiency of these elements. Phytic acid is partially degraded during the rising and baking of bread, but in some countries consumption of unleavened bread as a food staple is associated with secondary deficiencies of zinc, calcium, and possibly other elements.[17] Swine fed corn-soybean rations containing excess calcium may develop a zinc deficiency as the result of formation of a calcium zinc phytate complex in the gut.[18]

The serum calcium lowering effect of intravenously administered phosphate salts has been utilized in the clinical treatment of hypercalcemias of various etiologies. However, long-term administration of sodium and potassium salts of phosphate creates a danger of metastatic calcification which limits the usefulness of this procedure.[19]

Sodium phosphates are common ingredients of laxatives and enema preparations. Administration of phosphate enemas to subjects with renal insufficiency entails a risk of hyperphosphatemia, hypocalcemia, hypomagnesemia, and tetany. Since the kidney is the prime route of phosphorus excretion, diminished renal function reduces the capacity to excrete excess circulating phosphate. Caution has been recommended in the use of phosphate enemas prior to renal radiography in patients with severely limited creatinine clearance.[20]

REFERENCES

1. Warnet, J.-M., Claude, J.-R., and Truhaunt, R., Toxicologie biologique expérimentale du phosphore blanc, *Eur. J. Toxicol,* 6, 57–64, 1973.
2. Simon, F. A. and Pickering, L. K., Acute yellow phosphorus poisoning. "Smoking Stool Syndrome." *J.A.M.A.,* 235, 1343–1344, 1976.
3. Fletcher, G. L., The acute toxicity of a yellow phosphorus contaminated diet to brook trout (*Salvelinus fontinalis*), *Bull. Environ. Contam. Toxicol.,* 10, 123–128, 1973.
4. Specifications for the Identity and Purity of Food Additives and Their Toxicological Evaluation: Emulsifiers, Stabilizers, Bleaching and Maturing Agents, *W.H.O. Tech. Rep. Ser.,* No. 281, 1964.
5. Ellinger, R. H., Phosphates in food processing, in *Handbook of Food Additives,* Furia, T. E., Ed., CRC Press, Cleveland, 1972, 617–780.
6. Krook, L., Dietary calcium-phosphorus and lameness in the horse, *Cornell Vet.,* Suppl., 1, 58, 59–73, 1968.
7. Sie, T.-L., Draper, H. H., and Bell, R. Raines, Hypocalcemia, hyperparathyroidium and bone resorption in rats induced by dietary phosphate, *J. Nutr.,* 104, 1195–1201, 1974.
8. O'Dell, B. L., Morris, E. R., and Regan, W. O., Magnesium requirement of guinea pigs and rats, *J. Nutr.,* 70, 103–111, 1960.
9. Draper, H. H., Sie, T-L., and Bergan, J. G., Osteoporosis in aging rats induced by high phosphorus diets, *J. Nutr.,* 102, 1133–1142, 1972.
10. Blazhejevich, N. V., Spirichev, V. B., Pozdnyakov, A. L., Isaeva, V. A., and Gazdarov, A. K., Development of the kidneys and aortic calcinosis in rats with different rations of Ca and P in their ration, *Vopr. Pitan.,* No. 5, 28–34, 1974 (in Russian).
11. Reiss, E., Canterberry, J. M., Bercovitz, M. A., and Kaplan, E. L., The role of phosphate in the secretion of parathyroid hormone in man, *J. Clin. Invest.,* 49, 2146–2149, 1970.
12. Oppé, T. E. and Redstone, D., Calcium and phosphorus levels in healthy newborn infants given various types of milk, *Lancet* 1, 1045–1052, 1968.
13. Mazess, R. B. and Mather, W., Bone mineral content of North Alaskan Eskimo, *Am. J. Clin. Nutr.* 27, 916–925, 1974.

14. Bell, R. Raines, Draper, H. H., Tzeng, D. Y. M., Shin, H. K., and Schmidt, G. R., Physiological responses of human adults to foods containing phosphate additives, *J. Nutr.,* 107, 42–50, 1977.

15. Chapman, D. G. and Pugsley, L. I., Public health consideration of the use of phosphates in food processing, in *Symposium: Phosphates in Food Processing,* DeMan, J. M. and Melnychyn, P., Eds., Avi Publishing, Westport, Conn., 1971, 113–234.

16. Food and Agriculture Organization, *Evaluation of a Number of Antimicrobials and Anti-oxidants, FAO Nutrition Meeting Report, Series 31,* Rome, 1961.

17. Reinhold, J. G., Faradji, B., Abadi, P., and Ismail-Beigi, F., Decreased absorption of calcium, magnesium, zinc and phosphorus by humans due to increased fiber and phosphorus consumption as wheat bread, *J. Nutr.,* 106, 493–503, 1976.

18. Lewis, P. K., Jr., Hoekstra, W. G., and Grummer, R. H., Restricted calcium feeding versus zinc supplementation for the control of parakeratosis in swine, *J. Anim. Sci.,* 16, 578–587, 1957.

19. Lindeman, R. D. and Papper, S., Therapy of fluid and electrolyte disorders, *Ann. Intern. Med.,* 82, 64–70, 1975.

20. Chesney, R. W. and Haughton, P. B., Tetany following phosphate enemas in chronic renal disease, *Am. J. Dis. Child.* 127, 584–586, 1974.

NUTRIENT TOXICITIES IN ANIMALS AND MAN: MAGNESIUM

P. J. Lipsitz

INTRODUCTION

Magnesium is present in the earth's crust and is a major component of sea water. The cation has essential biologic functions in plants, animals, and humans. The magnesium salts have long been known for their healing properties.

Magnesium is widely distributed in foods; animal products are intermediate in the content of magnesium. It has been suggested that the American diet may be marginal in its magnesium content, but it would be very unlikely for deficiency states to develop under normal conditions. It has also been suggested that men require more magnesium than women to remain in equilibrium.[1]

Animal studies of the toxicology of magnesium are limited and pertain mainly to experimental work. Experimental and clinical reports ascribe toxicity of magnesium in man to supplementation of diets in renal disease and parenteral or oral administration of magnesium salts.

The physiology of magnesium in man has only been well studied over the last two decades, as a result of advances in analytic methods. The cation is the fourth most abundant in the human body. Half the total body content is in bone and half in the intracellular compartments. The extracellular fluid contains only about 1% of the total body magnesium. Homeostasis is maintained by hormonal influence, probably para-thyroid hormone. Serum levels range from 1.5 to 1.8 meq/l. Approximately two thirds of ingested magnesium remains unabsorbed and is excreted in the feces, and the other third is excreted in the urine. It is likely that calcium and magnesium have a common transport system from the intestinal tract and renal tubules. The pharmaceutical products of various magnesium salts are used as antacids or laxatives. Parenteral salts are used as potential anticonvulsants and anesthetic agents.

ANIMAL STUDIES

Magnesium is one of the major cations present in the Precambrian seas from which life evolved. This common ancestry and origin accounts for similar concentrations of magnesium and potassium in most tissues of different species. In the 1930s, magnesium was identified as a dietary essential for the rat.

No reports of toxicity of dietary magnesium have been published in animal studies. In the experimental model of rats on high fat diets, Vitale et al.[2] showed reduction of atherosclerotic changes in the animals supplemented with magnesium oxide. This action was produced by the decreased lipid deposits in the blood vessels, although the serum cholesterol level was not decreased. Neal and Neal[3] also showed that magnesium sulfate in distilled water gave complete protection against rabbit atherosclerosis. Bushman et al.[4] showed a reciprocal relationship between urinary excretion of magnesium and phos-phorus in lambs. Magnesium oxide in the diet of the lambs reduced the incidence of urolithiasis.

Renal function studies during infusion of magnesium in the rat suggest that the mode of excretion is by tubular secretion.[5] In the dog, chronic hypermagnesemia causes renal vasodilatation as well as systemic vasodilatation with decreased renal blood flow and filtration rate.[6] Acute severe hypermagnesemia in renal ligated dogs inhibits the cellular transport systems for electrolytes.

In hibernating animals, the serum magnesium rises. It has been suggested that the action of magnesium on the central nervous system maintains hibernation. In acute infusion experiments with magnesium chloride in dogs, there is a rapid fall in body temperature.[7]

In the cat, direct application of magnesium to the central nervous tissues blocks central synaptic transmission and may cause general anesthesia.[8] Feldberg[9] was able to produce an anesthetic-like condition in the cat by injection of magnesium chloride in the cerebral ventricles. The condition lasted about 1 hr and was associated with hyperexcitability. In the animal experiments of Aldrete et al.,[10] magnesium decreased the electrical activity of the cerebral cortex when there was associated respiratory insufficiency. A similar decrease in electrical activity was observed when artificial respiration was used, and there was associated circulatory failure.

Pharmacologic doses of magnesium have a curare-form action on the neuromuscular junction, presumably by interfering with the release of acetylcholine from motor nerve terminals.[11] In studies of general and coronary hemodynamics in the dog, Maxwell et al.[12] showed a tachycardia and a fall in the systemic blood pressure with injection of magnesium chloride. It was postulated that the mechanism of action was most likely due to reduced acetylcholine activity interfering with the release of adrenal catecholamines. Cardiac arrest had been observed in animals when the serum level of magnesium exceeded 25 meq/l; these animals were resuscitated by the injection of calcium chloride.

In the newborn lamb, it has been demonstrated that magnesium sulfate enema will cause toxicity and death of the animal.[13] Dunne et al.[14] studied the effect of magnesium sulfate administered i.p. to newborn rabbits. The animals were made anoxic, and the gasping time was noted to be increased with improvement of the survival rate.

In general, it can be stated that no adverse toxicologic effects of ingested magnesium can be demonstrated in animals. Excess magnesium salts in the diet of the animals appear to be protective for atherosclerotic changes and urolithiasis. Toxicity and death in young animals can be produced by magnesium sulfate enemas.

CLINICAL STUDIES

The absorption of magnesium salts from orally administered antacids and purgatives in the human is dependent on the salt used, e.g., 5% of magnesium in magnesium trisilicate and 21 to 28% of magnesium sulfate is absorbed.[15] With normal renal function, the excess magnesium is excreted. Stevens et al.[16] found magnesium sulfate to be poorly absorbed from the gut, and its action as a purgative is due to its osmotic effect. Toxicity of absorbed magnesium was observed to have a central effect.

The study of Somjen et al.[17] could not demonstrate anesthesia in human subjects by i.v. administration of magnesium sulfate. Rapid i.v. injection of magnesium sulfate can cause nausea and vomiting.[18] It induces peripheral vasodilatation and warmth in the skin.

Pharmacologic toxicity may be produced by excess intake of magnesium when there is marked renal failure, either acute or chronic.[19] The serum magnesium elevation seems to correlate with the degree of nitrogen retention.[20] Parenteral injection of magnesium salts is most dangerous in patients with renal failure. It is common to observe cardiotoxic effects, as manifest by EKG changes (see Table 1) in these patients. The most common cause of hypermagnesemia in patients with chronic renal disease is related to the use of magnesium salts as antacids and laxatives.[21] This may produce a state of dehydration with renal decompensation and a superimposed hypermagnesemia. It is noted that elevated serum magnesium levels in uremic patients has no clinical significance; however, a large intake of magnesium will produce pharmacologic toxicity. Table 1 illustrates the clinical observations related to increased serum magnesium levels.[19,22,23]

Magnesium sulfate is used as a central depressant primarily for the prevention of seizures in the hypertensive crisis of acute glomerulonephritis[10,24] and toxemia of pregnancy.[25-28] Magnesium sulfate was found to be safe and of some value in the preven-

Table 1
THE CLINICAL OBSERVATIONS OF
INCREASED SERUM MAGNESIUM LEVELS

Serum concentration (meq/l)	Clinical observation
4–6	Hypotension
>5.4	Difficulty in urination
6–8	Central nervous system depression
8–12	Loss of deep tendon reflexes
	Drowsiness
	Ataxia
	Slurred speech
	EKG changes
	Increase in PR interval
	Prolonged QRS
	Increased height of T wave
12–18	Respiratory depression
13–17	Coma
19–20	Cardiac arrest in diastole

tion and control of nephritic convulsion in adults, but its effects were neither certain nor dramatic.[24] Using magnesium sulfate in the acute hypertensive stage of acute glomerulonephritis in children, Harris et al.[29] found it to be effective by increasing glomerular filtration rate and increasing the effective renal plasma flow. The use of newer antihypertensive agents in acute nephritis has decreased the usefulness of magnesium sulfate.[20]

In the treatment of toxemia,[25-28] various regimens for administration of i.v. and/or i.m. magnesium sulfate have been proposed. The patient must be observed carefully and the serum levels checked repeatedly. The aim of therapy is to depress the central nervous system and avoid convulsions. Magnesium sulfate would not be expected to cause a fall in the blood pressure. During therapy, the patellar reflex should be present, urinary flow should be at least 100 ml in the previous 4-hr period, and respiration should not be depressed.[25] Renal function and the state of hydration should be monitored closely.

In 1967, research indicated that when toxemic mothers were treated with magnesium sulfate given by continuous i.v. infusion, their newborns exhibited severe depression;[30] these clinical findings were similar to those observed in the adult with renal failure and hypermagnesemia.[19] Further study[31] relating to this problem in the newborn indicated that there was no correlation between the serum magnesium level and symptomatology, but there was a strong correlation between the i.v. administration of magnesium sulfate for greater than 24 hr and depression of the newborn. The analysis of the study also indicated a trend to low Apgar scores when the magnesium sulfate was administered for more than 24 hr. Serial serum levels in the newborns indicated a slow renal excretion of magnesium and recovery of the newborn in 48 to 72 hr. Pritchard and Pritchard[32] could not demonstrate magnesium toxicity in the newborn after i.v. administration of a loading dose of magnesium sulfate followed by i.m. administration every 4 hr.

Treatment of toxicity in the newborn may be accomplished by maintaining hydration and normal acid-base balance. The urinary excretion of magnesium is enhanced by increased urinary flow. Calcium salts have been found to be antagonistic to the central and peripheral depression of magnesium; however, its use in the newborn has given equivocal results. Exchange transfusion in the severely depressed newborn has been found to be effective in reversing the symptomatology of excess magnesium.[33]

Hypermagnesemia has occurred in patients receiving enemas of magnesium sulfate.[34] Outerbridge et al.[35] reported a fatality from a magnesium sulfate enema in a newborn.

The only humans with magnesium toxicity reported thus far are patients with renal failure, an occasional toxemic mother receiving parenteral magnesium sulfate, and the newborn of such a mother. Enemas of magnesium sulfate should never be used in the newborn and are probably never indicated for adult administration.

CONCLUSION

Although magnesium is an important cation with numerous biologic functions in all species, signs of toxicity are rarely noted. However, toxic manifestation may be produced in the animal and human by parenteral administration of magnesium salts. Various stages of toxicity can be observed progressing from hypotension to respiratory and cardiac arrest.

In animals and humans, magnesium has been used to produce a "sleep-like" state similar to anesthesia. Peripheral neuromuscular paralysis and central depression can be reversed by calcium. Iatrogenic disease in man can be produced when there is renal damage or failure and orally ingested magnesium salts are used as laxatives or purgatives. The recommended daily requirement of magnesium is dependent on age and sex.

REFERENCES

1. **Seelig, M. S.,** The requirement of magnesium by the normal adult, *Am. J. Clin. Nutr.,* 14, 342–390, 1964.
2. **Vitale, J. J., White, P. L., Nakamura, M., Hegsted, D. M., Zamcheck, N., and Hellerstein, E. E.,** Inter-relationships between experimental hypercholesteremia, magnesium requirement and experimental atherosclerosis, *J. Exp. Med.,* 106, 757–766, 1957.
3. **Neal, J. B. and Neal, M.,** Effects of hard water and $MgSO_4$ on rabbit atherosclerosis, *Arch. Pathol.,* 73, 400–403, 1962.
4. **Bushman, D. H., Emerick, R. J., and Embry, L. B.,** Experimentally induced ovine phosphatic urolithiasis: relationships involving dietary calcium, phosphorous and magnesium, *J. Nutr.,* 87, 499–504, 1965.
5. **Heaton, F. W.,** The kidney and magnesium homeostasis, *Ann. N.Y. Acad. Sci.,* 162, 775–785, 1969.
6. **Randall, R. E., Jr.,** Magnesium metabolism in chronic renal disease, *Ann. N.Y. Acad. Sci.,* 162, 831–846, 1969.
7. **Heagy, F. C. and Burton, A. C.,** Effects of intravenous injection of magnesium chloride on body temperature of unanesthetized dog with some observations on magnesium levels and body temperature in man, *Am. J. Physiol.,* 152, 407–416, 1948.
8. **Rubin, M. A., Hoff, H. E., Winkler, A. W., and Smith, P. K.,** Intravenous potassium, calcium and magnesium and cortical electrogram of cat, *J. Neurophysiol.,* 6, 23–28, 1943.
9. **Feldberg, W.,** Anesthesia and sleep-like conditions produced by injections into the cerebral ventricles of the cat, *J. Physiol.,* (London), 140, 20P–21P, 1958.
10. **Aldrete, J. A., Barnes, D. R., and Aikawa, J. K.,** Does magnesium produce anesthesia? Evaluation of its effects on the cardiovascular and neurologic systems, *Anesth. Analg.* (Cleveland), 47, 428–433, 1968.
11. **Goodman, L. S. and Gilman, A.,** *The Pharmacologic Basis of Therapeutics,* 5th ed., Macmillan, New York, 1975, 787–791.
12. **Maxwell, G. M., Elliot, R. B., and Burnell, R. H.,** Effects of hypermagnesemia on general and coronary hemodynamics of the dog, *Am. J. Physiol.,* 208, 158–161, 1965.
13. **Andrews, B. F., Campbell, D. R., and Thomas, P.,** Effects of hypertonic magnesium sulfate enemas on newborn and young lambs, *Lancet,* 2, 64–65, 1965.
14. **Dunne, J. T., Milligan, J. E., and Thomas, B. W.,** The effect of magnesium sulfate on anoxia and resuscitation in the neonate, *Am. J. Obstet. Gynecol.,* 109, 369–374, 1971.
15. *Nutr. Rev.,* (Editorial), 26, 12–15, 1968.
16. **Stevens, A. R. and Wolff, H. G.,** Magnesium intoxication: absorption from the intact gastrointestinal tract, *Arch. Neurol. Psychiatry,* 63, 749–759, 1950.

17. Somjen, G., Hilmy, M., and Stephen, C. R., Failure to anesthetize human subjects by intravenous administraton of magnesium sulfate, *J. Pharmacol. Exp. Ther.,* 154, 652–659, 1966.

18. Kelly, H. G., Cross, H. C., Turton, M. R., and Hatcher, J. D., Renal and cardiovascular effects induced by intravenous infusion of magnesium sulfate, *Can. Med. Assoc. J.,* 82, 866–871, 1960.

19. Randall, R. E., Jr., Cohen, M. D., Spray, C. C., and Rossmeisl, E. C., Hypermagnesium in renal failure, *Ann. Intern. Med.,* 61, 73–88, 1964.

20. Wacker, W. E. C. and Parisi, A. F., Magnesium metabolism, *N. Engl. J. Med.,* 278, 658–663; 712–717; 772–776, 1968.

21. Hoff, H. E., Smith, P. K., and Winkler, A. W., Effects of magnesium on nervous system in relation to its concentration in serum, *Am. J. Physiol.,* 130, 292–297, 1940.

22. Wacker, W. E. C. and Vallee, B. L., Magnesium metabolism, *N. Engl. J. Med.,* 259, 431–438; 475–482, 1958.

23. Fishman, R. A., Neurological aspects of magnesium metabolism, *Arch. Neurol.,* (Chicago), 12, 562–569, 1965.

24. Winkler, A. W., Smith, P. K., and Hoff, H. E., Intravenous magnesium sulfate in the treatment of nephritic convulsions in adults, *J. Clin. Invest.,* 21, 207–216, 1942.

25. Pritchard, J. A., The use of the magnesium ion in the management of eclamptogenic toxemias, *Surg. Gynecol. Obstet.,* 100, 131–140, 1955.

26. Chesley, L. C. and Tepper, I., Plasma levels of magnesium attained in magnesium sulfate therapy for pre-eclampsia and eclampsia, *Surg. Clin. North Am.,* 37, 353–367, 1957.

27. Flowers, C. E., Jr., Easterling, W. E., Jr., and White, F. D., Magnesium sulfate in toxemia of pregnancy new dose schedule on body weight, *Obstet. Gynecol.,* 19, 315–327, 1962.

28. Flowers, C. E., Jr., Magnesium sulfate in obstetrics. A study of magnesium in plasma, urine and muscle, *Am. J. Obstet. Gynecol.,* 91, 763–776, 1965.

29. Harris, J. S. and De Maria, W. J. A., Effect of magnesium sulfate on renal dynamics in acute glomerulonephritis in children, *Pediatrics,* 11, 191–206, 1953.

30. Lipsitz, P. J. and English, I. C., Hypermagnesemia in the newborn infant, *Pediatrics,* 40, 856–862, 1967.

31. Lipsitz, P. J., The clinical and biochemical effects of excess magnesium in the newborn, *Pediatrics,* 47, 501–509, 1971.

32. Pritchard, J. A. and Pritchard, S. A., Standardized treatment of 154 consecutive cases of eclampsia, *Am. J. Obstet. Gynecol.,* 123, 543–549, 1975.

33. Brady, J. P. and Williams, H. C., Magnesium intoxication in a premature infant, *Pediatrics,* 40, 100–103, 1967.

34. Fawcett, D. W. and Gens, J. P., Magnesium poisoning following enema of epsom salt, *JAMA,* 123, 1028–1029, 1943.

35. Outerbridge, E. W., Papageorgiou, A., and Stern, L., Magnesium sulfate enema in a newborn, *JAMA,* 224, 1392–1393, 1973.

NUTRIENT TOXICITIES IN ANIMAL AND MAN: SODIUM

H. D. Battarbee and G. R. Meneely

SODIUM CHLORIDE

Until recently, views that a substance in such common use as ordinary table salt might be noxious when consumed in amounts dictated by dietary custom and taste have been received by many scientists with persistent skepticism. These doubts are not altogether unwarranted, for studies of chronic sodium toxicity in man and animals have dealt primarily with the difficult subject of essential hypertension, with the exception of a few studies on cardiovascular lesions, stroke[1-4] and gastric cancer.[5-8] The effects of a high sodium diet are insidious, genetic differences in susceptibility exist, and dietary potassium has been found to exert a protective effect.[9-10] Calcium may likewise have a protective effect.[11] Consequently, studies have often yielded equivocal results. Further, the task of sorting through the voluminous literature on the subject in order to reach some logical conclusion is a formidable one.

Evidence that chronic sodium excess might have some toxic effects derives from four basic types of studies, each of which will be topically discussed later:

1. Epidemiological studies involving geographically or ethnically isolated populations consuming various amounts of sodium.
2. Intrapopulation studies of individuals consuming various quantities of salt.
3. Clinical information from normal and abnormal subjects treated with drugs such as diuretics and/or salt restriction or supplements.
4. Animal studies.

Results from the first two types of studies have been conflicting for a host of reasons and are always difficult to interpret. The latter two types of studies have yielded fairly conclusive results, especially the animal studies. In animal studies, chronic sodium excess leads to an increased incidence of hypertension and increased morbidity and mortality in virtually all vertebrate species tested. With recent advances in the genetics of hypertension and a better understanding of cardiovascular physiology, it is now possible to resolve some of the observed differences in experimental and epidemiological studies and to formulate new hypotheses relating to sodium toxicity.

Since the work of Janeway[12] in 1913, it has been known that death tends to occur sooner in persons with higher arterial pressures. At any level of elevated pressure, the younger the age, the greater is the reduction in life expectancy.[13] Of the American population aged 18 to 79 years, 16% have systolic blood pressures of 160 mm Hg or greater or diastolic pressures of 95 mm Hg or greater or both,[14] and the prevalence of hypertension ranges up to 40% in older age groups.[15,16] When the diastolic pressure is 105 mm Hg or more, the risk of a first major coronary event is four times greater than when it is less than 85 mm Hg.[17] Life insurance underwriters have never had any illusions about the prognostic significance of hypertension.[18] There are good reasons to believe that excessive dietary sodium and deficient dietary potassium play a cardinal role in the genesis of hypertension among a large faction of our population. Further evidence indicates that the noxious effect begins in childhood where the habit of excess salt consumption is acquired at the family table.[19,20] Meanwhile, the present methods of preparation of food, both in the home and before it arrives there, deplete the protective potassium.

The failure of some investigators to give attention to potassium consumption and excretion is undoubtedly responsible for some of the confusion that exists in the current literature relating to sodium toxicity. A relationship between dietary sodium and potassium was reported as early as 1894 by von Bunge[21,22] and has more recently been investigated by Miller.[23] Both elements are essential for normal growth and survival. They are usually consumed together in foodstuffs and extremes of intake can vary the ratio, sodium to potassium, in the bodies of mice and dogs from 1.5 to 2.4.[24] The maximum and minimum tolerances of these electrolytes are interdependent and vary with water consumption and renal excretory efficiency. Acute toxicity, visible edema, occurs in the healthy adult man with 35 to 40 g of sodium chloride per day when he is consuming an ordinary diet containing 3.7 to 7.4 g of potassium chloride per day. Chronic sodium toxicity occurs in many humans, but not all, with much lower intakes.

PHYSIOLOGY OF SODIUM AND POTASSIUM

There is reason to believe that, early in the evolution of life, the potassium content of the Precambrian seas was much higher with respect to the sodium concentration than later came to be the case.[25] This is adduced as a reason why potassium is the principal intracellular electrolyte. At the time of "emergence from the sea," the sodium concentration in the sea had become relatively high. This latter is thought to be the basis for the electrolyte pattern of the extracellular compartment, with its relatively much higher concentration of sodium.

Intra- and Extracellular Osmolarity

Sodium and potassium are important contributors to intra- and extracellular osmolarity. Mammalian interstitial fluid is an almost protein-free filtrate containing sodium as its chief cation. Under steady state conditions, this solution is in equilibrium with the intracellular environment, that is, the effective osmotic pressures of the two compartments are equal because the cell membrane is freely permeable to water. The intracellular fluid, more complex, has several chief contributors to its osmotic pressure: calcium, magnesium, protein, phosphate, and potassium. A significant number of these intracellular ions are not present as osmotically active substances but are sequestered by various structures within the cell and are regulated by and regulators of cell activity.

Cells have no mechanism for directly controlling their own individual osmotic pressure. They rely upon the osmotic control of the extracellular fluid. Any substantial departure from normal intracellular fluid osmotic pressure cannot be tolerated by the organism. The osmotic pressure of the extracellular fluid is closely regulated by one of the most complex homeostatic devices.

Electrogenic Pump

Sodium and potassium are also involved in the electrophysiology of cells. Early studies[26,27] revealed that the permeability characteristics of cell membranes are not sufficient to account for the disparity in the composition of fluids inside and outside the cells. It is necessary to postulate the existence of an "electrogenic pump." This pump is energy dependent, deriving its energy from the metabolic processes of the cell.[28] It transports sodium from the intracellular compartment to the extracellular fluid and potassium into the interior of the cell. The two ions are transported in a loosely coupled manner.[29] The diffusion properties of each of the two ions with respect to the cell membrane is such that a membrane potential is generated by the diffusion of potassium down its electrochemical gradient, leaving a net intracellular negative charge.[28] This electrogenic pump has been shown to be sensitive to intracellular sodium concentration and extracellular potassium concentration. Increases in either or both concentrations increase the rate of the pump.[29] The result is a pump with its own system of feedback control related directly to the concentration of the transported substances. Changes in the permeability of the cell membrane can lead to marked changes in the cell's membrane potential. The diffusion of sodium and potassium down their electrochemical gradients after a change in permeability is the factor responsible for the genesis of the propagated action potential observed in excitable tissues.[28]

Homeostatic Mechanisms

Sodium and potassium are also indispensable components of the homeostatic machinery of the whole organism. Both ions are integral parts of the control mechanism of acid-base balance, renal tubular sodium and potassium being appropriately exchanged for hydrogen ions or reabsorbed depending upon the needs of the organism. Also, the ions are intimately involved in the regulation of extracellular fluid volumes and the maintenance of blood pressure. Small sustained changes in the sodium concentration of the extracellular fluid can have profound effects upon arterial blood pressures.[30] Small changes in blood potassium concentration have a profound effect on cardiac performance. The ions are likewise involved in carbohydrate metabolism in that the active transport of glucose across the intestinal mucosa requires the movement of sodium in the same direction,[31] the peripheral uptake of glucose by cells is associated with a shift of potassium from the extracellular fluid into cells, and some enzymes of glycolysis and oxidative phosporylation are potassium-dependent.[32] Potassium depletion also inhibits the secretion of insulin by pancreatic islet cells in response to a hyperglycemic stimulus. Protein and amino acid metabolism are likewise affected by these two electrolytes.[31]

Because of the profound influences that even subtle changes in sodium and potassium balance have upon cell metabolism and the organism, it is not surprising to find that amazingly complex systems have evolved to maintain the balance of these electrolytes within critical tolerances. The importance of this balance is underscored by the redundance of controls that have developed. A failure in any one component of this control system can often be compensated for by the remaining parts so that the organism does not perish.

Salt and Evolution

Physiophyly concerns the evolution of bodily function as morphogenesis concerns the evolution of form. It takes time for a new or altered function to evolve. Life began 2 billion years ago, emergence from the sea was 360 million years ago, and manlike creatures came into being 2 million years ago. By then, many physiologic mechanisms were firmly encoded in the genes. Dahl[33] and Meneely[34] pointed out that a primal herbivorous people probably consumed at the most 10 mEq/day of sodium, 0.6 g as sodium chloride. A strictly carnivorous man might have consumed 60 mEq (3.5 g as NaCl) on successful hunting days.

Now, for a mere 2000 years or less, "civilized" man has added 6 to 18 g/day of salt to his food. There has hardly been enough time for the genetic material to alter in a favorable adaptation. Further, hypertension as defined today usually manifests itself after the procreative age and, therefore, is not naturally selected out of the species as are most other maladies.

EVIDENCE OF SODIUM TOXICITY

Epidemiological Studies

Epidemiological studies on sodium toxicity as it relates to hypertension have not been enthusiastically received by clinicians or even epidemiologists themselves. There have been enough conflicting reports[35-38] to confound the issue of sodium toxicity. Some explanations of the source of these differing results can be found in the light of recent

Table 1
POPULATIONS IN WHICH THE BLOOD PRESSURE INCREASES LITTLE OR NOT AT ALL WITH AGE (AVERAGE OF BOTH SEXES, IN DIFFERENT AGE GROUPS)

Population	Systolic pressure (mm Hg)			Dietary salt	Ref.
	20–29 years	40–49 years	60–69 years		
Brazil (Carajas)	107	100	109	No salt: use lyes of vegetable ash (K salts)	171
New Guinea (Murapins)	126	126	123	0.6 g (24-hr urine)	172
Botswana (!Kung bushmen)	119	116	122	2.0 g (24-hr urine)	173
Cook Islands (Pukapukas)	113	116	125	2.9–4.1 g (24-hr urine)	123

From Joosens, J. V., *Triangle* (published by Sandor, Ltd., Basle, Switzerland), 12(1), 9–16, 1973. With permission.

Table 2
POPULATIONS IN WHICH BLOOD PRESSURE INCREASES MARKEDLY WITH AGE (AVERAGE OF BOTH SEXES, IN DIFFERENT AGE GROUPS)

Population	Systolic pressure (mm Hg)			Dietary salt	Ref.
	20–29 years	40–49 years	60–69 years		
United States	119	130	149	±10 g/24 hr (DAHL)	99
Portugal	126	134	155	Not measured	174
Sweden	125	138	159	Not measured	175
Belgium	132	143	163	4–20 g (24-hr urine)	176
Cook Island (Raotongas)	124	151	165	7.0–8.2 g (24-hr urine)	123
Norway	130	141	167	Not measured	177
Wales	120	138	169	8.0 g (24-hr urine)	35
Bahama Islands (Negroes)	129	154	176	15–30 g (24-hr urine)	178

From Joosens, J. V., *Triangle* (published by Sandor, Ltd., Basle, Switzerland), 12(1), 9–16, 1973. With permission.

research confirming some earlier suspicions of modifying influences. In spite of these conflicting reports, a few general statements can be made about hypertension and sodium toxicity. It can be said that studies of different so called "primitive" peoples have revealed that populations exist in which blood pressure does not increase with age and in which essential hypertension is rare. Other groups exist, such as the populus of the United States and most "developed" countries, in which blood pressure advances with age and in which essential hypertension affects 15 to 30% of the general population. The incidence of the malady approaches 30 to 40% of the members of some ethnic groups. The common factor in those populations without hypertension is the low sodium content of their diet. In contrast, the blood pressure of those consuming greater amounts of salt (4 to 40 g/day), whether they are primitive or developed, increases with age (see Tables 1 and 2).

Table 3
CHANGES IN SODIUM AND POTASSIUM
CONTENT OF PEAS

Food (100 g edible portion)	Na (mg)	K (mg)
Fresh peas	0.9	380
Frozen peas	100	160
Canned peas, liquid poured off	230	180
Add salt, serve with salted butter	?	?

One source of the disparate results observed in epidemiological studies is exemplified in a study by Dahl in 1954. In this study, a dietary history was taken from subjects. Sodium consumption was estimated on the basis of whether salt was not added by the individual at the table, was added after tasting the food, or was added before tasting the food. A significant correlation was found between estimated sodium consumption and the incidence of hypertension in this small group in Brookhaven.[39] A subsequent larger study utilizing individuals selected from the general population did not confirm these results. Thus the Framingham study,[38] one of the larger intrapopulational studies, yielded negative results. Later, it was demonstrated that subjective estimates of the amount of salt consumed by the individual correlate very poorly with actual measurements of 24-hr urinary sodium excretion. Subjects consuming salt in amounts comparable to the range of the average American diet (8 to 15 g/day) do not sense large differences in the salinity of food very well.[33] In addition, much of the salt added to foodstuffs is added by commercial food preparers and packagers. Since convenience foods are eaten in ever increasing amounts, it is the commercial preparer who determines salt consumption. It has been reported that Americans only exercise voluntary control over 30% or less of their sodium intake.[40]

In addition to inaccuracies in subjective estimates of salt consumption, the collection and measurement of urinary sodium excretion also presents some difficulties. The compliance of subjects in the collection of 24-hr urine samples oftentimes leaves a lot to be desired, and even complete collections to assess sodium consumption may not be ideal. Sodium excretion of individuals consuming constant daily amounts of the element varies surprisingly from day to day, and studies have shown that correlations between 24-hr urinary sodium and actual consumption are rather poor. It is doubtful, in this slowly oscillating system of sodium balance, whether a 24-hr urine specimen accurately reflects one day of sodium intake. Six days of collection are required to reveal sodium excretion accurately even under steady state conditions.[41] The collection of 24-hr urine specimens in an epidemiological study is difficult, the collection of 6-day samples is nearly impossible. Most studies of sodium consumption and hypertension have utilized 24-hr sodium excretion and, despite the caveats, show a significant correlation between the two. A few recent reports suggest that the collection of an overnight urine specimen (first morning specimen) can be useful in estimating sodium consumption.[42]

Another factor largely ignored by most studies to date is dietary potassium. Human studies strongly suggest that potassium has a protective effect in sodium loading[41,43-50] and animal studies confirm this.[9,51-55] No mammals other than man cook food, and as man has become more urbanized, he has turned increasingly to preserved and convenience foods. Both the preservation and cooking of foods tend to deplete the potassium content of food and usually add large amounts of sodium. Table 3 exemplifies the extent to which potassium is depleted and sodium increased in a processed 100-g serving of peas. Canned peas, drained and before butter and salt are added for service at the table, contain

Table 4
SODIUM AND POTASSIUM CONTENT
OF SEVERAL FOODS

Food (100 g edible portion)	Na (mg)	K (mg)
Olives	2400	55
White bread	507	105
Cornflakes	660	160
Cheddar cheese	700	82
Dried nonfat milk	525	1335
Bacon	1700	225
Chipped beef	4300	200
Smoked ham, raw	2530	248
Frankfurter	1100	230
Salami	1260	302
Canned crab meat	1000	110
Canned salmon	540	330

250 times as much sodium as the fresh product and more than half the potassium is gone. The intake of sodium is thereby increased and that of potassium reduced. Some other convenience foods show an even greater rearrangement of elemental composition (see Table 4). Thus man in the processing of food has reduced the amount of protective potassium consumed and loaded his body with an excess of sodium.

Intrapopulation Studies

Intrapopulational studies attempting to correlate sodium consumption and hypertension have met with little success. Only a few have revealed significant correlations.[39,42,56-60] Many of the problems that plague the interpopulational studies also pertain to the intrapopulational ones, but the major difficulty appears to stem from what Meneely[61] has termed the "saturation effect." There is a large body of clinical and experimental evidence that essential hypertension has a large genetic component.[62,63] The condition has been thought to be transmitted by a dominant gene or genes. It is far from clear whether a single gene with incomplete penetrance or several genes might be involved.[64] Another facet of the problem is that, to judge from animal models, many forms of hypertension are genetic in origin but only in a conditional sense. Thus Dahl developed a strain of rats that became hypertensive only if they received more sodium than was required for normal growth and development.[62,63] Studies within populations consuming large amounts of salt reveal that some individuals consume tremendous amounts with relative impunity, while others develop severe hypertension with small amounts, and still others are affected in some intermediate fashion. Attempts to correlate the incidence or degree of hypertension with sodium consumption under these conditions is almost futile. The dynamic range of sodium consumption is not likely to be large enough to bring out a clear correlation. These individuals are saturated with the element. Only those humans who are genetically predestined to have salt-induced hypertension will manifest it under conditions of excess sodium intake. Those lacking the salt-sensitive trait will remain normotensive in spite of an excess sodium intake and there will be no distinction between the salt intake (or urinary sodium excretion) of normotensive and hypertensive subjects. Perhaps more meaningful intrapopulation studies could be conducted by sorting out families predisposed to hypertension from those without the trait and correlating salt consumption and hypertension in the two subpopulations. Recently, such studies have yielded some affirmative results.[42,59]

In addition to the above factors that confuse the interpretation of epidemiological studies, there exists another which is very difficult to deal with. Hypertension is a

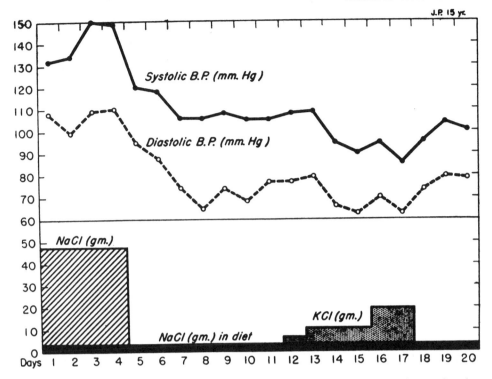

FIGURE 1.   Hypertensigenic effect of sodium chloride and protective effect of potassium in a 15-year-old salt-eating diabetic girl. BP = blood pressure. (Reprinted from McQuarrie, I., Thompson, W. H., and Anderson, J. A., *J. Nutr.*, 11, 77—101, 1936. With permission.)

self-perpetuating entity. Human[65] and animal[66] evidence suggests that once sustained hypertension exists, it persists even though the conditions that brought about the malady are removed. Thus, once a chronic sodium dependent hypertension manifests itself, a train of events takes place that sustains an elevated blood pressure even though salt consumption may subsequently be quite low. Attempts to correlate salt consumption and hypertension in individuals in whom the original insult may have occurred substantially displaced in time are obviously impossible.

Clinical Information and Animal Studies

It is obvious that sodium chloride increases blood pressure in some[67,68] and does not in others[69,70] and equally obvious that restriction of dietary sodium reduces high blood pressure in some and not in others.[71-73] The genetic explanation of this has already been noted. Saluretic diuretic agents reduce blood pressure in most patients with hypertension. It has been shown that patients with essential hypertension who respond to diuretic drugs respond equally well to low sodium diets.[56,74] The only common denominator among these drug preparations is natriureses.

The ability of potassium in the short term to reduce blood pressure may be seen in Figure 1 from work by McQuarrie et al.[44] Although the patient's condition was not normal (she had diabetes), her blood pressure increased with excess ingestion of sodium chloride, and it was reduced after reduction of sodium chloride. McDonough performed a similar study in a normal subject and produced similar results.[67] Shaper et al.[75] observed a rapid rise in the blood pressure of young Samburus (who customarily do not eat salt), when given 15 g of salt per day during military service. Tobian[76] and Kirkendall[77] also reported hypertension that was due to salt intake in previously healthy individuals. Even more impressive is the case shown in Figure 2, also from McQuarrie et al.[44] Extra

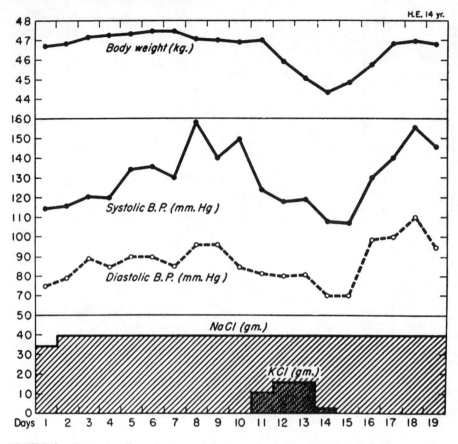

FIGURE 2.   Dramatic effect of extra potassium on the sodium chloride induced hypertension of a 14-year-old salt-eating diabetic patient. (Adapted from McQuarrie, I., Thompson, W. H., and Anderson, J. A., *J. Nutr.*, 11, 77–101, 1936. With permission.)

potassium chloride produced an impressive reduction in blood pressure in spite of continued ingestion of excess dietary salt.

Recent animal studies have yielded good evidence to explain the pleomorphic human epidemiological observations. Alexander et al.[78,79] may well have been the first (1954) to develop a "spontaneously" hypertensive animal, a strain of rabbits. Dahl and Schackow[80] credit Smirk and his group[81] with the first report of a colony of rats with genetically determined elevations in blood pressure. In 1962, a sodium chloride sensitive strain of rats was developed by Dahl et al.[62,63] In 1963, Okamoto,[2-4] Aoki et al.[1] and Yoshitoshi[5] isolated a Wistar-derived strain of spontaneously hypertensive rats. Louis et al.[84] found that this strain continued to become hypertensive with limited salt intake and further explored the genetic component.[85] Hypertension in all these "spontaneously" hypertensive animals is brought on earlier and made much worse by extra dietary sodium chloride.[5] Meanwhile Schlager and Weibust[86-88] found that variations of blood pressure among mice were genetically determined and concluded that several genes were involved. An important paper by Aoki et al.[1] sheds a good deal of light. Their careful analysis of substrains of their hypertensive rats showed genetic independence of "sensitivity" to salt, i.e., its propensity to induce hypertension, and of "toxicity," by which they meant a propensity of salt to induce vascular lesions and to accelerate mortality. They present a fourfold genetic categorization: high sensitivity, high toxicity; high sensitivity, low toxicity; low sensitivity, high toxicity; and low sensitivity, low toxicity.

Meneely et al.[9] demonstrated that rats of Sprague-Dawley derivation would develop

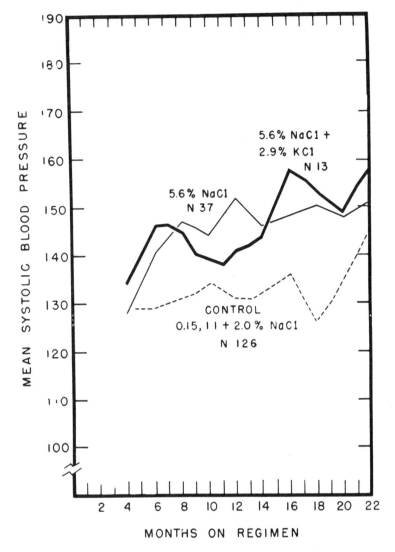

FIGURE 3. Rats eating intermediate levels of extra sodium chloride developed a moderate hypertension with or without the addition of protective potassium chloride. (Adapted from Meneely, G. R., Ball, C. O. T., and Youmans, J. B., *Ann. Intern. Med.,* 47, 263–273, 1957. With permission.)

hypertension when fed levels of salt corresponding to a human diet of perhaps 14 to 28 g/day with other dietary constituents held constant (Figure 3). Dahl et al.[62,63] found that long-term consumption of excess sodium chloride may induce pathologic lesions without direct relation to blood pressure. Further, it was demonstrated that if a similar sodium chloride intake was provided but extra potassium chloride was also fed throughout life, the level of blood pressure manifested by these rats was not significantly lower than those not receiving the extra potassium, but there was a very large increase in the median duration of life.[9] Figure 4 presents a composite of two separate experiments, a "pilot" study of 13 rats and a follow-up study of 30 rats, each with concurrent groups of "unprotected" rats. An almost identical experiment performed later produced similar results,[51,52] Coleman et al,[89] comment that one of the animal models of hypertension, the dog, is ". . . salt induced hypertension, where excessive salt is combined with reduced renal function. . ." It is true that it is almost impossible to induce hypertension in the

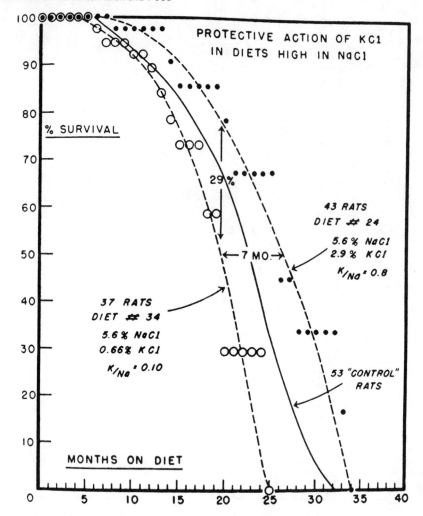

FIGURE 4.   Despite the lack of significant differences in blood pressure between the two groups, extra dietary potassium chloride resulted in a prolongation of the median duration of life of 7 months. (Adapted from Meneely, G. R., Ball, C. O. T., and Youmans, J. B., *Ann. Intern. Med.*, 47, 263–273, 1957. With permission.)

dog[90] unless one does "reduce" renal function.[91] This is not so in the Sprague-Dawley derived rat. The majority of these animals will manifest hypertension with any intake of salt much in excess of that of the Hubbel-Mendel-Wakeman mineral mix[92] fed at the 2% level. The sodium allowance in this mix is 0.14% as sodium chloride in the final diet. Ordinary "laboratory chow" contains seven times more sodium chloride, 1% or higher. Laboratory chow is a high sodium diet. This is further evidenced by the fact that rats eating laboratory chow manifest fully suppressed aldosterone secretion. When they are fed a diet with much lower sodium content, aldosterone secretion increases briskly in a very few days.[93]

Dahl et al.[62,63] note that they have repeatedly seen lessening of vascular disease, clearing of retinopathy, for example, even when a patient's pressure failed to decrease with a low salt diet. Priddle[94] reported similar findings. Kempner[95] reported the same findings earlier, and more recently Okamoto and co-workers[3,4] explored the effect of extra sodium chloride on the genetic susceptibility to stroke in their hypertensive rats. Thus, there is both animal and human evidence that elevation of blood pressure is not the only toxic effect of excess sodium chloride. Studies in animals and clinical observations in humans reveal a genetic dissociation of several independent noxious effects.

EXCESS SODIUM IN MAN

Salt Loading

Aside from the rather uncertain matter of treks to salt licks, no terrestrial mammals except man add salt to their food in the wild. Feral man was probably confronted with a dietary sodium load that rarely exceeded 3.5 g/day, and dietary potassium was plentiful. Man in fact is admirably equipped to cope with a harsh, dry terrestrial environment in which sodium is usually in short supply, and potassium ingested with foodstuffs is found in excess. The metanephric kidney, endocrine systems, and an integument that is relatively impermeable to water have evolved to conserve the precious liquid along with sodium, while excreting a potassium load. Conversely, there are no efficient physiologic mechanisms for conserving potassium and eliminating excess sodium, particularly chronic sodium excess.

Although it now appears that the minimum daily requirement for sodium is approximately a few hundred milligrams, surely no more than a gram, the average consumption of this substance is from 10 to 35 times this amount among most "civilized" people. Most reference works cite the Food and Nutrition Board Recommended Dietary Allowance of 1 g of sodium chloride per kilogram of water intake (in food or fluid). Water consumption is actually a function of the amount of salt in the diet as noted by Pliny the Elder.[96] In the experimental animal, water consumption varies directly with dietary salt.[97] Since the average daily turnover of water is about 2,750 ml for a 70 kg adult, this recommendation would suggest about 3 g/day of sodium chloride. Much evidence indicates this is more than enough.

Aside from the evolutionary considerations of sodium excess and potassium deficit, recent data upon renin and aldosterone secretion further support the hypothesis that civilized man consumes too much sodium. Aldosterone serves two bodily functions. It is secreted to promote the excretion of potassium under conditions of excess, and it is necessary for the renal retention of sodium under conditions of sodium deprivation. Sealey et al.[98] have clearly shown in their studies of the relationship between urinary aldosterone excretion and urinary sodium excretion that normal man consuming in excess of 5 g/day of sodium chloride manifests a greatly suppressed aldosterone secretion (see Figure 5). In other words, such individuals operate with one portion of their homeostatic machinery suppressed to the extent that it is practically inoperable unless a potassium excess intervenes. Most biological control systems do not ordinarily function at the extremes of their function curves. To do so renders them ineffective and insensitive. Instead, they usually operate near the midpoint or in the more sensitive portion of their function curve. The midpoint for the urinary aldosterone and sodium curve is about 3 g of sodium chloride per day, suggesting that this amount might represent an appropriate daily consumption of salt.

Epidemiological studies further support the hypothesis that a daily consumption of 3 g/day of sodium chloride, perhaps less, might represent an appropriate maximum intake (see Tables 1 and 2). Populations consuming this amount or less show very little or no increase in blood pressure with age. It is possible that the "accepted norms" for blood pressure in the American population may be the usual pressure rather than the normal pressure for *Homosapiens recens*. Blood pressures far lower than those found after age 50 among men and women in the U.S.[99] have been found in the sixth, seventh, and eighth decades of life in Mayan Indians,[100] Andes natives,[101] Cook Island Polynesians,[102] African bushmen,[103] clerks in Calcutta,[104] female rural Northern Indians,[105] German contemplative monks,[106] Kenyan Samburu elders,[75] Buddist farmers in Thailand,[107]

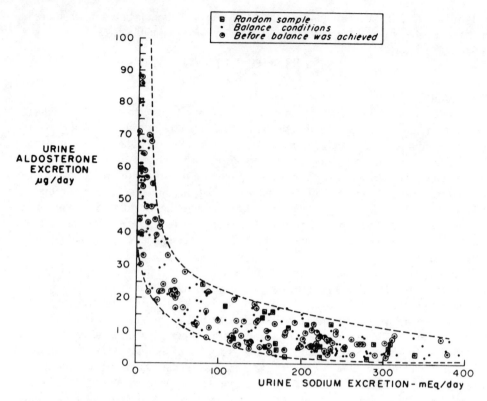

FIGURE 5.   Relationship of urinary aldosterone excretion to daily urine sodium excretion. (Adapted from Sealey, J. E., Bühler, F. R., Laragh, J. H., Manning, E. L., and Brunner, H. R., *Circ. Res.*, 31, 367–378, 1972. With permission.)

Japanese farmers of the Tottori culture,[108] New Guinea natives,[109] Masai Warriors,[110] Somali nomads,[111] Eskimos,[112,113] U.S. Naval aviators,[114] American doctors and nurses moving to China in the 1920s,[115-117] inhabitants of Norielsk, a Russian Arctic industrial city,[118] and isolated Navajos.[119] In most of these populations, blood pressure remains at a level characteristic of persons in the United States at about age 25, increasing little if at all with advancing age. Newer reports continue to provide instances of populations whose systolic blood pressure remains below 120, 125, or 130 throughout adult life.[120-123] One cannot, therefore, dismiss the possibility that the American data on blood pressure reflect merely the usual, not the normal, levels.

Taste for Salt

Most human beings have a taste for salt, but there is good evidence that this is due to its properties as a condiment rather than some mysterious physiologic need. Lapicque[124] contributed an invaluable anthropologic observation concerning a tribe in Africa that lixiviated its salt from the ashes of certain plants. Analysis showed it to be fairly pure potassium chloride. Sodium chloride was available and cheap in the area inhabited by this vegetarian people, but they used it only when their own potassium chloride was scarce. Even then, they would mix the trader's salt with some of their own potassium chloride. Lapicque thus convincingly documented the use of salt purely as a condiment similar to the use of pepper, curry powder, paprika (capsicum), pimento, and numerous other widely used flavorings. Potassium is readily available in nature away from the sea, sodium is not.[125] The lixivium from most plant ashes is rich in potassium carbonate which, because of its disagreeable taste, earned the name "salt of wormwood." It was by chance that Lapicque's inhabitants of the basins of the Ogoone and the Sangha "by the

Table 5

SODIUM INTAKE OF INFANTS AT 6 MONTHS OF AGE

	Total mEq/day	Dietary salt intake		Na:K
		/kg	/100 kcal	
Infant-human milk	6−8	1	1.0	0.6
Puyau average	45 ± 12	6	6	0.7−1.5
6-month-old infant				
Purvis				
1969 survey	40−45	6	6	1.15
1972 survey	25−30	3−4	3−4	0.89
Adult	150−200	2−3	6−8	1.2−2.0

From *Pediatrics,* 53, 115−121, 1974. With permission.

unconscious experimentation of successive generations" came upon a particular species of plant whose ashes yielded a nearly pure potassium chloride with a good sharp saline taste. Lapicque drew attention to the extraordinary ability of human beings, just by experimentation, to identify other gratifying substances of widely differing plant origins. Thus, caffeine was found by the Abyssinians and the Arabians in coffee, by the Chinese in tea, by the Paraguyans in maté and by the natives of the Amazon region in Guarana. Lapicque thus effectively disproved von Bunge's thesis[21] that a vegetarian diet required extra sodium chloride. One of the authors found some years ago that potassium chloride could be admixed with sodium chloride, up to 50% for some, without a noticeable change in taste[126] and Frank and Mickelsen[127] recently made a similar observation. A product with such a mixture has recently appeared on the market.

Dahl[19] found in patients consuming long-term diets containing up to 250 meq per day of sodium chloride that the addition of 5 or 10 g could go unnoticed. It was concluded that salt bears no necessary relation to requirement. Wotman et al.[128] confirmed an elevated salt taste threshold as characteristic of patients with hypertension. The salt preference of animals may represent an effort to maintain sodium homeostasis[129,130] since sodium is scarce in nature away from the sea,[125] whereas salt appetite is acquired by human beings primarily as a result of social and dietary custom and usage.[33,112] Schechter et al.[131] found no difference in taste threshold between hypertensive and normotensive subjects but, when offered distilled water or saline solution by free choice, the hypertensive patients consumed more than four times as much sodium chloride as did normotensive control subjects.

Children's Diet

Dahl[19] drew attention to the unenviable infant mortality rate in the U.S.[132] believing that high sodium chloride levels in the infant diet were partly to blame. He pointed out that extra salt is neither necessary nor useful. Certain baby foods he tested in the late 1960s were lethal for rats because of their high sodium chloride content.[20] The Committee on Nutrition of the American Academy of Pediatrics concluded at that time that the salt intake of infants was several times larger than the requirement and recommended that the salt content of infant foods be reduced. Subsequent surveys have shown that the manufacturers have complied. Some parents needlessly add salt to their infant's food in accordance with their own taste preference. This has the effect of giving unnecessary salt to the child and cultivating in him a taste for excess salt.[19,20] Table 5 shows the results of one survey. Before 1969, some children were consuming an adult equivalent of 25 g/day. Epidemiological studies clearly show that such a large intake of sodium has devastating long-term effects. Many parents are still not aware that the salt

Table 6

SOURCE OF SALT IN THE U.S. INFANTS DIET — 1972 SURVEY
OF 374 INFANTS

Age (months)	mEq of Na per day					
	2	4	6	8	10	12
From						
Formula	8.0	4.5	1.5	1.0	1.0	0
Milk	2.0	7.0	12.5	13.5	12.5	12
Baby food	3.5	7.5	12.0	15.0	13.5	12
Table food[a]	0	0	3.0	9.5	24.0	39
Total	13.5	19.0	29.0	38.0	51.0	63
Estimated sodium of an infant receiving only human milk	6	7	8	9	10	12

[a] Sodium content estimated from Handbook #8, U.S. Dept. of Agriculture, December 1963. Does not include table salt.

From *Pediatrics*, 53, 115—121, 1974. With permission.

requirement of infants and children is low and that excesses are noxious. The sodium content of breast milk is very low, 7.4 mEq/l, less than 0.5 g as sodium chloride. This amount is more than adequate for normal growth and development. Since breast feeding is no longer in vogue, cow's milk is now used for most infant formuli. The salt content of cow's milk is four to five times that of breast milk. Table 6 presents some recent revealing information about salt consumption in children. The table very clearly shows a tremendous increment in salt consumption as a child ages and his dietary habits change from one consisting primarily of breast milk to one primarily of table food. The figure does not take account of any additional salting that occurs at the dinner table. The estimated salt intake of 12-month-old children in this survey exceeded an equivalent of 30 g/day for an adult. This is more than is consumed in Northern Japan where the incidence of hypertension is nearly 40% of the general population.

MECHANISMS OF SALT INDUCED HYPERTENSION

Guyton and co-workers[89,133-135] have pioneered many aspects of body fluid regulation and in the past few years have developed many important facets of this complex system. Their systems analysis approach to the control of arterial blood pressure and hypertension provides a great deal of quantitative insight to hypertension.

Extracellular Fluid Volume and Sodium

Increased sodium intake in man and in animals affects electrolyte and fluid balance to a considerable extent. Studies of sodium loading reveal an increase in body weight, exchangeable sodium, extracellular fluid volume, plasma volume, and blood volume. Under these conditions, the urinary output of sodium, potassium, and calcium increases, the plasma concentration of sodium increases slightly, and total body potassium and plasma potassium concentration reportedly decrease.[136] In such preparations, the cardiac output is increased due to increased venous return and greater stroke volume, but the heart rate reflexly declines. The total peripheral resistance measured on the 9th and 24th days in experimental animals is normal, but since the hematocrit value and blood viscosity are decreased, this suggests a reduction in average blood vessel diameter.[137] Dogs with a reduced renal mass have been shown to have a rapid increase in extracellular fluid volume, blood volume, right atrial pressure, stroke volume, cardiac output, and blood pressure within a few days after increased sodium intake. In these animals the heart rate is reflexly decreased, the serum sodium concentration is increased, and the total peripheral resistance is increased .... further contributing to the increase in blood pressure. Cardiac output soon returns to normal, leaving an animal with an elevated blood pressure and increased total peripheral resistance as a result of these two autoregulatory systems working at cross purposes.

Local Tissue Autoregulation

It has been shown that the process involved in the increase in total peripheral resistance in high cardiac output-volume expansion hypertension is due to a purely local phenomenon referred to as autoregulation. This control of blood flow is relatively independent of neural control and vasoactive substances that enter the vascular bed via the arterial blood supply[136] and is responsible for the relatively constant blood flow through tissues in spite of significant changes in arterial pressure. Such intrinsic regulation of blood flow is thought to occur as a result of smooth muscle activation as more oxygen is delivered and by the washing out of vasodilator substances.[136,138,139] It is thought that under circumstances of a high salt intake there is an increase in extracellular fluid volume and blood volume that promotes an increase in cardiac stroke volume via the Frank-Starling mechanism[140] and a consequent increase in blood pressure. Tissue perfusion likewise increases and as the tissues "perceive" that blood flow exceeds the needs of that particular tissue, there is a constriction of the vascular resistance vessels (arterioles) that reduces the flow to a level commensurate with the needs of the tissue at that particular time. After a period of time, these elevations in blood pressure tend to become permanent, even when the animals are taken off of the high sodium diet.[65,66]

Permanent Hypertension

The changes that are thought to be responsible for the development of a permanent type of hypertension are the subject of much controversy. Some investigators feel that passive changes in the caliber of the resistance vessels are at least in part responsible for the perpetuation of an elevated blood pressure. It has been hypothesized that these changes are due to sodium retention and potassium depletion by such vessels, vessel wall edema, and vessel wall hypertrophy. Aortic sodium content and possibly water content have been shown to increase with an increased dietary sodium load,[141,142] and it is possible that an increase in water content initially limits the distensibility of resistance vessels and also narrows their lumena.[76] The locus of such extra sodium and water has not been established, but it appears to be intracellular. Sodium could accumulate inside the smooth muscle cell as a result of an increased sodium influx brought about by greater diffusion down an electrochemical gradient, and/or due to greater cell membrane permeability and decreased sodium efflux due to slowing of the sodium pump. Also, smooth muscle cells in the larger resistance vessels are under increased stretch, and this is known to enhance the number of "spike" action potentials in some muscles.[142] Hypertension itself leads to changes in the sodium and water content of vessels. Experimentally induced and clinical coarctation of the aorta have been shown to lead to increased intracellular sodium and water content with unchanged or decreased potassium content in the wall of the proximal segment of the vessel.[143,144] Thus, some "water logging" seen during dietary sodium loading could be due to stretch in the presence of an increased sodium gradient and/or due to slowing of the sodium pump brought about by a decrease in extracellular fluid potassium concentration. Greater potassium excretion occurs under these conditions (discussed in following section). Such changes as these might be prevented by the feeding of potassium chloride to prospective hypertensives. In fact, many of the changes seen in sodium loading, increased exchangeable sodium and decreased total body potassium, can be prevented by the simultaneous feeding of potassium chloride and this also ameliorates the detrimental effects of such hypertension.[10,51]

The differential ion concentration of the intracellular versus the extracellular compartments is thought to be generated by means of an ATP dependent "pump" that removes sodium from the cell and transports potassium into the cell. The diffusion of potassium down its concentration gradient from the interior of the cell to the extracellular fluid is thought to leave a net negative charge relative to the exterior of the cell. Such electrogenic pumps are common throughout the body, being found characteristically in such excitable tissues as nerve and striated muscle. In muscle, changes in membrane potential (action potentials) are responsible for the initiation of contractile events. It has been found that the electrogenic pump can be stimulated by the addition of potassium to the extracellular fluid and sodium to the intracellular fluid, and that it can be inhibited by lowering the extracellular fluid potassium concentration or intracellular sodium concentration. Such changes in the rate of the electrogenic pump are reflected in the membrane potential: increasing its rate hyperpolarizes the membrane and decreasing its rate hypopolarizes it. A similar pump has been demonstrated in vascular smooth muscle. This pump is much more potassium dependent than either of the more familiar pumps of skeletal muscle or cardiac muscle or those of nervous tissue.[145] In general, it is very sensitive to its ionic environment.[146,147] It has been shown that a reduction in the potassium concentration in an intact vascular bed results in an almost immediate contraction and increase in resistance[148,149] that can be blocked by ouabain.[150] Some investigators feel that sodium loading and the consequent increase in potassium excretion may lead to hypokalemic depolarization of vascular smooth muscle and an increase in vasoconstriction of such tissue. Such increased activity could be at least in part responsible for the vascular smooth muscle hypertrophy seen in salt loaded hypertension.

Exchangeable Sodium and Total Body Potassium

Despite the fact that electrolytes, particularly sodium and potassium, play an important role in hypertension, dynamic studies have gone lacking due to the technological difficulties imposed in attempting to ascertain their distribution and quantities — the only way to measure total body potassium is to homogenize the animal unless whole body potassium counting facilities are available. Exchangeable sodium lends itself more readily to quantitation. Meneely et al. performed such studies[10,51] and revealed that the inclusion of potassium chloride in the diets of salt-fed hypertensive rats did in fact ameliorate the detrimental effects of such hypertension. Similar relationships between potassium chloride and hypertension have been shown to exist in man.[84] There is evidence that the exchangeable sodium is markedly affected by the amount of potassium chloride included in the diet. In Meneely's study, those animals eating excess sodium whose exchangeable sodium was increased had a proportional increase in systolic blood pressure. The inclusion of potassium chloride in the diets of these animals lowered their exchangeable sodium, and they had a proportionately lower blood pressure and greatly enhanced survival.

Vascular Hyperreactivity

Mendlowitz[151] has reviewed the role of vascular reactivity in systemic arterial hypertension and summarized the extensive contributions he and his associates have made. He concludes that "In most systemic vascular beds and in the systemic circulation as a whole, vascular reactivity to vasoactive substances acting directly or indirectly on the sympathetic neural and vascular receptor complex is increased in hereditary essential hypertension." Further, he states, " . . . such increased reactivity is present early in the course of the disease and even in essential prehypertension . . . ." Our own interest in this centers on the isolation and characterization of a humoral factor present in hypertensive patients and salt-eating hypertensive rats that sensitizes bioassay animals to exogenous norepinephrine.[152] Other investigators have shown that there is increased secretion and decreased storage of catecholamines in salt loaded animals.[153]

Sensitizing Factor to Catecholamines

Lately there has been a renewed interest in a "sensitizing factor" to catecholamines in hypertension, but because of the diverse results of both older and newer studies and the frequent utilization of inadequate controls, reports have been conflicting[151,154-165] Several investigators have found evidence for a vasopressor agent potentiating factor among hypertensive subjects,[160,166-168] and a transmissible humoral factor has been suggested by a study by Dahl et al.[169] utilizing the parabiotic union of a strain of rats genetically susceptible to the development of hypertension.

Recently, studies by Mizukoshi[170] and from the authors' laboratories[152] resulted in reports of a significant increase in vasopressor sensitivity to norepinephrine and angiotensin after the administration of small quantities of serum from hypertensive humans or animals to bioassay animals. This increase was thought to be due to a heat-stable, nondialyzable substance that was produced and released by the ischemic kidneys of patients with renovascular hypertension. It was found in the plasma of patients with essential and malignant hypertension and some normotensive subjects. Sodium restriction in these hypertensive subjects unequivocally reduced the plasma activity of this substance.

SUMMARY AND CONCLUSIONS

Sodium toxicity in animals and most particularly in man has been a highly controversial subject for a long period of time. There is no question that the element can be deleterious, even fatal, when consumed acutely in very large amounts, and the train of events is such that there is little doubt as to cause and effect. In contrast, the long-term effects of chronic daily ingestion of "moderate" or what some individuals consider small amounts of sodium are much more covert. This has led to a great deal of difficulty in interpreting many studies. Nevertheless, epidemiological, clinical, and animal studies strongly suggest that chronic sodium excess is pernicious and insidious. It appears that under most circumstances, approximately 1/2 g of sodium chloride per day, certainly no more than a gram, would represent an adequate intake. The peoples of most "civilized" countries consume 10 to 35 times this amount, and it exacts a dreadful toll.

The aforementioned difficulty in interpretation of results is due in part to the fact that studies on chronic excess sodium deal principally with the complex subject of essential hypertension. There is much evidence to indicate that other manifestations of toxicity are at least equally important. Many modifying influences have been shown to affect the manifestations of sodium excess, and the failure of earlier studies to recognize these influences is largely responsible for the diversity of experimental results observed. Differences in genetic susceptibility, modifying elements, the insensitivity of methods, and the extreme complexity of the subject confound the efforts of investigators. Difficulty arises even with the identification of hypertensive subjects, for it is now apparent from epidemiological studies that what was previously thought to be a normal pressure was more likely the usual pressure in many developed countries. Recent advancements in each of these troublesome facets allow recognition of the shortcomings of many earlier studies and hopefully will lead to future better controlled studies.

Although it may not be possible to predict exactly how a given individual or animal will respond to a given sodium load, recognition of the toxic properties of sodium and knowledge of the mechanisms involved in its toxicity offer great possibilities in the area of preventive medicine. It may be possible, by the familial sorting out of hypertension-prone subjects and dietary and/or drug intervention, to prevent or minimize the development of this self-perpetuating disease state and its sequelae in susceptible individuals. This says nothing of the prospects in other areas of sodium toxicity, of which we are largely ignorant. In addition, it would be wise to reconsider customs in the preservation, marketing, and preparation of foods. These latter, although they would reap the greatest benefit, appear to be the least practical. Man responds to restricted salt intake not unlike one addicted to drug. Wars have been fought, wives and children have been sold into slavery, and many have been dismembered or put to death for dealing in the material as lucrative contraband to insure a bountiful supply of this substance. A massive reeducation program would be necessary to implement a large-scale reduction in salt consumption, and even then a great deal of noncompliance would have to be expected, especially among the youth where it would do the most good.

ACKNOWLEDGMENT

Much of the material pertinent to sodium and potassium nutrition was assembled for Chapter 26 (Sodium and Potassium) in *Present Knowledge in Nutrition,* 4th ed., Hegsted, D. M. et al., Eds., published by the Nutrition Foundation, Inc., Washington, D.C., 1976. We are grateful to Dr. William J. Darby, President of the Foundation and Dr. D. M. Hegsted, Editor, *Nutrition Reviews* for their generous policy which permitted us to quote freely from that chapter. We are also grateful to Dr. Simon Dack, Editor, and the Yorke Publishing Company, Publisher, of *The American Journal of Cardiology* for permission to quote from a recent article in that journal.

REFERENCES

1. Aoki, K., Yamori, Y., Ooshima, A., and Okamoto, K., *Jpn. Circ. J.,* 36, 539—545, 1972.
2. Okamoto, K. and Aoki, K., *Jpn. Circ. J.,* 27, 282—293, 1963.
3. Okamoto, K., *Int. Rev. Exp. Pathol.,* 7, 227—270, 1969.
4. Okamoto, K., Yamori, Y., and Nagaoka, A., *Circ. Res., Suppl.,* 34, 35, 143—153, 1974.
5. Yoshitoshi, Y., *Proc. 3rd World Congr. of Gastroenterology,* Tokyo, 1966, 1, 179.
6. Nagaoka, A., Iwatsuka, H., Suzuoki, A., and Okamoto, K., *Am. J. Physiol.,* 230, 1354—1359, 1976.
7. Sato, T., Fukuyama, T., Suzuki, T., Takayanagi, J., Murakami, T., Shiotscurki, N., Tanaka, R., and Tsuji, R., *Bull. Inst. Public Health,* Japan, 8(4), 187, 1959.
8. Hirayama, T., *UICC Monograph Ser.* (Proc. 9th Int. Cancer Congress, Tokyo, 1966), Vol. 10, Springer, Berlin, 1967, 37.
9. Meneely, G. R., Ball, C. O. T., and Youmans, J. B., *Ann. Intern. Med.,* 47, 263—273, 1957.
10. Meneely, G. R. and Ball, C. O. T., *Am. J. Med.,* 25, 713—725, 1958.
11. Langford, H. G. and Watson, R. L., *Trans. Am. Clin. Climatol. Assoc.,* 83, 125—133, 1972.
12. Janeway, T. C., *Arch. Intern. Med.,* 12, 755—798, 1913.
13. Society of Actuaries, *Build and Blood Pressure Study,* Vol. 1, Society of Actuaries, Chicago, 1959.
14. *Blood Pressure of Adults by Age and Sex, United States, 1960—1962,* U.S. Dept. of Health, Education and Welfare, Public Health Service, National Center for Health Statistics (Series II, No. 4), 1960—1962, 9.
15. *National Center for Health Statistics: Hypertension and Hypertensive Heart Disease in Adults, U.S., 1960—62,* U.S. Dept. of Health, Education and Welfare Vital and Health Statistics (Series II, No. 13), 1966.
16. *Health Examination Survey, United States, 1960—1962, Hypertension and Hypertensive Heart Disease in Adults,* U.S. Dept. of Health, Education and Welfare, Publication 1000 (Series II, No. 13), 1966.
17. *Primary Prevention of Atherosclerotic Diseases. Report of Inter-Society Commission for Heart Disease Resources* (Circulation 42), A55—95, 1970.
18. *The Underwriting Significance of Hypertension for the Life Insurance Industry,* Dept. of Health, Education and Welfare, U.S. Public Health Service, Bethesda, Md., Publication V (NIH) 74—426, 1974.

19. Dahl, L. K., *Am. J. Clin. Nutr.,* 21, 787–792, 1968.
20. Dahl, L. K., Heine, M., Leite, G., and Tassinari, L., *Proc. Soc. Exp. Biol. Med.,* 133, 1405–1408, 1970.
21. von Bunge, G., *Z. Biol.,* 10, 111, 1874.
22. von Bunge, G., in *Lehrbuch der Physiologischen und Pathologischen Chemie,* 3rd ed., Verlag von F. C. W. Vogen, Leipiz, 1894.
23. Miller, H. G., *J. Biol. Chem.,* 67, 71–77, 1926.
24. Miller, H. G., *J. Biol. Chem.,* 70, 593–598, 1926.
25. MaCallum, A. B., *Physiol. Rev.,* 6, 316–357, 1926.
26. Hodgekin, A. L., *The Conduction of Nervous Impulse,* Charles C Thomas, Springfield, Ill., 1964.
27. Ling, G. and Gerard, R. W., *J. Cell Comp. Physiol.,* 34, 383–396, 1949.
28. Woodbury, J. W., in *Physiology and Biophysics: Biophysics of the Cell Membrane,* Ruch, T. C. and Patton, H. D., Eds., W. B. Saunders, Philadelphia, 1966, 1–72.
29. Skou, J. C., *Q. Rev. Biophys.,* 4, 401–434, 1975.
30. Coleman, T. G., Manning, R. D., Jr., Norman, R. A., Granger, H. J., and Guyton, A. C., *Am. J. Med. Sci.,* 264, 103–110, 1972.
31. Lehninger, A. L., in *Biochemistry: The Molecular Basis of Cell Structures and Function,* Worth Publishers, New York, 1971, 619–622.
32. White, A., Handler, P., and Smith, E. L., in *Principles of Biochemistry,* McGraw-Hill, New York, 1964, 699.
33. Dahl, L. K., *N. Engl. J. Med.,* 258, 1152–1205, 1958.
34. Meneely, G. R. and Dahl, L. K., *Med. Clin. North Am.,* 45, 271–283, 1961.
35. Miall, W. E., *Br. Med. J.,* 2, 1204–1210, 1959.
36. Malthora, S. L., *Am. J. Clin. Nutr.,* 23, 1353–1363, 1970.
37. Paul, O., in *The Epidemiology of Hypertension: The Natural History of Hypertension,* Stamler, J., Stamler, R., and Pullman, T., Eds., Grune & Stratton, New York, 1967, 365.
38. Dawber, T. R., Kannel, W. B., Kagan, A., Donabediaru, R. K., McNamara, P. M., and Pearson, G., in *The Epidemiology of Hypertension: Environmental Factors in Hypertension,* Stamler, J., Stamler, R., and Pullman, T. N., Eds., Grune & Stratton, 1967, 255.
39. Dahl, L. K. and Love, R. A., *Arch. Intern. Med.,* 94, 525–531, 1954.
40. Pietenin, P. I., Wong, O., and Altschul, A. M., personal communication.
41. Langford, H. and Watson, R. L., in *The Epidemiology and Control of Hypertension: Electrolytes and Hypertension,* Paul, O., Ed., Stratton Intercontinental Medical Book Corp., New York, 1975, 119–130.
42. Steinbach, M. et al., *Rev. Roum. Med. Interne,* 13, 261–263, 1975.
43. Addison, W., *Can. Med. Assoc. J.,* 18, 281–285, 1928.
44. McQuarrie, I., Thompson, W. H., and Anderson, J. A., *J. Nutr.,* 11, 77–101, 1936.
45. Thompson, W. H. and McQuarrie, I., *Proc. Soc. Exp. Biol. Med.,* 31, 907–909, 1933–1934.
46. Priddle, W. W., *Can. Med. Assoc. J.,* 25, 5–8, 1931.
47. DeWesslow, O. L. V. S. and Thomson, W. A. R., *Quart. J. Med.,* 8, 361–374, 1939.
48. McQuarrie, I., in *Hypertension: A Discussion of Chapman, C. B.,* University of Minnesota Press, Minneapolis, 1950, 517–523.
49. Bartorelli, C., Gargano, N., and Leonett, G., in *Antihypertensive Therapy, An International Symposium,* Gross, F., Ed., Springer-Verlag, Berlin, 1966, 442–435.
50. Sasaki, N., Mitsuhashi, T., and Fukushi, S., *Igaku To Seibutsugaku,* 51, 103–105, 1959.
51. Meneely, G. R., Lemley-Stone, J., and Darby, W. J., *Am. J. Cardiol.,* 8, 527–532, 1961.
52. Lemley-Stone, J., *Am. J. Cardiol.,* 8, 748–853, 1961.
53. Dahl, L. K., *J. Exp. Med.,* 136, 318–328, 1972.
54. Unsigned Summary, *Nutr. Rev.,* 20, 195–197, 1962.
55. Medical News, *J.A.M.A.,* 235, 785–786, February 23, 1976.
56. Parijs, J., Joosens, J. V., Van der Linden, L., Verstreken, G., and Amery, A. K. P. C., *Am. Heart J.,* 85, 22–34, 1973.
57. Haragus, S. T., Zagreanu, I., Straus, H., and Galea, V., *Sixth World Congress of Cardiology,* British Medical Association, 1970, 180.
58. Joosens, P. V., *Triangle,* 12, 9–16, 1973.
59. Watson, R. L. and Langford, H. G., *Circulation,* 36, 264, 1967.
60. Steinbach, M. et al., *Rev. Roum. Med. Interne,* 12, 3–6, 1974.
61. Meneely, G. R. and Battarbee, H. D., *Am. J. Cardiol.,* 38, 768–781, 1976.
62. Dahl, L. K., Heine, M., and Tassinari, L., *J. Exp. Med.,* 115, 1173–1190, 1966.
63. Dahl, L. K., Heine, M., and Tassinari, L., *J. Exp. Med.,* 118, 605–611, 1963.
64. McKusick, V. A., *Circulation,* 22, 857–863, 1960.
65. Tobian, L., Ishii, M., and Duke, M., *J. Lab. Clin. Med.,* 73, 309–319, 1969.

66. Dahl, L. K., *J. Exp. Med.*, 114, 231–236, 1961.
67. McDonough, J. and Wilhemj, C. M., *Am. J. Dig, Dis.*, 21, 180–181, 1954.
68. Brown, W. J., Jr., Brown, K., and Krishian, I., *Circulation*, 43, 508–519, 1971.
69. Grant, H. and Reischman, F., *Am. Heart J.*, 32, 704–712, 1946.
70. Gross, G., Weller, J. M., and Hoobler, S. W., *Am. J. Clin. Nutr.*, 24, 605–608, 1971.
71. Chapman, C. B., Gibbons, T., and Henschel, A., *N. Engl. J. Med.*, 243, 899–905, 1950.
72. Grollman, A., Harrison, T. R., Mason, M. F., Baxter, J., Crampton, J., and Reichsman, F., *J.A.M.A.*, 129, 533–537, 1945.
73. Kempner, W., *N.C. Med. J.*, 6, 61–117, 1945.
74. Dustan, H. P., Bravo, E. L., and Tarazi, R. C., *Am. J. Cardiol.*, 31, 606–615, 1973.
75. Shaper, A. G., *Am. Heart J.*, 63, 437–442, 1962.
76. Tobian, L., *Phys. Rev.*, 40, 280–312, 1960.
77. Kirkendall, W. M., *Clin. Res.*, 20, 89, 1972.
78. Alexander, N., Hinshaw, L. B., and Drury, D. R., *Proc. Soc. Exp. Biol. Med.*, 86, 855–861, 1954.
79. Alexander, N., Hinshaw, L. B., and Drury, D. R., *Proc. Soc. Exp. Biol. Med.*, 92, 249–253, 1956.
80. Dahl, L. K. and Schackow, E., *Can. Med. Assoc. J.*, 90, 155–160, 1964.
81. Smirk, F. H. and Hall, W. H., *Nature*, 182, 727–728, 1959.
82. Smirk, F. H. and Eryctishir, I., *Proc. Univ. Otago Med. Sch.*, 37, 28, 1959.
83. Smirk, F. H., in *The Epidemiology of Hypertension*, Grune & Stratton, New York, 1967, 39–55.
84. Louis, W. J., Tabei, R., and Spector, S., *Lancet*, 2, 1283–1288, 1971.
85. Louis, W. J., Tabei, R., Spector, S., and Sjoerdsma, A., *Circ. Res.*, Suppl. 1, 1–93, 1969.
86. Schlager, G., *J. Hered.*, 56, 278, 1965.
87. Schlager, G. and Weibust, R. S., *Genetics*, 55, 497–506, 1967.
88. Schlager, G., *Can. J. Genet. Cytol.*, 10, 853–864, 1968.
89. Coleman, T. G., Cowley, A. W., Jr., and Guyton, A. C., in *MTP International Review of Science*, Physiology Series One, Vol. 1: Cardiovascular Physiology, Guyton, A. C. and Jones, C. E., Eds., University Park Press, Baltimore, 1974, 259–297.
90. Vogel, J. A., *Am. J. Physiol.*, 210, 186–190, 1973.
91. Langston, J. B., Guyton, A. C., Douglas, B. H., and Dorsett, P. E., *Circ. Res.*, 12, 508–513, 1963.
92. Hubbell, R. B., Mendel, L. B., and Wakeman, A. J., *J. Nutr.*, 14, 273–285, 1937.
93. Battarbee, H. D., Farrar, G. E., and Braley, J. K., *J. Appl. Physiol.*, April 1977, to be published.
94. Priddle, W. W., *Can. Med. Assoc. J.*, 86, 1–9, 1962.
95. Kempner, W., *Am. J. Med.*, 4, 545–577, 1948.
96. Piny the Elder (Gaius Pinius Secundus), *Naturalis Historia*, circa 65–75 A.D.
97. Meneely, G. R., Tucker, R. G., and Darby, W. J., *J. Nutr.*, 48, 489–498, 1952.
98. Sealey, J. E., Bühler, F. R., Laragh, J. H., Manning, E. L., and Brunner, H. R., *Circ. Res.*, 31, 367–378, 1972.
99. Gordon, T., in *National Center for Health Statistics: Blood Pressure of Adults by Age and Sex*, Public Health Service Pub. No. 1000 (Series 11, No. 4), 1964.
100. Hoobler, S. W., Tejada, C., Guzman, M., and Pardo, A., *Circulation*, 32, Suppl. II, 11–116, 1965.
101. Ponce Barreda, L. A., *Sanitoria*, 14, 42–48, 1959.
102. Murphy, W., *N.Z. Med. J.*, 54, 64–73, 1955.
103. Kaminer, B. and Lutz, W. P., *Circulation*, 22, 289–295, 1960.
104. Das, B. C., *Clin. Sci.*, 19, 439–448, 1960.
105. Padmovati, S. and Gupta, S., *Circulation*, 19, 395–405, 1959.
106. Seile, F., *Med. Klin.*, 26, 929–931, 1930.
107. *Thailand, Nutrition Survey of the Armed Forces*, A report by the Interdepartmental Committee on Nutrition for the National Defense (ICNND), U.S. Government Printing Office, Washington, D.C., 1960.
108. Sasaki, N., *Jpn. Heart J.*, 3, 313–324, 1962.
109. Whyte, H. M., Graham, I. A., and deWolf, M. S., *Aust. Ann. Med.*, 7, 328–335, 1958.
110. Orr, T. B. and Gilks, J. L., *Studies on Nutrition: Physique and Health of Two African Tribes* (Special Rep., Series 155), Medical Research Council, 1931.
111. Lapiccorella, V., Lapiccirella, R., Abboni, F., and Liotta, S., *Bull. W.H.O.*, 27, 681–697, 1964.
112. Meneely, G. R. and Dahl, L. K., *Med. Clin. North Am.*, 45, 271–283, 1961.
113. Thomas, W. A., *J.A.M.A.*, 88, 1559–1560, 1927.
114. Oberman, A. et al., *Circulation*, 36, 812–822, 1967.
115. Foster, J. H., *Arch. Intern. Med.*, 40, 38–45, 1927.

116. Tung, C. L., *Arch. Intern. Med.*, 40, 153–158, 1927.
117. Foster, J. H., *N. Engl. J. Med.*, 203, 1073–1076, 1930.
118. Zubri, H. L. and Klienov, E. A., *Govetakaia Meditsina*, 27, 135–137, 1964.
119. Fulmer, H. S. and Roberts, R. W., *Ann. Intern. Med.*, 59, 740–764, 1963.
120. MacFarland, W. V., Howard, B., and Scoggins, B. A., *Proc. 24th Int. Congr. Physiol. Sci.*, Washington (1968), 7, 274, 1971.
121. Sinnett, P. F., *J. Chronic Dis.*, 265–290, 1973.
122. Oliver, W. J., Cohen, E. L., and Neel, J. V., *Circulation*, 2, 146–151, 1975.
123. Prior, I. A. M., Evans, J. G., Harvey, H. P. B., Davidson, F., and Lindsey, M., *N. Engl. J. Med.*, 279, 515–520, 1968.
124. Lapicque, L., *Anthropologie*, 7, 35–45, 1896.
125. Blair-West, J. R., Coghlan, J. P., Denton, D. A., Funder, J. W., Nelson, J., Scoggins, B. A., and Wright, R. D., *Circ. Res. Suppl. II*, 26, 27, 11–251, 1970.
126. Ball, C. O. T. and Meneely, G. R., *J. Am. Diet. Assoc.*, 33, 366–370, 1957.
127. Frank, R. L. and Mickelsen, O., *Am. J. Clin. Nutr.*, 22, 464–479, 1969.
128. Wotman, S., Mandel, I. D., Thompson, R. H., and Laragh, J. H., *J. Chronic Dis.*, 20, 833–840, 1967.
129. Quartermain, D., Miller, N. E., and Wolfe, G., *J. Comp. Physiol. Psychol.*, 63, 417–420, 1967.
130. Denton, D. A., *Nutr. Abstr. Rev.*, 39, 1043–1049, 1969.
131. Schecter, P. J., Horwitz, D., and Kenkin, I., *JAMA*, 225, 1311–1315, 1953.
132. *Infant Mortality in the United States and Abroad*, Metropolitan Life Insurance Co., Stat. Bull., 18, 3, 1967.
133. Guyton, A. C., Coleman, T. G., Cowley, A. W., Manning, R. D., Jr., Norman, R. A., and Ferguson, J. D., *Circ. Res.*, 35, 159–176, 1974.
134. Guyton, A. C., Coleman, T. G., Cowley, A. W., Scheel, K. W., Manning, R. D., Jr., and Norman R. A., *Am. J. Med.*, 584–594, 1972.
135. Guyton, A. C. et al., *Ann. Rev. Physiol.*, 34, 13–46, 1972.
136. Haddy, F. J., *Physiol. Rev.*, 48, 688–707, 1968.
137. Haddy, F. J., *Arch. Intern. Med.*, 133, 916–931, 1974.
138. Haddy, F. J., *Am. J. Physiol.*, 195, 111–119, 1958.
139. Haddy, F. J., *Circ. Res.*, 15 (Suppl. I), 49–59, 1964.
140. Sarnoff, S. J. and Mitchell, J. H., in *Handbook of Physiology: The Control of the Function of the Heart*, Hamilton, W. F. and Dow, P., Eds., Williams & Wilkins, Baltimore, 1962, chap. 15.
141. Redleaf, P. D., *Circ. Res.*, 6, 343–351, 1958.
142. Koletsky, S., *Am. J. Cardiol.*, 8, 576–581, 1961.
143. Hollander, W. J., *Clin. Invest.*, 47, 1221–1229, 1968.
144. Villamil, M. F., *Clin. Res.*, 19, 198, 1971.
145. Garrahan, P., *Am. J. Physiol.*, 209, 955–960, 1965.
146. Haddy, F. J., in *Electrolytes and Cardiovascular Disease*, S. Karger, Bard, Switzerland, 1965, 383–400.
147. Scott, J. B., *Fed. Proc.*, 27, 1403–1407, 1968.
148. Anderson, D. K., *Circ. Res.*, 31, 165–173, 1972.
149. Haddy, F. J., *Am. J. Physiol.*, 204, 202–212, 1963.
150. Chen, W. T., *Proc. Soc. Exp. Biol. Med.*, 140, 820–824, 1972.
151. Mendlowitz, M., *Am. Heart J.*, 85, 252–259, 1973.
152. Self, L. E., Battarbee, H. D., Gaar, K., Jr., and Meneely, G. R., *Proc. Soc. Exp. Biol. Med.*, 153, 7–12, 1976.
153. deChamplain, J., Krakoff, L., and Axelrod, J., *Circ. Res.*, 24, 25 (*Suppl. I*); 75–92, 1969.
154. Redleaf, P. D. and Tobian, L., *Circ. Res.*, 6, 185–198, 1958.
155. McQueen, E. G., *Clin. Sci.*, 21, 133–140, 1961.
156. Hinke, J. A. M., *Circ. Res.*, 17, 359–371, 1965.
157. McGregor, D. D. and Smirk, F. H., *Am. J. Physiol.*, 214, 1429–1433, 1968.
158. Gordon, D. B. and Nogueira, A., *Circ. Res.*, 10, 269–273, 1962.
159. Daly, J. J. and Duff, R. S., *Clin. Sci.*, 19, 457–463, 1960.
160. Doyle, A. E. and Fraser, J. R. E., *Circ. Res.*, 9, 755–761, 1961.
161. Mendlowitz, M., *Am. Heart J.*, 73, 121–128, 1967.
162. Kaplan, N. M. and Shilah, J. G., *J. Clin. Invest.*, 43, 659, 669, 1964.
163. Bohr, D. R., *Fed. Proc.*, 33, 127–132, 1974.
164. Bandick, N. R. and Sparks, H. W., *Am. J. Physiol.*, 219, 340–344, 1970.
165. Hollback, M., Lundgren, Y., and Weiss, L., *Acta Physiol. Scand.*, 81, 176–181, 1971.
166. Page, I. H., Taylor, R. D., and Prince, R., *Am. J. Physiol.*, 159, 440–456, 1949.
167. Gordon, D. B., Drury, D. R., and Shapiro, S., *Am. J. Physiol.*, 175, 123–128, 1953.

168. Lapu, A. N., Maxwell, M. H., and White, F. N., *Proc. Soc. Exp. Biol. Med.,* 134, 617—620, 1970.
169. Dahl, L. K., Knudsen, K. D., Heine, M., and Leite, G., *J. Exp. Med.,* 126, 687—699, 1967.
170. Mizukoshi, H., *J. Clin. Endocrinol.,* 34, 1016—1024, 1972.
171. Lowenstein, F. W., Blood pressure in relation to age and sex in the tropics and sub-tropics, *Lancet,* 1, 389—392, 1961.
172. Goldrick, R. B., Sinnett, P. F., and Whyte, H. M., in *Atherosclerosis,* Proceedings of the Second International Symposium, "An Assessment of Coronary Heart Disease and Coronary Risk Factors in a New Guinea Highland Population," Jones, R. J., Ed., Springer-Verlag, Berlin, 1970, 366—368.
173. Truswell, A. S., Kennelly, B. M., Hansen, J. D. L., and Lee, R. B., Blood pressures of !Kung bushmen in Borthern Botswana, *Am. Heart J.,* 84, 5—12, 1972.
174. Das Neves Almeida, F., Pression sanguine and rapport avec age, sexe et profession, in Abstr. VI Congr. Int. Hyg. Med. Prev., Madrid, 1971, 70.
175. Carlson, L. A. and Lindstet, S., The Stockholm prospective study, *Acta Med. Scand.,* Suppl. 493, 1—135, 1968.
176. Joossens, J. V., Williams, J., Claessens, J., Claes, J., and Lissens, W., in *Nutrition and Cardiovascular Diseases,* "Sodium and Hypertension," Morgagni Edizione Scientifiche, Rome, 1972, 91—110.
177. Bøe, J., Humefelt, S., and Wedervang, F., The blood pressure in a population: blood pressure readings and height and weight determinations in the adult population of the City of Bergen, *Acta Med. Scand.,* Suppl. 321, 1—336, 1957.
178. Moser, M., Morgan, R., Hale, M., Hoobler, S. W., Remington, R., Dodge, H. J., and Macaulay, A. I., Epidemiology of hypertension with particular reference to the Bahamas. I. Preliminary report of blood pressures and review of possible etiologic factors, *Am. Heart J.,* 4, 727—733, 1959.

TOXIC AND NONTOXIC EFFECTS
OF CHLORIDE IN ANIMALS AND MAN

H. Kaunitz

During the early part of the century, French authors[1] believed that exclusion of NaCl from the diet was associated with a reduction of elevated blood pressures. Probably because it was relatively easy to determine chloride at that time and sodium assays were complicated and not very reliable, chloride was assumed to be the culprit, and for many years, so-called "chloride-low" diets were prescribed. A similar confusion of the effects of chloride and NaCl can be seen in the work of plant physiologists even as late as the 1940s, when certain properties were assigned to chloride and it was actually the effects of NaCl that were being studied.[2] Only recently have some plant physiologists[3] separated the effects of sodium and chloride and found that chloride has a toxic effect on certain bacteria.

During the 1930s and 1940s, the development of the flame photometer permitted easier and more accurate Na^+ and K^+ determinations, which led to the accumulation of new data of interest to nephrologists. The impression grew that changes in the "milieu interieur" were dominated by the movements of Na^+ and K^+. Studies of Cl^- metabolism were carried out only infrequently as typified by the fact that, even now, the chapter on circulation in Volume 3 of the *Handbook of Physiology*[4] has 80 entries for Na and only 14 for Cl. This viewpoint was expressed by Homer Smith.[5] Schwartz et al.[6] summarized the prevailing ideas of that time: "Anions in general, and chloride in particular, have been considered 'physiologically indifferent' and thus devoid of any important role in determining electrolyte equilibrium." These authors go on to say "it is now clear that this view is not correct and that under many physiologic and clinical conditions anion characteristics have considerable influence on both the reabsorption of bicarbonate and the secretion of potassium."

Despite the tendency in the past to ignore or discount the physiologic effects of anions individually; in certain cases, specific chloride effects have been demonstrated. One obvious example is HCl in gastric juice. In addition, the chloride ion plays a specific part in the "chloride shift" which occurs during oxygenation of hemoglobin in the lungs, when Cl^- leaves erythrocytes after oxygenation of hemoglobin to balance electrically the inward movement of bicarbonate ions from the plasma. The latter move inward as more carbonic acid is formed within the erythrocyte in the presence of the more acidic oxyhemoglobin. And finally, the Cl ion is excluded from parenchymatous cells to an even greater extent than is the Na ion. This fact only became known with the help of NMR analyses.

In searching the literature for comparisons of anions with regard to some specific biological function, one is struck by the paucity of such material. In one early study, dogs with bladder fistulae were infused with Na or K salts of Cl^-, HCO_3^-, and $H_2PO_4^{-7}$. As was expected, more of the infused water was excreted when the infusion contained a K salt; moreover, infusion of $NaHCO_3$ or NaH_2PO_4 brought about marked chloride retention compared to the same K salts. Furthermore, this was even more pronounced than the water retention.

The importance of the Cl ion can be seen in a number of clinical conditions associated with a loss of Cl and K ions and a rising blood HCO_3^- level, such as that occurring after severe vomiting or gastric drainage, intake of various diuretics and adrenal corticoids, or hypercapnia. Schwartz and his collaborators have carried out experiments to separate and clarify some of the processes involved.[6] When gastric HCl is lost, the Cl ion in blood and intercellular fluid is replaced by HCO_3 ion and is reflected in the glomerular filtrate. The

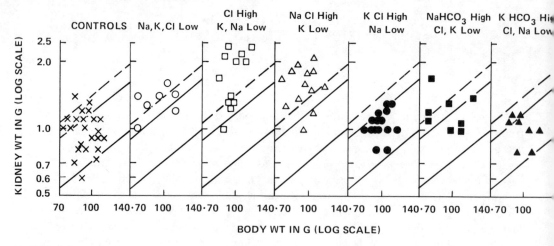

FIGURE 1.   Relation of kidney weight to body weight of male rats with different levels of Na, K, and Cl. Food in was restricted to just enough to allow the rats to maintain their weight. (From Kaunitz, H., *Lab. Invest.*, 5, 132, 1 With permission.)

kidney, lacking reabsorbable Cl ions, exchanges H and K ions for Na ions, resulting in acidification of the urine and loss of K ions. Bicarbonate ions are released from the renal epithelial cells and further increase plasma HCO_3^- levels. This condition of metabolic alkalosis was originally thought to be a consequence of K ion deficiency, and the role of Cl ions was not appreciated. It has now been shown that correction of the Cl deficit alone will restore normal HCO_3^- levels, despite loss of K ion.[8] The chloride ion plays a similar part in metabolic alkalosis induced by diuretics and the posthypercapnic state.

Interaction between K and Cl was observed in experiments in which rats were fed purified diets supplemented with combinations of Na, K, and Cl ions. In the first experiments (carried out in 1956),[9] the body weights of rats were kept constant by restricted feeding to permit direct comparison of organ weights. Figure 1 shows that all K-deficient groups tended to have heavy kidneys with those given Cl (alone or as NaCl) having significantly heavier kidneys than the other K-deficient groups (no supplement or Na only). In 1972, a similar experiment was carried out with ad libitum fed rats.[10] In these experiments (Table 1), the groups fed NaCl had by far, the heaviest kidneys — even heavier than in the control rats, which weighed nearly three times as much in terminal body weight. The kidneys of all K-deficient rats were examined histologically for signs of K deficiency[11] and graded on a scale of 1+ to 4+. Severe lesions were seen in 50% of the rats fed either no supplement, Cl ion, or NaCl. Only one of the 15 rats fed only Na ion had a 4+ lesion. By and large, these results again indicate that Cl ion in the absence of K ion damages the kidney.

The effect of Cl ion in the heart and testes was different from that in the kidney. Rats fed only Na ion had the heaviest hearts, far exceeding those in the rats fed Cl or NaCl. They were even heavier than the hearts of the much heavier rats fed K ion or KCl. Testicular weights and histology (Figure 2) revealed that in the K-deficient rats, exclusion of Cl ion from the diet led to severely degenerated, small testes. Cl supplementation (with or without Na) induced heavier testes without lesions.

During the last 25 years, the role of Cl ion in the transport of Na ion through isolated amphibian skin has been studied in many laboratories, especially in Ussing's laboratory in Copenhagen. In such studies, the ionic compositions of the solutions bathing opposite sides of a frog skin could be varied, and resulting changes in electrical phenomena could be measured.[12] With their technique, other investigators found that Cl ion on the inside of the skin leads to stimulation of Na ion influx,[13,14] an effect not seen with other

Table 1

EFFECTS OF FEEDING ELECTROLYTE-DEFICIENT DIETS TO YOUNG MALE RATS FOR 5 WEEKS

Experimental conditions	Survival	Food intake (g/day)	Terminal body wt (g)	Kidneys Wt (mg)	Kidneys 4+ Lesions	Hearts Wt (mg)	Hearts 4+ Lesions	Testes Wt (mg)	Testes 4+ Lesions
Experiment I									
K-, Na-, Cl-deficient	10/12	7.0 ± 0.50	94 ± 3.5	1790 ± 65	6/10	368 ± 18	2/10	1080 ± 108	6/9
+ Na	7/12	6.7 ± 0.20	104 ± 3.3	1940 ± 90	1/7	520 ± 30	1/6	620 ± 86	5/5
+ Na + Cl	6/12	6.7 ± 0.25	105 ± 5.0	2720 ± 270	4/5	399 ± 34	1/5	1260 ± 146	1/5
+ Cl	10/12	6.7 ± 0.69	89 ± 3.0	1840 ± 86	7/10	289 ± 16	0/10	1600 ± 80	0/10
+ K + Cl	11/12	12.4 ± 0.60	154 ± 6.8	1580 ± 177		474 ± 21		2700 ± 77	
Controls (+ K, Na, Cl)	9/10	13.2 ± 0.50	289 ± 14.9	2020 ± 78		843 ± 40		2950 ± 91	
Experiment II									
K-, Na-, Cl-deficient	8/8		103 ± 2.8	2200 ± 144	1/4	433 ± 24	1/4	697 ± 48	4/4
+ Na	10/16	6.8 ± 0.36	105 ± 4.4	1840 ± 102	0/8	529 ± 33	2/10	605 ± 49	10/10
+ Na + Cl	11/16	6.2 ± 0.17	121 ± 4.4	2730 ± 155	8/11	433 ± 22	2/11	1340 ± 116	0/11
+ Cl	8/8		100 ± 2.2	1950 ± 82	2/8	304 ± 8	0/8	1470 ± 95	0/8
+ K	8/8		174 ± 6.3	1720 ± 45	0/8	476 ± 16	0/8	2610 ± 147	0/8
+ K + Cl	8/8		167 ± 6.0	1690 ± 57	0/8	466 ± 17	0/8	2640 ± 164	0/8
Controls (+ K, Na, Cl)	6/6		302 ± 7.6	2310 ± 79	0/6	857 ± 25	0/6	3000 ± 50	0/6

From Kaunitz, H. and Johnson, R. E., *Proc. Soc. Exp. Biol. Med.*, 141, 875, 1972. With permission.

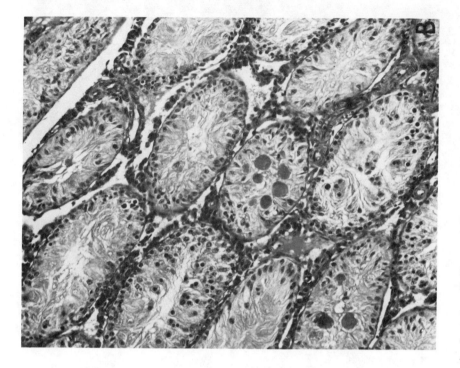

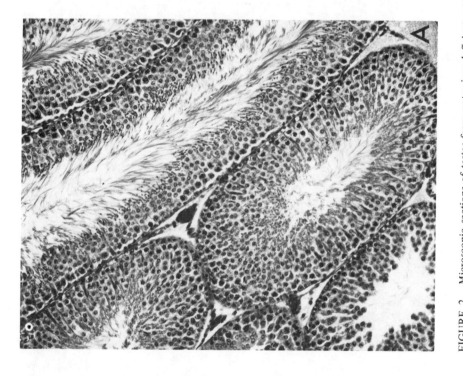

FIGURE 2.   Microscopic sections of testes from potassium-deficient rats. A. Testis of rat given NaCl; essentially normal. B. Testis of rat given NaHCO₃; advanced atrophy of seminiferous tubules; spermatogenesis absent; ghosts of spermatozoa and degenerating giant cells in tubular lumen. (Magnification × 160.) (From Kaunitz, H. and Johnson, R. E., *Proc. Soc. Exp. Biol. Med.*, 141, 875, 1972. With permission.)

anions. It was also found that, in contrast to the active transport of Na ion, only 10% of Cl ion transfer is active.[14] The mode of transfer has not been elucidated. Ussing and Windhager[15] suggested the existence of a transcellular pathway for Na ion and Cl ion, but Cereijido and Rotuno[16] favored an extracellular pathway along the cell surface. These mechanisms are not confined to amphibian skin but are of some importance for the transport of Na and Cl ion through epithelial structures in general.[17]

While the above examples demonstrate some of the more specific important effects of Cl ion and, undoubtedly, Na ion has its own specific effects, one must not lose sight of the fact that, by and large, the two ions are coupled in their action most of the time: They occur in equivalent amounts in blood and intercellular fluid; about four fifths of the Na and Cl ions filtered by the glomeruli is reabsorbed together; and finally, our taste buds insist on NaCl.

REFERENCES

1. Achard, C. and Loeper, M., Sur la rétention dans les tissus au cours de certain états morbids, *C. R. Seances Soc. Biol. Paris,* 53, 346, 1901.
2. Haas, A. R. C., Influence of chlorine on plants, *Soil Sci.,* 60, 53, 1945.
3. Steinborn, J. and Roughly, R. J., Toxicity of sodium and chloride ions to *Rhizobium* spp. in broth and peat culture, *J. Appl. Bacteriol.,* 39, 133, 1975.
4. Hamilton, W. F., Ed., *Handbook of Physiology,* Vol. III, Section 2, American Physiological Society, Washington, D.C., 1963.
5. Smith, H. W., *The Kidney: Structure and Function in Health and Disease,* Oxford, New York, 1951, 305.
6. Schwartz, W. B., van Ypersele de Strihou, C., and Kassirer, J. P., Role of anions in metabolic alkalosis and potassium deficiency, *N. Engl. J. Med.,* 279, 630, 1968.
7. Kaunitz, H., Über die Mineralstoffauscheidung nach Belastung mit verschiedenen Salzen, *Biochem. Z.,* 293, 142, 1937.
8. Kassirer, J. P., Berkman, P. M., Lawrenz, D. R., and Schwartz, W. B., Critical role of chloride in correction of hypokalemic alkalosis in man, *Am. J. Med.,* 38, 172, 1965.
9. Kaunitz, H., Relation of chloride, sodium, and potassium intake to renal and adrenal size, *Lab. Invest.,* 5, 132, 1956.
10. Kaunitz, H. and Johnson, R. E., Chloride: effect on morphology of heart, kidney, and testes in potassium-deficient rats, *Proc. Soc. Exp. Biol. Med.,* 141, 875, 1972.
11. Follis, R. H., Orent-Keiles, E., and McCollum, E. V., *Am. J. Pathol.,* 18, 29, 1942.
12. Ussing, H. H. and Zerahn, K., Active transport of sodium as the source of electric current in the short-circuited, isolated frog skin, *Acta Physiol. Scand.,* 23, 110, 1951.
13. Ferreira, K. T. G., Anionic dependence of sodium transport in the frog skin, *Biochim. Biophys. Acta,* 150, 587, 1968.
14. Huf, E. G., The role of Cl⁻ and other anions in active Na⁺ transport in isolated frog skin, *Acta Physiol. Scand.,* 84, 366, 1972.
15. Ussing, H. H. and Windhager, E. E., Nature of shunt path and active transport path through frog skin epithelium, *Acta Physiol. Scand.,* 61, 484, 1964.
16. Cereijido, M. and Rotuno, C. A., Fluxes and distribution of sodium in frog skin, *J. Gen. Physiol.,* 51, 280s, 1968.
17. Watlington, C. O., Regulation of NaCl transport – relation to chloride conductance, *Biochim. Biophys. Acta,* 249, 339, 1971.

NUTRIENT TOXICITIES IN ANIMALS AND MAN: POTASSIUM*

Philip O. Ettinger

Elemental potassium does not exist in nature because of its intense affinity for combination with other elements. A silvery grey metal, it reacts immediately with water to form potassium hydroxide, so that if ingested it would induce severe burns of the mouth, pharynx, esophagus, and stomach.

Acute toxicity from orally administered potassium compounds in healthy man is unusual. Large doses of potassium salts are considered "moderately toxic"[1] although they usually induce vomiting[1,2] and most of the absorbed potassium is rapidly excreted by the normal kidney. Toxicity is predominantly due to cardiac electrical effects and/or respiratory muscle paralysis.

In the experimental anesthetized animal (Table 1), elevation of plasma potassium ion induces a biphasic, reproducible sequence of initial increase, then profound decrease in electrical impulse formation and conduction in all cardiac tissues.[3-10] These changes result ultimately in cardiac asystole in the majority; in a few, ventricular fibrillation may occur. Some of these alterations, especially the conduction disturbances and arrhythmias, are superficially similar to those induced by quinidine intoxication,[11] although the cellular electrophysiologic responses to these substances differ in important ways. When potassium salts are administered orally, animals develop flaccid muscular paralysis and may die of cardiac or respiratory failure,[12-15] occasionally following convulsions. Sometimes diarrhea, organ dehydration, and early renal tubular necrosis are seen. If the toxin administration is discontinued, complete recovery usually follows, although anorexia, thirst, fever, weakness, excessive urination, convulsions, and gastrointestinal disturbances may persist for 24 to 48 hr. Hyperkalemia may occur in the following ways:

1. In acute or chronic renal insufficiency of any etiology, particularly when trauma is present.

2. After extensive tissue trauma, which results in the liberation of quantities of previously intracellular K^+ ion.

3. During intravenous administration of large amounts of banked blood, where hemolysis has liberated K^+ ion from red blood cells.

4. During administration of large amounts of potassium containing foods or medicines, including certain penicillins which are supplied as potassium salts.[16,17]

5. In metabolic acidosis, where hydrogen ion penetration of cells causes potassium ion extrusion from those cells, thus raising plasma K^+.

6. During involution of a large postpartum uterus.

7. When normally well-tolerated amounts of potassium salts are administered concomitantly with certain diuretics (spironolactone, triamterene) whose action depends upon K^+ retention by the distal renal tubule.

8. After surgery, particularly cardiac surgery,[18] usually associated with oliguria, extravascular accumulations of blood, multiple transfusions, and acidosis.

9. Rarely, when digitalis glycosides are ingested or injected in toxic quantities with suicidal intent.[19]

10. In acute adrenal insufficiency.

In addition to the above, iatrogenic hyperkalemia is a potential complication which

* This investigation was supported in part by the United States Public Health Service National Heart and Lung Institute Research Grant No. HL 09914 and Postgraduate Training Grant No. HL 05510 and a Research Grant, No. 75-BER-1, from the Bergen County Heart Association, New Jersey.

Table 1

CELLULAR ELECTROPHYSIOLOGIC AND EKG CHANGES
IN HYPERKALEMIA[a]

Plasma K[+b] (mEq/l)	Cardiac transmembrane potential change	Resultant cardiac or EKG change
5.0–5.5	Mechanism unknown	Slightly increased excitability, shorter PR interval
5.5–6.5 (7.0)	Normal or minimally lowered resting membrane potential (RMP)	Usually no or minimal QRS widening
	Shortened action potential duration (APD), steep phase 3	Shortened QT, tall T waves
7.0–9.0	Lowered RMP	Decreased cardiac excitability
	Slower rise of phase O	Decreased conduction velocity; PR and QRS prolong
	Shortened APD	Short (QT-QRS)
	Atrial muscle totally depolarized at K = 9.0	P waves disappear; "sinoventricular conduction"; A-V block occurs and junctional rhythms begin
>9.0	Further lowering of RMP	Ventricular muscle becomes inexcitable; ultimate asystole in most
	Further slowing of phase O	Slow conduction; wide, bizarre QRS
	Shortened APD	Short (QT-QRS)
	Irregular firing of junctional pacemakers; atrium silent	No P waves, irregular QRS rhythm
	Unknown mechanism	In some, ventricular fibrillation

[a] Experimental animal studies, in vivo and in vitro.
[b] Plasma K[+] levels are approximate.

may develop wherever physicians have attempted to use potassium salts therapeutically, usually for treatment in the following circumstances:[2]

1. For the treatment of potassium deficiency states:

    a. Loss of potassium ion in gastrointestinal fluids, as in vomiting, diarrhea, or during intestinal drainage via an indwelling tube.

    b. Loss of potassium in urine caused by mercurial, thiazide, or newer diuretics, such as furosemide and ethacrynic acid.

    c. Reduced plasma and extracellular potassium caused by electrolyte shifts (typically, following early treatment of diabetic acidosis).

    d. In a rare neuromuscular disorder, familial periodic paralysis.

2. For the treatment of certain cardiac disturbances in the absence of potassium deficiency:

  a. In ventricular tachycardias of digitalis and nondigitalis origin.[20-29] Such use is rare today as less toxic agents are available.

  b. In paroxysmal atrial tachycardia with block, an arrhythmia often associated with digitalis intoxication.[27]

  c. As a diagnostic agent for distinguishing between supposed "organic" and "functional" ST-T changes in the EKG.[30-31] Such use is rare today because of toxic sequelae.

Because most clinical instances of hyperkalemia are observed in acute or chronic kidney diseases, an understanding of the mode of renal excretion of potassium ion is essential. Potassium ion is filtered by the glomerulus and completely reabsorbed; all potassium appearing in the urine is derived from tubular excretion of this ion in exchange for sodium at distal exchange sites.[32] As a general rule, an intravenous infusion rate of 20 mEq/hr is well tolerated in normal man. Surprisingly, because of an apparent utilization of latent secretory capacity, even the failing kidney may continue to excrete potassium relatively independent of the amount filtered at least until terminal azotemia supervenes.[33] In such advanced renal failure, potassium ion clearance may be as much as two to six times inulin clearance.[32-34] However, the latent renal tubular capacity for potassium secretion may be at or near maximal in renal failure, so that hyperkalemia in renal failure may rapidly develop in three circumstances. First, minor increases in potassium load at this time could overtax the secretory capacity. Second, minor reductions in the sodium load presented to the kidney, such as caused by the institution of a low salt diet or salt depletion caused by vomiting or diarrhea, could prevent normal potassium-sodium exchange and allow rapid rises in plasma potassium. Third, the development of acidosis can induce hyperkalemia, as an increase of 0.4 to 1.2 mEq of potassium ion per liter occurs with each decline of 0.1 pH unit in extracellular fluid or blood.[35-37]

While the effects of acute hyperkalemia alone are reproducible in the experimental animal, chronic sustained hyperkalemia is extremely difficult to reproduce. In addition, hyperkalemic effects on cardiac and other tissues in man and other animals are probably greatly modified by other acute metabolic alterations occurring simultaneously, especially acidosis, hyponatremia, and hypocalcemia. Thus, extrapolation of known experimental data into the clinical sphere may not provide an entirely accurate view of clinical K^+ excess. In addition, it has been suggested that the rapid, as opposed to slow, development of hyperkalemia may have differing effects.[5]

Reduction of the force of cardiac muscle contraction in hyperkalemia has been suggested by several in vitro studies.[38,39] However, other studies in the intact animal are conflicting, indicating instead that maintenance of cardiac performance is jeopardized only after severe asynchrony of contraction occurs as a result of cardiac conduction disturbances.[8,40,41]

When oral potassium chloride is administered as a liquid, gastric irritation is common.[42] However, even enteric coating does not entirely prevent gastric ulceration.[43] More recently, enteric coated potassium chloride tablets have been implicated as a cause of ulceration of the small intestine and several fatal cases have been reported.[44] The hazard of such therapy is small, however.

Treatment of hyperkalemia (Table 2) is directed against both the elevated K^+ ion and the underlying cause, if possible.[45,46] With plasma K^+ in the 5.5 to 7.0 mEq range and associated with sinus rhythm and narrow QRS, sodium polystyrene sulfonate resin (Kayexalate®) may be administered. Widened QRS and/or arrhythmia requires urgent therapy. Asystole and bradycardia require temporary transvenous pacing of the ventricle,

Table 2
TREATMENT OF HYPERKALEMIA[45,46]

Plasma K⁺ (mEq/l)	EKG abnormality	Emergency therapy	Subsequent therapy
5.5–7.0 mEq/l	Narrow QRS ± tall T waves	None required	Sodium polystyrene sulfonate resin 12–16 g p.o., q.i.d., *or* Sodium sulfate, 3–5 g p.o. daily
>7.0 mEq/l	Widened QRS, ± arrhythmia	$NaHCO_3$, 40–100 mEq i.v., *or* $CaCl_2$ or gluconate, 10 ml of 10% solution (1 g) i.v.	500–1000 ml 10% glucose solution i.v., with 2.0 g Ca gluconate and 25 U regular insulin i.v., *or* 500–1000 ml 1/6 molar Na lactate solution i.v., with 100 g glucose, 2.0 g Ca gluconate and 25 U regular insulin i.v.
>7–8 mEq/l	Asystole *or* bradycardia *or* A-V block	Temporary transvenous pacing *and* $NaHCO_3$ or $CaCl_2$ or gluconate, 10 ml of 10% solution (1 g) i.v.	500–1000 ml 10% glucose solution i.v., with 2.0 g Ca gluconate and 25 U regular insulin i.v., *or* 500–1000 ml 1/6 molar NA lactate solution i.v., with 100 g glucose, 2.0 g Ca gluconate and 25 U regular insulin i.v.

as does atrioventricular block. The prompt administration of sodium or calcium ion is best when rapid tachyarrhythmias are caused by K⁺ excess. Sodium bicarbonate solution or even saline tends to reverse the EKG abnormality, as do molar sodium lactate or calcium gluconate or chloride. As sodium and calcium ions do not lower K⁺ but only reverse the cardiac electrical abnormality, an intravenous solution of 10% glucose (possibly with the addition of regular insulin intravenously) should be begun to induce intracellular movement of extracellular K⁺. Oliguria and acidosis should be corrected (see below), and any responsible drugs or biologicals (see above) removed from the therapeutic regimen. Peritoneal and/or hemodialysis may be required. Careful monitoring of K⁺ to prevent hypokalemia is recommended.

In renal disease with acidosis, other measures may be required.[45] Dietary protein (a source of K⁺) should be restricted. Alkali may be administered in the form of Shohl's solution (sodium citrate and citric acid). If acidosis is not present, sodium sulfate can be given orally, 3 to 5 g daily, and the sodium content of the diet should not be restricted. Strauss and Welt[45] suggest infusion of a liter of 10% glucose solution containing 2.0 g of calcium gluconate, or a liter of sixth molar sodium lactate solution to which has been added 100 g of glucose and 2.0 g of calcium lactate. It is not clear if exogenous insulin is useful in the nondiabetic patient when glucose is given intravenously.

While not of direct relevance to potassium intoxication when induced orally or by intravenous administration, experimental perfusion of local regions of myocardium with potassium salts (local myocardial K^+ excess) induces electrocardiographic changes and arrhythmias simulating acute ischemia.[47] The mechanism of arrhythmia appears to be local conduction delay and reentry, and may be similar to changes occurring in acute myocardial infarction in man.

REFERENCES

1. Gleason, M. N., Gosselin, R. E., Hodge, H. C., and Smith, R. P., *Clinical Toxicology of Commercial Products,* 3rd ed., Williams & Wilkins, Baltimore, 1969, 118.

2. Goodman, L. S. and Gilman, A., *The Pharmacologic Basis of Therapeutics,* 5th ed., Macmillan, New York, 1975, 776–780.

3. Winkler, A. W., Hoff, H. E., and Smith, P. K., Electrocardiographic changes and concentrations of potassium in serum following intravenous injection of potassium chloride, *Am. J. Physiol.,* 124, 478, 1938.

4. Vassalle, M., Greenspan, K., Jomain, S., and Hoffman, B. F., Effects of potassium on automaticity and conduction of canine hearts, *Am. J. Physiol.,* 207, 334, 1964.

5. Surawicz, B., Chlebus, H., and Mazzoleni, A., Hemodynamic and electrocardiographic effects of hyperpotassemia. Differences in response to slow and rapid increases in concentration of plasma K^+, *Am. Heart J.,* 73, 647, 1967.

6. Fisch, C., Feigenbaum, H., and Bowers, J. A., The effects of potassium on atrioventricular conduction of normal dogs, *Am. J. Cardiol.,* 11, 487, 1963.

7. Cohen, H. C., Gozo, E. G., Jr., and Pick, A., The nature and type of arrhythmias in acute experimental hyperkalemia in the intact dog, *Am. Heart J.,* 82, 777, 1971.

8. Ettinger, P. O., Regan, T. J., Oldewurtel, H. A., and Khan, M. I., Ventricular conduction delay and asystole during systemic hyperkalemia, *Am. J. Cardiol.,* 33, 876, 1974.

9. Fisch, C., Knoebel, S. B., and Feigenbaum, H., Potassium and the monophasic action potential, electrocardiogram, conduction and arrhythmias, *Prog. Cardiovasc. Dis.,* 8, 387, 1966.

10. Fisch, C., Greenspan, K., and Edwards, R. E., Complete atrioventricular block due to potassium, *Circ. Res.,* 19, 373, 1966.

11. Gleason, M. N., Gosselin, R. E., Hodge, H. C., and Smith, R. P., *Clinical Toxicology of Commercial Products,* 3rd ed., Williams & Wilkins, Baltimore, 1969, 202.

12. Anberg, S. and Helmholtz, H. F., The fatal dose of various substances on intravenous injection in the guinea pig, *J. Pharmacol. Exp. Ther.,* 6, 595, 1915.

13. Loeser, D. and Konwiser, A. L., A study of the toxicity of strontium and comparison with other cations employed in therapeutics, *J. Lab. Clin. Med.,* 15, 35, 1929.

14. Ulrich, J. L. and Shternov, V. A., The comparative action of hypertonic solutions of the chlorates and chlorides of potassium, sodium, calcium and magnesium, *J. Pharmacol. Exp. Ther.,* 35, 1, 1929.

15. Boyd, E. M. and Shanas, M. N., The acute oral toxicity of potassium chloride, *Arch. Int. Pharmacodyn.,* 133, 275, 1961.

16. Pearson, R. E. and Fish, K. H., Jr., Potassium content of selected medicines, foods and salt substitutes, *Hosp. Pharm.,* 6, 6, 1971.

17. Mercer, C. W. and Logic, J. R., Cardiac arrest due to hyperkalemia following intravenous penicillin administration, *Chest,* 64, 359, 1973.

18. Williams, J. R., Jr., Morrow, A. G., and Braunwald, E., Incidence and management of "medical" complications following cardiac operations, *Circulation,* 32, 608, 1965.

19. Smith, T. W., Haber, E., Yeatman, L., and Butler, V. P., Jr., Reversal of advanced digoxin intoxication with Fab fragments of digoxinspecific antibodies, *N. Engl. J. Med.,* 294, 797, 1976.

20. Sampson, J. J. and Anderson, E. M., The therapeutic use of potassium in certain arrhythmias, *Proc. Soc. Exp. Biol. Med.,* 28, 163, 1930.

21. Sampson, J. J. and Anderson, E. M., Treatment of certain cardiac arrhythmias with potassium salts, *J.A.M.A.,* 99, 2257, 1932.

22. Sampson, J. J., Albertson, E. C., and Kondo, B., The effects of man on potassium administration in relation to digitalis glycosides, with special reference to blood serum potassium, the electrocardiogram, and ectopic beats, *Am. Heart J.,* 26, 164, 1943.

23. Casteden, L. I. M., Effect of potassium salts on cardiac irregularities, *Br. Med. J.,* 1, 7, 1941.

24. Enselberg, C. D., Simmons, H. G., and Mintz, A. A., The effects of potassium upon the heart, with special reference to the possibility of treatment of toxic arrhythmias due to digitalis, *Am. Heart J.,* 39, 713, 1950.

25. **Stempien, S. J. and Katz, K. H.,** Quinidine and potassium in the treatment of refractory paroxysmal ventricular tachycardia, *Am. Heart J.,* 24, 555, 1942.

26. **Lown, B., Weller, J. M., Wyatt, N., Hoigue, R., and Merrill, J. P.,** Effects of alterations of body potassium on digitalis toxicity, *J. Clin. Invest.,* 31, 648, 1952.

27. **Lown, B., Wyatt, N. F., Crocker, A. T., Goodale, W. T., and Levine, S. A.,** Interrelationship of digitalis and potassium in auricular tachycardia with block, *Am. Heart J.,* 45, 589, 1953.

28. **Bettinger, J. C., Surawicz, B., Bryfogle, J. W., Anderson, B. N., Jr., and Bellet, S.,** The effect of intravenous administration of potassium chloride on ectopic rhythm, ectopic beats and disturbances in A-V conduction, *Am. J. Med.,* 21, 521, 1956.

29. **Vassalle, M. and Greenspan, K.,** Effects of potassium on ouabain-induced arrhythmias, *Am. J. Cardiol.,* 12, 692, 1963.

30. **Wasserburger, R. H. and Corliss, R. J.,** Value of oral potassium salts in differentiation of functional and organic T-wave changes, *Am. J. Cardiol.,* 10, 673, 1962.

31. **Dodge, H. T., Grant, R. P., and Seavey, P. W.,** The effect of induced hyperkalemia on the normal and abnormal electrocardiogram, *Am. Heart J.,* 45, 725, 1953.

32. **Leaf, A. and Camara, A. A.,** Renal tubular secretion of potassium in man, *J. Clin. Invest.,* 28, 1526, 1949.

33. **Mudge, G. H.,** Disorders of renal tubular function, *Am. J. Med.,* 20, 448, 1956.

34. **Platt, R.,** Sodium and potassium excretion in chronic renal failure, *Clin. Sci.,* 9, 367, 1950.

35. **Berliner, R. W., Kennedy, T. T., Jr., and Hilton, J. G.,** Renal mechanisms for excretion of potassium, *Am. J. Physiol.,* 162, 348, 1950.

36. **Berliner, R. W., Kennedy, T. T., Jr., and Oreoff, J.,** Relationship between acidification of urine and potassium metabolism, *Am. J. Med.,* 11, 274, 1951.

37. **Burnell, J. M., Villamil, M. F., Uyeno, B. T., and Scribner, B. H.,** Effect in humans of extracellular pH change on the relationship between serum potassium concentration and intracellular potassium, *J. Clin. Invest.,* 35, 935, 1956.

38. **Garb, S.,** The effects of potassium, ammonium, calcium, strontium and magnesium on the electrogram of mammalian heart muscle, *J. Pharmacol. Exp. Ther.,* 101, 317, 1951.

39. **LeVeen, H. H., Schatman, B., and Moskowitz, H.,** Effect of digitalis on potassium toxicity in isolated turtle heart, *Proc. Soc. Exp. Biol. Med.,* 100, 538, 1959.

40. **Leight, L., Roush, G., Rafi, E., and McGaff, C. J.,** The effect of intravenous potassium on myocardial contractility and cardiac dynamics, *Am. J. Cardiol.,* 12, 686, 1963.

41. **Goodyer, A. V. N., Goodkind, M. J., and Stanley, E. J.,** The effects of abnormal concentrations on the serum electrolytes on left ventricular function in the intact animal, *Am. Heart J.,* 67, 779, 1964.

42. **Bockus, H. L.,** *Gastroenterology,* 3rd ed., W. B. Saunders, Philadelphia, 1974, 510.

43. **Jacobs, E. and Pringot, J.,** Gastric ulcers due to the intake of potassium chloride, *Am. J. Dig. Dis.,* 18, 289, 1973.

44. Hazards of enteric coated potassium chloride tablets, Medical Letter, 7, 26, 1965.

45. **Strauss, M. B. and Welt, L. G.,** *Diseases of the Kidney,* 2nd ed., Little, Brown & Co., Boston, 1971, 244—246.

46. **Ettinger, P. O., Regan, T. J., and Oldewurtel, H. A.,** Hyperkalemia, cardiac conduction, and the electrocardiogram. A Review, *Am. Heart J.,* 88, 360, 1974.

47. **Ettinger, P. O., Regan, T. J., Oldewurtel, H. A., and Khan, M. I.,** Ventricular conduction delay and arrhythmias during regional hyperkalemia in the dog, *Circ. Res.,* 33, 521, 1973.

EFFECT OF NUTRIENT TOXICITIES IN ANIMALS AND MAN: SULFUR

P. R. Bird

This review covers a range of compounds containing sulfur that may be ingested and that may in some circumstances impair productive performance or health. Passing reference is also given to sulfur compounds that are adventitiously imbibed in fluids, or inhaled, or given parenterally.

INORGANIC SULFUR COMPOUNDS

Elemental Sulfur

This element has long been used as a curative for various maladies of humans and other animals. It has been taken orally by humans as a laxative,[1] given to farm animals as a purgative and to cure mange,[5] and ringworm,[2,19] given subcutaneously as a fungicide, parasiticide, and keratolytic agent, and parenterally for conditions such as rheumatoid arthritis.[1,2] Sulfur is adventitiously inhaled with dust from some industrial processes but is apparently not injurious to the lungs.[3] However, sulfur aerosols can cause eye irritation at a threshold as low as 0.2 ppm.[1] The lethal dose of colloidal sulfur given parenterally is reported to be 5 to 10 mg/kg body weight.[4]

Deleterious effects of ingested sulfur have been found when given in large amounts to livestock with the objective of curing skin ailments. In one instance ewes were each given on average an estimated 45 to 70 g of sublimed sulfur mixed with barley and 140 of the 480 died on the first day.[6] Other evidence was given that even 7 g of sulfur per day killed two pregnant ewes within 17 days but two others survived and lambed normally 5 months later.[6] Symptoms observed were labored respiration, strong smell of H_2S on the breath, severe abdominal pain, temperature, inability to walk, diarrhea, intense inflammation of the abomasum and intestines, and haemorrhage throughout the viscera and musculature.

Cattle deaths, too, have resulted from feeding sulfur with silage to cure ringworm[2,20] and lice infestation.[7] In the latter case the affected steers could each have received 50 g of sulfur or more. The symptoms were as described above together with muscular twitching, apparent blindness and paralysis in one beast, blackening of the muscles of the loin and back, complete inappetence, and apparent lack of muscular contractility of the rumen and intestines of those steers that survived beyond the first day.

The use of large amounts of sulfur when ensiling can cause diarrhea and inappetence in cattle fed that material.[8] The appetite depressing effect of sulfur has been used to restrict the intake of a fattening ration fed to lambs, thus apparently preventing lamb deaths from enterotoxemia.[9]

Elemental sulfur has been added to ruminant rations deficient in sulfur but containing adequate nitrogen, thereby improving the utilization of nitrogen, intake and digestibility of the feed, and growth of the sheep.[10-14] While 0.2 to 0.4% of elemental sulfur in the semipurified diets gave results comparable to those using other forms of sulfur, higher percentages were less effective and feed refusals were found.[11,13] Holstein bull calves weighing about 120 kg were fed semipurified diets containing 1.68% elemental sulfur for 77 days without sign of any adverse effect on feed intake, weight gain, or nitrogen retention.[15]

The available evidence indicates that 0.4% or less elemental sulfur, which is not used as effectively as Na_2SO_4 or methionine sulfur,[11,12,43] can be successfully and safely used for both sheep and cattle when rations contain adequate nitrogen but are deficient in sulfur.

The mechanism of the toxic effect of elemental sulfur is presumably twofold:

1. Astringent action of high concentrations of sulfur per se on the intestinal mucosa and on body tissues following absorption if this occurs;
2. Formation of H_2S by the action of bacterial enzymes within the rumen and gut, and possibly also by other enzymes within the tissues.[16] Hydrogen sulfide may be produced by direct reduction of elemental sulfur by assimilatory sulfate-reducing bacteria[16] but also via oxidation to $SO_3^=$ [17] and subsequent reduction.[18]

Inorganic Sulfate

Terrestrial sulfate is derived mainly from weathering of sedimentary rocks (FeS_2, $CaSO_4$), cyclic sulfur from the oceans, industrial pollution, volcanic emissions, and microbial oxidation of soil organic matter and sulfate esters.[19,20] Ground water may contain sulfate varying from virtually zero up to at least 1330 ppm SO_4-S.[21] Sulfate is used by yeast, fungi, microorganisms, and plants for the synthesis of organic compounds containing sulfur, but it is an end product of sulfur metabolism in animals. Plants absorb sulfur from the soil in the form of inorganic sulfate — and some SO_2 from the air — but where the requirement of sulfur for the synthesis of methionine and cysteine has not been exceeded then the plant tissues generally do not accumulate much sulfate.[22] The inorganic SO_4-S content of some pastures in England[23] and Australia[24] ranged from 0.07 to 0.4% and 0.03 to 0.3%, respectively. However, accumulation of significant amounts of sulfate occurs in certain plant species such as halophytes from coastal sand plains[25] which contained, on an average, 0.6% of the dry matter as sulfate sulfur. Moreover, when plants are grown in the presence of adequate sulfur but inadequate nitrogen[26] or are fertilized with sulfate[27,28] the ratio of nitrogen/sulfur will fall due to storage of inorganic sulfate.

It is clear that grazing animals — and animals and man that consume water containing sulfates — will sometimes be exposed to a relatively high intake of sulfate. While there appears to be little experimental evidence upon which standards can be set for any species there is some evidence that sulfate can be deleterious despite the claim that, "It is one of the least toxic anions known, surpassing even the chloride ion. . ."[2]

The feed intake and growth of lambs were found to be less when 0.54 or 0.69% SO_4-S was given in the diet than when 0.25 or 0.40% of SO_4-S was used.[12] The daily intake of SO_4-S would have been about 5.4 g at the highest level and about 2.6 g at the lowest level, which gave the best liveweight response. Other work has shown no effect of high intake of sulfate on appetite. Pregnant ewes were given up to 21 g SO_4-S/day with the only tangible result being the aborting of nine ewes and evidence of nutritional muscular dystrophy in those lambs.[29]

Substantial amounts of sulfate may be taken in with drinking water. The provision of 4.5 g/day of Na_2SO_4-S (1130 mg/l) or of 2.65 g SO_4-S from $CaSO_4$ and from $MgSO_4$ (each 38 mg S/l) in the drinking water of adult wethers over a 15-month period did not adversely affect water intake, feed intake, weight gain, or wool production.[31,32] These sheep were adapted to these salts by gradually increasing the concentration in the water over a period of 5 to 9 weeks. In short-term experiments with Hereford heifers the inclusion of 1125 mg SO_4-S/l in the drinking water decreased the intake of water by 35%, hay by 30%, and decreased the apparent growth rate from 0.6 kg/day (tap water containing 37 ppm SO_4-S) to −0.5 kg/day.[21] Growing Hereford heifers ate 12% less hay and grew more slowly when 930 mg SO_4-S/l was added to their drinking water.[33] These heifers received 35 g of sulfur per day from this source. The effect of 480 ppm SO_4-S was also to produce a small decrease in feed intake and an apparent growth depression, but no effect on water intake. In a choice situation these animals discriminated against water containing 480 ppm SO_4-S and rejected water containing 710 ppm SO_4-S.[33] Yearling

Angus steers discriminated against water containing 670 mg SO_4-S/l and rejected water with 1100 mg SO_4-S/l.[34]

These data for cattle differ substantially from earlier data[35] where definite effects of sulfate toxicity were seen only when 2250 mg SO_4-S/l was given, but no adverse effects on intake or growth were found when 1575 mg SO_4-S/l was given. The $SO_4^=$ anion was found to be more potent than Cl^-.[21,35] The variance between experimental results makes it difficult to determine safe limits for livestock. No doubt variation in intake of water, sex, and age of animal, breed, type of feed offered, amount of sulfur ingested in the diet, stress factors, and adaptation all influence the tolerance to sulfate waters shown in the examples cited above. However, it seems clear that the recommendation[30] that water containing less than 165 mg SO_4-S/l would not be detrimental to stock has substance. In fact ruminants could probably tolerate at least 500 mg SO_4-S/l without adverse long-term effects on productivity. Sheep may be more tolerant than cattle, but more evidence is needed to define limits for all species.

Ammonium bisulfate has been used to preserve silage, and this has led to poor intake and growth of cattle.[36,37] However, since an equivalent amount of Na_2SO_4 (0.83% S) did not produce the same effect $SO_4^=$ per se was thought not to be responsible.

The inclusion of 1% sulfur, as H_2SO_4 or as Na_2SO_4, in the diet of mature sheep significantly depressed intake.[38] Intake of grass hay and liveweight change were affected when 0.5, 1.0, or 1.5% Na_2SO_4-S was included in the ration.[39] There was no effect of putting 5.3% NaCl in the ration. With the sulfate treatments feed intake was progressively reduced during the first 4 days but stabilized thereafter. A rapid and complete recovery was made following the removal of supplemental sulfur from the rations. This pattern was repeated when H_2SO_4 or Na_2SO_4 was given either in the feed or intraruminally.[40] Sulfuric acid given intraruminally had only a slightly larger effect on intake than did Na_2SO_4 given either in the feed or per rumen, but H_2SO_4 in the diet had much more effect. Palatability was not a factor with Na_2SO_4.

Sulfates have had some usage as an intake depressant. In one instance the inclusion of 0.84 g Na_2SO_4-S, together with 12 g urea, in a drought supplement of 255 g wheat fed to sheep delayed the intake of this meal from 10 to 15 min without Na_2SO_4 to 30 min and this had some effect in reducing mortality from urea toxicity.[4] This rapid effect of Na_2SO_4 seems to have been due to a palatability effect. Gypsum has been used to regulate the intake of concentrates by cattle.[218] Daily intraruminal doses of 700 g of gypsum caused complete inanition in steers,[217] but 450 g/day given in the feed to lactating cows did not affect intake.[216] However, intake was reduced when gypsum was used to increase the concentration of sulfur in the diet to 0.3%.[216]

In rats[52,55] and chicks[53,54,57] and presumably all other animals, sulfate can spare the requirement of organic sulfur, since sulfate is required for the formation of mucopolysaccharides of cartilage, bone connective tissue, and mucous secretions[16,60] and must be supplied from cysteine or methionine degradation when no exogenous sulfate is supplied. When added to diets otherwise adequate in sulfur amino acids, sulfate can stimulate growth.[54,56] High levels of dietary sulfate have also been used to counteract molybdenum toxicity,[59] but growth rates of chicks fed 0.076% or 0.135% S as $Na_2SO_4 - K_2SO_4$, or as $CaSO_4$, was depressed compared with purified rations not containing sulfate.[58] With practical broiler rations the inclusion of up to 0.03%S as Na_2SO_4 stimulated growth and improved feed conversion efficiency.[54] This level of sulfate in chick rations seems acceptable.

The mechanism of sulfate toxicity is still rather obscure. Depending upon the animal species and the composition of the diet there are several possible explanations for poor productive performance. The likely mechanisms are considered below:

1. Palatability — As discussed above, there is evidence that animals discriminate

against high and potentially toxic concentrations of sulfate in drinking water, but will accept it if no alternative is provided. There is less evidence to suggest that feed is refused, or that it is eaten less readily, when it contains sulfate.

2. Digestive disturbance — Feed refusals by ruminants cannot be attributed to a depression in digestibility, rumen pH, or volatile fatty acid production.[40] The digestion of cellulose and of starch by ruminal microorganisms in vitro is unaffected by Na_2SO_4 in concentrations less than 1000 $\mu g/ml$.[61,62] Concentrations of $S^=$ up to 1000 $\mu g/ml$ did not affect microbial growth in vitro[42] although higher concentrations may impair fermentation activity under in vitro conditions.[63] However, high concentrations of $S^=$ in vivo are not maintained[42] and it is unlikely that microbial fermentative activity is impaired by ingested sulfate per se. Gypsum decreased the rate of passage of feed but without affecting the in vitro digestibility of alfalfa cellulose in rumen liquor taken from the dosed steers.[217] Rumen motility appeared to be unaffected but the criteria did not include an estimate of the intensity of contraction and the measurements were of brief duration.

3. Electrolyte imbalance — No evidence has been provided that high intakes of Na_2SO_4 or K_2SO_4 result in substantial alteration in electrolyte balance in rumen fluid or plasma.[39,40] However, an excessive intake of $CaSO_4$ produced a degree of acidosis in steers.[217]

4. Mineral balance — Calcium retention was affected by the inclusion of 0.2% Na_2SO_4-S (but not by 0.2% elemental sulfur) in the ration of sheep.[43] The effect was not found in cattle[21] nor sheep[31] given drinking water containing high levels of Na_2SO_4. The concentration of calcium, magnesium, and phosphorus in plasma and rumen liquor of sheep was unaffected by Na_2SO_4.[39,40]

Sulfate is known to influence the storage and retention of several trace elements.[64] Perhaps the most dramatic interaction is that of copper-molybdenum-sulfur. Molybdenosis in sheep can be remitted by increasing the intake of copper and/or $SO_4^=$.[64] Conversely, molybdenum deficiency in nonruminants may be a result of excessive intake of sulfate.[68] Copper deficiency in ruminants and nonruminants may occur as a result of excessive intakes of molybdenum and $SO_4^=$.[64,68,75,77] Sulfate added without molybdenum is usually considered to be ineffective in reducing liver copper storage in ruminants,[65,67] but there is evidence that sulfur alone can slightly reduce copper availability.[68-71] The mode of action may be by formation in the rumen of an insoluble CuS product[134] and/or a cupric thiomolybdate complex.[68] Ingested $SO_4^=$ — along with organic sulfur — is largely reduced to sulfides in the rumen,[44,49,135] and one should therefore consider the total sulfur content of a feedstuff rather than just its SO_4-S content when assessing its potential effect on copper or molybdenum status.[77,134] However, the intake of grass rich in sulfate has been implicated in cases of conditioned copper deficiency in lambs,[76] and where the molybdenum intake also is high, the possibility of induced copper deficiency must be considered.

The interaction between sulfur and selenium is also relevant in that nutritional muscular dystrophy (NMD) is associated with a deficiency of selenium that may be induced by a high intake of $SO_4^=$.[64] In some instances adding $SO_4^=$ to the diet of sheep can produce cardiac lesions or NMD[72-74,79] but not in all cases.[136,137] For example, increasing the total sulfur content of a ration to give 0.5% increased the incidence of NMD in lambs prior to weaning,[72] but in other experiments higher intakes of sulfur did not.[29]

These differences are hard to explain, but may be due to different forms and levels of sulfur and selenium in the basal diets. Thus the excretion of ^{75}Se in sheep was greater when the diet sulfur content was 0.24% than when 0.05%.[78] Ruminal microbes can convert inorganic selenium into organic forms[80,81] — selenate competitively inhibits sulfate activation within bacterial and mammalian tissues.[82,83] Thus more inorganic selenium may be incorporated — and less excreted — when the intake of sulfur is small.

The selenium content of pastures can be substantially depressed by application of sulfates[79,84] or sulfur bearing fertilizer,[85] and this may lead to levels of selenium which are insufficient for the grazing animal.

While excess intake of sulfate — and other forms of sulfur — may lead to a deficiency of copper, selenium, and possibly other elements such as lead[87] and zinc, the resultant effect on productivity would be a long-term consequence and not an immediate effect on intake and growth.

e. Production of H_2S — It has been shown that the infusion of up to 6 g SO_4-S/day into the rumen of sheep produced a cessation of intake whereas the intraduodenal route of infusion did not.[42] The ruminal microflora rapidly adapt to $SO_4^=$[44] due to the build-up of anaerobic dissimilatory SO_4-reducing bacteria such as *Desulphovibrio* species[88] that can employ $SO_4^=$ instead of O_2 as the terminal e^1 acceptor in the respiratory chain. H_2S results from ingestion or ruminal infusion of $SO_4^=$,[44,135] and it is this that causes an immediate effect on intake (see Sulfide) and perhaps contributes to the longer term induction of heavy metal deficiencies by formation of insoluble sulfides.[25,134]

It has been suggested that the daily intake by sheep of sulfur in any form should not and need not exceed 4 g, or that the concentration of sulfur in the diet of 0.2% DM should permit efficient usage of dietary energy and nitrogen for growth and production by sheep and cattle without reducing intake.[42] In the absence of adequate dietary nitrogen even small additions of sulfate to the ration (final concentration of sulfur 0.25%) may decrease nitrogen retention and tend to decrease intake and digestibility of the ration by sheep and cattle.[50] The addition of urea to a similar diet markedly improved the performance of both species.[50] The recommended relationship of dietary nitrogen/ sulfur is about 13 for sheep and 15 for cattle,[50,51] but these ratios need to be narrowed when elemental sulfur is used as a supplement or when the dietary sulfur is in a relatively indigestible form.[216]

When supplementing rations with sulfur account must be taken of the form of sulfur, the concentration of sulfur and nitrogen in the basal ration, and the amount of sulfur received in the drinking water so that the appropriate intake of digestible sulfur and ratio of nitrogen/sulfur is obtained.

Sulfide

Experimental evidence with small mammals indicates that sulfide may be ingested or H_2S inhaled, or even absorbed through the skin in quantities sufficient to cause toxicity.[47] The evolution of H_2S gas from sewers or marshes, due to the activity of dissimilatory sulfur reducing bacteria,[16] may give rise to $S^=$ concentrations in the air exceeding 10,000 ppm, and human fatalities have occurred since 1000 ppm is lethal.[47] One might speculate on the fate of the "Assyrian horde" (Kings 2:19) which camped near a canal outside of Jerusalem". . . . and it came to pass that night that the angel of the Lord went out and smote in the camp an hundred and four score and five thousand; and when they arose early in the morning, behold, they were all dead corpses" (Lord Byron used this event in his poem "The Destruction of Sennacherib"). Certainly lethal and sublethal poisoning of humans by H_2S in industry is well-known.[2,89-92] The odor can be detected when present in the air in less than 1 ppm, but paralysis of the olfactory nerve occurs when concentrations exceed 150 ppm so that lethal concentrations of H_2S may be encountered without adequate warning.[47,90,92]

Fish, too, are susceptible to $S^=$ released from industry[93] or emanating from bacterial action on sludge.[94] Concentrations as low as 0.3 mg/l of water were toxic to juvenile salmon.

Alkali metal sulfides when absorbed give rise to H_2S or $S^=$ with the same effect as if free sulfide had been absorbed.[133] Tetraphosphorus trisulfide (P_4S_3), which is used in safety matches, has a low toxicity — doses of 15 mg/kg body weight produce no ill effects in humans and rabbits survived doses of 170 mg/kg.[131] Toxic symptoms are those of phosphorus. Phosphorus pentasulfide is potentially more toxic since it reacts immediately with water to give H_2S. Antimony trisulfide and pentasulfides are also components of matches and fireworks but are relatively insoluble and nontoxic.[132]

The toxic effect of sulfide is exerted primarily through paralysis of the nervous system with resulting respiratory failure.[47,92] Unless death occurs the adverse effects are usually but not always reversible.[2,42,47]

Intraruminal infusion of H_2S may cause sheep to collapse.[42,45,46] This result is achieved through eructation of ruminal gases and consequent absorption of H_2S from the lungs.[45] The liver and blood can detoxify $S^=$[47,95] at a rate estimated to exceed that of $S^=$ absorption from the rumen,[42] but the respiratory route affords a direct approach to the respiratory-circulatory control centers within the brain. Ruminants are thus rendered much more susceptible to sulfur poisoning than are monogastric animals which do not possess a microbial fermentation capacity in the stomach — and hence do not have the same ability to produce H_2S from other forms of sulfur — nor the propensity to discharge gaseous products of digestion via the lungs. The susceptibility of the ruminant to H_2S intoxication could well depend upon the concentration of H_2S in the rumen gas phase. Conditions that may determine this are

1. Rate of eating — This will determine the supply of sulfur available for bacterial degradation
2. Dietary nitrogen and energy — Readily available forms should permit rapid utilization of $S^=$ for bacterial growth,[111] thereby limiting the accumulation of free H_2S in the rumen.
3. Molybdenum intake — While high levels of dietary molybdenum may decrease the rate of sulfate reduction in vivo[211] it also inhibits $S^=$ absorption from the rumen to the extent that the ruminal concentration of $S^=$ markedly increases.[109]
4. Rumen pH — The higher the pH the slower the rate of absorption of $S^=$ from the rumen.[108,110]

The effect of sulfide on the intake may be due to paralysis of the appetite center of the brain, or may result from the inhibition of ruminal contractions.[42] These contractions are necessary to maintain rumen fermentation, the flow of digesta from the rumen, and hence continuity of intake.

Sulfur dioxide, sulfite, bisulfite and metabisulfite

SO_2 has a long history as a fumigant, an antiseptic agent, and a preservative.[2,96-98] Sulfurous acid (H_2SO_3) forming compounds such as sodium or potassium sulfite, bisulfite and metabisulfite when dissolved in solution form sulfite ($SO_3^=$) and bisulfite (HSO_3^-) ions, the proportions depending on pH.[98] The principal functions of these compounds in the food industry[96] are as follows:

1. To inhibit nonenzymic browning of canned, heat processed fish, meat, fruit, or vegetables; dehydrated meat, milk, fruit products; frozen food; wines, cordials; foods stored at room temperature.
2. To inhibit enzymic-browning and other reactions during storage of dehydrated freeze-dried fruit and vegetables and hydrated vegetables and fruit.
3. To prevent oxidation of ascorbic acid, lipids, oils, and carotenoids in vegetables, meat products, fruit juices, and beer and to act as a reducing agent, thus preventing discoloration of minced meat products when exposed to the air.

4. To inhibit microbial spoilage of soft fruits, meat products, wines, and fruit juices stored at room or refrigerator temperature.

SO_2 is also widely used in industry and is a waste product from sulfide ores, coals, and petroleum. It has been incriminated as a toxic principle in smog, chiefly because of its effect on mucous membranes which predisposes to respiratory infections,[2] but another view is that at ambient levels of 0.6 ppm or less SO_2 is not injurious.[100]

The major concern over the use of SO_2 (and SO_2 forming agents) in human diets is centered on possible mutagenic effects.[98] Thus, bisulfite can modify the nucleic acid residues uracil and cytosine in yeasts, E. coli, and phage cells,[113,120] Possible alternatives to the use of SO_2 in foodstuffs have been described.[96]

In the U.S. the dietary intake of sulfite has been estimated as 0.1 mg/kg, or 7 mg/day.[99] An estimate for adults in Belgium is 16 mg/day,[114] the difference being due to a greater consumption of SO_2 in beverages.

Rats, mice, and monkeys dosed daily with $NaHSO_3$ to give an intake of up to 200 mg SO_2/kg showed no ill effects and did not excrete free sulfite during the 5- or 30-day periods.[99] Sulfite is rapidly oxidized to $SO_4^=$ in the tissues due to the action of sulfite oxidases[2,99] and this accounts for the relatively low toxicity of sulfite. However, a congenital deficiency of this enzyme results in the accumulation of endogenous $SO_3^=$ from the catabolism of amino acids containing sulfur, and this can produce pathological changes.[102]

The inclusion of 0.1% of sodium metabisulfite ($S_2O_5^=$) in the diet of growing rats had no effect on net protein utilization.[115] However, rats are affected when the diet contains 0.1% sodium bisulfite,[101] but this is due, in part at least, to an SO_2 — induced thiamine deficiency.

The LD_{50} values for parenteral dosage of sulfite range from 1 to 7 mM/kg, depending on the species of animal and the route of administration.[2]

Ruminants are susceptible to sulfite toxicity,[103,104] but feeding dry pulp containing up to 1000 ppm SO_2 (0.05%S) is considered safe.[105] SO_2 is used in drying fodder beet but no adverse effects are found when cows are fed beet containing 400 mg SO_2/kg.[221] In one instance giving 60 to 202 g Na_2SO_3 in the diet of cows progressively decreased feed intake to zero,[104] but this effect was not found later when 160 g Na_2SO_3 was given daily.[106] A dosage of 80 to 160 g of metabisulfite or of sulfur dioxide gas per day will produce anorexia in cattle.[103] Metabisulfite is used to preserve silage but if used to the extent of around 3.6 kg per tonne of fresh herbage, and if cattle are given 50 kg or less of silage daily, the residual sulfite is insufficient to adversely affect them.[103] Severe intoxication is unlikely since silage containing excess sulfite is not readily consumed. However, daily dosing with more than 120 g metabisulfite or sulfur dioxide resulted in anorexia, weight loss, constipation then dysentery, lesions of the trachea and other organs, severe depression, and death.[103] Sheep given 8 g/day of Na_2SO_3 as a ruminal infusion were unaffected, except that propionate concentrations in the rumen increased,[112], as found earlier,[104,106] and acetate and methane production was inhibited, presumably because of a toxic effect of $SO_3^=$ on methanogenic bacteria.

Since sulfite is an intermediate in the reduction of sulfate,[107] it is possible that in the ruminant the toxic effect of sulfite is expressed via the formation of H_2S. Alternatively, inappetence might be due to an aversive taste.

Thiosulfate and Polythionates

Thiosulfate is relatively nontoxic when given orally or parenterally. With humans, at least, 18 to 24 g doses of sodium thiosulfate produce no ill effects.[116,121] Thiosulfate is mostly excreted unchanged in the urine and has been used for the determination of extracellular space[117] and glomerular filtration rate.[118]

The parenteral dosage of 3 to 4 g $Na_2S_2O_3$/kg body weight produces mortality in dogs.[119] Adult humans are given about 0.2 g $Na_2S_2O_3$/kg, or approximately 12 g doses, for treatment of cyanide poisoning.[120]

Polythionates are used in dermatological preparations for application to the skin and as an antidote against cyanide.[2] Tetrathionate when injected to the extent of 0.1 g/kg body weight is lethal to rabbits and 10 mg/kg causes necrosis of the kidneys.[122,123]

Persulfates

Potassium monopersulfate ($KHSO_5$) is used as an oxidizing, bleaching, and germicidal preparation. In rats the lethal oral dose is about 2 g/kg body weight.[124]

Ammonium persulfate [$(NH_4)S_2O_8$] is used in the commercial preparation of flour, but at the concentrations used, it has no adverse effects.[125] Rats can tolerate 6.5 g/kg body weight.[126] The LD_{50} for mice given intraperitoneal injections of sodium persulfate is 226 mg/kg.[127]

Thiocyanates

These compounds have had a pharmacological use as an antihypertensive agent and an experimental use in inhibiting gastric secretion, thyroid function, and determination of extracellular fluid volume.[2]

The LD_{50} in mice given potassium thiocyanate orally is about 0.6 g/kg body weight.[128] Renal excretion is slow due to efficient reabsorption of filtered thiocyanate,[129] and the compound accumulates when repeated doses are given, with toxic consequences.[130]

Thiocyanates are found in cruciferous plants[146] and are potentially goitrogenic.[145] Hydrolysis of plant glucosinolates releases thiocyanate and isothiocyanate. Further discussion of these compounds is found below.

ORGANIC SULPHUR COMPOUNDS

Glucosinolates

These thioglucosides are derived from a host of cruciferous plants, such as turnips, cabbages, kale, radish, and mustard. Some 65 glucosinolates have been isolated; among these are sinigrin, sinalbin, and progoitrin.[138] Glucosinolates are described by the formula

$$R-\underset{\underset{N-O-SO_2O^-}{\|}}{C}-S-C_6H_{11}O_5$$

where the R group is allyl (sinigrin), P–HO benzyl (sinalbin), 2–HO–3–butenyl (progoitrin), etc.

The unique taste and odor of members of the Cruciferae family are due to the presence of such compounds.

Goitrogenic compounds are present in the seeds, roots, and leaves of many of the genera, particularly Brassica. The seeds of rutabaga, rape, turnip, and cabbage are particularly rich in progoitrin glucosinolates. Turnip seed may contain 0 to 0.1% progoitrin; cabbage seed may contain 0 to 0.7% progoitrin and up to 2.6% sinigrin. The edible portion of cabbage may contain 0.035% (350 ppm) of sinigrin and at least 0.5 ppm progoitrin.[138]

The inclusion of seed meal containing 8 to 10% of glucosinolates to provide 10% of the diet of rats for long periods has resulted in some weight loss arising from the production of goitrin.[139] The glucosinolates are hydrolyzed by the enzyme thioglucosidase to form glucose, sulfate, and variable proportions of thiocyanate, isothiocyanate, nitrites, and sulfur. Even where thioglucosidase has been inactivated by cooking, some goitrins may be produced due to the enzymic activity of intestinal microflora.[140] The proportion of goitrins formed during hydrolysis depends on the pH, temperature, and concentration of meal and upon pretreatment storage conditions.[138]

Human goiter problems are largely due to a deficiency of dietary iodine,[141] but other conditioning factors may be implicated in some instances. Thus, the consumption of excessive amounts of cabbage[142] or kale[143] is believed to contribute to the problem. Evidence that milk from cows fed kale or other crucifers can cause goiter in humans is conflicting.[138] It seems that the transfer of goitrogen into milk is insufficient to cause harm, but the iodine content of the milk may be substantially depressed,[144] and this could conceivably result in goiter in the milk-fed offspring.[138]

Thiocyanate, a product of glucosinolate hydrolyses, is a recognized goitrogen in that it inhibits the uptake of iodine by the thyroid.[145] Fresh cabbage may contain 0.7 to 10.2 mg thiocyanate/100 g tissue;[146] however, unless the diet contains very little iodine the consequences of ingesting large amounts of such material is probably negligible. In some instances defatted crucifer seed meal could contain progoitrin equivalent to 2.7% goitrin — compared with about 0.1 ppm goitrin from cabbage — and clearly then there may be adverse effects on animal performance.[138] Protein isolates from rapeseed and other cruciferous oilseeds are increasingly being incorporated into livestock rations.

Methionine and Analogues of Methionine

In mammals biosynthesis of methionine from cysteine is not possible, since the cystathionine synthase reaction (homocysteine + serine → cystathionine) is irreversible. Dietary cyst(e)ine can reduce the animal's requirement for methionine by satisfying the requirement for cysteine, glutathione, taurine, and ester sulfate, but it cannot entirely replace it.[148,166] This, and the fact that methionine is frequently the growth limiting amino acid in human and animal diets,[149,167,168] has led to an extensive use of synthetic methionine in nutrition.

Methionine has been described as the most toxic of the nutritionally important amino acids.[150] The range between amounts of methionine which give benefit and those which are toxic is relatively small.[151] Results from many experiments with rats show that increasing the methionine content of a low protein diet much above 1.5%, or roughly three times the requirement, depresses growth and feed intake.[150] Pathological and biochemical lesions of the internal tissues and organs also occur[152,153,173,175] and the reduction in feed intake accounts only for part of the decreased weight gains.[152,169, 172,175] With a diet supplying 10% casein and about 0.3% methionine the addition of 0.3% methionine gave a marked stimulation to growth, but the addition of more than 1.4% methionine resulted in growth rates markedly less than those obtained with the 0.3% supplement.[150]

Many factors can influence the response to supplemental methionine:

1. Arginine, ethanolamine, glycine, threonine, and serine can alleviate methionine toxicity;[150,152,154,155,171,172,192]

2. The growth depression from a given level of methionine decreases as the protein content of the diet is increased.[156,157] Relatively small changes in the concentration of dietary protein can markedly affect responses to methionine.[150]

3. Pyridoxine deficient rats are more susceptible to methionine toxicity[158] presumably because cystathionine clearance is then inhibited.[191]

4. D-methionine is usually regarded as less toxic than either L-methionine or D,L-methionine,[159,160,169,170] although it is apparently less well utilized by humans[165] and other monogastric animals.[54]

5. Susceptibility decreases with age.[160]

6. Adaptation may occur. Pathological changes in rats were reversed after 2 months of feeding diets containing up to 4% methionine, although body weights were still much depressed.[160] Partial recovery of growth rates were found after only 5 days of feeding a ration containing 3% methionine[172] and after giving 2% methionine for 4 weeks[175] or 8 weeks,[172] indicating that metabolic adaptation had to some extent occurred.

7. Animal species differences. Guinea pigs are much more sensitive to methionine than are rats. Thus, a dose of 10 mM/kg/day caused complete inanition and death within 65 hr, but rats were unaffected after 21 days.[155] As little as 1 mM of methionine/kg taken orally by humans can elicit nausea, vomiting, and prostration.[164] Differences between species in biochemical effects are also evident since guinea pigs[155] and rabbits[161] develop an acute hypoglycemia when given excess methionine, but the rat does not.[155] The symptoms of methionine intoxication in humans seems similar to those in guinea pigs or rabbits.[155]

8. Patients with cirrhosis of the liver are rendered more susceptible to methionine toxicity[174] and were found to tolerate less than 100 mg/kg/day,[162] whereas methionine intoxication in healthy humans may require a dosage of up to 500 mg/kg.[163]

9. Taste or olfactory discrimination — Some individuals may reject feed on this basis, and fail to thrive as a result.

In the rat, excess methionine severely disorganizes nitrogen metabolism, increases urinary excretion of creatinine and amino acids, and decreases nitrogen retention.[171] Hypertrophy of the liver and kidneys, splenic hemosiderosis, pancreatic degeneration, and fat deposition in the liver occurs, while body fat diminishes.[160]

Methionine induced toxicity is possibly mediated via an hepatic insufficiency of ATP,[155] defective metabolism of the labile methyl group of methionine,[159] a secondary deficiency of threonine,[173,192] altered metabolism of tryptophan,[193] and/or the accumulation of toxic amounts of homocysteine or cystathionine.[169] The precise mechanisms are obscure.

Many "inborn errors" of sulfur metabolism have been detected, and most of these result in neurological disturbance in humans.[200] Homocystinuria is a condition where a deficiency of cystathionine synthase results in the accumulation in the plasma and urine of homocystine, homocysteine, methionine, methionine sulfoxide, cystathionine, and many other metabolites of methionine.[200] Mental retardation and convulsions are features of the disease.

The difference between species in dosage effects and biochemical lesions produced make extrapolation from animal experiments to human application difficult.[155] Man seems to be rather intolerant of excess methionine.

The symptoms in patients given 30 g methionine intravenously were nausea, vomiting, hypotension, tachycardia, fever, disorientation, and liver disfunction.[163] Man is more sensitive to methionine given orally than intravenously.[174] Furthermore, chlortetracycline either partially or completely offsets the effect of oral doses of methionine on neurological perturbation, implying that the toxic agent was a product of methionine degradation in the gut.[174] With rats intestinal antibiotics did not affect the overall effect of excess dietary methionine.[193]

Chicks consuming semisynthetic diets containing 0.35% of D,L-methionine and 0.35% of cystine were adversely affected by adding as little as 0.5% methionine to the ration during the 8- to 14-day posthatching stage,[169] The addition of 0.1% methionine to practical broiler rations may be beneficial.[54]

D,L- and hydroxy analogue of methionine have been given postruminally to sheep in order to stimulate wool growth.[176,177,179,180] The maximum response to D,L-methionine appears to occur with 1 to 2 g/day. In sheep methionine is the limiting amino acid at the tissue level,[181],[182] and there is a general anabolic effect of absorbed supplemental methionine.[176],[178] Doses in excess of 2.5 g/day gave a smaller response in wool growth, down to zero at 10 g/day.[177] Mild scouring and inappetence were observed in one case.[176] When 15 g/day of D,L-methionine was given in the diet there was a trend towards decreased intake, liveweight gain, and nitrogen retention.[178] Adding 1% of methionine to a corn-alfalfa ration did not adversely affect the growth of lambs.[209] The inclusion of up to 0.5% methionine in semipurified rations fed to fattening lambs increased the growth rate, but higher percentages were not as effective.[11],[12] However, 1.1% or 1.5% of methionine in similar rations improved nitrogen retention.[186],[188] This supplement would have supplied up to 10 g methionine per day. D,L- methionine given subcutaneously or intraperitoneally to sheep in doses exceeding 13 g/week, or per month, caused lost weight relative to controls.[179]

Methionine hydroxy analogue (MHA) has been used instead of methionine in experiments with sheep because it is cheaper, less subject to ruminal demethiolation,[184],[185] and is equally effective in increasing wool growth.[176],[183] However, MHA given intraperitoneally in amounts as little as 1 g/day severely depresses wool growth and body weight gains.[179] Up to 80 g MHA/day given to cows increased the production of milk and milk fat, apparently without any adverse effects,[187],[213] although in one instance milk production declined due to feed refusals.[214]

MHA can be substituted for methionine in the rations of chicks[189] and rats[115],[190] and it is about as active as methionine for promoting growth in chicks and rats.[54],[115] MHA is less toxic than D,L or D,L-methionine; almost twice as much MHA (2.84%) was needed to depress growth to the same extent as did 1.25% D,L-methionine.[169]

Ethionine at one tenth the concentration of methionine was more toxic to chicks.[169] With guinea pigs the situation is probably similar.[155] While excess methionine and ethionine induce rather similar biochemical and histological changes in the liver and pancreas[155],[160],[175] with ethionine adaptation does not occur.[160]

Other analogues of methionine, such as methionine hydantoin, methionine nitrate sulfate, 4-thia pentan-1-al-NaHSO$_3$, and methionine ethyl ester HCl did not affect net protein utilization of chicks when included at the level of 0.1% of the diet, but methyl methionine sulfonium HSO$_4$ markedly inhibited NPU.[115]

Oxidation Products of Methionine

Compounds such as methionine sulfoxide, methionine sulfone, and methionine sulfoximine can be found during food processing when heat treatment is involved.

When given to rats to the extent of 0.1% of the diet the sulfoxide may have slightly improved NPU, but the sulfone and certainly the sulfoximine was deleterious.[115]

Methionine sulfoxide, but not methionine sulfone, is reported to be capable of replacing methionine in the diet of rats.[195] Methionine sulfoxide was reported to be nontoxic to dogs[196] and rats.[197]

Methyl cysteine is a breakdown product of methionine. S-methyl-L-cysteine is as toxic as L-methionine.[198] Thiomethane, and finally sulfate, is produced from methyl cysteine in the intact rat.[199] Information on the effects of CH$_3$SH is scanty, but mercaptans might contribute to hypotension in patients with liver disease, where CH$_3$SH may accumulate.[174]

Cyst(e)ine and Homocyst(e)ine

While dietary methionine is essential for the growth of monogastric animals, homocystine (plus methylating agents) can entirely replace it,[207] and cystine can partially replace it, to the extent of around 80% in man.[212] With monogastrics the methionine requirement is usually assessed in terms of the total concentration of amino acids containing sulfur in the diet, and while this may be regarded as optimal, an excessive proportion of cystine relative to methionine causes an imbalance, with reduced intake, reduced weight gains, and lower feed efficiency in rats,[208] chicks,[57] and pigs.[166] Conversely, an appropriate combination of methionine and cystine gives better growth than when methionine provides all of the organic sulfur.[57,201,202]

The critical proportion of cystine/methionine depends on many factors:

1. Age and weight of the animal — Mature animals fed at maintenance levels can tolerate a greater proportion of cystine[203,204] because the requirement for methionine per se is then relatively small.

2. Presence of inorganic sulfur in the diet — The optimal percentage of cystine in the diet is reduced by adding sulfate.[57]

3. Concentration of sulfur amino acids on the diet — If this is excessive, then more cystine can be tolerated, until methionine becomes limiting.[57] Conversely, with a low concentration of sulfur amino acids a high proportion of cystine markedly reduces appetite.[208]

4. Presence of choline in the diet — With choline added rats can tolerate a higher proportion of cystine,[208] presumably because synthesis of choline from methionine is spared.

5. Species of animal — Differences between species of monogastric animals probably exist. Young pigs could tolerate at least 56% of cystine,[166] whereas the corresponding proportion for chicks was 48%.[57] Ruminants can use inorganic sulfate as effectively as cystine for synthesis of microbial protein[147] and can exist comfortably without any source of dietary sulfur amino acids.[205,206]

A direct excess of cystine in the diet of rats can cause growth failure, inappetence, and sometimes death.[156] With a diet containing 6% casein the inclusion of 5% cystine resulted in growth failure and many deaths. When the casein content of the ration was increased growth rates were initially unaffected, but most rats died abruptly towards the end of the 4-week period. Lactalbumen, in place of casein, overcame the deleterious effect of excess cystine. Despite the obvious toxicity of cystine it is relatively safer than excess methionine.[156]

With chicks, cystine and cysteine fed at a level isosulfurous to 1.25% methionine depressed rate of gain and feed efficiency equally, but less than did 1.25% methionine.[169] Excess cysteine, equivalent to 4% methionine on a molar basis, did not greatly affect the growth of rats.[193] The intraperitoneal injection of 5.5 mM cysteine per kg body weight produced no adverse effects in rats.[210]

With sheep, cystine or cysteine infused into the abomasum or intravenously is as effective as methionine in stimulating wool growth,[176,177] but cyst(e)ine is much less toxic.[177] Cystine or cysteine is rapidly and extensively degraded in the rumen,[49] a factor which further eliminates the possibility of cyst(e)ine toxicity per se, but in exceptional circumstances, this could lead to H_2S intoxication (see Sulfide).

Cystinuria in man is a relatively common disease, characterized by a defective renal transport of cystine and other dibasic amino acids and sometimes the formation of cystine calculi in the urinary tract.[200] Cystinosis is a rare condition in which cystine crystals are deposited in the lysosomes of eye, bone marrow, leucocytes, and kidney so that renal failure and early death occur.[200]

With rats, homocystine (2.4%) was less toxic than 3% of L-methionine,[150] but with chicks homocystine and homocysteine (1.13%) equally depressed rate of gain and feed efficiency[169] to an extent equivalent to 1.25% methionine. Another experiment with rats has shown that excess dietary homocysteine depressed weight increase more than did 4% methionine.[193]

Homocysteine has a convulsant activity which is not shared by homocystine, methionine, cysteine, or their intermediary metabolites.[194] This property of homocysteine may account for convulsions observed in the condition of homocystinuria.[210] While 5.5 mM of D,L-homocysteine per kilogram produced convulsions and some deaths in rats, homocystine and methionine did not, although there appeared to be a delayed lethal effect with homocystine,[210] due presumably to the conversion of homocystine to homocysteine.

Homocystine toxicity may be partially alleviated by glycine or serine[150] presumably because this results in the faster metabolism of homocysteine via the synthesis of cystathionine. Protection against homocysteine is given by prior doses of homoserine, serine, glycine, betaine, or glucose.[210]

Cysteine Sulfoxides and Metabolites

These compounds are precursors of many volatile compounds such as disulfides from the Liliaceae family, notably the genus Allium. Onions, garlic, and chives in particular contain S-substituted cysteine sulfoxides and products of this such as allylthiosulfinate (allicin) in garlic, and various sulfides in onions and garlic.[138] These compounds have, variously, flavor, odor, lachrymatory, or antibiotic properties.

These is little evidence that any of these compounds is harmful when eaten, but they may contribute to goiter in some areas where onions are consumed in large amounts.[138]

The inclusion of dried onions as 10% of the diet of rats, however, did not adversely affect net protein utilization.

TOLERANCE OF SOME ANIMAL SPECIES TO SULFUR[a]

Sulfur compound	Response	Man	Cattle	Sheep	Rat	Chick	Pig	Guinea pig	Dog
Elemental-S	Adverse		50 g (7)[d]	7–70 g (6)[d]	1% (115)				
	Safe		1.7% (15)						
Sulfate-S	Adverse	3.9 g (30)[d]; 0.13 g/l (30)[f]	2.25 g/l (35)[f]; 1.13 g/l (21)[f]; 0.5–0.9 g/l (33)[f]; 0.3% (216); 130 g/d (217)[d]	0.5–1.5% (38,39); 0.54% (12); 6 g/d (42)[f]		0.08–0.14% (58); 1.7 g/l (30)[f]			0.8 g/l (30)[f]
	Safe	0.2–0.4 g/l (30)[f]	0.8% (36); 0.2% (50); 1.58 g/l (35)[f]	0.4% (12); 0.2% (50); 1.13 g/l (31,32)[f]	3.4 g/l (30)[f]; 0.6 g/l (30)[f]	0.03% (54)			
Sulfur dioxide-S (also sulfite bisulfite metabisulfite)	Adverse	25 ppm (1)[e]	15–51 g/d (104); 26–80 g/d (103)		0.03% (101)				
	Safe	2.5 ppm (3)[e]	41 g/d (106); 0.05% (105); 0.02% (221)	2 g/d (112)	0.03% (115); 0.06 g/d (99)[d]; 3.8 g/l (30)[f]				
Thiosulfate-S	Adverse	10 g (121)[c]						1.2 g/kgW (126)[d]	1.2 g/kg W (119)[d]
	Safe	7 g (116)[d]							0.6 g/kg W (119)[d]
Persulfate-S	Adverse				0.42 g/kgW (124)[d]				
	Safe				2 g/kgW (126)[d]				
Hydrogen sulfide-S	Adverse	50 ppm (2,92)[e]		2.9 g/d (42)[d]					
	Safe	19 ppm (3)[e]							

a Values in the table are, for comparative purposes, expressed in terms of elemental sulfur. The data given must be regarded with caution, since many factors can influence the response of an animal to a given concentration of a particular nutrient in the diet (see text) and in many cases adequate data are not available. The numbers in parentheses are reference citations.

b Subcutaneous or intraperitoneal route

c Abomasal or intravenous route

d Oral dose or infusion

e Given in air

f Given in drinking water

TOLERANCE OF SOME ANIMAL SPECIES TO SULFUR[a] (continued)

Sulfur compound	Response	Man	Cattle	Sheep	Rat	Chick	Pig	Guinea pig	Dog
Methionine-S	Adverse	0.03 g/kgW (164) 0.02 g/kgW (162) 0.1 g/kgW (163)		2 g/d (177,180)[c] 3 g/d (179)[c] 3 g/d (178)	0.4% (150,156) (160, 172)	0.18% (169)		0.3 g/kgW/d (155)[d]	
	Safe		0.05% (215) 5 g/d (215)[c]	0.5 g/d (177)[c] 1.7 g/d (178) 0.1% (11,12) 0.2–0.3% (186,188)	0.17% (193) 0.1% (150,175,208) 0.3 g/kgW/d (155)[d]	0.07% (169)	0.08% (166)		
MHA-S	Adverse		10–17 g/d (214)	0.2 g/d (179)[b]		0.5% (169)			
	Safe		5–10 g/d (214) 17 g/d (187,213)	2.6 g/d (180)[c] 1.1 g/d (183)		0.26% (169)			
Cyst(e)ine-S	Adverse				1.3% (156)	0.1% (57)			
	Safe			2 g/d (177)[c]	0.8% (193) 1.3 g/kg W (210)[b]	0.26% (169) 0.08% (57)	0.07% (166)		
Homocys-t(e)ine-S	Adverse				0.8% (193)	0.27% (169)			
	Safe				0.6% (150) 0.7 g/kg W (210)[b]				

REFERENCES

1. Sollmann, T., in *A Manual of Pharmacology and Its Application to Therapeutics and Toxicology,* 3rd ed., W. B. Saunders, Philadelphia, 1957.
2. Sorbo, B., The pharmacology and toxicology of inorganic sulfur compounds, in *Sulfur in Organic and Inorganic Chemistry,* Vol. 2, Senning, A., Ed., Marcel Dekker, New York, 1972, chap. 16.
3. Patty, F. A., *Industrial Hygiene and Toxicology,* 2nd ed., Vol. 2, Interscience, New York, 1962.
4. Greengard, H. and Woolley, J. R., Studies on colloidal sulfur-polysulfide mixture. I. Toxicity, *J. Am. Pharmacol. Ass.,* 29, 289–292, 1940.
5. Hoare, W., *Veterinary Materia Medica,* Bailliere, Tindall and Cox, London, 1924.
6. White, J. B., Sulphur poisoning in ewes, *Vet. Rec.,* 76, 278–279, 1964.
7. Coghlin, C. L., Hydrogen sulphide poisoning in cattle, *Can. J. Comp. Med.,* 8, 111–113, 1944.
8. Lampila, M., *J. Sci. Agric. Soc. Finl.,* 32, 169, 1960.
9. Christensen, J. F., Deem, A. W., Esplin, A. L., and Cross, F., The use of sulphur in reducing death losses from enterotoxaemia in feeder lambs, *J. Am. Vet. Med. Assoc.,* 111, 144–148, 1947.
10. Starks, P. B., Hale, W. H., Garrigus, U. S., and Forbes, R. M., The utilization of feed nitrogen by lambs as affected by elemental sulfur, *J. Anim. Sci.,* 12, 480–291, 1953.
11. Starks, P. B., Hale, W. H., Garrigus, U. S., Forbes, R. M., and James, M. F., Response of lambs fed sulfur, sulfate sulfur and methionine, *J. Anim. Sci.,* 13, 249–258, 1954.
12. Albert, W. A., Garrigus, U. S., Forbes, R. M., and Norton, H. W., The sulfur requirement of growing-fattening lambs in terms of methionine, sodium sulfate, and elemental sulfur, *J. Anim. Sci.,* 15, 559–569, 1956.
13. Albert, W. A., Garrigus, U. S., Forbes, R. M., and Hale, W. H., Modified urea supplements with corn silage for wintering ewe lambs, *J. Anim. Sci.,* 14, 143–152, 1955.
14. Sokolowski, J. H., Hatfield, E. E., and Garrigus, U. S., Effect of inorganic sulfur on potassium nitrate utilization by lambs, *J. Anim. Sci.,* 28, 391–396, 1969.
15. Chalupa, W., Oltjen, R. R., Slyter, L. L., and Dinius, D. A., Sulfur deficiency and tolerance in bull calves, *J. Anim. Sci.,* 33, 278, 1971.
16. Roy, A. B. and Trudinger, P. A., *The Biochemistry of Inorganic Compounds of Sulphur,* Cambridge University Press, New York, 1970.
17. Suzuki, I. and Silver, M., The initial product and properties of the sulfur-oxidising enzyme of thiobacilli, *Biochem. Biophys. Acta,* 122, 22–23, 1966.
18. Peck, H. D., Sulfur requirements and metabolism of micro organisms, in *Symposium: Sulfur in Nutrition,* Muth, O. H. and Oldfield, J. E., Eds., Avi Publishing, Westport, Conn., 1970, chap. 6.
19. Berner, R. A., Worldwide sulfur pollution of rivers, *J. Geophys. Res.,* 76, 6597–6600, 1971.
20. Freney, J. R., Sulfur-containing organics, in *Soil Biochemistry,* McLaren, A. D. and Peterson, G. H., Eds., Arnold and Dekker, London 1967, chap. 10.
21. Weeth, H. J. and Hunter, J. E., Drinking of sulfate-water by cattle, *J. Anim. Sci.,* 32, 277–281, 1977.
22. Thompson, J. F., Smith, I. K., and Moore, D. P., Sulfur requirement and metabolism in *Symposium: Sulfur in Nutrition,* Muth, O. H. and Oldfield, J. E., Eds., Avi Publishing, Westport, Conn., 1970, chap. 7.
23. Allcroft, R. and Lewis, G., Copper nutrition in ruminants. Disorders associated with copper-molybdenum-sulphate content of feeding stuffs, *J. Sci. Food Agric.,* 8, Suppl., 96–104, 1957.
24. Beck, A. B., The levels of copper, molybdenum and inorganic sulphate in some Western Australian pastures, *Aust. J. Exp. Agric. Anim. Husb.,* 2, 40–45, 1962.
25. Spais, A. G., Lazaridis, T. K., and Agiannidis, A. K., Studies on sulphur metabolism in sheep in association with copper deficiency, *Res. Vet. Sci.,* 9, 337–344, 1968.
26. Stewart, B. A. and Porter, L. K., Nitrogen-sulfur relationships in wheat (*Triticum aestivum* L.), corn (*Zea mays*) and beans (*Phaseolus vulgaris*), *Agron. J.,* 61, 267–271, 1969.
27. Jones, R. K., Robinson, P. J., Haydock, K. P., and Megarrity, R. G., Sulphur – nitrogen relationships in the tropical legume *Stylosanthes humilis, Aust. J. Agric. Res.,* 22, 885–894, 1971.
28. Eppendorfer, W., The effects of nitrogen and sulphur on total-N and -S concentrations in grain and straw, and on cystine and methionine contents of grain, in *Coppenhagen Royal Veterinary Agricultural University Yearbook,* 1969, 100–116.
29. Boyazoglu, P. A., Jordan, R. M., and Meade, R. J., Sulfur-selenium-vitamin E interrelations in ovine nutrition, *J. Anim. Sci.,* 26, 1390–1396, 1967.

30. McKee, J. E. and Wolf, H. W., in *Water Quality Criteria*, 2nd ed., The Resources Agency of California, State Water Quality Control Board, Sacramento, California, Publ. No. 3A, 1963, p. 154, 213, 270–279.

31. Peirce, A. W., Studies on salt tolerance of sheep. III. The tolerance of sheep for mixtures of sodium chloride and sodium sulphate in the drinking water, *Aust. J. Agric. Res.*, 11, 548–556, 1960.

32. Peirce, A. W., Studies on salt tolerance of sheep. VI. The tolerance of wethers in pens for drinking waters of the type obtained from underground sources in Australia, *Aust. J. Agric. Res.*, 17, 209–218, 1966.

33. Weeth, H. J. and Capps, D. L., Tolerance of growing cattle for sulfate-water, *J. Anim. Sci.*, 34, 256–260, 1972.

34. Weeth, H. J. and Digesti, R. D., Taste response of cattle to sodium chloride and sodium sulfate, *J. Anim. Sci.*, 34, 906, 1972.

35. Embry, L. B., Hoelscher, M. A., Wahlstrom, R. C., Carlson, C. W., Krista, L. M., Brosz, W. R., Gastler, G. F., and Olson, G. E., Salinity and livestock water quality, in *S. D. Agric. Exp. Stn. Bull.*, 1959, 481.

36. McCarrick, R. B., Keane, E., and Tobin, J., The nutritive values of ammonium bisulphate and molassed silages. I. Comparisons of acceptability and feeding value for beef and dairy cattle, *Ir. J. Agric. Res.*, 4, 115–123, 1965.

37. McCarrick, R. B., Poole, D. R. B., and Maguire, M. F., The nutritive values of ammonium bisulphate and molassed silages. II − Effect of sulphate intake on performance of growing and mature cattle, *Ir. J. Agric. Res.*, 4, 125–133, 1965.

38. L'Estrange, J. L., Clarke, J. J., and McAleese, D. M., Studies on high intake of various sulphate salts and sulphuric acid in sheep. I. Effects on voluntary feed intake, digestibility and acid-base balance, *Ir. J. Agric. Res.*, 8, 133–150, 1969.

39. L'Estrange, J. L., Upton, P. K., and McAleese, D. M., Effects of dietary sulphate on voluntary feed intake and metabolism of sheep. 1. A comparison between different levels of sodium sulphate and sodium chloride, *Ir. J. Agric. Res.*, 11, 127-144, 1972.

40. Upton, P. K., L'Estrange, J. L., and McAleese, D. M., Effect of high intakes of dietary sulphate on voluntary feed intake and metabolism of sheep. 2. Palatability and metabolic effects of sodium sulphate and sulphuric acid, *Ir. J. Agric. Res.*, 11, 145–158, 1972.

41. McBarron, E. J. and McInnes, P., Observations on urea toxicity in sheep, *Aust. Vet. J.*, 44, 90–96, 1968.

42. Bird, P. R., Sulphur metabolism and excretion studies in ruminants. X. Sulphide toxicity in sheep, *Aust. J. Biol. Sci.*, 25, 1087–1098, 1972.

43. Goodrich, R. D. and Tillman, A. D., Effects of sulfur and nitrogen sources and copper levels on the metabolism of certain minerals by sheep, *J. Anim. Sci.*, 25, 484–491, 1966.

44. Bird, P. R. and Moir, R. J., Sulphur metabolism and excretion studies in ruminants. 1. The absorption of sulphate in sheep after intraruminal or intraduodenal infusions of sodium sulphate, *Aust. J. Biol. Sci.*, 24, 1319–1328, 1971.

45. Dougherty, R. W., Mullenax, C. H., and Allison, M. J., Physiological phenomena associated with eructation, in *Physiology of Digestion in the Ruminant*, Dougherty, R. W., Ed., Butterworths, Washington, D.C., 1965, 159–170.

46. Bray, A. C., Sulphur metabolism in sheep. II. The absorption of inorganic sulphate and inorganic sulphide from the sheep's rumen, *Aust. J. Agric. Res.*, 20, 739–748, 1969.

47. Evans, C. L., The toxicity of hydrogen sulphide and other sulphides, *Q. J. Exp. Physiol.*, 52, 231–248, 1967.

48. Anderson, C. M., The metabolism of sulphur in the rumen of sheep, *N.Z. J. Sci. Technol.*, 37, 379–394, 1956.

49. Bird, P. R., Sulphur metabolism and excretion studies in ruminants. V. Ruminal desulphuration of methionine and cyst(e)ine, *Aust. J. Biol. Sci.*, 25, 185–193, 1972.

50. Bird, P. R., Sulphur metabolism and excretion studies in ruminants. XIII. Intake and utilization of wheat straw by sheep and cattle, *Aust. J. Agric. Res.*, 25, 631–642, 1974.

51. Bird, P. R., Sulphur metabolism and excretion studies in ruminants. IX. Sulphur, nitrogen and energy utilization by sheep fed a sulphur deficient and a sulphate supplemented roughage-based diet, *Aust. J. Biol. Sci.*, 25, 1073–1085, 1972.

52. Wellers, G., Boelle, G., and Chevan, J., Investigations on sulphur metabolism. IV. Partial fulfillment of the requirement for endogenous sulphur by sodium sulphate in the adult male rat, *J. Physiol. (Paris)*, 52, 903–910, 1960.

53. Gordon, R. S. and Sizer, I. W., The biological equivalence of methionine hydroxy analogue, *Poult. Sci.*, 34, 1198, 1955.

54. Almquist, H. J., Sulfur nutrition of non-ruminant species, in *Symposium: Sulfur in Nutrition,* Muth, O. H. and Oldfield, J. E., Eds., Avi Publishing, Westport, Conn., 1970, chap. 13.

55. Michels, F. G. and Smith, J. T., A comparison of the utilization of organic and inorganic sulfur by the rat, *J. Nutr.,* 87, 217–220, 1965.

56. Martin, W. G., The neglected nutrient: sulphur, *Sulphur Inst. J.,* 4, 5, 1968.

57. Sasse, C. E. and Baker, D. H., Sulfur utilization by the chick with emphasis on the effect of inorganic sulfate on the cystine-methionine interrelationship, *J. Nutr.,* 104, 244–251, 1974.

58. Leach, R. M., Zeigler, T. R., and Norris, L. C., The effect of dietary sulfate on the growth rate of chicks fed a purified diet, *Poult. Sci.,* 39, 1577–1578, 1960.

59. Miller, E. C. and Denton, C. A., Molybdenum-sulfate interrelationship in the growing chick, *Poult. Sci.,* 38, 910–916, 1959.

60. Dziewiatkowski, D. D., Metabolism of sulfate eaters, in *Symposium: Sulfur in Nutrition,* Muth, O. H. and Oldfield, J. E., Eds., Avi Publishing, Westport, Conn., 1970, chap. 8.

61. Hubbert, F., Cheng, E., and Burroughs, W., Mineral requirements of rumen microorganisms for cellulose digestion in vitro, *J. Anim. Sci.,* 17, 559–568, 1958.

62. Kennedy, L. G., Mitchell, G. E., and Little, C. O., Influence of sulfur on in vitro starch digestion by rumen micro organisms, *J. Anim. Sci.,* 32, 359–363, 1971.

63. Kawashima, R., Uesaka, S., and Toyama, T., Studies on importance of trace elements in farm animal feeding. XXXVI. Effects of sulphate, molybdate and phosphorus on sulphide production by rumen bacteria, *Bull. Res. Inst. Food Sci. Kyoto University,* 31, 7–12, 1968.

64. Underwood, E. J., *Trace Elements in Human and Animal Nutrition,* 3rd ed., Academic Press, New York, 1970, chap. 3–4.

65. Dick, A. T., The control of copper storage in the liver of sheep by inorganic sulphate and molybdenum, *Aust. Vet. J.,* 29, 233–239, 1953.

66. Dick, A. T., Studies on the assimilation and storage of copper in crossbred sheep, *Aust. J. Agric. Res.,* 5, 511–544, 1954.

67. Marcilese, N. A., Ammerman, C. B., Valsecchi, R. M., Dunavant, B. G., and Davis, G. K., Effect of dietary molybdenum and sulfate upon copper metabolism in sheep, *J. Nutr.,* 99, 177–183, 1969.

68. Suttle, N. F., Recent studies of the copper-molybdenum antagonism, *Proc. Nutr. Soc.,* 33, 299–305, 1974.

69. Evans, J. L. and Davis, G. K., Mineral interrelationships (copper, molybdenum, sulfur and phosphorus) in the nutrition of the rat, in *Proc. 6th Int. Congr. Nutr.,* E. and S. Livingstone Ltd., Edinburgh, Scotland, 1964, 520–521.

70. Wynne, K. W. and McClymont, G. L., Copper-molybdenum-sulphate interactions in induction of ovine hypercuperaemia and hypocuprosis, *Aust. J. Agric. Res.,* 7, 45–56, 1956.

71. Bingley, J. B., Effects of high doses of molybdenum and sulphate on the distribution of copper in plasma and in blood of sheep, *Aust. J. Agric. Res.,* 25, 467–474, 1974.

72. Hintz, H. F. and Hogue, D. E., Effect of selenium, sulfur and sulfur amino acids on nutritional muscular dystrophy in the lamb, *J. Nutr.,* 82, 495–498, 1964.

73. Whanger, P. D., Muth, O. H., Oldfield, J. E., and Weswig, P. H., Influence of sulfur on incidence of white muscle disease in lambs, *J. Nutr.,* 97, 553–562, 1969.

74. Whanger, P. D., Weswig, P. H., Muth, O. H., and Oldfield, J. E., Selenium and white muscle disease: Effect of sulfate and energy levels on plasma enzymes and ruminal microbes, *J., Am. Vet. Med. Assoc.,* 31, 965–972, 1970.

75. Mills, C., Comparative studies of copper, molybdenum and sulphur metabolism in the ruminant and the rat, *Proc. Nutr. Soc.,* 19, 162–169, 1960.

76. Spais, A., Les oligoéléments et specialement le cuirire en nutrition et pathologie animales, *An. Med. Vet.,* 3, 131, 1962.

77. Suttle, N. F., Effects of molybdenum and sulphur at concentrations commonly found in ruminant diets on the availability of copper to sheep, in *Trace Element Metabolism in Animals 2,* Hoekstra, W. G., Suttie, J. W., Ganther, H. E., and Mertz, W., Eds., University Park Press, Baltimore, 1974, p. 612–617.

78. Pope, A. L., Moir, R. J., Somers, M., and Underwood, E. J., Effect of sulphur on [75] selenium absorption in sheep, *J. Anim. Sci.,* 27, 1771, 1968.

79. Schubert, J. R., Muth, O. H., Oldfield, J. E., and Remmert, L. F., Experimental results with selenium in white muscle disease of lambs and calves, *Fed. Proc.,* 20, 689–694, 1961.

80. Rosenfield, I., Biosynthesis of seleno-compounds from inorganic selenium by sheep, *Fed. Proc.,* 20, 10, 1961.

81. **Paulsen, G. D., Baumann, C. A., and Pope, A. L.,** Metabolism of ^{75}Se-selenite, ^{75}Se-selenate, ^{75}Se-selenomethionine and ^{35}S-sulfate by rumen micro-organisms in vitro, *J. Anim. Sci., 27,* 497–504, 1968.

82. **Postgate, J. R.,** Competitive inhibition of sulphate reduction by selenate, *Nature (London),* 164, 670–671, 1949.

83. **Shrift, A.,** Biological activities of selenium compounds, *Bot. Rev.,* 24, 550–583, 1958.

84. **Pratley, J. E. and McFarlane, J. D.,** The effect of sulphate on the selenium content of pasture plants, *Aust. J. Exp. Agric. Anim. Husb.,* 14, 533–538, 1974.

85. **Muth, O. H.,** White muscle disease (myopathy) in lambs and calves. 1. Occurrence and nature of the disease under Oregon conditions, *J. Am. Vet. Med. Assoc.,* 126, 355–361, 1955.

86. **Gissel-Nielsen, G.,** Uptake and distribution of added selenate and selenite by barley and red clover as influenced by sulphur, *J. Sci. Food Agric.,* 24, 649–655, 1973.

87. **Morrison, J. N., Quartermain, J., Humphries, W. R., and Mills, C. F.,** The influence of dietary sulphate on the toxicity of lead to sheep, *Proc. Nutr. Soc.,* 34, 77A, 1975.

88. **Peck, H. D.,** Comparative metabolism of inorganic sulphur compounds, *Bacteriol. Rev.,* 26, 67–94, 1962.

89. **Thoman, M.,** Sewer gas: Hydrogen sulfide intoxication, *Clin. Toxicol.,* 2, 383–386, 1970.

90. **Adelson, L. and Sunshine, I.,** Fatal hydrogen sulfide intoxication, *Arch. Pathol.,* 81, 375–380, 1966.

91. **Poda, G. A.,** Hydrogen sulphide can be handled safely, *Arch. Environ. Health,* 12, 795–800, 1966.

92. **Milby, J. H.,** Hydrogen sulfide intoxication. Review of the literature and report of unusual accident resulting in two cases of non-fatal poisoning, *J. Occup. Med.,* 4, 431–437, 1962.

93. **Chevalier, J. R.,** Toxicity of sodium sulfide to common shiners – dynamic bioassay, *Tappi,* 56, 135–136, 1973.

94. **Ziebell, C. D., Pine, R. E., Mills, A. D., and Cunningham, R. K.,** Field toxicity studies and juvenile salmon distribution in Port Angeles harbor, Washington, *J. Water Pollut. Control Fed.,* 42, 229–236, 1970.

95. **Curtis, C. G., Bartholomew, T. C., Rose, F. A., and Dodgson, K. S.,** Detoxification of sodium ^{35}S-sulphide in the rat, *Biochem. Pharmacol.,* 21, 2313–2321, 1972.

96. **Roberts, A. C. and McWeeny, D. J.,** The uses of sulphur dioxide in the food industry, *J. Food Technol.,* 7, 221–238, 1972.

97. **Schroeter, L. C.,** *Sulphur Dioxide,* Permagon Press, Oxford, 1966.

98. **Smith, G. and Stevens, M. F. G.,** Sulphur dioxide – preservative and potential nitrogen, *Pharm. J.,* 570–572, Dec. 16, 1972.

99. **Gibson, W. B. and Strong, F. M.,** Metabolism and elimination of sulphite by rats, mice and monkeys, *Food Cosmet. Toxicol.,* 11, 185–198, 1973.

100. **Schimmel, H. and Murawski, T. J.,** SO_2 – Harmful pollutant or air quality indicator? *J. Air Pollut. Control Assoc.,* 25, 739–740, 1975.

101. **Frank, N. R., Yoder, R. E., Yokohama, E., and Speizer, F. E.,** The diffusion of $^{35}SO_2$ from tissue fluids into the lungs following exposure of dogs to $^{35}SO_2$, *Health Phys.,* 13, 31–38, 1967.

102. **Mudd, S. H., Irreverre, F., and Laster, I.,** Sulfite oxidase deficiency in man: Demonstration of the enzymatic effect, *Science,* 156, 1599–1602, 1967.

103. **Luedke, A. J., Bratzler, J. W., and Dunne, H. W.,** Sodium metabisulfite and sulfur dioxide gas (Silage Preservative) poisoning in cattle, *Am. J. Vet. Res.,* 20, 690–696, 1959.

104. **Alhassan, W. S. and Satter, L. D.,** Observations on sodium sulfite administration to the ruminant, *J. Dairy Sci.,* 51, 981, 1968.

105. **Kaemmerer, Von K.,** Zur toxikologischen bedeutung von sulfit beim Wiederkäuer, *Zucker,* 25, 123–127, 1972.

106. **Alhassan, W. S., Krabill, L. F., and Satter, L. D.,** Manipulation of the ruminal fermentation. 1. Effect of sodium sulfite on bovine ruminal fatty acid concentration and milk composition, *J. Dairy Sci.,* 52, 376–379, 1969.

107. **Henderickx, H.,** The incorporation of sulfate in the ruminal proteins, *Arch. Int. Physiol. Biochim.,* 69, 449–458, 1961.

108. **Moir, R. J.,** Implication of the N:S ratio and differential recycling, in *Symposium: Sulfur in Nutrition,* Muth, O. H. and Oldfield, J. E., Eds., Avi Publishing, Westport, Conn., 1970, chap. 11.

109. **Gawthorne, J. M. and Nader, C. J.,** The effect of molybdenum on the conversion of sulphate to sulphide and microbial protein-sulphur in the rumen of sheep, *Br. J. Nutr.,* 35, 11–23, 1976.

110. **Bray, A. C. and Till, A. R.,** Metabolism of sulphur in the gastro-intestinal tract, in *Digestion and Metabolism in the Ruminant,* McDonald, I. W. and Warner, A. C. I., Eds., The University of New England Publishing Unit, Armidale, 1975, 243–260.

111. **Lampila, M.,** Nutritional factors promoting the in vitro growth of rumen bacteria of the cow fed on a purified protein-free diet, *Ann. Agric. Fenn.,* 6, 14–29, 1967.

112. **Van Nevel, C. J., Demeyer, D. I., Cottyn, B. G., and Henderickx, H. K.,** Effect of sodium sulfite on methane and propionate in the rumen, *Z. Tierphysiol. Tierernaehr. Futtermittelkd.,* 26, 91–100, 1970.

113. **Shapiro, R., Cohen, B. I., and Servis, R. E.,** Specific deamination of RNA by sodium bisulphite, *Nature (London),* 227, 1047, 1970.

114. **Bigwood, E. J.,** Consommation probable de SO_2 en Belgique, *Arch. Belg. Med. Soc.,* 26, 473, 1968.

115. **Miller, D. S. and Samuel, P. D.,** Effects of the addition of sulphur compounds to the diet on utilization of protein in young growing rats, *J. Sci. Food Agric.,* 21, 616–618, 1970.

116. **Nyiri, W.,** Über das schicksal von Natriumthiosulfat im Organismus, *Biochem. Z.,* 141, 160–165, 1923.

117. **Cardozo, R. H. and Edelman, I. S.,** The volume of distribution of sodium thiosulfate as a measure of the extracellular fluid space, *J. Clin. Invest.,* 31, 280–290, 1952.

118. **Brun, C.,** Thiosulfate as a measure of the glomerular filtration rate in normal and diseased human kidneys, *Acta Med. Scand. Suppl.,* ccxxxiv 63–70, 1949.

119. **Dennis, D. L. and Fletcher, W. S.,** Toxicity of sodium thiosulfate (NSC-45624), a nitrogen mustard antagonist, in the dog, *Cancer Chemother. Rep.,* 50, 255–257, 1966.

120. **Hayatsu, H. and Miura, A.,** The mutagenic action of sodium bisulfite, *Biochem. Biophys. Res. Commun.,* 39, 156–160, 1970.

121. **Litwins, J., Boyd, L. J., and Greenwald, L.,** The action of sodium thiosulphate on the blood, *Exp. Med. Surg.,* 1, 252–259, 1943.

122. **Gilman, A., Philips, F. S., Koelle, E. S., Allen, R. P., and St. John, E.,** The metabolic reduction and nephrotic action of tetrathionate in relation to a possible interaction with sulfhydryl compounds, *Am. J. Physiol.,* 147, 115–126, 1946.

123. **Saunders, J. P. and Wills, J. H.,** The nephrotoxic action of sodium tetrathionate, *J. Pharmacol. Exp. Ther.,* 112, 197–201, 1954.

124. **Gleason, M. N., Gosselin, R. E., Hodge, H. C., and Smith, R. P.,** *Clinical Toxicology of Commercial Products,* 3rd ed., Williams & Wilkins, Baltimore, 1969, 118.

125. **Arnold, A. and Goble, F. C.,** Studies with dogs fed flour treated with ammonium persulphate, *Cereal Chem.,* 27, 375–382, 1950.

126. **Eichler, O.,** *Handbuch der Experimentellen Pharmakoligie,* Vol. X, Springer, Berlin, 1950.

127. **Nofre, C., Dufour, H., and Cier, A.,** Toxicité générale comparée des anions mineraux chez la Souris, *C. R. Acad. Sci.,* 257, 791–794, 1963.

128. **Anderson, R. C. and Chen, K. K.,** Absorption and toxicity of sodium and potassium thiocyanates, *J. Am. Pharmacol. Assoc.,* 29, 152–161, 1940.

129. **Pullman, T. N. and McClure, W. W.,** The renal tubular reabsorption of thiocyanate in normal man, *J. Lab. Clin. Med.,* 43, 815–823, 1954.

130. **Garvin, C. F.,** The fatal toxic manifestations of the thiocyanates, *J.A.M.A.,* 112, 1125–1127, 1939.

131. **Santesson, C. G. and Malmgren, R.,** Ueber die Wirkung des Phosphorsesquisulfides (P_4S_3), *Scand. Arch. Physiol.,* 15, 259–327, 1903.

132. *The Merck Index of Chemicals and Drugs,* 7th ed., Merck, Rahway, 1960, 88.

133. **Laug, E. P. and Draize, J. H.,** The percutaneous absorption of ammonium hydrogen sulfide and hydrogen sulfide, *J. Pharmacol. Exp. Ther.,* 76, 179–188, 1942.

134. **Bird, P. R.,** Sulphur metabolism and excretion studies in ruminants. III. The effect of sulphur intake on the availability of copper in sheep, *Proc. Aust. Soc. Anim. Prod.,* 8, 212–218, 1970.

135. **Bird, P. R. and Hume, I. D.,** Sulphur metabolism and excretion studies in ruminants. IV. Cystine and sulphate effects upon the flow of sulphur from the rumen and upon sulphur excretion by sheep, *Aust. J. Agric. Res.,* 22, 443–452, 1971.

136. **Boyazoglu, P. A., Jordan, R. M., and Meade, R. J.,** Sulfur – Selenium – Vit. E inter-relationships in sheep, *J. Anim. Sci.,* 23, 1207, 1964.

137. **Boyazoglu, P. A. and Jordan, R. M.,** Sulfur influence on ovine white muscle disease, *J. Anim. Sci.,* 21, 988, 1962.

138. **Van Etten, C. H. and Wolff, I. A.,** Natural sulphur compounds, in *Toxicants Occurring Naturally in Food,* 2nd ed., 1973, chap. 10.

139. **Van Etten, C. H., Gagne, W. E., Robbins, D. J., Booth, A. N., Daxenbichler, M. E., and Wolff, I. A.,** Biological evaluation of crambe seed meals and derived products by rat feeding, *Cereal Chem.,* 46, 145–155, 1969.

140. **Oginsky, E. L., Stein, A. E., and Greer, M. A.,** Myrosinase activity in bacteria as demonstrated by the conversion of progoitrin to goitrin, *Proc. Soc. Exp. Biol. Med.,* 119, 360–364, 1965.

141. **Greer, M. A.,** The natural occurrence of goitrogenic agents, *Recent Prog. Horm. Res.,* 18, 187–219, 1962.

142. **Kelly, F. C. and Snedden, W. W.,** Prevalence and geographical distribution of endemic goitre, in *Endemic Goitre, W.H.O. Monogr. Ser.* No. 44, World Health Organization, Geneva, 1960, 91.

143. **Michajlovskij, N., Sedlak, J., Jusic, M., and Buzina, R.,** Goitrogenic substances of kale and their possible relations to the endemic goitre on the Island of Krk (Yugoslavia), *Endocrinol. Exp.,* 3, 65–72, 1969.

144. **Piironen, E. and Virtanen, A. I.,** The effect of thiocyanate in nutrition on the iodine content of cow's milk, *Z. Ernaehrungswiss,* 3, 140, 1963.

145. **VanderLaan, J. E. and VanderLaan, W. P.,** The iodine concentrating mechanism of the rat thyroid and its inhibition by thiocyanate, *Endocrinology,* 40, 403–416, 1947.

146. **Michajlovskij, N. and Langer, P.,** The relation between thiocyanate formation and the goitrogenic effect of foods. I. The preformed thiocyanate content of some foods, *Hoppe-Seyler's Z. Physiol. Chem.,* 312, 26–30, 1958.

147. **Hume, I. D. and Bird, P. R.,** Synthesis of microbial protein in the rumen. IV. The influence of the level and form of dietary sulphur, *Aust. J. Agric. Res.,* 21, 315–322, 1970.

148. **Finklestein, J. D.,** Control of sulfur metabolism in mammals, in *Symposium: Sulfur in Nutrition,* Muth, O. H. and Oldfield, J. E., Eds., Avi Publishing, Westport, Conn., 1970, chap. 5.

149. **Miller, D. S. and Donoso, G.,** Relationship between the sulphur/nitrogen ratio and the protein value of diets, *J. Sci. Food Agric.,* 14, 345–349, 1963.

150. **Benevenga, N. J. and Harper, A. E.,** Alleviation of methionine and homocystine toxicity in the rat, *J. Nutr.,* 93, 44–52, 1967.

151. **Kies, C. and Fox, H. M.,** Effect of amino acid supplementation of dehydrated potato flakes on protein nutritive value for human adults, *J. Food Sci.,* 37, 378–380, 1972.

152. **Cohen, H. P., Choitz, H. C., and Berg, C. P.,** Response of rats to diets high in methionine and related compounds, *J. Nutr.,* 64, 555–569, 1958.

153. **Klavins, J. V. and Peacocke, I. L.,** Pathology of amino acid excess. III. Effects of administration of excessive amounts of sulphur amino acids: methionine with equimolar amounts of glycine and arginine, *Br. J. Exp. Pathol.,* 45, 535–547, 1964.

154. **Brown, J. H. and Allison, J. B.,** Effects of excess dietary DL-methionine and/or L-arginine on rats, *Proc. Soc. Exp. Biol. Med.,* 69, 196–198, 1948.

155. **Hardwick, D. F., Applegarth, D. A., Cockroft, D. M., Ross, P. M., and Calder, R. J.,** Pathogenesis of methionine-induced toxicity, *Metabolism,* 19, 381–391, 1970.

156. **Sauberlich, H. E.,** Studies on the toxicity and antagonism of amino acids for weanling rats, *J. Nutr.,* 75, 61–72, 1961.

157. **Grau, C. R. and Kamei, M.,** Amino acid imbalance and the growth requirements for lysine and methionine, *J. Nutr.,* 41, 89–101, 1950.

158. **De Bey, H. J., Snell, E. E., and Baumann, C. A.,** Studies on the inter-relationship between methionine and vitamin B6, *J. Nutr.,* 46, 203–214, 1952.

159. **Benevenga, N. J.,** Toxicities of methionine and other amino acids, *J. Agr. Food Chem.,* 22, 2–9, 1974.

160. **Stekol, J. A. and Szaran, J.,** Pathological effects of excessive methionine in the diet of growing rats, *J. Nutr.,* 77, 81–90, 1962.

161. **Jeanjean, M. and Taper, H.,** Toxicité de la méthionine chez la lapin, *Pathol. Eur.,* 2, 93–104, 1967.

162. **Kinsell, L. W., Harper, H. A., Giese, G. K., Margen, S., McCallie, D. P., and Hess, J. R.,** Studies in methionine metabolism. II. Fasting plasma methionine levels in normal and hepatopathic individuals in response to daily methionine ingestion, *J. Clin. Invest.,* 28, 1439–1450, 1949.

163. **Floyd, J. S., Fajans, S. S., Conn, J. W., Knopf, R. F., and Rull, J.,** Stimulation of insulin secretion by amino acids, *J. Clin. Invest.,* 45, 1487–1501, 1966.

164. **Perry, T. L., Hardwick, D. F., Dixon, G. H., Dolman, C. G., and Hansen, S.,** Hypermethioninaemia: A metabolic disorder associated with cirrhosis, islet cell hyperplasia and renal tubular degeneration, *Paediatrics,* 36, 236–250, 1965.

165. **Kies, C., Fox, H., and Aprahamian, S.,** Comparative value of L-, DL- and D- methionine supplementation of an oat-based diet for humans, *J. Nutr.,* 105, 809–814, 1975.

166. **Baker, D. H., Clausing, W., Harmon, B. G., Jensen, A. H., and Becker, D. E.,** Replacement value of cystine for methionine for the young pig, *J. Anim. Sci.,* 29, 581–584, 1969.

167. **Walker, D. M. and Kirk, R. D.,** Methionine supplementation of milk protein for pre-ruminant lambs. 1. Effect of protein concentration and source of carbohydrate on nitrogen balance, *Aust. J. Agric. Res.,* 26, 673–679, 1975.

168. **Berry, T. H., Becker, D. D., Rasmussen, O. G., Jensen, A. H., and Norton, H. W.,** The limiting amino acids in soybean protein, *J. Anim. Sci.,* 21, 558–561, 1962.

169. **Katz, R. S. and Baker, D. H.,** Toxicities of various organic sulfur compounds for chicks fed crystalline amino acid diets containing threonine and glycine at their minimal dietary requirements for maximal growth, *J. Anim. Sci.,* 41, 1355-1361, 1975.

170. **Harper, A. E., Benevenga, N. J., and Wohlhueter, R. M.,** Effects of ingestion of disproportionate amounts of amino acids, *Physiol. Rev.,* 50, 428−558, 1970.

171. **Klain, G. J., Vaughan, D. A., and Vaughan, L. N.,** Some metabolic effects of methionine toxicity in the rat, *J. Nutr.,* 80, 337−341, 1963.

172. **Daniel, R. G. and Waisman, H. A.,** Adaptation of the weanling rat to diets containing excess methionine, *J. Nutr.,* 99, 229−306, 1970.

173. **Girard-Globa, A., Robin, P. and Forestier, M.,** Long term adaptation of weaning rats to high dietary levels of methionine and serine, *J. Nutr.,* 102, 209−217, 1972.

174. **Phear, E. A., Ruebner, B., Sherlock, S., and Summerskill, W. H.,** Methionine toxicity in liver disease and its prevention by chlorotetracycline, *J. Clin. Sci.,* 15, 93−117, 1956.

175. **Klavins, J. V., Kinney, T. D., and Kaufman, N.,** Histopathologic changes in methionine excess, *Arch. Pathol.,* 75, 661−673, 1963.

176. **Reis, P. J.,** The growth and composition of wool. IV. The differential response of growth and of sulphur content of wool to the level of sulphur containing amino acids given per abomasum, *Aust. J. Biol. Sci.,* 20, 809−825, 1967.

177. **Reis, P. J., Tunks, D. A., and Downes, A. M.,** The influence of abomasal and intravenous supplements of sulphur-containing amino acids on wool growth rate, *Aust. J. Biol. Sci.,* 26, 249−258, 1973.

178. **Doyle, P. T. and Bird, P. R.,** The influence of dietary supplements of DL-methionine on the growth rate of wool, *Aust. J. Agric. Res.,* 26, 337−342, 1975.

179. **Langlands, J. P.,** Efficiency of wool production of growing sheep. 3. The use of sulphur-containing amino acids to stimulate wool growth, *Aust. J. Exp. Agric. Anim. Husb.,* 10, 665−671, 1970.

180. **Reis, P. J.,** The influence of abomasal supplements of some amino and sulphur containing compounds on wool growth rate, *Aust. J. Biol. Sci.,* 23, 441−446, 1970.

181. **Nimrick, K., Hatfield, E. E., Kaminski, J., and Owens, F. N.,** Qualitative assessment of supplemental amino acid needs for growing lambs fed urea as the sole nitrogen source, *J. Nutr.,* 100, 1293−1300, 1970.

182. **Wakeling, A. E., Lewis, D., and Annison, E. F.,** The amino acid requirements of ruminants, *Proc. Nutr. Soc.,* 29, 60A, 1970.

183. **Langlands, J. P.,** Methionine hydroxy analogue as a dietary supplement for sheep, *Proc. Aust. Soc. Anim. Prod.,* 9, 321−325, 1972.

184. **Salsbury, R. L., Marvel, D. K., Woodmansee, C. W., and Haenlein, G. F.,** Utilization of methionine and methionine hydroxy analog by rumen micro organisms in vitro, *J. Dairy Sci.,* 54, 390−396, 1971.

185. **Langar, P. N., Buttery, P. J., and Lewis, D.,** N-steroyl-DL-methionine − a new form of protected methionine for ruminant feeds, *Proc. Nutr. Soc.,* 32, 86A, 1973.

186. **McLaren, G. A., Anderson, G. C., and Barth, K. M.,** Influence of methionine and tryptophan on nitrogen utilization by lambs fed high levels of non-protein nitrogen, *J. Anim. Sci.,* 24, 231−234, 1965.

187. **Griel, L. C., Patton, R. A., McCarthy, R. D., and Chandler, P. T.,** Milk production response to feeding methionine hydroxy analog to lactating dairy cows, *J. Dairy Sci.,* 51, 1866−1868, 1968.

188. **Loosli, J. K. and Harris, L. E.,** Methionine increases the value of urea for lambs, *J. Anim. Sci.,* 4, 435−437, 1945.

189. **Machlin, L. J. and Gordon, R. S.,** Effect of protein level on the requirement of the chick for non-specific nitrogen, glycine and the sulfur amino acids, *Poult. Sci.,* 36, 1137, 1957.

190. **Chow, K. and Walser, M.,** Effects of substitution of methionine, leucine, phenylalanine, or valine by their hydroxy-analogs in the diet of rats, *J. Nutr.,* 105, 372−378, 1975.

191. **Finklestein, J. D. and Mudd, S. H.,** Trans-sulfuration in mammals. The methionine-sparing effect of cystine, *J. Biol. Chem.,* 242, 873−880, 1967.

192. **Katz, R. S. and Baker, D. H.,** Methionine toxicity in the chick: Nutritional and metabolic implications, *J. Nutr.,* 105, 1168, 1975.

193. **Sprince, H.,** Metabolic aspects and inter-relationships of methionine and tryptophan: behavioural implications, *Nutr. Rep. Int.,* 1, 243−261, 1970.

194. **Sprince, H.,** Congeners of gammo-hydroxybutyrate in relation to depressant and convulsant activity, *Biol. Psychiatry,* 1, 301−315, 1969.

195. **Njaa, L. R.,** Utilization of methionine sulphoxide and methionine sulphone by the young rat, *Br. J. Nutr.,* 16, 571−577, 1962.

196. **Virtue, R. W. and Doster-Virtue, M. E.**, Studies on the production of taurocholic acid in the dog. V. Methionine sulfoxide, *J. Biol. Chem.*, 137, 227-231, 1941.

197. **Bennett, M. A.**, The replaceability of D,L-methionine in the diet of albino rats with its partially oxidized derivative D,L-methionine sulphoxide, *Biochem. J.*, 33, 1794–1797, 1939.

198. **Case, G. L. and Benevenga, N. J.**, S-methylcysteine as a methionine analogue in methionine toxicity studies, *Fed. Proc.*, 31, 715, 1972.

199. **Canellakis, E. S. and Tarver, H.**, The metabolism of methyl mercaptan in the intact animal, *Arch. Biochem. Biophys*, 42, 446–455, 1953.

200. **Mudd, S. H.**, Errors of sulfur metabolism, in *Symposium: Sulfur in Nutrition*, Muth, O. H. and Oldfield, J. E., Eds., Avi Publishing Westport, Conn., 1970, Chap. 15.

201. **Graber, G. and Baker, D. H.**, Sulfur amino and nutrition of the growing chick: quantitative aspects concerning the efficacy of dietary methionine, cysteine and cystine, *J. Anim. Sci.*, 33, 1005–1011, 1971.

202. **Sowers, J. E., Stockland, W. L., and Meade, R. J.**, L-methionine and L-cysteine requirements of the growing rat, *J. Anim. Sci.*, 35, 782–788, 1972.

203. **Baker, D. H., Becker, D. E., Norton, H. W., Jensen, A. H., and Harmon, B. G.**, Quantitative evaluation of the tryptophan, methionine and lysine needs of adult swine for maintenance, *J. Nutr.*, 89, 441–447, 1966.

204. **Graber, G., Scott, H. M., and Baker, D. H.**, Sulfur amino acid nutrition of the growing chick: effect of age on the capacity of cystine to spare dietary methionine, *Poult. Sci.*, 50, 1450–1455, 1971.

205. **Thomas, W. E., Loosli, J. K., Williams, H. H., and Maynard, L. A.**, The utilization of inorganic sulfates and urea nitrogen by lambs, *J. Nutr.*, 43, 515–523, 1951.

206. **Whanger, P. D. and Matrone, G.**, Effect of sulfur deficiency on metabolism in sheep, in *Symposium: Sulfur in Nutrition*, Muth, O. H. and Oldfield, J. E., Eds., Avi, Publishing Westport, Conn., 1970, Chap. 10.

207. **Du Vigneaud, V. and Rachele, J. R.**, The concept of transmethylation in mammalian metabolism and its establishment by isotopic labeling through in viro experimentation, in *Transmethylation and Methionine Biosynthesis*, Shapiro, S. K. and Schlenk, F., Eds., University of Chicago Press, Chicago, 1965, Chap. 1.

208. **Byington, M. H., Howe, J. M., and Clark, H. E.**, Effects of different levels and proportions of methionine, cystine, choline and inorganic sulfur on growth and body composition of young rats, *J. Nutr.*, 102, 219–228, 1972.

209. **Fulk, D. W.**, Effects of methionine supplementation on wool and body growth and various blood constituents in nursing and weaned lambs, *Diss. Abstr. Int.*, 32, 3719–B, 1972.

210. **Sprince, H., Parker, C. M., and Josephs, J. A.**, Homocysteine-induced convulsions in the rat: Protection by homoserine, serine, betaine, glycine and glucose, *Agents Actions*, 1, 9–13, 1969.

211. **Huisingh, J., Milholland, D. C., and Matrone, G.**, Effect of molybdate on sulfide production from methionine and sulfate by ruminal micro-organisms of sheep, *J. Nutr.*, 105, 1199–1205, 1975.

212. **Rose, W. C. and Wixom, R. L.**, Amino acid requirements of man. Sparing effect of cystine on methionine requirement, *J. Biol. Chem.*, 216, 763–773, 1955.

213. **Patton, R. A., McCarthy, R. D., and Griel, L. C.**, Observations on rumen fluid blood serum and milk lipids of cows fed methionine hydroxy analog, *J. Dairy Sci.*, 53, 776–780, 1970.

214. **Polan, C. E., Chandler, P. T., and Miller, C. N.**, Methionine: Varying levels for lactating cows, *J. Dairy Sci.*, 53, 607–610, 1970.

215. **Steinacker, G., Devlin, T. J., and Ingalls, J. R.**, Effect of methionine supplementation posterior to the rumen on nitrogen utilization and sulfur balance of steers on a high roughage ration, *Can. J. Anim. Sci.*, 50, 319–324, 1970.

216. **Bouchard, R. and Conrad, H. R.**, Sulfur requirements of lactating dairy cows. II. Utilization of sulfates, molasses and lignin-sulfonate, *J. Dairy Sci.*, 56, 1429–1434, 1973.

217. **Kroger, D. and Carroll, F. D.**, Possible mechanism of agricultural gypsum in regulating appetite. *J. Anim. Sci.*, 23, 1011–1015, 1964.

218. **Wagnon, K. A.**, Agricultural gypsum as a regulator for self-feeding cottenseed meal to cattle on the range, *J. Range Manage.*, 13, 279–280, 1960.

219. **Mal'tsev, V. S.**, Profilaktika i lechenie trikhofitii krupnogo rogotogo skota putem skarmlivaniya sery. *Probl. Veterinarnoi Sanitarii.*, 42, 119–122, 1972.

220. **Julien, R. J. and Harrison, K. B.**, Sulphur poisoning in cattle, *Can. Vet. J.*, 16, 28–29, 1975.

221. **Weigand, E., Kirchgessner, M., Granzer, W., and Ranfft, K.**, Zur Fütterunghoher Sulfitmengen an Milchkühe. SO_2 – Verträglichkeit and SO_2 – Gehalt der Milch, *Zentralbl. Veterinaermed.*, 19A, 490–501, 1972.

THE EFFECT OF NUTRIENT TOXICITIES
IN ANIMALS AND MAN: IRON*

William H. Crosby

IRON METABOLISM[1-4]

Its pervasiveness on our planet together with the chemical flexibility of its atom involve the element iron in many common inorganic compounds and many kinds of organic reactions. Properly harnessed, the reactivity of iron in biologic systems is essential for the movement of oxygen and to several enzymes of intermediary metabolism. Unharnessed ionic iron is toxic, the severity of its toxicity depending, of course, upon the dose. Large doses cause death of both plants and animals. To control or contain its toxicity, iron is normally moved through biological systems escorted by some sort of carrier; biologically uncommitted iron is normally held in cells by some sort of biologically neutral storage mechanism. It is well recognized, however, that excessive amounts of iron cannot be safely stored. Long-term presence of excessive amounts of iron causes damage and disease of the storage organs.

Distribution of Iron in the Body

Mammalian iron metabolism has been extensively investigated. Iron bound by porphyrin in combination with various proteins provides a number of important pigments which manipulate oxygen within the body. For example, the iron in hemoglobin transports oxygen in the blood. The iron in myoglobin stores oxygen in muscle. The iron in the enzymes catalase and peroxidase removes the oxygen from hydrogen peroxide, thereby converting a dangerously oxidative compound into water. These are examples of physiologically committed iron. Uncommitted iron is in transit or in storage. In the plasma, iron is bound by a carrier protein, transferrin. In tissues, excess iron is deposited in the storage protein ferritin. Excesses of storage iron may overflow the ferritin to form microscopically visible aggregates of hemosiderin.

The total amount of iron in a normal adult human is about 4 g. Most of it is invested in hemoglobin. Up to 1 g is in storage, much of which is in the liver and the rest in the reticuloendothelial system. The remaining small amounts of iron are physiologically committed in every cell of the body (Table 1).

Obligatory Metabolic Loss[5]

Any cell lost from the body carries with it a tiny decrement of iron. Desquamation of surface epithelial cells from skin, intestine, and urinary system loses about 1 mg of iron per day, an obligatory metabolic loss. Even iron-deficient people must lose this iron.

Table 1
IRON COMPARTMENTS IN NORMAL 70-Kg MAN

	mg
Hemoglobin	2250
Storage	500–1000
Myoglobin	130
Marrow and spleen	100 (hemoglobin production and destruction)
Enzymes	8
Transferrin	3

* Supported by NIH Contract AM-16452 from the National Institutes of Health.

Because of the high concentration of iron in hemoglobin, the shedding of 1 ml of red blood cells loses the same amount of iron, 1 mg. Thus, bleeding is, by far, the easiest avenue for the loss of iron and the most common cause of iron deficiency. Controlled blood loss is the easiest way to treat iron overload.

Control of Iron Absorption[6,7]

Our environment contains an abundance of iron, and everything we eat or drink contains some iron. It is estimated that the American diet contains about 6 mg of iron per 1000 cal, which means that most Americans ingest 10 to 20 mg of iron per day in food. Their requirement is considerably less. Normal men require about 1 mg to offset the obligatory 1 mg lost with cells that are shed. Normal women require, on the average, about 2 mg/day because of the additional loss of hemoglobin at menstrual periods. Metabolic balance of essential nutrient minerals other than iron (sodium, calcium, and copper, for example) is maintained by excreting whatever excess amounts may be absorbed. This is not true for iron because there is no effective excretory mechanism. When our bodies have too much iron, the excess cannot be excreted and it must be placed in storage. The mechanism for maintaining iron balance resides in the ability of the intestine to refrain from absorbing available dietary iron that is not needed. Most iron absorption occurs in the duodenum and upper jejunum. The absorptive ability of these mucosal surfaces is somehow, by some mechanism that we do not understand, tuned to the body's requirement. The need for iron is increased during periods of growth when blood volume and hemoglobin mass are expanding. Iron requirement is increased during pregnancy. A normal fetus, together with the blood loss at delivery, requires about 700 mg of iron — an extra 2 or 3 mg/day during the pregnancy. Iron requirement is also increased by blood loss. A unit of blood contributed to a blood bank (480 ml) contains 200 to 250 mg of iron. The donor becomes temporarily iron deficient by this amount, but after the donation he absorbs more than the usual amount of iron from his diet and restores his iron status to normal in 2 to 3 months.

Lost Control of Iron Absorption[7,8]

The normal small intestine is informed about the body's requirements for iron and is somehow able to translate the information into appropriate behavior, absorbing more iron from the diet when it is needed and rejecting available iron which is not needed. Any prolonged inability of the intestine to reject available, unneeded iron in a normal diet results in the absorption of iron which cannot be excreted. This excess is deposited in storage organs. The deposition of the iron pigment eventually results in damage to these organs, causing complications such as diabetes, cirrhosis, sterility, and heart disease. This is the sort of disease caused by chronic iron intoxication.

Iron Excretion[2,5]

The ability of the normal human to excrete excess iron is limited, but, nevertheless, slightly more than the obligatory metabolic loss can be excreted. For example, some of the glandular epithelial cells are able to accept more iron than they require for their internal metabolism and the excess is stored as ferritin and hemosiderin. These siderotic cells have a finite life span; when they die and fall off the surface of the body, their accumulated iron is lost with them. The tissue macrophages are self-propelled scavenger cells which can accumulate, among other debris, excess amounts of iron. When these cells become heavily loaded with iron, they come to the intestinal mucosa and crawl into the lumen, carrying with them their bits of hemosiderin. These two populations of cells provide a means whereby excess iron in small amounts can be removed from the body. There may be other yet undiscovered small pathways. It is suspected that the maximum total capacity of the excretory mechanisms is of the order of 2 to 5 mg/day.[9]

FERRITIN [10,11]

Organisms in the biosphere of our planet lead a precarious existence with respect to iron. Because iron is a protean constituent of our environment, a problem common to all plants and animals is to accept enough iron from the environment, but not too much. However, should an organism receive too much iron, it has the means, within limits, to neutralize the excess by forming ferritin.

Ferritin is a protein with a molecular weight of about 500,000 which is capable of accepting for storage more than 20% of its weight in iron. Ferritin is a spherical molecule composed of 24 spherical subunits arranged around a core of ferric iron. The ability to synthesize ferritin is common to all forms of cellular life. It is not synthesized unless it is needed, and the presence of excess iron is the stimulus that switches on ferritin synthesis.

In some cellular systems the stored iron can be withdrawn, the ferritin thereby functioning as a reservoir; however, in many cells the ferritin iron remains in storage until the cell dies. Ferritin is a metabolic sink within which iron not needed by the cell is neutralized. Its widespread employment in all cellular types indicates that the ability to synthesize ferritin must have been achieved very early in the evolution of life. The ferritin system is a primitive device, essential for survival, and obviously a biological success.

The ferritins in different organisms vary, distinguishable by physical and immunologic differences. Even in one species, the ferritin extracted from different organs is not the same. Furthermore ferritin from the same organ of one animal can sometimes be separated into several isoferritins.[12] The basis for these subtle physicochemical differences has not yet been established.

HEMOSIDERIN AND SIDEROSIS [13,14]

"Haemosiderin is a substance occurring in the form of intra or extracellular deposits that (1) are visible in the light microscope as brown granules (2) contain trivalent iron as demonstrated by histochemical tests such as Perls' test, and (3) often contain variable amounts of ferritin as revealed by electron microscopy."[15]

Normal storage iron, most of which is held by ferritin in the liver, is thinly spread and invisible in tissue sections examined under the light microscope. When the load of storage iron within a cell increases beyond the normal capacity of the ferritin system, it forms aggregates which become large enough to be visible by light microscopy. This visible aggregation of storage iron is called hemosiderin. Hemosiderin may consist of large amounts of ferritin or it may be composed of ferritin plus accumulations of ferric hydroxides which form when the ferritin mechanism becomes massively overloaded in the iron storage diseases. Most hemosiderin is intracellular, but in heavily iron-loaded tissues siderotic cells may die and then their ferruginous contents then coalesce to form large or small extracellular deposits. The presence of hemosiderin in a cell or tissue is called hemosiderosis or siderosis.

Causes of Siderosis
Siderosis, the accumulation as hemosiderin of excess amounts of iron in tissues, occurs as a consequence of several kinds of abnormal situations.

Failure to Control Absorption of Iron[16]
Hereditary hemochromatosis — Hemochromatosis is characteristic of a group of hereditary abnormalities whereby the intestine lacks the normal ability to keep out available but unneeded dietary iron. Unusable iron enters the body, and because of limited excretory capacity the excess must be placed in storage.
Acquired hemochromatosis — A similiar intestinal abnormality sometimes occurs with

acquired diseases. People with Laennec's alcoholic cirrhosis may accumulate abnormal amounts of iron from a normal diet. The same is true of some patients with primary acquired sideroblastic anemia, a neoplastic disorder of the bone marrow characterized by the presence of erythroblasts containing microscopically visible granules of iron which are often arranged in a ring around the nucleus, "ring sideroblasts."

Excessive Quantities of Ingested Iron

Medicinal iron[17-19] — Because iron deficiency is the most common cause of anemia, some physicians indiscriminately treat every case of anemia, whatever its cause, with iron. Furthermore, many people who do not need iron ingest excessive amounts because urged to do so by advertising of iron-loaded nostrums. Some of these people who are not iron deficient accumulate large quantities of the medicinal iron. There are some recorded cases of patients who, early in life, visited a doctor and, acting upon his advice and without further inquiry, continued to take iron pills indefinitely. It goes without saying that iron medication can benefit only those anemias that are caused by iron deficiency. Further, iron is not a harmless pill to be used as a placebo.

Dietary iron[20,21] — It is possible to fortify the diet with iron to such an extent as to cause siderosis. The best known example of this is among the indigenous Bantu-speaking Blacks of South Africa. The people of this culture have been studied extensively by hematologists of the University of Witwatersrand in Johannesburg. Excess iron in the Bantu diet is a consequence of preparing food and drink in iron vessels in such a way that relatively large amounts of iron leach into the porridge (cooked in soft iron pots) and beer (brewed in sheet-steel cans). Both of these staples are acidic; the beer is only mildly alcoholic (1.5%). The average amount of iron ingested daily is about 100 mg, but in some cases it is as great as 200 mg. The acidity of the food tends to retain the iron in an ionic form which, together with the effects of alcohol, facilitates the absorption of iron. The unintentional fortification of the diet results in widespread iron-storage disease among the Bantu. There is concern in the U.S. that intentional, indiscriminate fortification of the national diet may result in an unintentional increase of the incidence of iron-storage disease among Americans.

In Ethiopia, the native diet contains iron in excess of 300 to 400 mg/day. This is due to contamination with dirt of the grain used for basic food. The dirt that gets into the grain during threshing is rich in iron, but it is present as unabsorbable oxides which, therefore, are neither nutrient nor toxic.

Combination of Genetic Predisposition and Excess Ingested Iron

Clinical experience and nutritional surveys indicate a range of differences in ability to withstand iron overdose. Twenty percent of the overdosed Bantu do *not* accumulate excessive storage iron. In the U.S., some people have taken iron pills every day for 20 years without developing siderosis. On the other hand, patients have developed severe hemochromatosis on what must be considered an iron-deficient diet. In considering the risks of siderotic disease, the likely victim of a moderately iron-loaded diet would be the person with some genetic debility of his control of iron absorption and iron balance. In this type of person, a moderate excess of dietary iron could result in positive iron balance. On a fortified diet, he would accumulate iron; on an unfortified diet, he might not.

Injections of Therapeutic Iron

Patients with anemia who do not respond to orally administered iron are often and erroneously suspected of inability to absorb iron. (Failure to respond is almost always because the anemic patient is not iron deficient; the anemia has another basis.) These patients are sometimes treated by injections of iron. The iron is not needed, and it cannot be excreted. Placed into storage, it causes siderosis.

Transfusion

The hemoglobin of a unit of blood contains 200 to 250 mg of iron. Some anemic patients require repeated transfusions in order to survive. When the transfused red cells are ultimately destroyed, the hemoglobin iron is not reutilized if the anemia is not due to blood loss. Therefore, it goes into storage, causing siderosis.

FOOD FORTIFICATION [22-24]

In the 1960s the American diet was estimated to contain 6 mg of iron per 1000 cal. Extrapolation of this estimate indicated that the average American woman received 12 mg of iron per day, while the average man received 18 mg or more. Since that time, iron fortification of foods has become a preoccupation of the food industry. It is advertised as a virtue that a single serving of fortified cereal or breakfast beverage contains the complete daily requirements of iron. This increased fortification of foods has occurred at the same time that the additives used to fortify foods were changed from iron phosphate compounds, which are almost totally unassimilable. Much of the iron is currently added as ferrous sulfate and reduced iron, which are easily assimilable. Thus, the total of assimilable iron in the American diet has been increased on two counts: the total amount of supplementation and the absorbability of the supplement. Further increases in the intensity of fortification are contemplated. The food industry and the Food and Drug Administration have seriously proposed that the amount of iron *required* to be added to all enriched flour and bread be increased 300 to 400%. This alone would bring the daily intake of some citizens into the range of 100 mg of elemental iron per day, the amount ingested by South African Blacks who develop siderosis.

RESPONSES TO LARGE AMOUNTS OF IRON

Treatment of Iron Deficiency Anemia[25]

Iron deficiency anemia is usually treated with large doses of iron salts. Ferrous sulfate, for example, in 300-mg tablets provides 60 mg of elemental iron. One to four of these tablets per day may be prescribed. A patient with moderate to severe iron deficiency anemia can absorb about 35 mg of a 250-mg daily dose of iron and incorporate the iron into new hemoglobin, thereby enlarging the red cell mass, day by day, and correcting the anemia. Patients with mild iron deficiency absorb less than patients with severe anemia.

Unneeded Therapy with Iron

A normal person given a single large dose of ferrous sulfate absorbs a good amount of it, much more than is required to offset the obligatory 1-mg daily metabolic loss.[26] This one-shot event does not, however, represent what must happen when the large dose is repeated every day for many months or years. In an experiment by Finch et al.,[27] medical students were given three tablets of 300-mg ferrous sulfate each day for 6 months to 1 year. The load of iron in their storage organs was then measured and was found not to be in excess of the normal 1 g. One must conclude that the repetition of the large dose of iron conditioned the intestine so that it did not absorb more iron than the body required. The same is true of some patients with anemia. In my practice I have encountered patients who, early in their lives, were found to have anemia; on the orders of physicians, they took therapeutic iron daily for many years thereafter. In each case, there was no history or indication of blood loss. For example, a man with hereditary spherocytosis had taken iron for 37 years. He had consumed bottle after bottle of ferrous sulfate tablets. When I examined him, the serum iron and iron-binding capacity were normal and he had a flat iron tolerance test: the oral dosing with 120 mg of iron as ferrous sulfate did not produce any increase of serum iron concentration. How can we reconcile these two phenomena: the inability of normal people to reject iron from a single large dose and their ability to take large repeated doses without accumulating iron? Iron balance is not maintained by excretion of excess iron. Therefore, we surmise that the repeated dose of iron results in the creation of an intestinal barrier to the absorption of unneeded iron even when massive amounts are ingested.[28] This adaptation occurs in some people, but not all.

On the other hand, some patients who have taken unneeded iron tablets for many years did develop hemochromatosis. Wallerstein and Robbins reported the case of a man with hereditary spherocytosis who, after taking iron pills for 10 years, was found to have hemochromatosis.[18] It seems likely that such patients do not have a normal intestinal mechanism for the control of iron absorption. When the anemia for which the iron is mistakenly prescribed is an "iron-loading anemia," the administration of therapeutic iron accelerates the rate at which the iron accumulates.

Large Doses of Dietary Iron

Among the South African Bantu-speaking Blacks, dietary intake of iron may be 100 mg or more per day.[20] Of these people, 80% show some excessive storage iron postmortem; 20% have a severe excess, and 4% have siderosis in a distribution and amount which is indistinguishable from hereditary hemochromatosis.[29]

The reason for the variability of the intestine to control absorption of dietary iron involves many factors.

1. The dose of iron is not the same for all.[20]
2. Alcohol, taken with food, facilitates the absorption of iron. The habitual use of alcohol by a person with the hemochromatosis gene may tip the balance so that he absorbs more iron that he can excrete.[30]
3. Some develop alcoholic cirrhosis, and patients with cirrhosis often lose the ability to control absorption of iron.[31]
4. Other dietary deficiencies or habits and other diseases may affect the control of iron balance. For example, orange juice, taken with a meal, facilitates food iron absorption because it contains ascorbic acid.[32]
5. People with the gene of hemochromatosis may not develop the disease when the amount of iron in the diet is normal. They may absorb a small excess of iron, but are able to excrete it. When the dietary load is very large, however, they may absorb more than they can excrete and iron-storage disease results.
6. Fully expressed hereditary hemochromatosis results in accumulation of excessive iron from a normal diet.[33]

Any of these pathogenetic mechanisms may play a part among the Bantu. The dose of dietary iron, the use and abuse of alcohol, and the presence or absence of the hemochromatosis gene and its varied expressivity may all contribute to the diversity of iron-storage disease. On the one end of the spectrum, 20% of the Bantu have no excess of storage iron while, on the other end, 20% have heavily iron-laden livers together with cirrhosis.

In any population, African or American, the presence of the hemochromatosis gene in combination with environmental factors such as heavily iron-fortified food or the abuse of alcohol or both can result in variable rates of clinically recognizable complications of iron-storage disease.

An example of the consequences of food fortification with iron may be forthcoming from Sweden, a country where for 30 years national policy has required a level of enrichment that increases the intake of iron over that in the natural food by 42%. Recently, a survey was conducted in a small community on all persons aged 30 through 39.[34] Of 197 men, 9 had persistently high serum iron levels and abnormal indices of saturation (>50%). Of these, eight were given a desferal iron excretion test and four had grossly abnormal results. Liver biopsy demonstrated all four to have excessive iron stores.

HEMOCHROMATOSIS[16,35-37]

Hemochromatosis is the iron-storage disease that results from the inability of the intestine to keep out unneeded iron; the iron accumulates, causing siderosis and, ultimately, damage of the storage organs. Hemochromatosis thus defined is a complication of a number of different hereditary abnormalities. The mechanism that keeps out unnecessary iron is complex and probably can be disrupted at a number of different points. Disruption at any point might result in the absorption of unnecessary iron.

Several varieties of hereditary anemia are complicated by hemochromatosis.[38]

Thalassemia major (Cooley's anemia) — This is the homozygous form of β thalassemia. In addition to a severe hypochromic anemia, the affected children absorb unneeded iron. The requirement for transfusions adds to the burden of iron-storage disease, but the children develop hemochromatosis even with no transfusions,[39] and it is the hemochromatosis that kills them. Most die of myocardial siderosis. People with heterozygous β thalassemia (thalassemia minor) do not accumulate iron from ordinary diets.

Atransferrinemia — This rare congenital absence of transferrin (the plasma's iron transporting protein) produces hemochromatosis in association with a severe iron-deficiency anemia.[40] The intestine absorbs an increased amount of iron which is put into storage and cannot be recovered from storage because of the absence of the transport protein.

Hereditary hemolytic anemia — Hereditary spherocytosis and hereditary nonspherocytic hemolytic anemias are usually not complicated by hemochromatosis. In some families, however, iron-storage disease is associated with the hemolytic disease.[18,19,41-43]

Sex-linked hypochromic anemia — This moderate to severe anemia may be associated with severe hemochromatosis in young men.[44] In some families the anemia is partially responsive to large doses of pyridoxine.[46]

It is evident that all of the above diseases are genetic in etiology and are easily distinguishable from one another. We surmise, therefore, that several different genetic mechanisms can result in the syndrome of hemochromatosis.

Acquired hemochromatosis — Disruption of the intestinal control of iron balance is, in some conditions, acquired. In neoplastic disorders of the bone marrow characterized by sideroblastic anemia, the ability to keep out unneeded iron may be lost.[47] This is also true of 5 to 10% of patients who develop alcoholic cirrhosis.[31]

Idiopathic hemochromatosis[33,48] — This commonly used term is a misnomer. Hemochromatosis, as it usually occurs, unassociated with anemia, is hereditary, not idiopathic. It is usually discovered in middle life, or later, when the massive accumulation of storage iron has caused a serious disease that brings the patient to medical attention. This end stage of the disease may be called florid hemochromatosis. Because of the hereditary nature of the disease, the family of each patient should be investigated to discover other cases, often in people who are not yet ill (preclinical hemochromatosis). These people can be treated and serious illness prevented. Family studies have demonstrated several patterns of relationship. Siderosis has been demonstrated in successive generations, an affected parent transmitting the fault to a child who, in some cases, has transmitted it to the third generation. However, when first-degree relatives are examined, it is more common to find the disease in siblings than in parents or children. When both parents of affected offspring show no evidence of iron-storage disease, we may suspect that one or both are genetic carriers of hemochromatosis. The differences in the patterns of diagnosable disease have resulted in confusion and argument about the genetic transmission of hemochromatosis. The involvement of successive generations is evidence for an autosomal dominant pattern of inheritance. The demonstration of two unaffected parents of severely affected children is evidence for an autosomal recessive pattern. Penetrance, expressivity, and environmental stresses also modify the one criterion by which we can recognize the presence of the disease: siderosis, the accumulation of abnormal amounts of storage iron. We have, as yet, no genetic marker by which we can identify unaffected carriers of the hemochromatosis gene.

Adding to the confusion is the probability that hereditary hemochromatosis is not a single genetic disease. What we recognize in the affected person is siderosis, the consequence of a failure of a complex control mechanism to keep out available unneeded iron. The complex system might fail at any one of five or ten or more places. Each of these points of potential failure may be under separate genetic control. Conceived thus, hemochromatosis is a family of hereditary diseases. There may be several patterns of inheritance.

Hemochromatosis in Women

It has long been recognized that hemochromatosis is more common in men than in women by a factor of perhaps 10 to 1.[35] There are various ways in which this may come about.

1. Some genetic variants may have a sex-linked pattern of inheritance.
2. Expression of the gene may be inhibited in women.
3. Women eat less food than men; therefore, they eat less iron.
4. Women lose iron in menstruation, which helps to correct any positive metabolic balance.

Identifying Hemochromatosis

We have no marker that can identify every person who carries the gene of hereditary hemochromatosis. To diagnose the disease, we look for abnormalities of iron metabolism. However, an abnormal test has no valid interpretation, other than to arouse suspicion, unless it indicates or is associated with abnormal amounts of iron in tissue stores.

Serum Iron, Total Iron-binding Capacity (TIBC), Saturation Index[49]

In most cases of hemochromatosis, serum iron is high, outside a normal upper limit of 160 μg/dl. However, serum iron may be in the normal range when, for example, the patient with hemochromatosis suffers from an inflammatory disease. TIBC of the transferrin in serum is often low or low normal (less than 300 μg/dl). Saturation index, the percentage of TIBC that contains iron, is high, greater than 50 to 60% and usually close to 90%. Even when serum iron is low, saturation is high. Among these three indicators, the saturation index is the most significant. Because of the diurnal variation of serum iron (normally higher in the morning than at night) and other natural variations, a single abnormal test should not be relied upon.

Serum Ferritin[50,51]

The function in the circulating blood of a trace amount of the iron-storage protein ferritin is not yet known, nor are its origin, destination, and half-life. The concentration does, however, serve as a useful index of the amount of storage iron in tissues. Thus, in iron deficiency the serum ferritin is uniformly low. In most people with iron-storage disease, the ferritin concentration is elevated. It was hoped that the measurement of serum ferritin might provide a reliable screening device to search out cases of siderotic hemochromatosis during epidemiologic nutritional surveys. It has been demonstrated, however, that in some families, people with preclinical hemochromatosis (heavily loaded with iron but, as yet, without resulting complications, e.g., diabetes or cirrhosis) have normal serum ferritin levels.[52] Thus, when serum ferritin level is high, it represents iron-storage disease, but iron stores may be substantially increased without elevation of the serum ferritin. In the surveys, the test will find some cases of hemochromatosis, but not all of them. The approximate normal range of serum ferritin is 10 to 100 ng/ml for women and 20 to 300 ng/ml for men.

Deferoxamine (DF) Iron-excretion Test

DF is an iron chelator of microbial origin. Each molecule can combine tightly with one atom of iron. Injected DF combines with available iron to form feroxamine, which is then excreted into the urine. DF is employed to treat acute iron poisoning. Its use in the treatment of transfusion-induced iron-storage disease is still experimental. For the treatment of hemochromatosis without anemia, it is vastly inferior to phlebotomy. However, intramuscular injection of 500 mg of DF has been employed as an indicator of the amount of storage iron. After the injection, all urine is collected for 24 hours and the amount of iron excreted is measured. The test is without value for diagnosing iron deficiency because of overlap between the normal group and the iron deficient. When iron storage disease is present, iron excretion usually exceeds the normal range considerably. Ascorbic acid deficiency may interfere with the efficiency of DF, and, for reasons yet undiscovered, some people known to have iron-storage disease do not excrete an abnormal amount of iron after DF injection.[54] Following DF, most normal people excrete in their urine 0.1 to 1.5 mg in 24 hr; most iron-deficient subjects excrete 0 to 1.0 mg and most iron-loaded, ascorbate-replete subjects excrete 3.5 to 20 mg.

Marrow Aspiration[55]

Normal marrow contains a small amount of stainable storage iron (hemosiderin) in its reticuloendothelial cells. It is common practice to examine smears for storage iron whenever a marrow biopsy is done. In iron deficiency, the amount in the marrow is diminished or absent. In iron storage disease, it is increased, except in hereditary hemochromatosis where, more often than not, the amount of visible iron in the marrow is within normal limits. Bone marrow aspiration studies have been performed on small groups of normal young women, and iron deficiency has been diagnosed in 20 to 25% who had less than a "normal" amount of storage iron. It should be understood that an otherwise normal person with any storage iron cannot be regarded as iron deficient in the sense that the body lacks iron for physiologic requirements. Normal storage iron is iron that remains after all physiologic requirements have been met, and any storage iron represents "money in the bank." Marrow aspirates are graded on a microscopist's estimate with a range of 0 to 4+. Zero and trace amounts may be regarded as representing iron deficiency, or impending iron deficiency, while 3 to 4+ represents iron-storage disease.

Liver Biopsy[56]

There is little or no stainable iron in normal liver tissue, even though the liver contains most of the body's storage iron. The normal storage iron is diffusely held as ferritin in aggregates too small to be seen by light microscopy. In iron-storage disease, the amount of visible storage iron, hemosiderin, is variably increased. A fair correlation exists between visual estimates and chemical analysis. The liver stores are contained in epithelial hepatic cells and reticuloendothelial cells (Kupffer cells). In hereditary hemochromatosis, the heavier concentration of iron is in the hepatic cells. In transfusion siderosis, the Kupffer cells are more heavily loaded. Liver biopsy is frequently used to confirm a diagnosis of hemochromatosis when that diagnosis has been suggested by other circumstantial tests such as high serum saturation index, high serum ferritin, or the excretion of large amounts of urinary iron after injection of deferoxamine. Even when the diagnosis of hemochromatosis seems a certainty, it is important to learn if the liver has become fibrotic or cirrhotic.

Iron Absorption Test[57]

Iron deficiency is normally repaired by increasing the amount of iron absorbed from the diet. Even the mild deficiency induced by the donation of a unit of blood has been demonstrated to increase the absorption of a test dose of radioiron, increasing it from about 10% to about 40%. The measurement of iron absorption, although cumbersome, is a most sensitive index of iron deficiency. It may also be a sensitive indicator of hemochromatosis. People with hemochromatosis absorb more iron than normal. This should be true of those with preclinical hemochromatosis and, perhaps, it may be true also of those with compensated hemochromatosis. Research upon this subject is indicated.

Patients heavily iron-loaded with florid hemochromatosis may absorb only a normal amount of a challenge dose of radio iron.[58] However, after a *part* of the excess storage iron has been removed, the same patient then absorbs an abnormally large amount of the radioiron.[59]

Cobalt Absorption Test

The intestine of iron-deficient animals not only absorbs more iron than the normal intestine, but it also absorbs more cobalt and more manganese.[60] This is true of iron-deficient humans who absorb a more-than-normal amount of cobalt and excrete it in their urine. The same is true of patients with hereditary hemochromatosis who, in spite of their iron-loaded condition, absorb and excrete an increased amount of radiocobalt.[61,62] This test has not been widely employed.

The Incidence of Hemochromatosis

Until a test is developed to identify all carriers of the hemochromatosis gene, the incidence of the disorder cannot be determined. We lack a marker with which to identify all people who possess the gene and who, therefore, are liable to develop and transmit the disease. In sickle-cell disorders, the healthy carrier does not have sickle-cell anemia, but his red cells contain a clinically insignificant but easily demonstrable amount of hemoglobin S. The healthy woman who transmits hemophilia has a demonstrably but insignificantly diminished amount of antihemophilic factor in her blood plasma. We have as yet no such indicator of the existence of hemochromatosis in the healthy carrier.

Case Finding

The existence of hemochromatosis must be recognized by identifying abnormalities of iron metabolism or the damage done thereby. Cases of florid hemochromatosis are found among the groups of patients with cirrhosis, diabetes, refractory heart disease, arthritis, sterility, and impotence. Many cases may be overlooked because hemochromatosis is an uncommon cause of any of these diseases, and the physician, even the specialist, who may encounter one case of hemochromatosis in a year is unlikely to remember "hemochromatosis" on the day when that one patient appears. A patient's wife once told me of a conversation with a doctor. Following his examination of the patient, he said to the wife, "Your husband has a peculiar pigmentation of his skin. One might almost think he had hemochromatosis" — almost but not quite. It was another 5 years before the tests were performed and the diagnosis established.

Finding hemochromatosis in any patient must lead to a search for the disease in other family members (blood relatives). In this way, we can find early and unsuspected cases. Hemochromatosis is one of the small number of hereditary disorders which, once diagnosed, are amenable to preventive measures. People with preclinical disease can be treated, and the severe late complications can be prevented.

The number of cases of hemochromatosis in the U.S. has been estimated from records of the number of cases found at autopsy. These studies were done prior to 1955 and include only those cases which demonstrated the late results of severe iron-storage disease. The figures collected from several large general hospitals indicated a rate of 1 in 7000.[36] This was not the real incidence. There are mild cases, there are people who die young of unrelated diseases, there are "compensated" cases where iron excretion keeps even with a slightly increased absorption, and there are carriers of the gene with no discernible evidence of metabolic disorder. Based upon the crude evidence of 1 in 7000 at postmortem examination, it has been estimated that the gene frequency in the U.S. may be as high as 1 in 50.[33] This figure involves an assumption that "florid" hemochromatosis represents the homozygous state, the gene received from both parents. From Glasgow, a postmortem study extending from 1900 to 1970 indicates the incidence of florid hemochromatosis to be 1 in 500.[63] In Johannesburg, South Africa, the postmortem incidence of the hereditary pattern of iron overload among the Bantu is 4 in 100.[29] These differences may reflect differences in gene frequency among the populations, or they may represent interactions of the gene with environment, involving heavier doses of iron in the diet, more intensive use of alcohol, or other matters of diet or malnutrition not yet linked to the problems of iron metabolism.

INJURY CAUSED BY IRON-STORAGE DISEASES

The persistent presence of a large excess of iron results in injury to certain of the storage organs. Late in the course of hemochromatosis, the liver is often cirrhotic. When the degree of siderosis becomes severe in young people, myocardial disease is a common cause of death; impotence may occur in young men, amenorrhea may occur in young women, and sterility may occur in both. These sexual problems may be secondary to iron loading of the anterior pituitary. Diabetes may be of either pituitary or pancreatic origin. Iron loading of skeletal muscle is not associated with incapacity, and the excess pigment present in thyroid and adrenal glands usually causes no incapacity.

When iron-storage disease is caused by repeated blood transfusions over months and years, the complications of secondary hemochromatosis are not as commonplace. One hundred transfusions given to an anemic patient who is not losing blood adds 20 to 25 g of iron to his body load: when the transfused red cells in each unit of blood reach the end of their life span, they are destroyed and the iron goes into storage. Twenty grams of iron is equivalent to the amount often recovered by phlebotomy from patients with hereditary hemochromatosis; yet, the patient with transfusion siderosis often has no complications of iron-storage disease.[64] Myocardial siderosis provides an exception, particularly in children. Severe heart disease is often associated with excess iron derived from transfusions. The reasons for the difference in pathogenicity of storage iron in the two varieties of disease are not known, but they can be speculated upon.

1. In transfusion siderosis, most of the storage iron is found in reticuloendothelial cells of the liver, spleen, and marrow. In hereditary hemochromatosis, the heavier load is in the epithelial hepatic cells, while the load in the Kupffer cells of the liver and the reticuloendothelial cells in the marrow and spleen is relatively light.[64] The difference in location may diminish the toxicity of the iron. The reticuloendothelial cells may be mobile and capable of moving to the surface of the body (e.g., intestinal mucosa) where the cell and its iron can exit from the body. The hepatic epithelial cell, on the other hand, dies *in situ,* and its load of iron is left to accumulate in the hepatic lobule or at its margin, where it may stimulate the formation of fibrous scar tissue.

2. In most cases of transfusion siderosis, duration of time in storage is briefer. People who depend upon blood transfusion to survive do not, in general, survive for long.

Patients with hereditary hemochromatosis have had a lifetime of excessive iron stores. Duration and magnitude of iron stores must be of some importance.[65]

3. The mechanism for the storage of iron may differ. We do not know the nature of the fundamental fault (or faults) in hereditary hemochromatosis. There may be some abnormality in the control or synthesis of ferritin or in the manner whereby the iron gains access to the storage cells. These differences may result in injury to the cells.

Other Factors Relating to Injury by Siderosis

Exposure to other toxic agents, especially alcohol, may contribute to the pathogenicity of stored iron.[66] Hemosiderin deposited in tissues in large amounts contains a vast excess of iron, but iron is not the only metal entrapped in the pigment. Analysis has demonstrated that siderotic hepatic tissue also contains abnormally large concentrations of other metals, including copper, lead, zinc, and manganese.[67] One of these, rather than the iron itself, may be the pathogen that results in injury to the storage cells.[68] The pattern of nonferrous metals in transfusion siderosis has not yet been studied.

CANCER AND SIDEROSIS

Hepatoma, primary cancer of the liver, has become the most common cause of death among patients with hemochromatosis.[69] It occurs in those patients whose livers have been loaded with iron pigment for many years, and in most cases the cancer has developed while the organ is still siderotic. It remains to be learned what will happen in those patients whose bodies have been emptied of excess storage iron and kept clean of siderosis by persistent preventive treatment. It is noteworthy that hepatoma, a relatively rare tumor, is one of the leading types of cancer encountered among the Bantu of South Africa.[70] Cancers other than hepatoma have been reported to occur with unexpected frequency in patients with hemochromatosis.[69] Intramuscular injection of therapeutic iron-dextran compounds has, in rare instances, been followed by the development of sarcoma in the site of injection.[71,72]

INFECTION IN IRON-STORAGE DISEASE

Bacteria require iron to thrive. It has been argued that a superabundance of iron provides invasive bacteria a better-than-average opportunity in people with iron-storage diseases.[73] Many people with transfusion siderosis are susceptible to infections, but it remains to be proved that this is a function of the iron and not the underlying disease. People with uncomplicated "idiopathic" hemochromatosis seem not to be at any increased risk.

On the other hand, it has also been proposed that people with iron deficiency are often at increased risk of infection.[74,75] This, too, is a complex problem, and unclouded conclusions are difficult to achieve.

TREATMENT OF IRON STORAGE DISEASES

Phlebotomy Therapy[36,49,76]

Removal of the patient's blood, one unit at a time, is by far the easiest, cheapest, safest, and most effective way to accomplish the reduction of siderosis. A unit of blood is removed into a plastic bag or a bottle as though the patient were donating the blood for transfusion. We remove 300 to 600 ml each time, the volume depending upon such factors as the patient's size, age, and demonstrated tolerance. Before each phlebotomy, the hemoglobin concentration of his blood is measured to ascertain that he is not anemic.

The concentration of iron in hemoglobin is 0.347%. The concentration of hemoglobin in red blood cells is 33%. Therefore, each milliliter of red cells contains approximately 1 mg of iron. A 500-ml unit of blood contains 250 mg of iron when the concentration of hemoglobin is 15 g/dl, or 165 mg when the hemoglobin concentration is 10 g/dl. Removing two units per week, each containing 200 mg of iron, does, in the course of a year, remove 20 g of iron. This averages about 55 mg/day.

Iron-chelating Agents

The reason for this mathematical exercise is to demonstrate the effectiveness of phlebotomy therapy as compared with the next best, the chelating agents. Deferoxamine is the best and least toxic of the presently available chelators. It is expensive and must be given by intramuscular injection, which is uncomfortable. Each injection results in the excretion of 5 to 15 mg of iron, a small amount compared to the amount removed by a single phlebotomy. Patients who require transfusion to remain alive cannot be treated by phlebotomy. Therefore, effort is being made to achieve improved means of inducing iron excretion by the use of chelating agents.[77,78]

Hemochromatosis with Anemia

Several hereditary anemias are complicated by the syndrome of hemochromatosis. In some patients, the anemia is severe enough to require blood transfusions. When anemia is not so severe, it is possible to remove stored iron by phlebotomy without detriment to the patient.[45,46] When anemia requires transfusion, therapeutic phlebotomy is out of the question. For such patients, improvement in the effectivensss of iron chelators or the means of employing them might improve health and prolong life. Care should be taken that anemic patients with hemochromatosis are not abused by exposure to excessive amounts of unneeded iron.[79,80] For example, therapeutic iron should not be prescribed and iron supplements should not be used. Proposals to increase the compulsory fortification of bread and flour with iron have met emphatic opposition from the Cooley's Anemia Society because the increased amount of iron in food would hasten the accumulation of the iron burden in all people with hemochromatosis, including children with Cooley's anemia.

Dietary Restriction

Iron-storage disease is best treated by removal of iron and not by dietary restriction to iron-deficient foods. However, when patients know that they have iron-storage disease, they should avoid heavily iron-enriched foods and, of course, should not take supplemental iron (in vitamin mixtures, for example) or therapeutic iron.

REFERENCES

1. Bothwell, T. H. and Finch, C. A. *Iron Metabolism*, Little, Brown & Co., Boston, 1962.
2. Crosby, W. H., Iron absorption, in *Handbook of Physiology*, Vol. 3, Field, J., Ed., Williams & Wilkins, Baltimore, 1968, 1553–1570.
3. Fairbanks, V. F., Fahey, J. L., and Beutler, E., *Clinical Disorders of Iron Metabolism*, 2nd ed., Grune & Stratton, New York, 1971.
4. Jacobs, A. and Worwood, M., Eds., *Iron in Biochemistry and Medicine*, Academic Press, New York, 1974.
5. Green, R., Charlton, R., Seftel, H., Bothwell, T., Mayet, F., Adams, B., Finch, C., and Layrisse, M., Body iron excretion in man, *Am. J. Med.*, 45, 336–353, 1968.
6. Crosby, W. H., Intestinal response to the body's requirement for iron, *JAMA*, 208, 347, 1969.
7. Crosby, W. H., Mucosal block. An evaluation of concepts relating to control of iron absorption, *Semin. Hematol.*, 3, 299–313, 1966.
8. Crosby, W. H., Regulation of iron metabolism, in *Regulation of Hematopoiesis*, Gordon, A. L., Ed., Appleton-Century-Crofts, New York, 1970, 519–538.
9. Crosby, W. H., Conrad, C. E., and Wheby, M. S., The rate of iron accumulation in iron storage disease, *Blood*, 22, 429–440, 1963.
10. Chrichton, R. R., Ed., *Proteins of Iron Storage and Transport in Biochemistry and Medicine*, Elsiever, New York, 1975.
11. Harrison, P. M., Ferritin and haemosiderin, in *Iron Metabolism*, Gross, F., Ed., Springer-Verlag, Berlin, 1964, 40–56.
12. Crichton, R. R., Millar, J. R., Cumming, R. L. C., and Bryce, C. F. A., The organ specificity of ferritin in human and horse liver spleen, *Biochem. J.*, 131, 51–59, 1973.
13. Richter, G. W. and Bessis, M. C., Commentary on hemosiderin, *Blood*, 25, 370–374, 1965.
14. Sturgeon, P. and Shoden, A., Mechanisms of iron storage, in *Iron Metabolism*, Gross, F., Ed., Springer-Verlag, Berlin, 1964, 121–146.
15. Richter, G. W., The cellular transformation of injected colloidal iron into ferritin and hemosiderin in experimental animals, *J. Exp. Med.*, 109, 197, 1959.
16. Fairbanks, V. F., Fahey, J. L., and Beutler, E., *Clinical Disorders of Iron Metabolism*, 2nd ed., Grune & Stratton, New York, 1971, 399–477.
17. Castleman, B. and Towne, V. W., Case records of the Massachusetts General Hospital: Case number 38512, *N. Engl. J. Med.*, 247, 992–995, 1952.
18. Wallerstein, R. O. and Robbins, S. L., Hemochromatosis after prolonged iron therapy in a patient with chronic hemolytic anemia, *Am. J. Med.*, 14, 256–260, 1953.
19. Pletcher, W. D., Broday, G. L., and Meyers, M. C., Hemochromatosis following prolonged iron therapy in a patient with hereditary nonspherocytic hemolytic anemia, *Am. J. Med. Sci.*, 246, 27–34, 1963.
20. Charlton, R. W., Bothwell, T. H., and Seftel, H. C., Dietary iron overload, *Clin. Haematol.*, 2, 383–403, 1973.
21. Hofvander, Y., Hematological investigations in Ethiopia with special reference to a high iron intake, *Acta Med. Scand. Suppl.*, 494, 1–73, 1968.
22. Aaron, H., Increased iron fortification of foods, *Med. Lett.*, 14, 81–83, 1972.
23. Crosby, W. H., The iron-enrichment-now brouhaha, *JAMA*, 231, 1054–1056, 1975.
24. Swiss, L. D. and Beaton, G. H., A prediction of the effects of iron fortification, *Am. J. Clin. Nutr.*, 27, 373–379, 1974.
25. Pritchard, J. A., The response to iron in iron deficiency, *JAMA*, 175, 478–482, 1961.
26. Smith, M. D. and Pannacciulli, I. M., Absorption of inorganic iron from graded doses, *Br. J. Haematol.*, 4, 428–434, 1958.
27. Finch, S., Haskins, D., and Finch, C. A., Iron metabolism. Hemopoiesis following phlebotomy. Iron as a limiting factor, *J. Clin. Invest.*, 29, 1078–1086, 1950.
28. Weintraub, L. R., Conrad, M. E., and Crosby, W. H., Control of iron absorption. An evaluation of the concept of mucosal block, *Blood*, 26, 887, 1965.
29. Bothwell, T. H. and Bradlow, B. A., Siderosis in the Bantu. Histopathological and chemical study, *Arch. Pathol.*, 70, 279–292, 1960.
30. Charlton, R. W., Jacobs, P., Seftel, H., and Bothwell, T. H., Effect of alcohol on iron absorption, *Br. Med. J.*, 2, 1427–1429, 1964.
31. Conrad, M. E., Berman, A., and Crosby, W. H., Iron kinetics in Laennec's cirrhosis, *Gastroenterology*, 43, 385–390, 1962.
32. Sayers, M. H., Lynch, S. R., Jacobs, P., Charlton, R. W., Bothwell, T. H., Walker, R. B., and Mayet, F., The effects of ascorbic acid supplementation on the absorption of iron in maize, wheat and soya, *Br. J. Haematol.*, 24, 209–218, 1973.

33. **Scheinberg, H.,** The genetics of hemochromatosis, *Arch. Intern. Med.,* 132, 126–128, 1973; 133, 1073, 1974; 135, 1269, 1975.
34. **Olsson, K. S., Heedman, P. A., and Stangaard, F.,** *JAMA,* in press.
35. **Sheldon, J. H.,** *Haemochromatosis,* Oxford University Press, London, 1935, 94–99 and 164–165.
36. **Finch, S. C. and Finch, C. A.,** Idiopathic hemochromatosis, an iron storage disease, *Medicine,* 34, 381–430, 1955.
37. **Crosby, W. H.,** Hemochromatosis. The unsolved problems, *Semin. Hematol.,* 14, 135–143, 1977.
38. **Crosby, W. H.,** Hereditary of hemochromatosis, in *Controversy in Internal Medicine,* Ingelfinger, F. J., Relman, A. S., and Finland, M., Eds., W. B. Saunders, Philadelphia, 1966, 271–284.
39. **Ellis, J. T., Schulman, I., and Smith, C. H.,** Generalized siderosis with fibrosis of liver and pancreas in Cooley's (Mediterranean) anemia, *Am. J. Pathol.,* 30, 287–294, 1954.
40. **Goya, N., Miyazaki, S., Kodate, S. et al.,** A family of congenital atransferrinemia, *Blood,* 40, 239–245, 1972.
41. **Barry, M., Scheuer, P. J., Sherlock, S., Ross, C. F., and Williams, R.,** Hereditary spherocytosis with secondary hemochromatosis, *Lancet,* 2, 481–485, 1968.
42. **Gelpe, A. P. and Ende, N.,** An hereditary anemia with hemochromatosis, *Am. J. Med.,* 25, 303–314, 1958.
43. **Canfield, C. J., Herman, Y. F., and Herman, R. H.,** Refractory hemolytic disease associated with glucose-6-phosphate hemolytic disease, *J. Lab. Clin. Med.,* 66, 96–106, 1965.
44. **Losowsky, M. S. and Hall, R.,** Hereditary sideroblastic anaemia, *Br. J. Haematol.,* 11, 70–85, 1965.
45. **Crosby, W. H. and Sheehy, T. W.,** Hypochromic iron-loading anaemia. Studies of iron and haemoglobin metabolism by means of vigorous phlebotomy, *Br. J. Haematol.,* 6, 56–65, 1960.
46. **Weintraub, L. R., Conrad, M. E., and Crosby, W. H.,** Iron-loading anemia. Treatment with repeated phlebotomies and pyridoxine, *N. Engl. J. Med.,* 275, 169–176, 1966.
47. **Dacie, J. V. and Mollin, D. L.,** Siderocytes, sideroblasts and sideroblastic anaemia, *Acta Med. Scand. Suppl.,* 445, 237–248, 1966.
48. **Crosby, W. H.,** Hemochromatosis, *Arch. Intern. Med.,* 133, 1072, 1974; 135, 1269, 1975.
49. **Barry, M.,** Iron overload. Clinical aspects, evaluation, and treatment, *Clin. Haematol.,* 2, 405–426, 1973.
50. **Lipschitz, D. A., Cook, J. D., and Finch, C. A.,** A clinical evaluation of serum ferritin as an index of iron stores, *N. Engl. J. Med.,* 290, 1213–1216, 1974.
51. **Jacobs, A. and Worwood, M.,** Ferritin in serum. Clinical and biochemical implications, *N. Engl. J. Med.,* 292, 951–956, 1975.
52. **Wands, J. R., Rowe, J. A., Mezey, S. E. et al.,** Normal serum ferritin concentration in precirrhotic hemochromatosis, *N. Engl. J. Med.,* 294, 302–305, 1967.
53. **Harker, L. A., Funk, D. D., and Finch, C. A.,** Evaluation of storage iron by chelates, *Am. J. Med.,* 45, 105–115, 1968.
54. **Wapnick, A. A., Lynch, S. R., Charlton, R. W., Seftel, H. C., and Bothwell, T. H.,** The effect of ascorbic acid deficiency on desferrioxamine-induced urinary iron excretion, *Br. J. Haematol.,* 17, 563–568, 1969.
55. **Rath, C. E. and Finch, C. A.,** Sternal marrow hemosiderin. A method for the determination of available iron stores in man, *J. Lab. Clin. Med.,* 33, 81–86, 1948.
56. **Scheuer, P. J., Williams, R., and Muir, A. R.,** Hepatic pathology in relatives of patients with haemochromatosis, *J. Pathol. Bacteriol.,* 84, 53–64, 1962.
57. **Kuhn, I. N., Monsen, E. R., Cook, J. D., and Finch, C. A.,** Iron absorption in man, *J. Lab. Clin. Med.,* 71, 715–721, 1968.
58. **Chodos, R. B., Ross, J. F., Apt, L., Pollycove, M., and Halkett, J. A. E.,** The absorption of radioiron labeled foods and iron salts in normal and iron-deficient subjects and in idiopathic hemochromatosis, *J. Clin. Invest.,* 36, 314–326, 1957.
59. **Smith, P. M., Godfrey, B. E., and Williams R.,** Iron absorption in idiopathic hemochromatosis and its measurement using a whole-body counter, *Clin. Sci.,* 37, 519–521, 1969.
60. **Pollack, S., George, J. N., Reba, R. C., and Crosby, W. H.,** The absorption of nonferrous metals in iron deficiency, *J. Clin. Invest.,* 44, 1470–1473, 1965.
61. **Olatunbosun, D., Corbett, W. E. N., Ludwig, J. et al.,** Alteration of cobalt absorption in portal cirrhosis and idiopathic hemochromatosis, *J. Lab. Clin. Med.,* 75, 754–762, 1970.
62. **Sorbie, J., Olatunbosun, D., Corbett, W. E. N., and Valberg, L. S.,** Cobalt excretion test for the assessment of body iron stores, *Can. Med. Assoc. J.,* 104, 777–782, 1971.
63. **MacSween, R. N. M. and Scott, R.,** Hepatic cirrhosis. A clinicopathological review, *J. Clin. Pathol.,* 26, 936–942, 1973.

64. **Bothwell, T. H.,** The relationship of transfusional haemosiderosis to idiopathic haemochromatosis, *S. Afri. J. Clin. Sci.,* 4, 53—70, 1953.

65. **Barry, M.,** Iron and the liver, *Gut,* 15, 324—334, 1974.

66. **Powell, L. W.,** Tissue damage in haemochromatosis. An analysis of the roles of iron and alcoholism, *Gut,* 11, 980, 1970.

67. **Butt, E. M., Nusbaum, R. E., Gilmour, T. C., and Didio, S. L.,** Trace metal patterns in disease states. I. Hemochromatosis and refractory anemia, *Am. J. Clin. Pathol.,* 26, 225—242, 1956.

68. **Alstatt, L. B., Pollack, S., Feldman, M. H., and Crosby, W. H.,** Liver manganese in hemochromatosis, *Proc. Soc. Exp. Biol. Med.,* 124, 353—355, 1967.

69. **Bomford, A., Walker, R. J., and Williams, R.,** Treatment of iron overload including results in a personal series of 85 patients with idiopathic haemochromatosis, in *Iron Metabolism and Its Disorders,* Kief, H., Ed., Elsevier, New York, 1975, 324—331.

70. **Geddes, E. W. and Falkson, G.,** Malignant hepatoma in the Bantu, *Cancer,* 25, 1271—1278, 1970.

71. **Greenberg, G.,** Sarcoma after intramuscular iron injection, *Br. Med. J.,* 1, 1508-1509, 1976.

72. **Abramowicz, M.,** Adverse effects of parenteral iron, *Med. Lett.,* in press.

73. **Weinberg, E. D.,** Iron and susceptibility to infectious disease, *Science,* 184, 952—956, 1974.

74. **Abramson, J. H., Sacks, T. C., Flug, D., Elishkovsky, R., and Cohen, R.,** Bacteruria and hemoglobin levels in pregnancy, *JAMA,* 215, 1631—1637, 1971.

75. **Chandra, R. K.,** Iron-deficiency anaemia and immunological responses, *Lancet,* 2, 1200—1201, 1976.

76. **Sherlock, S.,** Hemochromatosis, course and treatment, *Annu. Rev. Med.,* 27, 143—149, 1976.

77. **Propper, R. D., Shurin, S. B., and Nathan, D. B.,** Reassessment of desferrioxamine B in iron overload, *N. Engl. J. Med.,* 294, 1421—1423, 1976.

78. **Anderson, W. F. and Hiller, M. C.,** Eds., Development of Iron Chelates for Clinical Use, Publ. No. (NIH) 76-994, Department of Health, Education and Welfare, Bethesda, Md., 1976.

79. **Crosby, W. H.,** Iron enrichment. One's food, another's poison, *Arch. Intern. Med.,* 126, 911—913, 1970.

80. **AMA Committee on Iron Deficiency,** Iron deficiency in the United States, *JAMA,* 203, 407—412, 1968.

Trace Elements

EFFECT OF NUTRIENT TOXICITIES IN ANIMALS AND MAN: COPPER

A. E. Moffitt, Jr.

TOXICITY OF COPPER IN MAN

Historical

Early discussions of the biologic actions of copper refer to its medicinal properties rather than its toxic effects. For example, in 77 A.D. Pliny[1] stated that he gave many prescriptions for the use of copper as a healing agent for ulcers. It is difficult to estimate the first discovery of the toxic effects of copper. However, Davenport[2] has suggested that Moses in 1200 B.C. must have been aware of the symptoms produced by food prepared in copper vessels since he recommended that they should be kept absolutely clean. In 1567, Paracelsus[3] classified copper as a cause of "miners' diseases," i.e., various disturbances of the lungs, stomach, and intestines of individuals engaged in mining or smelting. In 1700, Ramazzini was reported to have remarked on the green color of the complexion, beard, and hair of workmen engaged in the melting and hammering of copper and tin.[2] From the 18th until the early 20th century there were numerous reports of apparent copper toxicity.[4-7] As Lewin observed in 1900, in almost all early reported cases of copper poisoning, some other metal or combination of metals, especially lead, was involved and may have been a more appropriate cause of the observed symptomatology.[9]

Nonindustrial Copper Poisoning

Acute Toxicity

Acute copper poisoning in man is a rare occurrence. When a case is identified, it is usually the result of the accidental or deliberate ingestion of several grams of copper sulfate ($CuSO_4$) or the consumption of acidic food or drink which has had prolonged contact with copper-containing vessels.[9,10] Hall[10] has summarized several episodes of acute copper poisoning, including 9 cases with vomiting and diarrhea following the ingestion of apples boiled in a copper vessel at a school lunch kitchen; 15 cases characterized by nausea, vomiting, cramps, or diarrhea after ingestion of a punch prepared in a three-tiered fountain punch bowl made of chrome-plated copper; and 10 cases from a party at which cocktails were prepared in metal shakers with worn inner copper platings. Several instances of nausea and vomiting have been reported following ingestion of carbonated beverages from postmix vending machines with defective check valves and copper water lines.[11] In England, copper hot-water boilers have been employed to heat water for tea. Nicholas[12] has reported that 20 workers exhibited nausea, vomiting, and diarrhea after drinking their morning tea prepared with water which had been allowed to stand in a boiler for more than 48 hr. Subsequent inspection of the boiler showed corrosion and a sample of tea from the brew pan contained more than 30 ppm copper.

In India, ingestion of copper sulfate has been a favorite method of attempted suicide. Chuttani et al.[13] have reported that acute copper sulfate poisoning constituted approximately 33% of all cases of poisoning admitted to a major hospital in New Delhi during 1961. Of 48 cases studied in particular detail, all were characterized by metallic taste, nausea, vomiting, and burning in the epigastrium and 20 to 30% exhibited diarrhea, jaundice, hemoglobinuria, and anuria. According to statements obtained from the patients, the amount of copper ingested in these cases varied from 1 to 100 g dissolved in water. Of 12 fatal cases, 7 died apparently from shock and hypotension or from subsequent hepatic or renal damage with coma and uremia. Pathological reports showed gastric and intestinal ulcerations, liver-cell necrosis (centrilobular), bile thrombi, and renal damage characterized by glomerular congestion and swelling or necrosis of tubular cells.

Although often indistinguishable from other forms of food poisoning,[14] the acute oral toxicity of copper and its salts appears to be dependent upon the amount of ionic copper retained in the stomach rather than on the amounts actually ingested.[2,10] Because of their irritant effects, even small (milligram) amounts of soluble copper salts cause vomiting and diarrhea; indeed, these clinical features probably protect the patient from the serious systemic toxic effects, such as hemolysis, liver damage, gastrointestinal bleeding, oleguria, hemoglobinuria, hematuria, proteinuria, hypotension, tachycardia, convulsions, coma, or death.[9,13] When larger quantities (a gram or more) of copper salts are ingested, as in Chuttani's cases above,[13] ulcerations of the gastric mucosa, jaundice, hemolyses, hepatic necrosis, and renal damage from deposition of hemoglobin and/or copper are the pathlogical findings.[9]

Holtzman et al.[16] have observed hemolysis in burn patients treated with copper sulfate topically. The same phenomenom has been reported as a result of the use of copper-containing hemodialysis equipment.[17] In the latter case, copper introduced into the circulation during hemodialysis can elicit febrile reactions strikingly similar to the classical metal fume fever experienced by copper smelter workers.[17,18] Hemolytic anemia is not an unexpected complication of copper poisoning since copper denatures hemoglobin, damages cell membranes, and inhibits the activity of glucose-6-phosphate-dehydrogenase and glutathione reductase.[15] Hemolytic anemia is also a common clinical feature of Wilson's disease, which is the disease associated with the most striking abnormality of copper retention in man.[19]

Chronic Toxicity

Nonindustrial chronic toxicity is rarely encountered. Salmon and Wright[20] have suggested that excessive copper in drinking water may have caused acrodynia (pink disease) in a 15-month-old infant. The present U.S. drinking water standard for copper of 1 mg/l was established to prevent the undesirable taste imparted to water at high copper concentrations.[9] Nevertheless, there have not been any reports of human or animal toxicity associated with this level. The use of copper alloys in dental and other prostheses has been associated with gingivitis[21] and eczematous dermatitis.[22] Copper-containing intrauterine contraceptive devices and copper-supplemented animal feeds constitute two other situations of potential chronic toxicities.[9]

Industrial Copper Poisoning

Browning has observed that there is very little evidence that copper presents a serious industrial health hazard, either from acute or chronic exposure.[18]

Acute Toxicity

Exposures to copper fume (finely divided particles of oxides of copper) occur in copper and brass plants and in welding copper-containing metals.[18,23] Health effects in workers employed in these processes have included irritation of the upper respiratory tract, metallic or sweet taste, nausea, metal fume fever, and, in some cases, discoloration of skin or hair.[23]

Acute intoxication in the form of metal-fume fever (brass chills or brass foundry workers' ague) can occur following inhalation of copper fume or dust.[9,18] Metal-fume fever is typically a 24- to 48-hr illness characterized by chills, fever, aching muscles, dryness in the mouth and throat, and headaches. There have been reports of apparent metal-fume fever among workers welding copper tanks,[25] men handling copper oxide powder in a paint factory,[9,26] and workers exposed to metallic copper dust produced during the polishing of copper plates.[27]

Inhalation of dusts and mists of copper salts in high concentrations can result in congestion of nasal mucous membranes and, on occasion, ulceration with perforation of the nasal septum.[23] If copper salts in sufficient concentration reach the gastrointestinal tract (following expectoration and swallowing), they act as irritants producing salivation, nausea, vomiting, gastric pain, and diarrhea as described above for acute nonindustrial toxicity.

Contact dermatitis associated with copper has been reported,[26,28] but few cases of industrial dermatitis caused by copper or copper salts have been documented.[9,18] Eye contact with copper salts has been reported as a cause of conjunctivitis and edema of the eyelids.[26,29] In severe cases, turbidity and/or ulceration of the cornea can occur. In his recent review of the literature, Cohen[26] found no reports of toxic eye reactions due to copper fume or dust.

The Occupational Safety and Health Administration, U.S. Department of Labor, has adopted separate permissible exposure limits for copper dust and fume. The 8-hr time-weighted average limit for copper dust is 1 mg/m^3 air; for copper fume, the limit is 0.2 mg/m^3. Both limits are derived from the threshold limit values adopted by the American Conference of Governmental Industrial Hygienists.[23]

Chronic Toxicity

As observed by Browning,[18] the question of whether chronic copper poisoning exists as a separate occupational disease entity has been disputed since the 19th century. Lewin's studies of workers engaged in the extraction of metallic copper and the manufacture of bronze led him to the conclusion that chronic copper poisoning did not exist and that reported cases were probably due to contamination with other metals.[8] The discoloration of skin, hair, and teeth often seen in copper workers is considered a surface patina composed of copper carbonate and is not apparently associated with intoxication or ill health.[8,18]

Pimentel and Marques[30] have described a disease called vineyard sprayer's lung which is characterized by pulmonary copper deposition, fibrosis, and blue coloration of the lungs of certain vineyard workers who have had many years of exposure to Bordeaux mixture, a 1 to 2% solution of copper sulfate neutralized with hydrated lime. This mixture is widely used in certain European countries as an antimildew agent on grape vines. In a more recent report, granulomas of undetermined etiology have appeared in the livers of some of these workers.[31] Despite years of exposures to insoluble copper dusts, Chilean copper miners have not shown any unusual disease patterns or abnormal liver or blood copper concentrations.[9]

Although an effect of excess dietary copper on the central nervous system has been observed in animals,[32] there does not appear to be any conclusive evidence for the existence of a form of copper poisoning affecting the central or peripheral nervous system in man.[18]

Various researchers have found an increased incidence of respiratory cancer among workers in copper smelters in the U.S.[33,34] and Japan.[35] In all cases, however, the evidence suggests that copper itself did not play any etiologic role in the observed cancer deaths.

The National Academy of Sciences[9] has recently offered the following summary opinion on copper as an industrial health hazard:

Although copper can act as a toxic agent in an occupational setting, it is benign under ordinary circumstances. However, if workers are exposed to excessive concentrations of the metal in any of its forms, there may be undesirable health effects. Because of the absence of reports on significant environmental effects from airborne copper, copper and its compounds as dusts or fumes dispersed into the atmosphere have not been considered hazardous.

TOXICITY OF COPPER IN ANIMALS

Occurrence

Cases of acute poisoning with copper are rare and generally accidental in animals. Chronic copper poisoning in ruminant grazing animals (sheep and cattle), however, is fairly common in Australia, Canada, and in the Midwest and Great Plains states.[9,36] In contrast to copper deficiency, sheep are more susceptible than cattle to copper toxicoses and may be affected at any age, although most reported cases have been in mature animals. In Australia, two types of chronic copper poisoning of sheep have been recognized: (1) a phytogenous type, which occurs when weather conditions result in the early growth of subterranean clover (*Trifolium subterraneum*) containing up to 20 ppm (dry weight) of copper with negligible amounts of molybdenum, and (2) a hepatogenous type, which follows the consumption of weeds (mainly *Heliotropum europaeum,* but also some species of *Senecio*) containing hepatotoxic alkaloids, which increase the ability of the liver to accumulate copper[37] by impairing normal liver metabolic and excretory processes.

In the U.S., dietary imbalances of copper and molybdenum appear to be the major contributor to copper toxicoses in sheep.[9] Buck,[38] for example, has reported over 20 episodes of chronic copper toxicosis among sheep in Iowa from June 1968 through June 1970. In this state as well as in the other Great Plains and upper Midwest states, grains and forages contain adequate levels of copper but low levels of molybdenum. The U.S. Food and Drug Administration generally recognizes as safe up to 15 ppm copper as a livestock feed additive, whereas molybdenum is prohibited as a feed ingredient.[9] These regulations, however, fail to recognize the dietary requirements of sheep for a proper balance between copper and molybdenum (6 to 10 parts Cu/1 part Mo). An adequate dietary intake of molybdenum has been shown to prevent copper toxicoses in sheep because these elements form an in vivo complex which inhibits copper accumulation in the ruminant liver.[39,40] Moreover, molybdenum, in conjunction with dietary sulfate, increases the urinary and biliary excretion of copper.[41] Accordingly, hepatic accumulation of copper and resulting copper toxicity in sheep are commonly observed when vitamin-mineral preparations containing copper but not molybdenum are added to feedstuffs with an existing copper:molybdenum imbalance. It has been suggested that this problem could be readily solved through the use of computerized feed-control systems in feed mills wherein chemical analysis of mineral ingredients in feeds for each animal species would be under close surveillance.[9]

Other possible causes of copper toxicoses in ruminants include: the consumption of grain or forage contaminated by copper-containing pesticides, such as Bordeaux mixture which contains 1 to 3% copper sulfate; the use of copper sulfate to control helminthiasis and infectious pododermatitis in cattle and sheep; the use of calcium-copper EDTA as an injectable source of copper in countries where sheep are prone to copper deficiency; maintaining sheep indoors in winter pens with no access to green forage containing sufficient molybdenum; and contamination of soils or vegetation in the vicinity of smelting or mining operations.[9,42]

Symptoms of Intoxication – Ruminant Animals

In all animals, the continued ingestion of copper in excess of nutritional requirements leads to some accumulation in the tissues, especially in the liver. The capacity for hepatic copper storage and the hepatic copper levels that can be tolerated without adverse effects vary greatly among animal species.[42] Monogastric (nonruminant) animals appear to have a higher resistance to excessive dietary copper levels than multigastric (ruminant) animals.

Acute copper poisoning in sheep, albeit a rare event, may be produced by large oral doses of copper salts. Signs of acute poisoning include vomiting, abdominal pain, excessive salivation, and diarrhea, followed within 24 to 48 hr by collapse and/or death.[9,42]

In chronic copper poisoning, sheep may passively accumulate copper over a period varying from a few weeks to more than a year as a result of the regular consumption of small but excessive amounts of copper in the diet, especially when the dietary copper:molybdenum ratio is greater than 10:1. During this preclinical phase, no signs of toxicity are evident until a critical level of hepatic copper is attained (usually 3 to 15 times the normal level or about 150 ppm). At this point, a second (hemolytic) phase of the disease develops as a result of the rapid liberation of high concentrations of copper from the liver into the circulation. The animals abruptly become weak, tremble, and lose their appetite. Hemoglobinuria, hemoglobinemia, jaundice, and death usually result in a few days after the onset of the second phase. Although morbidity is usually less than 5%, mortality in diseased animals generally exceeds 75%.[9]

Necropsy of poisoned sheep reveals generalized icterus. The liver is slightly enlarged, yellow, and friable; the gall bladder is distended with thick greenish-brown bile; the kidneys are enlarged, friable, and show hemorrhagic mottling; the spleen is enlarged with discoloration of the parenchyma; and there may be excess pericardial fluid and some epicardial hemorrhage.[36]

The changes that occur in the livers of copper-treated sheep prior to the development of the hemolytic crisis have been studied by Ishmael et al.[43] in lambs given a dose of copper sulfate (1 g) as a drench on 5 days/week. Hemolysis and subsequently death occurred between 31 days and 10 weeks from the start of the treatment. Some 6 weeks prior, liver biopsies showed necrosis of isolated parenchymal cells and swollen paraaminosalicyclic acid-positive, copper-containing Kupffer cells rich in acid phosphatase. At a later stage, many swollen parenchymal cells showed fatty change, karyomegaly, and nuclear vacuolation. At the time of the hemolytic crisis, extensive focal necrosis of the liver was seen; afterwards, large amounts of bile pigments were present in the small bile ducts and periportal fibrosis occurred. Activities of adenosine triphosphatase, nonspecific esterase, and succinic and glutamic dehydrogenases were depressed. The levels of copper in the livers of these lambs at autopsy ranged from 1641 to 4064 ppm (dry weight).

Other investigators[44,45] have reported a sharp reduction of glutathione concentration and an accumulation of methemoglobin in the blood during the hemolytic crisis of copper poisoning apparently as a result of the sudden release of copper from the liver. It has been suggested that death may result from kidney failure pursuant to a blockage of the kidneys by hemoglobin.

Symptoms of Intoxication — Nonruminant Animals

As previously stated, nonruminants can tolerate toxic levels of copper for ruminants (20 to 50 ppm).[9,42,46] Dietary concentrations greater than 250 ppm are required to produce toxicity in swine[47,48] and rats.[49] Hemolytic anemia and jaundice have not been reported in rats or rabbits suffering from chronic copper poisoning or in birds.[42,49] Although jaundice has been observed in pigs fed toxic levels (250 to 500 ppm) of copper, this condition is not a marked feature of copper toxicosis in pigs. Adult birds exposed to high intakes of copper exhibit a marked loss of body weight and anemia, with no evidence of intravascular hemolysis.[50] Wiederanders[51] subjected turkeys to copper loading through subcutaneous copper injections, yet copper toxicity and hemolysis were not produced. The author suggested that turkeys and other fowl may have metabolic and excretory pathways for copper that differ from those of mammals. For chickens, pigeons, and ducks, Pullar[52] found the minimum lethal dose of copper to vary from 300 to 1500 mg/kg body weight, depending on the form of copper fed.

Rats appear to be extremely tolerant of high intakes. Boyden et al.[49] have shown that normal growth and health were maintained on diets containing 500 ppm copper, or about 100 times normal, despite a 14-fold increase in liver copper. Whereas sheep accumulate copper in the liver in proportion to the dietary intake, rats maintain normal hepatic copper levels until a critical dietary level of copper is reached.[53,54]

In respect of the mechanism of toxic action of copper in mammals, studies with rats and mice have revealed that copper accumulates in liver lysosomes.[55] Accordingly, it has been postulated that in copper poisoning, acid hydroxylases capable of producing cellular injury are released, thereby causing hepatic damage.[56]

Miscellaneous Toxic Effects

Copper has long been known to be a component of many essential enzymes in biological systems, such as the cytochrome oxidases of the cellular electron transport system and monoamine oxidase which oxidizes amines to aldehydes for cross-linkage in the formation of collagen and for deamination of norepinephrine and serotonin.[42,57] Therefore, alterations in copper intake, notably copper deficiency, may be associated with changes in copper-dependent enzyme activity[9,42,46] yielding toxic manifestations.

Moffitt and Murphy[58] have shown that copper loading can alter the metabolism of certain foreign chemicals by hepatic microsomal enzymes. In this study, rats were fed 450 ppm copper sulfate in the drinking water for 30 days and subsequently were challenged with various toxicants. Copper loading appeared to protect against the induction of methemoglobinemia by 3,4-dichloroaniline and inhibited the hepatic metabolism of aniline in vitro.

Aquatic Organisms

Copper is toxic to many aquatic organisms.[9,46] Local high concentrations of copper in streams may result from influxes of copper-containing fertilizers, industrial discharges, or use of copper salts as molluscicides for the control of aquatic vegetation.[9] The gills of many fish apparently lack barriers to the absorption of copper and efficient excretory mechanisms, because toxic levels for fish are similar to those for algae, plankton, insects, and invertebrates.[46] Toxic levels in water are functions of the species, the age of the individual organism, concentrations of organic and inorganic constituents, water temperature, and the chemical form of copper.[9] The toxicity of copper to fish, however, is inversely related to the hardness or the calcium content of the water, which appears to be the case for other aquatic organisms, with a ratio of 40:1 to 70:1 in hard and soft waters.[59]

Based on data from a number of studies,[60,64] it has been concluded that acute toxicosis in fish is unusual if the copper concentration in fresh water is less than 0.025 ppm.[9] In soft fresh water, however, 0.01 to 0.02 ppm has been found to be toxic.[59,62]

REFERENCES

1. **Bailey, K. C.,** Ed., *The Elder Pliny's Chapters on Chemical Subjects, Part II,* Arnold and Co., London, 1932, 25–55.
2. **Davenport, S. J.,** Review of Literature on the Health Hazards of Metals. I. Copper, 7666, U.S. Department of the Interior, Bureau of Mines, Washington, D.C., Inf. Circ., November 1953.
3. **Koelsch, F.,** Theophrastus vonHohenheim, Paracelsus, (On Miners' Phthisis and Other Miners' Diseases), Deut-Gesell, Gewerbeh., Frankfurt, a.M., Series No. 12, 1925, 69.
4. **Galtier, G. P.,** Treatise on General Toxicology (or Poisons and Poisoning in General), Chamerot, Paris, 1855, 372.
5. **Chevallier, A.,** Note on workers who prepare verdigris, *Ann. Hyg. Publique. Med. Leg.,* 37, 392–399, 1847.

6. **Bochefontaine,** Cholera, smallpox, thyphoid fever, and anthrax among copperworkers of Villedieu, *C.R. Soc. Biol.,* Paris, 97, 1375–1376, 1883.
7. **Bochefontaine,** Contribution to the study of action of copper on persons in continued contact with this metal, *C.R. Soc. Biol.,* Paris, 36, 79–83, 1884.
8. **Lewin, L.,** Investigations in copper workers, *Dtsch. Med. Wochenschr.,* 26, 689–694, 1900.
9. *Copper,* National Academy of Sciences, Washington, D.C., 1977, 115.
10. **Hall, R.,** Copper Containers for Food and Drink, National Clearinghouse for Poison Control Centers Bulletin, March to April 1970, 1–7.
11. **Hopper, S. H. and Adams, H. S.,** Copper poisoning from vending machines, *Public Health Rep.,* 73, 910–914, 1958.
12. **Nicholas, P. O.,** Food poisoning due to copper in the morning tea, *Lancet,* 2, 40–42, 1968.
13. **Chuttani, H. K., Gupta, P. S., Gulati, S., and Gupta, D. N.,** Acute copper sulfate poisoning, *Am. J. Med.,* 39, 849–854, 1965.
14. **Wyllie, J.,** Copper poisoning at a cocktail party, *Am. J. Public Health,* 47, 617, 1957.
15. **Louria, D. B., Joselow, M. M., and Browder, A. A.,** The human toxicity of certain trace elements, *Ann. Intern. Med.,* 76, 307–319, 1972.
16. **Holtzman, N. A., Elliott, D. A., and Heller, R. H.,** Copper intoxication. Report of a case with observations on ceruloplasmin, *N. Engl. J. Med.,* 275, 347–352, 1966.
17. **Manzler, A. D. and Schreiner, A. W.,** Copper-induced acute hemolytic anemia: a new complication of hemodialysis, *Ann. Intern. Med.,* 73, 409–412, 1970.
18. **Browning, E.,** Copper in *Toxicity of Industrial Metals,* 2nd ed., Appleton-Century-Crofts, London, 1969, 145–153.
19. **Evans, G. W.,** Copper homeostasis in the mammalian system, *Physiol. Rev.,* 53, 535–570, 1973.
20. **Salmon, M. A. and Wright, T.,** Chronic copper poisoning presenting as pink disease, *Arch. Dis. Child.,* 46, 108–110, 1971.
21. **Trachtenberg, D. I.,** Allergic response to copper – its possible gingival implications, *J. Periodontol.,* 43, 705–707, 1972.
22. **Barranco, V. P.,** Eczematous dermatitis caused by internal exposure to copper, *Arch. Dermatol.,* 106, 386–387, 1972.
23. Documentation for Threshold Limit Values, Am. Conf. Governmental Industrial Hygienists, Cincinnati, 1974. Stokinger, H. E., personal communication, 1976.
24. **Hamilton, A. and Hardy, H.,** *Industrial Toxicology,* 3rd ed., Publishing Sciences Group, Acton, Mass., 1974, 575.
25. **Fishburn, C. W. and Zenz, C.,** Metal fume fever: a report of a case, *J. Occup. Med.,* 11, 142–144, 1969.
26. **Cohen, S. R.,** A review of the health hazards of copper exposure, *J. Occup. Med.,* 16, 621–624, 1974.
27. **Gleason, R. P.,** Exposure to copper dust, *Am. Ind. Hyg. Assoc. J.,* 29, 461–462, 1968.
28. **Saltzer, E. I. and Wilson, J. W.,** Allergic contact dermatitis due to copper, *Arch. Dermatol.,* 98, 375–376, 1968.
29. **Stokinger, H. E.,** Copper, in *Industrial Hygiene and Toxicology,* Vol. 2, 2nd ed., Fassett, D. W. and Irish, D. A., Eds., Interscience, New York, 1963, 1033–1037.
30. **Pimentel, J. C. and Marques, F.,** "Vineyard sprayer's lung": a new occupational disease, *Thorax,* 24, 678–688, 1969.
31. **Pimentel, J. C. and Menezas, A. P.,** Liver granulomas containing copper in vineyard sprayer's lung. A new etiology of hepatic granulomatosis, *Am. Rev. Respir. Dis.,* 111, 189–195, 1975.
32. **Bowler, K. and Duncan, C. J.,** The effect of copper on membrane enzymes, *Biochim. Biophys. Acta,* 196, 116–119, 1970.
33. **Lee, A. M. and Fraumeni, J. F., Jr.,** Arsenic and respiratory cancer in man: an occupational study, *J. Natl. Cancer Inst.,* 42, 1045–1052, 1969.
34. **Milham, S., Jr. and Strong, T.,** Human arsenic exposure in relation to a copper smelter, *Environ. Res.,* 7, 176–182, 1974.
35. **Kuratsune, M., Tokudome, S., Shirakusa, T., Yoshida, M., Tokumitsu, Y., Hayano, T., and Seita, M.,** Occupational lung cancer among copper smelters, *Int. J. Cancer,* 13, 552–558, 1974.
36. **Gallagher, C. H.,** *Nutritional Deficiencies and Enzymological Disturbances,* J. B. Lippincott, Philadelphia, 1964, 97–100.
37. **Todd, J. R.,** Chronic copper toxicity of ruminants, *Proc. Nutr. Soc.,* 28, 189–197, 1969.
38. **Buck, W. B.,** Diagnosis of feed-related toxicoses, *J. Am. Vet. Med. Assoc.,* 155, 1434–1443, 1970.
39. **Dick, A. T.,** The control of copper storage in the liver of sheep by inorganic sulfate and molybdenum, *Aust. Vet. J.,* 29, 233–239, 1953.

40. Dowdy, R. P. and Matrone, G., Copper-molybdenum interaction in sheep and chicks, *J. Nutr.*, 95, 191–196, 1968; A copper-molybdenum complex: its effects and movement in the piglet and sheep, *J. Nutr.*, 95, 197–201, 1968.

41. Marcilese, N. A., Ammerman, C. B., Valsecchi, R. M., Dunavant, B. G., and Davis, G. K., Effect of dietary molybdenum and sulfate upon copper metabolism in sheep, *J. Nutr.*, 99, 177–183, 1969.

42. Underwood, E. J., Copper, in *Trace Elements in Human and Animal Nutrition*, 3rd ed., Academic Press, New York, 1971, 57–115, 543.

43. Ishmael, J., Gopinath, C., and Howell, J. M., Experimental chronic copper toxicity in sheep. Histological and histochemical changes during the development of lesions in the liver, *Res. Vet. Sci.*, 12, 358–366, 1971.

44. Todd, J. R. and Thompson, R. H., Studies on chronic copper poisoning. II. Biochemical studies on the blood of sheep during the hemolytic crisis, *Br. Vet. J.*, 119, 161–173, 1963.

45. Todd, J. R. and Thompson, R. H., Studies on chronic copper poisoning. III. Effects of copper acetate injected into the blood stream of sheep, *J. Comp. Path. Ther.*, 74, 542–551, 1964.

46. Schroeder, H. A., Nason, A. P., Tipton, I. H., and Balassa, J. J., Essential trace elements in man: copper, *J. Chronic Dis.*, 19, 1007–1034, 1966.

47. Suttle, N. F. and Mills, C. F., Studies of the toxicity of copper to pigs. I. Effects of oral supplements of zinc and iron salts on the development of copper toxicoses, *Br. J. Nutr.*, 20, 125–148, 1966.

48. Suttle, N. F. and Mills, C. F., Studies on the toxicity of copper to pigs. II. Effect of protein source and other dietary components on the response to high and moderate intakes of copper, *Br. J. Nutr.*, 20, 149–161, 1966.

49. Boyden, R., Potter, V. R., and Elvehjem, C. A., Effect of feeding high levels of copper to albino rats, *J. Nutr.*, 15, 397–402, 1938.

50. Goldberg, A., Williams, C. B., Jones, R. S., Yamagita, M., Cartwright, G. E., and Wintrobe, M. M., Studies on copper metabolism. XXII. Hemolytic anemia in chickens induced by the administration of copper, *J. Lab. Clin. Med.*, 48, 442–453, 1956.

51. Wiederanders, R. E., Copper loading in the turkey, *Proc. Soc. Exp. Biol. Med.*, 128, 627–629, 1968.

52. Pullar, E. M., The toxicity of various copper compounds and mixtures for domesticated birds, *Aust. Vet. J.*, 16, 147–162, 1940.

53. Milne, D. B. and Weswig, P. H., Effect of supplementary copper on blood and liver copper-containing fractions in rats, *J. Nutr.*, 95, 429–433, 1968.

54. Lal, S. and Sourkes, T. L., Deposition of copper in rat tissues – the effect of dose and duration of administration of copper sulfate, *Toxicol. Appl. Pharmacol.*, 20, 269–283, 1971.

55. Goldfischer, S., Demonstration of copper and acid phosphatase activity in hepatocyte lysosomes in experimental copper toxicity, *Nature*, 215, 74–75, 1967.

56. Lindquist, R. R., Studies on the pathogenesis of hepatolenticular degeneration. III. The effect of copper on rat liver lysosomes, *Am. J. Pathol.*, 53, 903–927, 1968.

57. Fisher, G. L., Function and homeostasis of copper and zinc in mammals, *Sci. Total Environ.*, 4, 373–412, 1975.

58. Moffitt, A. E., Jr. and Murphy, S. D., Effect of excess and deficient copper intake on hepatic microsomal metabolism and toxicity of foreign chemicals, in *Trace Substances In Environmental Health*, Vol. 7, Hemphill, D. D., Ed., University of Missouri, Columbia, 1974, 205–210.

59. Pickering, Q. H. and Henderson, C., The acute toxicity of some heavy metals to different species of warm water fishes, *Air Water Pollut.*, 10, 453–463, 1966.

60. Doudoroff, P. and Katz, M., Critical review of the literature on the toxicity of industrial wastes and their components to fish. II. The metals, as salts, *Sewage Ind. Wastes*, 25, 802–839, 1953.

61. Mount, D. I., Chronic toxicity of copper to fathead minnows. (*Pimephales promelas*, rafinesque), *Water Res.*, 2, 215–223, 1968.

62. Mount, D. I. and Stephan, C. E., Chronic toxicity of copper to the fathead minnow (*Pimephales promelas*) in soft water, *J. Fish Res. Board Can.*, 26, 2449–2457, 1969.

63. Sprague, J. B., Promising anti-pollutant: chelating agent NTA protects fish from copper and zinc, *Nature*, 220, 1345–1346, 1968.

64. O'Hara, J., Alterations in oxygen consumption by bluegills exposed to sublethal treatment with copper, *Water Res.*, 5, 321–327, 1971.

COBALT: AN ESSENTIAL TRACE ELEMENT

James F. Sullivan and Robert E. Burch

INTRODUCTION

The presence of cobalt in plants and animals was recognized some 50 years ago.[1,2] In 1929 the first clear indication of a biologic activity for cobalt followed the demonstration that rats ingesting cobalt developed polycythemia.[3] In 1926 Minot showed that the ingestion of liver by humans cured or alleviated pernicious anemia.[4] The recognition of cobalt as the essential metal in vitamin B_{12}, a nutrient necessary for life in animals and in many bacteria, followed.[5]

The origin of the name "cobalt" is of interest. One theory is that the term is derived from the German word *Kobold,* meaning a hobgoblin or gnome.[6] A second theory implicates the Greek word *cobalos,*[7] which means mine. The first derivation implies a mischievous spirit who produced skin lesions in miners of arsenical silver — cobalt ores. The second derivation, also related to mines, may refer to noxious fumes arising from the smelting of cobalt arsenates. Cobalt blue was used to produce colors ranging from dark blue to various shades of green in the production of colored pottery, porcelain, and glass as early as 1450 B.C.[8]

The purpose of this review is to outline the effect of the cobalt ion in metabolism with only a brief review of its essential role in vitamin B_{12} activity.

CHEMISTRY

Cobalt is quite different from other essential trace elements in that as an ion or in a simple protein complex, it is not required by any organism.[9] Cobalt in the specific complex of vitamin B_{12} is essential for animal life. On a weight basis, vitamin B_{12} is the most potent of the known vitamins. It is also unique in that neither it nor any precursor is produced by plants. Thus, the only known source of vitamin B_{12} in nature is ultimately its synthesis by micro-organisms. The metabolic functions of vitamin B_{12} are numerous. In essence, where nucleic acid synthesis or reproduction of cells occur, vitamin B_{12} is required.

Coenzyme B_{12} is involved in fat, carbohydrate, and protein metabolism. It is necessary for the hydrogen transfer and isomerization whereby methylmalonyl-CoA is converted to succinyl-CoA. Propionyl-CoA formed from odd-chain fatty acid oxidation, is converted to methyl-malonyl CoA, and is metabolized further by the above reaction.

Methylcobalamin is required for the conversion of hymocysteine to methionine in animals. This reaction also requires folic acid and seems necessary for the prevention of megaloblastic anemia. Vitamin B_{12} also seems to be active in the maintenance of sulfhydryl groups in the reduced form necessary for many biochemical reactions. Numerous excellent reviews of the biochemistry and functions of vitamin B_{12} have appeared[10,11] which outline these aspects in greater depth.

EFFECT OF COBALT ON THE PANCREAS AND THYROID

The intravenous injection of cobaltous chloride in a number of animals has produced elevations of blood sugar, glucagon, and lipids. Changes in pancreatic islet histology involving both alpha and beta cells have been described. Some investigators have been unable to corroborate these histologic findings. Species differences may explain this disparity. However, the mechanism by which alterations of blood glucose and hyperlipemia are produced and the relationship of these abnormalities to pancreatic cellular changes remains unclear. There is no apparent relationship of these observations to known disease in humans at this time.

In 1951, it was demonstrated that parenteral administration of cobalt chloride caused an elevation of blood sugar in guinea pigs that was associated with degranulation of the alpha cells of the islets of Langerhans.[12] These observations led to a hypothesis that cobalt-released glucagon following cobalt administration to guinea pigs.[13] Cobalt-treated rats also developed hyperglycemia accompanied by increases in blood glucagon. Histologic changes in the pancreas were not found.[14] Other workers failed to confirm the described findings in the pancreas of guinea pigs and of rabbits.[15,16]

Continued research into the presumed action of cobalt on pancreatic alpha cells resulted in the observation that, in rabbits, the increase in blood sugar was associated with a transient elevation in serum cholesterol.[17]

A similar study indicated that the hyperglycemia produced by cobalt also caused an increase in serum cholesterol, while other drugs producing hyperglycemia did not cause cholesterol elevation. If dihydroergotamine was given to block adrenalin at the time of cobalt injection, the hyperglycemia was absent while cholesterol elevation was seen.[18] Stimulated by the observation of hypertriglyceridemia in a patient receiving cobalt as a medication, serum lipid changes after intravenous or intraperitoneal cobalt were studies. After cobalt injection in the rabbit, the serum component showing the greatest increase was the triglyceride fraction.[19] Concurrent changes in the alpha-2 and beta globulins were also observed. Thus, considerable evidence accumulated suggesting that cobalt ingestion would produce hyperglycemia, increases in serum glucagon, and serum lipids.

Other studies approached the problem from a different aspect. They described the presence of a lipid-mobilizing hormone whose activity was stimulated by cobalt.[20] An increase in beta lipoproteins was found, as was a decrease in heparin-activated lipoprotein lipase. Further investigation of changes in serum proteins in the cobalt-induced hyperlipemia disclosed a decrease in both serum albumin and total proteins. It was postulated that the serum protein changes were secondary to the hyperlipemia, although the mechanism was unclear.[21] Observations of changes in serum proteins led to the hypothesis that the cobalt-stimulated production of lipid-carrying proteins, having a pattern consistent with very low-density hyperlipemia, was the essential change induced by cobalt when given to rats.[22] A biologic effect of insulin, in terms of glucose utilization and antilipolytic activity (seven times normal) was noted. Although serum insulin levels were not increased, the pattern of hepatic gluconeogenesis suggested an excess of insulin-like activity. The latter observations were made in vitro with hepatic slices. Recent in vitro studies have suggested that cobalt inhibits serum-immunoreactive insulin levels and that this effect is produced by prevention of calcium reentry into beta islet cells by cobalt, thus inhibiting immunoreactive insulin from these cells.[23]

Only tentative hypotheses may be advanced to relate these various observations. Cobalt injection appears to increase glucagon release from pancreatic alpha cells. This may be accompanied by a decrease in insulin as measured by serum immunoreactive methods with an enhanced insulin-like effect on glucose and lipolytic activity.[22,23] Each factor might work singly or concurrently to mobilize fatty acids with increased hepatic lipid synthesis and subsequent elevations of serum triglyceride levels. The morphology of alpha and beta cells of the pancreas is controversial and a consensus of opinion is not available.

The appearance of thyroid hyperplasia and hypothyroidism in patients was first noted following the widespread use of cobalt as an erythropoietic stimulant. The initial recognition was made in a 1954 case report.[24] Isolated case studies and other clinical descriptions clearly related cobalt to thyroid hypofunction and goiter[25-27] but mechanisms were unknown in these early studies.

Five cases were reported in which thyroid hyperplasia occurred. Three showed a visible goiter; laboratory evidence of decreased thyroid function was demonstrable in three who showed decreased uptake of radioactive iodine and decreased protein-bound iodine in the serum. One patient developed serious myxedema. All evidence of thyroid dysfunction disappeared when cobalt was discontinued.[28]

In the succeeding years, other reports of thyroid dysfunction followed. However, some authors have administered cobalt at a maximum dose of 6 mg/kg daily for 10 weeks without evidence of thyroid toxicity.[29]

It is now believed that there is definite evidence of production of thyroid dysfunction and goiter following cobalt therapy. This, along with other adverse effects, has resulted in its very restricted use as a hematinic.[30] Since cobalt-induced thyroid goiter and hypothyroidism are unlikely to be seen in patients, the interest in the cause of the disorder has waned.

The early studies of thyroid function seemed to validate the observation that cobalt blocked tyrosine iodinase and prevented iodine uptake.[28,29] Undoubtedly, studies of thyroid function by the more sophisticated methods available today would be of great interest.

COBALT AND EXPERIMENTAL EPILEPSY

Cobalt applied locally to the brain has the ability to produce epileptic seizures in experimental animals. Repeated studies with various species have also shown convulsive activity with the local application of cobalt. The mechanism of seizure induction is unknown. Early studies suggested that this phenomenon related to inflammatory response or to scar formation. Subsequent studies have failed to support this hypothesis.

In 1960, cobalt was first used in a study of epilepsy. Cobalt powder, when applied to the cerebral cortex of mice, was found to result in seizures accompanied by abnormal electroencephalograms.[31] These findings were confirmed in rats and amplified by the demonstration of increased seizure reactivity to photo- and audio-stimulation. The nature of lesions in the cortex where cobalt was applied were initially a coagulation necrosis, then an inflammatory reaction, and subsequently scar formation.[32] Similar epileptic foci were produced in monkeys[33] and cats.[34] These early studies produced definite cortical lesions when cobalt was applied to the surface or injected into the cortex. Extradural, subdural, and even extracranial placement of cobalt also produced an increased tendency to epilepsy.[35] These studies were repeated by many workers,[36-41] and cobalt was thought to be an excellent method of producing epilepsy in various animals. Areas of the brain showing the cobalt-induced epilepsy included the thalamus, hippocampus, and amygdaloid areas. Cobalt applied to the cerebellum inhibited convulsions originating from a cortical lesion.[41,42] The inhibition of the cerebellar lesion was found to be more effective in the rat than in the cat.

Adverse seizures, those in which the head turned away from the side of the cortical stimulation, were noted for the first time after cobalt implants in the frontal lobes of rats. However, lesions in the cerebellum in this experiment did not alter seizure activity.[43]

Determinations of amino acid content in the cerebral cortex of humans with epilepsy showed a higher glycine and lower taurine content.[44] Aspartic acid, GABA, and glutamic acid concentrations were also decreased. Lesions produced by cobalt in the cerebrum showed the same abnormality. Taurine consistently decreased the convulsive threshold and corrected the amino acid abnormalities previously noted.[44]

A detailed and extensive study of brain histopathology was carried out in rats with cobalt-induced epilepsy. Necrosis of tissue, inflammatory reaction, and scar formations were accompanied by increased levels of calcium.[45] Introduction of a new method of studying cobalt-induced epilepsy consisted of deposition of the pia of cobalt in minute amounts by use of iontophoresis. This method was not accompanied by any cortical lesion. Nevertheless, epilepsy was induced, and the cobalt was found in the upper three levels of the cortex.[46]

Choline acetyltransferase and cholinesterase activities were depressed in areas of brain subjected to cobalt in rats.[47] These enzyme changes occurred early; the enzyme activity was normal again in 3 weeks but the lowered seizure threshold persisted. These findings suggest that the activities of these two enzymes were not related to enhanced seizure activity.

These investigations into the nature of cobalt-induced epilepsy clearly indicate that:

1. It is reproducible in most animal species.
2. It is not dependent upon tissue damage.
3. It can be inhibited by cerebellar lesions and that the cobalt is confined to the three outer layers of the cortex.

Although the pathogenesis remains uncertain, the amino acid abnormalities and the corrective action of taurine would seem to be an area in which further research would be most fruitful.

COBALT AND TUMOR PRODUCTION

Cobalt as well as nickel and cadmium have the potentiality of producing tumors when injected into the skeletal muscle of experimental animals, especially rhabdomyosarcomas.[48] These tumors are readily transplanted into other animals of the same species, however the inducing metal need not be present in the transplanted tumor.[49] The reason that such tumors are produced is unknown and only fragmentary evidence of altered biochemistry has been observed.

In the early 1950s a series of papers described induction of rhabdomyosarcomas by injection of cobalt into the thigh muscle of rats. Nickel and cadmium produced similar effects.[50-56] With cobalt injections, tumors other than rhabdomyosarcomas were rare. Injections of cobalt into cardiac muscle through the chest wall produced massive rhabdomyosarcomas in which the cellular origin appeared to be derived, at least in part, from the heart.[57]

Dust from a nickel refinery showed possible carcinogenic activity when injected locally in rats. The tumors were sarcomas arising from skeletal muscle and injection of cobalt produced similar neoplasias.[58]

The oral injection of sodium cobalt/nitrate and sodium nitrate by mice exerted a protective effect against the carcinogen, methylcholanthrene. The mechanism was thought to be dependent upon methemoglobinemic hypoxia.[59] This work was repeated using cobalt chloride, and a similar degree of protection was observed indicating that the cobalt *per se* was probably the important safety factor.[60] Thus, cobalt seemed to function as an inducer and as a protector.

A further investigation into the cobalt-induced tumors revealed that the highest metal concentration was in the center of the tumor and that the metal is eliminated as the tumor ages.[49] The primary localization of the metal was intracellular and the metal seemed to be bound by nucleic acids; smaller amounts of cobalt were in the mitochondria. A study of the mitochondrial respiration in these metal-induced rhabdomyosarcomas indicated that the oxidative activity was decreased by cobalt, cadmium, and nickel in that order. Oxidation of pyruvate was altered the most, and cobalt showed the lowest pyruvate oxidation of the three metals.[61]

Using cell cultures in studying the carcinogenic aspect of cobalt, myoblasts and fibroblasts were exposed to ionic cobalt. At high concentration the cells were killed; at lower cobalt concentrations, no effect was noted.[62] Upon allowing the cobalt to dissolve slowly in horse serum prior to injection, cytological changes resembling premalignant cells were seen in the cell cultures.

The study of cobalt-produced tumor has demonstrated that injection into muscle will produce rhabdomyosarcoma and the tumor may be transplanted without the inducing metal. There is some evidence of altered tissue enzyme activity. Otherwise, there is little other knowledge in this area.

HEMATOPOIETIC EFFECT OF COBALT

Excluding the role of cobalt in vitamin B_{12}, cobalt appears to have the ability to increase red cell production in most animal species as well as in man. The nature of this stimulus is not entirely clear. Bone marrow hypoxia, increased erythropoietic levels, and a combination of the two have been considered. The use of cobalt in the treatment of various anemias has been discontinued because of poor therapeutic response and associated complications secondary to cobalt usage.

The ability of cobalt to stimulate red blood cell formation, other than its role in vitamin B_{12}, has been known for some time.[63] Originally thought to be due to tissue anoxia, this concept was considered to be inadequate,[64] and attention was turned to increased production of erythropoietin.[65]

The erythropoietic activity in plasma of rats treated with cobalt was compared to that of rats chronically bled. Although plasma of cobalt-treated animals showed ability to stimulate reticulocyte response, the activity was only a third of that produced by the anemic animals.[66] The activity of hypoxia and cobalt were compared using lactic dehydrogenase in the kidney as an indicator of hypoxia.[67] As increased production of red cells correlated with lactic dehydrogenase changes, the association of erythropoietic response with anoxia of renal tissue was assumed. Injection of cobalt into rabbits indicated that the increased red cell production was accompanied by definite and severe depression of oxidative phosphorylation in the kidney, bone marrow, and liver. This tissue hypoxia was thought to be the effect produced by cobalt which resulted in increased red cell production.[68] Renal perfusion, comparing hypoxic blood and blood containing cobalt, resulted in similar increases of erythropoietin activity in dogs.[69] Repetition of the combined effects of cobalt and testosterone in perfused dog kidney indicated minimal change in renal histology, ruling out cell destruction as a source of increased plasma erythropoietin.[70] Cobalt and testosterone were seen as producing a greater red cell increase acting in concert, and plasma erythropoietin was increased to a greater level than either agent acting alone.[71]

A novel approach to this problem using anephric rats indicated that cobalt and testosterone caused increased plasma erythropoietin. The authors concluded that nonrenal sources of erythropoietin existed and that the action of cobalt was secondary to cellular hypoxia.[72] Recent work has indicated increased cyclic AMP activity in the rat kidney is accompanied by increased erythropoietin when cobalt is injected.[73] Cobalt administration was shown to produce increased blood lactate followed by an increase in renal cortical cyclic AMP. Subsequent to these changes plasma erythropoietin was increased.[74]

In an extension of this experiment, cobalt was injected into rats and changes in the pH of blood were correlated with erythropoietin activity. Alkalosis, secondary to cobalt injection, was found to increase affinity of hemoglobin for oxygen, and increased levels of erythropoietin resulted. Nephrectemized rats developed acidosis. Erythropoietin could not be found in the blood following cobalt chloride treatment. The blood pH and relative affinity for O_2 with accompanying tissue oxygen was thought to be of importance in erythropoietin production.[75]

The use of cobalt as an effective and safe stimulator of erythropoiesis is contra-indicated in clinical medicine. The mechanisms by which cobalt induces an increase in red blood cell mass is unclear. Blood pH changes with accompanying changes in red cell affinity for oxygen, altered lactate production, increased cyclic AMP, tissue hypoxia, and increased erythropoietin production all appear to be interrelated. The key to cobalt-induced erythropoiesis continues to be a fascinating biomedical puzzle.

COBALT/BACTERIA ENZYMES

The only recognized essential role of cobalt is that of a component of vitamin B_{12}. This vitamin is synthesized by microorganisms living in a symbiotic relationship with certain plants. These plants are then the source of vitamin B_{12} for most animals, and ultimately for man.

Measurement of vitamin B_{12} using bacterial assays has demonstrated that such plants as alfalfa[76] and turnip greens contained vitamin B_{12} but were unable to synthesize it.[77] The actual synthesis of vitamin B_{12} has been demonstrated in certain algae, which also have the capacity to fix atmospheric nitrogen.[78,79] Those algae lacking the nitrogen fixation capacity appear to be unable to synthesize vitamin B_{12} but require it for their growth.[80,81] Confirmatory studies with a variety of plants and bacteria have demonstrated the symbiotic relationship and have affirmed the utilization of cobalt in the synthesis of vitamin B_{12}.

Ruminants differ from most other animals in that they produce vitamin B_{12} by a microbial synthesis within the rumen when cobalt is adequate in their diet. Both the synthesis of vitamin B_{12} and its absorption in ruminants are very inefficient. In contrast, nonruminant animals absorb ingested vitamin B_{12} quite efficiently.

The demonstration of a wasting disease occurring in cattle and sheep grazing on pasture having a low cobalt content was the initial important observation relating cobalt to disease. The beneficial response to cobalt in these animals was convincing evidence of a disease resulting from cobalt deficiency.[82-85] Because of the known requirement for cobalt to synthesize vitamin B_{12} in the symbiotic bacteria associated with plants, it was suggested that rumen bacteria could not synthesize vitamin B_{12} in the absence of cobalt. Thus, a deficiency of vitamin B_{12} was postulated as the cause of this disease in sheep and cattle.[84,86-88]

The capacity to synthesize vitamin B_{12} again appears to be strongly associated with the fixation of atmospheric nitrogen.[89] The exact biochemical steps through which microorganisms produce vitamin B_{12} are not known.[90,91] The salient features in these complex relationships indicate that microorganisms in symbiosis with certain plants utilize cobalt in the synthesis of vitamin B_{12}.

Higher plants and animals are dependent upon this process for vitamin B_{12} requirements. Ruminants synthesize vitamin B_{12} through the activity of intraruminal microorganisms when cobalt is available.

Cobalt as a divalent cation has been shown to alter enzymatic activities and cellular physiology in various bacteria and animals. These changes in enzyme activity are diverse in nature and are inexplicable by any single quality of the cobalt cation. At present, they seem to be without apparent significant physiologic significance. The best-studied of these reactions are briefly listed in Table 1.

Cobalt has also been used to study various metalloenzymes by substitution for the naturally occurring metal. Zinc-containing alcohol dehydrogenase[92,93] has been so studied, as has the alkaline phosphatase derived from *Escherichia coli*.[94] The geometric configuration of the phosphoglucomatase of rabbit muscle has been extensively studied.[95-98] The crystalline structure and clarification of spatial and geometric relationships of these metalloenzymes has resulted.

Table 1
COBALT AND ENZYMATIC ACTIVITY

Organism	Activity or effect
Pseudomonas aeruginosa	Oxidative enzymes inhibited[99]
Aspergillus niger Neurospora crassa	Toxic effect associated with altered iron metabolism[100,101]
Escherichia coli	Inhibition of growth[102,103] Cobalt containing dipeptidase[104]
Propionibacteria	Cobalt containing transcarboxylase[105]
Rat	Cobalt induced changes in plasma kiniogen[106]
Rat liver	Hemeoxidase induced[107]
	Cytochrome P-450 decreased[108]

COBALT DEFICIENCY IN ANIMALS

As with many other trace metals, actual cobalt deficiency was first recognized in animals.[109] Sheep and cattle in Australia developed a disease characterized by debilitation and wasting with a high mortality rate. The response to cobalt ingestion clearly indicated that cobalt deficiency was an important element in the pathogenesis of this disease.[110]

Research in ruminant animals indicated that vitamin B_{12} deficiency resulted from insufficient dietary cobalt. The pasture upon which these animals fed was low in cobalt. Synthesis of vitamin B_{12} by microorganisms in the rumen was greatly decreased leading to the vitamin deficiency.

The recognition that pernicious anemia in humans and wasting disease in cattle appeared to have an identical pathogenesis, the lack of vitamin B_{12}, stimulated research in relation to vitamin B_{12} and cobalt. The metabolism of cobalt in man seems to have significant differences from that in ruminant animals. These animals show poor absorption of oral cobalt, which is excreted primarily in the feces. With injection of cobalt into animals, the excretion is high in the urine and less in the feces. There appears to be a small amount of ingested cobalt which is absorbed, accumulated in the liver, and excreted via the biliary tract.[111,112] The rate of absorption in man is quite high, and the main excretory pathway is through the kidneys.

Cobalt is found in most organs of the body in man and animals, with the highest concentration being found in the liver. Similar concentrations of cobalt are found in the organs of most species studied.[113,114] A comparison of tissue levels appears in Table 2.

In vitamin B_{12}-deficient animals, the tissue content of cobalt is decreased. The early stages of cobalt (and hence vitamin B_{12}) insufficiency is difficult to diagnose. Tissue levels are not dependable and assays for serum B_{12} levels appear to be the most valuable means for early diagnosis.[115]

Ruminants use acetate and propionic acid as primary sources of energy. The discovery that methylmalonyl isomerase (a vitamin B_{12}-dependent enzyme) catalyzes the conversion of methylmalonyl-CoA to succinyl-CoA gave definite evidence of one vitamin B_{12} reaction. In vitamin B_{12} deficiency, the activity of this enzyme in the liver is quite low. Thus, slower removal of intravenously injected propionic acid occurs.[116]

Deficiencies of folic acid or vitamin B_{12} result in increased urinary excretion of

Table 2
COBALT CONTENT OF KIDNEY, LIVER, AND HEART

Cobalt in	Animal tissues (wet weight in $\mu g/g$)	Man (wet weight in $\mu g/g$)
Kidney	0.1–0.36	0.05–0.11
Liver	0.05–0.76	0.12–0.28
Heart	0.13–0.28	0.17–0.23

Table 3
COBALT REQUIREMENTS

	Cobalt in feed	Co	Vitamin B_{12} in feed
Lambs	0.11 ppm	0.1–1.0 mm	11.1 μg/day B_{12}
Sheep	0.11 ppm	0.1–1.0 mm	200 μg/day B_{12}
Calves	0.11 ppm	0.1–1.0 mm	20–40 μg B_{12} kg
Cattle	0.11 ppm	0.1–1.0 mm	20–40 μg B_{12} kg
Men	0.6–1.2 μg/day	0.1 mg–1.0 mg/day	1 μg/day

Table 4
COBALT CONTENT OF VARIOUS FOODS

Leafy green vegetables	0.2–0.6 ppm co/dry weight
Organ meats (liver, kidney)	0.15–0.25 ppm
Muscle	0.08–0.12 ppm
Cow milk	0.4–1.1 μg/l
Dairy products, refined cereals, sugar	Very low

formiminoglutamic acid in humans and in some experimental animals.[117,118] Increased urinary excretion of formiminoglutamic acid has been found in vitamin B_{12}-deficient sheep, suggesting that the methyltetrahydrofolate-homocysteine transmethylase reaction that determines the availability of tetrahydrofolic acid and, indirectly, the excretion of formiminoglutamic acid may be impaired.[119]

The tryptamine alkaloids present in a grass (*Phalaris tuberosa*) are responsible for an acute and chronic disorder in sheep and, to a lesser degree, in cattle.[120] The acute form rapidly leads to death; the chronic form is characterized by gait disturbance, muscle tremor, apparent difficulty in breathing, and a marked tachycardia particularly on exertion. Cobalt or vitamin B_{12} will not aid the acute form; however, if the pasture area is dosed with cobalt the chronic disease does not occur.[121] The reason for this paradox is unknown.

The daily requirements of vitamin B_{12} and cobalt as reflected in the literature appear in Table 3. The relative content of vitamin B_{12} in various foods is shown in Table 4.

COBALT METABOLISM

The absorption of vitamin B_{12} and its transport in plasma have been carefully investigated and will not be included in this review. The absorption of cobalt itself, however, is of interest because of possible toxicity. In addition, the mechanism of cobalt absorption appears to be related to that of other divalent cations — particularly iron.

Orally administered cobalt is excreted primarily in the feces in most animals, with small amounts occurring in the urine. A limited amount of ingested cobalt is absorbed and appears in the liver. In humans, cobalt absorption differs from most animals because the efficiency of absorption is quite high. Urinary excretion is the major route of loss from the body.[110]

The absorption of iron in humans is similar to that of cobalt in animals. Iron absorption is apparently controlled in part by the mucosal cells of the small intestine. This mucosal block depends upon the activity of the protein apoferritin. This iron complex may be absorbed directly into the blood stream, with any excess being bound to the mucosal cells and eventually exfoliated into the intestine. A large excess of dietary iron beyond that normally ingested appears to overload the mucosal block and is absorbed by passive diffusion.[110] Iron absorption is increased by iron deficiency, decreased iron stores, or increased erythropoietin. Obviously, many steps in the absorption of iron remain uncertain. Studies of cobalt and iron absorption may aid in identifying factors important to each.

Vitamin D has been shown to increase both cobalt and iron absorption in chicks receiving a low-calcium diet.[122] Similar studies indicated that vitamin D also increased the absorption of calcium, strontium, beryllium, magnesium, barium, zinc, cadmium, and cobalt.[123-128] A mucosal transfer factor common to most divalent cations and influenced by vitamin D offered the best hypothesis for these observations. However, in rats made anemic by bleeding, only increased absorption of iron, manganese, and cobalt could be demonstrated.[129] This seems quite significant in that a common carrier of divalent cations was not apparent and each metal with increased absorption has been shown to produce chronic liver disease.[130-133]

A study of iron and cobalt absorption in iron deficiency, cirrhosis, and hema-chromatosis showed a direct correlation in the absorption of iron and cobalt. In patients with iron overload, iron absorption (but not cobalt) was decreased.[134]

A number of investigators have shown that physiologic changes which increase iron absorption also increase cobalt absorption. Cobalt absorption, however, is not affected in conditions in which iron absorption is decreased.[135-140]

Very little is known about cobalt carriers. One study has suggested that transferrin may have binding sites accessible to cobalt and manganese as well as iron.[141] It is apparent that knowledge of cobalt absorption and transfer are poorly understood.

TOXICITY OF COBALT

Much of the information concerning adverse effects of cobalt has been acquired by using cobalt and iron as hematinics. Nausea and vomiting, eighth nerve deafness, thyroid hyperplasia, myxedema, and skin hypersensitivity reactions have been described following the use of large doses of cobalt.[142,143] Toxic effects of cobalt have been fatal in a few reported cases.[144,145] Some interstitial pulmonary fibrosis and contact dermatitis have been reported to be related to cobalt.[146,147]

Cobalt has also been involved in an entity referred to as cobalt-beer drinker's cardiomyopathy. Appearing in such widely separated areas as Quebec, Canada,[148] Leuvain, Belgium,[149] and Minneapolis[150] and Omaha[151] in the U.S.A., the cases shared a common history of:

1. Ingestion of moderate to large amounts of beer containing cobalt
2. Fulminant right-sided heart failure
3. Cyanosis
4. Marked lactic acidosis
5. Hemorrhagic pericardial effusion

A relatively high mortality rate occurred in Canada and the U.S. while there was minimal mortality in Belgium. Marked urinary excretion of zinc and magnesium was found,[152] and deposition of increased cobalt in the myocardial tissue was demonstrated.[153] The heart appeared to be the organ showing most significant pathology. The myocardium was pale and flabby; there were extensive areas of myocardial necrosis without evidence of inflammatory reaction. Marked congestion of the liver with necrosis was present.

Removal of cobalt from beer resulted in a dramatic disappearance of the entity. The pathogenesis of the myocardial failure remains unclear. The presence of an inadequate diet (particularly protein inadequacy) rendering the patient susceptible to cobalt has been stressed.[149] The myocardium of rats oxidizes long chain fatty acids by preference and cobalt has been shown to inhibit utilization of the fatty acids, thus producing cardiac dysfunction.[154] Efforts to combine cobalt ingestion, beer drinking, and thiamine deficiency failed to reproduce the cardiac pathology.[155] The biochemical importance of cobalt's ability to block the oxidation of pyruvate to acetyl CoA (thus interfering with oxidative phosphorylation) has been emphasized.[150] A more complete description of this entity appears in the *Canadian Medical Association Journal*[156] and the *Annals of the New York Academy of Sciences.*[157]

It is obvious that cobalt was important in the pathogenesis of this syndrome since the entity disappeared after cobalt was removed from the beer. Why cobalt resulted in this characteristic syndrome is as much an enigma today as it was when this syndrome was first discovered.

REFERENCES

1. **Bertrand, G. and Machebouef, M.,** On the relatively high content of Ni and Co in the pancreas, *C. R. Acad. Sci.,* 182, 1305–1307, 1926.
2. **Dutoit, P. and Zbinden, C.,** Analyse spectrographique des cendres de sang et d'organes, *C. R. Acad. Sci.,* 188, 1628–1672, 1929.
3. **Waltner, K. and Waltner, K.,** Cobalt and blood, *Klin. Wochenschr.,* 7, 313–318, 1929.
4. **Minot, G. R. and Murphy, W. P.,** Treatment of pernicious anemia by special diet, *JAMA,* 87, 470, 1926.
5. **Florkin, M. and Stotz, E. H.,** *Comprehensive Biochemistry,* Elsevier, Amsterdam, 1963, 103.
6. U.S. Bureau of Mines Mineral Facts and Problems 1965, Bulletin 630, U.S. Department of the Interior, Bureau of Mines, Washington, D. C., 1965, 241.
7. **Hampel, C. A.,** *Rare Metals Handbook,* 2nd ed., Reinhold, New York, 1961, 114.
8. **Schroder, H. A., Nason, P. A., and Tipton, I. H.,** Essential trace metals in man: cobalt, *J. Chronic. Dis.,* 20, 869–890, 1967.
9. **Florkin, M. and Stotz, E. H.,** *Comprehensive Biochemistry,* Vol. 21, Elsevier, Amsterdam, 1970, 217–223.
10. **Castle, W. A. et al.,** Vitamins B$_{12}$ and folate, *Am. J. Med.,* 48, 539–617, 1970.
11. **Herbert, V.,** *The Pharmacological Basis of Therapeutics,* 5th ed., MacMillan, New York, 1975, 1325–1338.
12. **Van Campenhout, E. and Cornelis, G.,** Destruction experimentales des cellules alpha des ilots endourines du pancreas chez le cobaye, *C. R. Soc. Biol.,* 145, 933–935, 1951.
13. **Vuylsteke, C. A., Cornelis, G., deDuve, C,** Influence du traitement au cobalt sur le contenu en facteur H-G du pancreas de cobaye, *Arch. Int. Physiol.,* 60, 128–131, 1952.
14. **Lochner, J. D. V., Eisentraut, A. M., and Unger, R. H.,** The effect of CoCl-2 or glucagon levels in plasma and pancreas of rat, *Metabolism,* 13, 868–874, 1964.
15. **Korp, W. and LeCompte, P. M.,** Nature and function of alpha cells of pancreas; their possible role in production of glucagon, *Diabetes,* 4, 347–366, 1955.
16. **Creutzfeldt, W.,** Alpha cell cytotoxins; their influence on carbohydrate metabolism and the effect of the oral blood glucose reducing sulfonamides on the islet cells, *Diabetes,* 6, 135–145, 1957.
17. **Caren, R. and Carbo, L. J.,** Pancreatic alpha-cell function in relation to cholesterol metabolism, *J. Clin. Endocrinol.,* 16, 507–516, 1956.
18. **Boyd, G. S. and Maclean, N.,** Observations on the metabolic and histologic effects of cobalt chloride in the rabbit, with particular reference to cobalt-induced hypercholesterolaemia, *J. Exp. Physiol.* 44, 394–403, 1959.

19. **Caplan, R. M. and Block, W. D.,** Experimental production of hyperlipemia in rabbits by cobaltous chloride, *J. Invest. Dermatol.,* 40, 199–203, 1965.

20. **Zarafonetis, C. J. D., Bartlett, R. H., and Brody, G. L.,** Lipid mobilizer hormone in cobalt chloride hyperlipemia, *JAMA,* 191, 235–237, 1965.

21. **Zarafonetis, C. J. D., Dabich, L., and Brody, G. L.,** Plasma protein changes consequent to hyperlipemia induced by cobaltous chloride or triton, WR-1339, *Am. J. Med. Sci.,* 254, 506–512, 1967.

22. **Eaton, R. P.,** Cobalt chloride-induced hyperlipemia in the rat: effects of intermediary metabolism, *Am. J. Physiol.,* 222, 1550–1557, 1972.

23. **Henquin, J. C. and Lambert, A. E.,** Cobalt inhibition of insulin secretion and calcium uptake by isolated rat eyelets, *Am. J. Physiol.,* 228, 1669–1677, 1975.

24. **Gross, R. T., Kriss, J. P., and Spaet, T. H.,** Hematopoietic and goitrogenic effects of cobaltous chloride in patients with sickle-cell anemia, *Am. J. Dis. Child.,* 88, 503, 1954.

25. **Washburn, T. C. and Kaplan, E.,** Cobalt therapy and goiter, *Clinic Pediatr.,* 3, 89–92, 1964.

26. **Breidahl, H. and Fraser, R.,** Cobalt goiter, *Proc. Soc. Exp. Biol. Med.,* 48, 1026, 1955.

27. **Keitel, H. G.,** Cobalt and thyroid dysfunction, *JAMA,* 158, 1390, 1955.

28. **Kriss, J. P., Carnes, W. H., and Gross, R. T.,** Hypothyroidism and thyroid hyperplasia in patients with cobalt, *JAMA,* 157, 117–121, 1955.

29. **Jaimet, C. H. and Thode, H. G.,** Thyroid function studies on children receiving cobalt therapy, *JAMA,* 158, 1353–1355, 1955.

30. **Crosby, W. H.,** in The use of cobalt and cobalt-iron preparation in the therapy of anemia, *Blood,* 10, 852–861, 1955.

31. **Kopeloff, L. M.,** Experimental epilepsy in the mouse, *Proc. Soc. Exp. Biol. Med.,* 104, 500–503, 1960.

32. **Dow, R. S., Fernandez-Guardiola, A., and Manni, E.,** The production of cobalt experimental epilepsy in the rat, *Electroencephalogr. Clin. Neurophysiol.,* 14, 399–407, 1962.

33. **Chusid, J. G. and Kopeloff, L. M.,** Epileptogenic effects of pure metals implanted in the motor cortex of monkeys, *J. Appl. Physiol.,* 17, 697–700, 1962.

34. **Henjyoji, E. Y. and Dow, R. S.,** Cobalt-induced seizures in the cat, *Electroencephalogr. Clin. Neurophysiol.,* 19, 152–161, 1965.

35. **Payan, H. M., Strebel, R., and Levine, S.,** Epileptogenic effect of extradural and extracranial cobalt, *Nature,* 208, 792–793, 1965.

36. **Cesa-Bianchi, M. G., Mancia, M., and Mutani, R.,** Experimental epilepsy induced by cobalt powder in lower brain-stem and thalamic structures, *Electroencephalogr., Clin. Neurophysiol.,* 22, 525–536, 1967.

37. **Payan, H. M.,** Cerebral lesions produced in rats by various implants: epileptogenic effect of cobalt, *J. Neurosurg.,* 27, 146–152, 1967.

38. **Mutani, R.,** Cobalt experimental hippocampal epilepsy in the cat, *Epilepsia,* 8, 223–240, 1967.

39. **Mutani, R.,** Cobalt experimental amygdaloid epilepsy in the cat, *Epilepsia,* 8, 73–92, 1967.

40. **Mancia, M. and Lucioni, R.,** EEG and behavioral changes induced by subcortical introduction of cobalt powder in chronic cats, *Epilepsia,* 7, 308–317, 1966.

41. **Reimer, G. R., Grimm, R. J., and Dow, R. S.,** Effects of cerebellar stimulation on cobalt-induced epilepsy in the cat, *Electroencephalogr. Clin. Neurophysiol.,* 23, 456–462, 1967.

42. **Payan H. M., Levine, S., and Strebel, R.,** Inhibition of experimental epilepsy by chemical stimulation of cerebellum, *Neurology,* 16, 573–576, 1966.

43. **Cereghino, J. J. and Dow, R. S.,** Effect of cobalt applied to the cerebellum on cobalt-experimental epilepsy in the cat, *Epilepsia,* 11, 413–421, 1970.

44. **Van Gelder, N. M.,** Antagonism by taurine of cobalt-induced epilepsy in cat and mouse, *Brain Res.,* 47, 157–165, 1972.

45. **Payan, H. M.,** Morphology of cobalt-experimental epilepsy in rats, *Exp. Mol. Pathol.,* 15, 312–319, 1971.

46. **Willmore, L. J., Fuller, P. M., Butler A. B., and Bass, N. H.,** Neuronal compartmentation of ionic cobalt in rat cerebral cortex during initiation of epileptiform activity, *Exp. Neurol.* 47, 280–289, 1975.

47. **Goldberg, A. M., Pollock, J. J., and Hartman, E. R.,** Alterations in cholinergic enzymes during the development of cobalt-induced epilepsy in the rat, *Neuropharmacology,* 11, 253–259, 1972.

48. **Heath, J. C.,** The production of malignant tumors by cobalt in the rat, *Br. J. Cancer,* 10, 668–673, 1956.

49. Heath, J. C. and Webb, M., Content and intracellular distribution of the inducing metal in the primary rhabdomyosarcomata induced in the rat by cobalt, nickel and cadmium, *Br. J. Cancer,* 21, 768–779, 1067.

50. Heath, J. C., Cobalt as a carcinogen, *Nature,* 173, 822–833, 1954.

51. Heath, J. C. and Daniel, M. R., The production of malignant tumours by nickel in the rat, *Br. J. Cancer,* 18, 261–264, 1964.

52. Heath, J. C. and Daniel, M. R., The production of malignant tumours by cadmium in the rat, *Br. J. Cancer,* 18, 124–129, 1964.

53. Gilbert, I. G. F. and Radley, J. M., A procedure for the isolation of cell nuclei for trace metal studies, *Biochim. Biophys. Acta,* 82, 618–621, 1964.

54. Dingle, J. T., Heath, J. C., Webb, M., and Daniel, M. R., The biological action of cobalt and other metals, *Biochim. Biophys. Acta,* 65, 34–38, 1962.

55. Daniel, M. R., Strain differences in the response of rats to the injection of nickel sulphide, *Br. J. Cancer,* 20, 886–895, 1966.

56. Heath, J. C., Cadmium as a carcinogen, *Nature,* 193, 592–593, 1962.

57. Heath, J. C. and Daniel, M. R., The production of malignant tumours by cobalt in the rat: intra-thoracic tumours, *Br. J. Cancer,* 16, 473–478, 1962.

58. Gilman, J. P. W. and Ruckerbauer, G. M., Metal carcinogenesis 1. Observations on the carcino-genicity of a refinery dust, cobalt oxide, and colloidal thorium dioxide, *Cancer Res.,* 22, 152–157, 1962.

59. Thompson, R. S., Gautieri, R. F., and Mann, D. E., Effect of chronic oral administration of sodium cobaltinitrate and sodium nitrite on the minimal carcinogenic dose$_{50}$ of methyl-cholanthrene in albino mice, *J. Pharm. Sci.,* 54, 595–598, 1965.

60. Kasirsky, G., Gautieri, R. F., and Mann, D. E., Effect of cobaltous chloride on the minimal carcinogenic dose of methylcholanthrene in albino mice, *J. Pharm. Sci.,* 54, 491–493, 1965.

61. Heath, J. C. and Webb, M., Content and intracellular distribution of the inducing metal in the primary rhabdomyosarcomata induced in the rat by cobalt, nickel and cadmium, *Br. J. Cancer,* 21, 768–779, 1967.

62. Heath, J. C., Webb, M., and Caffrey, M., The interaction of carcinogenic metals with tissues and body fluids: cobalt and horse serum, *Br. J. Cancer.,* 23, 153–166, 1969.

63. Grant, W. C. and Root, W. S., Fundamental stimulus for erythropoiesis, *Physiol. Rev.,* 32, 449, 1952.

64. Warren, C. O., Schubmehl, O. C., and Wood, I. R., Studies with mechanism of cobalt polycythemia, *Am. J. Physiol.,* 142, 173, 1944.

65. Goldwasser, E., Jacobson, L. O., Fried, W., and Plzak, L., Mechanism of the erythropoietic effect of cobalt, *Science,* 125, 1085, 1957.

66. Brown, T. E. and Meineke, H. A., Presence of an active erythropoietic factor (erythropoietin) in plasma of rats after prolonged cobalt therapy, *Proc. Soc. Exp. Biol. Med.,* 99, 435, 1958.

67. Jensen, K. and Thorling, E. B., The effect of anaemia and cobalt of lactic dehydrogenase isoenzymes in kidney tissues of rabbits and its possible relation to the erythropoietin production, *Acta Pathol. Microbiol. Scand.,* 63, 385–390, 1965.

68. Yastrebov, A. P., Mechanism of cobalt action on erythropoiesis, *Fed. Proc.* (Transl. Suppl.), 25, 630–632, 1966.

69. Fisher, J. and Langston, J. W., The influence of hypoxiemia and cobalt on erythropoietin production in the isolated perfused dog kidney, *Blood,* 29, 115–125, 1967.

70. Fisher, J. and Langston, J. W., Sites of erythropoietin production and appearance in tissue fluids and extracts, *Ann. N.Y. Acad. Sci.* 149, 75–87, 1968.

71. Janda, W. E., Fried, W., and Gurney, C. W., Combined effect of cobalt and testosterone on erythropoiesis, *Proc. Soc. Exp. Biol. Med.,* 120, 443–446, 1965.

72. Fried, W. and Kilbridge, T., Effect of testosterone and of cobalt on erythropoietin production by anephric rats, *J. Lab. Clin. Med.,* 74, 623–629, 1969.

73. Rodgers, G. M., George, W. J., and Fisher, J. W., Increased kidney cyclic AMP levels and erythropoietin production following cobalt administration, *Proc. Soc. Exp. Biol. Med.,* 140, 977–981, 1972.

74. Miller, M. E., Howard, D., Stohlman, F., and Flanagan, P., Mechanism of erythropoietin production by cobaltous chloride, *Blood,* 44, 339–346, 1974.

75. Rodgers, G. M., Fisher, J. W., and George, W. J., Lactate stimulation of renal cortical adenylate cyclase: a mechanism for erythropoietin production following cobalt treatment of hypoxia, *J. Pharmacol. Exp. Ther.,* 190, 542–550, 1974.

76. Bickoff, E. M., Livingston, A. L., and Snell, N. S., The occurrence of vitamin B_{12} and other growth factors in alfalfa, *Arch. Biochem.,* 28, 242–252, 1950.

77. Gray, L. F. and Daniel, L. F., Studies of vitamin B$_{12}$ in turnip greens, *J. Nutr.*, 67, 623–634, 1958.
78. Fogg, G. E., *The Metabolism of Algae*, Methuen, London, 1953.
79. Holm-Hansen, O., Gerloff, G. C., and Folke, S., Cobalt as an essential element for blue-green algae, *Physiol. Plant.*, 7, 665–675, 1953.
80. Pintner, I. J. and Provasoli, L., Artificial cultivation of a red-pigmented marine blue-green alga, *Phormidium persicinum, J. Gen. Microbiol.*, 18, 190–197, 1958.
81. Provasoli, L., Nutrition and ecology of protozoa and algae, *Annu. Rev. Microbiol.*, 12, 279–308, 1958.
82. Marston, H. R., Cobalt, copper and molybdenum in the nutrition of animals and plants, *Physiol. Rev.*, 32, 66–121, 1952.
83. Underwood, E. J., *Trace Elements in Human and Animal Nutrition*, Academic Press, New York, 1956.
84. Young, R. S., Cobalt in biology and biochemistry, *Sci. Prog.* (London), 44, 16–37, 1956.
85. Johnson, R. R. and Bentley, O. G., Cobalt and the synthesis of vitamin B$_{12}$-like substances by rumen microorganisms, in *Trace Elements*, Lamb, C. A., Bentley, O. G., and Beattie, J. M., Eds., Academic Press, New York, 1958, 213–225.
86. Ford, J. E. and Hutner S. H., Role of vitamin B$_{12}$ in the metabolism of microorganisms, *Vitam. Horm.*, 13, 101–136, 1955.
87. Hunter, S. H., Aaronson, S., Nathan, H. A., Baker, H., Scher, S., and Cury, A., Trace elements in microorganisms: the temperature factor approach, in *Trace Elements*, Lamb, C. A., Bentley, O. G., and Beattie, J. M., Eds., Academic Press, New York, 1958, 47–65.
88. Robbins, W. J., Harvey, A., and Stebbins, M. E., Studies on euglena and vitamin B$_{12}$, *Bull. Torrey Bot. Club*, 77, 423–441, 1950.
89. Ahmed, S. and Evans, H. J., Effect of cobalt on the growth of soybeans in the absence of supplied nitrogen, *Biochem. Biophys. Res. Commun.*, 1, 271–275, 1959.
90. Residenauer, H. M., Cobalt in nitrogen fixation by a legume, *Nature*, 183, 375–376, 1959.
91. Evans, H. J. and Kliewer, M., Vitamin B$_{12}$ compounds in relation to the requirements of cobalt for higher plants and nitrogen-fixing organisms, *Ann. N.Y. Acad. Sci.*, 112, 735–755, 1964.
92. Sloan, D. L., Young, J. M., and Mildvan, A. S., Nuclear magnetic resonance studies of substrate interaction with cobalt substituted alcohol dehydrogenase from liver, *Biochemistry*, 14, 1998–2008, 1975.
93. Shore, J. D. and Santiago, D., The role of metal in liver alcohol dehydrogenase catalysis, *J. Biol. Chem.*, 250, 2008–2012, 1975.
94. Simpson, R. T. and Vallee, B. L., Zinc and cobalt alkaline phosphatases, *Ann. N.Y. Acad. Sci.*, 166, 670–695, 1969.
95. Ray, K. J., Jr. and Multani, J. S., Characterization of the metal binding site of phosphoglucomutase by spectral studies of its cobalt (II) and nickel (II) complexes, *Biochemistry*, 11, 2805–2812, 1972.
96. Ray, W. J., Jr., Role of bivalent cations in the phosphoglucomutase system, *J. Biol. Chem.*, 244, 3740–3747, 1969.
97. Ray, W. J., Jr. and Goodin, D. S., Ng, L., Cobalt (II) and nickel (II) complexes of phosphoglucomutase, *Biochemistry*, 11, 2800–2804, 1972.
98. Peck, E. J., Jr. and Ray, W. J., Jr., Role of bivalent cations in the phosphoglucomutase system, *J. Biol. Chem.*, 244, 3748–3753, 1969.
99. DeTurk, W. E. and Bernheim, F., The inhibition of enzyme induction and ammonia assimilation in *Pseudomonas aeruginosa* by sulfhydryl compounds and by cobalt, and its reversal by iron, *Arch. Biochem.*, 90, 218–223, 1960.
100. Adiga, P. R., Sastry, K., Venkatasubramanyam, V. et al, *Aspergillus niger, Biochem. J.*, 81, 545, 1961.
101. Padmanaban, G. and Sarma, P. S., Cobalt toxicity and iron metabolism in *Neurospora crassa, Biochem. J.*, 98, 330–334, 1966.
102. Blundell, M. R. and Wild, D. G., Inhibition of bacterial growth by metal salts, *Biochem. J.*, 115, 207–212, 1969.
103. Blundell, M. R. and Wild, D. G., Inhibition of bacterial growth by metal salts, *Biochem. J.*, 115, 213–223, 1969.
104. Hayman, S., Gatmaitan, J. S., and Patterson, E. K., The relationship of extrinsic and intrinsic metal ions to the specificity of a dipeptidase from *Escherichia coli* B, *Biochemistry*, 13, 3386–4494, 1974.
105. Northrop, D. B. and Wood, H. G., Transcarboxylase, *J. Biol. Chem.*, 244, 5801–5807, 1969.
106. Smith, R. J. and Contrera, J. F., Cobalt-induced alterations in plasma proteins, proteases and kinin system of the rat, *Biochem. Paramacol.*, 23, 1095–1103, 1974.

107. **Maines, M. D. and Kappas, A.,** Study of the developmental pattern of heme catabolism in liver and the effects of cobalt on cytochrome P-450 and the rate of heme oxidation during the neonatal period, *J. Exp. Med.,* 141, 1400–1410, 1975.

108. **Tephly, T. R. and Ribbein, P.,** The effect of cobalt chloride administration on the synthesis of hepatic microsomal cytochrome P-450, *Biochem. Biophys. Res. Commun.,* 42, 589–595, 1971.

109. **Marston, H. R.,** Problems associated with "coast disease" in South Australia, *J. Counc. Sci. Ind. Res.,* 8, 111–116, 1935.

110. **Underwood, E. J. and Filmer, J. F.,** Enzootic marasmus: the determination of the biologically potent element (cobalt) in limonite, *Aust Vet. J.,* 11, 84–92, 1935.

111. **Marston, H. R. and Lee, H. J.,** Primary site of the action of cobalt in ruminants, *Nature,* 164, 529–530, 1949.

112. **Underwood, E. J.,** *Trace Elements in Human and Animal Nutrition,* 3rd ed., Academic Press, New York, 1971, 141–165.

113. **Comar, C. L., Davis, G. K., and Taylor, R. F.,** Cobalt metabolism studies radioaction cobalt procedures with rats and cattle, *Arch. Biochem.,* 9, 149–158, 1946.

114. **Comar, C. L. and Davis, G. K.,** Cobalt metabolism studies, tissue distribution of radioactive cobalt administered to rabbits, swine, and young calves, *Arch. Biochem.,* 12, 257–266, 1947.

115. **Dawbarn, M. C., Hine, D. C., and Smith, J.,** The determination of vitamin B_{12} activity in the organs and excreta of sheep without the effect of cobalt deficiency on the vitamin B_{12} content of the blood plasma, *Aust. J. Exp. Biol. Med. Sci.,* 35, 273, 1957.

116. **Somers, M. and Gawthorne, J. M.,** The effect of dietary cobalt intake in the plasma vitamin B_{12} concentration of sheep, *Aust. J. Exp. Biol. Med. Sci.,* 47, 227–233, 1969.

117. **Beck, W. S., Flavin, M., and Ochoa, S.,** Metabolism of propionic acid in animal tissues. I. enzymatic conversion of propionate to succinate, *J. Biol. Chem.,* 229, 997, 1957.

118. **Beck, W. S. and Ochoa, S.,** Metabolism of propionic acid in animal tissues. IV. Further studies on the enzymatic isomerization of methylmalonyl ioenzymes, *J. Biol. Chem.,* 232, 931, 1958.

119. **Gawthorne, J. M.,** The excretion of methylmalonic acid and formiminoglutomic acids during the induction and remission of vitamin B_{12} deficiency in sheep, *Aust. J. Biol. Sci.,* 21, 789, 1968.

120. **Lee, H. J. and Kuchel, R. E.,** The aetiology of Phalaris staggers in sheep. I. Preliminary observations on the preventive role of cobalt, *Aust. J. Agric. Res.,* 4, 88–99, 1953.

121. **Gallagher, C. H., Koch, J. H., Moore, R. M., and Steel, J. D.,** Toxicity of *Phalaris tuberosa* for sheep, *Nature,* 204, 542, 1964.

122. **Masuhara, T. and Migicovsky, B. B.,** Vitamin D and the intestinal absorption of iron and cobalt, *J. Nutr.,* 80, 332–336, 1963.

123. **Sobel, A. E. and Burger, M.,** Calcification. XIII. The influence of calcium, phosphorus, and vitamin D on the removal of lead from blood and bone, *J. Biol. Chem.,* 212, 105–110, 1955.

124. **Meintzer, R. B. and Steenbock, H.,** Vitamin D and magnesium absorption, *J. Nutr.,* 56, 285–294, 1955.

125. **Greenberg, D. M.,** Studies in mineral metabolism with the aid of artificial radioactive isotopes. VIII. Tracer experiments with radioactive calcium and strontium on the mechanism of vitamin D action in rachitic rats, *J. Biol. Chem.,* 157, 99–104, 1945.

126. **Mrza, F. R. and Bacon, J. A.,** Influence of excessive amounts of vitamin D Sr^{89} metabolism in the rat, *Proc. Soc. Exp. Biol. Med.,* 104, 1–3, 1960.

127. **Worker, N. A. and Migicovsky, B. B.,** Effect of vitamin D on the utilization of beryllium, magnesium, calcium, strontium and barium in the chick, *J. Nutr.,* 74, 490–494, 1961.

128. **Worker, N. A. and Migicovsky, B. B.,** Effect of vitamin D on the utilization of zinc, cadmium and mercury in the chick, *J. Nutr.,* 75, 222–224, 1961.

129. **Pollack, S., George, J. N., Reba, R. C., Kaufman, R. M., and Crosby, W. H.,** The absorption of nonferrous metals in iron deficiency, *J. Clin. Invest,* 44, 1470–1473, 1965.

130. **Villaret, M., Bertrand, I., Justin-Besancon, L., and Even, R.,** Les cirrhoses cobaltiques, *C. R. Seances Soc. Biol.* (Paris), 108, 956, 1931.

131. **Cajano, A.,** Sulle lesioni del fegato nella intossicazione cronica sperimentale con cobalto, *Folia Med.* (Naples), 34, 8, 1951.

132. **Hurst, E. W. and Hurst, P. E.,** The etiology of hepatolenticular degeneration, poisoning with manganese, chloroform, phenylhydrazine, bile and guanidine, *J. Pathol. Bacteriol.,* 31, 303, 1928.

133. **Findlay, G. M.,** The experimental production of biliary cirrhosis by salts of manganese, *Br. J. Exp. Pathol.,* 5, 92, 1924.

134. **Valberg, L. S., Ludwig, J., and Olatunbosun, D.,** Alteration in cobalt absorption in patients with disorders of iron metabolism, *Gastroenterology,* 56, 241–251, 1969.

135. **Aisen, P., Aasa, R., and Redfield, A. G.,** The chromium, manganese, and cobalt complexes of transferrin, *J. Biol. Chem.,* 244, 4628–4633, 1969.

136. Olatunbosun, D., Corbett, W. E. N., Ludwig, J., and Valberg, L. S., Alteration of cobalt absorption in portal cirrhosis and idiopathic hemochromatosis, *J. Lab. Clin. Med.,* 75, 754–762, 1970.

137. Schade, S. G., Felsher, B. F., Bernier, G., and Conrad, M. E., Interrelationship of cobalt and iron absorption, *J. Lab. Clin. Med.,* 75, 435–441, 1970.

138. Thomson, A. B. R. and Valber, L. S., Kinetics of intestinal iron absorption in the rat: effect of cobalt, *Am. J. Physiol.,* 220, 1080–1085, 1971.

139. Thomson, A. B. R., Valberg, L. S., and Sinclair, D. G., Competitive nature of the intestinal transport mechanism for cobalt and iron in the rat, *J. Clin. Invest.,* 50, 2384–2391, 1971.

140. Thomson, A. B. R., Shaver, C., and Lee, D. J., Effect of varying iron stores on site of intestinal absorption of cobalt and iron, *Am. J. Physiol.,* 220, 674–678, 1971.

141. Sheehan, R. G., Interrelationships of iron and cobalt absorption: mucosal distribution of cobalt during absorption, *Proc. Soc. Exp. Biol. Med.,* 146, 993–996, 1974.

142. Gardner, F. H., Use of cobaltous chloride in the anemia associated with chronic renal disease, *J. Lab. Clin. Med.,* 41, 56–64, 1953.

143. Kriss, J. P., Carnes, W. H., and Gross, R. T., Hypothyroidism and thyroid hyperplasia in patients treated with cobalt, *JAMA,* 157, 155–161, 1955.

144. Schirrmacher, U. O. E., Case of cobalt poisoning, *Br. Med. J.,* 1, 544–545, 1967.

145. Jacobziner, H. and Raybin, H. W., Poison control: accidental cobalt poisoning, *Arch. Pediatr.,* 78, 200–205, 1961.

146. Coates, E. O. and Watson, J. H. L., Diffuse interstitial lung disease in tungsten carbide workers, *Ann. Intern. Med.,* 75, 709–716, 1971.

147. Camarasa, J. M. G., Cobalt contact dermatitis, *Acta Derm. Venereol.,* 47, 287–292, 1967.

148. Morin, Y. L., Foley, A. R., and Martineau, G., Quebec beer drinkers cardiomyopathy: forty-eight cases, *Can. Med. Assoc. J.,* 97, 881–883, 1967.

149. Kesteloot, H., Roelandt, J., and Willems, J., An enquiry into the role of cobalt in the heart disease of chronic beer drinkers, *Circulation,* 37, 854–864, 1968.

150. Alexander, C. S., Cobalt-beer cardiomyopathy: a clinical and pathologic study of twenty-eight cases, *Am. J. Med.,* 53, 395–417, 1972.

151. McDermott, P. H., Delaney, R. L., Egan, J. D., and Sullivan, J. F., Myocardosis and cardiac failure in men, *J.A.M.A.,* 198, 253–256, 1966.

152. Sullivan, J. F., Egan, J. D., and George, R. P., A distinctive myocardiopathy occurring in Omaha, Nebraska: clinical aspects, *Ann. N.Y. Acad. Sci.,* 156, 526–543, 1969.

153. Sullivan, J., Parker, M., and Carson, S. B., Tissue cobalt content in "beer drinkers' myocardiopathy," *J. Lab. Clin. Med.,* 71, 893–896, 1968.

154. Grice, H. C., Goodman, I., and Munro, I. C., Myocardial toxicity of cobalt in the rat, *Ann. N.Y. Acad. Sci.,* 156, 189–194, 1969.

155. Burch, R. E., Williams, R. V., and Sullivan, J. F., Effect of cobalt, beer, and thiamin-deficient diets in pigs, *Am. J. Clin. Nutr.,* 26, 403–408, 1973.

156. Morin, Y. and Daniel, P., Quebec beer-drinkers' cardiomyopathy: etiological considerations, *Can. Med. Assoc. J.,* 97, 926–928, 1967.

157. Sullivan, J. F., Egan, J. D., and George, R. P., Experimental "metabolic" cardiopathies and their relationship to human heart disease. VI. Primary cardiomyopathies and congestive heart failure in clinical medicine, *Ann. N.Y. Acad. Sci.,* 156, 526–542, 1969.

IODINE TOXICITY IN MAN AND ANIMALS

G. I. Vidor

INTRODUCTION

Iodine* has long been recognized as an essential nutrient in man and animals and is the most intensively studied nutritional trace element. It is an integral part of the thyroid hormones, and the relationship of iodine deficiency to endemic goiter is well established. The minimal requirement for goiter prophylaxis has been recommended by the World Health Organization. However, the upper "safe" limit of iodine intake, beyond which adverse reactions may occur, remains to be established.

While in the past public health authorities were concerned with eradication of iodine deficiency, in recent years concern has been growing about the rapid increases in dietary and environmental iodine. Rapid changes in agronomical practice, animal husbandry, dairy processing and technology, and food processing, preservation, packaging, and marketing, coupled with improved transport, have resulted in marked increases in dietary iodine, which is indicated by falling thyroid radioiodine uptakes and rising urinary iodine excretions. Iodine intakes in America, Australia, and many developed countries are significantly greater than recommended or required for adequate nutrition.

The adverse effects of iodine include toxic, idiosyncratic, and hypersensitivity reactions. Iodide goiter, hypothyroidism, and thyrotoxicosis have been reported in predisposed persons. Goiter has also been reported in animals in response to excess iodine. Toxic reactions to pharmacological doses of iodine and iodine-containing contrast media have been well documented. However, although reporting physicians were convinced and temporal relationships between dietary iodine intake and adverse reactions were apparent, scientific proof of a causal relationship is largely lacking. There is some laboratory and clinical evidence to suggest that iodine and iodine-containing foods and drugs may be haptenic, but there is no consensus on the role of immunological and nonimmunological mechanisms in untoward reactions to iodine.

SOURCES OF IODINE FOR MAN

Atmosphere

By evaporation the oceans are the major source of atmospheric iodine, which is, then, deposited on land and vegetation by rain and snow.[1] Seacoast air may contain up to 400 $\mu g/m^3$.[2] Vought et al. report ambient air contained 0.74 μg I per cubic meter.[1,3] The iodine content of air is increased by the pollution from combustion of gasoline and oil.[1]

Iodophors reduce the vapor pressure of iodine leading to the slow release of iodine into the atmosphere.[3,4] When these sanitizers are used, significantly increased air concentrations of iodine may occur.[3] Weather modification with artificial nucleating agents, usually silver iodide, used for hail suppression and rain or snow augmentation, may add to total atmospheric iodine as well as provide the iodine which enters rivers, soil and ecological food chains, and the human diet.

Human respiratory exposure to iodine has been calculated to be 5 $\mu g/day$ with an atmospheric concentration of 0.7 μg I per cubic meter and 100 μg of I per day if total absorption occurs with concentrations of 5 μg I per cubic meter.[5] However, the contribution of atmospheric iodine to total iodine intake is unknown.

Water and Beverages

The iodine content of water is variable. Fresh water from mountain rivers has a lower

---

* Iodine in this report is used in the generic sense to denote biologically available iodine, whether it be bound to nutrients or in drugs, dyes, or organic complexes.

iodine content than coastal rivers and lakes. Contamination of water by dry sewage, which may contain 7 μg I per kilogram, or treated effluent, containing as much as 1910 μg I per liter, may raise its iodine content appreciably.[1] The iodine content of urban water supplies may vary from 0 to 8.7 μg I per liter or, at times, higher. In some areas where domestic water supplies are iodinated, it may be 500 μg I per liter. In any case, water is not normally a significant source of iodine for man,[3-10] except in areas where water supplies are artificially iodinated. Sea water, however, contains higher concentrations of iodine (50 μg/kg); accordingly, this iodine is only available in animal feed and human diets containing fish, fish meal, or sea weed.

The iodine content of beer and wine is probably variable. Koutras[11] reports iodine contents of 43 to 48 μg/kg for beer and 8 to 32 μg/kg for wine.

Plants and Plant Products

Plants such as vegetables, fruits, and cereals absorb iodine from the soil by their root system and from the atmosphere.[12] Numerous factors, such as geological soil type, past glaciation,[13] the use of iodine-containing fertilizers, and atmospheric and water iodine, influence the iodine content of these foods. Consequently, significant variations in iodine content of vegetables, fruits, and cereals, grown in different localities, are likely to occur. The extreme variations, which may occur, are well illustrated in the data reported by the Chilean Iodine Information Bureau[14] (Table 1), Vought and London[15] (Table 2), and Koutras et al.[10] (Table 3). Nor must it be assumed, with modern methods of food processing and preservation and improved transportation, that populations are dependent on locally grown products. It is probable that extrapolation of data in Tables 1, 2, and 3 to different localities is invalid.

Though these foods are significant contributors of iodine in the diet, commercially baked bread is now the major source of this element. Large bakeries, employing the automated, continuous mix process, add calcium or potassium iodate, 12.5 to 20 μg/g of flour, as a conditioner and oxidant to dough.[16] The addition of iodized salt may further contribute to the iodine content of this bread. In conventionally baked bread, an iodate-containing yeast food is often employed. These products contain 0.1% potassium iodate and are used at the rate of 3 to 6 μg/g of flour.[16] These differing baking methods result in wide variations in iodine content of bread, ranging from 1 to 8 μg I per gram of bread (40 to 263 μg iodine per slice).[17] Bread, however, may cease to be a major source of dietary iodine if the present trends to replace iodate with a combination of bromate and ascorbic acid or azodicarbonamide[18] become universal.

Animals and Animal Products

Milk, milk products, eggs, meat, and fish contain larger quantities of iodine than "plant foods" (Table 1).[2,14-16] Milk and its products and meat are the main sources of iodine in the diet; fish, although it contains more iodine (Table 1), is not an important item of diet in developed countries but may make a significant contribution to dietary iodine[20] for coastal and island dwellers. Milk and milk products are probably the largest single source of dietary iodine at present. Iodine, which is concentrated and secreted by the mammary glands in milk,[21,22] increases following dietary iodine supplements in feed, salt blocks, or iodine-containing veterinary medication.[23-32]

In Australia iodine supplementation of livestock is not prevalent, and the high iodine content of milk was traced to iodophors, which have wide application in the dairy industry from farm to processor.[33] Surveys indicated that 87.1% of farms and 59.1% of dairy factories used iodophors for one or more sanitizing tasks. Iodine accumulates in the milk with each exposure to iodophors so that its concentration rises arithmetically, since no significant losses occur under normal handling.[33] The iodine content of manufactured products depends upon the initial iodine content of milk, the protein content of the

Table 1
AVERAGE IODINE CONTENT OF FOODS AND COMMODITIES USED IN PREPARED FOODS[a]

Food or product	Iodine content (μg/kg) Fresh basis	Dry basis	Food or product	Iodine content (μg/kg) Fresh basis	Dry basis
Cereal grains			Legumes		
Rice	22	39	Peas	23	223
Maize	27	43	Beans	36	245
Wheat	37	44	Mean	30	234
Flour	42	—			
Bread	58	—	Meat		
Barley	58	92	Mutton	27	—
Oats	60	91	Beef	28	—
Rye	72	84	Veal	28	—
Mean	47	65	Pork	45	—
			Bacon	77	—
Eggs			Lard	97	—
Mean	93	—	Mean	50	—
Fish			Milk		
Anadromous			Cow's milk	35	—
Sea trout	320	1,028	Cheese	51	—
Salmon	341	1,030	Butter	56	—
Mean	340	1,029	Mean	47	—
Freshwater					
Carp	17	68	Shell fish		
River bass	30	115	Crab and crabmeat	308	1,292
Lake trout	31	88	Oysters	577	4,712
River perch	40	194	Clams	783	3,595
Mean	30	116	Lobster	1,020	4,744
Marine			Shrimps	1,300	4,987
Sole	163	1,072	Mean	798	3,866
Sea bass	250	471			
Sardines	284	745	Vegetables		
Mackerel	371	1,031	Gourds, pumpkins,	12	600
Halibut	520	2,225	and marrow		
Herring	520	1,358	Cauliflower	12	221
Sea perch	742	3,105	Beets	21	233
Cod	1,463	7,493	Onions	22	204
Haddock	3,180	15,941	Cucumber	25	400
Mean	832	3,715	Lettuce	26	668
			Carrots	38	202
Fruits			Turnips	40	343
Pears	10	62	Asparagus	42	1,102
Tomatoes	17	196	Potatoes	45	197
Apples	16	277	Cabbage	52	260
Cranberries	29	100	Spinach	201	1,636
Mean	18	159	Mean	29	385

Note: Data compiled up to 1951.

[a] The iodine content of foods may vary in different localities.

From Fisher, K. D. and Carr, C. J., Eds., Iodine in Foods: Chemical Methodology and Sources of Iodine in the Human Diet, FDA 71-294, Life Sciences Research Office, Federation of American Societies for Experimental Biology, Bethesda, Md., 1974, 46. With permission.

Table 2
IODINE CONTENTS OF FOODS

		Iodine (μg/wet kg)	
Food category	No. of samples	Mean ± SE	Median
Seafoods	7	660 ± 180	540
Vegetables	13	320 ± 100	280
Meat products	12	260 ± 70	175
Eggs	11	260 ± 80	145
Dairy products	18	130 ± 10	139
Bread and cereal	18	100 ± 20	105
Fruits	18	40 ± 20	18

From Vought, R. L. and London, W. T., *Am. J. Clin. Nutr.,* 14, 186, 1964. With permission.

Table 3
IODINE CONTENT OF WATER, MILK, AND FOOD ITEMS IN TWO AREAS OF GREECE[a]

	Athens (goiter-free area)		Thessalia (endemic area)	
Food category	No. of samples	Iodine[b] (μg/wet kg)	No. of samples	Iodine[b] (μg/wet kg)
Drinking water	12	4.7 ± 0.3	163	2.4 ± 0.2
Milk				
Cow	12	41.5 ± 3.6	68	25.0 ± 2.2
Goat	—	—	56	22.0 ± 3.3
Sheep	—	—	59	94.0 ± 9.3
Soft cheese	15	151.2 ± 18.5	15	84.8 ± 12.2
Chicken dishes[c]	16	30.1 ± 10.2	16	5.7 ± 2.5
Meat dishes[c]	10	1.6 ± 0.5	16	0.8 ± 0.2
Fish dishes[c]	9	14.2 ± 6.6	—	—
Legume dishes[c]	14	0.9 ± 0.2	16	0.6 ± 0.3
Bread	12	15.6 ± 12.0	21	5.4 ± 2.5

[a] Data recalculated from Koutras et al.[10]
[b] Mean and standard error of mean.
[c] Estimated mean values based on average portion size.

From Fisher, K. D. and Carr, C. J., Eds., Iodine in Foods: Chemical Methodology and Sources of Iodine in the Human Diet, FDA 71-294, Life Sciencies Research Office, Federation of American Societies for Experimental Biology, Bethesda, Md., 1974, 46. With permission.

product, and whether the product has been heat treated. The iodine content of high-protein products, cheese and milk powders, is markedly higher than the iodine content of high-fat products, butter and reduced cream.[33]

Recent studies by Dunsmore and others in New South Wales indicate that contamination of milk and milk products in Australia is much higher than previously reported or suspected[33-38] (Table 4). The Australian experience may be relevant to other countries or areas where iodophors are widely used. In these studies gross contamination of milk with iodine was often evident although there was immense variability in the

Table 4
THE IODINE CONTENT OF MANUFACTURED DAIRY PRODUCTS IN NEW SOUTH WALES AND TASMANIA

Type	Item	No. of samples	I content (μg/kg) Min	I content (μg/kg) Max	Mean (A) (μg/kg)	Normal milk (N) (μg/kg)	Ratio A/N
		New South Wales					
ese	Cheddar	18	152	1,207	593	54	11.0
	Cottage	8	391	1,026	668	61	11.0
	Fetta	4	549	1,710	999	—	11.0
	Mozzarella	6	246	1,995	1,001	—	
	Ricotta	5	317	2,336	897	—	
	Other fancy varieties	9	119	1,971	714	—	
	Processed cheddar	4	241	1,011	677	—	
ter	Salted	23	20	865	156	56	2.8
	Salted and washed with iodinated water	1			1,116	—	
	Unsalted	2			98	—	
vders	Skim milk:						
	Roller-dried	5	2,237	3,985	2,985	81	35.6
	Spray-dried	3	4,522	8.216	6,730	646	10.4
	Whole milk	8	691	6,729	3,317	320	10.3
	Calf feeds	2			4,214	304	13.9
	Buttermilk	1			2,997	55	54.4
	Canned infant milk powder	7	100	3,613	1,512	—	
ssert products	Yoghurt	15	466	2,010	853	—	
	Sour cream	5	310	650	438	41	11.0
	Cultured buttermilk	2			574	—	
	Ice cream	9	316	706	525	35	15.0
ined goods	Infant liquid preparation	1			132	—	
	Milk beverages	4	308	1,506	817	—	
	Evaporated milk	5	111	540	250	—	
	Condensed milk	4	97	573	257	—	
	Reduced cream	2			60	—	
nfections	Chocolate	6	123	882	435	—	
		Tasmania					
eese		12	197	972	592		
tter		4	<20	220	86		
wders		8	2,408	11,480	4,808		

ta taken from Dunsmore, D. G., *Aust. J. Dairy Technol.*, 31, 125, 1976, and Tasmanian State Thyroid Advisory mmittee, unpublished information, 1975. With permission.

iodine content of milk. The monthly means ranged from 480 to 1075 μg I per liter for raw milk with extremes of 270 and 1600 μg I per liter and those for pasteurized milk, from 512 to 1100 μg I per liter with extremes of 264 and 1220 μg I per liter. The mean for the whole year was 727 μg I per liter for raw milk and 760 μg I per liter for pasteurized milk. A seasonal peak during the low-production winter period was noted[33] (Table 5).

Powdered milk products have a high iodine content with spray-dried skim milk (6370 μg I per kilogram) having double the iodine content of roller-dried milk (2985 μg I per kilogram), a result in accord with reports on uncontaminated milk[29,33] (Table 4). The means for cheese ranged from 593 to 1001 μg I per kilogram; noniodated butter, 98 to

Table 5
THE IODINE CONTENT OF WHOLE MILK AT
THREE SITES ALONG THE PRODUCTION LINE

Sampling date	Description	No. of samples	I conc (μg/l) Mean	I conc (μg/l) Range
March through April	Cow sample uncontaminated by iodophors	20	37	1—160
March through April	Factory raw bulk sample	20	570	370—1090
May	Bottled pasteurized milk	19	706	320—1170

From Dunsmore, D. G., *Aust. J. Dairy Technol.*, 31, 125, 1976. With permission.

156 μg I per kilogram; milk powders, 1512 to 6730 μg I per kilogram; dessert products, 438 to 853 μg I per kilogram; and canned goods, 60 to 817 μg I per kilogram. These data show a marked level of contamination with figures significantly higher than previously reported.[7,35-40] The iodine contents of ice cream (525 μg I per kilogram) and chocolate (435 μg I per kilogram) are similar to other reported figures.[7]

Tests commissioned by the Tasmanian State Advisory Committee showed that the mean iodine content of exported Tasmanian cheese was 592 μg/kg (range, 197 to 972 μg/kg); butter, 86 μg/kg (range, less than 20 to 220 μg/kg); milk powders, 4808 μg/kg (range, 2408 to 11,480 μg/kg), all of which approximate Dunsmore's findings in New South Wales.[41] These concentrations are well below the taste threshold of 2000 μg I per liter in milk,[42] below the minimum concentration of iodine 12,500 to 16,000 μg I per liter which has been reported to affect a bacteriological keeping quality test,[33,43,44] and below the minimum concentrations, 12,500 μg I per liter,[44] reported to affect acid development in lactic acid fermentation. The Tasmanian State Thyroid Advisory Committee has recommended a ceiling of 300 μg I per liter in milk,[45] and Switzerland has imposed a limit of 500 μg I per liter.[35]

Eggs accumulate systemic iodine and may contain high levels of it as a result of iodine compounds in feed supplements. Iodophor misting in poultry sheds can increase the atmospheric iodine content markedly; the importance or significance of this in relation to the iodine content of eggs and poultry is unknown. The iodine content of eggs is dependent on the level of iodine in feed and feed supplements and may, consequently, vary enormously[13-15] (Tables 1 and 2). When high concentrations, 100 to 500 mg/day, of iodine are fed to laying hens, the iodine content of eggs may reach 50 to 120 mg/kg or 5 to 12 mg/100 g.[46] Such high levels would not normally be reached because of established limitations (12.5 mg per kilogram or liter) on feed and water.[32]

Marine fish and shell fish are excellent sources of dietary iodine (Tables 1 and 2). Fresh water fish, however, contain only relatively low concentrations of iodine (Table 1).

Prepared and Processed Foods

In the United States, Australia, and probably many other affluent countries or communities, there is an increased trend toward the use of processed or prepared foods.[16] Intentional addition of iodine compounds is limited to salt, dietary supplements, infant formulas, and meat substitutes.[48] In Australia iodophors are used increasingly in the poultry, processed food, small goods, and meat industry.[45] The iodine content of such processed foods will therefore be variable. Significant differences may be observed in similar foods sampled in different localities. Factors which influence iodine content of processed and prepared foods are source and iodine content of the raw materials, use of iodized or noniodized salt, use of iodophor sanitizers, and application of heat or freezing. Data on the iodine content of processed and prepared foods are scanty[49,50] but indicate that these foods may be a significant source of dietary iodine. Examination of 12 types of frozen chicken dinners showed a range of 35 to 4605 μg I per package,[51] and analyses of products from a national "prepared food chain" showed a fairly constant 400 to 500 μg I per kilogram of product[52] (Table 6). These findings made in one locality cannot be considered applicable to other localities.

Salt, Food Additives, and Food Colors
Salt

The sale of iodized salt is increasing in the U.S. and many other parts of the developed world though per capita salt consumption in the U.S. is declining.[53] The Select Committee on "Generally Recognized as Safe" Substances (1974) estimated iodine intake from salt to be 270 μg/person/day.[16,54,55]

Accepted Iodinated Food Additives

Cuprous iodide and potassium iodide — They serve as nutrients or dietary supplements in table salt (max 0.01%).[16] Potassium iodide is also used in various food categories as shown in Table 7.

Potassium iodate and calcium iodate — They are used in the continuous mix baking process (max allowable 75 μg/kg, usually 12.5 to 20 μg/kg used).[16]

Alginates — They contain 9 mg I per kilogram but are used in very small quantities in processed foods as emulsifiers, stabilizers, and thickeners. The iodine intake from this source is estimated to be very small.[16]

Brown algae and kelp — These plants are sources of iodine in foods for special dietary use. The code of the U.S. Federal Register restricts intake from these foods to 150 μg I

Table 6
IODINE CONTENT OF FOODS FROM McDONALD CORPORATION AT OAK BROOK, ILLINOIS

Product	Iodine content (μg/kg product)
Egg McMuffin®	220
Hamburger	490
Cheeseburger	430
0.25-lb Hamburger	420
0.25-lb Cheeseburger	450
Big Mac®	420
Filet of fish	840
French fries	200
Chocolate shake	470
Vanilla shake	470
Strawberry shake	490

Note: This study was conducted in 1973, and the data were analyzed by Binnert's method.

From Nutritional Analysis of Foods Served at McDonald's Restaurants, WARF, Inc., Madison, Wis., 1973, 14. With permission.

Table 7
USE OF POTASSIUM IODIDE IN FOODS

Food categories	Usual use (%)[a]	Maximal use (%)[a]
Seasonings and flavorings	0.00866	0.00926
Imitation dairy products	0.00043	0.0043
Baby formulas	0.00013	0.00015
Processed fruit	0.00010	0.00010
Milk products	0.00002	0.00007

[a] Figures from the Subcommittee on Review of the GRAS List (Phase II), 1972.

From Fisher, K. D. and Carr, C. J., Eds., Iodine in Foods: Chemical Methodology and Sources of Iodine in the Human Diet, FDA 71-294, Life Sciences Research Office, Federation of American Societies for Experimental Biology, Bethesda, Md., 1974, 33. With permission.

Table 8
THERAPEUTIC DRUG PREPARATIONS CONTAINING
IODINE IN VARIOUS FORMS

Type of drug	Example	No. of products[a]
Thyroid preparations	Thyroglobulin	30
Radiopaque substances for roentgenography	Acetrizoate	53
Radioactive iodine	[131]Iodine	23
Anti-infective agents	Diiodohydroxyquin	21
Inorganic iodine salts	Potassium iodide	14
Antineoplastic agents	Sodium [131]iodide	4
Kidney function tests	Sodium iodohippurate	2
Expectorants	Potassium iodide	14[b]

[a] Determined from a computerized listing of OTC (over-the-counter) Drug Products supplied by the Bureau of Drugs, Food and Drug Administration, from an unpublished compilation, 1973.

[b] Determined from number of products listed in the American Hospital Formulary Service, American Society of Hospital Pharmacists, Washington, D.C., 1973.

From Fisher, K. D. and Carr, C. J., Eds., Iodine in Foods: Chemical Methodology and Sources of Iodine in the Human Diet, FDA 71-294, Life Sciences Research Office, Federation of American Societies for Experimental Biology, Bethesda, Md., 1974, 33. With permission.

per day.[56] The iodine content of algae varies with a range of 0.8 to 4.5 g per dry kilogram.[4]

Coloring Substances

Erythrosine (tetraiodofluorescein), which is partially degraded in the alimentary tract, is widely used as a red coloring in food and drugs and has become a significant source of dietary iodine.[58-60] Concentrations of iodine, up to 175 μg/g in dry cereals, fruit cocktails, maraschino cherries, and cake mixes, presumably due to erythrosine,[58] have been reported.

Medications

Many commonly used medications have a significant iodine content either as a part of the active ingredients or as a result of coloring with erythrosine.[58,61] Drugs with iodine content are hormonal preparations containing thyroid hormones and X-ray contrast media, some of which are in oily vehicles (Lipiodol) and may release iodine for months or even years; others are antibacterial agents (iodinated hydroxyquinolines, topical or parenteral), anticholinergic drugs (visceralgin and tyramide), and expectorants containing potassium iodide. Results of a survey of drugs with an iodine content are tabulated. Erythrosine-containing dyes in tablets and syrups may produce variable iodine contents in medications ranging from 1 to 3 μg I per tablet, and some pediatric mixtures of penicillin and tetracycline contain 100 μg I per milliliter (Table 8 and Appendix). In addition, many proprietary over-the-counter remedies, ranging from vitamin supplements to rectal suppositories, have a significant iodine content.[62,63]

Iodophors

These substances are combinations of iodine with a carrier, usually polyvinylpyrollidone. They can be combined with detergents and are commonly used sanitizers and disinfectants. Their antibacterial activity is due to iodine, which is slowly released from the carrier.[16] These substances are efficient, short contact-time sanitizers. They are relatively stable in the concentrate, nonstaining, color indicating, and noncorrosive to stainless steel.[34]

As was mentioned previously, iodophors have found extensive use in the dairy industry [33,34,64-69] and are being increasingly used in the poultry,[45,70] small goods, and prepared food industries,[45] laundries,[4] and hospitals.[15,34,39,71] In the dairy industry iodophors are used in udder cleaning and sanitizing, specifically premilking udder drying cloths, postmilking teat dipping, and postmilking udder creams. Iodophors are also used on dairy farms in the cleaning of milking machines and refrigerated farm vats. The uses of iodophors in the dairy factory include: washing and sanitizing bulk-milk, road-transport tankers and sanitizing raw milk vats, milk pipe lines, pasteurized milk vats, and rubber bottle-sealers on filling machines.[34] Owing to the many uses of iodophors in the dairy industry, serious contamination of milk and milk products may occur. (See section on Prepared and Processed Foods.) Dunsmore and Wheeler estimate that where bad practices are exercised and iodophors are used exclusively for sanitizing, contamination could reach 1018 μg I per liter of milk. They further state that with good practice contamination need only add 154.5 μg I per liter to milk[34] (Tables 4, 5, and 9).

Iodophors used by the poultry industry are dispersed by misting in poultry sheds to limit the spread of communicable diseases.[45,70] This practice increases atmospheric concentrations of iodine as well as deposition of iodine on walls, perches, feed belts, etc. The full extent of usage of these compounds in the poultry and other food industries as sanitizers is unknown, and their effects on iodine nutrition in animals and man remain undetermined.

The use of iodophors has been reported in laundries[15,71] and in hospitals as sanitizers for the cleaning of walls and floors,[4] mists in operating theatres, and surgical scrub.[39] Connolly has shown increased urinary iodine excretion in theatre staff following exposure to iodophors.

SOURCES OF IODINE FOR ANIMALS

Iodine, present in soil, air, and water, becomes a constituent of plants and animals used as food by other animals. Additional iodine may be acquired by deliberate supplementation or accidental contamination or adulteration. Iodine from air and water contributes to the total iodine intake of man and animal. Similarly, iodine-containing drugs and disinfectants may be incorporated in food or beverages and ingested.

Forage and Feeds

The iodine content of plants and plant products is influenced by many factors, including atmospheric, soil and water content of iodine, soil type, soil pH, and fertilizer practices.[72,73] Therefore, iodine content of forage and feeds will vary in different localities[30,32] (Tables 10–12).

Feed Supplements

The U.S. National Research Council (N.R.C.) recommends[74-78] that iodine should be incorporated in vitamin and mineral supplements made available in salt blocks or mineral mixtures in free-choice rations. Iodine deficiency can be prevented by feeding salt containing 0.0007% "stable" iodine. The suggested 800 μg/day allowance for a mature animal is normally exceeded by approximately double the amount, because feeding iodine-fortified salt at a rate of 0.25% of dry diet provides 0.19 mg/kg iodine.[5] Since iodine provided for foraging animals as salt blocks is subject to atmospheric conditions, a stable form of iodine has to be used. Currently, calcium and potassium iodate,

Table 9
CONTRIBUTION OF FARM AND FACTORY USES OF IODOPHORS
TO THE IODINE CONCENTRATION OF MARKET MILK

Iodophor use	Avg. iodine conc. applied (mg/l)	Bad practice		Good practice		Ref.
		Expected milk contaminate (µg/l I)	Comments	Expected milk contaminate (µg/l I)	Comments	
Cow hygiene						
Nonwiped premilking	50	34.1		0	Eliminate this step	66
Sanitizing premilking udder-drying cloth	50	0		0		66
Postmilking teat dipping	5000	196.6	Bad agricultural practice	57.0	Preventative mastitis treatment; good agricultural practice	66
Postmilking iodophor udder cream		53.6	Water prewash	0	Soap and towel prewash	66
Farm equipment cleaning						
Milking machine	25	243.4	µg/l/half vat	95.4	(µg/l/half vat)	67
Refrigerated farm vat	25	296.0		2.1		65
Factory equipment sanitizing						
Road transport tanker	25 soln	45.0		0		68
Raw milk vats	25 soln or 75 fog	0		0		68
Pasteurized milk vats	25 soln or 75 fog	25.0		0		68
Bottle filler and rubber seal sanitizing	25	125	Mean initial bottles	0	Eliminate this process	68
Net total		1018.7		154.5		
Allowance for physiological iodine		50.0		50.0		
Gross total		1068.7		204.5		

From Dunsmore, D. G. and Wheeler, A. M., *Aust. J. Dairy Technol.*, in press. With permission.

Table 10
IODINE CONTENT OF SOME ANIMAL FEEDS

Location of study	Hay (mg/kg)	Cow silage (mg/kg)	Feed concentrate (mg/kg)	Ref.
Ohio and Indiana	0.102	0.050	0.002	30
Illinois	0.62–1.02	0.34–0.70		32
Mean	0.82	0.64		
Maryland	1.31–2.54	1.00–1.87		32
Mean	1.87	1.64		

Table 11
IODINE CONTENT OF SOME
ANIMAL FEEDS

Type of feed	Average iodine content (μg/kg)	
	Fresh wt basis	Dry wt basis
Green forage		
Alfalfa	28	313
Mixed pasture	60	553
Hay		
Alfalfa	188	274
Meadow	194	368
Oat	219	—
Soybean	225	526
Sweet clover	—	320
Timothy	80	262
Oil seed meals		
Cottonseed	—	149
Groundnut	—	200
Linseed	—	110
Soybean	—	170
Straw		
Barley	—	630
Oat	—	468
Rye	—	490
Wheat	—	419

From Fisher, K. D. and Carr, C. J., Eds., Iodine in Foods: Chemical Methodology and Sources of Iodine in the Human Diet, FDA 71-294, Life Sciences Research Office, Federation of American Societies for Experimental Biology, Bethesda, Md., 1974, 46. With permission.

pentacalcium orthoperiodate, cuprous iodide, and thymol diiodide are used because of their stability and the bioavailability of the iodine.[11,13,24,25,79,80] 3, 5-Diiodosalicylic acid (DIS) has also been used, but cattle and poultry have less ability to use this source than other species, such as the rat.[81,82,235]

Ethylenediamine dihydroiodide (EDDI), a feed supplement, is readily absorbed and appears to be better concentrated and retained longer in the tissues than an equivalent amount of potassium iodide.[23]

Calcium iodobehenate, potassium iodide, sodium iodate, sodium iodide, and thymol

Table 12
NATURALLY OCCURRING IODINE IN PLANTS
AND PLANT PRODUCTS USED IN
COMMERCIAL ANIMAL FEEDS

Plant or plant part	Iodine content dry wt basis (μg/kg)	Ref.
Alfalfa (Medicago sativa)		
aerial parts, dehydrated, ground,	129	74
minimum 15% protein	120	75
aerial parts, dehydrated, ground,	161	74
minimum 17% protein	150	75, 76
aerial parts, dehydrated, ground,	140	75, 76
minimum 20% protein		
aerial parts, dehydrated, ground	200	75
minimum 22% protein		
Bermuda grass (Cynodon dactylon)		
suncured hay	115	74

From Fisher, K. D. and Carr, C. J., Eds., Iodine in Foods: Chemical Methodology and Sources of Iodine in the Human Diet, FDA 71-294, Life Sciences Research Office, Federation of American Societies for Experimental Biology, Bethesda, Md., 1974, 46. With permission.

diiodide are also used as dietary supplements of trace minerals.[16] The effect of potassium iodide in preventing goiter in several species of animals fed goitrogenic diets is well documented.[79]

Medications

Veterinary medications may add to iodine nutrition. Potassium iodide is used in fungal infections. EDDI is used in high dietary doses to prevent or treat footrot and soft-tissue lumpy-jaw in cattle.[27,83] It was suggested that EDDI was metabolized differently in the ruminant,[83] but subsequently, it was shown that iodine was converted to and absorbed as iodide.[23] At a level of 50 mg/cow/day, serum iodide and milk iodine levels are significantly elevated over control animals.[23] At therapeutic levels of 200 mg/cow/day, serum iodide levels were raised tenfold over controls, and milk iodine levels of 1559 ± 771 μg/l were found.[23,84] Experimental feeding of 1000 mg EDDI (50 mg/kg/day iodine) caused marked rises in serum iodide levels, 1971 ± 295 μg/l and milk iodine, 2393 ± 379 μg/l. No signs of iodine toxicity were noted at these levels in lactating cows.[23]

Meat and dairy products from supplemented animals reflect iodine intake. Miller et al.[26] found iodide in all tissues of dairy cows after administration of 300 to 600 μCi of ^{125}I or ^{131}I as sodium iodide.

IODINE REQUIREMENTS IN MAN AND ANIMALS

Man

Iodine intake is extremely variable with large fluctuations in intake occurring from person to person and from day to day in the same person. Intake is also influenced by socioeconomic and geographic factors. The daily requirement of iodine to prevent goiter in adult humans is 50 to 75 μg/day (1 μg/kg).[85] Pregnant and lactating women and growing children require more iodine (2 to 3 μg/kg). Iodine intakes between 100 and 300 μg/day are desirable;[85] intakes between 50 to 500 μg/day are safe, while intakes between

Table 13
ESTIMATED AVERAGE IODINE INTAKE FOR
ADULTS IN U.S.

Calculated from Daily Food Category Intakes
and Iodine Content Data

Food category	Average daily consumption (g/day)[a]		Average daily iodine intake (μg/day)[a,b]	
	Male	Female	Male	Female
Milk and milk products	397	269	51.6	35.0
Eggs	55	31	14.3	8.1
Meat and meat products	325	192	84.5	49.9
Seafood	14	9	9.2	5.9
Legumes	40	24	—[c]	—[c]
Grain and cereal products	122	81	12.2	8.1
Yellow and green vegetables	104	88	33.3	28.2
Other vegetables and fruits	96	56	3.8	2.2
Sugar and sweets	44	35	—[c]	—[c]
Beverages (excluding milk)	749	739	3.0	3.0
Estimated salt intake[d]	3.42	3.42	142.0	142.0
Iodine in food as additives[e]	—	—	100.1	100.1
Total			454.0	382.5

[a] Data from Household Food Consumption Survey 1965 to 1966, Report No. 71, U.S. Department of Agriculture, Agricultural Research Service, 1972, on 20- to 34-year olds.

[b] Calculated from figures on iodine content of foods, from Vought, R. L. and London, W. T., *Am. J. Clin. Nutr.,* 14, 186, 1964, and Vought, R. L. et al., *Arch. Environ. Health,* 20, 516, 1970.

[c] No figures available for iodine content.

[d] From Wood, *Summary of a Conference: Iodine Nutriture in the U.S.,* National Academy of Sciences, Washington, D.C., 1970, 30. Mean U.S. daily salt intake of 3.42 g/day @ 0.0076% I = 260 μg/day but 1968 figure of 54.8% of all salt iodized; therefore 260 μg/day × 54.8% = 142 μg/day.

[e] Based on the estimated mean daily potassium iodide intake for adults; 131 μg KI per day = 100.1 μg iodine per day; does not include iodine from other food additives for which estimates are unavailable (Select Committee on GRAS Substances, Evaluation of the Health Aspects of Potassium Iodide as a Food Additive, SCOGS 30, Life Sciences Research Office, Fed. Am. Soc. Exp. Biol., Bethesda, Md., 1974).

From Fisher, K. D. and Carr, C. J., Eds., Iodine in Foods: Chemical Methodology and Sources of Iodine in the Human Diet, FDA 71-294, Life Sciences Research Office, Federation of American Societies for Experimental Biology, Bethesda, Md., 1974, 46. With permission.

500 and 1000 μg/day are probably safer. The upper safe limit of iodine intake, the level beyond which thyroid inhibition occurs, remains to be established. Tables 13—15 depict the estimated iodine intake in adults and the urinary iodine excretion patterns in the U.S.

Animals

There is no general accord about the iodine requirements of animals. Recommendations by various authorities show marked differences[74-78,86-91] (Tables 16—17a).

Table 14
RANGE OF URINARY IODINE EXCRETION IN ADULTS IN THE UNITED STATES[a]

Iodine values[b] (µg/g creatinine)	Number of individuals					Percent of pop. sample
	Low income states		High income states			
	Male	Female	Male	Female	Total	
0—24	2	18	7	6	33	0.6
25—49	16	34	35	49	134	2.3
50—99	54	142	232	344	772	13.0
100—199	158	438	410	795	1801	30.5
200—299	150	297	222	484	1153	19.5
300—399	107	235	93	218	653	11.1
400—499	69	150	43	150	412	7.0
500—599	48	120	34	79	281	4.8
600—699	36	74	16	51	177	3.0
700—799	19	41	13	27	100	1.7
800+	66	206	43	70	385	6.5
Subtotals						
White	260	463	882	1478	3083	
Black	269	803	156	544	1772	
Spanish American	196	489	110	251	1046	
Grand totals	725	1755	1148	2273	5901	

a Cumulative number of individuals in iodine value classes, pooled by race and income, aged 17 to 59 years (U.S. Department of Health, Education, and Welfare, 1972).

b 34% intake > 300 µg I per day; 16% intake > 500 µg I per day.

From Fisher, K. D. and Carr, C. J., Eds., Iodine in Foods: Chemical Methodology and Sources of Iodine in the Human Diet, FDA 71-294, Life Sciences Research Office, Federation of American Societies for Experimental Biology, Bethesda, Md., 1974, 46. With permission.

Table 15
DIETARY IODINE INTAKE MEASURED BY CHEMICAL ANALYSIS OF IODINE CONTENT OF DIETS

Subject characteristics	Iodine intake (µg/day)		Ref.
	Mean	Range	
Metabolic diets, euthyroid, hospitalized	89	15—219	15
Cafeteria meals, euthyroid, hospitalized	175	65—529	71
Euthyroid, nonhospitalized	395	18—154	
Eastern Kentucky, goiter present	64	17—408	313
Eastern Kentucky, goiter absent	96	30—817	
Northern Virginia, goiter present	379	17—1579	1
Northern Virginia, goiter absent	230	4—877	
Hospitalized, euthyroid	533	274—482	315
Hospitalized, euthyroid, iodized salt used	677	595—713	

Modified from Fisher, K. D. and Carr, C. J., Eds., Iodine in Foods: Chemical Methodology and Sources of Iodine in the Human Diet, FDA 71—294, Life Sciences Research Office, Federation of American Societies for Experimental Biology, Bethesda, Md., 1974, 46. With permission.

Pregnant sows require 200 μg I per 100 lb body weight or 4 μg/kg of iodine per day, growing pigs, somewhat less.[76] For mature animals, probably 2 μg/kg/day will suffice. The use of stabilized iodized salt containing 0.007% I, which is incorporated at 0.5% of the given ration of fed free choice, will normally prove adequate. Iodine should be added at a rate of 200 μg/kg of feed.

In dairy cattle, a daily intake of 800 to 1600 μg of iodine is regarded as adequate and can be achieved by use of iodized salt containing 0.015% of iodine incorporated at a 1% level into grain rations.[75] When iodized salt is stabilized to retard loss, a product containing 0.0076% iodine would probably supply the needed iodine supplement. For beef cattle, a daily intake of 400 to 800 μg of iodine is recommended by the U.S.N.R.C.[74]

On the other hand, the British Agricultural Research Council recommends the following dietary concentrations of iodine:[91]

1. On diets free of goitrogens, pregnant and lactating animals should receive 800 μg I per kilogram of feed.

2. On diets free of goitrogens, nonpregnant animals should receive 120 μg I per kilogram of feed.

3. On diets containing goitrogens, pregnant and lactating animals should receive 2000 μg I per kilogram of feed.

4. On diets containing goitrogens, other animals should receive 1300 μg I per kilogram of feed.

The recommended iodine requirements in some instances would appear to be excessively high.

Mason[92] working on housed pregnant ewes found that iodine intake of 300 μg/day prevented goiter development in lambs, even when 500 mg sodium thiocyanate (= 362.5 SCN$^-$) was given daily. These findings were confirmed by Statham.[93] It would therefore seem reasonable to recommend a minimum basal intake of 1 μg/kg body weight and a diet containing 2 to 3 μg I per kilogram of iodine. In pregnant and lactating animals, however, an intake of 5 to 6 μg/kg would seem advisable. These allowances are probably sufficient to balance the effects of goitrogens in the diet.

IODINE KINETICS

Absorption, Distribution, and Excretion of Iodine
Man

Iodide and iodine are readily absorbed in the lungs, and significant absorption of iodine through skin also occurs following application of tincture of iodine, iodophors, or organic iodine-containing compounds.[7,19,45,94] Ingested iodine, whether taken as iodide or in complex organic form, is converted, if necessary, and absorbed as iodide or iodated amino acids in the intestine. Absorption is virtually complete with little dietary iodine excreted in the feces.[95]

Following absorption, iodine rapidly equilibrates with the extracellular fluid compartment. With the exception of the red cell, cell membranes are impervious to iodine, and an active transport mechanism is required for entry of iodide into cells.[96] All body cells contain iodide, but iodide is specifically taken up and concentrated by the thyroid and salivary glands, gastric mucosa, choroid plexus, skin, hair, mammary glands, and placenta.[97-99] In some mammals, the placenta has an active transport mechanism to

Table 16
RECOMMENDED IODINE INTAKES IN MICROGRAMS FOR ANIMALS[14,74-78,86-90]

	Poultry	Pregnant ewes	Sheep	Pigs	Pregnant sows	Dairy cattle	Beef cattle	Cattle	Dog	Cat
Orr and Leitch			400–2,000					3,000–30,000		
Scott									500	100
Norris				90						
Underwood	5–9		50–100	80–160		800–1,600	400–800	400–800	10–15 per kg	10–15 per kg
Chilean Iodine Educational Bureau	50–100		500–1,000	500–1,000	1,000–2,000			1,000–5,000		
U.S. National Research Council	11–37				200 per 100 lb.					
Mason		300								

Table 17
RECOMMENDED IODINE INTAKES FOR POULTRY

Bird wt (g)	Recommended intakes	
	Daily feed intakes (g)	I/day/bird (μg)
250	35	11
500	57	20
750	73	25
1000	84	29
1500	100	35
2500		
Maintenance	87	
Laying	125	37
Breeding	125	37

Table 17a
RECOMMENDED FEED SUPPLEMENTS FOR POULTRY

	Age (weeks)	I (μg/kg feed)
Starting chicks	0–8	350
Growing chicks	8–18	350
Laying and breeding hens	18+	300

Data from NRC Committee on Animal Nutrition, *Recommended Nutrient Allowances for Domestic Animals*, Part 1, 6th ed., National Academy of Sciences, Washington, D.C., 1971. 54.

Data from NRC Committee on Animal Nutrition, *Recommended Nutrient Allowances for Domestic Animals*, Part 1, 6th ed., National Academy of Sciences, Washington, D.C., 1971, 54.

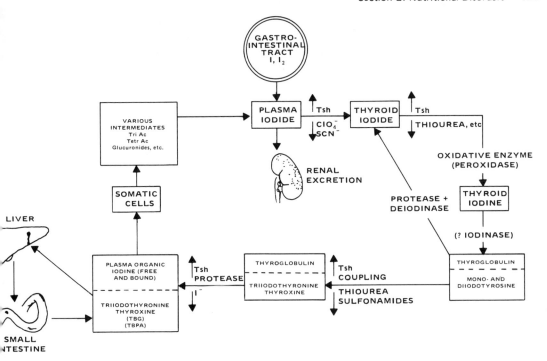

FIGURE 1.   Schematic diagram of iodine metabolism in euthyroid individuals. TriAc and TetrAc are the acetic acid analogues of tri- and tetraiodothyronine, respectively. TBG, thyroxine-binding globulin; TBPA, thyroxine-binding prealbumin. (From Stanbury, J. B., Wyngaarden, J. B., and Frederickson, D. S., *The Metabolic Basis of Inherited Disease,* 3rd ed., McGraw-Hill, N.Y., 1972, 1788. With permission.)

transport iodide from the maternal to the fetal circulation.[97] Iodide secreted into the digestive juices by the salivary glands, stomach, and intestines is completely reabsorbed and recycled.

The thyroid gland contains the body's main iodine pool; in man, it is estimated to be 7000 to 8000 μg of iodine. Most of the circulating iodine is hormonal thyroxine, of which 30 to 70 μg is protein bound and 0.5 μg is free thyroxine. Normally the plasma inorganic iodine levels are between 0.5 and 1.5 μg %.[96,97]

The main route of excretion is through the kidneys. The iodine is eliminated by glomerular filtration, and partial reabsorption occurs by passive diffusion. The renal iodide clearance is 30 to 40 ml/min.[95] The daily urinary output of iodide approximates intake when persons are in iodine balance.

Thyroid hormones inactivated by conjugation in the liver are excreted in bile. These conjugated hormones are partially split in the intestine leading to some reabsorption, the remainder, 10 to 15 μg I per day, being excreted in feces in organic form.[95] Negligible amounts of iodine may be lost through the lungs and sweat.

Animals

Dietary iodine and the various forms of supplemental iodine are absorbed as iodide in the rumen and distal small bowel of ruminants and the small bowel of monogastric animals. Absorption of iodide is virtually complete, and after absorption, iodide is distributed through all tissue fluids. Iodide is concentrated by the thyroid and actively secreted by the mammary and salivary glands, the abomasa of ruminants, and the stomachs of monogastric animals. Endogenous iodide secreted by the salivary glands is reabsorbed in the rumen and that secreted by the abomasa and duodena of ruminants or the stomachs of monogastric animals is reabsorbed in the distal small gut. The secretion of iodide into the saliva and abomasal or gastric juices maintains a significant iodine pool in the alimentary tract. The presence of this pool, which is continuously recycled, conserves iodine by preventing its excretion in milk or urine. Administration of thiocyanate or perchlorate inhibits secretion of iodide into saliva, into gastric and abomasal fluids, and into milk by the mammary gland and causes a sharp rise in urinary excretion of iodide.[247,259,260] This increased excretion of iodide has been attributed to a blockade of active renal iodide reabsorption,[25] but elimination of the alimentary iodine pool could also explain these observations. Mason[92] has found no correlation between renal thiocyanate and renal iodide clearances.

Active placental iodide transport has been reported in some mammals.[93,261-264] Fetal plasma was shown to have five times the iodide concentration of maternal plasma in cows.[264] Although the kidneys are the chief route of iodine excretion in most animals, negligible amounts of inorganic iodide and small quantities of organic iodine, usually derived from hepatically conjugated thyroid hormones, are lost in feces. Fecal loss of iodine, however, may increase when fecal output is increased and goitrogens are present in the diet.[266] In goats, the feces may be an important route of iodine excretion.[265]

Iodine and the Thyroid

The thyroid gland holds the major iodine compartment in the body. The thyroid develops early in fetal life. Within 12 weeks in man, the gland is able to concentrate and organify iodine.[96] Thyroglobulin is present in the follicles, and thyroxine synthesis occurs. The ability to concentrate and organify iodine and thyroxine synthesis appears concurrently in man and monkeys, but sequentially in lower species.[101]

The hypothalamic-pituitary-thyroid axis matures in the 18- to 22-week fetus.[102-104] When fetal thyrotropin (TSH) rises, it results in a marked increase in iodine uptake[105] by the thyroid followed by a rise in circulating thyroxine.[102] In lower mammals, rats, mice, rabbits, guinea pigs, and hamsters, who have short gestation periods, the hypothalamic-pituitary-thyroid axis does not mature until after birth.[102]

Iodide is taken up by the cell against a concentration gradient and against a negative intracellular potential. The energy requirement for this iodine uptake is supplied by the high-energy bonds of adenosine triphosphate (ATP), produced in mitochondria by oxidative phosphorylation or anaerobic glycolysis.[106]

Uncouplers of oxidative phosphorylation inhibit iodine uptake and lower the concentration of ATP. High concentrations of iodide have been reported as uncouplers of oxidative phosphorylation; in rats, injection of a large dose of iodide results in a marked reduction of thyroid ATP which persists for some time.[107] Thiocyanates, which inhibit iodine uptake, are also believed to be uncouplers of oxidative phosphorylation.[108]

After entry into the cell, iodide migrates to the luminal edge of the cell where iodination of tyrosyl radicals already incorporated into thyroglobulin occurs in the presence of thyroid peroxidase, which is also required for the coupling of the iodinated tyrosyl groups to form the iodothyronines.[109-112] The antithyroid drugs, carbimazole, methimazole, and propylthiouracil, inhibit thyroid peroxidase and consequently block organification of iodide. Increasing iodine concentration can partially surmount this inhibition.[111]

The availability of iodine influences thyroid synthetic functions and four degrees of excess intake are described:[114]

1. Slight excess, leading to increased absolute iodine uptake and increased formation of iodotyrosines and iodothyronines with increased stores of thyroid hormones resulting from positive iodine balance

2. Moderate excess of iodine which can inhibit release of hormone from hyperactive thyroids and those under stimulation by excess TSH

3. Large excess of iodine which inhibits organic binding of iodine — the Wolff-Chaikoff effect[114-116]

4. Gross excess of iodine which floods and saturates the transport system, although the inhibitory effects of iodide are evident long before this level of iodide is attained

Large doses of iodide have been shown to inhibit organification of iodide (Wolff-Chaikoff effect).[114-116] The exact mechanism of inhibition of organification of the iodide and the block in coupling of tyrosyl groups to form iodothyronines by excess iodide is unknown. However, normal thyroid glands will adapt to the excessive levels of iodide with reestablishment of organification and iodothyronine synthesis. The work of Braverman and Ingbar[118] suggests that in adaptation, reduced iodide transport into the cell results in intracellular iodide concentrations below critical levels, above which inhibition of organification would occur. The exact mechanism for the reduced iodine uptake by the cell is uncertain, though reduced efficiency in extraction of iodide from the blood is evident and may possibly be due to iodide uncoupling of oxidative phosphorylation and reduction of ATP.[107]

The normal thyroid gland is under hypothalamic-pituitary control. Hypothalamic thyrotrophin-releasing hormone (TRH) acts on the pituitary leading to the release of TSH. Thyrotrophin attaches to the thyroid-cell membrane receptor sites causing induction of AMP cyclase and an increase in cellular cyclic AMP which results in accelerated organification of iodide and increased hydrolysis of thyroglobulin with a consequent increase in iodothyronine release into circulation. Homeostasis is maintained by a negative feedback mechanism; rising serum levels of thyroxine and triodothyromine inhibit further release of TSH by the pituitary.

As a result of hydrolysis of thyroglobulin, the iodotyrosines, monoiodotyrosine (MIT) and diiodotyrosine (DIT), are also liberated; some may enter the circulation, but most are deiodinated; the released iodine is partially reorganified, and the remainder leaks back into the extracellular tissue fluids to be recycled.[96] High doses of iodide inhibit hydrolysis and release of iodothyronines in overactive glands but not in normal glands. The reason for this is not understood.[119] Release of iodothyronines follows hydrolysis of thyroglobulin by proteases in response to TSH. Iodothyronines migrate from the luminal edge of the cell and are secreted into circulation.

It is well known that additional iodine will overcome the effects of goitrogens in the diet and prevent thyroid enlargement;[92,267] however, gland histology is not necessarily normalized.[267] Dvoskin[267] reported that following the parenteral administration of both iodide and elemental iodine, but not iodide alone, thyroid histology returned to normal despite continued administration of uracil.

Administration of large doses of iodide has been reported to produce resumption of growth in thyroidectomized rats.[267,268] Restoration of normal growth rates has also been observed in thyroidectomized rats and goats after feeding of iodinated proteins.[269,270] Metamorphosis has been observed in hypophysectomized or thyroidectomized oxalotle following parenteral, but not oral, administration of iodide.[267]

Feeding of the iodinated proteins, casein and ardein, produced thyroxine-like effects with features of thyrotoxicosis in sheep and cattle.[271,272] The mechanism whereby iodide produces thyroxine-like effects in thyroidectomized animals is unknown, but peripheral synthesis of thyroxine is a possibility. DIT has been isolated from iodinated albumin,[273] and thyroxine has been isolated from albumin, globulin, casein, and other iodinated proteins.[274,275]

Nonendocrine Biological Effects of Iodine

A number of nonendocrine biological effects of iodine have been reviewed by Stone.[120] The relevance of in vitro studies in the living organism, however, remains debatable. In certain instances, the observed biological effect, particularly in relation to granulomatous tissue, still requires explanation.

Iodine, along with myeloperoxidase, hydrogen peroxide, and a halide, appears to increase the movement of polymorph leukocytes into areas of inflammation; bacteriocidal action is enhanced and phagocytosis is increased.[120-122] The proteolytic effect of leukocytes is also increased.[120] No inhibitory effect, comparable to the Wolff-Chaikoff phenomenon, is evoked by increasing concentrations of iodide.[123] In guinea pigs administration of iodides increased the cellular inflammatory response to induced inflammation.[124] Iodide in large doses also causes marked increases in salivary and bronchial glandular secretions in man and animals. This effect has been utilized for over 100 years by incorporating iodide in asthma medications and expectorants. Iodides also increase the activity of released protease in purulent sputum by interaction with a substrate.[120]

In addition, iodides have been used to treat gummas and granulomas empirically because of their recognized efficacy. Experimental work has confirmed the effects of iodide in resolving or preventing the development of granulomas.[120,125] Iodides concentrate at the sites of infection, granulomas, and tumors. This concentration was noted in infective granulation tissue but did not occur when large doses of iodide were used.[120] The reason for its concentration in granulomas is uncertain.

There are abnormalities of iodide distribution in humans and animals with tumors, and plasma iodide levels are usually elevated. Where small tumors were implanted subcutaneously in animals, a higher than normal ^{131}I uptake was noted in surrounding skin.[120] When larger tumors were implanted, ^{131}I uptake in skin doubled. Rats with large tumors showed progressive falls in thyroid radioiodine uptake and lower urinary ^{131}I excretion.

Injection of 5-hydroxytryptamine or histamine results in increased iodine distribution to skin and increasing numbers of mast cells.[120] Large injected doses of iodide in rats induced generalized erythema and muscular damage due to rupture of mast cells. It has been reported that iodides in high dosage may uncouple oxidative phosphorylation; thus, this therapy may interfere with ATPase systems.[107,146]

ADVERSE REACTIONS TO IODINE

It is impossible to assess the incidence or frequency of adverse reactions to iodine. Although temporal correlations exist between ingestion of iodine and the occurrence of untoward effects, it is often difficult to demonstrate causal relationship.

Dietary and environmental iodine have risen sharply in the last 10 to 15 years without a significant increase in reports of adverse effects, apart from an increase in thyrotoxicosis in Tasmania and possibly Central Europe and the Balkans. Although it is likely that no marked increase in untoward effects has occurred, it is invalid to assume that it is necessarily so. Mild adverse reactions may not be reported; even moderate or severe effects may not be presented, because few clinicians are aware of the ubiquity of iodine sources in the diet and environment. Another important factor possibly responsible for the sparsity of recent reports of adverse reactions is the difficulty in proving or establishing that iodine was responsible. Nevertheless, numerous, diverse untoward reactions have been documented.

Much is known of the biological effects, both endocrine and nonendocrine, of iodine, but many of the actual mechanisms of these effects remain obscure. The clinical features of acute and chronic overdosage and toxicity are well documented, but while the mechanisms of some adverse effects are known, the underlying cause of other manifestations remains uncertain. When adverse reactions occur to minute quantities of iodine, quantities normally regarded as harmless, in food and drugs, multiple factors may contribute to the reaction.

These reactions may be caused by immunological or nonimmunological mechanisms. Local chemical irritation, toxic manifestations, idiosyncrasy, genetic factors, enzyme deficiencies, and hypersensitivity may influence the nature and clinical features of the reaction. Local chemical irritation and toxic manifestations are often dose related while idiosyncratic and hypersensitivity reactions may follow minimal exposure to iodine. Iodine may act as an irritant on skin and epithelial surfaces, particularly the mucosa of the alimentary tract. On the skin, contact dermatoses can occur; in the alimentary tract, burning and soreness may result in the mouth and throat and along the esophagus, with nausea, vomiting, and diarrhea.

If the amount of a potentially toxic substance exceeds the body's tolerance, toxic reactions independent of immune mechanisms occur. Such reactions are inevitable after overdosage but may occur after minimal exposure in predisposed individuals. Toxic manifestations occurring after minimal exposure to iodine are due to idiosyncrasy. Idiosyncratic reactions may be determined by genetic factors or enzyme deficiencies or mediated by immune mechanisms.

Toxic reactions may be diagnosed on the basis of symptoms and signs, but it is not usually possible to distinguish idiosyncrasy from hypersensitivity on clinical grounds. Further diagnostic problems arise, because symptoms and signs, classically examples of hypersensitivity type reactions, may occur without a demonstrable immune response. This may explain the paucity of proven cases of iodine hypersensitivity despite frequent reports of allergic type reactions.

It is generally believed that hypersensitivity reactions to normally harmless quantities of iodine are due to interaction, directly or indirectly, with circulating antibodies or sensitized lymphocytes. Detrimental effects resulting from antigen-antibody interaction are called hypersensitivity, sensitivity, or allergic reactions. Four types of hypersensitivity reaction are identifiable:[126,127] type I, anaphylactic reactions; type II, cytotoxic reactions; type III, toxic immune-complex reactions; and type IV, cellular hypersensitivity. The features of these reactions are tabulated in Table 18.

The simpler traditional classification of hypersensitivity, immediate and delayed reactions, has some merit.[128-130] Immediate hypersensitivity reactions begin within a few minutes to a few hours after exposure and are caused by interaction of antigen with circulating antibodies. Examples of this type of reaction are anaphylaxis and hemolytic blood transfusion reactions. Delayed hypersensitivity reactions develop over 12 to 48 hr and are characterized by local or diffuse inflammatory exudates containing lymphocytes, eosinophils, and macrophages. These reactions are mediated by sensitized T lymphocytes which can react with a specific antigen. Contact dermatoses, graft rejection, and the tuberculin skin test are examples.[128,130]

Despite the introduction of numerous techniques, confirmation of hypersensitivity reactions to iodine remains difficult. The most reliable method is elimination of the suspected food or drug followed by careful observation for the disappearance of symptoms and signs. After resolution of the adverse effects, provocative reintroduction of the suspected substances is undertaken. Meticulous medical and dietary histories with expert clinical judgment are essential.[131-136] Because of the incidence of both false positive and false negative results, the various mucous membrane and skin sensitivity tests have limited application. The intradermal skin test and the intravenous sensitivity test may provoke severe, at times dangerous, reactions and yet produce false results. The "provocative subcutaneous food test" technique, which is distinct from "provocative dietary reintroduction," is the subject of controversy and requires further evaluation.[137-139] The passive transfer test risks transmission of hepatitis B in man. The radioallergosorbent test (RAST), an in vitro radioimmunoassay of allergen-specific

Table 18

MAJOR TYPES OF HYPERSENSITIVITY

Reaction	Mediated by	Onset	Example of action sites	Example of typical reactions
Type I — anaphylactic (allergy or atopy)	Circulating or cell-bound reaginic antibody, predominantly IgE	Immediate (minutes to a few hours)	Smooth muscle of bronchi and arteries; respiratory and gastrointestinal mucosae	Bronchial asthma, hay fever, systemic anaphylaxis, certain food and drug allergies
Type II — cytotoxic or cytolytic (complement-dependent)	Circulating antibody reacts with cell-bound antigen; secondary complement fixation	Immediate	Cell membranes in various tissues	Transfusion reactions, hemolytic disease of the newborn; acute allograft rejection, certain glomerulonephritides
Type III — toxic-complex (soluble complex or immune complex hypersensitivity)	Large doses of antigen; soluble antigen-antibody complexes; secondary complement fixation	Variable in time	Perivascular tissues; vascular endothelium; glomerular basement membrane	Acute inflammatory reactions, vasculitis, Arthus reaction, serum sickness, certain drug reactions
Type IV — cellular or delayed	"Sensitized" small lymphocytes reacting with antigen; production of transfer factor; macrophage-inhibiting factor; specific cytolytic ability	Delayed (24—72 hours)	Skin; parenchymatous organs	Contact dermatitis, tuberculin reactions, allograft rejection

From Talbot, J. M., Fisher, K. D., and Carr, C. J., Eds., A Review of the Significance of Untoward Reactions to Iodine in Foods, 71-294, Life Sciences Research Office, Federation of American Societies for Experimental Biology, Bethesda, Md., 1974, 1. With permission.

immunoglobulin E antibody, shows promise in the diagnosis of food allergy.[140] Current research techniques of estimating circulating antibody, leukocyte histamine release, and basophil degranulation and lymphocyte transformation tests have shown promise as practical tools. Halpern et al.[141] demonstrated in vitro positive lymphocyte transformation tests in four patients who had immediate-type hypersensitivity reactions to iodide.

Evidence suggests that iodine may be a hapten, though this concept is not universally accepted. A hapten is a substance which alone will not provoke an immune response but may serve as part of an antigen when bound to a carrier molecule, usually a protein. Individuals, sensitized to the hapten-carrier complex, produce antibodies which may interact with the hapten-carrier complex, the carrier protein alone, or the uncoupled hapten. Nickel can act as a hapten when combined with the structural protein of skin. The nickel-protein complex, the carrier protein, or free uncoupled nickel may provide reactions in sensitized subjects.

Treatment with iodine alters the serological specificity and antigenicity of tyrosine; the resultant antigen may stimulate antibody formation, and in a susceptible person, development of a hypersensitivity state. Guinea pigs can be anaphylactically sensitized by a mixture of iodine and guinea pig serum.[142] Sensitization can occur when homologous serum is iodinated with Lugol's iodine in conditions comparable to those occurring in iodine antisepsis. This type of sensitization may be a factor in human idiosyncrasy. In fact, Rosenberg et al.[143] believed that the vegetating iododerma in their patient was due to hypersensitivity, the potassium iodide taken for asthma acting as a hapten which combined with serum protein.

Many immunological studies using proteins treated with iodine as antigens have reported a periarteritis nodosa type of hypersensitivity reaction featuring rashes, eosinophilia, lymphadenopathy, arthralgia, and submucosal hemorrhages. These features have been attributed to a vasculitis caused by circulating antigen-antibody complexes.

Horn and Kabins[144] reported iodine sensitivity reactions consisting of high fever, rigors, and leukocytosis which occurred in three of their patients and mimicked an acute bacterial infection. Jacob et al.[145] reported the occurrence of fever, swelling of the submandibular glands, and extreme eosinophilia (white cell count of 70,000 with 85% eosinophils) in a patient, who was exposed to iodine before ingestion of an iodine-containing cough mixture. Using a skin-window technique, they found eosinophil exudation followed injection of isotonic potassium iodide into the area. The reaction was specific for the suspected allergen, providing confirmation that the eosinophilia was part of the iodide hypersensitivity reaction.

Acute Poisoning or Iodism

Oral administration of iodide may cause local gastrointestinal irritation. Acute poisoning even from large doses is rare, though occasional individuals may show a marked sensitivity. Onset of adverse symptoms may occur immediately or be delayed for some hours.[95]

Angio-edema is the commonest feature, and edema of the larynx may cause suffocation. Multiple cutaneous hemorrhages ranging from petechiae to larger ecchymosis develop, and later manifestations of the serum sickness type of hypersensitivity with fever, arthralgia, lymph node enlargement, and eosinophilia may occur. Fatal cases of polyarteritis nodosa have been reported.[95]

Chronic Poisoning or Iodism

This is more common than acute poisoning. There is no way to predict which patient will react unfavorably, and indeed, an individual may vary in his sensitivity to the drug from time to time. Eventually, iodism will occur in everyone if the dose is high enough.

Onset begins with an unpleasant taste and burning sensation in the mouth and throat as well as soreness of the teeth and gums. Salivation is increased. Coryza with watery nasal discharge and sneezing is common. Irritation of the eyes with lacrymation and swelling of the eyelids is not unusual. Mild iodism simulates a head cold with a severe headache originating in the paranasal sinuses and irritation of the mucous glands of the respiratory tract causes a productive cough. In addition the salivary glands — the parotids and submandibular glands — may be swollen and tender, and it may be mistaken for the mumps. The pharynx, tonsils, and larynx may be inflamed and painful. Skin lesions are common and vary in type and intensity. Acneiform lesions in the seborrheal areas are common. Eczema, pustules, vesicles, bullae, erythema, urticaria, petechiae, purpura, papular or nodular lesions, fungating lesions, erythema multiforme and erythema nodosum have all been reported. Fatal iododerma with fungating lesions may follow prolonged use of iodides. Symptoms of gastric irritation are common, usually manifested as nausea and vomiting with diarrhea which may be bloody. Depression of mood is also common.[95]

ADVERSE EFFECTS OF IODINE IN MAN

Deaths

The most common cause of death is anaphylactic shock which may follow even minute doses of iodine in a sensitized subject.[95] Death has also been recorded from periarteritis nodosa[143,147-151] and after thrombotic thrombocytopenic purpura[152] probably due to toxic immune complexes. In some reports, however, the actual cause of death was not always identifiable, particularly when iododerma was the predominating feature.[153] Neonatal deaths from airway obstruction by large goiters with or without associated hypothyroidism have also been recorded in infants born to mothers taking iodine-containing medications.[154]

Dermatological Effects

Cutaneous manifestations, alone or combined with systemic manifestations, are common; 25 types of skin lesion have been described where iodine may be an initiating or aggravating factor in the disorder. Skin eruptions are usually exanthematic.[155] The more common forms seen are acneiform papules and pustules — usually in relation to hair follicles and resembling acne vulgaris but more inflamed and widely distributed. Thick-walled bullae, containing clear fluid at first but becoming hemorrhagic or pustular, arise on plaques of erythema or on normal skin. Often few in number, they are most commonly found on the face, forearms, neck, and flexures but may also occur in the mouth. If iodine is continued, the bullae may be replaced by vegetating masses.[155] These vegetating or tuberous iododermas which are heaped up masses of hypertrophic epithelium with many pustules arise on the sites of bullae or pustules and may simulate granulomatous infection or other skin disorders, such as pemphigus vegetans.[155] Erythema, urticaria, purpura, acne and acneiform rashes, papular eruptions, and iododerma have been recorded following normal or supplemented dietary intake of iodine.[120]

Topical iodine may also produce eczematous reactions, acneiform or pustular lesions.[120] Iodine may also exaggerate or aggravate but rarely precipitate the lesions of acne, dermatitis herpetiformis, pyoderma gangrenosum, pustular psoriasis, pyoderma, and erythema nodosum[120] (Table 19).

Adverse Effects on the Thyroid

All current evidence indicates that in man iodine has no harmful effects on the normal thyroid gland which can adapt without ill effects to relatively massive doses of iodine (often greater than 1000 times the daily requirement), even when administered over prolonged periods. However, in the abnormal gland even modest increases in dietary

Table 19
REPORTED LESIONS

Lesion	Ref.	Lesion	Ref.
hema[a]	120, 157	Pustules	120
cles	120	Fixed eruptions	167
ae	120, 157–160	Fungating	147, 156
ucose	157	Papular[a]	120, 147, 159
eiform[a,b]	120, 143, 147, 148, 157–162	Nodular	120
or-like	120	Urticaria[a]	120, 148, 156, 160
ma	120	Exanthematic	143
hema multiforme	120	Erythema nodosum[b]	120, 160
natosis	163	Contact dermatosis	120, 143
ura[a]	120, 143, 147, 152, 163	Generalized pustular	120, 168
derma[a]	144, 147, 153, 157–159, 163, 165, 166	psoriasis[b]	
derma gangrenosum[b]	120	Dermatitis herpetiformis[b]	120, 169–171
se thinning of hair	154	Exfolliative dermatitis	143, 147

ecipitated by dietary iodine.
sions aggravated by iodine.

iodine may precipitate thyrotoxicosis, while in other instances larger quantities of ingested iodine may lead to goiter formation, hypothyroidism, or possibly lymphocytic thyroiditis.

Thyrotoxicosis

Most reports of iodine-induced thyrotoxicosis have arisen following therapeutic dosages and uses of iodine in the treatment of goiter.[172-176] Thyrotoxicosis has also been precipitated during iodine repletion tests,[177] when iodine has been administered to block thyroid uptake of radioiodine in the ^{125}I fibrinogen test,[178] and after iodine administration in the course of clinical research.[179,180] Transient episodes of iodine-induced thyrotoxicosis have been reported following the use of radiological contrast media and brief administration of iodine-containing drugs. In these patients, symptoms usually subsided spontaneously over a period of weeks or months.[179,198] It has been reported in epidemic form from endemic goiter areas when additional iodine entered the diet in salt[181-183] or bread.[6,7,39,184-187] Until recently, iodine-induced thyrotoxicosis was considered rare; reports from the mid-Western U.S. in the 1920's and Holland in the late 1940's were discounted, because statistical evaluation was impossible.

Tasmania is an iodine-deficient island with endemic goiters in a population of 400,000 people, of whom about 20% are over 50 years of age, making it an excellent place to observe and study iodine-induced thyrotoxicosis following iodine supplementation of the diet. In northern Tasmania, the incidence of thyrotoxicosis doubled in 1964, and there was a further trebling of the incidence from 1966 to 1967. A third rise in thyrotoxicosis was noted during 1971.[187] In each instance the increase followed an increase in available dietary iodine. The first increase, due to iodine residues in milk, followed the introduction of iodophors on dairy farms. The second rise followed the deliberate addition of iodate to bread improver as part of a universal iodine supplementation program for eradication of endemic goiter,[6] and the third rise was related to extension of the usage of iodophors to bulk-milk road tankers and processing plants. On each occasion, patients over the age of 40 years with long-standing goiter were chiefly affected. No increase in the incidence of endocrine ophthalmopathy was noted.[184]

In the first series of experiments by the author et al. to study these thyrotoxic patients,[185] the incidence of thyroid autoantibodies was lower than previously reported. Mouse thyroid-stimulating antibodies (LATS) were detected in only 36% of cases after

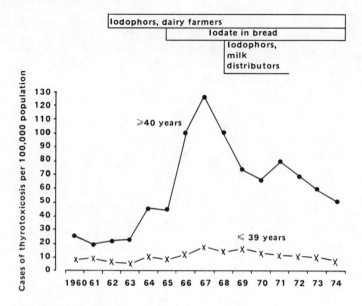

FIGURE 2. Age-specific incidence of thyrotoxicosis in northern Tasmania 1960–74. Bars indicate duration, but not magnitude, of various factors adding to dietary iodine. (From Stewart, J. C. and Vidor, G. I., *Br. Med. J.*, 1, 214, 1976. With permission.)

IgG concentration. In the second series from 1971 to 1972,[186] mouse thyroid-stimulating antibodies (MTS a/b, LATS) and human thyroid-stimulating antibodies (HTS a/b, LATS protector) could be detected in only 50% of the patients. All patients were shown to have either autoimmune thyrotoxicosis (Graves' disease) or autonomous thyrotoxicosis (Plummer's disease, toxic adenoma, and hot nodules), the latter being more common. No evidence of a third pathogenic mechanism was found.[186] As a result of these studies, it was concluded that iodine-induced thyrotoxicosis occurs only in people whose thyroid activity has escaped from the normal hypothalamic-pituitary control because of Graves' disease or autonomous nodules.[186,187]

Autonomous tissue is common in elderly euthyroid patients with long-standing goiter,[188] and the hormone output of such tissue is regulated by the availability of iodine and presumably the mass of the tissue. From this, it may be predicted that a small increase in dietary iodine will precipitate hyperthyroidism in patients harboring a large mass of autonomous tissue, while larger quantities of iodine would be required to produce the same effect in patients with smaller masses of unregulated tissue[187] (Figure 2).

Iodine-induced thyrotoxicosis is inevitable in any endemic goiter area when iodine nutrition improves through chance or design. The frequency with which this complication occurs will depend on the age structure of the population; the greater the proportion of elderly patients, the more common will iodine-induced thyrotoxicosis be. Iodine-induced hyperthyroidism in response to extra iodine can also occur in goitrous subjects living in iodine-sufficient areas.[180]

The occurrence of some cases of iodine-induced thyrotoxicosis should not deter public health authorities from implementing goiter prophylaxis programs, provided adequate facilities exist for early detection and treatment. It is strongly recommended that the state of iodine nutrition be monitored in all places, particularly in endemic goiter areas. As a result of the proliferation of dietary iodine sources resulting from technological changes, improved food preservation techniques, and rapid transport, spontaneous unsuspected changes in iodine nutrition are probably occurring in many areas traditionally considered iodine deficient.[187]

Response to Therapy

A few patients, usually with autoimmune thyrotoxicosis, respond poorly or incompletely to antithyroid drugs necessitating increased doses of propylthiouracil or carbimazole. In some instances, response to therapy is improved when iodine intake is reduced by eliminating bread, milk, and iodized salt from the diet.[189] Presumably an excess of iodine in these patients partially overcomes inhibition of peroxidase and organification of iodine[113] by antithyroid drugs.

Results of Therapy

A reduction in the remission rate of Graves' disease has been attributed to increases in ambient iodine nutrition, and an inverse relationship has been postulated. A 50 to 80% remission rate was expected following medical treatment in the early 1960s but had dropped to 15 to 30% by the early 1970s.[190-197] The deterioration in results corresponded to the period during which marked changes in the iodine environment occurred.[197] An increased relapse rate in medically treated thyrotoxics who receive

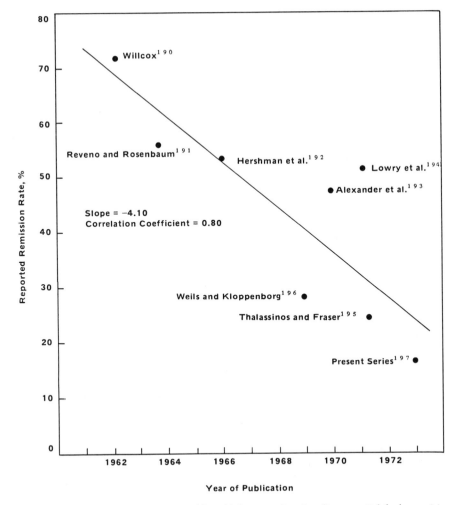

FIGURE 3. Remission rates after antithyroid therapy of toxic goiter reported during past ten years. (From Wartofsky, L., *J. Am. Med. Assoc.*, 226, 1087, 1973. With permission.)

additional iodine has also been reported[193] but was not substantiated by Thalassinos and Fraser.[195]

Iodide Goiter With and Without Hypothyroidism

Iodide goiter can occur at any age and is associated with hypothyroidism in 44% of the reported cases.[114] In the newborn, it is the sequel to prolonged ingestion of inorganic iodide or iodide-generating organic compounds during pregnancy.[154,199–205] In children and adults, it is most commonly seen in patients with chronic respiratory disease but may follow iodine medication for other indications. It has also been reported after contrast radiography using oily iodinated compounds, such as Lipiodol, which release iodine over prolonged periods.[206–209] Goiter may precede development of hypothyroidism or appear concurrently.[210–212]

Prolonged continuous ingestion of 10 to 100 times (or more) of the daily requirement of iodine in organic or inorganic form appears necessary for the development of goiter and/or myxedema.[114] Iodide goiter has, however, been reported to follow the intake of as little as 400 to 500 μg of iodine a day.[213] Goiter may develop within a few weeks of starting medication;[213–215] more often, symptoms develop after some months[214, 216,217] of medication and occasionally in patients who have taken iodides for years before the onset of goiter or hypothyroidism.[210,215,218]

Predisposing factors are uncertain but parenchymatous thyroid damage[114] due to physical agents such as radiation or inflammatory or autoimmune factors has been suggested.[114] Patients with autoimmune thyrotoxicosis treated with radioiodine and sufferers from chronic lymphocytic thyroiditis (Hashimoto's disease, autoimmune thyroiditis) are particularly susceptible to the effects of iodides and are prone to develop hypothyroidism.[219–221] Thyroid damage must be sufficiently slight to permit a return to the normal, which usually follows the withdrawal of iodine.

Genetic influences have not been sufficiently considered nor adequately evaluated in iodide goiter. Further study of this aspect and epidemiological observation would be highly desirable since only about 4% of patients receiving iodides for pulmonary disease develop goiter.[220] Trowbridge et al.[222,223] report an overall goiter incidence of 6.8% in American school children and 3.1% in adults on above-average iodine intakes, an incidence similar to both the 6 to 12% goiter rate seen in endemic coast goiter in Japan[114] and the 4% incidence reported in subjects taking iodine-containing medications.[220]

Although several mechanistic theories have been proposed to explain iodide goiter formation, proof for any of them is lacking. There is no evidence that iodide has a peripheral antithyroxine effect,[224,225] that it inhibits synthesis and release of or response to TSH, or that it inactivates TSH.[114] It has been variously suggested that iodides cause their effect by: 1. inflammatory changes in the thyroid parenchyma, 2. a "structure-breaking effect on some crucial enzyme," 3. inhibition of carbohydrate or some other intermediary metabolism resulting in reduced formation of pyridine dinucleotides, which, in turn, causes a reduction of substrate for thyroid peroxidase and consequent inhibition of organification, and 4. chemical mechanisms operating via the tri-iodide equilibrium and/or H_2OI^+ formation.[114]

The current evidence suggests that the most likely mechanism is a defect in organification. Here genetic factors, which have received inadequate consideration in the past, could operate by either of two mechanisms: 1. Partial or incomplete adaptation to the excess iodine load due to genetically determined enzyme deficiency or enzyme defect which may ultimately lead to complete failure of adaptation, and 2. Delayed, incomplete, or partial adaptation due to a genetically determined enzyme deficiency or a defect in response to the Wolff-Chaikoff effect, causing lowered blood hormone levels which produce increased TSH secretion and thyroid enlargement; iodide inhibition of hormone release in the TSH-stimulated gland leads occasionally to the gradual development of hypothyroidism.

The question of cogoitrogens has been considered following the observation of Purves and Griesbach[226] that administration of thiourea and iodine produced full "thyroidectomy cells" in the pituitary. It was also observed that iodides potentiate the goitrogenic effects of sulfonamides and sulfonylureas causing inhibition of organification and synthesis of iodothyronines, an effect which shows no "escape" over a period of 2 weeks.[227,228]

Most goitrous patients have marked thyroid enlargement; when hypothyroidism is associated with it, the usual clinical features of this disease are apparent. Serum thyroxine may be normal or low; TSH is invariably elevated; thyroid radioiodine uptake is usually high; plasma inorganic iodide is elevated, while the iodide clearance, which may be normal or high, results in an increased absolute iodine uptake. Urinary iodine excretion is high.[114,212,218,229-231] Iodine discharge by perchlorate and thiocyanates indicates an organification defect. An increase in the ratios of MIT to DIT and iodotyrosines to iodothyronines further indicates deficient organification and hormone synthesis.

The increased radioiodine uptake is TSH dependent and can be suppressed by triodothyronine though at times suppression takes longer than 7 days to develop and full suppression may not be apparent for 14 days or more.[232] That thyroid enlargement is due to TSH is indicated by regression in goiter size which can be produced by thyroxine medication. Withdrawal of iodine results in remission of symptoms and signs and regression of thyroid enlargement over several weeks. Reintroduction of iodine produces a recurrence of goiter and hypothyroidism, often quite rapidly.

Controversy exists over the typical histological appearance of the thyroid.[218] The majority of cases show thyroid hyperplasia[208,216,231,233,234] in man and mouse. The histologic appearance of Hashimoto's disease has also been described[218,236] in humans and hamsters. Whether these appearances in man are primarily due to excess iodine or to iodine-induced hypothyroidism in a gland already affected by autoimmune thyroiditis is uncertain.

Iodide Goiter in the Newborn

The fetal thyroid takes up iodine by the 15th week of gestation and the hypothalmic-pituitary-thyroid axis is functioning by the 22nd week of pregnancy. While iodides readily reach the fetal circulation possibly by active transport across the placenta,[99] there is no transfer of maternal thyroid hormones. Prolonged ingestion of iodides in the latter half of pregnancy carries the risk of development of iodide goiter in the fetus. However, the incidence of iodide goiter in fetuses of mothers taking iodides is unknown. The presence of thyroid enlargement in the fetus may cause extension of the neck and abnormal presentation at birth. If thyroid enlargement is marked, obstruction of labor may occur.

The neonatal mortality in these infants is high, particularly if hypothyroidism is also present. The major neonatal problem is usually respiratory obstruction due to the goiter.[154,199-205] Spontaneous complete recovery usually occurs in infants that survive the neonatal period though surgery is often required to relieve the respiratory obstruction. The fact that maternal milk will contain high concentrations of iodine and commercial whole milk or milk powders may be heavily contaminated by iodine must be considered in the management of these goitrous infants.

Iodide Goiter in Children and Adults

Goiter with or without hypothyroidism follows prolonged iodine intake. The iodine intake level may vary greatly from a few milligrams a day up to 1 to 2 g/day. Intakes of less than 1 mg/day have generally been regarded as harmless, but iodide goiters have been reported following intakes of 400 to 500 μg and approximately 750 μg/day. Relief of symptoms and regression of goiter follow withdrawal of iodine or, if medication is continued, administration of thyroxine.

Occasionally, a marked thyroid enlargement may cause pressure symptoms and respiratory obstruction. Of particular concern are the goiter surveys in American adults and children conducted by Trowbridge et al.[222,223] They report an overall goiter incidence of 3.1% in adults and 6.8% in children. Furthermore, adults and children with goiter and areas with high goiter prevalence tended to have higher rather than lower iodine excretion (Table 19a). Significant goiter incidences were noted in areas, not traditionally recognized as endemic goiter areas. The incidence reported by Trowbridge et al. compares unfavorably with that more recently seen in Tasmanian children (Table 19b). Urinary iodine excretion of children in Tasmania ranged from 90 to 650 μg/day. The mean daily iodine excretion was 225 μg/day, and iodine excretion exceeded 400 μg/day in very few subjects.[238] In the 1950s, goiter incidence in school children in the U.S. was reported to be 5%, but Trowbridge found incidences of up to 9.8% in some areas and an incidence of 4% in Georgia, which is not normally considered an endemic goiter area.[223] This increased goiter incidence occurred at a time when iodine intakes rose substantially above the necessary dietary requirements, and the possibility that this increased goiter incidence reflects the occurrence of iodide goiter requires careful evaluation.

Japanese Endemic Coast Goiter

This goiter is seen mainly among seaweed fishermen in villages along the coast of Hokkaido in northern Japan. The usual incidence is 6 to 12%[114] but in some villages may reach 20 to 25% of the population.[239] The condition is due to consumption of Laminariacaea japonica, L. ochotensis, and L. longissima which contain between 0.8 to 4.5 g iodine per kilogram.[57]

Girls are more often affected than boys. Histological appearances are those of colloid goiters. The cardinal difference between endemic coast goiter and sporadic iodide goiter is that hypothyroidism has not been reported. The sole problems encountered are cosmetic. The goiter regresses on stopping iodine consumption.

Table 19a
GOITER PREVALENCE AND IODINE EXCRETION

				Urinary iodine excretion			
				μg/g of creatinine		μg/day[b]	
Study area	No. examined	Percent with goiter	No. tested	Mean[a]	SD	Mean[a]	SD
Tecumseh, Mich.	1147	9.8	212	535	193	567	210
Breathitt County, Ky.	1267	8.4	164	497	231	472	208
Newton County, Tex.	809	8.3	58	392	203	381	191
Guadalupe County, Tex.	1,967	6.6	150	433	185	436	224
Savannah, Ga.	2595	4.4	170	344	213	358	350
All areas	7785	6.8	754	452	219	459	249

[a] Differences between means were significant (P < .005).
[b] Calculated valve.

From Trowbridge, F. L. et al., *Pediatrics,* 56, 82, 1975. With permission.

Table 19b

PERCENTAGE PREVALENCE OF PALPABLE GOITER IN TASMANIAN CHILDREN

	5 years		6–8 years		9–11 years		12–14 years		15–17 years	
	Male	Female	Male	Female	Male	Female	Male	Female	Male	Female
	23.5	20.1	27.9	28.8	34.8	41.5	38.3	44.7	37.5	49.0
	24.2	24.9	30.4	30.0	29.7	26.2	31.1	33.3	27.2	29.1
	22.1	25.9	23.2	26.0	22.1	25.5	19.3	24.3	18.5	20.0
a	10.1	11.6	14.4	15.6	17.3	18.2	16.6	18.7	13.8	20.0
b	8.2	8.6	11.0	12.7	14.3	17.4	15.6	18.1	9.2	16.7
c	1.4	0.9	1.2	2.2	3.8	4.4	4.3[d]	6.5[d,e]	4.0[d]	6.7[d,e]

e: Not less than 7000 children were examined in each survey year, e.g., just over 7000 in 1949 (the lowest number) and about 11,000 in 1976.

dophors introduced in dairy industry in 1963.

niversal iodation of bread in 1966.

ean 24-hr urinary iodine excretion = 238 μg I (range, 76–420 μg I per 24 hr)

ay contain a group of children with fixed goiter developed prior to universal iodation of bread.

ontribution from pubertal goiter.

a compiled from information supplied by Dr. Heather Gibson, Senior Medical Officer of the Tasmanian School Medical ices and Tasmanian State Thyroid Advisory Committee.

Thyroiditis

Acute thyroiditis has not been reported in response to increases in dietary iodine, but some cases have been reported following large doses of iodine (1 to 1.5 g/day). Symptoms develop within days of starting iodine medication. There is swelling of the thyroid gland associated with intense pain, both of which will subside within 48 to 72 hours after iodine is stopped.[240,241] It is unknown whether iodide can precipitate chronic lymphocytic thyroiditis though rising dietary intake of iodine has been linked with the increased incidence of lymphocytic infiltration noted in surgically removed thyroids.[242,243]

Miscellaneous Adverse Effects

Numerous, diverse effects on the various systems of the body have been recorded. Some reactions are due to the direct, irritant effects of excess iodide; others are undoubtedly idiosyncratic. Toxicity may also occur. In many instances, the immunological process may be suspected by analogy and extrapolation, but direct proof is lacking. With modern radioimmunoassay techniques, there is now some prospect that the position will be clarified.

The role of immune responses is difficult to prove. However, a number of reported manifestations of iodine toxicity defy explanation unless such responses are invoked. Anaphylaxis, with angio-edema,[120,156,164,246] bronchospasm,[143,148,244,245] hypotension,[244] and vomiting and diarrhea[160,164,244] are probably immunologically mediated. Circulating toxic immune complexes are probably responsible for some of the skeletal,[95,143,147,156,164] renal,[143,255,256] hepatic,[254] vascular,[120,143,147,148,150-152,156,244,246,257] and hematological[120,143,147,152,156,164,253,254,256,257] toxic manifestations reported (Table 20).

Central nervous system — The mechanisms involved in the development of irritability and depression are unknown, nor is it known how iodohydroxyquinolines and iodates produce retinal damage.

Table 19c
UNTOWARD REACTIONS TO IODINE IN FOODS

Type of reaction/circumstances	Iodine source	Number of cases	Verified
Iododerma, fatal (mother on iodine ℞)	Mother's milk	1	?
Acne, severe, adult	Iodized salt	7	2
Iododerma, fatal (iodized salt for years; therapeutic KI for 4 months)[a]	Iodized salt + KI	1	?
Iododerma, bullous, severe (iodized salt for months; 3 g NaI in 1 week)	Iodized salt + NaI	1	No
Iododerma, papular, indurated (iodized salt 2 months; none 12 months; ℞ colloid I for 2 days)	Iodized salt + colloid I	1	No
Acne, severe, adult	Iodized salt	1	No
Urticaria, chronic	Iodized salt	1	Yes
Acne, adult	Iodized salt	1	No
Iododerma, papular, indurated (presumptive diagnosis)	Iodized salt	1	No
Iododerma, bullous (iodized salt for years; KI cough mix for 4 days)	Iodized salt + KI	1	Yes
Iododerma, papular, indurated, chronic	Iodized salt	1	No
Iododerma, bullous, chronic (presumptive diagnosis)	Iodized salt	1	No
Purpura, thrombocytopenic, thrombotic	Kelp	1	No
Parotitis	Iodized salt	1	Probable
Dermatosis of lip	Seafood	1	No

[a] May have been complicated by neoplasms.

From Talbot, J. M., Fisher, K. D., and Carr, C. J., A Review of the Significance of Untoward Reactions to Iodine in Fo
FDA 71-294, Life Sciences Research Office, Federation of American Societies for Experimental Biology, Bethesda, N
1974, 1. With permission.

Table 20
SOURCES OF REPORTED ADVERSE EFFECTS

Adverse effects	Ref.
Central Nervous System	
Nervous irritability	95, 120
Depression	95, 244, 246
Respiratory System	
Nasal congestion[a]	95, 160
Sneezing[a]	244
Coryza[a]	95, 156
Pharyngitis	164
Asthma[c]	143, 148, 244, 245
Alimentary System	
Nausea[b]	55, 160, 244
Gingivitis	156
Salivary gland involvement[a,b]	120, 143, 147, 152, 156, 160, 164, 246, 250–2
Epigastric distress[b]	156
Vomiting[b]	160, 244
Abdominal pain[b]	244
Diarrhea[b]	164
Jaundice[c]	254

Table 20 (continued)
SOURCES OF REPORTED ADVERSE EFFECTS

Adverse effects	Ref.
Musculo-skeletal System	
ritis/arthralgia[c]	143, 147, 156, 164
rs[b]	147, 244
Genito-urinary System	
naturia[c]	143
hritis[c]	255
einuria[c]	143, 256
Cardiovascular System	
io-edema (eyelids, lips, tongue, tonsils, larynx)[a,b]	120, 156, 164, 246
al congestion[b]	160
junctival injection[b]	95, 156–159, 164, 167
arteritis nodosa[c]	143, 147, 148, 150, 151
chial hemorrhage and purpura[a,c]	143, 147, 151, 164, 257
culitis[c]	120
nonary edema[b,c]	244, 246
cardial infarction syndrome[b]	244
otension[b,c]	244
Hematological System	
inophilia[c]	143, 150, 156, 184, 253, 256
norrhages – submucosal	143, 147
ombocytopenia and/or purpura[a,c]	120, 143, 147, 152, 164, 257
kocytosis[c]	143, 147
nphadenopathy	143, 164, 246, 254
ombotic thrombocytopenic purpura[c]	152
Miscellaneous Unclassified	
um sickness[c]	95, 156
er[b]	120, 156, 160, 164, 250, 254
ngrene	120
lls and rigors[b]	143
sting cachexia	156, 246

Reported following dietary iodine ingestion.
Reported following contrast radiology with iodinated media.
Possibly due to immune mechanisms.

Respiratory system — Most of the effects described are probably due to the increased secretory activity produced by iodides. Pulmonary edema may be a sequel to excessive secretions by the bronchial mucosa in response to iodide or a result of direct alveolar capillary damage by circulating toxic immune complexes. Asthma and bronchospasm are usually seen in acute anaphylactic type reactions and usually associated with angio-edema of the eyelids, lips, tongue, and larynx, hypotension, nausea, vomiting, and at times, diarrhea.

Alimentary system — Nausea, vomiting, epigastric distress, and diarrhea may be due to the direct, irritant, local effect of iodides or iodine-containing medication. These manifestations may also be mediated by central mechanisms or immune responses as part of the anaphylactic symptom complex due to vasculitis caused by circulating immune complexes. Since iodide is not normally hepatotoxic, jaundice is difficult to explain without involving immune responses.

Enlargement of salivary glands, either the parotids or submandibular glands, may follow a single exposure to iodide during contrast radiography or more prolonged iodide medication. The salivary glands usually swell symmetrically and may be painful or quite painless. If painful, the condition may be mistaken for the mumps; however, it subsides rapidly on withdrawal of iodine.

Musculoskeletal system — Arthritis and arthralgia are difficult to explain without invoking possible immune disorder. Polyarthritis is often a feature of serum sickness, which is a classical example of an immune response due to toxic immune complexes.

Genitourinary system — There is no evidence that iodide is nephrotoxic. Renal manifestations are probably explicable on an immunological basis, possibly due to glomerular damage by immune complexes.

Cardiovascular system — As with other systems, iodide does not appear to have a direct toxic effect on the heart or blood vessels. Angio-edema and hypotension are part of the anaphylaxis symptom complex of cell-mediated immediate immune response. Periarteritis nodosa, vasculitis, and thrombotic thrombocytopenic purpura (micro-angiopathy with consumptive coagulopathy) are recognized as autoimmune disorders due to toxic immune complexes. Petechial hemorrhages and pulmonary edema can occur as a result of vasculitis and immune-complex damage to vessels.

Hematological system — Most of the hematological changes reported are probably best explained on the basis of immune responses though scientific proof is lacking and the evidence is presumptive based on extrapolation from observed facts. The initial response to immune-complex tissue damage is usually a neutrophil leukocytosis, which is followed by an eosinophil response. Thrombocytopenia may occur in autoimmune diseases due to immune-complex formation and may also possibly occur when iodinated compounds trigger an immune response. The thrombocytopenia in thrombotic thrombocytopenic purpura is probably secondary to the microangiopathy, in which damage to small vessels may be mediated by immune complexes, causing thrombotic phenomena with consumption of platelets and clotting factors, resulting in thrombocytopenia and hemorrhagic diathesis.

An increase in eosinophils in the circulation and tissues is seen in a number of conditions associated with high serum levels of IgE, such as allergic rhinitis, atopic asthma, parasitic infections, and disorders in which circulating antigen-antibody complexes capable of activating complement are present, e.g., periarteritis nodosa and rheumatoid arthritis. The functions of the eosinophil in anaphylactic reactions are probably diverse; they may have a regulatory function at all stages of the allergic response, namely, mediator release, mediator inactivation, and mediator replenishment. IgE, or tissue-sensitizing antibody, sensitizes mast cells to release the chemical mediators of anaphylaxis which include histamine, slow-reacting substance of anaphylaxis (SRS-A), and eosinophil chemotactic factor of anaphylaxis (ECF-A). ECF-A selectively attracts eosinophils from the leukocyte population. Histamine may contribute to the localization of eosinophils which release an inhibitor of histamine release, probably a prostaglandin. Eosinophil histaminase and arylsulfatase are then available for histamine and SRS-A inactivation.[258] The role of immune mechanisms in adverse reactions to iodine remains conjectural and must await scientific proof.

ADVERSE EFFECTS OF IODINE IN ANIMALS

Knowledge of the effects of excess iodine in animals is fragmented and the picture is still essentially obscure, though some species differences in response to excess iodine have

MEDIATOR RELEASE   **MEDIATOR INACTIVATION**   **MEDIATOR REPLENISHMENT**

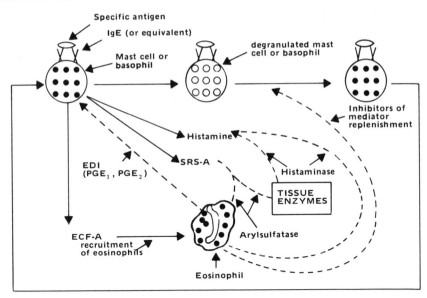

FIGURE 4.   Diagrammatic representation of the functions of the eosinophil during immediate-type (anaphylactic) hypersensitivity. The "inhibitory" effects of the eosinophil are represented by the interrupted lines. (From Kay, A. B., *Br. J. Haematol.,* 33, 316, 1976. With permission.)

been noted. Data in numerous reports on record are frequently conflicting, rendering evaluation difficult. Assessment is often complicated by the design of the experiments, since experimental conditions and diets are often artificial, and by the doses of iodine, which are frequently grossly excessive and unlikely to be encountered under field conditions except perhaps following the therapeutic use of iodine. Interpretation is further hampered by the mode of presentation of the data, since many reports feature the concentration of iodine in feeds, and in the absence of consumption data, actual iodine intake cannot be assessed.

Acute and Chronic Iodine Toxicity

Data on the acute, subacute, and chronic toxicity of iodides and iodates are available, at least in some species, but similar observations on other dietary iodine supplementing compounds are lacking. The toxicities of sodium and potassium iodides and iodates were extensively studied by Webster[276-279] and Highman[280] and their co-workers. They found that sodium iodide was well tolerated in animals after oral, intravenous, or intraperitoneal administration, while toxicity of the potassium ion rather than the iodide moiety was the limiting factor in potassium iodide dosimetry.[276] In mice after intraperitoneal injection of sodium iodide, the median lethal dose (LD_{50}) was 1690 ± 85 mg/kg, and for potassium iodide, the LD_{50} was lower, 1117 ± 30 mg/kg. No deaths were observed following intraperitoneal doses of up to 700 mg/kg of sodium iodide.[276]

Iodates were 8 to 10 times more toxic than iodides and better tolerated after oral administration than after parenteral dosing. The oral LD_{50} was approximately six times the intraperitoneal LD_{50} in mice. Larger doses of iodates were tolerated by feeding animals than by fasting ones, presumably due to slower absorption caused by the gastric contents.[276] Species variations in tolerance were noted between mice, rabbits, guinea pigs, and dogs, and larger animals generally showing greater tolerance than the smaller species. Sodium and potassium iodate were equally tolerated; the LD_{50} in mice was reported to be 108 mg/kg and in dogs, 200 mg/kg.[279] Amounts of iodate exceeding the single oral LD_{50} were well tolerated with few side effects when the dose of iodate was administered in divided doses over 24 hr.[277]

Iodate, a strong oxidizing agent, was seen to be a gastrointestinal irritant and hemolytic agent.[279] Neurological manifestations were observed after high doses of iodate; retinal degeneration was found to be dose dependent and occurred only after large intraperitoneal or intravenous doses. The general patterns of observed toxic manifestations after either iodates or iodides were similar, but with iodates, symptoms appeared earlier, progressed faster, and were observed at much lower dose levels. No mortality occurred in mice with intraperitoneal doses of 70 mg/kg of iodate; the LD_{50} for sodium iodate was 108 ± 4 mg/kg and for potassium iodate, 136 ± 5 mg/kg[276] (Table 21).

After administration of toxic doses of iodate, vomiting was seen early, followed sequentially by alternating periods of hyperactivity and lassitude, then weakness, prostration, and dyspnea. Paresis of the hind legs usually preceded convulsions, respiratory arrest, and death. Hemoglobinuria was observed in some animals after iodates, but not iodides.[276] Similar toxic manifestations were observed after massive doses of iodide, though symptoms appeared later and progressed more slowly. Deaths usually occurred days instead of hours later. Diarrhea was common after oral administration of toxic doses of iodates and less frequent after iodides.

Autopsies revealed necrosis of gastric parietal cells, hemorrhagic necrosis of the gastric mucosa, and patchy ulceration of the colonic and bladder mucosa. The lungs were congested with patchy consolidation evident. Fatty changes in the liver, kidneys, and myocardium were often associated with necrosis of the proximal renal convoluted tubules and hemoglobin casts in the kidneys. Retinal degeneration with pigmentary changes and absence of rods and cones, associated with atrophy of the inner and outer nuclear layers, was seen only after parenteral administration of toxic doses of iodate. Adrenal lipid deposits were depleted.[279,280]

In subacute toxicity studies with iodates, weight loss was common.[279] Reticulocytosis and a rise in nonprotein nitrogen of the blood were frequently noted. Methemoglobinemia and hemoglobinuria were observed in some animals. After administration of iodate, a marked rise in urinary iodide excretion occurred and iodate appeared in the urine only briefly, soon after dosing. Autopsies on sacrificed animals indicated that parietal cell necrosis and gastric mucosal ulceration after iodates were reversible. Fatty changes were apparent in the liver and kidneys, and hemosiderin deposits were noted in the spleen, renal tubules, and the Kupffer cells of the liver.[279]

In oral, chronic iodate toxicity studies, vomiting, which was initially marked, diminished after continued dosing with iodate. Few ill effects were noted in experimental animals apart from lowering of hemoglobin with an accompanying reticulocytosis.

Table 21
COMPARATIVE TOXICITIES OF ORAL IODATE AND IODIDE

Potassium salts	Not fasted		Fasted on saw dust		Fasted on screens	
	Iodate	Iodide	Iodate	Iodide	Iodate	Iodide
LD_{50} dose (mg/kg)	1177 ± 61	2068 ± 140	815 ± 29	1928 ± 90	531 ± 21	1862 ± 10
Highest tolerated dose without death (mg/kg)	819	1780	513	1420	359	1260

From Webster, S. H., Rice, M. E., Highman, B., and Von Oettingen, W. F., *J. Pharmacol. Exp. Ther.*, 120, 171, 1957. W permission.

Autopsies revealed hemosiderin deposits only in reticuloendothelial cells and renal tubules.[276]

These data indicate that iodide has low toxicity and that the greater toxicity of iodate is probably due to the powerful oxidative properties of this ion. Concentrations of iodide and iodate presently used for iodine supplementation appear safe. Present evidence also suggests that iodine from currently used supplemental compounds is absorbed and utilized adequately as iodide. The toxicities of these compounds depend on the biological availability and activity of the carrier anions and cations. Predictions concerning the toxicities of iodine-supplementing compounds other than iodate and iodide are not warranted. Further toxicity studies, relating to other approved and used compounds, are needed.

Iodism

The occurrence of iodism is dose dependent; symptoms may appear after varying periods, often quite brief periods, of excess iodine intake. Profuse nasal discharge, salivation, and cough are cardinal features, which may also be associated with enlargement of salivary glands and diarrhea.[281-283] Following prolonged administration of excessive iodine, thyroid enlargement may develop but hypothyroidism has not been documented. Feed intake is usually reduced causing depressed performance, manifested chiefly by absent or reduced weight gain. Hemoglobin may be lowered, and increased susceptibility to infection, especially of the respiratory tract, has been reported.[282,283] Central nervous system depression and lethargy are common.[282,283]

Excess Iodine Intake and the Thyroid

Despite an awareness of the possible occurrence of iodine-induced thyrotoxicosis following administration of iodine in feeds or medication, no conclusive proof of its occurrence has been adduced. Webster and Chesney have reported the occurrence of thyrotoxicosis in rabbits when iodine was administered to animals with hyperplastic glands;[284] their report, however, remains uncorroborated. Following administration of EDDI to sheep, elevation of body temperature and raised PBI levels were measured as evidence of hyperthyroidism by McCauley et al.[281] However, since PBI levels become unreliable indicators of blood thyroid-hormone levels when the system is flooded with excess iodine and direct thyroid hormone assays were not performed, their conclusions are questionable.

Manifestations similar to those seen after administration of thyroxine with many features of thyrotoxicosis have been observed in cattle and sheep, after ingestion of the iodinated proteins, casein and ardein, but not iodides.[271,272] In both sheep and cattle, a rise was noted in body temperature, while respiratory and pulse rates were elevated. Cardiac arrhythmia was also observed.[271,272] In cattle, milk output rose, and milk solids and fat were increased. Feed intake usually decreased with weight loss, emaciation, and depletion of fat stores. Muscular weakness and coughing fits were often noted, and death was usually associated with pulmonary congestion and consolidation. Cardiac hypertrophy and dilation were also observed at necropsy.[271] The mechanisms whereby these manifestations are produced after intake of iodinated protein are unknown. However, isolation of thyroxine from casein and other iodinated proteins has been reported, so the autopsy finding of "regression" in the thyroid[271] supports the hypothesis of extra-thyroid generation of thyroid hormone from these iodinated proteins.

Iodide Goiter

In most species, excessive iodine intake causes enlargement of the thyroid gland, but hypothyroidism has not been documented.[275,285-288] In most reports, massive doses of iodine on the order of 100 mg/kg of feed were administered resulting in thyroid enlargement and accumulation of colloid in the gland, but in the absence of reports of hypothyroidism, satisfactory adaptation of the thyroid to such doses must be assumed. The goiter incidence in horses could be directly linked with excess iodine intake in doses between 48 and 347 mg I per day.[285] When iodine was withdrawn, goiters regressed.

Thyroiditis

Large doses of iodine, administered to iodine-replete animals, may produce thyroid enlargement, but no inflammatory responses appear to be evoked. A single large dose of iodine administered to iodine-depleted or iodine-deficient dogs or hamsters has been observed to cause discharge of inorganic and organic iodine from the thyroid with necrosis of follicular cells.[288-292] In hamsters, follicles were devoid of colloid, follicular cells were desquamated, and polymorphonuclear infiltration of follicles and stroma was noted. The acute inflammatory changes subsided over two weeks with replacement of the polymorph infiltrate by round-cell infiltration. Repair with a return-to-normal histological appearance and without apparent, permanent damage occurred over a period of several weeks.[288,289] Administration of a small, preliminary priming dose of iodine before administration of the large dose appeared to protect the thyroid and prevented occurrence of acute changes.[292]

Species Responses to Excess Iodine

Many diverse adverse reactions occur in various species in response to excess iodine. The minimum levels at which adverse reactions may occur have not been established in most species. Impaired performance, increased susceptibility to infection, and the possibility of increased mortality among the offspring could be important economic factors to farmers, and even national productivity. While iodine supplementation in areas of iodine deficiency is logical and prudent, the practice remains questionable in areas of iodine sufficiency, at least until satisfactory information concerning its safety is presented.

Poultry

The recommended daily intake of iodine for poultry at various stages of development and at maturity ranges from 11 to 37 μg/g I per hen per day[78] and can be achieved by an iodine content of 300 to 350 μg I per kilogram of feed (Table 17). For optimal growth, 75 μg I per kilogram of feed is required by young chicks, while iodine contents of less than 300 μg I per kilogram of feed may result in thyroid enlargement in mature hens.[293] All currently accepted iodine-supplementing compounds are effective, but some evidence suggests that in poultry iodine from DIS is not as readily assimilated as other compounds.[235]

Iodine intake bears a direct linear relationship to the iodine content of eggs.[48] Feeding of 100 mg I per kilogram of feed caused a progressive rise in the iodine content of eggs to 3 mg I per egg over a 10-day period.[48] After feedings of 500 mg I per kilogram of feed, iodine contents of eggs rose progressively to 8 mg per egg until egg production stopped on the eighth day.[48] Egg production appeared to be inversely related to the level of iodine intake; in experiments it diminished progressively as level of intake rose above the mandatory iodine requirement. Though ova continued to develop, involution occurred beyond a threshold level of iodine and egg production stopped.[48,235,294] Egg production resumed a week after withdrawal of iodine.

In response to excess iodine intake, not only did egg production diminish and ultimately stop, but feed consumption and the weight of birds fell.[235] Studies have shown that excess iodine did not impair the fertility of eggs but embryonic death rate was high, hatchability reduced, and delayed hatching common. The newly hatched chicks were weak, their weights were lower than those of controls, and the perinatal death rate was high.[293-295]

Feeding 2.5 to 5.0 g I per kilogram of feed to 7-day-old chicks caused marked growth inhibition, accompanied by a 75% mortality rate at the higher level of iodine feeding.[293] The mechanisms producing these adverse effects are unknown.

Cattle

The optimum iodine intake for cattle remains uncertain, since recommended daily intakes range from 0.4 to 20 mg.[14,74,75,91] The calculated requirement was 1 mg/cow/day but the U.S. NRC Committee on Animal Nutrition recommended an intake 3.5 times the calculated figure; such levels of intake can be attained by feeding products containing 0.0076% stabilized iodine. An intake of 800 to 1600 μg I per day is regarded adequate for dairy cattle, while the requirement of beef cattle should be met by intakes of 400 to 800 μg I per day. The recommendations of the British Agricultural Research Council appear in the section on the Iodine Requirements of Animals. They recommend variations in the levels of iodine supplementation dependent on pregnancy, lactation, or goitrogen content of diet. The Chilean Iodine Information Bureau recommended iodine intakes of 5 to 20 mg/day.[14] Generally, iodine intakes of 2 to 3 μg I per kilogram of animal weight should be adequate for nonpregnant, nonlactating animals, while 4 to 6 μg I per kilogram animal weight should meet the needs of pregnant and lactating beasts.

All the currently used iodine supplementing compounds have proved to be adequate and suitable vehicles for supplementation.[24] Tests have indicated that DIS iodine is not as readily available as iodine from other compounds and is rapidly excreted,[25] while EDDI is absorbed better and retained in tissues longer than potassium iodide.[23]

Dietary and endogenous iodine secreted by the salivary glands is absorbed in the rumen while iodine secreted by the abomasal glands and duodenum is almost totally absorbed in the distal small bowel.[296-298] Organic iodine compounds, chiefly thyroid hormones, are conjugated in the liver and excreted in bile. These conjugated organic iodine compounds are then partially split and reabsorbed with a small unsplit fraction excreted in the feces. Although fecal loss of iodine is usually small, it may become substantial in the presence of large, bulky stools, diarrhea, or goitrogens in the diet.[299] However, the main routes of excretion of iodide are milk and urine. When thiocyanates inhibit secretion of iodide into saliva, abomasal fluid, and milk, urinary excretion of iodide rises sharply.[247,259,260]

The iodine content of milk is related to plasma iodine concentration and, consequently, iodine intake.[30,299] Studies show that in the absence of iodine supplementation, iodine levels in milk are low, 8 μg I per liter, and following supplements of 6.8 and 68 mg of iodine, it rises to 81 μg I per liter and 694 μg/l, respectively.[30] It has been noted that as milk output rises, milk iodine concentration falls but total iodine output in milk remains relatively unchanged.[25] Swanson's report,[32] that milk yield from iodine-supplemented cows during a 28-week study period was lower than in a matched control group, conflicts with Hemken's observation[30,299] that iodine supplements stimulated milk production.

Administration of thyroxine increases milk production in cows and produces a rise in the butter fat and solid content of milk.[25] The iodine concentration of milk falls, but total iodine output remains unchanged. Features of thyrotoxicosis, reduced feed consumption, weight loss, and diarrhea, associated with raised pulse and respiratory rates and rises in body temperature, may occur.[25] Features similar to those of thyroxine administration follow feeding of iodinated proteins to cattle, when an increase in the iodine and iodotyrosine content of milk may also be seen.[271,272] In mature animals the level of iodine intake at which adverse effects may be anticipated has not been adequately established; indeed, significant variations in tolerance may depend on the nature of the supplementing compound, the presence or absence of goitrogens in the diet, and whether the animal was pregnant or lactating.

Administration of 312 to 500 mg EDDI to cattle produced anorexia, lethargy, hyperthermia, marked salivation, and coughing. When these treated animals were stressed with bovine respiratory disease complex, they failed to respond to therapy; mortality exceeded 50% and survivors performed poorly.[281]

Calves fed 6.6 mg I per kilogram of body weight had poor appetites, abnormal feces, and poor appearances.[281] When iodine supplements exceeded 25 mg/kg of feed, toxic manifestations of increasing severity were seen as supplementation rose. Feed intake fell progressively; weight gain per unit of feed was reduced, and growth rate depressed; iodism with profuse nasal discharge, coughing, and diarrhea was common. The performance of calves was depressed at all levels of supplementation above 25 mg I per kilogram of feed. At higher levels of supplementation, 100 to 200 mg I per kilogram of feed, falls in hemoglobin and serum calcium levels were also observed.[282,283] While iodine supplementation in recognized areas of iodine deficiency is logical and sound practice, the need for, and indeed the wisdom of, iodine supplementation in other areas remains debatable.

Sheep

A marked diversity of opinion on the daily iodine requirement of sheep is apparent since recommendations range from 50 to 2000 μg I per day per animal.[14,86,89,90,92] Mason[92] reported that feeding pregnant housed ewes 250 to 300 μg I per day prevented goiter in newborn lambs despite a goitrogenic diet containing thiocyanates. Iodine kinetics in sheep are akin to those seen in other ruminants. Active renal iodide reabsorption, susceptible to thiocyanate blockade, has been postulated but requires confirmation.

Feeding of iodinated casein produced features of thyrotoxicosis; animals became sluggish, anorexic, weak, and emaciated. Respiratory and pulse rates were elevated, body temperature was raised, and profuse nasal discharge and coughing were noted. At necropsy, pulmonary congestion and consolidation were found. Similar changes were not observed in sheep fed an amount of iodide equivalent to that in casein.[271]

McCauley et al.[281] fed lambs either EDDI or potassium iodide in dose ranges of 94 to 750 mg/day and 197 to 785 mg/day, respectively. When doses of either compound exceeded 375 mg/day, central nervous system depression, anorexia, coughing, and elevated body temperature were observed. Reduced weight gain and, in some animals, actual weight loss were noted. A mortality rate of 20% was observed in the high iodine dose ranges. At autopsy half the lambs that died had pneumonia.[281] These workers suggested that iodide was an uncoupler of oxidative phosphorylation[146] which resulted in reduced ATP formation and caused increased susceptibility to infection in the presence of increased metabolic demand. They also noted loss of appetite and poor weight gain in lambs fed 7.2 mg I per kilogram body weight per day.[281]

No detrimental effects on growth, food consumption, wool production, or mortality were noted in ewes fed 20 mg to 60 mg of iodine a day.[300] However, in the second lambing season Malan et al. reported that though weight gain and patterns, wool production, and occurrence, periodicity, and duration of estrus were unchanged and though gestation periods and birth weights of lambs were normal, ewes lost weight after lambing and their milk supply was inadequate, causing a high mortality rate among lambs. In light of prevailing drought conditions and an unintended vitamin A deficiency in the diets of experimental animals, the authors questioned the role or contribution of the iodine supplements to the results observed.[300]

In the Derwent Valley of Tasmania, neonatal goiter in lambs is prevalent. Injection of pregnant ewes with 1 to 2 ml of lipiodol was, for a time, widely practiced. Reduced wool production in ewes and their lambs was noted the following year,[301] an observation in conflict with Malan's report.[300] In the absence of iodine deficiency and at the present state of our knowledge, indiscriminate iodine supplementation may be unwise.

Swine

The daily iodine requirements for swine have not been clearly established, and recommendations range from 80 to 2000 μg I per day. Iodine intakes of 2 to 3 μg/kg body weight in nonpregnant animals and 4 to 6 μg/kg body weight in pregnant sows appear to be sufficient.[14,76,86,89,90] Experimental data on iodine absorption, kinetics, and excretion are lacking, but minor species differences apart, iodine metabolism probably does not differ materially from other better studied species. Additional experimental information in this area would be desirable. Conflicting reports about levels of iodine supplementation, which cause adverse effects, render evaluation difficult. Iodine supplements of 1.5 to 2.5 g/kg of feed have been reported to cause no adverse effects in pregnant sows. Farrowing is normal and the progress of suckling pigs is satisfactor,[302] yet depressed performance was noted in young pigs after supplements of 4 mg/kg of feed.[303] Newton and Clawson[304] observed iodism, reduced feed intakes, depressed growth rates, and enlarged thyroid and salivary glands with lowered hemoglobin levels when iodine supplements exceeded 400 μg/kg of feed.[304] Conversely, Weiser and Zaitschek[305] reported no adverse effects in pigs fed 83 mg I per day, and Forbes[306] reported no ill effects from feeding 1.1 mg I per kg body weight.

Goats

The daily iodine requirements of goats have not been established. Theoretically, iodine intakes of 2 to 3 mg/kg body weight would be reasonable, while lactating and pregnant animals may need more, 4 to 6 mg I per kilogram of body weight.

Concentrations of radioactive iodine were noted in the abomasum, duodenum, cecum, and mammary glands at varying intervals after administration of a tracer dose. Peak thyroid uptake of radioiodine occurred at 72 hours. In the goat, the feces have been reported to be the main route of iodine excretion;[265] less than 50% of an iodine load could be recovered from urine.[247] Further studies of iodine metabolism and kinetics in the goat seem desirable.

The mammary gland actively secretes iodine. Milk to plasma ratios range from 7.1 to 24.1 and rise with increased milk output due to the higher concentration gradients maintained by the highly active mammary gland.[307] However, Lengeman reported that while small doses of iodine increased this ratio, large doses reduced it by a factor of 10, possibly due to the saturation of transport mechanisms.[247] Lengeman's report[247] conflicts with Reineke's observation[307] that excess iodine did not inhibit iodine concentration in milk.

Toxic levels of iodine supplementation, able to produce adverse effects in goats, have not been established. The suggestion that iodine secretion into milk may cause iodine depletion, which may adversely affect thyroid function and result in reduced milk production, requires evaluation so that optimal iodine intake can be maintained during lactation.[307,308]

Horses

Little is known of the daily iodine requirements and iodine metabolisms of horses, and little more of the adverse effects of excess iodine. Estimates based on iodine requirements of man and various other species suggest that intakes of iodine of 2 to 6 μg/kg body weight may be sufficient. Equine goiter has been reported in areas of iodine deficiency. No goiter was observed in horses receiving less than 7 mg of iodine a day, but iodide goiter was seen following iodine intakes from 48 to 432 mg/day in mares, foals, and yearlings. Elevated TSH levels were found in goitrous animals. Autopsy revealed only thyroid enlargement in the one foal that died.[285]

Dogs

The daily iodine requirement per kilogram of body weight is higher in dogs than in man since dogs do not conserve thyroid iodide and peripheral hormone efficiently. The daily requirement for an adult beagle was stated as 140 μg I per day.[290] In the dog, a greater release of nonhormonal iodine (MIT and DIT) and inefficient reutilization of released iodide were noted as well as considerable fecal losses of thyroxine and triiodothyronine.[290,291] Thyroid turnover of iodide, thyroxine, and triiodothyronine was more rapid in the dog than in man, and all these exchange rates were reduced by excess iodine. Triiodothyronine, 37 to 60% of which was derived by peripheral deiodination of thyroxine, appears to be the main regulator of metabolic function in the dog.[290]

Marked fluctuations in the iodine content of commercial brands of dog food ranging from 800 to 5100 μg/kg were noted; significant differences were seen between brands and even in the same brand from time to time. A 2- to 7-fold increase in the iodine content of commercial dog foods was observed between 1969 and 1975.[290] Iodate toxicity studies indicated that the minimum lethal oral dose of iodate was about 200 mg/kg, that the LD50 lay between 200 and 250 mg/kg, and that 100% mortality occurred with doses of 250 mg of iodate per kilogram.[279] With doses of 200 mg/kg and over, anorexia preceded coma and death.[279] Methemoglobinemia and hemoglobinuria were noted; iodate and iodide were present in urine. During autopsy, patchy hemorrhagic necrosis of gastric, colonic, and bladder mucosa was common, and irregular necrosis of hepatic lobules with fatty changes in the liver and renal tubules was noted. Retinal changes were present, and congestion and consolidation were seen in the lungs.[279]

In subacute toxicity studies with doses up to 100 mg of iodate per kilogram, apart from anorexia, emesis, a lowered red cell count, and transient weight loss, no significant toxic effects were noted. After sacrifice, mild fatty changes were observed in the liver and renal tubules, together with hemosiderin deposits in the spleen, Kupffer cells, and renal tubules. Dogs fed 30 to 45 mg of iodate per kilogram for one year showed no ill effects; food consumption, weight patterns, and general physical condition remained normal. Most animals showed a fall in their red cell count accompanied by a mild reticulocytosis. Methemoglobinemia was not encountered. After sacrifice, mild inflammation of some parts of the alimentary tract was commonly found and hemosiderin deposits were noted in Kupffer cells.[278] No retinal damage occurred. In view of the greater toxicity of iodate, significantly higher doses of iodide would have been tolerated before detrimental effects developed.

In iodine-deficient dogs, but not in iodine-replete dogs, administration of excess iodine caused acute thyroiditis with necrosis and desquamation of follicular cells and discharge of inorganic and organic iodine from the thyroid. Administration of a small preliminary dose of iodine appeared to prevent the acute response.[292]

Laboratory Animals

Recommendations on iodine nutrition for laboratory animals have been prepared by the U.S. NRC Committee on Animal Nutrition.[309] Exact iodine requirements for these animals have not been established, but iodine requirements of mice, rats, hamsters, and rabbits probably do not exceed 10 to 20 μg I per day.

Mice — The toxic effects of excess iodide and iodate have been described in a preceding section. Iodide goiter was produced by chronic administration of 100 mg I per day. Thyroid follicles were distended by colloid and lined by flat acinar epithelium.

Thyroid protease activity and iodothyronine release from the gland were reduced, but hypothyroidism was not described.[310]

Rat — High doses of iodine block organification in the rat, but adaptation was noted after 16 hr. Following chronic excess administration of iodide, inhibition disappeared after 3 to 4 days but iodine binding fell over the ensuing weeks. Though the thyroids enlarged, hypothyroidism was not documented.[311]

Rats fed diets containing iodine supplements of 0 to 2500 mg I per kilogram feed with resultant iodine intakes of up to 150 mg I per kilogram body weight ovulated normally; fertility was unaffected. Implantation, gestation periods, and the numbers in and birth weights of the litters were similar to control animals. However, mortality rates in the neonates increased commensurately with rising iodine intakes, i.e., mortality in the young was proportional to iodine intake. Low levels of iodine supplementation caused mortalities of less than 20%, but after supplements of 2000 mg/kg of feed, mortality in the litters rose to 85% and approached 100% when supplements of 2500 mg I per kilogram were used. Incomplete parturition, failure of lactation, and lack of mothering were also observed in high iodine-intake rats.[312] Maternal feed intakes were reduced at all levels of supplementation and probably contributed to the poor lactation and reduced weaning weights of the survivors from the litters. After withdrawal of iodine, littering and litter survival returned to normal.

Rabbits — Iodine supplementation of rabbits resulted in a significantly increased mortality rate in the young, even at the lowest level of supplementation, 250 mg I per kilogram of feed. It reached 100% when supplements of 500 mg/kg feed with resulting iodine intakes of up to 90 mg/kg body weight were used in the experiments. Although lactation was normal, there was no evidence of nursing.[312] Iodine-induced thyrotoxicosis has been reported after administration of iodine to rabbits with hyperplastic glands,[284] but further corroboration of this report is required.

Hamsters — Iodine intakes of up to 160 mg/kg body weight in hamsters resulted only in a slightly increased mortality among the young, but survivals were markedly better than those seen in rats and rabbits fed corresponding amounts of iodine.[312] Acute thyroiditis followed administration of large doses of iodine to iodine-deficient hamsters. Polymorphonuclear infiltration of follicles was associated with depletion of colloid with necrosis and desquamation of follicular cells, though little inflammatory reaction was observed in interstitial tissues of the gland in early stages. By the end of the first week, however, there was marked follicular and interstitial infiltration with polymorphonuclear and round cells. The acute inflammatory response subsided in the second week so that a mononuclear infiltration of the gland was predominant. Normal morphological appearances were reestablished over a period of 6 weeks. No inflammatory reaction followed administration of large doses of iodine to normal iodine-replete hamsters.[288,289]

METHODS TO CONTROL DIETARY IODINE

There has been no increase in the number of reports of adverse reactions to iodine during the last 10 years of rising intake of iodine, so there is no clear-cut evidence that the present level of iodine in the diet is harmful. No doubt exists that additional iodine can precipitate thyrotoxicosis in endemic goiter areas[187] and that goitrous subjects in iodine-replete areas also risk developing iodine-induced thyrotoxicosis with increasing dietary iodine levels.[180] Of concern is the high goiter incidence in American adults and school children and the appearance of a significant goiter incidence in areas, not traditionally recognized as endemic goiter zones. The most noteworthy fact in these surveys was that adults and children with goiter and areas with highest goiter prevalence tended to be those with high iodine intakes.[222,223]

Finally, one should not be lulled into a false sense of security, because reports of adverse reactions to iodine are not increasing. Few clinicians are aware of the state of iodine nutrition, and many possible adverse reactions may remain unrecognized. In other instances, when an adverse reaction to iodine is suspected, the case often remains unreported, because proof is lacking or the physician is disinterested or too indolent to report the case. In a limited survey on adverse reactions to drugs conducted by the author, it was found that doctors reported less than 10% of minor reactions, 20 to 25% of moderate reactions, and only 40% of severe adverse drug reactions to the Australian Health Department, which has a register for these reactions. The iodine content of the diet has risen steadily over the past decade, and there is no evidence yet that a plateau has been reached. The present level of mean intake is estimated at about 500 μg/day in America,[96] a figure markedly in excess of the level required for adequate nutrition.

Whereas in the past, public health authorities were concerned with goiter prophylaxis and maintenance of adequate nutritional intakes, their role as far as iodine is concerned is changing. Iodine intake is excessive and rising due to the adventitious presence of iodine in many foods and medications. The time has come to exercise some controls and limit the entry of iodine into the diet. Iodine nutrition is influenced by many factors and differs from region to region. For controls to be applied effectively and rationally, the state of iodine nutrition has to be accurately established in each locale. In Tasmania this is the responsibility of the multidisciplinary State Thyroid Advisory Committee. Members of the committee are drawn from a wide spectrum of disciplines — academicians from the Medical School at the University of Tasmania, an epidemiologist, clinicians, a veterinarian, a dietitian, a chemist, a dairy officer, and representatives from the public health and school medical services and the Department of Agriculture. The committee is charged with monitoring iodine nutrition, goiter incidences, thyrotoxicosis, and other adverse reactions. It has proved most successful in its functions. In areas where iodine intakes significantly exceed requirements, a reduction in iodine intake is desirable. This can be achieved without jeopardizing the economy of the dairy or cereal industry by educational programs in the proper use of iodophor sanitizers or by the use of alternative preparations coupled with the judicious application of controls.

Iodates and Bread

Potassium iodate is extensively used as a dough conditioner in continuous mix baking as well as in yeast food for conventional sponge-dough bread baking. As suitable, cheap, effective alternative products — bromate and ascorbic acid or azodicarbonamide — are available, there is now no need or justification for the continued use of iodate in baking. The practice should be abandoned and/or banned.

Erythrosine

The red iodine-containing dye, erythrosine, is used extensively to color cereals, foods, sweets, and medications and can make significant contributions to iodine intake. Its use in the food and drug industries should be abandoned or strictly controlled.

Iodophors

The introduction and widespread use of iodophors in the dairy industry has resulted in significant iodine residues in milk which, in turn, have led to considerable iodine contents in processed or manufactured dairy products. The increasing use of iodophors in other sectors of the prepared food industry is causing iodine contamination of prepared foods and small goods. These sanitizers are easy to use and effective. They are of economic importance in the dairy industry and possibly the poultry industry, where suitable alternatives are not readily available. The economic importance and the availability of suitable alternatives to iodophors in other sectors should be investigated, however, and usage of iodophors should be restricted and controlled, particularly when suitable alternative noniodine-containing products exist.

Iodophors in Tasmania initially caused iodine levels of 800 to 1000 μg I per liter, but now levels of less than 300 μg I per liter can be consistently achieved. The work of Dunsmore* and others shows that with good practice, iodine contamination due to iodophors can be reduced to 150 μg I per liter, a completely acceptable iodine content for milk. By careful education programs, contamination of Tasmanian milk by iodophors has been reduced to a level which the State Thyroid Advisory Committee finds satisfactory. Tasmanians, traditionally iodine-deficient, now have a mean iodine intake of 270 μg/day.

Therapeutic Drugs

There is probably little scope for reducing iodine intake in this field, once the use of erythrosine as a coloring agent is abandoned or restricted. The value and benefits of iodides in respiratory diseases are probably greatly overrated; they are probably used more frequently than justified. It is also likely that thyroid hormones are included unnecessarily in some patent slimming cures and hormone preparations which are probably overprescribed or given on inadequate or wrong indications. The iodohydroxyquinolines, because of their retinal toxicity, should be eliminated, since suitable alternative medication exists. Many patent or proprietary vitamin-mineral preparations have significant iodine contents which should be eliminated from them. A number of anticholinergic drugs are iodine salts, the use of which should be avoided because suitable noniodine-containing alternatives exist.

Animals

The currently recommended iodine requirements for domesticated animals, in most instances, appear overgenerous. Furthermore, iodine supplementation is practiced very extensively, often unnecessarily. The degree of need for supplementation requires assessment, because it will vary from area to area. Indiscriminate iodine supplementation may produce harmful or undesirable effects.

The question of iodine nutrition is a difficult and unstable problem. Iodine now enters the diet from many diverse sources, often through contamination of food. Complacency about the current levels of iodine in the diet is not justified. The latitude between mandatory dietary intake and the levels at which some suffer adverse effects is probably far narrower than suspected. Further studies of the epidemiology of iodide goiter and adverse reactions to iodine nutrition are needed. It may well be that public health authorities, who spent the first half of this century devising ways and means to increase dietary iodine, will spend the latter part of this century trying to reduce iodine intake.

* Attention is drawn to Dunsmore and Wheeler's excellent papers on the application of iodophors in the dairy industry and the methods whereby contamination can adequately be limited[3,4] (Table 10).

APPENDIX

IODINE-CONTAINING MEDICATIONS — 1976

This comprehensive list of iodine-containing medications includes topical, oral, and parenteral preparations as well as radiographic contrast media. This list, though comprehensive, must not be regarded as a complete list of iodine-containing medications.

The iodine in these preparations may be part of the active ingredients in inorganic form or in organic complexes or occur adventitiously due to the coloring agent, erythrosine. The iodine content, derived from erythrosine, may range from 1 to 3 μg per tablet to 500 μg per dose of some mixtures.

Drugs are listed both by generic and trade names. Where iodine content is due solely to erythrosine, only the trade name is listed:

　　＊　Iodine content due to erythrosine
　　†　Radiographic contrast medium

This list was compiled from the "Therapeutic Index of Iodine Compounds and Preparations Used in Human Medicine" in *Iodine Pharmaceuticals,* published by the Chilean Iodine Educational Bureau, London, 1952, and information supplied by pharmaceutical companies.

Abdec®	Anayodin
† Abrodil	Androcalcion
Absorbine®	Angioyodina
Accommin capsules 2	Ankylosyl
Acetyl iodosalicylic acid	Anning's liniment
Achorm II	An-Nu-Tol
Acnosil	Antimony triiodide
Acodine	Antiphlogistine®
Aconite-iodine	Antiustio®
Activated iodine	Anusol-HC® ointment and suppositories
Activin®	Anusol® ointment and suppositories
A-D-S iodine	Aphco®
Aero Klen	Aphidine
Aero-Klenz	Appliodine
Agontan Knoll®	＊ Apresoline®–50-mg tablet
Airol	Aqueous iodine solution
Alb-Argentum	Aretus
＊ Alert® tablets	Argentide
Algodin	Argoid
Alival	Arlef® capsules
Aljodan	Arsenic iodide
Alkylamine hydroiodide	Arsenic mercuric iodide
Allyl iodide	Arsiodon
Alogeral	Arteriodin
Alosiorin	Arthene
Alphaden®	Arthol
Alum-iodine compound	Arthritone
Aluminum iodide	Arthrytin
Amend's solution	A.S.A.®
Aminojod	Ascoleine
Aminophylline potassium iodide	＊ Ascoxal–24-mg tablet
Aminural	Asmocid
Amiodoxyl benzoate	Asmotil
Ammonium iodide	Aspriodine
Amoebindon	＊ At-10
＊ Am-O®-lin	Atomodine
Am®	Aubeliode
Amsa	Auricol
Anaesodine	Barium iodide
Anaesthiform	Bellapurin
＊ Analeptine	＊ Bellergal® retard 5-mg tablet
＊ Ananese	＊ Benadryl® capsules, syrup, and expectorant mixtu

Benzojodal
Betadine®
Betnovate C preparations
B.F.I.®
Bi-iodamine
Biliselectan
Bilitrast
Biloptin
Biniodide
Bi-Prin tablets
Bis-Hyd
Bismjol
Bismoid
Bismuth emetine iodide
Bismuth formic iodide
Bismuth iodide
Bismuth iodoform paste
Bismuth iodosalicylate
Bismuth iodosubgallate
Bismuth-loretin
Bismuth oxyiodide
Bismuth oxyodopyrogallate
Bismuth potassium iodide
Bismuth sodium iodide
Bismuth suspension
Bismuth thymol iodide
Bis-thiodide
Biyo-arseno
Biyoquinol
Blancard pills and syrup
Boecoform
Boric acid-iodine powder
Borniosal
Bracodin®
Brojosan
Brominated thyroid
Broncaseptina B
Broncholysin
Bronsyl
Brovonex
Brozanthine
Brufen tablets
Bryant's sherry
Burnham soluble iodine
Bursoline
Cacodyne
Cadior
Cadmium iodide
Caffeine iodide
Caffe-iodine
Cajote
Calcidin®
Calcidrine® syrup
Calciodoxyl benzoate
Calciodum
Calcium iodate
Calcium iodobehenate
Calcium iodostearate
Calmitol®
Calot's paste
Caloxodine
Calsiod
Calvita tablets®

† Campiodol
Capsiode
Carbalga
* Carbitral capsules
Carbosan
† Cardio-Conray®
Carson's® paint
Carvacrol iodide
Caside®
* Cedilanid® tablets
* Celontin® capsules
Ceptan
Cerbelaud's syrup
Channing's solution
Cher-Iomine
Chiniofon
Chloral hydrate with iodine solution
Chloroiodine
Chloroiodohydroxyquinoline
Chlorosane
† Cholepulvis
† Choleselectan
† Choloselectan
† Cholatrast®
Churchill's® inhalation
Churchill's caustic iodine solution
Cinagill
Cinchonine iodosulphate
* Clistin
Cocaine hydriodide
Coldsore cream
† Coleradiagnost
Colliodine
Collodium iodatum
Colloidal iodine
Colloidal silver iodide
Collosol iodine
Collthesalin
Collutorium iodo-tannicum
Combex® capsules
Complan
Complevite
† Conray® 280, 420, and 480
† Contrastil
Cortiode
Cosmiod
Cremo-Quin 1% and 2%
Crinotensyl
Criodin
Crookes emulsion with calcium and iodine
Cuprodine®
Cuprous iodide
Curadine
Decamethonium iodide
Decolorized iodine solution
Deka®
Dental cement
Depurol
Dermalgen
Dermycin®
Descleran
Desencin
Desicol® capsules

† Diaginol viscous
† Diaplac tablets
Diasthmalyt
Dietary supplement B.D.H.
Diethanolamine diiodopyridone acetate
Dihydrotachysterol solution
Diiodoaniline
Diiodohydroxybenzyl alcohol
Diiodohydroxyquinoline
Diiodophenol sulphonic acid
Diiodopuridone acetate
Diiodoquinoline
Diiodosalicylic acid
Diiodothyronine
† Diiodotyrosine and gelatin
Diiodyl
* Dijene tablets
Dijozol Seife
† Dikol
Dilodyl
* Dimetane® L.A.® tablets
Dimethyltubocurarine iodine
† Diodone
Diodoquin® ointment and tablets
† Diodrast
† Dionosil® (propyliodone injection)
Diphenyliodonium chloride
† Disodium diiodopyridone dicarboxylate
† Disodium methyldiiodopyridone dicarboxylate
† Disodium phenoltetraiodophthalein
† Disodium tetraiodophenolphthalein
Disoquin
† Ditopax
Dolorsan
Donovan's solution
* Doxin oral suspension
Dragee's® depuriode
† Duodon Leo®
Duozal
Duretal-iodo-calcique
Eddec capsules
E-D-P
Elactono
* Elixir Ephedrine
Elixir Rodi
Elixir Sibec
Elixophyllin-KI® Elixir
Emanal
Emetine periodide
Emetrol®
Emoleo® iodized
* E.M.U.V. capsules (erythromycin) tablets
Endoyodolen
Enkide
Entero-quintol
Entero-Vioform® tablets
Enterozol
Entodon®
Entrokin
Ephedrine-calcium iodized compound
Erosyn
Esjodin 1 and II
* Eskornade Spansule® capsules
Esoban U–F–I ointment

* Estopen
Eteiothalein sodium
Ethereal inhalation of iodine
† Ethiodan
Ethidol
† Ethyl diiodobrassidate
Ethyl diiodosalicylate
† Ethyl diiodostearate
Ethyl iodide
† Ethyl iodophenylundecylate
† Ethyl linolenate trihydriodide
† Ethyl triiodostearate
Ethylene tetraiodide
Eulissen
Euphedrine iodee
Eupinal
Eupnine Vernade
Eupnogene
Europhen
Euteberol
Evans® dermal powder
Falk-salt med jod
Fealil medoyol
* Fefol Spansule capsules
Fejoprot
* Felsol®
Ferrarsil
Ferrjodose
Ferrous iodide
Ferrous iodide syrup
* Fespan Spansule capsules
Filibon® capsules and tablets
First aid powder
Fixojod
Flaxedil®
Floraquin®
Floraquin vaginal tablets
Foille®
† Foriodo
Fornax
Frasex
French mixture
Furmethide iodide
Gadil yodico Wasserman
Gallae cum jodo
Gamaform
Gargaline
† Gastro-Conray
Geajod
Gehes Jod-Dragees
Geriflex capsules
Germidol
Gevrabon® supplement
Gevral® capsules
Gevral protein
Gevral T capsules
Glacia iodized table salt
Globaline
Gluciode
Glycerinum jodi
† Glyceryl diiodostearate
Glyco-111
Glyco-HI

Glyco-iosote
Goitron
Gram's iodine solution
Guaiacol-iodoform
Guttajod
Haliod
Halophine
Hamameliode
Hamiodine
Hamolen
Hemato-iodine "sténé"
Hepatoselectan
Heptuna® Plus
Hexamethonium iodide
Hexamethyldiaminoisopropanol diiodide
Hipotensor
Hippodin
Hippuran®
Hydriodic acid
Hyodin®
Hypaque® 25%, 50%, 85%
Hypaque oral liquid
Hypoloid® diodone
Hypoloid iodoxyl
Iba
Icoll
I-C-Z
Idocol
Idosobenzoic acid
Immetal
Inhajod
Inhaletten
Inhalitus
Inorganic iodides
Intron®
Iocamfen®
Iocapral®
Iocylate
Iodacadil
Iodalbacid
Iodalbin
Iodalosa
Iodalose
Iodalphionic acid
Iodamelis
Iodaminal
Iodanisol
Iodargol
Iodaseptine
Iodatol
Iodatum
Iodazine midi
Iod-calcium diuretin
Iod-calcium theozorin
Iod-calcotheobromine
Iod-cholegnostyl
Iodeikon
Iodeine Montagu
Iodel
Iodeopirine
Iodermiol
Iod-ethamine
Iodex® liquid

Iodex methyl salicylate
Iodex ointment
Iodex plain
Iodex pessaries
Iodex suppositories
Iodhema
Iodic acid
Iodicin
† Iodinol
Iodine adsorbate
Iodine-boric acid
Iodine-glycol
Iodine collodion
Iodine, colorless
Iodine cream (colorless)
Iodine diasporal
Iodine douche
Iodine lotion
Iodine medol
Iodine menthol spray
Iodine monochloride
Iodine nonstaining ointment
Iodine oil
Iodine ointment
Iodine paint
Iodine paper
Iodine petroxolin
Iodine-phenol spray
Iodine solutions
Iodine tincture
Iodine tribromide
Iodine trichloride
Iodinol
† Iodipin
Iodipinum solidum
Iodithesin
Iodival
Iodized balmosa
Iodized benzocaine compound
Iodized blood mixture
Iodized ethyl chaulmoograte
Iodized moolgrol
Iodized oils
† Iodized oil (veterinary)
Iodized phenol
Iodized tincture of guaiacol
Iodized throat tablets
Iodoacetanilide
† Iodoalphionic acid
Iodoaminobenzene
Iodoantipyrine
† Iodoaspartic acid
† Iodobenzoic esters
Iodobisman
Iodobismitol
† Iodobrassid
Iodocaffeine elixir
Iodocal
Iodocasein
Iodoceane
Iodochin
† Iodochloral
Iodocholeate
Iodocitin

Iodocitrane
Iodocresol
Iodoform
Iodoformal
Iodoformin
Iodogal
† Iodogalactose
Iodogen
Iodoguaiacol
Iodo-Hamamelis
Iodohemol
† Iodohydroxybenzene sulfonic acid
Iodohydroxyquinoline
Iodohydroxyquinoline sulfuric acid
Iodoichthyol
Iodo-isopropyl alcohol
Iodol®
Iodolake-S
Iodolate
Iodolecithin
Iodolep
† Iodolipin
Iodolysin
Iodomagnesium
Iodomagnesium iode
† Iodomandelic acid
† Iodomethyl glucoside
Iodone
Iodonium compounds
Iodopeptona
Iodophenal
Iodophenol
† Iodophenyl quinolinic acids
† Iodophthalein
Iodophor pre-op sponges
Iodophosphol
Iodoplastine
Iodoprotein
† Iodopyracet
† Iodo-ray
Iodorgan
† Iodosol
Iodosalicylic acid
Iodosan
Iodo-Scilline
Iodose
Iodosolfina
Iodostarin
† Iodosugars
Iodosyl
Iodotab
Iodothiophen
Iodothiouracil
Iodotropon
Iodovaleryl glycolurea
Iodovical
† Iodoxyl
Iodoxybenzoic acid
Iodrargil
Iodress
Iodsam
Iodurum

Iodyl
Ioform
† Iopax
Iopnoi
Iopnol
† Iopyracyl
Iosal
Iosyn
Iotanagen
Iothesin
Ioxanin
Ioxantin
Iozanthal
Ipocardiol
Isapogen
Isoban U-F-I ointment
† Isobutyl diiodoeurucate
* Isogel
† Iso-iodeikon
Isopropamide iodide
Itrumil
Jarabe® bago
Jecoferpon cum jodo
Jobichin
Jobramag
Jodafen
† Jodairal
† Jodaten
Jodbensin
Jodbiozyme
Jodbismol
Joddermasan
Joddiasporal
Jod-eigon-naurium
Jodella
Jodferratose
Jodferroplex
Jodfortan
Jodglidine
Jod-hythermine
Jodixod
† Jodobilan
† Jodolja
Jodorgan
Jodotannico
† Jodsol
† Jodtetragnost
Jodtheobromine
Jodtheocalcin
Jodtheosalin
† Joduron
Jodtheodural
Jodtheodurin
Jozo
Katiodin
† Keraphen
Kiophyllin
Klidina
† Kontrast "U"®
Kurchi bismuth iodide
Lead iodide
Leojodin
† Linseed oil, iodized
† Lipidol Ultrafluid

Lipidol Viscous
Lipiodol
Lipoiodine®
Lipoiodine diagnostic
Liquor alphidine
Lita
Lithium iodate
Lithium iodide
Little Liver pills
Lobiodrin
Locacorten® Vioform® cream, ointment, paste, and ear drops
Losetrin
L-Triiodothyronine injection
Lugol's solution
Luma antirheumatic compound
Lumoxyd
Lysanthine
Lysantine
Lytinol
Maalox® suspension
Madden's balsam
Magendi's ether
Magnesium iodide
Magsan
Maize oil, iodized
Malt-arsil iodado
Mandelamine® suspension
Mandl's paint
Mandrax capsules
Manganese iodide
Marinol
Materna tablets
Mecojod
Mecolod
Medi pulv.
Memban
Merai kacide
Merck® iodine powder
Mercuric diiodosalicylate
Mercuric iodate
Mercuric iodide
Mercuric iodide red
Mercuric iodide yellow
Mercuro-iodohemol
Mercurous iodide
Mercury iodide green
Mercury sozoiodolate
Merjodin
Merodicein® solution
Mesantoin® tablets
Methenamine tetraiodide
Methiodal
Methyl aspriodine
Methyldioxoiodotetrahydropyridineacetic acid
Methyl iodide
Metufan
Micelliode
Microcyclin tablets
Milontin® capsules
Mina-vita
Mineralin
Ministry of Food vitamin tablets

Mirion®
Misidine
Misr-iode
Misr-sulphaiode
Mixture of potassium iodide and stramonium
† Morpholine diiodopyridone
Morucalcose
† Morujudol
Mucargol
Multivitamin Forte tablets
* Mutabon® D and F
Myadec® capsules
* Mylicon® drops
† Myodil
† Myodil contrast medium for X-rays
Naicol
Naiodine
Nalotol
Nasciodine
* Navidrex tablets
N.Cl.
Negatol®
† Neo-hydriol
† Neo-iodipin
† Neo-iopax
Neokathesin
Neo-riodine
Neo-silvol®
† Neo-skiodan
† Neo-tenebryl
Niglin
N-methyl berbamine iodide
N-methyl cepharanthine iodide
* Norlestrin® tablets
* Norlutin®-A tablets
† Nosylan®
Nov-afecin
Novalene®
Novasmol
Nyal iodized throat tablets
Obron®
Oleo-iode
† Oleum iodatum
† Oleum iodisatum
Oliodatum
† Oliolase
† Opacin
† Opacol
† Optojod
Oridine®
Oroxine tablets
Ostoco
Ottonia
Oxo-ate B
Oxoid thyroid-iodoacetyl salicylate
Oxyliodide
P and B base iodine
Pafasiman
Paladac® with minerals tablets
Pamaquine hydriodide
Panalgyn
Panfluine

† Pantopaque®
Paraiode
Parasclerose
* Parnate® granules
* Parnate tablets
† Peanut oil, iodized
Pediculosis powder
Pentaphedrine
Penta-vite vitamin and mineral capsules
† Per-abrodil
† Perjodal
Perjoodtheodural
† Perurdil
Petroxolimu, iosi
* Pfipen oral suspension
Pharmacol lozenges
† Pheniodol
Phenodine
† Phentetiothalein sodium
Phenyl ethyl iodide
† Phenyl hydroxydiiodophenyl propionic acid
Phospholine Iodide®
Pillodines
Pilules de Blancard
Pneumodine
Pneumozyl iodado
Pommade de Dr. Tixier
* Ponstan® capsules
† Poppyseed oil, iodized
Potassium iodate
† Potassium iodide
Potassium iodide-theocalcin
Potassium iodine sustained-relief tablets
Potassium mercuric iodide
Pregl's solution
Presojod
† Priodax®
† Propyl triiodostearate
Protheonal
Protiodin
Providine iodine
Provocol
Prudine
Pulmocadil
Pulmocardine
PVP-iodine
† Pyelectan
† Pyelognost
† Pyelosil
† Pylumbrin
Pyroform
Quadronal
Quaternary ammonium iodides
Quimbo
Quinine biniodate
Quinine bismuth iodide
Quinine hydriodide
Quinine iodohydriodide
Quiniobine
Quinostab
Quinton's plasma
Rabanol

Raifort's iodized syrup
† Rapeseed oil, iodized
† Rayopake
Realdine
Reclus's pommade
* Redupon tablets
Reformina
† Renopac
† Renumbral
Resagill
Resorlan
† Retro-Conray
Rheumichthol
Rhinol
* Rhusal tablets
Ricord's ointment
Riodine
* Robinul® tablets
Rubyl
Ruphon
Sadisalyl iodee
Sadoiodyl
Saiodina
Saiodine
Salicitum
Salicivess
Salicylysin
Saliode
Saniodo B
Schnabel's ointment
Scleroton
* Segontin® tablets
† Selectan
Seloplastum
Sephcemine
Septicemine cortial
Septojod
† Sesame oil, iodized
Setajodum
† Shadocol
Showersan®
† Silver iodide
Sinarrea
* Sinoral
* Sinutab
* Sinutab® with codeine
* Sinuzets capsules
* Sinuzets forte caps
Siomine
† Skiadin
† Skiodan®
† Slipules iodo-ray
† Sodium diiodocyclohexane sulfonate
† Sodium diiodomethane sulfonate
Sodium diiodosalicylate
Sodium iodate
† Sodium iodide
Sodium iodide-guaiacol
† Sodium iodoacyl taurines
† Sodium iodohippurate
† Sodium iodohydroxybenzene sulfonate
† Sodium iodopyridone acetate
† Sodium monoiodomethane sulfonate

odium sozoiodolate
odium triiodobenzoyltaurine
ombrabil
ombrachol
orosil
ozoiodol Schnupfen-Pulver
pasmolen
pecifique bejean
pirobismol
pirone
pironine
.R.A. tablets
taniform
taniphur
tarch iodide
tearodine
telabid tablets
terules iodo-ray
tipolac
toxil® ophthalmic ointment
toxil ophthalmic solution
toxil topical ointment
trontium iodide
ulfiode
ulphodine
ulfur iodide
urgical soap
yncurine®
yniodin
yntagnost
aninol
aninol puro
araktogenol
artro-quiniobine
aylor's mixture
CP®
ebrocal
elepaque® tablets
emonium iodide
enebryl
ermosaline yodada
Terramycin® oral suspension
ertroxine tablets
etiothalein sodium
etraiodophenolphthalein
etraiodophthalic acids
etraiodophthalimidoethanol
etrandine iodides
etrodine
Thallium iodide
Thantis® lozenges
Theobarbide
Theobromine-potassium iodide-phenobarbital
Theomine
Theridol®
Thesodide
Thiodacaine midi
Thioderazine midi
Thiodine
Thionaidine
Thiphene diiodide
Thiosinamine ethyl iodide
Thomson's pills

Thycalsin
Thylodine
Thymol iodide
Thymoldine
Thyroid B.P.
Thyrologen
Thyrotren
Thyroxine
Thyroxine sodium
Tikajod
Tiodine cognet
† T-I-P
Tonavital
Toppin 22 tablets
Tossanol
* Trib® capsules
Tridine
Triiodophenoxyethylamines
Trilodine iodides
* Trimolets
Trisan
Trivan
Triyodo
Tubocurarine iodides
* Tuss-Ornade® Liquid
* Tuss-Ornade Spansule Capsules
Tyrimide coated capsules,   liquid, and tablets
† Tyrogel
U-F-I
Ulodine
Ulsanin
† Umbrenal
Unguentum bismuth oxyjodo-galeatis camphoratum
Uracel
Urandil
Urbital
† Urea-sodium iodide
† Uriodone
† Urokon®
† Uropac
† Uroselectan
† Uroselectan B
† Urumbrin
* Valium® syrup
* Valpin®
Vanier's syrup
† Vasiodone
Vasogen iodine
† Vasoselectan
Vasolimentum iodatum
† Vegetable oil, iodized
Verukol
* Vibramycin® oral suspension
Vi-brandt
Vijochin
Vi-mina
* Vincents powders and tablets
Vin iodotannique phosphate
Viodar
Vioform
Vioform cream, 1% and 3%
Vioform hydrocortisone, 3%, 1.0%, 0.5%

Vioform powder
Virol
Visceralgin
† Visco-Rayopake
Vi-syneral®
Vitamin and mineral tablets
Vitaminorum tablets
Viterra®
Viterra obstetric capsules
Viterra supplement capsules
Viterra therapeutic capsules
Vodine
Vulnojod
Waterbury's compound
Wyanoids®

Xancylate
Xaniomine
Xaniophen®
Yatren
Yodent
Yodo coradon
Yodolen
Yodopil
Yodo-teobronal
Yuvral vitamin and mineral capsules
* Zarontin® capsules and mixture
Zinc iodide
Zinc sozoiodolate
Zipp
Zisp

REFERENCES

1. **Vought, R. L., Brown, F. A., and London, W. T.,** Iodine in the environment, *Arch. Environ. Health,* 20, 516–522, 1970.
2. **Salter, W. T.,** Fluctations in body iodine, *Physiol. Rev.,* 20, 345–376, 1940.
3. **Schmidt, W. and Winicov, M.,** Preparation and properties of detergent iodine systems, *Chem. Spec. Manuf. Assoc. Proc. Mid-year Meet.,* 53, 128–130, 1967.
4. **Vought, R. L., London, W. T., Lutwak, L., and Dublin, T. D.,** Reliability of estimates of serum inorganic iodine and daily fecal and urinary iodine excretions from single casual specimens, *J. Clin. Endocrinol. Metab.,* 23, 1218–1228, 1963.
5. **Fisher, K. D. and Carr, C. J., Eds.,** Iodine in Foods: Chemical Methodology and Sources of Iodine in the Human Diet, FDA 71-294, Life Sciences Research Office, Federation of American Societies for Experimental Biology, Bethesda, Md., 1974, 46.
6. **Connolly, R. J., Vidor, G. I., and Stewart, J. C.,** Increase in thyrotoxicosis in endemic goiter after iodination of bread, *Lancet,* 1, 500–502, 1970.
7. **Connolly, R. J.,** The changing iodine environment of Tasmania, *Med. J. Aust.,* 2, 1191–1193, 1971.
8. **Olin, R. M.,** Iodine deficiency and prevalence of simple goiter in Michigan, *JAMA,* 82, 1328–1332, 1924.
9. **Matovinovic, J.,** Extent of iodine insufficiency in the United States, in *Summary of a Conference: Iodine Nutriture in the United States,* National Research Council Committee on Food Protection, National Academy of Science, Washington, D.C., 1970, 1–11.
10. **Koutras, D. A., Papapetrou, P. D., Yataganas, X., and Malamos, B.,** Dietary sources of iodine in areas with and without iodine-deficiency goiter, *Am. J. Clin. Nutr.,* 23, 870–874, 1970.
11. **Koutras, D. A.,** in *The Thyroid: A Fundamental and Clinical Text,* 3rd ed., Werner, S. C. and Ingbar, S. H., Eds., Harper & Row, New York, 1971, 409–423.
12. **Horn, F. P., Reid, J. L., and Jung, G. A.,** Iodine nutrition and thyroid function of ewes and lambs on orchard grass under different levels of nitrogen and micro-element fertilization, *J. Anim. Sci.,* 38, 968–974, 1974.
13. **Clements, F. W., Gibson, H. B., and Howeler-Coy, J. F.,** Goiter studies in Tasmania. 16 years' prophylaxis with iodide, *Bull. W.H.O.,* 38, 297–318, 1968.
14. Iodine Content of Foods: Automated Bibliography 1825–1951, Chilean Iodine Educational Bureau, London, 1952, 183.
15. **Vought, R. L. and London, W. T.,** Dietary sources of iodine, *Am. J. Clin. Nutr.,* 14, 186–192, 1964.
16. **Fisher, K. D. and Carr, C. J., Eds.,** Iodine in Foods: Chemical Methodology and Sources of Iodine in the Human Diet, FDA 71-294, Life Sciences Research Office, Federation of American Societies for Experimental Biology, Bethesda, Md., 1974, 33–46.

17. Sachs, B. A., Bread iodine content and thyroid radioiodine uptake: a tale of two cities, *Br. Med. J.*, 1, 79–81, 1972.

18. Am. Inst. Baking Lab. Sch. Baking, Azodicarbonamide as a substitute for iodates in continuous mix bread, *Am. Inst. Baking Bull.*, 127, 1–4, 1967.

19. Upjohn, A. C., Galbraith, H. J. B., and Solomons, B., Raised serum protein-bound iodine after topical clioquinol, *Postgrad. Med. J.*, 47, 515–516, 1971.

20. Food and Nutrient Intake of Individuals in the U.S., Spring 1965, Household Food Consumption Survey, 1965–1966, Report No. 11, U.S. Department of Agriculture, Gov. Print. Off., Washington, D.C., 1972, 291.

21. Anon., The secretion of iodine into milk of lactating goats, *Nutr. Rev.*, 20, 20–21, 1962.

22. Miller, J. K. and Swanson, E. W., Some factors affecting iodine secretion in milk, *J. Dairy Sci.*, 46, 927–932, 1963.

23. Miller, J. K. and Swanson, E. W., Metabolism of ethylenediamine-dihydriodide and sodium or potassium iodide by dairy cows, *J. Dairy Sci.*, 56, 378–384, 1973.

24. Miller, J. K., Moss, B. R., and Swanson, E. W., Calcium iodate and pentacalcium orthoperiodate as sources of supplemental iodine for cattle, *J. Dairy Sci.*, 51, 1831–1835, 1968.

25. Miller, J. K., Swanson, E. W., and Hansen, S. M., Effects of feeding potassium iodide, 3,5-diiodosalicylic acid, or L-thyroxine on iodine metabolism of lactating dairy cows, *J. Dairy Sci.*, 48, 888–894, 1965.

26. Miller, J. K., Swanson, E. W., and Lyke, W. A., Iodine concentration of nonthyroid tissues of cows, *J. Dairy Sci.*, 56, 1344–1346, 1973.

27. Miller, J. K. and Tillapaugh, K., Iodide medicated salt for beef cattle, Cornell Feed Service No. 62, Cooperative Extension Service, Cornell University, Ithaca, N.Y., 1966.

28. Mittler, S. and Benham, G. H., Nutritional availability of iodine from several insoluble iodine compounds, *J. Nutr.*, 53, 53–58, 1954.

29. Broadhead, G. D., Pearson, I. B., and Wilson, G. M., Seasonal changes in iodine metabolism. I. Iodine content of cow's milk, *Br. Med. J.*, 1, 343–348, 1965.

30. Hemken, R. W., Vandersall, J. H., Oskarsson, M. A., and Fryman, L. R., Iodine intake related to milk iodine and performance of dairy cattle, *J. Dairy Sci.*, 55, 931–934, 1972.

31. Iwarsson, K., Nyberg, J. A., and Ekman, L., Correlation between iodine supplementation and iodine concentration in milk from dairy cows, *Nord. Veterinaermed.*, 24, 559–566, 1972.

32. Swanson, E. W., Effect of dietary iodine on thyroxine secretion rate of lactating cows, *J. Dairy Sci.*, 55, 1763–1767, 1972.

33. Dunsmore, D. G., Iodophors and iodine in dairy products. I. The iodine content of Australian dairy products, *Aust. J. Dairy Technol.*, 31, 125–128, 1976.

34. Dunsmore, D. G. and Wheeler, A. M., *Aust. J. Dairy Technol.*, in press.

35. Baumgartner, H. and Muller, H., Disinfection of teats with iodophors and chlorine preparations, *Schweiz. Milchztg.*, 100, 81, 1974.

36. Iwarsson, K. and Ekman, L., Iodophor teat dipping and the iodine concentration in milk, *Nord. Veterinaermed.*, 26, 31–38, 1974.

37. Joerin, M. M. and Bowering, A., The total iodine content of cow's milk, *N. Z. J. Dairy Sci. Technol.*, 7, 155–158, 1972.

38. Pain, R. W., Unusual variants of hyperthyroidism, *South Aust. Clin.*, 6, 59–64, 1972.

39. Connolly, R. J., An increase in thyrotoxicoses in southern Tasmania after an increase in dietary iodine, *Med. J. Aust.*, 2, 1268–1271, 1971.

40. Twomey, A. and Joerin, M. M., Iodine in milk and milk products, *N. Z. J. Dairy Sci. Technol.*, 8, 75, 1973, abstr.

41. Tasmanian State Thyroid Advisory Committee, unpublished information.

42. Jensen, J. M., Trout, G. M., and Brunner, J. R., Iodophors. I. Effect on flavour of milk and other observations, *J. Dairy Sci.*, 46, 799–809, 1967.

43. Johns, C. K. and Berzins, I., Iodophors as preservatives for milk, *J. Milk Food Technol.*, 17, 313–319, 1954.

44. Zollikofer, E., Report of the Milchtechnische Institute, Zurich.

45. Tasmanian State Thyroid Advisory Committee, unpublished information, 1975.

46. Marcilese, N. A., Harms, R. H., Valsecchi, R. M., and Arrington, L. R., Iodine uptake by ova of hens given excess iodine and effect upon ova development, *J. Nutr.*, 94, 117–120, 1968.

47. Korp, J. H., Iodine detergent complex. Exemption for requirement of a tolerance, *Fed. Regist.*, 37, 6580, 1972.

48. Boehne, J. W., Safety and metabolism of iodine, in *Summary of a Conference: Iodine Nutriture in the United States,* National Research Council Committee on Food Protection, National Academy of Sciences, Washington, D.C., 1970, 20–29.

49. **Kuhajek, E. J. and Fiedelman, H. W.,** Nutritional iodine in processed foods, *Food Technol.* (Chicago), 27, 52–53, 1973.

50. **Reed, J. M.,** Problems in the use of iodized salt in processed foods, in *Summary of a Conference: Iodine Nutriture in the United States,* National Research Council Committee on Food Protection, National Academy of Sciences, Washington, D.C., 1970, 34–39.

51. **Anon.,** Frozen fried-chicken dinners, *Consumer Rep.,* 38, 402–405, 1973.

52. Nutritional Analysis of Foods Served at McDonald's Restaurants, WARF Institute, Inc., Madison, Wis., 1973, 1–14.

53. **Wood, F. O.,** Present usage of iodized salt in the United States – geographic differences, in *Summary of a Conference: Iodine Nutriture in the United States,* National Research Council Committee on Food Protection, National Academy of Sciences, Washington, D.C., 1970, 30–33.

54. Select Committee on GRAS Substances, Evaluation of the Health Aspects of Potassium Iodide as a Food Additive, SCOGS 30, Life Sciences Research Office, Federation of American Societies of Experimental Biology, Bethesda, Md., 1974.

55. Subcommittee on Review of GRAS List (Phase II), A Comprehensive Survey of Industry on the Use of Food Chemicals Generally Recognized as Safe (G.R.A.S.), D.H.E.W. Contract No. FDA 70-22, National Research Council Committee on Food Protection, Division of Biology and Agriculture, National Academy of Sciences, Washington, D.C., 1972.

56. Office of the Federal Register, General Services Administration, Kelp, Section 121, 1149 in Code of Federal Regulations Title 21, Food and Drugs, parts 10–129 revised, Gov. Print. Off., Washington, D.C., 1973.

57. **McClendon, J. F.,** Iodine and goiter with especial reference to the Far East, *J. Biol. Chem.,* 102, 91–99, 1933.

58. **Vought, R. L., Brown, F. A., and Wolff, J.,** Erythrosine: an adventitious source of iodine, *J. Clin. Endocrinol. Metab.,* 34, 747–752, 1972.

59. **Andersen, C. J., Reiding, N. R., and Nielsen, A. B.,** False evaluation of serum protein-bound-iodine caused by red colored drugs or food, *Scan. J. Clin. Lab. Invest.,* 16, 249, 1964.

60. **Bora, S. S., Radichevich, I., and Werner, S. C.,** Artifactual elevation of PBI from an iodinated dye used to stain medicinal capsules pink, *J. Clin. Endocrinol. Metab.,* 29, 1269–1271, 1969.

61. **Anon.,** Use of certified FD & C colors in food, *Food Technol.* (Chicago), 22, 14–17, 1968.

62. **Kline, O. L. and Boehne, W.,** Official fat soluble vitamins, in *Remington Pharmaceutical Sciences,* 14th ed., Mack Publishing, Easton, Pa., 1970, 1020–1023.

63. **Wayne, E. J., Koutras, D. A., and Alexander, W. D.,** *Clinical Aspects of Iodine Metabolism,* Blackwell Scientific, Oxford, 1964, 237–238.

64. **Dunsmore, D. G., Luckhurst, A. M., and Barnes, R. N.,** *Aust. J. Dairy Technol.,* in press.

65. **Dunsmore, D. G. and Nuzum, C.,** *Aust. J. Dairy Technol.,* in press.

66. **Dunsmore, D. G., Nuzum, C., and Dettman, E. B.,** *Aust. J. Dairy Technol.,* in press.

67. **Dunsmore, D. G., Nuzum, C., and Luckhurst, A. M.,** *Aust. J. Dairy Technol.,* in press.

68. **Dunsmore, D. G., Nuzum, C., and Scott, J. M.,** *Aust. J. Dairy Technol.,* in press.

69. **Dunsmore, D. G. and Cairncross, K. G.,** in preparation.

70. **Mason, R.,** Mt. Pleasant Veterinary Lab., Launceston, Tasmania, Australia, personal communication, 1970.

71. **Vought, R. L. and London, W. T.,** Iodine intake and excretion in healthy nonhospitalized subjects, *Am. J. Clin. Nutr.,* 15, 124–132, 1964.

72. **Alderman, G. and Jones, D. I. H.,** The iodine content of pastures, *J. Sci. Food Agric.,* 18, 197–199, 1967.

73. **Horn, F. P., Reid, R. L., and Jung, G. A.,** Iodine nutrition and thyroid function of ewes and lambs on orchard grass under different levels of nitrogen and micro-element fertilization, *J. Anim. Sci.,* 38, 968–974, 1974.

74. National Research Council Committee on Animal Nutrition, Recommended nutrient allowances for beef cattle, in *Recommended Nutrient Allowances for Domestic Animals,* 4th rev. ed., National Academy of Sciences, Washington, D.C., 1970, 55.

75. National Research Council Committee on Animal Nutrition, Nutrient requirements of dairy cattle, in *Recommended Nutrient Allowances for Domestic Animals,* 4th rev. ed., National Academy of Sciences, Washington, D.C., 1971, 54.

76. National Research Council Committee on Animal Nutrition, Nutrient requirements of swine, in *Recommended Nutrient Allowances for Domestic Animals,* 7th rev. ed., National Academy of Sciences, Washington, D.C., 1973, 56.

77. National Research Council Committee on Animal Nutrition, Nutrient requirements of sheep, in *Recommended Nutrient Allowance for Domestic Animals,* 4th rev. ed., National Academy of Sciences, Washington, D.C., 1968, 64.
78. National Research Council Committee on Animal Nutrition, Nutrient requirements of poultry, in *Recommended Nutrient Allowances for Domestic Animals,* 6th rev. ed., National Academy of Sciences, Washington, D.C., 1971, 54.
79. Ammerman, C. B. and Miller, S. M., Biological availability of minor mineral ions: a review, *J. Anim. Sci.,* 35, 681–694, 1972.
80. Moss, B. R. and Miller, J. K., Metabolism of sodium iodide, calcium iodate, and pentacalcium orthoperiodate initially placed in the bovine rumen or abomasum, *J. Dairy Sci.,* 53, 772–775, 1970.
81. Aschbacher, P. W., Miller, J. K., and Cragle, R. G., Metabolism of diiodosalicylic acid in dairy calves, *J. Dairy Sci.,* 46, 1114–1117, 1963.
82. Aschbacher, P. W. and Feil, V. J., Metabolism of 3,5-diiodosalicylic acid in cattle and rats, *J. Dairy Sci.,* 51, 762–766, 1971.
83. Herrick, J. B., What role for EDDI (ethylene diamine dihydroiodide) in bovine respiratory disease complex, *Vet. Med. Small Anim. Clin.,* 67, 480–482, 1972.
84. Long, J. F., Gilmore, L. O., and Hibbs, J. W., The effect of different levels of iodide feeding on serum inorganic and protein-bound iodine, with a note on the frequency of administration required to maintain a high level of serum inorganic iodide, *J. Dairy Sci.,* 39, 1323–1326, 1956.
85. National Research Council Committee on Dietary Allowances, *Recommended Dietary Allowances,* 8th ed., National Academy of Sciences, Washington, D.C., 1974, 128.
86. Orr, J. B. and Leitch, I., Iodine in Nutrition, Medical Research Council Special Report No. 123, Her Majesty's Stationary Off., London, 1929.
87. Scott, P. P., Greaves, J. P., and Scott, M. G., Nutrition of the cat. IV. Calcium and iodine deficiency on a milk diet, *Br. J. Nutr.,* 15, 35–51, 1961.
88. Norris, W. P., Fritz, T. E., and Taylor, J. A., Cycle of accomodation to restricted dietary iodide in the thyroid gland of beagle dog, *Am. J. Vet. Res.,* 31, 21–33, 1970.
89. Underwood, E. J., *The Mineral Nutrition of Livestock,* Commonwealth Agricultural Bureau, Aberdeen, Scotland, 1966, 107.
90. Underwood, E. J., *Trace Elements in Human and Animal Nutrition,* 2nd ed., Academic Press, New York, 1962, 218.
91. Anon., Nutrient requirements of farm livestock, No. 2: Ruminants. Technical Reviews and Summaries, Agricultural Research Council, Her Majesty's Stationary Off., London, 1965, 104.
92. Mason, R. W., Etiological Studies of Ovine Congenital Goitre in Tasmania with Particular Reference to the Cressy Longford Area, M.V. Sc. thesis, Melbourne University, Melbourne, Aust., 1976.
93. Statham, M. and Bray, A. C., Congenital goitre in sheep in southern Tasmania, *Aust. J. Agric. Res.,* 25, 751–768, 1975.
94. Reeve, R. S., Coupland, G. A. E., and Hales, I. B., The effect on serum iodine levels of painting tincture of iodine on the skin, *Med. J. Aust.,* 1, 891–892, 1973.
95. Welt, L. G. and Blythe, W. B., *The Pharmacological Basis of Therapeutics,* 4th ed., Goodman, L. S. and Gilman, A., Eds., Collier MacMillan, London, 1970, 821–825.
96. Ingbar, S. H. and Woeber, K. A., The thyroid, in *Text Book of Endocrinology,* 5th ed., Williams, R. H., Ed., W. B. Saunders, Philadelphia, 1974, 97–108.
97. Bastomsky, C. H., Thyroid iodide transport, in *Handbook of Physiology,* Vol. 3, Greep, R. O., Astwood, E. P., Greer, M. A., Solomon, D. H., and Geiger, S. R., Eds., American Physiology Society, Washington, D.C., 1974, 81–85.
98. Brown-Grant, K., *Physiol. Rev.,* 41, 189–213, 1961.
99. Wolff, J., *Physiol. Rev.,* 44, 45–90, 1964.
100. Stanbury, J. B., Wyngaarden, J. B., and Fredrickson, D. S., *The Metabolic Basis of Inherited Disease,* 3rd ed., McGraw-Hill, New York, 1972, 1778.
101. Boyd, J. D., Development of the human thyroid gland, in *The Thyroid Gland,* Pitt-Rivers, R. H. and Trotter, W. T., Eds., Butterworth, Washington, D.C., 1964.
102. Fisher, D. A. and Dussault, J. H., Development of the mammalian thyroid gland, in *Handbook of Physiology,* Vol. 3, Greep, R. O., Astwood, E. P., Greer, M. A., Solomon, D. H., and Geiger, S. R., American Physiology Society, Washington, D.C., 1974, 21–29.
103. Fisher, D. A., Hobel, C. J., Garza, R., and Pierce, C., *Pediatrics,* 46, 208–216, 1970.
104. Greenberg, A. H., Czernichow, P., Reba, R. C., Tyson, J., and Blizzard, R. M., *J. Clin. Invest.,* 49, 1790–1803, 1970.

105. Evans, T. C., Cretzchmar, R. M., Hodges, R. D., and Song, C. W., *J. Nucl. Med.,* 8, 157–165, 1967.
106. Tyler, D. D., Gonze, J., Lamy, F., and Dumond, J. E., *Biochem. J.,* 106, 123–133, 1968.
107. Maayan, M. L. and Ingbar, S. H., *Endocrinology,* 87, 1223-1226, 1970.
108. DeGroot, L. J. and Nagasaka, A., *Exerpta Med. Int. Congr. Ser.,* 238, 53, 1972.
109. Lissitzky, S., Codaclioni, J. L., Bismuth, J., and DePieds, G., *Biochem. Biophys. Res. Commun.,* 16, 249–263, 1964.
110. Maloof, F., Smith, S., and Soodak, M., Dissociation of iodination from protein synthesis in the rat thyroid, *Medicine,* 43, 375–378, 1964.
111. Taurog, A., Biosynthesis of iodoamino acids, in *Handbook of Physiology,* Vol. 3, Greep, R. O., Astwood, E. B., Greer, M. A., Solomon, D. H., and Geiger, S. R., Eds., American Physiology Society, Washington, D.C., 1974, 101–130.
112. Taurog, A., Thyroid peroxidase and thyroxine biosynthesis, *Recent Prog. Horm. Res.,* 26, 189–247, 1970.
113. Vought, R. L. and London, W. T., Iodine intake, excretion and thyroidal accumulation in healthy subjects, *J. Clin. Endocrinol. Metab.,* 27, 913–919, 1967.
114. Wolff, J., Iodide goiter and the pharmacological effects of excess iodide, *Am. J. Med.,* 47, 101–124, 1969.
115. Wolff, J. and Chaikoff, I. L., Inhibitory action of iodide upon organic binding of iodine by normal thyroid gland, *J. Biol. Chem.,* 172, 855–856, 1948.
116. Wolff, J. and Chaikoff, I. L., Plasma inorganic iodide as homeostatic regulator of thyroid function, *J. Biol. Chem.,* 174, 555–564, 1948.
117. Wolff, J. and Chaikoff, I. L., Plasma inorganic iodide, chemical regulator of normal thyroid function, *Endocrinology,* 42, 468–471, 1948.
118. Braverman, L. E. and Ingbar, S. H., Changes in thyroidal function during adaptation to large doses of iodide, *J. Clin. Invest.,* 42, 1216–1231, 1963.
119. Goldsmith, R. E. and Eisele, M. L., Effect of iodide on release of thyroid hormone in hyperthyroidism, *J. Clin. Endocrinol. Metab.,* 16, 130–137, 1956.
120. Stone, O. J., What are the non-endocrine biologic effects of iodides?, *Med. Times,* 99, 143–155, 1971.
121. Klebanoff, S. J., Myeloperoxidase-halide-hydrogen peroxide antibacterial systems, *J. Bacteriol.,* 95, 2131–2138, 1968.
122. Klebanoff, S. J., Iodination of bacteria: a bactericidal mechanism, *J. Exp. Med.,* 126, 1063–1078, 1967.
123. Stolc, V., Iodine metabolism in leukocytes: effect of graded iodide concentrations, *Biochem. Med.,* 10, 293–299, 1974.
124. Stone, O. J. and Willis, C. J., Iodide enhancement of inflammation – experimental with clinical correlation, *Tex. Rep. Biol. Med.,* 25, 205–213, 1967.
125. Mielens, Z. E., Rozitis, J., Jr., and Sansone, V. J., The effect of oral iodides on inflammation, *Tex. Rep. Biol. Med.,* 26, 117–121, 1968.
126. Lakin, J. D., Classification of hypersensitivity reactions, in *Allergic Diseases: Diagnosis and Management,* Patterson, R., Ed., J. P. Lippincott, Philadelphia, 1972, 1–30.
127. Parish, W. E., Allergy, in *Text Book of Dermatology,* Rook, A., Wilkinson, D. S., and Ebeling, F. J. G., Eds., Blackwell Scientific, Oxford, 1968, 158–185.
128. Austen, K. P., Introduction to immunology, in *Harrison's Principles of Internal Medicine,* 6th ed., Wintrobe, M. M., Thorn, G. W., Adams, R. D., Bennett, I. L., Jr., Braunwald, E., Isselbacher, K. J., and Petersdorf, R. G., Eds., McGraw-Hill, New York, 1970, 342–348.
129. Gordon, B. L. and Ford, D. K., *Essentials of Immunology,* F. A. Davis, Philadelphia, 1971, 82–91.
130. White, R. G. and Timbury, M. C., Types of allergic response and cell mediated immunity, in *Essentials of Immunology and Microbiology,* J. P. Lippincott, Philadelphia, 1973, 146, 154–164.
131. Almy, T. P., Gastro-intestinal allergy, in *Cecil – Loeb Test Book of Medicine,* 13th ed., Beeson, P. B. and McDermott, W. M., Eds., W. B. Saunders, Philadelphia, 1971, 1256–1257.
132. Bronsky, E. A. and Ellis, F. P., Diagnostic methods in allergic disease, in *Paediatrics,* 15th ed., Patterson, R. and Einhorn, A. H., Eds., Appleton-Century-Crofts, New York, 1972, 448–449.
133. DeSwarte, R. D., Drug allergy, in *Allergic Diseases: Diagnosis and Management,* Patterson, R., Ed., J. P. Lippincott, Philadelphia, 1972, 434–446.
134. Golbert, T. M., Food allergy and immunological diseases of the gastro-intestinal tract, in *Allergic Diseases: Diagnosis and Management,* Patterson, R., Ed., J. P. Lippincott, Philadelphia, 1972, 365–366.

135. Rosenoer, V. M., Diagnostic drugs, in *Side-effects of Drugs,* Meyler, L. and Herxheimer, A., Eds., Williams and Wilkins, Baltimore, 1968, 472.

136. Thomas, L., Drug allergy, in *Cecil — Loeb Text Book of Medicine,* 12th ed., Beeson, P. B. and McDermott, W. M., Eds., W. B. Saunders, Philadelphia, 1967, 442—444.

137. Chaplin, I., Committee on provocative food testing data does not strengthen validity of provocative subcutaneous food test technique, *Ann. Allergy,* 32, 52—55, 1974.

138. Sharp, M. C., Preliminary evaluation of subcutaneous provocative food testing experiment, *Ann. Allergy,* 32, 49—52, 1974.

139. Willoughby, J. W. and Sharp, M. C., Committee on provocative food testing data does not strengthen validity of provocative subcutaneous food test technique, *Ann. Allergy,* 32, 47—49, 1974.

140. Hoffman, D. R. and Haddad, Z. A., Diagnosis of multiple infant allergies in children by radioimmunoassay, *Pediatrics,* 54, 151—156, 1974.

141. Halpern, B., Ky, N. T., and Ameche, N., Diagnosis of drug allergy in vitro with the lymphocyte transformation test, *J. Allergy,* 40, 168—181, 1967.

142. Jacobs, J., Serological studies on iodinated sera, *J. Immunol.,* 23, 375—384, 1932.

143. Rosenberg, F. R., Einbinder, J., Walzer, R. A., and Nelson, C. T., Vegetating iododerma. An immunologic mechanism, *Arch. Dermatol.,* 105, 900—905, 1972.

144. Horn, B. and Kabins, S. A., Iodide fever, *Am. J. Med. Sci.,* 264, 467—471, 1972.

145. Jacob, H. S., Sidd, J. J., Greenberg, B. H., and Lingley, J. F., Extreme eosinophilia with iodide hypersensitivity. Report of a case with observations on the cellular composition of inflammatory exudates, *N. Engl. J. Med.,* 271, 1138—1140, 1964.

146. Middlebrook, M. and Szent-Gyorgyi, A., The action of iodide on oxidative phosphorylation, *Biochim. Biophys. Acta,* 18, 407—409, 1955.

147. Seymour, W. B., Jr., Poisoning from cutaneous application of iodine; rare aspect of its toxicologic properties, *Arch. Intern. Med.,* 59, 952—966, 1937.

148. Peacock, L. B. and Davison, H. M., Observations on iodide sensitivity, *Ann. Allergy,* 15, 158—164, 1957.

149. Rasmussen, H. J., Iodide hypersensitivity in the etiology of periarteritis nodosa, *J. Allergy,* 26, 394—407, 1955.

150. Wahlberg, J. E. and Wikstrom, K., A case of periarteritis nodosa with skin manifestations, probably provoked by iodide administration, *Acta Derm. Venereol.,* 43, 556—561, 1963.

151. Rich, A. R., Hypersensitivity to iodine as a cause of periarteritis nodosa, *Bull. Johns Hopkins Hosp.,* 77, 43—48, 1945.

152. Ehrich, W. E. and Seifter, J., Thrombotic thrombocytopenic purpura caused by iodine. Report of a case, *Arch. Pathol.,* 47, 446—449, 1949.

153. Eller, J. J. and Fox, E. C., Fatal iododerma, *Arch. Dermatol. Syphilol.,* 24, 745—757, 1931.

154. Galina, M. P., Avnet, M. L., and Einhorn, A., Iodides during pregnancy: an apparent cause of neonatal death, *N. Engl. J. Med.,* 267, 1124—1127, 1962.

155. Rook, A. and Rowell, N. R., Drug reactions: iodides and bromides, in *Text Book of Dermatology,* Rook, A., Wilkinson, D. S., and Ebeling, F. J. G., Eds., Blackwell Scientific, Oxford, 1968, 386.

156. Anon., Iodism, *South. Med. J.,* 54, 817—818, 1961.

157. Bechet, P. E., Iodized table salt as etiologic factor in iododerma, *Arch. Dermatol. Syphilol.,* 29, 529—536, 1934.

158. Bechet, P. E., Acneform iododerma (iodized salt), *Arch. Dermatol. Syphilol.,* 38, 966—967, 1938.

159. Bechet, P. E., The etiologic role of iodized table salt in iododerma, *J. Invest. Dermatol.,* 8, 408—417, 1947.

160. Falleroni, E., Asthma: management, in *Allergic Diseases: Diagnosis and Management,* Patterson, R., Ed., J. B. Lippincott, Philadelphia, 1972, 248—250.

161. Hitch, J. M. and Greenburg, B. G., Adolescent acne and dietary iodine, *Arch. Dermatol.,* 84, 898—911, 1961.

162. Shelmire, B., Acne from iodized salt, *JAMA,* 90, 1869—1870, 1928.

163. Mercantinie, S., Iododerma from sea food, *Can. Med. Assoc. J.,* 102, 759—761, 1970.

164. Bianco, R. P., Smith, P. J., Keen, R. R., and Jordan, J. E., Iodide intoxication: report of a case, *Oral Surg. Oral Med. Oral Pathol.,* 32, 876—880, 1971.

165. Chapman, R. S. and Main, R. A., Diffuse thinning of hair in iodide-induced hypothyroidism, *Br. J. Dermatol.,* 79, 103—105, 1967.

166. Pusey, W. A., Fatal iodism in a baby, *J. Cutaneous Dis. Syphilis,* 29, 309—310, 1911.

167. Baker, H., Fixed eruption due to iodide and antipyrine, *Br. J. Dermatol.,* 74, 310—316, 1962.

168. Shelley, W. B., Generalized pustular psoriasis induced by potassium iodide. A postulated role for dihydrofolic reductase, *JAMA,* 201, 133—138, 1967.

169. **Douglas, W. S. and Alexander, J. O. D.,** Dermatitis herpetiformis, iodine compounds and thyrotoxicosis, *Br. J. Dermatol.,* 92, 596–598, 1975.

170. **From, E. and Thomsen, K.,** Dermatitis herpetiformis. A case provoked by iodine, *Br. J. Dermatol.,* 91, 221–224, 1974.

171. **Warner, J., Brooks, S. E. H., James, W. P. T., and Louisy, S.,** Juvenile dermatitis herpetiformis in Jamaica: clinical and gastrointestinal features, *Br. J. Dermatol.,* 86, 226–237, 1972.

172. **Coindet, J. F.,** Decouverte d'un nouveau remede contre le goitre, *Ann. Chim. Phys.,* 15, 49–59, 1820.

173. **Breuer, R.,** Beitrag zur Aetiologie der Basedow'schen Krankheit und des Thyreoidismus, *Wien. Klin. Wochenschr.,* 13, 641–644, 1900.

174. **Kocher, T.,** Die Therapie des Kropfes, *Dtsch. Klin.,* 8, 1115–1184, 1904.

175. **Kocher, T.,** Ueber Iodbasedow, *Arch. Klin. Chir.,* 92, 1166–1193, 1910.

176. **Jackson, A. S.,** Iodin hyperthyroidism; an analysis of fifty cases, *Boston Med. Surg. J.,* 193, 1138–1140, 1925.

177. **Ek, B., Johnsson, S., and von Porat, B.,** Iodide repletion test in an endemic goitre area. Risk of iodine-induced hyperthyroidism, *Acta Med. Scand.,* 173, 341–348, 1963.

178. **Denham, M. J. and Himsworth, R. L.,** Hyperthyroidism induced by potassium iodide given in the course of ^{125}I-fibrinogen test, *Age Ageing,* 3, 221–225, 1974.

179. **Ermans, A. M. and Camus, M.,** Modification of thyroid function induced by chronic administration of iodide in the presence of "autonomous" thyroid tissue, *Acta Endocrinol. (Copenhagen),* 70, 463–475, 1972.

180. **Vagenakis, A. G., Wang, C. A., Burger, A., Maloof, F., Braverman, L. E., and Ingbar, S. H.,** Iodine-induced thyrotoxicosis in Boston, *N. Engl. J. Med.,* 287, 523–527, 1972.

181. **Hartsock, C. L.,** Iodized salt in the prevention of goiter: is it a safe measure for general use?, *JAMA,* 86, 1334–1338, 1926.

182. **Kimball, O. P.,** Induced hyperthyroidism, *JAMA,* 85, 1709–1710, 1925.

183. **McLure, R. D.,** Thyroid surgery as affected by the generalized use of iodized salt in an endemic goiter region – preventive surgery, *Am. Surg.,* 100, 924–932, 1934.

184. **Stewart, J. C., Vidor, G. I., Buttfield, I. H., and Hetzel, B. S.,** Epidemic thyrotoxicosis in northern Tasmania: studies of clinical features and iodine nutrition, *Aust. N. Z. J. Med.,* 1, 203–211, 1971.

185. **Vidor, G. I., Stewart, J. C., Wall, J. R., Wangel, A., and Hetzel, B. S.,** Pathogenesis of iodine-induced thyrotoxicosis: studies in northern Tasmania, *J. Clin. Endocrinol. Metab.,* 37, 901–909, 1973.

186. **Adams, D. D., Kennedy, T. H., Stewart, J. C., Utiger, R., and Vidor, G. I.,** Hyperthyroidism in Tasmania following iodide supplementation: measurement of thyroid-stimulating auto-antibodies and thyrotropin, *J. Clin. Endocrinol. Metab.,* 41, 221–228, 1975.

187. **Stewart, J. C. and Vidor, G. I.,** Thyrotoxicosis induced by iodine contamination of food – a common unrecognized condition?, *Br. Med. J.,* 1, 372–375, 1976.

188. **Miller, J. M. and Block, M. A.,** Functional autonomy in multinodular goiter, *JAMA,* 214, 535–539, 1970.

189. **Vidor, G. I.,** unpublished information, 1975.

190. **Willcox, P. H.,** Twelve years' experience of antithyroid treatment, *Postgrad. Med. J.,* 38, 275–280, 1962.

191. **Reveno, W. S. and Rosenbaum, H.,** Observations on the use of antithyroid drugs, *Ann. Intern. Med.,* 60, 982–989, 1964.

192. **Hershman, J. M., Givens, J. R., Cassidy, C. E., and Astwood, E. B.,** Long-term outcome of hyperthyroidism treated with antithyroid drugs, *J. Clin. Endocrinol. Metab.,* 26, 803–807, 1966.

193. **Alexander, W. D., Harden, R. M., Koutras, D. A., and Wayne, E.,** Influence of iodine intake after treatment with antithyroid drugs, *Lancet,* 2, 866–868, 1965.

194. **Lowry, R. C., Lowe, P., Hadden, D. R., Montgomery, D. A. D., and Weaver, J. A.,** Thyroid suppressibility: follow-up for two years after antithyroid treatment, *Br. Med., J.,* 2, 19–22, 1971.

195. **Thalassinos, N. C. and Fraser, T. R.,** Effect of potassium iodide on relapse-rate of thyrotoxicosis treated with antithyroid drugs, *Lancet,* 2, 183–184, 1971.

196. **Weils, J. A. and Kloppenburg, P. W. C.,** Incidence of remission of Grave's disease after treatment with thiouracil derivatives, *Ned. Tijdschr. Geneeskd.,* 113, 1637–1641, 1969.

197. **Wartofsky, L.,** Low remission after therapy for Grave's disease: possible relation of dietary iodine with antithyroid therapy results, *JAMA,* 226, 1083–1088, 1973.

198. **Nilsson, G.,** Self-limiting episodes of Jodbasedow, *Acta Endocrinol.,* 74, 475–482, 1973.

199. Studer, H. and Greer, M. A. A., A study of the mechanisms involved in the production of iodine-deficiency goiter, *Acta Endocrinol.,* 49, 610–628, 1965.

200. Martin, M. M. and Rento, R. D., Iodide goiter with hypothyroidism in two newborn infants, *J. Pediatr.,* 61, 94–99, 1962.

201. Croughs, W. and Visser, H. K. A., Familial iodide-induced goiter. Evidence for an abnormality in the pituitary-thyroid homeostatic control, *J. Pediatr.,* 67, 353–362, 1965.

202. Parmelee, A. H., Allen, E., Stein, I. F., and Buxbaum, H., Three cases of congenital goiter, *Am. J. Obstet. Gynecol.,* 40, 145–147, 1940.

203. Carswell, F., Kerr, M. M., and Hutchinson, J. H., Congenital goitre and hypothyroidism produced by maternal ingestion of iodides, *Lancet,* 1, 1241–1243, 1970.

204. Senior, B. and Chernoff, H. L., Iodide goiter in the newborn, *Pediatrics,* 47, 510–515, 1971.

205. Iancu, T., Boyanower, Y., and Laurian, N., Congenital goiter due to maternal ingestion of iodide, *Am. J. Dis. Child.,* 128, 528–530, 1974.

206. Savoie, J. C., deGennes, J. L., Michard, J. P., and Decourt, J., Study of two cases of acquired disorders of thyroid homogenesis following injections of iodized oil, *Ann. Endocrinol.,* 23, 293–310, 1962.

207. Decourt, J., Fauvert, R., deGennes, J. L. L., and Saucier, G., Myxedema with goitre caused by a disorder in the thyroidal synthesis of hormone appearing at a late stage in an adult, *Ann. Endocrinol.,* 19, 343–355, 1958.

208. Mornex, R., Peyrin, J. O., Pommateau, E., Berger, M., and Riffat, G., Myxedema with goitre caused by iodine saturation. A new type of defect of the process of thyroxinogenesis, *Ann. Endocrinol.,* 21, 704–716, 1960.

209. Raben, M. S., Endocrine conference, *J. Clin. Endocrinol. Metab.,* 13, 469–472, 1953.

210. Frey, H., Hypofunction of the thyroid gland, due to prolonged and excessive intake of potassium iodide, *Acta Endocrinol.* (Copenhagen), 47, 105–120, 1964.

211. Oppenheimer, J. H. and McPherson, H. J., The syndrome of iodide-induced goiter and myxedema, *Am. J. Med.,* 30, 281–288, 1961.

212. Rubenstein, H. M. and Oliner, J., Myxedema induced by prolonged iodine administration, *N. Engl. J. Med.,* 256, 47–52, 1957.

213. Dimitriadou, A. and Fraser, R., Iodide goitre, *Proc. R. Soc. Med.,* 54, 345–346, 1961.

214. Hurxthall, L. M., Myxedema following iodine administration for goiter in a girl aged six years, *Lahey Clin. Found. Bull.,* 4, 73–77. 1945.

215. Falliers, C. J., Goiter and thyroid dysfunction following the use of iodides in asthmatic children, *Am. J. Dis. Child.,* 99, 428–436, 1960.

216. Burrows, B., Niden, A. H., and Barclay, W. R., Goiter and myxedema due to iodide administration, *Ann. Intern. Med.,* 52, 858–870, 1960.

217. Taguchi, J. T. and Skillman, T., Iodine-induced myxedema: report of a case and review of the literature, *Am. J. Med. Sci.,* 239, 417–429, 1960.

218. Begg, T. B. and Hall, R., Iodide goitre and hypothyroidism, *Q. J. Med.,* 32, 351–362, 1963.

219. Braverman, L. E., Woeber, K. A., and Ingbar, S. H., Induction of myxedema by iodide in patients euthyroid after radio-iodine or surgical treatment of diffuse toxic goiter, *N. Engl. J. Med.,* 281, 816–821, 1969.

220. Braverman, L. E., Vegenakis, O. G., Wang, C., Maloof, F., and Ingbar, S. H., Studies on the pathogenesis of iodide myxedema, *Trans. Assoc. Am. Physicians,* 84, 130–138, 1971.

221. Braverman, L. E., Ingbar, S. H., Vagenakis, O. G., Adama, L., and Maloof, F., Enhanced susceptibility to iodide myxedema in patients with Hashimoto's disease, *J. Clin. Endocrinol. Metab.,* 32, 515–521, 1971.

222. Trowbridge, F. L., Matovinovic, J., McLaren, G. D., and Nichaman, M. Z., Iodine and goiter in children, *Pediatrics,* 56, 82–90, 1975.

223. Trowbridge, F. L., Hand, K. A., and Nichaman, M. Z., Findings relating to goiter and iodine in the ten-state nutrition survey, *Am. J. Clin. Nutr.,* 28, 712–716, 1975.

224. DeGroot, L. J., Action of potassium iodide on thyroxine metabolism, *J. Clin. Endocrinol. Metab.,* 26, 778–779, 1966.

225. Galton, V. A. and Ingbar, S. H., Effect of large doses of iodide on the peripheral metabolism of thyroxine in rats, *Endocrinology,* 81, 1439–1441, 1967.

226. Purves, H. D. and Griesbach, W. B., Observations on acidophil cell changes in pituitary in thyroxine deficiency states; acidophil degranulation in relation to goitrogenic agents and extrathyroidal thyroxine synthesis, *Br. J. Exp. Pathol.,* 27, 170–179, 1946.

227. Milne, K. and Greer, M., Clarification of the diverse actions of iodide on antithyroid effect of sulfonamides and thionamides, *Proc. Soc. Exp. Biol. Med.,* 109, 174–176, 1962.

228. **Milne, K. and Greer, M.,** Comparison of the effects of propylthiouracil and sulfadiazine on thyroidal biosynthesis and the manner in which they are influenced by supplemental iodine, *Endocrinology,* 71, 580–587, 1962.
229. **Morgans, M. E. and Trotter, W. R.,** Two cases of myxoedema attributed to iodide administration, *Lancet,* 2, 1335–1337, 1953.
230. **Goldner, M. G. and Adesman, J.,** Clinical conference on metabolic problems; iodide myxedema, *Metabolism,* 4, 545–551, 1955.
231. **Paris, J., McConnahey, W. M., Owen, C. A., Woolne, L. B., and Bahn, R. C.,** Iodide goiter, *J. Clin. Endocrinol. Metab.,* 20, 57–67, 1960.
232. **Murray, I. P. C. and Stewart, R. D. H.,** Iodide goitre, *Lancet,* 1, 922–925, 1967.
233. **Paley, K. R., Sobel, E. S., and Yalow, J.,** Some aspects of thyroidal iodine metabolism in a case of iodine-induced hypothyroidism, *J. Clin. Endocrinol. Metab.,* 18, 79–90, 1958.
234. **Gold, E.,** Iodide-induced goiter, *Surgery,* 45, 424–431, 1959.
235. **Asmundsen, V. S., Almquist, H. J., and Klose, A. A.,** Effect of different forms of iodine on laying hens, *J. Nutr.,* 12, 1–13, 1936.
236. **Turner, H. H.,** Goiter induced by iodide in children, *South. Med. J.,* 49, 1443–1451, 1956.
237. **Liewendahl, K. and Turula, M.,** Iodide-induced goiter and hypothyroidism in a patient with chronic lymphocytic thyroiditis, *Acta Endocrinol.* (Copenhagen), 71, 289–296, 1972.
238. **Tasmanian State Thyroid Advisory Committee,** unpublished data, 1974.
239. **Suzuki, H., Higuchi, T., Sawa, K., Ohtaki, S., and Horiochi, S.,** Endemic coast goiter in Hokkaido, Japan, *Acta Endocrinol.* (Copenhagen), 50, 161–176, 1965.
240. **Edmunds, H. T.,** Acute thyroiditis from potassium iodide, *Br. Med. J.,* 1, 354, 1955.
241. **Hellwig, C. A. and Wilkinson, P. M.,** Experimental production of chronic thyroiditis, *Growth,* 24, 169–177, 1960.
242. **Weaver, D. K., Batsakis, J. G., and Nishiyama, R. H.,** Relationship of iodine to "lymphocytic goiters," *Arch. Surg.,* 98, 183–186, 1969.
243. **Weaver, D. K., Nishiyama, R. H., Button, W. D., and Batsakis, J. G.,** Surgical thyroid disease: a survey before and after iodine prophylaxis, *Arch. Surg.,* 92, 796–801, 1966.
244. **Ansell, G.,** Adverse reactions to contrast agents. Scope of problem, *Invest. Radiol.,* 3, 374–391, 1970.
245. **Muller, C.,** Periarteritis nodosa – asthma bronchiale – iododerma tuberosum, *Acta Med. Scand.,* 136, 378–387, 1949.
246. **Sussman, R. M. and Miller, J.,** Iodide "mumps" after intravenous urography, *N. Engl. J. Med.,* 255, 432–434, 1956.
247. **Lengeman, F. W.,** Metabolism of radioiodide by lactating goats given iodine 131 for extended periods, *J. Dairy Sci.,* 53, 165–170, 1970.
248. **Anon.,** Nephrotoxic nephritis, *JAMA,* 184, 144, 1963.
249. **Clifton, L. and Makous, W.,** Iodate poisonings: early effect on regeneration of rhodopsin and the ERG, *Vision Res.,* 13, 919–924, 1973.
250. **Harden, R. M.,** Submandibular adenitis due to iodide administration, *Br. Med. J.,* 1, 160–161, 1961.
251. **Carter, J. E.,** Iodide "mumps," *N. Engl. J. Med.,* 264, 987–988, 1961.
252. **Editorial,** Iodism, *South. Med. J.,* 54, 817–818, 1961.
253. **Talner, L. B., Lang, J. H., Brasch, R. C., and Lasser, E. C.,** Elevated salivary iodine and salivary gland enlargement due to iodinated contrast media, *Am. J. Roentgenol. Radium Ther. Nucl. Med.,* 112, 380–382, 1971.
254. **Barker, W. H. and Wood, W. B., Jr.,** Severe febrile iodism during treatment of hyperthyroidism, *JAMA,* 114, 1029–1038, 1940.
255. **Leopold, I. H.,** Ocular complications of drugs, *JAMA,* 205, 631–633, 1968.
256. **Chapman, R. S. and Main, R. A.,** Diffuse thinning of hair in iodide-induced hypothyroidism, *Br. J. Dermatol.,* 79, 103–105, 1967.
257. **Stengel, A.,** Iodic purpura with fever, *Ther. Gaz.,* 26, 1–7, 1902.
258. **Kay, A. B.,** Functions of the eosinophil leukocyte, *Br. J. Haematol.,* 33, 313–318, 1976.
259. **Miller, J. K., Moss, B. R., Swanson, E. W., and Lyke, W. A.,** Effect of thyroid status and thiocyanate on absorption and excretion of iodine by cattle, *J. Dairy Sci.,* 58, 526–530, 1975.
260. **Moss, B. R., Hall, R. F., Miller, J. K., and Swanson, E. W.,** Effect of thiocynate on iodide concentrating systems of the calf, *Proc. Soc. Exp. Biol. Med.,* 129, 153–155, 1968.
261. **Wolff, J., Chaikoff, I. L., and Nicholls, C. W., Jr.,** Accumulation of thyrozine-like and other iodine compounds in fetal bovine thyroid, *Endocrinology,* 44, 510–519, 1949.
262. **Gorbman, A., Waterman, A., Barnes, C. M., and Bustard, L. K.,** Thyroidal function in fetal and pregnant sheep given chronic low level dosages of I^{131}, *J. Endocrinol.,* 60, 565–567, 1957.

263. Nathanielsz, P. W., Comline, R. S., Silver, M., and Thomas, A. L., Thyroid function in the fetal lamb during the last third of gestation, *J. Endocrinol.*, 58, 535–546, 1973.

264. Miller, J. K., Swanson, E. W., Aschbacher, P. W., and Cragle, R. G., Iodine transfer and concentration in the prepartum cow, fetus, and neonatal calf, *J. Dairy Sci.*, 50, 1301–1305, 1970.

265. Prakash, P. and Sharma, D., Thyroxine secretion rate and distribution and uptake of ^{131}I in goats, *Indian J. Exp. Biol.*, 10, 20–22, 1972.

266. Hemken, R. W., Iodine, *J. Dairy Sci.*, 53, 1138–1141, 1970.

267. Dvoskin, S., Thyroxine-like action of elemental iodine in rat and chick, *J. Endocrinol.*, 40, 334–352, 1947.

268. Evans, E. S., Taurog, A., Koneff, A. A., Potter, G. D., Chaikoff, I. L., and Simpson, M. E., Growth response of thyroidectomized rats to high levels of iodide, *Endocrinology*, 67, 619–634, 1963.

269. Rowlands, I. W., Preparation and biological effects of iodinated proteins; further experiments on restoration and maintenance of growth after thyroidectomy, *J. Endocrinol.*, 4, 305–311, 1945.

270. Reineke, E. P. and Turner, C. W., Growth response of thyroidectomized goats to artificially formed thyro-protein, *Endocrinology*, 29, 667–673, 1940.

271. Blaxter, K. L., Severe experimental hyperthyroidism in the ruminant. II. Physiological effects, *J. Agric. Sci.*, 38, 20–27, 1948.

272. Blaxter, K. L., Preparation and biological effects of iodinated proteins. 3. Effect of iodinated protein feeding on lactating cow. I. Effects of preparations of low activity and of iodinated ardein, *J. Endocrinol.*, 4, 237–265, 1945.

273. Oswald, A., Gewinnung von 3, 5-Dijodtyrosin aus Jodieweiss, *Z. Physiol. Chem.*, 70, 310-313, 1911.

274. Ludwig, W. and von Mutzenbecker, P., Uber die Einstellung von Thyroxin durch Jodierung von Eiweiss (Vorlaufige Mitterburg), *Z. Physiol. Chem.*, 244, 4, 1936.

275. Harrington, C. R. and Pitt-Rivers, R. V., Preparation of thyroxine from casein treated with iodine, *Nature*, 144, 205, 1939.

276. Webster, S. H., Rice, M. E., Highman, B., and von Oetingen, W. F., Toxicity of potassium and sodium iodates: acute toxicity in mice, *J. Pharmacol. Exp. Ther.*, 120, 171–178, 1957.

277. Webster, S. H., Rice, M. E., Highman, B., and Stohlman, E. F., The toxicology of potassium and sodium iodate. II. Subacute toxicity of potassium iodate in mice and guinea pigs, *Toxicol. Appl. Pharmacol.*, 1, 87–90, 1959.

278. Webster, S. H., Stohlman, E. F., and Highman, B., Chronic toxicity of potassium iodate in dogs, *Toxicol. Appl. Pharmacol.*, 6, 362, 1964.

279. Webster, S. H., Stohlman, E. F., and Highman, B., Toxicology of potassium and sodium iodate. III. Acute and subacute oral toxicity of potassium iodate in dogs, *Toxicol. Appl. Pharmacol.*, 8, 185–192, 1966.

280. Highman, B., Webster, S. H., and Rice, M. E., Degeneration of retinal and gastric parietal cells and other pathologic changes following administration of iodates, *Fed. Proc. Fed. Am. Soc. Exp. Biol.*, 14, 407, 1955.

281. McCauley, E. H., Linn, J. G., and Goodrich, R. D., Experimentally induced iodide toxicosis in lambs, *Am. J. Vet. Res.*, 34, 65–69, 1973.

282. Newton, G. L., Barrick, E. R., Harvey, R. W., and Wise, M. B., Effect of iodine on calf performance, *J. Anim. Sci.*, 34, 362, 1972.

283. Newton, G. L., Barrick, E. R., Harvey, R. W., and Wise, M. B., Iodine toxicity. Physiological effects of elevated dietary iodine on calves, *J. Anim. Sci.*, 38, 449–455, 1974.

284. Webster, B. and Chesney, A. M., Endemic goiter in rabbits: effect of administration of iodine, *Bull. Johns Hopkins Hosp.*, 43, 291–308, 1928.

285. Baker, H. J. and Lindsey, J. R., Equine goiter due to excess dietary iodine, *J. Am. Vet. Med. Assoc.*, 153, 1618–1630, 1968.

286. Wheeler, R. S. and Hoffman, E., Goiterogenic action of iodide and the etiology of goiters in chicks from thyroprotein-fed hens, *Proc. Soc. Exp. Biol. Med.*, 72, 250–254, 1949.

287. Correa, P. and Welsh, R. A., The effect of excessive iodine intake on the thyroid gland of the rat, *Arch. Pathol.*, 70, 247–251, 1960.

288. Follis, H. R., Jr., Thyroiditis resulting from administration of excess iodine to hamsters with hyperplastic goiters, *Proc. Soc. Exp. Biol. Med.*, 102, 425–429, 1959.

289. Follis, H. R., Jr., Further observations on thyroiditis and colloid accumulation in hyperplastic thyroid glands of hamsters receiving excess iodine, *Lab. Invest.*, 13, 1590–1599, 1964.

290. **Belshaw, B. E., Cooper, T. B., and Becker, D. V.,** The iodine requirement and influence of iodine intake on iodine metabolism and thyroid function in the adult beagle, *Endocrinology,* 96, 1280–1291, 1975.
291. **Belshaw, B. E., Barandes, M., Becker, D. V., and Berman, M.,** A model of iodine kinetics in the dog, *Endocrinology,* 95, 1078–1093, 1974.
292. **Belshaw, B. E. and Becker, D. V.,** Necrosis of follicular cells and discharge of thyroidal iodine induced by administering iodide to iodine-deficient dogs, *J. Clin. Endocrinol. Metab.,* 36, 466–474, 1973.
293. **Akerib, M.,** Iodine toxic to young animals, *World's Poult. Sci. J.,* 27, 35–37, 1975.
294. **Arrington, L. R., Santa Cruz, R. A., Harms, R. H., and Wilson, H. R.,** Effects of excess dietary iodine on pullets and laying hens, *J. Nutr.,* 92, 325–330, 1967.
295. **Perdomo, J. T., Harms, R. H., and Arrington, L. R.,** Effect of dietary iodine upon egg production, fertility and hatchability, *Proc. Soc. Exp. Biol. Med.,* 122, 758–760, 1966.
296. **Barua, J., Cragle, R. G., and Miller, J. K.,** Sites of gastrointestinal passage of iodide and thyroxine in young cattle, *J. Dairy Sci.,* 47, 539–541, 1964.
297. **Swanson, E. W. and Miller, J. K.,** Effect of feeding iodide on abomasal secretion of iodide, *J. Dairy Sci.,* 56, 671–672, 1973.
298. **Cragle, R. G.,** Dynamics of mineral elements in the digestive tract of ruminants, *Fed. Proc. Fed. Am. Soc. Exp. Biol.,* 32, 1910–1914, 1973.
299. **Hemken, R. W.,** Iodine, *J. Dairy Sci.,* 53, 1138–1141, 1970.
300. **Malan, A. I., du Toit, P. J., Grownewald, J. W., and Onderstepoort, J.,** Iodine in the nutrition of sheep. 2nd report, *J. Vet. Sci. Anim. Ind.,* 5, 189–200, 1935.
301. **Roche, D. and Harcicka, J. V.,** unpublished data, 1972–1974.
302. **Arrington, L. R., Taylor, R. N., Jr., Ammerman, C. B., and Shirley, R. L.,** Effects of excess dietary iodine upon rabbits, hamsters, rats and swine, *J. Nutr.,* 87, 394–398, 1965.
303. **Frape, D. L., Gage, J. W., Jr., Hays, V. W., Speer, V. C., and Catron, D. V.,** Studies on the iodine requirement and thyroid function of young pigs, *J. Anim. Sci.,* 17, 1225–1226, 1958.
304. **Newton, G. L. and Clawson, A. J.,** Iodine toxicity: physiological effects of elevated dietary iodine on pigs, *J. Anim. Sci.,* 39, 879–883, 1974.
305. **Weiser, S. and Zaitschek, A.,** Iodine tolerance experiments in young pigs, *Nutr. Abstr. Rev.,* 2, 885, 1933.
306. **Forbes, E. B.,** The value of iodine for livestock in central Pennsylvania, *J. Agric. Res.,* 45, 111, 1932.
307. **Reineke, E. P.,** Factors affecting the secretion of iodine-131 into milk of lactating goats, *J. Dairy Sci.,* 44, 937–942, 1961.
308. **Anon.,** The secretion of iodine into milk of lactating goats, *Nutr. Rev.,* 20, 20–21, 1962.
309. National Research Council Agricultural Board, *Nutrient Requirements of Laboratory Animals,* rev. ed., National Academy of Sciences, Washington, D.C., 1972.
310. **Ichikawa, A., Kawada, J., and Ito, Y.,** Iodide goiter in the mouse, *Endocrinol. Jap.,* 14, 333–341, 1967.
311. **Galton, V. A. and Pitt-Rivers, R. V.,** Effect of excessive iodine on the thyroid of the rat, *Endocrinology,* 64, 835–839, 1959.
312. **Ammerman, C. B., Arrington, L. R., Warnick, A. C., Edwards, J. L., Shirley, R. L., and Davis, G. K.,** Reproduction and lactation in rats fed excessive iodine, *J. Nutr.,* 84, 107–112, 1964.
313. **London, W. T., Vought, R. L., and Brown, F. A.,** Bread – a dietary source of large quantities of iodine, *N. Engl. J. Med.,* 273, 381, 1965.
314. **Pittman, J. A., Jr., Dailey, G. E., III, and Beschi, R. J.,** Changing normal values for thyroidal radioiodine uptake, *N. Engl. J. Med.,* 280, 1431–1434, 1969.

EFFECT OF NUTRIENT TOXICITY
IN ANIMALS AND MAN: MANGANESE

K. Nishiyama

Health disorders resulting from excessive manganese exposure can be classified into two types, according to the sites affected by manganese and the symptoms of the effect. The first is chronic manganese poisoning, in which symptoms are mainly caused by disturbances in the extrapyramidal system. The other type is composed of pulmonary disturbances which include pneumonia and pneumoconiosis. These health disorders have been reported to occur in workers at manganese mines, refineries, and plants manufacturing manganese steel, welding rods, or dry cell batteries. It has also been observed that the pulmonary disturbances occur in inhabitants in the vicinity of manganese factories. Most cases have been the result of the inhalation of fine dust of manganese compounds. Only one report in man has been the result of an excessive oral intake of manganese.[1]

CHRONIC MANGANESE POISONING

Environmental Conditions for its Occurrence

Before 1971, more than 800 cases of manganism had been reported and all had occurred in workers exposed to manganese dust and fumes.[2] It has been stated that the higher the oxidation state of the manganese compounds and the fresher and finer the crushed manganese dust, the easier is the occurrence of poisoning.[3] Even under similar exposure conditions, there are instances where symptoms fail to appear in some workers. While the latent period in affected workers varies greatly (from 2 months to 24 years), no definite relationship between the concentrations of manganese dust and the incidence of manganism has been observed. The difference in intestinal absorption of manganese, which increases with iron deficiency anemia, is considered to be one of the major factors in individual susceptibility to manganese.[4] In iron-deficient rats, the plasma-binding capacity of manganese is increased, and an increasing concentration of manganese in the brain has also been observed.[5]

The threshold limit value for manganese in the U.S. (a ceiling value), England, and Japan is 5 mg/m^3, but under environmental conditions with less than this value, some cases of chronic manganese poisoning have been produced. For example, Smyth et al.[6] reported cases which occurred in a ferromanganese factory where the concentration was in the range of 0.12 to 3.60 mg/m^3, with an average of 0.8 mg/m^3. The maximum permissible limit for manganese in the Soviet Union is 0.3 mg/m^3 as MnO_2.

Kawamura et al.[1] reported only one poisoning in humans due to an oral intake of manganese. This was the subacute manganese poisoning of 16 victims, which included two deaths (one suicide case) caused by drinking well-water which had been contaminated by manganese from waste dry cells buried underground. The period governing the contaminated water intake was thought to be 2 to 3 months and the manganese concentration of the well-water when the poisoning occurred was 14.34 ppm. These poisoning cases were diagnosed by the peculiar neurological symptoms presented by the patients, by histopathological changes in brain tissues, and by analysis of the manganese content of the liver obtained from the autopsy.

In addition to these main facets of manganese poisoning, there have been high and low occurrence regions of endemic goiter where the iodine concentration in well-water has been uniformly low. The manganese concentration in well-water in the high region ranged from 0.12 to 1.23 ppm, while in the low region it ranged from 0.05 to 0.11 ppm. From these findings, Manescu et al.[7] presumed that the ratio of iodine to manganese may possibly be associated with the incidence of goiter.

Animal Experiments

Many animal experiments using the oral administration of manganese have been carried out. However, in these experiments, despite an excessive administration of manganese, notable symptoms of poisoning have not always been found.

Suzuki[8] carried out free-water drinking experiments on mice using water containing a range of 20 to 5000 ppm of manganese for periods of 26 to 300 days. Based on results which showed a sudden accelerated increase of manganese accumulation in the organs and bad effects on body weight, death rate, and macroscopic changes in the organs, the lowest limit amount of oral intake in mice was assumed to be in the range of 500 to 1000 ppm. Cotzias and his co-workers gave mother mice milk containing various concentrations of manganese during labor and administered ^{54}Mn to newborn mice on the 6th day after birth. Those born from mothers given milk containing less than 280 ppm had no ^{54}Mn excretion until the 17th day while those born from mothers given milk containing more than 280 ppm had an ^{54}Mn excretion right after administration. In young mice born from mothers given milk containing more than 400 ppm, a sudden accelerated rate of loss of ^{54}Mn was observed during the 13th to 17th day of life. Body weights of young mice born from the group of mothers given milk with 400 ppm were 50% that of the control.[9] Although it is not appropriate to compare the above two reports, they suggest that the concentration limit for the oral intake of soluble manganese compounds which cause noticeable effects on mice, especially the destruction of the homeostasis of manganese in the body, may be several hundred parts per million.

Compared to oral administration, especially in experiments on monkeys, inhalation and injection produced manifestations of peculiar neurological symptoms. Data on manganese poisoning in the monkey experiments[10-17] are shown in Table 1. Suzuki et al.[16] reported that on subcutaneous administration of MnO_2 from 0.25 to 1 g once a week (a total amount of from 1.75 to 3 g), neurological symptoms including hyperexcitability, intention tremor, choreiform movement, disturbances of equilibration, and hand contracture were observed. The greater the amount administered at a time, the sooner the symptoms appeared. The manganese accumulation in the organs and tissues of these monkeys was largest in the salivary glands, thyroid gland, adrenal glands, basal ganglia, and lungs. It was also noted that the accumulations in these organs and tissues were higher than in the liver and kidneys. The accumulation in each organ increased proportionally with the dosage. There was no definite relationship between the severity of the neurological symptoms and the accumulations.

The absorption rate of deposited manganese in the lungs has not been determined. Morrow et al.[18] reported that the biological half time of absorbed MnO_2 in the lungs was in the range of 62 to 68 days in humans. Even insoluble manganese compounds can be absorbed by the lungs and the alimentary canal, leading to an elevation of the manganese level in the body. This elevation also occurs in the cerebrum and is especially remarkable in the basal ganglia. The biological half time in the brain is relatively long compared with that in the other main organs. The increased amount of manganese seems to influence the concentration of dopamine or serotonin in the brain.

On the administration of levodopa, the amount of both dopamine and manganese in the brain increases. This suggests that there is a close relationship between the two.[19] Cotzias and his co-workers observed increases in dopamine in the brains of mice 8 days after being given milk containing 1000 ppm of manganese. They also observed a high positive relationship between dopamine and manganese in the brains of newborn mice whose mothers had been given various amounts of manganese. These investigators supposed that the maturation of the dopaminergic apparatus would depend on

Table 1

REPORT ON EXPERIMENTAL MANGANESE POISONING IN MONKEYS

Reporter	Mn compound and its administration method	Item under observation and results
Mella[10]	$MnCl_2$, increase from 5 or 15 mg to 20 mg, daily or every other day, intraperitoneal	After 6–8 months, abnormal movements, histological changes in basal ganglia, Mn increase in brain
Van Bogaert et al.[11]	MnO_2, $NaClO_2$, NH_4 Cl mixture, 1 hr/day, for 95 days, inhalation	After 50 days, excitation, abnormal movements, atrophy of cortex
	$MnSO_4$ gradual increase from 10 mg to 50 mg, for 300 days, oral	No neurological symptoms, no histological changes
Pentschew et al.[12]	MnO_2, 2 g once, 1 month later 3.5 g once, intramuscular	After 9 months, extrapyramidal symptoms, histological change in the subthalamic nucleus and pallidum
Neff et al.[13]	MnO_2, 200 mg, 2 or 3 times/month subcutaneous	After 2 months, extrapyramidal symptoms, no histological changes in brain, but decrease in dopamine and serotonine in caudate
Osipova et al.[14]	$MnCl_2$, 2–4.5 mg single in the brain or 1–2 mg, 3–8 times, intracisternally	Typical parkinsonism, died after 5 days to 4 months
Dastur et al.[15]	MnO_2 or $MnCl_2$, repeatedly for several months, intraperitoneal	Extrapyramidal symptoms without histological changes, no relationship between symptoms and brain manganese content
	[54]Mn, single, intraperitoneal, killed after 6 hr–278 days	Increase of [54]Mn in central nerves, continuous level maintenance
Suzuki et al.[16]	MnO_2, 0.25–1 g/week for 9 weeks, subcutaneous	Appearance of neurotic symptoms, blood examination, Mn distribution in body
Nishiyama et al.[17]	MnO_2, 0.7 or 3 mg/m³ , 22 hr/day for 10 months, inhalation	Abnormality in chest X-rays and histological findings in lung, slight neurotic symptoms and serologic findings, marked Mn accumulation in basal ganglia, endocrine and exocrine glands

manganese.[9] However, as Neff et al.[13] and other investigators[20-22] observed in animals administered manganese, an excessive amount of manganese over a long period of administration seems to produce a remarkable decrease in dopamine, serotonin, norepinephrine, and homovanillic acid (HVA) levels in the brain. It is believed that human manganese poisoning, as described in the next section, starts a self-limited psychiatric phase, bearing similarities to mental effects of increased cerebral dopamine, followed by a permanent neurological phase with the extrapyramidal effects of decreased cerebral dopamine.[9] Excessive manganese seems to disturb the activity of tyrosine hydroxylase, but dopa-decarboxylase, monoamine oxidase (MAO), and catechol-O-methyltransferase (COMT) may not be influenced.[20]

Symptoms

Chronic manganese poisoning presents a clinical picture similar to that of Parkinson's disease. In its early stage, the prodromal period, the main symptoms are subjective; fatigue, languor, weakness, headache, and drowsiness sometime followed by insomnia, deterioration of memory, asthenia, and sexual excitement followed by impotence.[3] Salivation and excessive sweating are also seen. Mental disturbances, called locura manganica (manganese madness) in northern Chile (including unaccountable laughter, euphoria, impulsiveness, and hallucinations), may also be seen in the early stage.[4] Afterwards, with the more severe subjective symptoms given above, muscle rigidity, tremor, coordination disturbance, speech disturbance, micrographia, gait disturbance, and pro- or retropulsion may also appear and have been confirmed as symptoms of this poisoning.[3]

Laboratory Tests

The diagnosis of chronic manganese poisoning depends on a detailed examination of the manganese exposure history and neurological symptoms, but definite items for a laboratory test have yet to be established.

By measuring the manganese content of the blood, urine, hair, and other available biological specimens, the degree of manganese exposure and the body burden of manganese may be estimated in some cases. The examination of workers exposed to manganese has shown a very close relationship, with $r = 0.69$, between manganese concentrations in the air and in the urine.[23] In a balance study on manganese in normal subjects, a nearly linear relationship, with $r = 0.96$, between the oral intake of manganese and the amount of excreted manganese in the feces has also been observed.[24] Some cases have been reported where the feces contained more than 6 mg/100 g and the worker was evidently working under a manganese concentration in the air of 2 mg/m^3.[25] The degree of the body-burden of manganese may be estimated by the increased urinary excretion of manganese after the administration of ethylenediaminetetraacetic acid (EDTA). In examining a manganese mine worker presenting peculiar neurological symptoms, Rosenstock et al.[26] considered it strong evidence for the diagnosis of poisoning that the hair (29 ppm) and chest hair (109 ppm) contained high levels of manganese and that the urinary excretion of manganese increased on the administration of EDTA. Despite numerous reports on the amount of manganese in the biological specimens in manganism, no values for biological threshold limits have been determined. As the amounts of manganese in the tissues of most organs decrease rapidly after the cessation of manganese exposure, it is necessary to carefully consider the time relationship between the exposure and collection times of the specimens in order to evaluate the measurements. One distinctive feature of chronic manganese poisoning, which is said to differ from that of heavy metal poisonings, is that the chronically poisoned patients have a lower tissue manganese concentration as well as a slower turnover rate of ^{54}Mn than do healthy subjects under the same exposure conditions.

Following the administration of an excessive amount of manganese in the early stage, an increase in red blood cells and in hemoglobin content is observed, but these later show a decrease. In poison patients, an increase in calcium content and adenosine deaminase activity in their sera has been observed,[27] as well as an elevated basal metabolic rate.[28] To evaluate EEG changes, more cases with detailed data are necessary. In the human autopsy cases and in the animal experiments, the brain, especially the basal ganglia, showed histopathological changes such as the degeneration, destruction, or disappearance of the nerve cells. In rabbits administered manganese, remarkable tissue changes in the testicles were observed in the relatively early stage. Acid phosphatase, adenosine triphosphatase, and succinic dehydrogenase activities in the testicles were significantly reduced.[29] The reduced activities of succinic dehydrogenase and lactic dehydrogenase in the brain, liver, and testicles were also observed in manganese-treated rats.[30] However, these findings from human autopsies and animal experiments are not yet useful for the direct diagnosis of early poisoning.

Treatment

Isolation from manganese exposure is the most important treatment. The persistence of neurological symptoms makes it difficult to recover even after isolation, but at least the possibility of a progression of symptoms would be eliminated. An administration of chelate compounds, including EDTA, to eliminate body manganese would be useful before irreversible neurological changes occur.

Long-term, high-dosage L-dopa therapy has had beneficial effects in some cases of chronic manganism. One case with hypotonia presented an unfavorable response to L-dopa therapy, but later improved after the administration of 5-hydroxytryptophan.[31] Studies on transferring the effective dosage of L-dopa to the target organ without side effects have been carried out.[32]

PULMONARY DISTURBANCES

Animal experiments and the epidemiological studies on manganese workers clearly show that the inhalation of manganese dust causes disturbances in the respiratory organs. According to studies by Lloyd Davies,[33] the incidence of pneumonia in manganese workers was 2.6%, which was very high when compared to 0.073% for the control. He also stated, based on animal experiments, that manganese pneumonitis appears to be a more appropriate term than manganese pneumonia.

Elstad[34] reported that in Sauda, Norway the death rate of the inhabitants in the vicinity of a ferromanganese factory due to croupous pneumonia was 3.2% as compared to 0.4% for the whole country. Suzuki[35] also reported that there is a high incidence of pneumonia in Japan in those people living in the vicinity of a manganese factory.

In animal experiments which studied the inhalation of MnO_2 dust, inflammatory changes in the lungs were observed in most cases. Nishiyama et al.[17] conducted animal experiments in which monkeys and mice inhaled MnO_2 dust at concentrations of 0.7 and 3 mg/m^3 as manganese for 22 hr a day, for 10 continuous months. The monkeys developed abnormal shadows with time including granular, nodular, reticular, honeycomb shadows, and bulla which were observed in the pulmonary radiograms taken monthly. In the histopathological examination which was made 10 months later, they also showed abnormal changes including hyperplasia of their lymphoid tissue and papillary growth of the bronchial epithelium. The mice had slight inflammatory changes in their lungs after 0.5 to 1 month. Afterwards, these changes disappeared and pathological pictures similar to those of the monkeys appeared. At this stage, no tendency for definite fibrosis was observed.

In the experiments on mice, the inhalation of MnO_2 was shown to lead to an increase in the death rate due to infection by bacteria or virus, and the shortening of their lifespans.[36] From the examinations of manganese workers, a high rate of occurrence of manganoconiosis has been reported.[2]

MANGANESE TRICARBONYL COMPOUNDS

Manganese carbonyl compounds added to fuel oils are excellent smoke inhibitors and combustion improvers; when added to gasoline, they produce excellent antiknock effects. Manganese cyclopentadienyl tricarbonyl (MCT), produced in the Soviet Union, and methylcyclopentadienyl manganese tricarbonyl (MMT), produced in the U.S., are used to supplement or replace lead antiknock compounds.

MCT causes anaphylaxis of the muscle nerves and histological changes in the respiratory organs of rats, guinea pigs, and rabbits on repeated exposure to 0.001 mg/l. A 2-hr exposure of 0.1 mg/l is fatal. On administration of a fatal amount, animals show central nervous symptoms.[37] The oral LD_{50} of MMT in rats is 58 mg/kg.[38] A threshold limit value of 0.2 mg/m^3 of MMT was proposed in 1970 in the U.S. for industrial exposure.

Of the manganese produced by the combustion of MMT, 99.5% is Mn_3O_4. Moore et al.[39] carried out an experiment in which rats and hamsters inhaled diluted automobile exhaust gases produced by gasoline with MMT added. The animals were exposed to these gases at a concentration of 117 $\mu g/m^3$ as manganese, for 8 hr a day for 56 days. An increase in manganese levels in their tissues was seen, but no definite histopathological changes due to manganese or MMT occurred.

ENVIRONMENTAL POLLUTION BY MANGANESE

Manufacturing and production processes using manganese greatly influence the way this element acts in the ecosystem.[2] Air pollution caused by manganese should be a seriously regarded form of environmental pollution. Disturbances in the respiratory organs by manganese itself, its role in the exacerbation of bacterial or viral infection, and the oxidation catalytic effect on other chemical compounds should be considered in the future.[40] Oxidative compounds of manganese react with SO_2 or NO_2 and become soluble manganese compounds. $MnSO_4$ acts as an oxidative catalyst in the oxidation reaction of SO_2 to SO_3. According to Amdur et al.,[41] when guinea pigs inhaled SO_2 and Mn^{+2} simultaneously, the airway resistance increased three times as compared to the inhalation of only SO_2.

The ambient air quality standard for manganese is 10 $\mu g/m^3$ for an average 24-hr exposure and 30 $\mu g/m^3$ for a single exposure in the Soviet Union. Hickey et al.[42] investigated the relationship between various metal concentrations in the air at various sites in the U.S. and the death rate of the inhabitants due to renal, heart, cardiovascular system diseases and to respiratory system neoplasms and others, but no definite relationship between manganese and any of the above diseases was observed. Even though an increased amount of manganese in the air may be of little harm to adults, if one considers that in young rats the intestinal absorption of manganese is on the order of 70% and the entrance of manganese into the brain is fourfold that in adult rats, the effects of air manganese on the newborn or infants should not be disregarded.[5] Any increase in the manganese level in the air from any source should be under careful control.

REFERENCES

1. Kawamura, R., Ikuta, H., Fukuzumi, S., Yamada, R., Tsubaki, S., Kodama, T., and Kurata, S., Intoxication by manganese in well water, *Kitasato Arch. Exp. Med.,* 18, 145–169, 1941.
2. National Research Council, Committee on Medical and Biological Effects of Environmental Pollutants, *Manganese,* National Academy of Sciences, Washington, D.C., 101–103, 1973.
3. Rodier, J., Manganese poisoning in Moroccan miners, *Br. J. Ind. Med.,* 12, 21–35, 1955.
4. Mena, I., Horiuchi, K., Burke, K., and Cotzias, G. C., Chronic manganese poisoning. Individual susceptibility and absorption of iron, *Neurology,* 19, 1000–1006, 1969.
5. Mena, I., The role of manganese in human disease, *Ann. Clin. Lab. Sci.,* 4, 487–491, 1974.
6. Smyth, L. T., Ruhf, R. C., Whitman, N. E., and Dugan, T., Clinical manganism and exposure to manganese in the production and processing of ferromanganese alloy, *J. Occup. Med.,* 15, 101–109, 1973.
7. Manescu, S., Negoescu, I., Lupulescu, A., and Diaconescu, M., Investigations on the goitrogenic action of manganese, *Rom. Med. Rev.,* 5, 88–90, 1961.
8. Suzuki, Y., Studies on excessive oral intake of manganese. II. Minimum dose for manganese accumulation in mouse organ, *Shikoku Acta Med.,* 30, 32–45, 1974.
9. Cotzias, G. C., Miller, S. T., Papavasiliou, P. S., and Tang, L. C., Interactions between manganese and brain dopamine, *Med. Clin. North Am.,* 60, 729–738, 1976.
10. Mella, H., The experimental production of basal ganglion symptomatology in macacus rhesus, *Arch. Neurol. Psychiatry,* 11, 405–417, 1924.
11. Van Bogaert, L. and Dallemagne, M. J., Approches expérimentales des troubles nerveux du manganisme, *Monatsschr. Psychiatr. Neurol.,* 111, 60–89, 1945/46.
12. Pentschew, A., Ebner, F. F., and Kovatch, R. M., Experimental manganese encephalopathy in monkeys. A preliminary report, *J. Neuropath. Exp. Neurol.,* 22, 488–499, 1963.
13. Neff, N. H., Barrett, R. E., and Costa, E., Selective depletion of caudate nucleus dopamine and serotonin during chronic manganese dioxide administration to squirrel monkeys, *Experientia,* 25, 1140–1141, 1969.
14. Osipova, I. A., Dyakova, I. N., and Urmancheeva, T. G., Manganese-induced parkinsonism in rhesus monkeys, *Gig. Tr. Prof. Zabol.,* 13, 48–49, 1969.
15. Dastur, D. K., Manghani, D. K., and Raghavendran, K. V., Distribution and fate of ^{54}Mn in the monkey: studies of different parts of the central nervous system and other organs, *J. Clin. Invest.,* 50, 9–20, 1971.
16. Suzuki, Y., Mouri, T., Suzuki, Y., Nishiyama, K., Fujii, N., and Yano, H., Study of subacute toxicity of manganese dioxide in monkeys, *Tokushima J. Exp. Med.,* 22, 5–10, 1975.
17. Nishiyama, K., Suzuki, Y., Fujii, N., and Yano, H., Effects of Inhalation of Manganese Dixoide Dusts on Respiratory Organs in Monkeys and Mice, in Abstr. 18th Int. Congr. Occupational Health, Brighton, 1975, 429.
18. Morrow, P. E., Gibb, F. R., and Gazioglu, K., The clearance of dust from the lower respiratory tract of man. An experimental study, in *Inhaled Particles and Vapours,* Vol. 2, Davies, C. N., Ed., Pergamon Press, Oxford, 1967, 351-358.
19. Papavasiliou, P. S., Miller, S. T., Cotzias, G. C., Kraner, H. W., and Hsieh, R. S., Sequential analysis: manganese, catecholamines and L-DOPA induced dyskinesia, *J. Neurochem.,* 25, 215–219, 1975.
20. Bonella, E. and Diez-Ewald, M., Effect of L-dopa on brain concentration of dopamine and homovanillic acid in rats after chronic manganese chloride administration, *J. Neurochem.,* 22, 297–299, 1974.
21. Chandra, S. V. and Srivastava, S. P., Experimental production of early brain lesions in rats by parenteral administration of manganese chloride, *Acta Pharmacol. Toxicol.,* 28, 177–183, 1970.
22. Mustafa, S. J. and Chandra, S. V., Levels of 5-hydroxytriptamine, dopamine and norepinephrine in whole brain of rabbits in chronic manganese toxicity, *J. Neurochem.,* 18, 931–933, 1971.
23. Tanaka, S. and Lieben, J., Manganese poisoning and exposure in Pennsylvania, *Arch. Environ. Health,* 19, 674–684, 1969.
24. Suzuki, Y., Studies on excessive oral intake of manganese. I. Metabolic balance of manganese in the adult human, *Shikoku Acta Med.,* 29, 484–494, 1974.
25. Jindrichova, J., Anwendungsmöglichkeit der Manganbestimmung im Stuhl als Expositionstest, *Int. Arch. Gewerbepath. Gewerbehyg.,* 25, 347–359, 1969.
26. Rosenstock, H. A., Simons, D. G., and Meyer, J. S., Chronic manganism. Neurologic and laboratory studies during treatment with levodopa, *JAMA,* 217, 1354–1358, 1971.
27. Chandra, S. V., Seth, P. K., and Mankeschwar, J. K., Manganese poisoning. Clinical and biochemical observations, *Environ. Health,* 7, 374–380, 1974.

28. **Suzuki, Y., Nishiyama, K., Doi, M., Hirose, H., and Shibata, H.,** Studies on chronic manganese poisoning, *Tokushima J. Exp. Med.,* 7, 124–132, 1960.

29. **Chandra, S. V., Ara, R., Nagar, N., and Seth, P. K.,** Sterility in experimental manganese toxicity, *Acta Biol. Med. Ger.,* 30, 857–862, 1973.

30. **Singh, J., Husain, R., Tandon, S. K., Seth, P. K., and Chandra, S. V.,** Biochemical and histopathological alterations in early manganese toxicity in rats, *Environ. Physiol. Biochem.,* 4, 16–23, 1974.

31. **Mena, I., Court, J., Fuenzalida, S., Papavasiliou, P. S., and Cotzias, G. C.,** Modification of chronic manganese poisoning. Treatment with L-dopa or 5-OH tryptophane, *N. Engl. J. Med.,* 282, 5–10, 1970.

32. **Cotzias, G. C.,** Levodopa, manganese, and degenerations of the brain, *Harvey Lect.,* 68, 115–147, 1974.

33. **Lloyd Davies, T. A. and Harding, H. E.,** Manganese pneumonitis. Further clinical observations, *Br. J. Ind. Med.,* 6, 82–90, 1949.

34. **Elstad, G.,** Manganholdig fabrikkrek som medvirkende arsak ved pneumoniepidemier i en industrihygd., *Nord. Med.,* 3, 2527–2533, 1939.

35. **Suzuki, Y.,** Environmental contamination by manganese, *Jpn. J. Ind. Health,* 12, 529–533, 1970.

36. **Maigetter, R. Z., Ehrlich, R., Fenters, J. D., and Gardner, D. E.,** Potentiating effects of manganese dioxide on experimental respiratory infections, *Environ. Res.,* 11, 386–391, 1976.

37. **Arkhipova, O. G., Tolgskaya, M. S., and Kochetkova, T. A.,** Toxicity within a factory of the vapour of the new antiknock compound, manganese cyclopentadienyl tricarbonyl, *Hyg. Sanit.* (USSR), 30, 40–44, 1965.

38. National Environmental Research Center, *Scientific and Technical Assessment Report on Manganese,* Program Element No. 1AA001, ROAP No. 26AAA, EPA-600/6-74-002, Office of Research and Development, Office of Program Integration, U.S. Environmental Protection Agency, Washington, D.C., April 1975, 6-3.

39. **Moore, W., Hysell, D., Miller, R., Malanchuk, M., Hinners, R., Yang, Y., and Stara, J. F.,** Exposure of laboratory animals to atmospheric manganese from automotive emissions, *Environ. Res.,* 9, 274–284, 1975.

40. **Sullivan, R. J.,** *Preliminary Air Pollution Survey of Manganese and Its Compounds. A Literature Review,* U.S. Department of Education and Welfare, Washington, D.C., 1969.

41. **Amdur, M. O. and Underhill, D.,** The effect of various aerosols on the response of guinea pigs to sulfur dioxide, *Arch. Environ. Health,* 16, 460–468, 1968.

42. **Hickey, R. J., Schoff, E. P., and Clelland, R. C.,** Relationship between air pollution and certain chronic disease death rates, *Arch. Environ. Health,* 15, 728–738, 1967.

EFFECT OF SPECIFIC NUTRIENT TOXICITIES
IN ANIMALS AND MAN: ZINC

H.-J. Lantzsch and H. Schenkel

INTRODUCTION

Zn is a relatively nontoxic element to animals and man in comparison to other metals. The normal Zn concentration in major feeds and foods ranges from 20 to 80 ppm on a dry matter basis. These levels are far below those which have been found to produce deleterious effects.

Replacement of Zn-coated (galvanized) containers, used for preparation and storage of feed, by containers produced from plastic materials has eliminated a frequent source of Zn poisoning. Nowadays, main sources of Zn poisoning are ingestion of Zn-contaminated feeds grown in areas near Zn smelting plants and feeding of ready mixtures accidentally supplemented with toxic amounts of Zn.

ANIMAL STUDIES

Laboratory Animals

Growth and General Health

In rats, dietary Zn levels up to 1000 ppm were well tolerated without toxic effects on growth and health, even when given through several generations.[96,156,175] With dietary Zn levels between 2000 and 8400 ppm, variable results were obtained. In a series of experiments using Zn in the oxide, carbonate, chloride, or sulfate form, as Zn dust, or in a biologically bound form (oysters), no effect on growth and health was observed;[5,17,49,63,89,90,106,145,152,156,175,176] however, reduced growth or weight losses together with reduced food consumption were reported by others.[26,48,106,131-133,143,144,152,167] Growth inhibition is substantiated by findings of reduced nitrogen, phosphorus, and sulfur retention[132,133] and a lowered body fat content.[48,131]

These conflicting results may be caused in part by the different ages of the animals and the composition of the diets used. Younger animals are more susceptible to excess Zn than older ones.[102,146] The dietary protein level,[103] the dietary protein source,[103,106] special components like distiller's dried solubles, liver extract, or whole milk powder,[49,101,143,146] and dietary Ca and P levels[152] influence the toxic effect of Zn on growth.

In rats, dietary Zn levels above 10,000 ppm resulted in severe growth inhibition or weight losses and anorexia.[17,39,70,101,131-133,146,156,167,173,176] In most cases, death occurred within 4 weeks after being placed on the diets.[131,146,156,173,176] In Table 1, lethal doses are given for some Zn salts administered orally or by injection.

General clinical symptoms of highly Zn-toxified laboratory animals were diarrhea and vomiting, dyspnea and tremor (so-called "zinc-tremor"), convulsions with following paresis, hemorrhagic gastritis, enteritis and mucosal erosions owing to the caustic effect of, particularly, $ZnCl_2$, and a depigmentation of black hair.[19,26,46,54,79,106,163,176] In cats, fibrous changes of the acinar tissue of the pancreas were noted.[36,142]

Ingestion of up to 125 mg/day Zn from raw whole oysters caused no signs of toxicity in cats.[105] A daily intake of 200 to 240 mg Zn in the oxide form over a long period of time resulted in weight losses;[142] however, in another report higher daily doses of 480 to 800 mg were needed to produce weight loss.[36] In dogs, a daily intake of 400 to 800 mg Zn in the oxide form during 3 to 19 weeks was without effect on growth and health.[36]

Blood Findings

In rats, ingestion of large amounts of Zn induced a microcytic, hypochromic type

Table 1
TOXICITY OF SOME ZINC SALTS

Compound	Animal	Route	Toxic dose	Dosage expressed in mg of	
				compound/kg	metal/kg
Zinc acetate	Rabbit	Oral	MLD	1000—1800	356—641
Zinc chloride	Rat	i.v.	LD	60—90	28, 8—43, 2
Zinc sulfate	Rat	Oral	LD	2200	500
		s.c.	LD	330—440	75—100
		i.v.	LD	50—60	11, 4—13, 6
	Rabbit	Oral	LD	2000—2200	455—500
		s.c.	LD	300—440	68, 3—100
		i.v.	LD	44	10
	Dog	s.c.	LD	78	17, 7
		i.v.	LD	66	15

From Venugopal, B. and Luckey, T. P., in *Environmental Quality and Safety,* Suppl. Vol. 1, Coulston, F. and Corte, F., Eds., Georg Thieme Verlag, D-Stuttgart, 1975, 19. With permission.

anemia. While dietary Zn levels up to 2000 ppm are without effect on hemoglobin concentration or hematocrit value,[145,156] a slight reduction was seen with 2500 ppm Zn in a casein-based diet.[106] Dietary Zn levels between 4000 and 20,000 ppm caused a reduction in hemoglobin level ranging from 30 to 70%.[26,39,48,49,98,103, 106,115,118,143,144,146,156,167,173] Similar results were obtained with mice on diets containing 2500 or 5000 ppm Zn.[172] At lower dietary Zn levels (100 to 500 ppm) a Zn:Cu ratio of 50:1 or more caused anemia in meat-fed mice and rats.[52]

Hemoglobin reduction was accompanied by appearance of red blood cells of irregular size and shape,[144,146,156] shortening of the life span of the red blood cells,[144] reduction in the hematocrit value,[26,146] and later on by reduction in the red blood cell counts.[156,173] White blood cell counts were found to be increased[146,156] or uninfluenced.[49]

Smaller[29,44,101] or lacking effects[37] of high dietary Zn levels on hemoglobin value may be due in part to other dietary components or ingredients in the diets used. The severity of Zn-induced anemia in rats may be accentuated by animal protein (casein) instead of plant protein (soybean) and higher protein levels,[103,106] or counteracted by higher Ca,[52] Cu, and Fe levels[26,39,49,101,106,115,146] and the presence of special ingredients — for example, distiller's dried solubles or liver.[103] In cats and dogs, daily intakes of 140 to 180 mg and 400 to 800 mg, respectively, of Zn as oxide were without effects on hemoglobin values or red and white blood cell counts.[36]

Application of 30 mg Zn as chloride directly in the stomach of fasting rats caused a rise in the blood sugar level; a dose of 65 mg Zn as sulfate caused a rise in hemoglobin and red blood cell counts, concomitant with a fall in blood pressure.[157] Similar results were obtained with sheep,[122] indicating a hemoconcentration. After parenterally administering Zn as malate in rabbits, cats, and dogs, again a rise in blood sugar level accompanied by a glucosuria was observed.[134]

Reproduction

In rats, reproduction performance was uninfluenced by dietary Zn levels up to 2500 ppm given as oxide, carbonate, chloride, sulfate, or dust, even when given a long time before mating[63,85,139,156] and when given to several generations.[139,156] Feeding daily doses of 4.4 to 9.7 mg Zn per animal as acetate or citrate 29 weeks before mating and 2 to 38 mg Zn as acetate, malate, citrate, or oxide during pregnancy and lactation also had no effect upon health and fertility of the parents or health and early growth of the offspring.[160]

Dietary Zn levels of 4000 and 5000 ppm given during pregnancy caused a significant reduction in fetal growth, birth weight, occurrence of fetal resorption (4 to 29%), and stillbirths, but no anatomical malformations.[29,63,85,139] If Zn feeding was started before mating, all offspring were stillborn[156] or a 100% fetal resorption was noted.[139] This was not confirmed by others.[89] High dietary fat level (17%) may be responsible for the weakened effect of 7000 ppm Zn on reproduction performance.[44] After 5 months on a 5000 ppm Zn diet, females ceased to become pregnant. Reproduction became normal again after removal of Zn from the diet.[156] No reproduction occurred on a 10,000 ppm Zn diet.[156] Feeding of a 4000 ppm Zn diet to lactating rats resulted in growth reduction of the suckling young during the final 7 days of lactation.[21] Milk production of the mothers was unchanged. Milk Zn was found to be elevated threefold, while milk fat, lactose, Cu, Fe, and K were reduced.[21,85]

Enzyme Activities

Activity of various enzymes was found to be altered in tissues of rats fed excess Zn. As mentioned above, in rats, growth and hemoglobin were uninfluenced by dietary Zn levels of up to 1000 and 2000 ppm, respectively. However, activity of several enzymes such as carbonic anhydrase in lung, pituitary, brain, and red blood cells and lactic dehydrogenase in serum was found to be already reduced at this dietary Zn concentration, while the activity of the alkaline phosphatase in serum, kidneys, pancreas, and epiphyseal tibia was increased.[74]

Reduction of cytochrome oxidase activity in liver,[29,167] heart,[29,39,101] and brain,[21] xanthine oxidase in liver[28] and heart,[29] catalase in liver,[167] ceruloplasmin in serum,[29,98,115,174] and alkaline phosphatase in intestines[133] was observed in rats on diets of 4000 to 10,000 ppm Zn. However, if the Cu intake of the animals is inadequate, much lower dietary Zn levels are sufficient to produce a reduction in ceruloplasmin activity.[19] Supplemental dietary Cu or Cu injections restored cytochrome oxidase catalase and ceruloplasmin activity.[21,39,98,101,115,167,174]

Succinic dehydrogenase activity in liver and heart,[29] DPNase and isocitric dehydrogenase activity in liver,[167] and proteolytic activity in pancreas[70] were not influenced by dietary Zn levels between 4000 and 10,000 ppm. Peroxidation of polyunsaturated fatty acids in the livers of rats receiving a 1000 ppm Zn diet was found to be reduced.[22]

Zn Accumulation in Organs and Tissues

Zn toxicity symptoms described above indicate that increasing amounts of Zn enter the body after the ingestion of Zn-enriched food. In rats, increased Zn absorption is evidenced by a marked rise in the serum Zn level. This is already seen at 500 ppm dietary Zn,[91] much earlier than toxicity symptoms and tissue Zn accumulation occur.[4] With increasing dietary Zn levels (1000 to 7500 ppm) a continuous rise in the blood Zn levels was noted.[20,144,174] Concomitantly Zn accumulation occurs in many organs and tissues, especially in the liver,[5,20-22,26,73,91,96,101,103,106,143,175] the long bones,[5,71,78,97,143,145,152] pancreas,[73] and teeth,[78] and to a lesser extent in kidneys, spleen, heart, testes, brain, thymus, and intestinal tissues.[5,20,22,26,29,73,101] Similar results were reported for cats[36,142] and dogs.[36]

In the reports available, considerable differences are to be seen in the extent of Zn accumulation in relation to the dietary Zn level. These variations may be attributed in part to differences in the age of the animals, the duration of the feeding period, and the composition of the diets used. As seen in the rat femur, the accumulated Zn after exposure to excess Zn is rapidly released following removal of Zn from the diet.[97]

During embryonal development Zn accumulated in the liver and body of rat fetuses whose mothers were on diets containing 2000 ppm Zn or more, but to a lesser degree than their mothers themselves, suggesting a placental barrier to the transfer of abnormal amounts of Zn to the fetuses.[29,30,85,138,139] The transfer of Zn from the mammary gland to the milk seems to be limited, too. The Zn content of young rats suckling from mothers fed either 2000 or 5000 ppm Zn was not significantly different.[85]

In growing and adult rats, most of the accumulated liver Zn (60 to 80%) was found in the soluble fraction,[12,20,32,175] bound to a metallothionein-like substance.[20] This was also true for the pancreas.[33] It is assumed that this protein has a primary role in the temporary storage and detoxification of Zn which is similar to ferritin (or hemosiderin) in Fe storage. In livers of pregnant rats and their offspring, in addition to being found in the soluble fraction, Zn was also found deposited in nuclei, mitochondria, and microsomes.[30]

Competition of Zn with Other Elements

In growing and adult rats, chronic ingestion of large amounts of Zn antagonized the intestinal absorption of Cu and Fe. Intestinal absorption of both elements was reduced, as directly seen from absorption studies with ^{64}Cu[164,165] and ^{59}Fe[143] or by reduced blood,[19,26,48,98,174] liver,[19,21,26,29,30,48,85,98,101,103,106,139,175] and whole body Cu,[48] reduced blood[144] and liver Fe,[26,28,30,103,106,143] and reduced Fe content of some other organs.[143] In pregnant rats on high-Zn diets, lower amounts of Cu and Fe were transferred to the fetuses, evidenced by the diminished liver and body contents.[29,30,85,139] Similarly, in young rats suckling from mothers fed high-Zn diets, reduced liver, brain, and whole body Cu and Fe were noted.[21,85]

In growing rats the soluble liver fraction was mainly affected in Cu depletion by dietary Zn levels up to 2000 ppm; however, at higher levels the Cu content of the microsomes was significantly depressed.[175] In pregnant rats only the mitochondria-bound Cu was found to be reduced, whereas the Fe concentration of all four subcellular fractions (soluble fraction, nuclei, mitochondria, microsomes) was affected.[30] Likewise, in the fetal livers the Cu and Fe levels of all subcellular fractions were lowered.[30]

Besides reducing Cu and Fe absorption, it is suggested that Zn also antagonizes the Cu and Fe metabolism at a cellular level. The suggestion was substantiated by the findings of a decreased incorporation of absorbed ^{64}Cu in the liver in conjunction with an increased excretion in the urine,[101] a reduced Fe incorporation into liver ferritin accompanied by an increased uptake of Zn,[24,143] and an enhanced fecal excretion of i.p.-injected ^{59}Fe, owing to a more rapid turnover of the red blood cells caused by Zn.[144]

In rats, Ca, Mg, P, and S metabolism was influenced by ingestion of toxic amounts of Zn. The retention of these elements was markedly reduced due to a higher fecal excretion[132,133,152] or by a lowered transfer from the mothers to the fetuses.[29] Concomitantly, urinary P losses were decreased.[132,133,152] Development and mineralization of the bones was disturbed, particularly when excess Zn was associated with a low Ca intake.[45,71,78,131,152] More Ca accumulated in the heart and brain of pregnant rats and in the liver and brain of weanling rats whose mothers were fed diets with 4000 ppm Zn.[21,29]

Horses

Daily intakes of 800 to 8000 mg Zn as lactate by mares during pregnancy and lactation failed to induce symptoms of illness in either the mares or the suckling foals.[47] In fillies, growth rate decreased markedly after the intake exceeded 90 mg Zn given as oxide per kilogram of body weight per day. These amounts approximated 3600 ppm Zn in the feed on a dry matter basis.[177] As seen in one foal, a dose of 184 mg Zn/kg body weight/day resulted in anemia.[177] Clinical symptoms observed in these animals included swelling of the epiphyseal regions of the long bones, stiffness, and lameness. Similar symptoms were seen in fillies receiving Zn as oxide together with Pb, indicating that Pb is not responsible for these symptoms, which were also noted in foals reared on pastures adjacent to smelters emitting both Zn and Pb.[75]

In fillies, whole blood Zn levels closely paralleled Zn intake when dosages were higher than 60 mg/kg body weight. Organ-Zn levels also rose, especially in the liver but in kidneys, bones, and skeletal muscles as well. Brain and lung Zn levels were not affected.[177]

Ruminants
Dairy Cows

Feeding dietary Zn levels of 1279 ppm during 6 weeks,[110] or daily doses of 8000 mg Zn as oxide during 12 weeks to dairy cows[6] was without any toxic effect. Body weight, food intake, hemoglobin and hematocrit values, activity of lactic dehydrogenase and alkaline phosphatase in serum, and milk production were normal. Plasma Zn levels increased with increasing dietary Zn levels.[110]

In drinking water, 8 ppm Zn were reported to be toxic for dairy cows.[125] The cows suffered from chronic constipation accompanied by a loss in milk yield. After feeding dairy nuts accidentally contaminated by a concentrate producer with 20,000 ppm Zn as oxide, most cows of the herd became seriously ill.[1] The animals suffered from acute enteritis and milk production was dramatically reduced. In some cases death occurred. Post-mortem findings essentially were pulmonary emphysema and degenerative changes in the liver.

As with laboratory animals and horses, especially the livers of Zn-poisoned dairy cows accumulated Zn in large amounts.[1] Also, high Zn levels were noted in dairy cows grazing on pastures adjacent to Zn smelters;[169] concomitantly, liver Cu concentration decreased.[1,169] Increased Zn levels were also found in the kidneys[1] and the milk,[6,110,169] but Zn enrichment of the milk is limited and plateaus with dietary levels substantially higher than 700 ppm.[110]

It is interesting to note that under normal conditions rumen liquor of cows contained 300 to 700 μg Zn/l,[166] a level which was found to be toxic to cellulolytic rumen bacteria in vitro.[72] Like in man, inhalation of ZnO fumes by dairy cows resulted in fever, increased pulse and respiration frequency, dyspnea, and pulmonary emphysema. In acute cases death occurred.[68,82]

Calves

Dietary Zn levels up to 600 ppm as oxide or sulfate were without toxic effects on growth and food intake in calves.[87,111,121,147] Consumption of diets with 900 ppm or higher of Zn in the oxide or carbonate form resulted in growth reduction and lowered feed efficiency. On 3000 ppm diets calves lost weight,[43,120] but no clinical symptoms appeared. After 5 weeks on a 3000 ppm Zn diet hemoglobin and hematocrit values were not affected, but after 12 weeks a 1300 ppm diet was sufficient to produce a reduction.[123] Elevated blood glucose levels were established in calves after feeding them diets with 200 to 400 ppm Zn.[147] Serum Zn levels increased with increasing dietary Zn levels from 100 to 1700 ppm,[112,113,123,149] followed by a plateau at higher dietary concentrations.[123] The same effect was brought about by Zn in the sulfate and oxide forms.[111]

Much more Zn was retained in the body of calves than rats when fed with 200 and 600 ppm diets, as shown by balance studies[112,123] and organ analysis. Especially the pancreas, liver, and kidneys are organs with a striking accumulation capacity.[86,87,112,113,149] This was not confirmed by other calf studies.[43,123]

Like with rats, accumulated liver and pancreas Zn was mainly associated with a 10,000-mol wt protein in the soluble fraction[86,87] to which Cu is simultaneously bound.[10,11,42] These findings may be relevant to the mutual interaction between Zn and Cu. A moderate increase in the Zn concentration was found in gastrointestinal tissues,[86,112,113,149] hair, testes, and ribs.[112,123] The Zn concentrations in muscles, heart, spleen, lungs, and, surprisingly, tibia were not affected.[112,123] Six weeks after transfer to a normal Zn diet, liver Zn was still in progress toward a return to normal values.[123]

As with rats, ingestion of increasing amounts of Zn by calves caused a progressive decrease in serum Cu and Fe and liver Cu and an increase in liver Ca. In contrast to findings in rats, liver Fe was elevated.[86,123] Furthermore, there was a depletion of the liver P.[123]

Sheep

First signs of toxic effects on growth in sheep were caused by 1000 ppm Zn as oxide. At 1500 ppm food intake was affected; 2000 ppm resulted in marked inhibition of growth and at 4000 ppm the animals lost weight.[120] First cases of death occurred at 3000 ppm.[120]

Above 1000 ppm, the pancreas showed a replacement of acinar with connective tissues,[122] similar to the histopathological changes in the pancreas of Zn-fed cats.[36,142] While increasing daily oral doses from 200 to 1000 mg Zn as stearate given during 10 months as a suspension to young sheep and goats was without toxic effects on growth and health,[23] intraruminal doses of 2000 mg Zn as oxide during 11 days in addition to a 1000-ppm Zn diet caused total growth inhibition, and higher doses of 4000 or 6000 mg caused weight losses accompanied by marked decreases in feed and water consumption.[120] With 6000 mg one of two animals died.[120] A diet with 4000 ppm Zn was almost completely refused by sheep which were allowed free choice to feed with increasing Zn levels.[120]

After 7 to 10 weeks on Zn feeding a decline in blood hemoglobin was first seen at 2000 ppm dietary Zn and a reduction in the hematocrit values at 3000 ppm, but strongly marked anemic conditions were not attained with these levels, nor with higher levels up to 4000 ppm.[122] As with rats, an increase in the hemoglobin and hematocrit values was observed after intraruminal application of 4000 or more mg Zn due to a hemoconcentration.[122]

Daily oral doses of 460 mg Zn as stearate[23] or injections of 5 mg/kg body weight Zn as sulfate during pregnancy did not affect reproduction performance.[80]

In lambs, substantially increased Zn levels in serum, liver, pancreas, and kidneys occurred above 1500 ppm dietary Zn, while serum and liver Cu decreased.[122] In sheep on Cu-supplemented diets, a liver Cu depletion was already observed after ingestion of low daily doses of 100 mg Zn.[35] On the other hand, long continued intraruminal application of 2000 mg Zn/day as sulfate in a solved form to pregnant ewes did not change liver Cu, but raised kidney Cu.[64] Also in sheep most liver Cu and Zn is bound to the same low molecular weight proteins in the soluble liver tractions.[10,11] Serum Fe concentrations showed no consistent changes.[122] In contrast to rat findings and in accordance with results in calves, liver Fe increased at dietary Zn levels above 3500 ppm, in part due to a decrease in the organ weight.[122]

Like in rats and calves, Ca accumulated in the sheep liver at these high Zn levels (4000 ppm) but no effect on P, Mg, and Na concentrations was noted.[122] In lambs, Zn accumulated to a moderate degree in bones and only to a small degree in brain, lungs, eyes, and wool.[122] Muscle Zn concentrations were not influenced.[122]

In contrast to rat findings, no toxic effects of high dietary Zn levels up to 4000 ppm on bone mineralization were observed in lambs,[122] although in another experiment with lambs Ca and P retention was highly significantly reduced by 2000 or 4000 ppm Zn owing to a decreased Ca and P absorption and an increased endogenous fecal Ca loss.[159]

High levels of dietary Zn altered the rumen digestion, indicated by an increased pH of the rumen fluid, a drop in the production of total volatile fatty acids, and a lowered acetate:propionate ratio.[122] This is in accordance with the toxic effect of Zn on rumen microorganisms in vitro.[72]

Pigs

In growing pigs dietary Zn levels up to 1000 ppm were well tolerated without toxic effects on growth and feed intake.[15,96,99] With 2000 ppm Zn as carbonate a moderate growth inhibition and reduction in feed intake was seen,[15] but this was not confirmed by other studies using Zn as oxide or sulfate.[27,92] Likewise, a daily dose of 4300 mg Zn as lactate dissolved in skim milk or buttermilk did not affect weight gain.[135] Dietary levels higher than 2000 ppm were necessary to reduce growth and feed intake; however, 3400 and 4000 ppm Zn given as sulfate or oxide caused only moderate effects.[15,92] Above 7000 ppm the animals refused feed intake,[15,92] presumably because of the unpalatability of the feed and in consequence growth was markedly decreased. Feeding of 1000 ppm Zn (lactate) contained in whole milk or skim milk, corresponding to approximately 10,000 ppm on a dry matter basis, resulted in reduced weight gains and appetite.[51,135]

Typical clinical symptoms in Zn-poisoned pigs on diets with Zn concentrations of 2000 ppm or higher were stiffness and lameness caused by a nonspecific arthritis which was characterized by swollen joints. The animals got up on toes with a stilted gait. Post-mortem damages in the articular cartilage of the femur, tibia, and humerus, hemorrhages in the axillary space, ventricles of the brain, lymph nodes, spleen, and the viscera, gastritis, and catarrhal enteritis were observed.[15,50,51,135] Death occurred frequently within 3 weeks of treatment.

In contrast to findings in other mammals, Zn-poisoned pigs exhibited no anemia, as seen by the unchanged hemoglobin values.[15,27,92,135] Red and white blood cell counts were found to be in the normal range[92] or white cells were reduced.[135]

Substantially raised Zn levels in serum, liver, kidneys, pancreas, bones, and bristles of growing pigs and milk of lactating sows already occurred at the low dietary level of 100 ppm.[13,99,128,151] Most of the accumulated liver Zn was found in a low-molecular weight protein called Cu,Zn-thionein.[13,14] With increasing dietary levels, Zn accumulation proceeded in plasma, liver, pancreas, and bristles.[15,27,99] The Zn concentration of the skin[99] and the muscles[92] was not changed.

In contrast to results in rats, dairy cows, calves, and sheep, liver Cu of growing pigs was not reduced by low[13] or toxic dietary Zn concentrations;[27,96] however, the Cu distribution within the proteins of the soluble liver fraction was changed.[13] Serum and kidney Cu also were found to be in the normal range.[96] On Cu-supplemented diets (125 to 750 ppm) low dietary Zn concentrations (100 to 500 ppm) were sufficient to deplete increased liver Cu.[2,7,60,93,128,155] A Mo-induced Cu accumulation in liver, spleen, kidneys, femur, and blood plasma[96] and a Cu-induced parakeratosis with several symptoms of a chronic Cu intoxication[119] were prevented by 1000 and 4000 ppm Zn, respectively. In accordance with findings in rats and turkey poults and in contrast to results in calves and sheep, dietary Zn levels of 4000 ppm caused an Fe depletion of the liver.[27]

Poultry

As indicated in rats, the effect of excess dietary Zn intake in chickens clearly depends on the Cu status of the animals. In Cu-deficient chickens, low dietary Zn levels (50 to 400 ppm) are sufficient to produce growth inhibition, a microcytic, hypochromic anemia with reticulocytosis, reduction in heart cytochrome oxidase activity, subcuteaneous hemorrhages, leg abnormalities, degenerative changes in the elastic tissue of the aorta, and high mortality.[65,66,136] The toxic effects of these low dietary Zn levels were removed in part or completely by the addition of 10 to 50 ppm Cu alone or in combination with 100 to

200 ppm Fe,[65,66,136] whereas these Cu and Fe supplements were insufficient to overcome the toxic effects of higher dietary Zn levels in Cu-deficient diets.[65]

In adequately Cu-supplied growing chickens, dietary Zn had no deleterious effect up to 1000 ppm.[66,83,94,107,129] Dietary concentrations of 1500 ppm and above impaired growth, feed and water consumption, feed efficiency, and development of feathers and bones,[81,83,94,129,158] but recently it was found that 2400 ppm Zn in the oxide form was without effect on growth.[88]

In laying hens, addition of 2320 mg Zn/l of drinking water reduced water consumption and stopped egg production.[154] Growth and egg production returned to normal after removing the Zn from the diet.[83,154] Tolerance of growing chickens to various Zn compounds differs; Zn in the oxide form proved to be somewhat less toxic than in the carbonate and sulfate form,[129,158] and Zn in the acetate and carbonate form somewhat less toxic than in the proteinate form.[94] A sharp rise in chick mortality was noted at 2000 ppm Zn as sulfate,[81] 3000 ppm Zn as carbonate,[129] and 4000 ppm Zn as oxide.[83]

Recently it has become evident that dietary Zn levels of 2000 ppm or more induce a Se deficiency, indicated by the appearance of exudative diathesis, muscular dystrophy, and high mortality. The symptoms were completely prevented by the addition of 0.5 ppm Se to the diet.[81]

It appears that turkey poults on a practical diet tolerate higher dietary Zn levels than chickens. To produce a marked depression in weight gain, 4000 ppm Zn as oxide were necessary; even at a 10,000 ppm level no mortality occurred.[170], Growth inhibition in turkey poults was caused by 2300 ppm Zn as oxide in a purified diet.[171]

Excess Zn-fed chickens accumulated Zn in various organs and tissues, especially in the liver, tibia, and testes and, to a lower degree, in the kidneys and feathers.[83,88,158,171] The amount of Zn deposited in the liver and tibia increased with increasing dietary Zn levels[83,88,158,171] and amount of time on the diet.[88] Zn concentration in the liver and tibia was found to be somewhat lower after feeding Zn in the oxide rather than in the sulfate and carbonate form.[158] In breast muscle and spleen no Zn accumulated.[83,158] Liver Zn returned to normal 6 weeks after removing the Zn from the diet.[83]

Liver Cu of Cu-deficient growing chickens was not affected by 2500 ppm Zn in the diet;[65] 5000 ppm Zn were necessary in chickens on Cu-supplemented diets to cause a reduction.[83] Zn antagonized the intestinal Cu absorption and displaced Cu from a 10,000 mol wt duodenal protein.[15] In turkey poults the Cu level in liver, tibia, beak, heart, kidneys, and feathers was unaffected by 2000 ppm Zn,[171] while the spleen Cu concentration was already increased by 1000 ppm Zn.[83]

In growing chickens, liver Fe was not changed up to 3000 ppm Zn,[158] but on a low-Fe diet Fe storage from an iron supplement was prevented by 2500 ppm Zn.[65] In turkey poults Fe in liver, kidneys, and tibia was depleted by 1000 or 2000 ppm Zn,[171] while bone mineralization was not disturbed by 2000 ppm Zn.[171]

HUMAN STUDIES

Zinc poisoning in man may occur as a result of three different processes: ingestion of toxic amounts of Zn with foods and drinks, inhalation of dusts or fumes of Zn in fairly high amounts, and direct contact of Zn or Zn salts with the skin.[63,163,168] Within the topic of this chapter the first process is of the greatest interest. Zn intoxication by ingestion of high Zn amounts may be caused by intake of food with a high Zn content, intake of foods and drinks which were contaminated with Zn during preparation or storage, or intake of Zn salts in high doses.

The main dietary sources of Zn are meats, fish, and poultry with a contribution of 37% to the total dietary Zn intake. Dairy products and cereals, respectively, provide

20%.[104] An unusually high Zn content can be found in yeast, oysters, and mushrooms.[62] Mussels especially are able to accumulate high amounts of Zn.[161]

After consumption of a selective high-Zn diet prepared from yeast cake, oysters, and bran pudding, no toxic symptoms were reported. An approximate estimate indicated that each subject ingested at this one meal between 225 and 275 mg Zn.[38] In another experiment oysters containing Zn at a level of 1970 ppm on a fresh weight basis were given to four test persons. Only one of them, ingesting 345 mg oyster-zinc, suffered from nausea and diarrhea. But there was no evidence whether the amount of oysters eaten or the quantity of hot spiced sauce consumed with them was responsible for the effects noted.[105] Zn poisoning in man caused by a high Zn content in shellfish has not been reported.[105,161] Injuries by Zn residues in foodstuffs derived from plants and animals are not to be expected.[61,109]

Daily Zn intake from drinking water should also be noticed. The particularly high Zn content in natural water caused by industrial water pollution is largely removed by the plants which prepare drinking water. The real Zn intake from drinking water is determined by the material of the water pipes used: water in polyethylene pipes has the lowest and water in Zn-coated pipes the highest Zn content. In the latter, the accompanying ions like lead and cadmium are also of interest in relation to the heavy metal burden by water intake. It is recommended to take water from a water tap a few minutes after it has been turned on.[69,100,116]

Drinking water quality standards for European countries provide a Zn content not more than 5 mg/l.[3] Exceeding this value may result in an astringent effect, opalescent appearance, and a fine granular sediment. The testable threshold concentration for Zn is reported to be 2 ppm[100] and 15 ppm.[67] Water and drinks containing 7 to 8 ppm were perceived to be bitter.[141] It has been stated that consumption over long periods of time of drinking water containing Zn amounts up to 40 ppm caused no deleterious effects.[62] A Zn concentration of 675 to 2280 ppm is emetic.[67] A postulated carcinogenic effect of Zn in drinking water[55-57] and alcoholic drinks[108] has not been established.[172]

Zn poisoning by ingestion of toxic amounts of Zn sometimes occurred after the preparation or storage of acid food in Zn-coated (galvanized) iron containers.[62,100,163] Instances of mass Zn poisoning were reported after the consumption of a celebration menu,[16] stewed apples,[62] potato salad,[140] limeade,[18] and fruit punch.[16] In all those cases during preparation or storage toxic amounts of Zn were brought into solution. For example, acid rinsing of tubes used for food storage resulted in a solution containing about 2500 ppm Zn.[16] Sometimes, however, difficulties in interpretation arise from the fact that the specificity of the toxicity generally could not be established. It could be shown that toxic amounts of antimony were also brought into solution during the preparation of limeade.[18] Therefore the legislator endeavors to displace Zn in the production of requisites that are handled in direct contact with foodstuffs, whereas Zn alloys like brass and tombac are suited very well for the production of such requisites.[141]

The toxicity of various Zn compounds in man has not been studied systematically.[126] It is mentioned that the range between nutritionally required amounts of Zn and toxic levels seems to be wide.[58,178] If the content of available Zn is at 20%, 11 mg Zn are necessary in the daily diet of man.[178] Doses up to 220 mg Zn sulfate given three times daily to patients for acceleration of wound healing[126] and the treatment of leg ulcers[53,76,77] are well tolerated. No toxic effects on serum electrolytes, urea, full blood count, and liver function could be noted.[53,77] Undesirable side effects after the Zn sulfate treatment leading to discontinuation of therapy also occurred in placebo treatments.[53] Significant differences regarding the side effects, however, occurred when two different pharmaceutical forms of Zn sulfate were used.[53] Gastric irritation and diarrhea could often be avoided when Zn sulfate was given in the middle of the meal or half an hour after.[126,148]

There is also a marked difference between the ability of people to tolerate Zn sulfate therapy. Young patients tolerate it much more poorly.[127] After ingestion of 150 mg Zn vomiting occurs.[141] Other authors stated the emetic dose of Zn sulfate to be 1 to 2 g (225 to 450 mg Zn).[58,100,126]

As much as 12 g elemental Zn mixed with peanut butter was swallowed over 2 days by a 16-year-old boy. A pronounced lethargy developed 3 days after ingestion connected with light-headedness, slight staggering of gait, and difficulty in writing legibly, all suggestive of a cerebellar involvement. Raised serum amylase and lipase values indicated some pancreatic derangement.[117] In an earlier review a lethal dosage of 3 to 5 g Zn chloride and 7 to 8 g Zn sulfate was stated.[40]

The first symptoms of Zn toxicity in humans appeared 2 to 12 hr after ingestion[16,140,168] and were observed as fever, gastroenteric irritation, nausea, vomiting, diarrhea, exhaustion, headache, dizziness, and muscular pain.[16,18,31,34,58,140] An acute renal failure caused by Zn chloride poisoning was reported.[31] Dehydration and electrolyte imbalances are severe consequences of Zn intoxication.[34,58]

For its importance in Zn intoxication, metal fume fever should also be mentioned. Metal fume fever is an acute allergic industrial disease of short duration caused by inhalation of metal oxide fumes. Although the oxides of several metals have been reported to cause metal fume fever, Zn oxide is the most common offender.[124,130,137] The symptoms depend on the concentration of the metal oxide fumes inhaled, the duration of exposure, and the length of time thereafter. They were described as pharyngeal irritation, metallic or sweet taste, depression, nausea, lethargy, cough, fever, headache, salivation, chills, muscular pain, and gastrointestinal complaints associated with repeated episodes of emesis. Also, a high white cell count is typical.[34,59,124,130,137,153,163,168] A raised Zn concentration in blood cells, whole blood, and fasting gastric juice and a slightly higher urinary output of Zn could be observed.[59] In an older study no acute or chronic illness was found in men who had been exposed to Zn oxide at levels of about 3 to 15 mg Zn/m^3 air.[8] It was estimated that metal fume fever will not occur if the air breathed does not contain more than 15 mg Zn/m^3 air.[62,163] In newer reports a limit of 5 mg Zn/m^3 air at the factory is considered acceptable.[69,124]

Exposure to Zn chloride fumes caused a more severe poisoning. A description in more detail is available.[41] Damage to the mucous membranes of the nasopharynx and the respiratory tract were the main effects. Several persons died as a result of shock and a profuse outpouring of secretion water, which clogged the lungs.

Zn poisoning also may occur after thoughtless handling of bronze powder, cosmetics, and sprays containing Zn compounds. Prolonged inhalation can sometimes produce pneumonitis as well as fatal pneumonia.[34,84]

SUMMARY AND CONCLUSIONS

Zn is a relatively nontoxic element to animals and man in comparison to other metals. In animals on diets providing all other nutritional requirements, a wide margin of safety exists between the normal dietary Zn concentrations (40 to 50 ppm) and the threshold level above which first signs of toxicity occur (about 1000 ppm). This fact demonstrates an effective homeostatic control mechanism, operating by changes in intestinal absorption and endogenous excretion of Zn.[25,114] It is suggested, however, that the endogenous excretion is of greater importance in the homeostatic control system because it is an active, energy-dependent transport process.[95] The increasing Zn accumulation in various organs and tissues following the ingestion of diets with increasing Zn levels may be the result of a decreasing effectiveness of the excretory part of the homeostatic control system.

The Zn toxicity syndrome in animals is characterized by poor growth and a microcytic, hypochromic anemia. While growth inhibition is assumed to be due largely, but not entirely, to reduced food consumption probably caused by the unpalatability of the high Zn diets, the anemia results first from a conditioned Cu deficiency and second from an induced Fe deficiency brought about by an interference of high Zn intakes with the absorption and utilization of the metals.[162] Therefore the threshold dietary Zn concentration at which manifestation of Zn toxicity will occur depends on the Cu and Fe stores of the animals and the dietary Cu and Fe levels. The threshold dietary Zn concentration depends further on the dietary Ca, P, and phytic acid concentrations which interfere with the intestinal Zn absorption.[162] Some species-dependent differences in trace element and mineral metabolism changes are to be established between Zn-poisoned animals. Calves accumulate large amounts of Zn at dietary levels of 200 to 600 ppm which are without effect in rats. In rats and turkey poults the liver Fe concentration is reduced while it is increased in calves and sheep. Pigs do not exhibit an anemia, which does develop in other mammals and birds. Liver Cu concentration in pigs is not lowered on diets providing a normal amount of Cu, but only on high-Cu diets. In lambs and turkey poults, disturbances in bone development and mineralization as seen in Zn-poisoned rats are lacking.

In human beings, studies with chronic ingestion of excess Zn have not been carried out. Accidental ingestion of acute toxic amounts of Zn is usually accompanied by vomiting and diarrhea.

REFERENCES

1. **Allen, G. S.,** An outbreak of zinc poisoning in cattle, *Vet. Rec.,* 83, 8–9, 1968.
2. **Allen, M. M., Barber, R. S., Braude, R., and Mitchell, K. G.,** Copper and zinc supplements for fattening pigs, *Proc. Nutr. Soc.,* 17, xii, 1958.
3. **Anon.,** Einheitliche Anforderungen an die Beschaffenheit, Untersuchung und Beurteilung von Trinkwasser in Europa, *Schriften. Ver. Wasser Boden Lufthyg. Berlin Dahlem,* 14b, 1–50, 1971.
4. **Ansari, M. S., Miller, W. J., Lassiter, J. W., Neathery, M. W., and Gentry, R. P.,** Effects of high but nontoxic dietary zinc on zinc metabolism and adaptations in rats, *Proc. Soc. Exp. Biol. Med.,* 150, 534–536, 1975.
5. **Ansari, M. S., Miller, W. J., Neathery, M. W., Lassiter, J. W., Gentry, R. P., and Kincaid, R. L.,** Zinc metabolism and homeostasis in rats fed a wide range of high dietary zinc levels, *Proc. Soc. Exp. Biol. Med.,* 152, 192–194, 1976.
6. **Archibald, J. G.,** Zinc in cow's milk, *J. Dairy Sci.,* 27, 257–261, 1944.
7. **Barber, R. S., Braude, R., and Mitchell, K. G.,** Further studies on antibiotic, copper and zinc supplements for growing pigs, *Br. J. Nutr.,* 14, 499–508, 1960.
8. **Batchelor, R. P., Fehnel, J. W., Thomson, R. M., and Drinker, K. R.,** A clinical and laboratory investigation of the effect of metallic zinc, of zinc oxide, and of zinc sulphide upon the health of workmen, *J. Ind. Hyg.,* 8, 322–363, 1926.
9. **Becker, W. M. and Hoekstra, W. G.,** The intestinal absorption of zinc, in *Intestinal Absorption of Metal Ions, Trace Elements and Radionuclides,* Skoryna, S. C. and Waldron-Edward, D., Eds., Pergamon Press, Oxford, 1971, 229–256.
10. **Bremner, I. and Marshall, R. B.,** Hepatic copper- and zinc-binding proteins in ruminants. I. Distribution of Cu and Zn among soluble proteins of livers of varying Cu and Zn content, *Br. J. Nutr.,* 32, 283–291, 1974.
11. **Bremner, I. and Marshall, R. B.,** Hepatic copper- and zinc-binding proteins in ruminants. II. Relationship between Cu and Zn concentrations and the occurrence of a metallothionein-like fraction, *Br. J. Nutr.,* 32, 293–300, 1974.
12. **Bremner, I. and Davies, N. T.,** The induction of metallothionein in rat liver by zinc injection and restriction of food intake, *Biochem. J.,* 149, 733–738, 1975.
13. **Bremner, I.,** The relationship between the zinc status of pigs and the occurrence of copper- and zinc-binding proteins in liver, *Br. J. Nutr.,* 35, 245–252, 1976.
14. **Bremner, I. and Young, B. W.,** Isolation of (copper, zinc)-thioneins from pig liver, *Biochem. J.,* 155, 631–635, 1976.
15. **Brink, M. F., Becker, D. E., Terrill, S. W., and Jensen, A. H.,** Zinc toxicity in the weanling pig, *J. Anim. Sci.,* 18, 836–842, 1959.

16. **Brown, M. A., Thom, J. V., Orth, G. L., Cova, P., and Juarez, J.,** Food poisoning involving zinc contamination, *Arch. Environ. Health,* 8, 657–660, 1964.

17. **Burke, A. D., Woodson, F., and Heller, V. G.,** The possible toxicity of buttermilk soured in zinc containers, *J. Dairy Sci.,* 11, 79–88, 1928.

18. **Callender, G. R. and Gentzkow, C. J.,** Acute poisoning by the zinc and antimony content of limeade prepared in a galvanized iron can, *Mil. Surg.,* 80, 67–71, 1937.

19. **Campbell, J. K. and Mills, C. F.,** Effects of dietary cadmium and zinc on rats maintained on diets low in copper, *Proc. Nutr. Soc.,* 33, 15A-16A, 1974.

20. **Chen, R. W., Eakin, D. J., and Whanger, P. D.,** Biological function of metallothionein. II. Its role in zinc metabolism in the rat, *Nutr. Rep. Int.,* 10, 195–200, 1974.

21. **Chu, R. C. and Cox, D. H.,** Zinc, iron, copper, calcium, cytochrome oxidase, and phospholipid in rats of lactating mothers fed excess zinc, *Nutr. Rep. Int.,* 5, 61–66, 1972.

22. **Chvapil, M., Peng, Y. M., Aronson, A. L., and Zukoski, C.,** Effect of zinc on lipid peroxidation and metal content in some tissues of rats, *J. Nutr.,* 104, 434–443, 1974.

23. **Cohrs, P.,** Zinkvergiftung bei Haustieren?, *Dtsch. Tieraerztl. Wochenschr.,* 53, 33–34, 1946.

24. **Coleman, C. B. and Matrone, G.,** In vivo effect of zinc on iron-induced ferritin synthesis in rat liver, *Biochim. Biophys. Acta,* 177, 106–112, 1969.

25. **Cotzias, G. C., Borg, D. C., and Selleck, B.,** Specificity of zinc pathway through the body: turnover of Zn^{65} in the mouse, *Am. J. Physiol.,* 202, 359–363, 1962.

26. **Cox, D. H. and Harris, D. L.,** Effect of excess dietary zinc on iron and copper in the rat, *J. Nutr.,* 70, 514–520, 1960.

27. **Cox, D. H. and Hale, O. M.,** Liver iron depletion without copper loss in swine fed excess zinc, *J. Nutr.,* 77, 225–228, 1962.

28. **Cox, D. H. and Harris, D. L.,** Reduction of liver xanthine oxidase activity and iron storage proteins in rats fed excess zinc, *J. Nutr.,* 78, 415–418, 1962.

29. **Cox, D. H., Schlicker, S. A., and Chu, R. C.,** Excess dietary zinc for the maternal rat, and zinc, iron, copper, calcium, and magnesium content and enzyme activity in maternal and fetal tissues, *J. Nutr.,* 98, 459–466, 1969.

30. **Cox, D. H.,** Excess dietary zinc and subcellular changes in hepatic zinc, iron, and copper in maternal and fetal rats, *Nutr. Rep. Int.,* 5, 145–150, 1972.

31. **Csata, S., Gallays, F., and Toth, M.,** Akute Niereninsuffizienz als Folge einer Zinkchloridvergiftung, *Z. Urol.,* 61, 327–330, 1968.

32. **Davies, N. T., Bremner, I., and Mills, C. F.,** Studies on the induction of a low-molecular-weight zinc-binding protein in rat liver, *Biochem. Soc. Trans.,* 1, 985–988, 1973.

33. **Davies, N. T. and Bremner, I.,** Studies on the appearance of a zinc-binding protein in rat pancreas, *Biochem. Soc. Trans.,* 2, 654–656, 1974.

34. **Deichmann, W. B. and Gerarde, H. W.,** *Toxicology of Drugs and Chemicals,* Academic Press, New York, 1969, 142, 638.

35. **Dick, A. T.,** Studies on the assimilation of storage of copper in the crossbred sheep, *Aust. J. Agric. Res.,* 5, 511–544, 1954.

36. **Drinker, K. R., Thompson, P. K., and Marsh, M.,** An investigation of the effect of long-continued ingestion of zinc, in the form of zinc oxide, by cats and dogs, together with observations upon the excretion and the storage of zinc, *Am. J. Physiol.,* 80, 31–64, 1927.

37. **Drinker, K. R., Thompson, P. K., and Marsh, M.,** An investigation of the effect upon rats of long-continued ingestion of zinc compounds, with especial reference to the relation of zinc excretion to zinc intake, *Am. J. Physiol.,* 81, 284–306, 1927.

38. **Drinker, K. R., Fehnel, J. W., and Marsh, M.,** The normal excretion of zinc in the urine and feces of man, *J. Biol. Chem.,* 72, 375–383, 1927.

39. **Duncan, G. D., Gray, L. F., and Daniel, L. J.,** Effect of zinc on cytochrome oxidase activity, *Proc. Soc. Exp. Biol. Med.,* 83, 625–627, 1953.

40. **Esser, A.,** Klinisch-anatomische und spektrographische Untersuchungen des Zentralnervensystems bei akuten Metallvergiftungen unter besonderer Berücksichtigung ihrer Bedeutung für gerichtliche Medizin und Gewerbepathologie. I. Teil: Strontium, Barium, Magnesium-Aluminium, Thorium (radioaktive Stoffe), Thallium-Zink, Cadmium, Quecksilber, *Dtsch. Z. Gesamte Gerichtl. Med.,* 25, 239–317, 1936.

41. **Evans, E. H.,** Casualties following exposure to zinc chloride smoke, *Lancet,* 2, 368–370, 1945.

42. **Evans, G. W., Majors, P. F., and Cornatzer, W. E.,** Mechanism for cadmium and zinc antagonism of copper metabolism, *Biochem. Biophys. Res. Commun.,* 40, 1142–1148, 1970.

43. **Feaster, J. P., Hansard, S. L., McCall, J. T., Skipper, F. H., and Davis, G. K.,** Absorption and tissue distribution of radiozinc in steers fed high-zinc rations, *J. Anim. Sci.,* 13, 781–788, 1954.

44. **Feaster, J. P., van Middelem, C. H., and Davis, G. K.,** Zinc-DDT interrelationships in growth and reproduction in the rat, *J. Nutr.,* 102, 523–527, 1972.
45. **Ferguson, H. W. and Leaver, A. G.,** The effects of diets high in zinc at different levels of calcium and vitamin D on the rat humerus and incisor, *Calcif. Tissue Res.,* 8, 265–275, 1972.
46. **Gesswein, M.,** Über chronische Zinkvergiftung bei Ratten, *Virchows Arch. Pathol. Anat. Physiol.,* 332, 481–493, 1959.
47. **Graham, R., Sampson, J., and Hester, H. R.,** Results of feeding zinc to pregnant mares and to mares nursing foals, *J. Am. Vet. Med. Assoc.,* 97, 41–47, 1940.
48. **Grant-Frost, D. R. and Underwood, E. J.,** Zinc toxicity in the rat and its interrelation with copper, *Aust. J. Exp. Biol. Med. Sci.,* 36, 339–345, 1958.
49. **Gray, L. F. and Ellis, G. H.,** Some interrelationships of copper, molybdenum, zinc and lead in the nutrition of the rat, *J. Nutr.,* 40, 441–452, 1950.
50. **Grimmett, R. E. R. and McIntosh, I. G.,** Suspected zinc poisoning in pigs, *N.Z. J. Agric.,* 53, 34–37, 1936.
51. **Grimmett, R. E. R., McIntosh, I. G., Wall, E. M., and Hopkirk, C. S. M.,** Chronic zinc-poisoning of pigs, *N.Z. J. Agric.,* 54, 216–223, 1937.
52. **Guggenheim, K.,** The role of zinc, copper and calcium in the etiology of the "meat anemia", *Blood,* 23, 786–794, 1964.
53. **Haeger, K., Lanner, E., and Magnusson, P.-O.,** Oral zinc sulfate in the treatment of venous leg ulcers, in *Clinical Applications of Zinc Metabolism,* Pories, W. J., Strain, W. H., Hsu, J. M., and Woosley, R. L., Eds., Charles C Thomas, Springfield, Ill., 1974, 158–167.
54. **Hahn, F. and Schunk, R.,** Untersuchungen ueber die akute Zinkvergiftung, *Arch. Exp. Pathol. Pharmakol.,* 226, 424–434, 1955.
55. **Halme, E.,** On the carcinogenic influence of drinking water and foodstuffs containing zinc, *Vitalst. Zivilisationskr.,* 13, 179–182, 1968.
56. **Halme, E.,** On the carcinogenic influence of drinking water and foodstuffs containing zinc, *Vitalst. Zivilisationskr.,* 13, 230–232, 1968.
57. **Halme, E.,** Kancerogene Wirkung von zinkhaltigem Trinkwasser, *Staedtehygiene,* 20, 174–175 1969.
58. **Halsted, J. A., Smith, J. C., and Irwin, M. I.,** A conspectus of research on zinc requirements of man, *J. Nutr.,* 104, 345–378, 1974.
59. **Hamdi, E. A.,** Chronic exposure to zinc of furnace operators in a brass foundry, *Br. J. Ind. Med.,* 26, 126–134, 1969.
60. **Hanrahan, T. J. and O'Grady, J. F.,** Copper supplementation of pig diets. The effect of protein level and zinc supplementation on the response to added copper, *Anim. Prod.,* 10, 423–432, 1968.
61. **Hapke, H.-J.,** Wirkung und Schaeden durch Blei, Cadmium und Zink bei Nutztieren, *Staub Reinhalt. Luft,* 34, 8–10, 1974.
62. **Hegsted, D. M., McKibben, J. M., and Drinker, K. R.,** The biological, hygienic, and medical properties of zinc and zinc compounds, *Public Health Rep.,* Suppl. 179, 1–44, 1945.
63. **Heller, V. G. and Burke, A. D.,** Toxicity of zinc, *J. Biol. Chem.,* 74, 85–93, 1927.
64. **Hemingway, R. G., Inglis, J. S. S., and Brown, N. A.,** Effects of daily administration of lead acetate and zinc sulphate during pregnancy on the copper, lead and zinc status of ewes and their lambs, *Res. Vet. Sci.,* 5, 7–16, 1964.
65. **Hill, C. H. and Matrone, G.,** A study of copper and zinc interrelationships, in Proc. XIIth World's Poultry Congr., Sydney, Australia, 1962, 219–222.
66. **Hill, C. H., Matrone, G., Payne, W. L., and Barber, C. W.,** In vivo interactions of cadmium with copper, zinc and iron, *J. Nutr.,* 80, 227–235, 1963.
67. **Hinman, J. J., Jr.,** Desirable characteristics of a municipal water supply, *J. Am. Water Works Assoc.,* 30, 484–494, 1938.
68. **Hoffman, E.,** Nytt fall av zinkfoergiftning hos noet till foeljd av svetsningsarbete i ladugård, *Skand. Veterinartidskrift,* 33, 84–87, 1943.
69. **Holtmeier, H. J., Kuhn, M., and Rummel, K.,** *Zink – ein lebenswichtiges Mineral,* Wissenschaftl. Verlagsgesellschaft mbH, Stuttgart, 1976, 24–26, 51.
70. **Hove, E., Elvehjem, C. A., and Hart, E. B.,** Further studies on zinc deficiency in rats, *Am. J. Physiol.,* 124, 750–758, 1938.
71. **Hove, E., Elvehjem, C. A., and Hart, E. B.,** The effect of zinc on alkaline phosphatases, *J. Biol. Chem.,* 134, 425–442, 1940.
72. **Hubbert, F., Cheng, E., and Burroughs, W.,** Mineral requirements of rumen microorganisms for cellulose digestion in vitro, *J. Anim. Sci.,* 17, 559–568, 1958.
73. **Huber, A. M. and Gershoff, S. N.,** Effects of dietary zinc and calcium on the retention and distribution of zinc in rats fed semipurified diets, *J. Nutr.,* 100, 949–954, 1970.

74. **Huber, A. M. and Gershoff, S. N.,** Effects of dietary zinc on zinc enzymes in the rat, *J. Nutr.,* 103, 1175–1181, 1973.
75. **Hupka, E.,** Ueber Flugstaubvergiftungen in der Umgebung von Metallhuetten, *Wien. Tieraerztl. Monatsschr.,* 42, 763–775, 1955.
76. **Husain, S. L.,** Oral zinc sulphate in leg ulcers, *Lancet,* 1, 1069–1071, 1969.
77. **Husain, S. L. and Bessant, R. G.,** Oral zinc sulfate in the treatment of leg ulcers, in *Clinical Applications of Zinc Metabolism,* Pories, W. J., Strain, W. H., Hsu, J. M., and Woosley, R. L., Eds., Charles C Thomas, Springfield, Ill., 1974, 168–180.
78. **Huxley, H. G. and Leaver, A. G.,** The effect of different levels of dietary zinc and calcium upon the zinc concentration of the rat femur and incisor, *Arch. Oral Biol.,* 11, 1337–1344, 1966.
79. **Jaeger, H.,** Ueber Zinkwirkungen, *Arch. Exp. Pathol. Pharmakol.,* 159, 139–153, 1931.
80. **James, L. F., Lazar, V. A., and Binns, W.,** Effects of sublethal doses of certain minerals on pregnant ewes and fetal development, *Am. J. Vet. Res.,* 27, 132–135, 1966.
81. **Jensen, L. S.,** Precipitation of a selenium deficiency by high dietary levels of copper and zinc, *Proc. Soc. Exp. Biol. Med.,* 149, 113–116, 1975.
82. **Johansson, A.,** Fall av zinkfoergiftning hos noet till foeljd av svetsningsarbete i ladugård, *Svensk Veterinaertidskrift,* 32, 163–172, 1942.
83. **Johnson, D., Jr., Mehring, A. L., Jr., Savino, F. X., and Titus, H. W.,** The tolerance of growing chickens for dietary zinc, *Poult. Sci.,* 41, 311–317, 1962.
84. **Kemper, F. and Trautmann, A.,** Inhalation von Zinkoxid, *Dtsch. Med. Wochenschr.,* 97, 307, 1972.
85. **Ketcheson, M. R., Barron, G. P., and Cox, D. H.,** Relationship of maternal dietary zinc during gestation and lactation to development and zinc, iron and copper content of the postnatal rat, *J. Nutr.,* 98, 303–311, 1969.
86. **Kincaid, R. L., Miller, W. J., Gentry, R. P., Neathery, M. W., and Hampton, D. L.,** Intracellular distribution of zinc and zinc-65 in calves receiving high but nontoxic amounts of zinc, *J. Dairy Sci.,* 59, 552–555, 1976.
87. **Kincaid, R. L., Miller, W. J., Fowler, P. R., Gentry, R. P., Hampton, D. L., and Neathery, M. W.,** Effects of high dietary zinc upon zinc metabolism and intracellular distribution in cows and calves, *J. Dairy Sci.,* 59, 1580–1584, 1976.
88. **Kincaid, R. L., Miller, W. J., Jensen, L. S., Hampton, D. L., Neathery, M. W., and Gentry, R. P.,** Effect of high amounts of dietary zinc and age upon tissue zinc in young chicks, *Poult. Sci.,* 55, 1954–1957, 1976.
89. **Kinnamon, K. E.,** Some independent and combined effects of copper, molybdenum, and zinc on the placental transfer of zinc-65 in the rat, *J. Nutr.,* 81, 312–320, 1963.
90. **Kinnamon, K. E. and Bunce, G. E.,** Effects of copper, molybdenum, and zinc on zinc-65 tissue distribution and excretion in the rat, *J. Nutr.,* 86, 225–230, 1965.
91. **Kirchgessner, M. and Pallauf, J.,** Zinkrepletion in Serum und Leber wachsender Ratten. IV. Mitteilung. Zum Stoffwechsel des Zinks im tierischen Organismus, *Z. Tierphysiol. Tierernaehr. Futtermittelkd.,* 29, 77–85, 1972.
92. **Klaus, W., Michna, S., and Neumann, H.-J.,** Sind hoehere Zinksulfatgaben fuer Schweine schaedlich?, *Futter und Fuetterung,* 9, 73–75, 1958.
93. **Kline, R. D., Hays, V. W., and Cromwell, G. L.,** Related effects of copper, zinc and iron on performance, hematology and copper stores of pigs, *J. Anim. Sci.,* 34, 393–396, 1972.
94. **Klussendorf, R. C. and Pensack, J. M.,** Newer aspects of zinc metabolism, *J. Am. Vet. Med. Assoc.,* 132, 446–450, 1958.
95. **Kowarski, S., Blair-Stanek, C. S., and Schachter, D.,** Active transport of zinc and identification of zinc-binding protein in rat jejunal mucosa, *Am. J. Physiol.,* 226, 401–407, 1974.
96. **Kulwich, R., Hansard, S. L., Comar, C. L., and Davis, G. K.,** Copper, molybdenum and zinc interrelationships in rats and swine, *Proc. Soc. Exp. Biol. Med.,* 84, 487–491, 1953.
97. **Leaver, A. G.,** An effect of vitamin D upon the uptake and release of zinc by bone, *Arch. Oral Biol.,* 12, 773–775, 1967.
98. **Lee, D., Jr. and Matrone, G.,** Iron and copper effects on serum ceruloplasmin activity of rats with zinc-induced copper deficiency, *Proc. Soc. Exp. Biol. Med.,* 130, 1190–1194, 1969.
99. **Lewis, P. K., Jr., Hoekstra, W. G., and Grummer, R. H.,** Restricted calcium feeding versus zinc supplementation for the control of parakeratosis in swine, *J. Anim. Sci.,* 16, 578–588, 1957.
100. **Lindner, E.,** *Toxikologie der Nahrungsmittel,* Georg Thieme Verlag, Suttgart, 1974, 114–115.
101. **Magee, A. C. and Matrone, G.,** Studies on growth, copper metabolism and iron metabolism of rats fed high levels of zinc, *J. Nutr.,* 72, 233–242, 1960.
102. **Magee, A. C.,** Effect of zinc toxicity on rats fed various supplements, *Fed. Proc.,* 22, 437, 1963.
103. **Magee, A. C. and Spahr, S.,** Effects of dietary supplements on young rats fed high levels of zinc, *J. Nutr.,* 82, 209–216, 1964.

104. Mahaffey, K. R., Corneliussen, P. E., Jelinek, C. F., and Fiorino, J. A., Heavy metal exposure from foods, *Environ. Health Perspect.*, 12, 63–69, 1975.

105. Mannell, W. A., Effect of oysters with a high zinc content on cats and man, *Food Cosmet. Toxicol.*, 5, 841–842, 1967.

106. McCall, J. T., Mason, J. V., and Davis, G. K., Effect of source and level of dietary protein on the toxicity of zinc to the rat, *J. Nutr.*, 74, 51–57, 1961.

107. Mehring, A. L., Jr., Brumbaugh, J. H., and Titus, H. W., A comparison of the growth of chicks fed diets containing different quantities of zinc, *Poult. Sci.*, 35, 956–958, 1956.

108. McGlashan, N. D., Zinc and oesophageal cancer, *Lancet*, 1, 578, 1967.

109. Michels, S., Croessmann, G., and Scholl, G., Ueber die Kontamination von Nahrungsmitteln mit Schwermetallen, *Staub Reinhalt. Luft*, 34, 24–26, 1974.

110. Miller, W. J., Clifton, C. M., Fowler, P. R., and Perkins, H. F., Influence of high levels of dietary zinc on zinc in milk, performance and biochemistry of lactating cows, *J. Dairy Sci.*, 48, 450–453, 1965.

111. Miller, W. J., Blackmon, D. M., Hiers, J. M., Jr., Fowler, P. R., Clifton, C. M., and Gentry, R. P., Effects of adding two forms of supplemental zinc to a practical diet on skin regeneration in Holstein heifers and evaluation of a procedure for determining rate of wound healing, *J. Dairy Sci.*, 50, 715–721, 1967.

112. Miller, W. J., Blackmon, D. M., Gentry, R. P., and Pate, F. M., Effects of high but nontoxic levels of zinc in practical diets on ^{65}Zn and zinc metabolism in Holstein calves, *J. Nutr.*, 100, 893–902, 1970.

113. Miller, W. J., Wells, E. S., Gentry, R. P., and Neathery, M. W., Endogenous zinc excretion and ^{65}Zn metabolism in Holstein calves fed intermediate to high but nontoxic zinc levels in practical diets, *J. Nutr.*, 101, 1673–1681, 1971.

114. Miller, W. J., Dynamics of absorption rates, endogenous excretion, tissue turnover, and homeostatic control mechanisms of zinc, cadmium, manganese, and nickel in ruminants, *Fed. Proc.*, 32, 1915–1920, 1973.

115. Minato, A. and Ogiso, T., Studies on metalloproteins. IX. The effects of excessive dietary zinc on serum copper and oxidase activity of ceruloplasmin, *Yakugaku Zasshi*, 86, 521, 1966; *Chem. Abstr.*, 65, 9420, 1966.

116. Moeller, H. and Wilk, G., Der Gehalt von Spurenelementen in Saeuglingsnahrungen und Infusionsloesungen, in *Spurenelemente in der Entwicklung von Mensch und Tier*, Betke, K. and Bidlingmeier, F., Eds., Urban & Schwarzenberg, München, 1975, 137–148.

117. Murphy, J. V., Intoxication following ingestion of elemental zinc, *JAMA*, 212, 2119–2120, 1970.

118. Myers, V. C., Beard, H. H., and Barnes, B. O., Studies in the nutritional anemia of the rat. IV. The production of hemoglobinemia and polycythemia in normal animals by means of inorganic elements, *J. Biol. Chem.*, 94, 117–122, 1931–1932.

119. O'Hara, P. J., Newman, A. P., and Jackson, R., Parakeratosis and copper poisoning in pigs fed a copper supplement, *Aust. Vet. J.*, 36, 225–229, 1960.

120. Ott, E. A., Smith, W. H., Harrington, R. B., and Beeson, W. M., Zinc toxicity in ruminants. I. Effect of high levels of dietary zinc on grains, feed consumption and feed efficiency of lambs, *J. Anim. Sci.*, 25, 414–418, 1966.

121. Ott, E. A., Smith, W. H., Harrington, R. B., and Beeson, W. M., Zinc toxicity in ruminants. II. Effect of high levels of dietary zinc on gains, feed consumption and feed efficiency of beef cattle, *J. Anim. Sci.*, 25, 419–423, 1966.

122. Ott, E. A., Smith, W. H., Harrington, R. B., Stob, M., Parker, H. E., and Beeson, W. M., Zinc toxicity in ruminants. III. Physiological changes in tissues and alterations in rumen metabolism in lambs, *J. Anim. Sci.*, 25, 424–431, 1966.

123. Ott, E. A., Smith, W. H., Harrington, R. B., Parker, H. F., and Beeson, W. M., Zinc toxicity in ruminants. IV. Physiological changes in tissues of beef cattle, *J. Anim. Sci.*, 25, 432–438, 1966.

124. Papp, J. P., Metal fume fever, *Postgrad. Med.*, 43, 160–163, 1968.

125. Pickup, J., Worden, A. N., Bunyan, J., and Wood, E. C., Chronic constipation in dairy cattle associated with a high level of zinc in the water supply, *Vet. Rec.*, 66, 93–94, 1954.

126. Pories, W. J., Henzel, J. H., Rob, C. G., and Strain, W. H., Acceleration of wound healing in man with zinc sulphate given by mouth, *Lancet*, 1, 121–124, 1967.

127. Pories, W. J., Discussion remark, in *Clinical Applications of Zinc Metabolism*, Pories, W. J., Strain, W. H., Hsu, J. M., and Woosley, R. L., Eds., Charles C Thomas, Springfield, Ill., 1974, 179.

128. Ritchie, H. D., Luecke, R. W., Baltzer, B. V., Miller, E. R., Ullrey, D. E., and Hoefer, J. A., Copper and zinc interrelationships in the pig, *J. Nutr.*, 79, 117–123, 1963.

129. Roberson, R. H. and Schaible, P. J., The tolerance of growing chicks for high levels of different forms of zinc, *Poult. Sci.*, 39, 893–896, 1960.

130. **Rohrs, L. C.,** Metal-fume fever from inhaling zinc oxide, *Arch. Ind. Hyg. Occup. Med.,* 16, 42–47, 1957.
131. **Sadasivan, V.,** Studies on the biochemistry of zinc. I. Effect of feeding zinc on the liver and bones of rats, *Biochem. J.,* 48, 527–530, 1951.
132. **Sadasivan, V.,** Studies on the biochemistry of zinc. II. The effect of intake of zinc on the metabolism of rats maintained on a stock diet, *Biochem. J.,* 49, 186–191, 1951.
133. **Sadasivan, V.,** Studies on the biochemistry of zinc. III. Further investigation on the influence of zinc on metabolism, *Biochem. J.,* 52, 452–455, 1952.
134. **Salant, W. and Wise, L. E.,** The production of glycosuria by zinc salts, *J. Biol. Chem.,* 34, 447–462, 1918.
135. **Sampson, J., Graham, R., and Hester, H. R.,** Studies on feeding zinc to pigs, *Cornell Vet.,* 37, 225–236, 1942.
136. **Savage, J. E., Bird, D. W., and O'Dell, B. L.,** Accentuation of copper deficiency pathology by zinc, *Poult. Sci.,* 42, 1304, 1963.
137. **Sayers, R. R.,** Metal fume fever and its prevention, *Public Health Rep.,* 53, 1080–1086, 1938.
138. **Schlicker, S. A. and Cox, D. H.,** Maternal dietary zinc in excess, fetal development and iron and copper metabolism, *Fed. Proc.,* 26, 520, 1967.
139. **Schlicker, S. A. and Cox, D. H.,** Maternal dietary zinc, and development and zinc, iron, and copper content of the rat fetus, *J. Nutr.,* 95, 287–294, 1968.
140. **Schmidt-Lange, W.,** Zinkvergiftungen durch Kartoffelsalat, *Medizinische,* 1, 842–844, 1955.
141. **Schormueller, J.,** *Lehrbuch der Lebensmittelchemie,* 2nd ed., Springer-Verlag, Berlin, 1974, 118, 770.
142. **Scott, D. A. and Fisher, A. M.,** Studies on the pancreas and liver of normal and of zinc-fed cats, *Am. J. Physiol.,* 121, 253–260, 1938.
143. **Settlemire, C. T. and Matrone, G.,** In vivo interference of zinc with ferritin iron in the rat, *J. Nutr.,* 92, 153–158, 1967.
144. **Settlemire, C. T. and Matrone, G.,** In vivo effect of zinc on iron turnover in rats and life span of the erythrocyte, *J. Nutr.,* 92, 159–164, 1967.
145. **Shah, B. G., Meranger, J. C., Belonje, B., and Deshmukh, D. S.,** Zinc retention in young rats fed increasing levels of high-zinc oysters, *J. Fish. Res. Board Can.,* 28, 843–848, 1971.
146. **Smith, S. E. and Larson, E. J.,** Zinc toxicity in rats. Antagonistic effects of copper and liver, *J. Biol. Chem.,* 163, 29–38, 1946.
147. **Sommer, E., Zalewska, E., and Cakala, S.,** The effect of zinc on some immune parameters in beef calves, *Bull. Vet. Inst. Pulawy,* 19, 32–37, 1975.
148. **Spencer, H.,** Discussion remark, in *Clinical Applications of Zinc Metabolism,* Pories, W. J., Strain, W. H., Hsu, J. M., and Woosley, R. L., Eds., Charles C Thomas, Springfield, Ill., 1974, 179.
149. **Stake, P. E., Miller, W. J., Gentry, R. P., and Neathery, M. W.,** Zinc metabolic adaptations in calves fed a high but nontoxic zinc level for varying time periods, *J. Anim. Sci.,* 40, 132–137, 1975.
150. **Starcher, B. C.,** Studies on the mechanism of copper absorption in the chick, *J. Nutr.,* 97, 321–326, 1969.
151. **Stevenson, J. W. and Earle, I. P.,** Zinc in sow's colostrum and milk, *J. Anim. Sci.,* 23, 300, 1964.
152. **Stewart, A. K. and Magee, A. C.,** Effect of zinc toxicity on calcium, phosphorus and magnesium metabolism of young rats, *J. Nutr.,* 82, 287–295, 1964.
153. **Sturgis, C. C., Drinker, P., and Thomson, R. M.,** Metal fume fever. I. Clinical observations on the effect of the experimental inhalation of zinc oxide by two apparently normal persons, *J. Ind. Hyg.,* 9, 88–97, 1927.
154. **Sturkie, P. D.,** The effects of excess zinc on water consumption in chickens, *Poult. Sci.* 35, 1123–1124, 1956.
155. **Suttle, N. F. and Mills, C. F.,** Studies of the toxicity of copper to pigs. I. Effects of oral supplements of zinc and iron salts on the development of copper toxicosis, *Br. J. Nutr.,* 20, 135–148, 1966.
156. **Sutton, W. R. and Nelson, V. E.,** Studies on zinc, *Proc. Soc. Exp. Biol. Med.,* 36, 211–213, 1937.
157. **Sutton, W. R. and Nelson, V. E.,** Blood sugar changes in the rat produced by salts of beryllium, magnesium, and zinc with some observations on hemoglobin and red blood corpuscles, *Proc. Iowa Acad. Sci.,* 45, 115–121, 1938.
158. **Tahara, S., Iida, I., Shoji, K., and Aso, K.,** Effects of high levels of different forms of zinc on growth of chicks, *Jpn. J. Zootech. Sci.,* 33, 304–310, 1962.

159. **Thompson, A., Hansard, S. L., and Bell, M. C.,** The influence of aluminum and zinc upon the absorption and retention of calcium and phosphorus in lambs, *J. Anim. Sci.,* 18, 187–197, 1959.

160. **Thompson, P. K., Marsh, M., and Drinker, K. R.,** The effect of zinc administration upon reproduction and growth in the albino rat, together with a demonstration of the constant concentration of zinc in a given species, regardless of age, *Am. J. Physiol.,* 80, 65–74, 1927.

161. **Tiews, H.,** Welches Ausmass hat die Oelverschmutzung der Meere und deren Verunreinigung durch sonstige schaedliche Stoffe erreicht und welche Auswirkungen haben diese Verunreiningungen?, in *Zur Sache 3/71,* Presse- und Informationszentrum des Deutschen Bundestages, Bonn, 1971, 169–173.

162. **Underwood, E. J.,** *Trace Elements in Human and Animal Nutrition,* 3rd ed., Academic Press, New York, 1971, 218, 242.

163. **Vallee, B. L.,** Biochemistry, physiology and pathology of zinc, *Physiol. Rev.,* 30, 443–490, 1959.

164. **Van Campen, D. R.,** Effects of zinc, cadmium, silver and mercury on the absorption and distribution of copper-64 in rats, *J. Nutr.,* 88, 125–130, 1966.

165. **Van Campen, D. R. and Scaife, P. U.,** Zinc interference with copper absorption in rats, *J. Nutr.,* 91, 473–476, 1967.

166. **Van Koetsveld, E. E.,** Das Zink in der Tierernaehrung, besonders fuer Rindvieh, *Z. Tierphysiol. Tierernaehr. Futtermittelkd.,* 16, 318–322, 1961.

167. **Van Reen, R.,** Effects of excessive dietary zinc in the rat and the interrelationship with copper, *Arch. Biochem. Biophys.,* 46, 337–344, 1953.

168. **Van Reen, R.,** Zinc toxicity in man and experimental species, in *Zinc Metabolism,* Prasad, A. S., Ed., Charles C Thomas, Springfield, Ill., 1966, 411–426.

168a. **Venugopal, B. and Luckey, T. P.,** Toxicology of non-radioactive heavy metals and their salts, in *Environmental Quality and Safety,* Suppl. Vol. 1, Coulston, F. and Corte, F., Eds., Georg Thieme Verlag, Stuttgart, 1975, 4–73.

169. **Vetter, H. and Maehlhop, R.,** Untersuchungen ueber Blei-, Zink- und Fluor-Immissionen und dadurch verursachte Schaeden an Pflanzen und Tieren, *Landwirtsch. Forsch.,* 24, 294–315, 1971.

170. **Vohra, P. and Kratzer, F. H.,** Zinc, copper and manganese toxicities in turkey poults and their alleviation by EDTA, *Poult. Sci.,* 47, 699–704, 1968.

171. **Vohra, P., Gottfredson, G. D., and Kratzer, F. H.,** The effects of high levels of dietary EDTA, zinc or copper on the mineral contents of some tissue of turkey poults, *Poult. Sci.,* 47, 1334–1343, 1968.

172. **Walters, M. and Roe, F. J. C.,** A study of the effects of zinc and tin administration orally to mice over a prolonged period, *Food Cosmet. Toxicol.,* 3, 271–276, 1965.

173. **Waltner, K. and Waltner, K.,** Ueber die Wirkung einiger Metalle. II. Mitteilung., *Arch. Exp. Pathol. Pharmakol.,* 146, 310–312, 1929.

174. **Whanger, P. D. and Weswig, P. H.,** Effect of some copper antagonists on induction of ceruloplasmin in the rat, *J. Nutr.,* 100, 341–348, 1970.

175. **Whanger, P. D. and Weswig, P. H.,** Effect of supplementary zinc on the intracellular distribution of hepatic copper in rats, *J. Nutr,* 101, 1093–1097, 1971.

176. **Wilkins, J. H.,** A note on the toxicity of zinc chloride, *Vet. Rec.,* 60, 81–84, 1948.

177. **Willoughby, R. A., MacDonald, E., McSherry, B. J., and Brown, G.,** Lead and zinc poisoning and the interaction between Pb and Zn poisoning in the foal, *Can. J. Comp. Med.,* 36, 348–359, 1972.

178. **W.H.O.,** Trace elements in human nutrition. Report of a W.H.O. expert committee, *W.H.O. Tech. Rep. Ser.,* No. 532, 9–15, Geneva, 1973.

NUTRIENT TOXICITIES IN ANIMALS AND MAN: SELENIUM

John L. Martin

Although the inadequacy of selenium probably presents a far greater health problem to animals and man than an overabundance, the dangers of selenium toxicity must be thoroughly recognized.

The ingestion of seleniferous plants by livestock, with the consequent development of well-defined disorders, has been a problem to stockmen and farmers for many years. Perhaps the first report describing a disease syndrome resulting from the ingestion of seleniferous plants was that of Marco Polo during his famous travels to the Orient. In 1295 he[1] recorded in his journals that beasts of burden were afflicted by a peculiar disorder when they fed on a particular plant which grew in western China near the border of Turkestan and Tibet. One of the symptoms he described was that the hoofs of the affected animals dropped off. Only recently has it become evident that this disorder is characteristic of a type of chronic selenosis.

In 1860 Dr. T. C. Madison,[2] an army surgeon stationed at Fort Randall, (then located in the Nebraska Territory but now a part of South Dakota near the Nebraska border) published a sanitary report in which he described a fatal disease in horses grazing near the fort. This disease was of a chronic nature and began about 10 days after the horses were moved into the area. Losses continued for several months but ceased when other forage was provided. Again, one of the characteristic symptoms exhibited by the horses was extreme tenderness and inflammation of the feet. These symptoms were accompanied by loss of hair from the mane and tail. This account may not have represented the only instance in which cavalry horses suffered from selenium toxicosis. It has even been speculated that this ailment played an important role in the disaster suffered by General Custer and his troops at the Little Big Horn.[3] In this military engagement, a relief expedition failed to reach the beleaguered troops of General Custer in time to provide needed support. The officer in command of this expedition wrote in his official report that a peculiar sickness affected his horses and was responsible for the delay. This is a plausible explanation since the route taken by the relief expedition ran through seleniferous regions.

In the early 1930s, experiments by Franke[4] and Robinson[5] definitely established that selenium present in grains, grasses, and weeds was highly toxic to grazing livestock. Livestock ingesting seleniferous plants generally develop one of two types of selenosis. "Alkali disease," characterized by retarded growth, emaciation, deformed hoofs, loss of hair, arthritis, and eventual death, is a chronic disorder and results from animals consuming seleniferous grains and grasses, as in Figure 1. The selenium in such plants is present largely as selenoamino acids of the protein type, i.e., selenocystine and selenomethionine.[6-10] Selenium content in such plants seldom runs higher than 35 ppm.

"Blind staggers," the other type of selenosis, is a more acute disorder characterized by signs of central nervous system impairment. Affected animals stumble, bump, and push against objects in their path. Weakness and inappetence are also present. Unless there is recovery, affected animals become paralyzed prior to death, as in Figure 2. Blind staggers results from animals consuming seleniferous weeds such as certain species of the genus *Astragalus,* some of which contain extraordinarily high concentrations of selenium, sometimes reaching several thousand ppm. (See Figure 3.) The selenium is present in these plants primarily as selenoamino acids of the nonprotein type, i.e., selenocystathionine and Se-methylselenocysteine.[11-18]

Toxicity symptoms resulting from the administration of toxic levels of either selenite or selenate tend to resemble those of blind staggers rather than those of alkali disease.

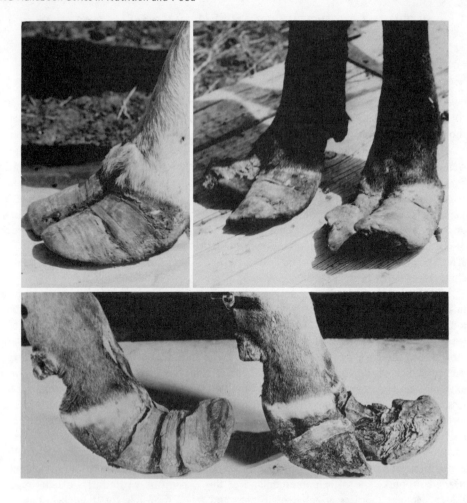

FIGURE 1.   Alkali disease in horses showing various degrees of hoof damage. Note the marked separation of the coronary band at the upper right hoof. (By courtesy of O. E. Olson and from Rosenfeld, I. and Beath, O. A., *Selenium,* Academic Press, New York, 1964, 157. With permission.)

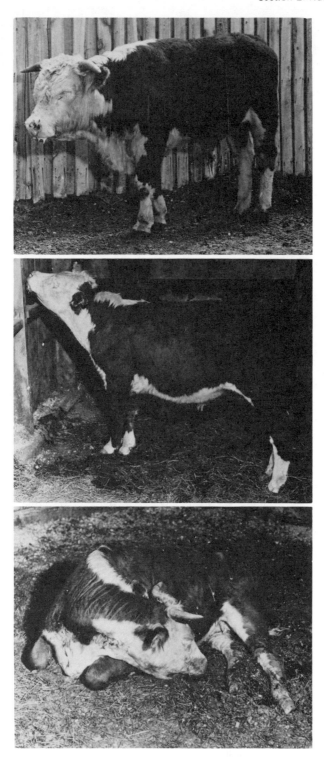

FIGURE 2.   Blind staggers in cattle showing the three stages of the
disease in increasing severity from top to bottom. Note salivation
and closure of eyes in all stages and emaciation and paralysis in the
final stage. (From Rosenfeld, I. and Beath, O. A., *Selenium,*
Academic Press, New York, 1964, 148. With permission.)

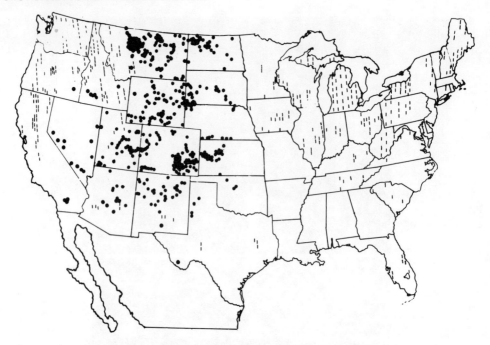

FIGURE 3.   Distribution of vegetation containing more than 50 ppm selenium, in relation to distribution of occurrence of white muscle disease in livestock. Dots represent seleniferous vegetation, lines represent white muscle disease occurrence. (From Rosenfeld, I. and Beath, O. A., *Selenium*, Academic Press, New York, 1964, 10. Base map copyright, American Map Co., Inc., New York. License no. 18196. Used with permission.)

This resemblance in toxicity symptoms would be expected if it is true that alkali disease is the result of the consumption of abnormally high quantities of selenoamino acids of the protein type. It is now accepted in most circles that animals in general are incapable of synthesizing selenoamino acids from either selenate or selenite.[19-21] However, in ruminants in which gastrointestinal microflora abound and which are capable of synthesizing selenoamino acids from inorganic selenium, it might be possible for bovine to exhibit symptoms of alkali disease following the administration of toxic levels of either selanate or selenite. Maag et al.[22] observed no clinical symptoms of alkali disease in any of the 13 Hereford steers administered toxic quantities of selenite. Recently, however, Olson and Embry[23] reported that of five Hereford heifers fed diets containing toxic levels of selenite, one animal developed selenium toxicity symptoms similar to those of alkali disease.

The quantity of selenium accumulated in tissues depends on the chemical form in which the selenium is administered. For instance, Smith et al.[24] found that the concentrations of selenium in various tissues of rabbits fed seleniferous oats were many times higher than those found in the tissues of rabbits fed comparable amounts of selenium as selenite. They found 7 times more selenium in the liver, 15 times more in the skin, and 29 times more in skeletal muscle, which comprises the bulk of the body protein.

In recent studies by Martin and Hurlbut,[25,26] mice were fed diets supplemented with toxic levels of selenite, Se-methylselenocysteine, selenocystine, or selenomethionine. The accumulation and retention of tissue selenium in mice fed diets supplemented with selenite was nearly identical to that of mice fed diets supplemented with either Se-methylselenocysteine or selenocystine — even though both the latter are organo-selenium compounds and selenocystine is the selenium analog of cystine, a protein amino acid. Conversely, both the accumulation and retention of tissue selenium was much

FIGURE 4.   Cirrhosis of liver (left). The external surface has a coarsely granular, hobnail appearance. Early cirrhosis (center) and normal liver (right). (By courtesy of O. E. Olson and from Rosenfeld, I. and Beath, O. A., *Selenium,* Academic Press, New York, 1964, 196. With permission.)

greater in mice fed selenomethionine-supplemented diets. This was particularly true in skeletal muscle in which there was negligible accumulation of selenium in mice fed diets supplemented with selenite, Se-methylselenocysteine, or selenocystine. The selenium concentration in the muscle of mice fed toxic levels of selenomethionine rose to 20 ppm, however. Muscle containing such a high level of selenium would have to be considered highly toxic if eaten by other animals. The point is that meat of livestock fed selenium-supplemented diets would be entirely safe for consumption even from animals supplemented with toxic levels of selenium unless supplemented with selenomethionine, which would be extremely unlikely.

Extensive tissue damage results from both acute and chronic selenoses and most investigators agree that the most severe and consistent degeneration occurs in those organs which concentrate selenium to the greatest extent, that is, the liver and kidneys. (See Figure 4.) The characteristic gross pathological changes in animals suffering from alkali disease, blind staggers, and experimental selenoses were summarized by Rosenfeld and Beath.[27]

The observations by Glenn et al.[28] are somewhat contradictory to other reports. These investigators fed toxic quantities of sodium selenate to sheep and observed that most tissue damage occurred in the myocardium, with only occasional degenerative changes in the liver and kidneys. Because of the structural and functional similarities between cardiac and skeletal muscle, the researchers also compared the pathological changes in these tissues. They found that cardiac muscle, which accumulated more than twice as much selenium as did skeletal muscle, suffered much more severe degenerative changes.

The toxicity of selenium depends on many factors such as chemical form, animal species, quality of diet, and the presence of certain additives. Except for hydrogen selenide, selenite is generally regarded as the most toxic chemical form of selenium. Of almost equal magnitude are the selenoamino acids and selenate. It is quite likely that the selenoamino acids are not toxic *per se* but become toxic only after their biological oxidation to higher oxidation state selenium compounds.[25,29] Organic selenium compounds whose sulfur analogs are not normal sulfur metabolites exhibit less toxicity. Dimethyl selenide, which results from the biological detoxification of selenite, is about 1/500 as toxic as selenite.[30] Selenium in its most toxic form is toxic when supplied at a level of 1 to 5 mg selenium per kilogram body weight.[31-33]

The body attempts to detoxify the selenite by reductive methylation which occurs in the liver. Under most conditions the most important excretory pathway is via the kidneys and the principal urinary excretory product is the trimethylselenonium ion.[34-35] When selenite is administered in doses approaching toxic levels, detoxification occurs by the formation of volatile dimethyl selenide that is eliminated by pulmonary excretion. In animals having functional rumen, fecal elimination is an important excretory pathway.[36]

Various agents have been shown to act in a protective manner against selenium toxicity. The more prominent agents are arsenic, linseed oil meal, and a combination of methionine and vitamin E.

Early investigations by Moxon[37] demonstrated that chronic and acute selenosis produced by seleniferous grains were minimized by the administration of arsenite. Until recently, the exact mechanism by which arsenic counteracted selenium toxicity was somewhat obscure. Ganther and Baumann[38] and Levander and Baumann[39,40] reported that arsenic increased the gastrointestinal excretion of selenium in rats. They further found that this increased excretion was accompanied by a corresponding decrease in the retention of selenium in the rats' livers. Moreover, these investigators observed that gastrointestinal excretion of selenium was elevated because there was an increase in its clearance from the liver into the bile.

Moxon[37] and Halverson[41] discovered that linseed oil meal provides protection against chronic selenosis. The active principle, a nonprotein factor, was found to reduce the incidence of liver injury caused by selenium but not to lower the concentration of selenium in that organ.[42] Experiments by Levander et al.[43] provided evidence that linseed oil meal apparently lessens the effects of chronic selenosis by binding the selenium in a less toxic form.

Experiments by Levander and Morris[44] demonstrated that neither methionine nor vitamin E alone give much protection against liver damage from excess selenium, but that combinations of methionine and vitamin E are active and the level of protection is roughly proportional to the level of E that is added to the diet. These researchers postulated that methionine detoxifies selenium by forming certain methylated derivatives of the element, which are then eliminated either via the breath or the urine, and that vitamin E and some fat-soluble antioxidants increase the availability of the methyl group of methionine for this process. It was recently reported[45] that combinations of vitamin E and vitamin C significantly reduce the toxicity of selenite selenium.

Numerous investigations have been carried out showing that selenium administered as inorganic selenium becomes rather firmly attached to all types of proteins from milk proteins[46,47] to keratins.[48,49] Until recently, however, it was not known whether administered selenite or selenate become bound to protein because they react with sulfhydryl groups or if they are reduced to selenoamino acids that are then incorporated into the proteins through peptide bonds. Although the biosynthesis of selenoamino acids from selenite in animal systems is postulated in the literature,[49-51] some investigators began to doubt if such synthesis could take place since, except perhaps in the cat family, there is no known pathway for the synthesis of sulfur amino acids from inorganic sulfur in monogastric animals.

Following the lead of Schwarz and Sweeney,[52] Cummins and Martin[19] were able to demonstrate that neither selenocystine nor selenomethionine was present in liver protein or urine of rabbits fed a selenite-supplemented diet. The results of Cummins and Martin were quickly confirmed by Jenkins et al.[20,21]

A brief summary of the metabolic fate of selenite as it acts as a toxicant is presented below. Selenite rapidly finds its way into the general circulation regardless of its mode of administration. In the general circulation it quickly but loosely combines with plasma albumin. As such it is carried throughout the body. Before it becomes more firmly attached to the various globular fractions it can freely detach itself from the albumin and easily enter the various cells of the body. In the cells it exerts its toxic action by catalyzing the oxidation of important sulfhydryl metabolites, including sulfhydryl-

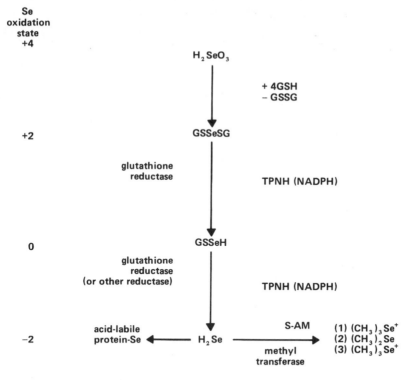

FIGURE 5.   Pathways for reductive methylation of selenite. (From Ganther, H. E. and Hsieh, H. S., *Trace Element Metabolism in Animals,* Vol. 2, University Park Press, Baltimore, 1974. With permission.)

containing enzymes, thus inactivating them. It has recently been shown that the enzymic activity of fatty acid synthetase, a sulfhydryl-containing enzyme, is inhibited in vivo by selenite.[53] Furthermore, the excessive requirement for NADPH, which is needed for selenite detoxification, may soon exceed the availability of this important reductant, particularly when unusually high quantities of selenite need to be detoxified. The depletion of NADPH would obviously lead to serious metabolic imbalances adding to the toxic nature of selenium. The metabolic scheme whereby selenite is reduced to dimethyl selenide and trimethyl selenonium in liver tissue has been elucidated by Ganther and Hsieh.[54] (See Figure 5.)

An interesting biochemical property of selenium is its interaction with such toxic metals as mercury and cadmium to significantly reduce their toxicity. Ganther et al.[55] reported that the selenium normally found in tuna significantly lessens the toxicity methylmercury which might be contaminating the tuna. Parizek et al.[56] have recently reviewed this entire area of interaction between selenium and toxic metals.

Although many selenium compounds are clearly toxic and care must be taken to prevent accidental overdose, selenium is readily excreted. Therefore there is no chance of its accumulation in tissues of animals administered subtoxic levels. In muscle tissues there is virtually no selenium deposition regardless of its level of supplementation, except if selenomethionine is supplemented, in which case selenium levels can become quite high.

REFERENCES

1. Polo, M., *The Travels of Marco Polo*, Komroff, M., Ed., Liveright, New York, 1926, 81.
2. Madison, T. C., *U.S. Congr., 36th, 1st Session, Senate Exch. Doc.*, 52, 37, 1860.
3. Selwood, P. W., *General Chemistry*, Henry Holt, New York, 1954, 236.
4. Franke, K. W., A new toxicant occurring in certain samples of plant foodstuffs. I. Results obtained in preliminary feeding trials, *J. Nutr.*, 8, 597, 1934.
5. Robinson, W. O., Determination of selenium in wheat and soils, *J. Assoc. Off. Agr. Chem.*, 16, 423, 1933.
6. Franke, K. W., A new toxicant occurring naturally in certain samples of plant foodstuffs. II. The occurrence of the toxicant in the protein fraction, *J. Nutr.*, 8, 609, 1934.
7. Painter, E. P. and Franke, K. W., Selenium in proteins from toxic foodstuffs. II. The effect of acid hydrolysis, *Cereal Chem.*, 13, 172, 1936.
8. Smith, A. L., Separation of the Selenium Compounds in Seleniferous Plant Protein Hydrolysates by Paper Partition Chromatography, M.S. thesis, South Dakota College of Agriculture, Brookings, S.D., 1949.
9. Peterson, P. J. and Butler, G. W., The uptake and assimilation of selenite by higher plants, *Aust. J. Biol. Sci.*, 15, 126, 1962.
10. Olson, O. E., Novacek, E. J., Whitehead, E. I., and Palmer, I. S., Investigations on selenium in wheat, *Phytochemistry*, 9, 1181, 1970.
11. Trelease, S. F., Disomma, A. A., and Jacobs, A. L., Seleno-amino acids found in *Astragalus bisulcatus*, *Science*, 132, 3427, 1960.
12. Shrift, A. and Virupaksha, T. K., Seleno-amino acids in selenium-containing accumulating plants, *Biochim. Biophys. Acta*, 100, 65, 1965.
13. Virupaksha, T. K. and Shrift, A., Biochemical differences between selenium accumulator and non-accumulator *Astragalus* species, *Biochim. Biophys. Acta*, 107, 69, 1965.
14. Horn, M. J. and Jones, D. B., Isolation from *Astragalus pectinatus* of a crystalline amino acid complex containing selenium and sulfur, *J. Biol. Chem.*, 139, 649, 1941.
15. Virupaksha, T. K. and Shrift, A., Biosynthesis of selenocystathionine from selenate in *Stanleya pinnata*, *Biochim. Biophys. Acta*, 74, 791, 1963.
16. Martin, J. L. and Gerlach, M. L., Separate elution by ion-exchange chromatography of some biologically important selenoamino acids, *Anal. Biochem.*, 29, 257, 1969.
17. Martin, J. L., Shrift, A., and Gerlach, M. L., Field use of ^{75}Se-Selenite for the study of selenium metabolism in *Astragalus*, *Phytochemistry*, 10, 945, 1971.
18. Nigam, S. N. and McConnell, W. B., Seleno amino compounds from *Astragalus bisulcatus* isolation and identification of a x-L-glutamyl-Se-methyl-L-cysteine and Se-methylseleno-L-cysteine, *Biochim. Biophys. Acta*, 192, 185, 1969.
19. Cummins, L. M. and Martin, J. L., Are selenocystine and selenomethionine synthesized in vivo from sodium selenite in mammals?, *Biochemistry*, 6, 3162, 1967.
20. Jenkins, K. J., Evidence for the absence of selenocystine and selenomethionine in the serum proteins of chicks administered selenite, *Can. J. Biochem.*, 46, 1417, 1968.
21. Jenkins, K. J., Hidirogblou, M., and Ryan, J. F., Intravascular transport of selenium by chick serum proteins, *Can. J. Physiol. Pharmacol.*, 47, 459, 1969.
22. Maag, D. D., Osborn, J. S., and Clopton, J. R., The effect of sodium selenite on cattle, *Am. J. Vet. Res.*, 21, 1049, 1960.
23. Olson, O. E. and Embry, L. B., Chronic selenite toxicity in cattle, *Proc. S. D., Acad. Sci.*, 52, 50–58, 1973.
24. Smith, M. I., Westfall, B. B., and Stohlman, E. F., Studies on the fate of selenium in the organisms, *U.S. Public Health Report*, 53, 1199, 1938.
25. Martin, J. L. and Hurlbut, J. A., Tissue selenium levels and growth responses of mice fed selenomethionine, Se-methylselenocysteine or sodium selenite, *Phosphorus and Sulfur*, 1, 295–300, 1976.
26. Martin, J. L. and Hurlbut, J. A., Difference between the metabolic fate of selenomethionine and selenocysteine, *Fed. Proc.*, 34(3), 924, (Abstract), 1975.
27. Rosenfeld, I. and Beath, O. A., *Selenium*, Academic Press, New York, 1964, 146.
28. Glenn, M. W., Martin, J. L., and Cummins, L. M., Sodium selenate toxicosis: the distribution of selenium within the body after prolonged feeding of toxic quantities of sodium selenate to sheep, *Am. J. Vet. Res.*, 25, 1495, 1964.
29. Martin, J. L. and Gerlach, M., Selenium metabolism in animals, *Ann. N.Y. Acad. Sci.*, 192, 193, 1972.
30. McConnell, K. P. and Portman, O. W., Toxicity of dimethylselenide in the rat and mouse, *Proc. Soc. Exp. Biol. Med.*, 79, 230, 1952.

31. Smith, M. I. and Lillie, R. D., The chronic toxicity of naturally occurring food selenium, *Nat. Inst. Health Bull.,* 174, 1, 1940.

32. Smith, M. I., Stohlman, E. F., and Lillie, R. D., The toxicity and pathology of selenium, *J. Pharmacol. Exp. Therap.,* 60, 449, 1937.

33. Franke, K. W. and Moxon, A. L., A comparison of the minimum fatal doses of selenium, tellurium, arsenic and vanadium, *J. Pharmacol. Exp. Therap.,* 58, 454, 1936.

34. Byard, J. L., Trimethyl selenide: a urinary metabolic of selenite, *Arch. Biochem. Biophys.,* 130, 556, 1969.

35. Palmer, I. S., Fischer, D. D., Halverson, A. W., and Olson, O. E., Identification of a major selenium excretory product in rat urine, *Biochim. Biophys. Acta,* 177, 336, 1969.

36. Cousins, F. B. and Cairney, I. M., Aspects of selenium metabolism in sheep, *Aust. J. Agric. Res.,* 12, 927, 1961.

37. Moxon, A. L., The effect of arsenic on the toxicity of seleniferous grains, *Science,* 88, 81, 1938.

38. Ganther, H. E. and Baumann, C. A., Selenium metabolism. I. Effects of diet, arsenic and cadmium, *J. Nutr.,* 77, 210, 1962.

39. Levander, O. A. and Baumann, C. A., Selenium metabolism. V. Studies on the distribution of selenium in rats given arsenic, *Toxicol. Appl. Pharmacol.,* 9, 98, 1966.

40. Levander, O. A. and Baumann, C. A., Selenium metabolism. VI. Effect of arsenic on the excretion of selenium in the bile, *Toxicol. Appl. Pharmacol.,* 9, 106, 1966.

41. Halverson, A. W., Hendrick, C. M., and Olson, O. E., Observations on the protective effect of linseed oil meal and some extracts against chronic selenium poisoning in rats, *J. Nutr.,* 56, 51, 1955.

42. Olson, O. E. and Halverson, A. W., Effect of linseed oil meal and arsenicals on selenium poisoning in the rat, *Proc. S. D. Acad. Sci.,* 33, 90, 1954.

43. Levander, O. A., Young, M. L., and Meeks, S. A., Studies on the binding of selenium by liver homogenates from rats fed diets containing either casein or casein plus linseed oil meal, *Toxicol. Appl. Pharmacol.,* 16, 79, 1970.

44. Levander, O. A. and Morris, V. C., Interactions of methionine, vitamin E and antioxidants in selenium toxicity in the rat, *J. Nutr.,* 100, 1111, 1970.

45. Hill, C. H., Interaction of Ascorbic Acid and vitamin E in alleviating selenium toxicity in chicks, *Fed. Proc.,* 35, 577 (abstract), 1976.

46. McConnell, K. P., Passage of selenium through the mammary glands of the white rat and the distribution of selenium in the milk proteins after subcutaneous injection of sodium selenate, *J. Biol. Chem.,* 173, 653, 1948.

47. Kiermeier, F. and Wigand, W., Selenium content of milk and milk powder, *Z. Lebensm. Unters. Forsch.,* 139, 205, 1969.

48. McConnell, K. P. and Kreamer, A. E., Incorporation of selenium-75 into dog hair, *Proc. Soc. Exp. Biol. Med.,* 105, 170, 1960.

49. Rosenfeld, I., Biosynthesis of seleno-compounds from inorganic selenium by sheep, *Proc. Soc. Exp. Biol. Med.,* 11, 670, 1962.

50. McConnell, K. P. and Wabnitz, C. H., Studies on the fixation of radioselenium in proteins, *J. Biol. Chem.,* 226, 765, 1957.

51. McConnell, K. P., Kreamer, A. E., and Roth, D. M., Presence of selenium-75 in the mercapturic fraction of dog urine, *J. Biol. Chem.,* 234, 2932, 1959.

52. Schwarz, K. and Sweeney, E., Selenite binding to sulfur amino acids, *Fed. Proc.,* 23, 421, 1964.

53. Donaldson, W. E., Selenium toxicity possible mechanism of action, *Fed. Proc.,* 35, 577, (abstract), 1976.

54. Ganther, H. E. and Hsieh, H. S., Mechanisms for the conversion of selenides in mammalian tissues, in *Trace Element Metabolism in Animals – 2,* Hoekstra, W., Ed., University Park Press, Baltimore, 1974, 339.

55. Ganther, H. E., Gaudie, C., Sunde, M. L., Kopecky, J. J., Wagner, P., Sang, H. O., and Hoekstra, W. G., Selenium relation to decreased toxicity of methylmercury added to diets containing tuna, *Science,* 175, 1122, 1972.

56. Parizek, J., Kalouskova, A., Babicky, J., Bennes, J., and Pavlik, L., Interaction of Selenium with Mercury, Cadmium, and Other Toxic Metals, in *Trace Element Metabolism in Animals – 2,* Hoekstra, W., Ed., University Park Press, Baltimore, 1974, 119–131.

NUTRIENT TOXICITIES IN ANIMALS AND MAN: CHROMIUM

G. L. Romoser

INTRODUCTION

Man is living in an environment which, for his continual well-being, is becoming more and more industrialized. Under proper and realistically implemented controls, this industrial progress (if it can be termed such) should be of no serious consequence. It should, indeed, be most beneficial to mankind.

An undesirable by-product of a disorderly, uncontrolled industrial development is environmental pollution, which can assume many forms. One of these is widespread dispersion of certain toxic elements such as mercury, cadmium, vanadium, lead, chromium, and others.

The ubiquity of chromium and its continuing importance in the furtherance of the needs of an advancing civilization makes this element an intriguing one. This element seems to possess properties which have biomedical applications of great benefit to man. On the other hand, under certain conditions and in certain chemical forms it seems to be detrimental to the well-being of certain animals and man.

In this section the nutritional and toxicological properties of this interesting element for certain species of animals and man will be explored.

DISTRIBUTION

Chromium is a metallic element which was discovered in 1797 by the French chemist Louis Nicholas Vauquelin (1763—1829). Vauquelin named the element chromium (from Greek *chroma*, meaning ''color'') because of the many different colors of its compounds.

This element, a common one which is estimated to make up about 1/3000 of the earth's crust, is surpassed in abundance by only seven other metals. The chromium industry in the United States uses primarily chromite ($FeOCr_2O_3$) for the production of alloys of chrome and various chemicals containing chrome. Although there are deposits of chromite in the U.S., located primarily in Utah, California, and Montana, the ore is no longer mined domestically. Economics are now such that the ore must be imported. High-grade ore deposits are mined in the U.S.S.R., South Africa, the Philippines, Turkey, and Southern Rhodesia.

During the past two decades, sophistication in the field of microanalysis has increased tremendously. For example, the development of such techniques as atomic absorption spectrophotometry, emission spectroscopy, neutron-activation analysis, and gas-liquid chromotography have made possible the detection of chromium at levels below 1 ppb (part per billion). Its presence has been detected in virtually all living matter as well as in the air, soil, marine and surface waters.

CHARACTERISTICS

Although chromium has oxidation states ranging from Cr^{-2} to Cr^{+6}, it most commonly occurs as Cr^0, Cr^{+2}, Cr^{+3}, and Cr^{+6}. Chromium or its various salts are responsible for the characteristic colors of some precious and semiprecious gems such as the ruby and the garnet.

Other chemical and physical properties of chromium are listed below.

A. Chemical
   Symbol   Cr
   Atomic Number   24
   Atomic weight   51.996
   Melting point   1890°C
   Boiling point   2482°C
   Specific gravity   7.18
B. Physical
   Lustrous
   Hard
   Steel grey color
   Metallic
   Resistant to cor-
     rosion and tarnish

USES

Chromium has many industrial uses, e.g., as pigments (in the bichromate or lead forms) in the leather tanning process, in mordant dyeing, metallurgy, and in the production of corrosion-resistant decorative trim used on automobiles and appliances. Chromium or its salts alone or in combination with other metals or their salts is also used as a catalyst in some chemical processes.

Of all of the chromite ore processed in the U.S., it is estimated that approximately half is used in metallurgical applications and the remainder in refactories and by the chemical industry. Most of the ore is used in the production of stainless steel.[1]

NUTRITIONAL VALUE OF CHROMIUM

Chromium, while not presently considered to be essential for the chick, has, nevertheless, been shown by Schwartz and Mertz[2] to play an essential role in improving the metabolism of glucose in rats fed certain types of diets. Its physiological role in the diet of man is still in question, but some response in glucose tolerance to dietary chromium by diabetics has been reported.

Four diabetic subjects were treated with chromium administered in daily doses of between 180 and 1,000 μg/day. Three showed marked improvement of oral glucose tolerance after several weeks of treatment. Two subjects were observed as outpatients for an extended period of time. One showed improvement from the chromium administration, whereas the other was unaffected. No reduction of hypoglycemic medication was permitted in any of the patients. Chromium supplementation was not considered to be a therapeutic agent in the treatment of diabetes.[3] In cases of impaired glucose tolerance in nondiabetic geriatric patients, four out of ten treated with 150 μg/day of chromium for a prolonged period of time responded positively to treatment.[4]

Positive responses to chromium supplementation have been observed in children in Jordan, Nigeria, and Turkey suffering from an impairment in glucose tolerance.[5,6] However, in a study conducted in Egypt, malnourished children suffering from impaired glucose tolerance failed to respond to chromium administration. It was found that they had had a high dietary intake of chromium and high blood chromium levels. Therefore, it was postulated, that the impaired glucose tolerance was due to some factor other than a chromium deficiency.[7]

The feeding of a torula yeast diet low in available chromium to rats impaired their tolerance to intravenously injected glucose. Glucose removal rates dropped approximately 50% within three weeks after the animals were placed on the experimental diets. The defect did not occur in the presence of dietary chromium and it could be reversed in approximately 8 hr when 20 to 50 μg of chromium were administered via stomach tube.[8] Likewise, the feeding of chromium-deficient laboratory chow to monkeys led to impairment of glucose tolerance. The addition of 10 μg of chromium per liter of drinking water prevented the impairment.[9]

Severe deficiencies of chromium have been produced in rats and mice by taking extreme precautions to preclude chance contamination. The addition of as little as 5 mg chromium per liter of water was sufficient to decrease the mortality of male mice. The life span of the last surviving 10% of experimental rats was increased 110% from chromium supplementation, and the treated animals grew at a more rapid rate than their counterparts receiving the chromium-deficient diets.[10,11]

In chromium-deficient rats, an elevation of circulating cholesterol was observed which could be reduced by the dietary administration of chromium. Furthermore, there was a significantly lower incidence of aortic plaques in rats which received chromium in the diet compared with the negative controls.[12]

Surprisingly little research has been done in poultry nutrition with this element, either as it relates to metabolic function or toxicity. In one of the few published reports on the subject it was observed that broiler strain chicks could tolerate levels of 30 to 100 ppm Cr as Na_2CrO_4 in a practical diet which was fed to the test animals from the 11th to the 32nd day of age.[13]

TOXICITY OF CHROMIUM IN ANIMALS

Hexavalent chromium (Cr VI) is a highly toxic form of chromium and is a strong oxidizing agent. It will penetrate cell membranes readily and causes skin ulcerations. On the other hand, the trivalent (Cr III) form of chromium is considerably less toxic for animals and appears to be limited in its toxicological aspects to parenteral administration.

The LD_{50} for Cr III for the rat when injected intravenously is 10 mg per kilogram body weight.[8] Oral levels (Cr VI) in the range of 50 ppm have depressed the growth rate and damaged the livers and kidneys of experimental animals.[14] No ill effects were observed in mice or rats fed levels of Cr III of 2 and 5 ppm for their complete life cycle. There was no increased incidence of tumors as compared to the control animals and, in fact, the females receiving the chromium supplement seemed to be more resistant to lung infection than the controls.[10,11]

In other animals it was found that Cr III in food or water had little effect. For example, dogs which were fed 10 to 15g of chromium as chromic chloride exhibited severe gastrointestinal distress (vomiting and diarrhea). However, they recovered rapidly.[15] In mice, high levels of Cr III salts are tolerated. Parenteral administration of 2.29 and 0.8 g/kg of chromic acetate and chromic chloride, respectively, were required to produce mortality.[16]

A summary of several important studies relating to the toxicology of various compounds of chromium for animals is presented in Table 1.

TOXICITY OF CHROMIUM IN MAN

Data on the nutritional toxicity of chromium and its dietary intake by man is limited. Furthermore, in view of differences in analytical techniques among laboratories, there have been variances in estimates of dietary intakes of chromium.

Table 1
TOXICOLOGY OF VARIOUS COMPOUNDS OF CHROMIUM FOR ANIMALS

Species	Compound tested	Level(s) used	Method of administration	Observations	Ref.
Dog	K bichromate	5.2 mg/kg as Cr	IV	Temporary decrease in renal filtration and reabsorption capacity during 2nd and 3rd hour after injection; prolonged decrease in nephron function between 8th and 24th hour after injection.	17
	Cr VI	5.2 mg/kg/day	IV	Fatal at 4—6 days Anuria, azotemia, hyperglycemia in terminal stage and hyalinosis and sclerosis of renal tissue	17
			OD	Fatal at 80—159 days Anuria, azotemia, hyperglycemia in terminal stage and hyalinosis and sclerosis of renal tissue	17
	Na di-chromate	1—10 mg/kg/day	OD	After 1—2 years — decreased levels of glycogen, SH groups, decreased activity of cytochrome oxidase, succinate dehydrogenase, NAD diaphorase, and choline esterase; increased activity of monoamine oxidase in the myocardium and myofibrosis and sclerosis of small branches of coronary arteries	18
	Potassium chromate	10 g	IV (single dose)	Fatal	19
	K di-chromate	3 mg/100 cc blood	IV (2 doses)	Marked renal damage	20
		210 mg as Cr	IV (single dose)	Rapid death	15
	Chromic chloride	0.8 g/kg	SC (single dose)	Fatal	15
Guinea pig	$CrCl_3$	1 mg/kg as Cr	OD	After 9 days — inhibition of cytochrome oxidase and succinate dehydrogenase in brain, kidney, muscle, liver, and thyroid gland	21
	$K_2C_2O_4$	0.5% solution	TA	Dermatitis, 12—14 days; decrease in monocytes, increase in lymphocytes	22
	$CrCl_3$	4.7 µg/kg/day	OD	After 9 days — increased oxygen demand by liver, kidney, brain, thyroid gland, and skeletal muscle	23
	Cr III salts	1.0 mg/kg/day	OD	After 9 days — decreased respiration by brain, kidney, and thyroid	23
			TA	Ability of Cr III salts to elicit allergic reaction in Cr-sensitive animals decreased gradually ($CrCl_3$ > $Cr(NO_3)_3$ > $Cr_2(SO_4)_3$ > $Cr(OAc)_3$ > Cr oxalate)	24
Mouse	Potassium di-chromate	10 mg	SC (single dose)	Fatal — renal damage	25
	Cr VI	5 ppm	DW	Administration throughout life span depressed growth rate regardless of sex; all tumors found in treated mice at necropsy were malignant	26
	FeCr catalyst	Various	IG	LD_{50} of 6.95 g/kg; histological examination revealed protein dystrophy in liver and kidneys; lower doses after 18 days revealed loss of weight, increase in oxygen consumption, increase in weight: body	26

TOXICOLOGY OF VARIOUS COMPOUNDS OF CHROMIUM FOR ANIMALS

Species	Compound tested	Level(s) used	Method of administration[a]	Observations	Ref.
	Zinc chromate	0.1 mg/dose	IV (1 dose/month for 10 months)	No apparent toxicity	27
	Barium chromate	0.75 mg	IV (single dose)	Fatal	27
		2.5 mg/dose	IV (9 doses at 6-week intervals)	No apparent toxicity	27
	Chromic chloride	0.8 g/kg	IV (single dose)	Minimum lethal dose	28
	Chromic acetate	2.29 g/kg	IV (single dose)	Minimum lethal dose	28
	Chromic sulfate	0.085 g/kg	IV (single dose)	Minimum lethal dose	29
Rabbit	Cr VI	1.5—5 mg/liter	DW	Decrease in hemoglobin, erythrocyte count, and transaminase activity in blood	30
	$Cr_2(SO_4)_3$	10—100 µg/kg as Cr	SC	Inhibition of blood catalase activity	31
	CrO_3	720 ppm as Cr VI	IO	Slight or no irritating effect on conjunctiva when administered in tap water	32
	Chromic chloride	0.52 g/kg	SC (single dose)	Fatal	15
	Potassium dichromate	1.5 cc of a 1% soln./kg body wt.	SC (single dose)	80% died — renal damage	33
		20 mg	SC (single dose)	Fatal — renal damage	26
		0.5—1 cc of a 0.5% soln./kg body wt.	SC (single dose)	Nephritis — renal damage	26
Rat	NH_4CrO_4	0.05 mg/m³	Inhalation (2 hr, alternate days for 6 months)	Proliferation of epithelium, bronchial papillary growth in bronchii, pneumonia	26
	$K_2Cr_2O_7$	13 mg/kg as Cr over a period of 90 days	DW	After 90 days — loss of weight, hair, hair luster, skin desquimation; after 180 days — depression of hemoglobin and leukocytes	34
		17 mg/kg as Cr	SC	Signs of intoxication (0.5 hr)	34
		25 mg/kg as Cr	SC	Lethal (0.5 hr)	34
	K bichromate	20 mg/kg as Cr	SC	After 180 days — significantly lowered ascorbic acid content of kidneys	34
		13 mg/kg as Cr	OD	After 180 days — significantly lowered ascorbic acid content of kidneys	34
	Alumino Cr (VI) catalyst	Dust	TA/IN	Ascorbic acid levels in adrenal glands decreased (more pronounced with treatment period [30, 60, 90 days])	34

Table 1 (continued)
TOXICOLOGY OF VARIOUS COMPOUNDS OF CHROMIUM FOR ANIMALS

Species	Compound tested	Level(s) used	Method of administration[a]	Observations	Ref.
	$K_2Cr_2O_7$	5 mg	SC	Increased uptake of p-amino hippurate by renal cortical slices 24 hr after injection	14
		25 mg	SC	Gluconeogenesis by kidney slices inhibited and blood urea nitrogen increased 24 hr after injection	14
		15 mg/kg	IC	Bimodal excretion of Tamm-Horsfall mucoprotein within 12 hr; renal tubular necrosis and cast formation at 24—72 hr	34
	Chrome-alum/ K_2CrO_4	400 µg as Cr	SC	Daily administration of Cr III or Cr VI produced hypertrophy of liver, kidney, and spleen; fatty liver, abscesses, and proteinuria; distribution of Cr VI in tissues greater than Cr III; Cr VI more toxic at lower levels than Cr III	34
	K bichromate	1 µg Cr/g	SC	Alternate daily administration for 25 days increased growth of transplanted sarcoma, but increased survival time; Cr content of blood, spleen, kidneys and muscles increased; significantly decreased in brain	34
	Cr VI	3.5 mg/l	DW	After 81 days (mean intake 114 µg/day) — spleen, kidneys, and liver tissue contained 1690, 230, and 76 µg Cr/g, respectively	34
	CrO_3	72 ppm as Cr VI	OD	Administration of single dose of 5—20 mg tap water containing 72 ppm as Cr VI had no effect on growth, organ weights, hematological or enzymatic activities, or tissue histology	34
	Potassium chromate	500 ppm	DW	Maximum nontoxic concentration	35
	Zinc chromate	1%	OD	Maximum nontoxic concentration	35

DW, drinking water; IC, intracutaneous; IG, intragastric; IN, inhalation; IO, intraoccular; IV, intravenous; OD, oral dose; SC, subcutaneous; TA, topical application.

Table 2
DAILY INTAKES OF CHROMIUM IN VARIOUS DIETS

Diet or food (quantity)	Chromium, μg	Area
Institutional	78	Vermont
Institutional	52	Syracuse, N.Y.
Hospital	101	Vermont
Ad libitum	65	Syracuse, N.Y.
Dinner(200 g)	64	Rome, Italy
Self-selected	130—140	Japan
Self-selected	200—400	Cincinnati, Ohio
Vegetarian	11—55	India
Sugar cane (500 g)	100	South Africa
Lentil soup (200 ml)	38	Egypt
Maize (1 l)	10	Nigeria
Milk powder (1 l)	18	Jordan

Reproduced from Medical and Biologic Effects of Environmental Pollutants; Chromium, page 11, with the permission of the National Academy of Sciences, Washington, D.C.

Table 3
CHROMIUM CONCENTRATION OF VARIOUS FOODS[29,38,39]

Food	No. samples	Cr,μg/g (wet wt) Mean	Range
Condiments	6	2.71	0.01—10
Meat	9	0.13	0.03—0.27
Fish	2	0.02	0.01—0.02
Vegetables	36	0.18	0—3.62
Grains	15	0.13	0—0.52
Cereal products	10	0.22	0.05—0.23
Vegetable oils	12	0.08	0.03—0.23
Fruits	9	0.01	0—0.2
Sugars and syrups	65	0.10	0—1.1
Frozen T.V. dinners	3	0.09	0.05—0.13

Dietary intakes of chromium have been estimated from very low values[16] to as much as 400 μg/day.[36] Calculated daily intakes of chromium from different areas have been published by many investigators. Some of these values are shown in Table 2.[37]

Foods have been reported to contain chromium in various quantities; condiments, meats, vegetables, and cereal products contain higher levels than fruits, vegetable oils, and fish. Table 3 shows some of the reported findings for chromium in different foods.

At the present time, little is known of the impact of chromium in the food chain or its dispersion in the environment from a nutritional standpoint. More information is needed relative to its detection and role in the physiology of man. This knowledge will then permit the nutritionist to make intelligent recommendations relative to the requirement for chromium and its most available and beneficial dietary sources.

While little is known about the "nutritional" toxicity of chromium, the toxicological effects of industrial exposure to different chromium compounds is well known. In man as well as in animals, metallic chromium is innocuous and biologically inert, whereas its compounds (especially in the hexavalent state) are the major causes of health problems. Such problems, reported to be direct results of chromium exposure, include a wide range of diseases, from mild dermatitis to lung cancer.

SUMMARY

Chromium is considered to be an essential element in the diet of man and also in several species of animals. In animals, deficiencies have been produced which resulted in an impairment in their tolerance to glucose. Similarly, certain diabetic patients and subjects with low levels of chromium in their diets have responded positively to orally and parenterally administered chromium.

Lack of standardized analytical techniques for chromium in foods and clinical specimens has, at the present time, precluded the establishment of dietary recommendations for this element.

Hexavalent chromium is more rapidly absorbed than other forms and is more corrosive and irritating than other valent states. In animals and man, extreme exposure to the salts of chromium, especially the hexavalent forms, is fatal. Lower levels can cause skin dermatitis, ulcers, perforated nasal septum, and kidney malfunction. Prolonged exposure (15 to 20 years) of industry workers to chromium salts has revealed a predisposition to lung cancer.

REFERENCES

1. National Research Council, Trends in Usage of Chromium. Report of the Panel on Chromium of the Committee on Technical Aspects of Critical and Strategic Materials, *National Materials Advisory Board Report* N.M.A.B. 256, National Academy of Sciences, Washington, D.C., 1970.
2. **Schwartz, K. and Mertz, W.,** Chromium (III) and the glucose factor, *Arch. Biochem. Biophys.,* 85, 292—295, 1959.
3. **Glinsman, W. H. and Mertz, W.,** Effect of trivalent chromium on glucose tolerance, *Metab. Clin. Exp.,* 15, 510, 1966.
4. **Levine, R. A., Streeten, D. H., and Doisy, R. J.,** Effects of all chromium supplementation on the glucose tolerance of elderly human subjects, *Metab. Clin. Exp.,* 17, 114, 1968.
5. **Hopkins, L. L., Jr., Ransome-Kuti, O., Majaj, A. S.,** Improvement in impaired carbohydrate metabolism by chromium (III) in malnourished infants, *Am. J. Clin. Nutr.,* 21, 203—211, 1968.
6. **Gurson, C. T. and Saner, G.,** Effect of chromium on glucose utilization in marasmic protein calorie nutrition, *Am. J. Clin. Nutr.,* 21, 1313—1319, 1971.
7. **Carter, J. P., Katteb, A., Abd-El-Hadi, K., Davis, J. T., El Gholmy, A., and Pathwardhan, Z. N.,** Chromium (III) in hypoglycemia and in impaired glucose utilization in kwashiorkor, *Am. J. Clin. Nutr.,* 21, 195-202, 1968.
8. **Mertz, W., Roginski, E. E., and Reba, R. C.,** Biological activity and fate of trace quantities of intravenous chromium in the rat, *Am. J. Physiol.,* 209, 489-494, 1965.
9. **Davidson, I. W. F. and Blackwell, W. L.,** Changes in carbohydrate metabolism in squirrel monkeys with chromium dietary supplementation, *Proc. Soc. Exp. Biol. Med.,* 127, 66-70, 1968.
10. **Schroeder, H. A., Balassa, J. J., and Vinton, W. H., Jr.,** Chromium, cadmium and lead in rats. Effects on life span, and tissue levels, *J. Nutr.,* 86:51-66, 1965.
11. **Schroeder, H. A., Balassa, J. J., and Vinton, W. H., Jr.,** Chromium, lead, cadmium, nickel and titanium in mice. Effect on mortality, tumors and tissue levels, *J. Nutr.,* 83, 239-250, 1964.
12. **Schroeder, H. A. and Balassa, J. J.,** Influence of chromium, cadmium and lead on rat aortic lipids and circulating cholesterol, *Am. J. Physiol.,* 209, 433—437, 1965.
13. **Romoser, G. L., Loveless, L., Machlin, L. J., and Gordon, R. S.,** Toxicity of vanadium and chromium for the growing chicken, *Poult. Sci.,* 39, 1288, 1960.
14. **MacKenzie, R. D., Byerrum, R. U., Decker, C. F., Hoppert, C. A., and Langham, R. F.,** Chronic toxicity studies. II. Hexavalent and trivalent chromium administered in drinking water to rats, *Arch. Ind. Health,* 18, 232-234, 1958.

15. **Brard, D.**, *Toxicologie du Chrome,* Hermann and Co., Paris, 1935.
16. **Schroeder, H. A., Balassa, J. J., and Tipton, I. H.**, Action of chromium on kidney filtration and reabsorption function, *J. Chronic Dis.,* 15, 941-964, 1962.
17. **Simavoryan, P. S.**, Experimental histological investigation of the pathomorphology of the myocardium in chronic chromium intoxication, *Tr. Erevan. Med. Inst.,* 15(1), 213-218, 1971.
18. **Shakhnazarov, A. M.**, Experiments on the effects of baryta, strontia, chrome, molybdenum, tungsten, tellurium, titanium, osmium, platinum, iridium, rhodium, palladium, nickel, cobalt, uranium, cerium, iron and manganese, on the animal system, *Arkh. Patol.,* 35(11), 67—73, 1973.
19. **Gmelin, C. G.**, Experimental nephropathies, IV Glycosuria in dogs poisoned with uranyn nitrate, mercury bichloride and potassium dichromate, *Edinburgh Med. J.,* 26, 131—139, 1826.
20. **Hepler, O. E. and Simonds, J. P.**, Changes in the activity of some enzyme systems of tissue respiration caused by different concentrations of chromium, *Arch. Pathol.,* 41, 42-29, 1946.
21. **Ergeshov, I.**, Hematological parallels in experimental sensitization by chromium compounds, *Zdravookhr. Turkm.,* 16(10), 3-5, 1972.
22. **Kalugin, V. V. and Sosonkin, I. E.**, Effect of chromium on the intensity of oxygen consumption by homogenates of guinea pig organs and tissues, *Vopr. Gig. Tr. Profpatol. Mater. Nauchn. Konf.;* 175-177, 1972.
23. **Ergeshov, I.**, Chromium complexes and allergic eczematous chromium dermatitis, *Zdravookhr. Turkm.;* 18(3), 19-21, 1974.
24. **Scheiner, D. M., Katz, S. A.**, Experimental nephritis in guinea pigs by subcutaneous injections of chromates, *Rev. Latinoam. Quim.,* 3(4), 160-2, 1973.
25. **Ophuls, W.**, Studies on the detoxicating hormone of the liver (yakriton). Contribution to the usage against experimental chromate nephritis, *Proc. Soc. Exp. Biol. Med.,* 9, 11-12, 1911.
26. **Ohta, S.**, Observations on the physiological action of chromium, *Tohoku J. Exp. Med.,* 39, 37-46, 1940.
27. **Priestley, J. J.**, *Anat. Physiol.,* 11, 285-301, 1877.
28. The Merck Index of Chemicals and Drugs, 7th ed., Merck & Co., Inc., Rahway, N.J., 1960.
29. **Schroeder, H. A.**, Level of chromium (VI) in water sources and its effect on the development of experimental atherosclerosis in warm-blooded animals, *Air Qual. Monogr.,* No. 70-15, American Petroleum Institute, Washington, D.C., 1970.
30. **Novakova, S., Mautner, G., and Dinoeva, S.**, Blood catalase activity of rabbits under the influence of cadmium sulfate, *Gig. Sanit.,* 5, 78-80, 1974.
31. **Kichina, M. M., Komeko, V. D., and Andryuk, A. K.**, Toxicological studies on chromium. Toxicity test for 72 ppm chromium (VI) solution, *Sb. Rab. Leningr. Vet. Inst.,* 38, 87-91, 1974.
32. **Fujii, T., Sakomoto, Y., Nakagawa, Y., Fukumori, N., Mikuriya, H., Fukuda, H., Abe, Y., Yano, N., Yuzawa, K., and Hiraga, K.**, Studies on the detoxifying hormone of the liver (yakriton). Effect of yakriton upon chronic nephritis, *Tokyo Toritsu Eisei Kenkyusho Kenkyu Nempo,* 25, 547-55, 1974.
33. **Hasegawa, M.**, Certain biochemical and morphological changes in kidneys during chromium poisoning and therapeutic effectiveness of unithiol, *Tohoku J. Exp. Med.,* 32, 163-176, 1938.
34. **Sarkisyan, A. A., Epremyan, G. A. and Simavoryan, P. S.**, Chromates in animal nutrition, *Zh. Eksp. Klin. Med.,* 11(5), 25-31, 1971.
35. **Gross, W. G. and Heller, V. G.**, Trace elements in diets and excreta, *J. Ind. Hyg. Toxicol.,* 28, 52-56, 1946.
36. **Tipton, I. H., Stewart, P. L., and Martin, P. G.**, *Health Phys.,* 12, 1683—1689, 1966.
37. Division of Medical Science, National Research Council, *Medical and Biological Effects of Environmental Pollutants: Chromium,* National Academy of Sciences, Washington, D.C., 1974.
38. **Schroeder, H. A., Nason, A. P., and Tipton, I. H.**, Chromium deficiency as a factor in atherosclerosis, *J. Chronic Dis.,* 23, 123-142, 1970.
39. **Sullivan, R. J.**, Preliminary Air Pollution Survey of Chromium and Its Compounds. *A Literature Review, Publ.* National Air Pollution Control Administration

Species	Chemical form	Route administered	Amounts administered	Gross, morphological, and biochemical changes	Ref.
Male albino rat	$NiCl_2$	Intratracheal injection	1 mg	Found in decreasing concentrations in kidney, lung, adrenal glands, liver, pancreas, spleen, and heart; rapid, transient increase in serum glucose, decrease in serum insulin, glycosuria; with concurrent administration of insulin, increase in serum glucose was prevented	1
Male guinea pig	$NiCl_2$	i.p. injection	8 mg/kg	Found in decreasing concentrations in kidney, pituitary, lung, liver, spleen, heart, adrenal gland, testes, and pancreas	1
Male albino rat	$NiCl_2$	Water	225 ppm	Decrease in growth, serum cholesterol, TG, excretion-urine volume, urinary Ca, Zn	1
Rabbit	—	Implanted stainless steel prostheses	—	At tissue near implant, Ni released by action of sweat and blood	2
Mice	Ni acetate	Water	5 ppm	Increased longevity	3
Rat	Ni acetate	Water	5 ppm	Rate of growth enhanced at four age intervals in males up to 6 months, two in females up to 18 months; Ni-fed rats smaller than controls; some decrease in Zn and Cu in lung; some decrease in Mn and Cu in spleen; more Cr in heart and spleen, more Mn in kidney than in rats not fed Ni; survival and longevity not affected; Ni not tumorigenic or carcinogenic; no accumulation in tissues	4
Fathead minnow	—	Water	0.38 mg/l	Did not adversely affect survival, growth, or reproduction; caused significant decrease in number of eggs, spawning, and hatchability	5
Rat	$NiSO_4$ NiS $NiCl_2$	i.p. or i.v. injection	5 mg	Especially in hydrosol compounds, Ni caused regularly anaphylactoid edema, which developed after a bilateral nephrectomy; cannot be ascribed to release of renal erythropoietic factor after parenteral administration	6
Chick	$NiCl_2$	Diet	3 ppm	Yellow lipochrome pigments/g of shank skin; other previously reported gross changes diminished or inconsistent; Ni deficiency decreased O_2 uptake in liver homogenates in presence of α-glycerophosphate; increased liver lipids; decreased liver phospholipids; ultrastructural changes in hepatocytes; no change in liver cholesterol	7

NUTRIENT TOXICITIES IN ANIMALS AND MAN: NICKEL (continued)

Species	Chemical form	Route administered	Amounts administered	Gross, morphological, and biochemical changes	Ref.
Rat	$NiCl_2$	Diet	—	Ultrastructural changes in hepatocytes; first generation Ni-deficient rat has 15% fetal loss at birth; no perinatal mortality in controls	7
Rat	$NiCl_2$	—	—	Decreased I_2 abs by thyroid at 0.5—5.0 mg/kg/day; I_2 abs not affected at 5 μg/kg/day	8
Male albino New Zealand rat	$NiCl_2$	i.v. injection	240 μg/kg 4.5 μg/kg	Sacrifice 2 hr after injection, highest concentration in kidney, pituitary, skin, and lung; after 24 hr of 34—48 consecutive daily injections (4.5 μg/kg/day), highest in kidney, pituitary, spleen, lung, and skin; low in brain and certain portions of the eye	9
Fischer 344 rat	$NiCl_2$	Oral intubation	—	With free access to food and water, regardless of quantity of Ni given, animals eliminated entire amount within 48hr; 3—6% in urine, remainder in feces; peak in urine at 4 hr	10
Fischer 344 rat	$NiCl_2$	i.p. injection	4, 16, 64 mg/kg body wt.	Almost all Ni that reached bloodstream eliminated quickly by kidney, preventing entrance into tissues; 1—2% in feces, urine peak at 3 hr; abs on G.I. tract slight; remainder passed intact through gut; 64 mg/kg orally — no effect; 64 mg/kg i.p. — 60% lethal	10
Male Wistar rat	NiS and NiS 3,4-benzpyrene	Intratracheal injection	5 mg NiS; 2 mg BP mixture	One out of ten developed hepatoma; increased pre-malignant lesions with NiS and 3,4-benzpyrene	11
Ohio State University brown rat	Ni acetate	Diet	0, 100, 500, 1000 ppm	High dietary levels of Ni: decreased weight gain, HBg, Hct, alkaline phosphatase (plasma); no significant influence on plasma ceruloplasmin or cytochrome oxidase activity; decrease in heart, not in liver; increasing Ni: increased plasma Ni, no influence on plasma Cu, increased plasma Zn, increased RBC Ni, no influence on RBC Zn, increased RBC Cu, Fe; increased Ni found in heart, liver, testes; increase in kidney and increased Fe in all renal and hepatic cellular fractions	12
Mice	—	Water	0.01 M	Ni increased mortality from encephalomyocarditis virus (ECMV) and inhibited Newcastle disease virus protection to ECMV mortality	13
Rhode Island red chick	$NiCl_2$	i.p. injection	10, 20, 40 mg/kg body wt. range	Transient hyperglycemia in fed birds, prolonged in starved birds; concomitant decrease in FFA in fed	14

Species	Chemical form	Route administered	Amounts administered	Gross, morphological, and biochemical changes	Ref.
Bovine	$NiCl_2$	—	0.1—0.2 ml of $NiCl_2$ solution	In vitro metal-induced augmentation of release of GH, TSH, LH, and ACTH shows decreasing potency — Cu > Zn > Ni; Ni and Cu caused increasing inhibition of prolactin release with increasing concentration	15
Adult human	—	Excretion	Men: 7.9 ± 4.0 μg/dl Women: 14.4 ± 6.3 μg/dl	Conclusion: sweating is a significant route of excretion; prolonged exposure to heat may deplete essential trace minerals	16
Wistar albino rat	$NiCl_2$	i.v. injection	Rat: 82 μg/kg body wt. Rabbit: 240 μg/kg body wt.	Ni rapidly cleared from plasma or serum during first 2 days after injection, slower rate at 3—7, days; urinary excretion averaged 78% of administered dose during first day after injection in rabbits and 78% during the 3 days after injection in rats; suggest Ni diluted within a volume composed two compartments; Ni eliminated by first order kinetics	17
New Zealand albino rabbit	—	Diet	—	On a normal diet, Ni in 3 days feces collections from ten persons averaged 3.3 μg/g wet weight, and 14.2 μg/g dry weight; fecal excretion of Ni averaged 258 μg/day; conclusion — fecal excretion major route	18
Healthy adult human	—	Diet	Ni deficiency 2.7 ppb, 3 ppm	Ni deficiency compared to Ni supplement: increased weight, decreased yellow lipochrome pigments/g shank skin, increased phospholipids in liver and brain, increased plasma cholesterol changes noted with addition of n-propyl uracil, epinephrine, and T_3; indicated through interaction, Ni may influence energy metabolism	19
Chick	—	Diet	44 ppb-deficient diet 3.4 ppm-supplemented diet	After 30 days, Ni-supplemented chicks showed no difference from Ni-deprived chicks in body weight, leg appearance, hematological parameters, concentration of serum cholesterol, or histological appearance of organs and tissue by light microscopy; electron microscope showed dilatation of perimitochondrial RER in hepatocytes of Ni-deprived chicks	20
Male White Rock chick	$NiCl_2$				
Human adult, Sprague-Dawley rat, Maine lobster, beagle dog, and New Zealand rabbit	—	—	—	Species variation between man, rabbit, and rats in the concentration of total serum Ni and the partitions of serum-ultrafiltrable and protein-bound Ni; Due to species difference in Ni binding properties of serum albumin	21

NUTRIENT TOXICITIES IN ANIMALS AND MAN: NICKEL (continued)

Species	Chemical form	Route administered	Amounts administered	Gross, morphological, and biochemical changes	Ref.
Sprague-Dawley rat	NiCO$_4$	i.v. injection	2.2 mg/100 g body wt.	Increase hepatic ATP; may be due to Ni inhibition of hepatic ATP-ase	22
Charles River-CD mice Long-Evansblu rat	—	Water	5 ppm	Size of litters decreased with each generation, with two breeding failures; few males in third generation; relative toxicity in feeding experiment: Pb > Cd > Se > Ni, As, Ti, Mo; breeding exposure: Pb = Cd > Se > Ni > Ti > Mo > As	23
Rat	—	Diet	5 mg/kg 0.5 mg/kg	Elimination of Ni from the body through G.I. tract; retention occurred with 5 mg/kg; no changes in functional activity of G.I. tract with 0.5 mg/kg	24
Holstein, Brown Swiss male calf	NiCO$_3$	Diet	0, 62.5, 250, 1000 ppm	Food intake and growth slightly retarded at 250—1000 ppm, but responded to posttreatment; N retention decreased by 1000 ppm; increased propionate, decreased butyrate in fluid from animals on higher level Ni supplementation; kidneys nephritic, degree of severity with Ni level; relative to body weight, fresh weight of lung, heart, spleen, liver, gall bladder, brain, and testes unaffected; ruminal, abomasal, duodenal, hepatic, and testicular tissues not affected histologically	25
White Rock chick	NiCl$_2$	Diet	Control < 0.08 ppm fed 5 ppm	Chicks fed low Ni showed bright orangish-yellow color in legs, slightly thickened legs, swollen hock joints, slightly abnormal gait and length, and decreased width ratio; Ni retained to greater extent in bone, kidney, liver, heart, and aorta at 6 and 24 hour intervals after dosage in low Ni chicks than in control	26
Man	Ni acetate	Diet	60 mg/day	Hypogeusia in patient with multiple myeloma with excessive levels of thiols; in patient treated with penicillin, increase in taste acuity with administration of Zn and Ni; proportionate decrease in detection and recognition thresholds for each of four taste qualities; hypothesis is that this followed shifts in distribution of metal ions among various metabolic pools via forced metalthiol interaction	27
Rat	—	Diet	50 µg	50 µg each of Co and Ni increased blood phagocytic activity during first week, after which, activity	28

Species	Chemical form	Route administered	Amounts administered	Gross, morphological, and biochemical changes	Ref.
Sprague-Dawley rat	$NiCO_4$	i.v. injection	2.2 mg Ni/100 g 6.6 mg Ni/100 g	Severe respiratory symptoms; diffuse dilations of RER at LD_{100}; mitochondria and other cytoplasmic organelles not affected; hepatocytic nucleolar changes within 2–24 hr after injection; ultrastructure returned to normal by sixth day in surviving rats of LD_{50} dose; LD_{50} death at 3–5 days; LD_{100} death by the third day	29
Sprague-Dawley male rat	$NiCO_4$	i.v. injection	5 µl/100 g body wt	30% of administered $NiCO_4$ exhaled without metabolic alteration within 6 hr after injection; remaining $NiCO_4$ underwent intracellular decomposition and oxidation to Ni^+ and CO; CO became bound to Hbg, with maximum saturation of 35% 2 hr after injection; within 6 hr, 49% administered ^{14}C exhaled as ^{14}CO and 1.1% exhaled as $^{14}CO_2$; less than 1% administered ^{14}C was excreted in urine during 24 hr after $NiCO_4$ injection	30
Webster Swiss mouse	Ni acetate	Diet	0, 1, 100, 1600 ppm	Number of pups weaned decreased by 1600 ppm; decrease in growth of male and female; decrease in female body wt at 1100 ppm; liver: decrease cytochrome oxidase, isocitric dehydrogenase, NADH cytochrome C reductase; heart: decrease cyctochrome oxidase and malic dehydrogenase; kidney: decrease malic dehydrogenase	31
Hubbard broiler chick	Ni acetate $NiSO_4$	Diet	0, 100, 300, 500, 700, 900, 1100, 1300 ppm	Decrease in growth of chick to 4 weeks of age with approximately 700 ppm; decreased nitrogen retention and energy metabolism	32
Swiss albino mice Fischer rat	Ni acetate Nickelocene	i.p. or oral gavage	i.p., 12.5–150 mg/kg body wt Oral, 200–800 mg/kg body wt	Ni acetate more toxic than nickelocene when given orally and Ni acetate more toxic when given i.p. rather than orally; i.p. toxicity shown by diarrhea, respiratory difficulty, and lethargy 2–3 hr after injection; initial fatalities occur 6–10 hr after treatment; surviving animals were alive on day 10 and appeared normal on day 35; necropsies of survivors showed no outstanding pathological features; animals injected by the i.p. method showed intestinal adhesions and whitish mucous layers covering spleen, liver, and kidneys; for orally treated, mortalities occurred 24–72 hr after treatment	33

Animal	Compound	Route	Dose	Effects	Ref.
Sprague-Dawley male rat	$NiCO_4$	i.v. injection	5 µl/100 g body wt	Lung as major excretory organ: 38% expired from body burden after 24 hr; decreasing concentrations in muscle and fat, bone and connective tissue, viscera and blood, and in the lung, liver, brain, and spinal cord; 83.9% was recovered in tissues, feces, and expired air after 24 hr; total serum $NiCO_4$: 80.5% bound to serum proteins, of which 81.1% was bound to albumin during the 4 days after injection; 31% of dose in urine	34
Sprague-Dawley male rat	$NiCO_4$	i.v. injection i.p. injection	0.57—4.52 mg/100 g	Immediate clinical observations: 5% died within first hour of administration of LD_{50} ; weakness, anorexia, decrease in H_2O intake, paralysis of hindleg, and decrease in weight seen in first 24 hr, recovery from symptoms by second day; delayed reactions on third; tachypnea, dyspnea, weight decrease, inactivity, ruffled fur, abdominal distention; pulmonary histological changes of diffuse hypertrophy, hyperplasia of alveolar living cells, increased mitotic activity, and cytologic abnormalities; pulmonary parenchyma was found to be primary target for $NiCO_4$, regardless of its route of administration	35
Male albino rat	$NiSO_4$	Subcutaneous injection	0.04 µmol/kg body wt	Shrinkage of testes tubules, spermatozoa in epididymis completely degenerated, hyperemia, and necrosis of ductuli is progressive; testiculate tissue and spermatogenesis is normal by 12 days	36
Wistar C3H rat Webster Swiss mouse	Ni_3S_2 NiO	i.m. injection	Rats, 20 mg Mice, 5 mg	Rats: 80% tumor incidence at sites injected with Ni_3S_2, with 150-day latent period; 41% NiO, with 302-day latent period; 89% of all exposed animals developed tumor where Ni_3S_2 was administered, 66% when NiO was administered; metastases occurred in almost all Ni_3S_2 animals Mice: no difference in tumorigenic activity of the two Ni compounds in mice, either in average latent period or number of tumor-bearing animals; progression time for NiO-induced tumors was shorter than with Ni_3S_2 ; strain difference in mice with susceptibility to Ni	37

REFERENCES

Clary, J., Nickel chloride-induced metabolic changes in the rat and guinea pig, *Toxicol. Appl. Pharmacol.*, 31, 55–66, 1975.

Samitz, M. and Katz, S., Nickel dermatitis hazards from prostheses, *in vivo* and *in vitro* solubilization studies, *Br. J. Dermatol.*, 92, 287–290, 1975.

Schroeder, H. and Mitchener, M., Life-term effects of mercury, methyl mercury and nine other trace metals on mice, *J. Nutr.*, 105, 452–458, 1975.

Schroeder, H., Mitchener, M., and Nason, A., Life-term effects of nickel in rats: survival, tumors, interactions with trace elements and tissue levels, *J. Nutr.*, 104, 239–243, 1974.

Pickering, Q., Chronic toxicity of nickel to the fathead minnow, *J. Water Pollut. Control Fed.*, 46, 760–765, 1974.

Jasmin, G., Anaphylactoid edema induced in rats by nickel and cobalt salts, *Proc. Soc. Exp. Biol. Med.*, 147, 289–292, 1974.

Nielsen, F. and Ollerich, D., Nickel: a new essential trace element, *Fed. Proc. Fed. Am. Soc. Exp. Biol.*, 33, 1767–1772, 1974.

Lestrovoi, A., Itskova, A., and Eliseev, I., Effect of nickel on the iodine fixation of the thyroid gland when administered perorally and by inhalation, *Gig. Sanit.*, 10, 105–106, 1974.

Parker, K. and Sunderman, F., Jr., Distribution of ^{63}Ni in rabbit tissues following intravenous injections of ^{63}NiCl$_2$, *Res. Commun. Chem. Pathol. Pharmacol.*, 7, 755–762, 1974.

Ho, W. and Furst, A., Nickel excretion by rats following a single treatment, *Proc. West. Pharmacol. Soc.*, 16, 245–248, 1973.

Kasprzak, K., Marchow, L., and Breboorowicz, J., Pathological reactions in rat lungs following intratracheal injection of nickel subsulfide and 3,4-benzpyrene, *Res. Commun. Chem. Pathol. Pharmacol.*, 6, 237–246, 1973.

Whanger, P., Effect of dietary nickel on enzyme activities and mineral content in rats, *Toxicol. Appl. Pharmacol.*, 25, 323–331, 1973.

Gainer, J., Effects of metals on viral infections in mice, *Environ. Health Perspect.*, 4, 98–99, 1973.

Freeman, B. and Langslow, D., Responses of plasma glucose, free-fatty acids, and glucagon to cobalt and nickel chlorides by *Gallus domesticus*, *Comp. Biochem. Physiol.*, 46, 427–436, 1973.

LaBella, F., Dular, R., Vivian, S., and Queen, G., Pituitary hormone releasing or inhibiting activity of metal ions present in hypothalamic extracts, *Biochem. Biophys. Res. Commun.*, 52, 786–791, 1973.

Hohnadel, D., Sunderman, F., Jr., Mechay, M., and Neely, M., Atomic absorption spectrometry of nickel, copper, zinc, and lead in sweat collected from health subjects during sauna bathing, *Clin. Chem.* (N.Y.), 19, 128–1292, 1973.

Onkelinx, C., Becker, J., and Sunderman, F., Jr., Compartmental analysis of the metabolism of ^{63}Ni (II) in rats and rabbits, *Res. Commun. Chem. Pathol. Pharmacol.*, 6, 664–676, 1973.

Horak, E. and Sunderman, F., Jr., Fecal nickel excretion by healthy adults, *Clin. Chem.* (N.Y.), 19, 429–430, 1973.

Nielsen, F., Effect of the dietary level of nickel on the responsiveness of chicks to changes in hormonal status, *Fed. Proc. Fed. Am. Soc. Exp. Biol.*, 31, 700, 1972.

Sunderman, F., Jr., Nomoto, S., Morang, R., Mechay, M., Burke, C., and Nielsen, S., Nickel deprivation in chicks, *J. Nutr.*, 102, 259–267, 1972.

Hendel, R. and Sunderman, F., Jr., Species variation in the proportions of ultrafiltrable and protein-bound serum nickel, *Res. Commun. Chem. Pathol. Pharmacol.*, 4, 141–146, 1972.

Sunderman, F., Jr., Effect of nickel carbonyl upon hepatic concentrations of adenosine triphosphate, *Res. Commun. Chem. Pathol. Pharmacol.*, 2, 545–551, 1971.

Schroeder, H. and Mitchener, M., Toxic effects of trace elements on the reproduction of mice and rats, *Arch. Environ. Health*, 23, 102–106, 1971.

Elakhovskaya, N., Study of metabolism of nickel entering the body with drinking water, *Gig. Sanit.*, 37, 20–22, 1972.

O'Dell, G., Miller, W., King, W., Moore, S., and Blackman, D., Nickel toxicity in the young bovine, *J. Nutr.*, 100, 1447–1454, 1970.

Nielsen, F. and Sauberlich, H., Evidence for a possible requirement for nickel by the chick, *Proc. Soc. Exp. Biol. Med.*, 134, 845–849, 1970.

Heakin, R. and Bradley, D., Hypogeusia corrected by Ni^{++} and Zn^{++}, *Life Sci.*, 9, 701–709, 1970.

Litvinenko, M., Effect of the trace nutrients cobalt and nickel on phagocytic activity, *Tr. Khar'k. Gos. Med. Inst.*, 79, 44–47, 1969.

Hackett, R. and Sunderman, F., Jr., Nickel carbonyl. Effects upon the ultrastructure of hepatic parenchymal cells, *Arch. Environ. Health*, 19, 337–343, 1969.

Kasprzak, K. and Sunderman, F., Jr., The metabolism of nickel carbonyl-^{14}C, *Toxicol. Appl. Pharmacol.*, 15, 295–303, 1969.

31.   **Weber, C. and Reid, B.,** Nickel toxicity in young growing mice, *J. Anim. Sci.,* 28, 620–623, 1969.

32.   **Weber, C. and Reid, B.,** Nickel toxicity in growing chicks, *J. Nutr.,* 95, 612–616, 1968.

33.   **Haro, R., Furst, A., and Falk, H.,** Studies on the acute toxicity of nickelocene, *Proc. West. Pharmacol. Soc.,* 39–42, 1968.

34.   **Sunderman, F., Jr. and Selin, C.,** The metabolism of nickel-63 carbonyl, *Toxicol. Appl. Pharmacol.,* 12, 207– 1968.

35.   **Hackett, R. and Sunderman, F., Jr.,** Acute pathological reactions to administration of nickel carbonyl, *Environ. Health,* 14, 604–613, 1967.

36.   **Hoey, M.,** The effects of metallic salts on the histology and functioning of the rat testis, *J. Reprod. Fertil.,* 461–471, 1966.

37.   **Gilman, J.,** Metal carcinogenesis. II. A study on the carcinogenic activity of cobalt, copper, iron, and n compounds, *Cancer Res.,* 22, 158–162, 1962.

EFFECT OF NUTRIENT TOXICITIES IN
ANIMALS AND MAN: SILICON

D. R. Bennett

INTRODUCTION

Although silicon has been known as an element since its isolation in 1810 by Berzelius,[51] the last decade has witnessed a marked change in how it is viewed biologically. Older biological roles of essentiality of silicon in some lower forms of life and only significant toxicity in some higher forms of life now include a role of essentiality in chickens and rats.

The initial portion of this section on oral-silicon toxicity is devoted to a brief survey of the chemical classification and physiological role of silicon compounds. An overview of the chemistry and biological activity of silicon compounds is essential to a basic understanding of their essential trace-element role[18-20,77,95,96] and oral toxicity as it relates to food and nutrition. General review articles and texts on the chemistry of silicon[24,28,47,85,99,105] are included for the biological investigator who needs such information on silicon compounds.

CHEMICAL CLASSIFICATION

The chemistry of silicon and carbon would be expected to have certain similarities as a result of their relative Group IV positions in the current periodic table. Their differences are more significant, however, when one describes their chemical reactivity.[16] For example, because of the electronegativity of silicon (1.8) compared to carbon (2.5) and hydrogen (2.1), the hydrogen on silicon has hydride reactivity unlike the hydrogen on carbon. Silicon bonding is both ionic and covalent in character. Thus, the Si—O bond is stronger, shorter, and less obtuse than the O—O bond. Silanols do not behave like alcohols but like acids of the phenol type; therefore, they rapidly lose water to form anhydrides and esters. Similarly, the silyl ethers, Si—O—Si, are chemically more similar to carboxylic acids than organic ethers because the former readily react with alcohol under mild acidic catalysis.

Germanium and silicon are mimes in that they have a similar distribution within certain organisms.[71] Germanium is actually an antagonist of silicon in at least one transport process across membranes;[104] thus, silicon is more similar to germanium than carbon in the Group IV series in this instance.

Although silicon is not considered metallic, its chemistry is usually indexed under organometallic chemistry; thus, some refer to it as one of the half-way or metalloid elements along with B, Be, As, Sb, Te, Po, and At.[22] Silicon is the only element in this group now considered biologically essential.

All biological roles for silicon are dependent on how silicon is chemically dressed, for the element is not found naturally even though it is the second most abundant element on earth. Its chemical wardrobe is extensive, and silicon compounds are classified into inorganic and organic categories. Organosilicon compounds are characterized by the presence of at least one Si—C bond.[100] Organosilicon chemistry[28] was extensively developed by Kipping in the first four decades of this century. There is little evidence that organosilicon compounds of the Si—C type are synthesized by or serve as a source of silicon or silicate for organisms. No compound in this category has been cited as a cause of a pneumoconiosis. Organosilicon chemistry is usually exemplified by the organo-siloxanes, better known as the commercially important, trivial-named silicones,[99] although there are numerous other examples. Organosiloxanes are often referred to as biologically inert; however, many organosilicon compounds have considerable promise for biological activity and potential toxicity.[11,12,32,35,102,103]

The inorganic compounds of silicon, best exemplified by the abundant silicates, contain no Si—C bonds. Organic silicates are molecules which have no Si—C bond, but they do possess at least one C—O—Si bond. Unfortunately, the term organic silicates has different meanings in the world's chemical literature, for the organic silicates are classified as a subsection of inorganic chemistry in most of the Western World, while, for example, Soviet chemists consider organic silicates organosilicon compounds. All inorganic silicon compounds, including the organic silicates, are not equal in oral bioavailability. Bioavailability is determined by molecular size, configuration, and the number of various polymeric species even in a substance like silicic acid which is often inappropriately perceived by biologists as nonpolymeric in solution.[10,15,47,58,105]

PHYSIOLOGICAL SIGNIFICANCE

The need for silicon in shell formation of diatoms, radiolarians (protozoa), and a few siliceous forms of *Porifera* was known early in this century.[25,61] Silicates are absorbed, accumulated, and metabolized in bacteria, but essentiality is not proven.[42] A few plant species of the *Gramineae* (barley, wheat, oats, rye, rice) seem to be more disease resistant and thrive better if silicon is present, but definite essentiality is not proven.[61] The rigorous evidence of silicon-deficient diets needed to establish silicon as an essential trace element in chickens and rats was accomplished technologically by Carlisle[19] and Schwarz[96] working independently in the last decade.

However, a physiological role for silicon in mammals had been suspected prior to the last decade by a few investigators based on the unequal distribution of silicon in tissues. These investigators did not view silicon as biologically insignificant contamination. Fregert has summarized this literature prior to 1960.[33]

Analytical methods employed for determination of silicon in biological materials include colorimetry[8,9,33,48,50,56,70] atomic absorption,[27,46] emission spectroscopy,[60,87] neutron activation,[74] and electron microprobe microanalysis.[72,73,78,91] The colorimetric method is widely employed, but problems of sensitivity, silicon contamination, and phosphorus interference in biological specimens make comparison of data among investigators difficult. Colorimetric data since 1955 are probably more reliable, since recent workers have attempted to control rigorously the problems of contamination and phosphorus interference.

Older quantitative data were characterized by presenting silicon as units of SiO_2 per gram of nitrogen, or wet or dry weight of tissue. At present, parts per million of silicon per gram of dried tissue is preferred. Table 1 has been adapted on that basis and summarizes the data available from mammals since 1956.

Data from Table 1 identify bone, skin, lung, and lymph nodes as organs of high-silicon density. The bone and skin in younger animals are especially high in silicon density. Further fractionation of skin reveals the highest density in epidermis, hair, and nails, i.e., areas of high keratin content. Silicon levels are especially high in sites of endochondral- and periosteal-active bone growth,[19] and, subcellularly, silicon is associated with the mitochondria of rat liver, kidney, and spleen[73] as well as rat osteoblasts.[20] Thus, the elevated bone and skin levels of silicon correlate with silicon's proposed roles in calcification of bone and keratinization respectively. The lungs of the primate and human are especially high in older individuals. This high level has been suggested to be associated with gradual deposition of higher molecular weight species of polymeric silicon compounds following chronic environmental exposure. A similar mechanism, following chronic exposure to silica and silicates in the diet and/or air, is proposed to account for the high silicon level in lymph nodes.

SILICON IN VARIOUS TISSUES

	Adult rat			Guinea pig[90]	Young rabbit[19]	Adult rabbit[19]	Cow[90]	Rhesus monkey[60]	Human		
	Young rat[70]	Ref. 59	Ref. 60						Ref. 33	Ref. 3[a]	Ref. 17
Whole femur	86.7 ± 11.4	115.2 ± 16.1	38.2 ± 19.2								
Femur shaft								42.9 ± 9.3			
Femur head								456.3 ± 71.0			
Rib	90.2 ± 20.2	91.8 ± 22.4	4.0 ± 4.0								
Skin	9.0 ± 3.0	3.6 ± 0.9			45	10			23.0 ± 8.7		
Epidermis									106.0 ± 2.7		
Dermis-sole									13.0 ± 4.6		
Dermis									19.0 ± 3.8		
Dermis-hairy									25.0 ± 5.2		
Hair				113.1					90.0 ± 15.5	26	41.3 ± 11.7
Nail	4.0 ± 1.7	7.3 ± 1.3	7.8 ± 6.8						56.0 ± 13.7		
Lung				59.1 ± 0.8				941.7 ± 889.3	653.0		
Lymph nodes			18.9 ± 25.3				56.5	100.9 ± 48.8			
Esophagus			33.1 ± 18.5								
Stomach, cardiac			13.9 ± 13.4								
Stomach, fundic			15.4 ± 12.7								
Duodenum			4.9 ± 2.7					6.3 ± 1.0			
Jejunum			4.3 ± 3.9					11.4 ± 1.7			
Ileum			5.1 ± 2.9								
Caecum			21.8 ± 13.3								
Colon			8.5 ± 5.1					12.1 ± 2.2			
Hypothalamus			3.0 ± 2.0					3.2 ± 1.5			
Pituitary			4.0 ± 4.8					3.6 ± 2.0			
Prostate			5.4 ± 4.8					20.0 ± 13.1			
Seminal vesicles			11.6 ± 9.1					12.8 ± 7.2			
Epidydymis			6.1 ± 4.9					6.5 ± 4.9			
Testes			8.1 ± 8.1					12.5 ± 7.5			
Thyroid			9.2 ± 12.4					9.6 ± 3.2			
Pancreas			5.0 ± 2.9					7.6 ± 5.3			
Adrenals			10.6 ± 7.0					5.3			
Brain[c]	2.0 ± 0.4	2.9 ± 0.2	4.0 ± 4.5					7.0 ± 3.5			
Tongue			2.3 ± 3.1								
Submaxillary			4.1 ± 4.1					17.0 ± 7.8			
Thymus			1.3 ± 0.6								
Spleen	3.8 ± 1.0	5.5 ± 1.3	3.4 ± 2.9		55	2		3.4 ± 1.7		2	
Heart	3.0 ± 0.9	4.3 ± 1.5	4.4 ± 5.7		8	14	13.9	4.2 ± 1.3		18	
Skeletal muscle			3.3 ± 2.6		4	12	17.9	4.4 ± 1.8	18.0 ± 5.0		
Liver	2.6 ± 1.1	4.7 ± 1.8	5.0 ± 4.7			4	18.2	3.7 ± 3.7		5	
Kidney	1.5 ± 0.4	4.6 ± 1.8	2.3 ± 3.2				29.3	7.4 ± 6.9	42.0 ± 15.2	17	
Depot fat			2.8 ± 3.0					5.6 ± 5.8			
Eye, lens								1.5 ± 0.6			
Tendon	4.0 ± 1.8	4.8 ± 1.9							28 ± 6.8		

a   ppm Si ± s.d. expressed per dried tissue except serum, blood, bile, and urine, which are expressed per wet weight tissue.

b   Data estimated from graph in original papers.

c   More detailed analysis of the human brain may be found in Reference 3.

Table 1 (continued)
PPM· SILICON IN VARIOUS TISSUES

	Young rat[70]	Adult rat		Guinea pig[70]	Young rabbit[19]	Adult rabbit[19]	Cow[90]	Rhesus monkey[60]	Human		
		Ref. 59	Ref. 60						Ref. 33	Ref. 3a	Ref. 17
Aorta						80	14			41.0 ± 12.6	
Serum			0.9 ± 0.3						1.9 ± 0.3		
Blood							2		1.1 ± 0.6		
Bile									3.7 ± 3.2		
Urine									7.0 ± 2.0	28	
Urine mg/24 hr			23.1 ± 9.1	4.7 ± 3.6						37.1 ± 8.1	

Even when total silicon is not especially high in tissues, fractionation procedures reveal that connective tissue and cartilage are sites of high-silicon density in such tissues.[19,96] Further detailed studies demonstrate that silicon is associated with acid mucopolysacchar- ides in such tissues.[21,94] Silicon is therefore proposed as a connective tissue cross-linking agent, presumably in the form of an organic silicate. No such endogenous compound has been identified to date, although the unconfirmed presence of the Si–O–C bond has been proposed in bacterial studies.[41]

In summary, silicon's essential role is related to calcification of bone, integrity of connective tissue, and keratinization. Thus, the skeletal role of silicon in the shell of diatoms has its mammalian counterpart in the integumentary and skeletal systems which also have supporting and covering functions. Selected localization of silicon in the senile plaques in the brain associated with Alzheimer's disease,[3,78] and the relative deficiency of silicon in atherosclerotic vessels, is provocative for further pathophysiological functions of silicon.[36,63-65,97]

PHARMACOLOGICAL-TOXICOLOGICAL SIGNIFICANCE

Inhalation

Inorganic silicates represent the largest commodity industry in the U.S. The major source of such silicon toxicity is chronic industrial exposure or environmental contamination, and the principal route of entry for such materials is by inhalation. Toxicity is rather specifically organotropic, i.e., directed to the lung and pleura, although kidney damage may also occur with chronic exposure.[57] The two major types of pulmonary pathology observed are fibrogenic pneumoconioses and carcinoma of the mesothelioma type.

The fibrogenic pneumoconiosis, silicosis, is caused only by certain forms of inorganic silica. The crystalline silicas, excepting stishovite, i.e., quartz, tridymite, cristobalite, keatite, and coesite are capable of inducing silicosis if other criteria are satisfied. Such criteria are morphology of the crystal (including physical and/or chemical conditions of manufacture which modify morphology), dose, and duration of exposure. The presence of trace elements such as Fe, Cr, Co, Ni, Sc, and/or contaminating oils, benzopyrene, etc., may modify the response. Morphological determinants include the type of lattice and electronic structure, particle size, $<3 \mu$, and filament structure.[37,38,84] Plant silica is a hydrated amorphous polymerized silicic acid gel which is referred to as opal. It is much more difficult, if not impossible, to produce silicosis with amorphous silicas such as opaline silica from plants or diatomaceous earths. This specificity of morphological etiology coupled with the chemical specificity of an antagonist of silica, viz. isotactic poly-(2-vinylpyridine-1-oxide),[13,45,69,92] has led to the concept that only silicas possessing a certain ordered arrangement of silanol groups on the silica surface are capable of inducing silicosis. Cellular mechanisms for the pathogenesis of silicosis-induced fibrogenesis are reviewed.[1,2,15,43]

Other than silicas, naturally occurring incombustible mineral silicates which include asbestos, glass, and ceramics comprise a second large group of inorganic silicon compounds. Asbestos, a hydrated magnesium silicate of many forms, is also capable of inducing a type of fibrogenic pneumoconiosis (asbestosis) as well as pleural meso- theliomas. Asbestos-induced fibrosis differs from silica-induced fibrosis functionally in that the former is not antagonized by poly-(2-vinylpyridine-1-oxide); but it is antagonized experimentally by ethylenediamine tetraacetic acid.[26,66]

Asbestos can be divided into two major morphological forms.[15,105] The amphiboles are fibers composed of solid tubes. Amphiboles includes crocidolite, amosite, antho- phyllite, tremolite, and actinolite — all of which comprise less than 10% of commercial asbestos in the U.S. The sheeted, rolled-up, hollow-tube fibers of chrsysolite represent over 90% of potential asbestos exposure. Talc is a flaky, granular, hydrated magnesium silicate and may contain up to 30% fibrous silicates as asbestos.[10] Asbestos has recently been extensively covered in an excellent review.[39]

Oral

The oral source of inorganic silicon compounds include diet, water, and medications. A number of foods contain silicon, especially the outer coverings of unrefined whole grain and certain seafoods.[97] Aluminosilicate, an anticaking or drying agent, is allowed to the extent of 0.01% in selected foods such as salt, seasoning, sugar, and baking powder.[34] Large amounts of silicates in water are only found in certain regions, e.g., the Balkan countries, as a result of erosion of silicate-magnetic rocks (plutonites)[67] or as a result of environmental pollution arising from industrial activities.[23]

Human dietary-silicon intake is considered to be an average of 7 ppm/day.[93] However, silica is added as a nonnutritive bulk constituent to animal feeds. Thus, a 150- to 250-g rat may consume a mean of 7 mg of silicon per day. The daily requirement for young animals is approximately 100 μg silicon/g diet for chicks[18,19] and 500 μg silicon/g diet for rats.[95,96] Silicon is generally not considered a dietary problem because so much silicon is available in the grasses — about 40% compared to 0.2% in legumes or less in seeds of cereal grains. However, individuals on highly processed grain or principally animal-protein diets may be somewhat marginal regarding requirement. Schwarz has recently proposed that individuals in certain low-silicon water regions may be more susceptible to atherosclerosis.[98]

Only 3 to 5% of dietary silicon is absorbed.[60,70,89] Policard et al.[83] have suggested that polymeric molecules of silicic acid containing up to 4 to 5 silicon-oxygen units characterize the transport form of silicic acid in blood. This molecular size correlates with the gastrointestinal absorption of simple polydimethyl trisiloxanes and tetra-siloxanes.[11,106] Tetraethoxysilane is said to be more bioavailable for growth[97] because it is absorbed in threefold greater amounts compared to sodium metasilicate in the guinea pig.[89]

Early studies using silicic acid labeled with Si^{31} (half-life, 2.6 hr) demonstrated very rapid renal clearance of silicic acid following parenteral administration[44,81] if amounts of silicic acid equivalent to dietary silicon intake were given. In cows[5] and man,[8,9] 25% tubular reabsorption of silicic acid, but not tubular secretion, was noted. Neither expired air nor bile has been examined as a route for silicon elimination. Generally, 90% of administered silica or sodium metasilicate is found in the feces, presumably representing unabsorbed material. Silica is present in the urine in amounts proportional to that in the diet, but there are considerable interindividual differences.

Pharmacological doses of ^{31}Si-labeled silicic acid temporarily increase the silicon tissue/plasma level especially in the kidney and, to a lesser degree, in the liver. Studies comparing parenteral to oral administration of silicic acid reveal that the kidney is capable of excreting much larger doses of silicic acid than are normally absorbed, although silica is deposited in the mitochondria of renal cells[83] at high oral-dose levels.

Drugs

A number of silicon compounds are present in pharmaceutical preparations. Colloidal silicon dioxide is a submicroscopic-fumed silica prepared by vapor-phase hydrolysis of a silicon compound. Colloidal silicon dioxide is not soluble in water or acid. It is used as a pharmaceutic aid (tablet diluent, suspending agent, thickening agent). Silica gels are amorphous polysilicic acids — essentially condensation polymers of orthosilicic acid which are used as pharmaceutic aids in desiccants, thickeners, and gelling agents. A number of insoluble-complex silicates are also used in pharmaceutical preparation: kaolin has adsorbent and demulcent properties, bentonite is a suspending agent, and talc is a dusting powder, while pumice is a dental abrasive. Magnesium trisilicate is a mixture of magnesium oxide and a soluble silicate. Magnesium oxide readily reacts with mineral acids; thus, magnesium trisilicate is used as an antacid. Although oral bioavailability of dietary inorganic silicon is limited, conversely the antacid-classifying terms of non-absorbable or nonsystemic for magnesium trisilicate are a relative and not an absolute description. Page and Heffner[40,79] showed in 1941 that a mean of 5.2% of this antacid is absorbed orally in man. This amount can alter urinary pH by one unit and thus influence the rate of renal elimination of numerous other drugs. The degree of oral bioavailability of inorganic silicon compounds like silicic acid, silica, and the silicates is dependent on their solubility, degree of polymerization, and polymeric form of the material.

Various forms of asbestos are also found in foods[31] and drugs.[31,76,86] Clinically significant gastrointestinal absorption of potentially toxic particles, which produce asbestosis and/or carcinoma on inhalation, is not likely to occur.[14,39] This is an extremely active area of investigation at present, and more data on longer periods of both exposure and induction time are required before a final decision can be reached. Bioavailability is as important as intrinsic activity when one determines the relevance of any silicon compound to silicon essentiality. The principle is identical for oral-silicon toxicity.

The principal target organ for toxicity of silicon compounds ingested orally is the kidney. Two major types of functional renal toxicity are observed: obstructive (urolithiasis) and toxic (nephropathy).

Urolithiasis

Urinary siliceous calculous disease is principally a problem with range animals who eat pasture plants containing a relatively high silicon content in the form of opaline silica.[50] Bovine urolithiasis is characterized by high silicon content, in contrast to man, where oxalates, urates, or phosphates play the predominant role.

The major constituents of siliceous calculi are silica and other elements, i.e., Mg, Ca, and P, along with considerable organic material — mostly glycoprotein. These calculi have been produced in rats with oral ingestion of magnesium trisilicate, sodium metasilicate, or water glass, but calculi were particularly evident with tetraethoxysilane,[53] which is absorbed to the greatest degree. A level of silicic acid in the urine above the saturation point does not in itself usually lead to silicon urolithiasis. The formation of silica urolithiasis is augmented by increased urinary protein in the form of glyco- or mucoproteins and increased urinary electrolytes, elevated urinary pH, possible elevated urinary phosphate, and lowered urinary output.[4,6,53-55] Complexes of silicon compounds with fatty acids and cholesterol may also possibly become a part of the matrix of the stone.[53]

A pelleted atherogenic feed containing relatively insoluble silicates such as 3% talc or the more soluble 3% magnesium silicate and 12% silicic acid gel as nonnutritive bulk constituents produced silicon urolithiasis in dogs.[35]

When magnesium trisilicate was introduced as an antacid in 1937, it was assumed to be orally nonabsorbed. As previously noted, a mean of 5.2% of a dose of magnesium trisilicate is excreted (and, therefore, presumably absorbed) in healthy young male humans.[40,79] Although this amount of urinary silicon excretion can alter urinary pH on short-term therapy, clinically significant calculous formation is not produced. Higher doses, or even recommended doses, when administered over longer periods of time, can lead to silicon urolithiasis in man. Joekes et al.[49] have reviewed the available literature on this subject and conclude that this phenomenon in man is probably under-reported because most hospital laboratories do not routinely do stone analysis.

Nephropathy

It was well established early that the organic silicate tetraethyl-ortho-silicate orally produced extensive tubular necrosis similar to mercury in rabbits, guinea pigs, and rats.[52] Newberne and Wilson[75] have recently exposed rats and dogs orally for 30 days to sodium silicate, magnesium trisilicate, silica, and aluminum silicate. The dose level was equivalent to 0.8 g/kg/day for all compounds. No pathology was noted for any compound in rats; however, renal lesions were produced in beagle dogs for the first two compounds. Significant clinical abnormalities included polydipsia, polyuria, and soft feces in both species. However, renal focal subcapsular hemorrhages were noted grossly only in dogs. Microscopically, focal hypertrophy of tubular epithelium with or without degenerative changes, inflammatory cell infiltration into the interstitium, and dilatation and collapse of some tubules occurred. Glomeruli were not damaged. No evidence of urolithiasis or other type of mechanical blockage was noted. These early microscopic changes were not associated with alteration of urine-specific gravity or blood-urea nitrogen measurements. Long-term studies of feeding silica and silicates to rats in amounts comparable to antacid medication have revealed no nephropathy. The dog may be more sensitive to renal toxicity induced by silicic acid, or the rat may absorb lesser amounts orally of silicic acid; other investigators have demonstrated such changes in rats with parenteral silicic acid[81,82] and with a quartz suspension orally in guinea pigs.[68]

A moderate increased frequency of renal pathology in individuals suffering from silicosis has been noted.[57] An in-depth anatomical and renal-function evaluation of such renal pathology in an individual associated with long-term chronic occupational exposure to silicon dusts has been conducted.[87] Albuminuria and hypertension were observed clinically. The major changes on renal biopsy were a focal membranoproliferative type of glomerulonephritis and degenerative changes in the proximal tubules associated with deposition of electron-dense particles in cytosomes. Biopsy also revealed an eight- to tenfold increase over the mean silicon content of individuals with chronic renal failure from multiple etiologies other than silicon-dust exposure. The proximal tubular nephropathy is similar to that induced by a number of heavy metals. Interestingly, the renal pathology was not associated with biochemical alterations in proximal tubular function induced by certain heavy metals. The authors suggest that this is due to the fact that, unlike heavy metal-induced alteration in function, silicon does not affect the transport enzyme, Na-K-ATPase.

Silicon has been implicated as a causative factor in the etiology of Balkan or endemic nephropathy which affects individuals living along the Danube and other rivers in Bulgaria, Yugoslavia, and Romania. It is characterized by a chronic persistent malignant degeneration of the renal parenchyma (chronic interstitial nephropathy). Polyps, papillocarcinoma, and carcinoma of the upper urinary tract are documented in one third of the cases at autopsy. There is considerable research being devoted to the etiology of this condition. One of the etiologies considered is silicate minerals leached into the drinking water of the region.[67,68] Other etiologies, infection[7] and immune,[88] are also potential candidates as causative agents of this phenomenon. A recent editorial[29] summarizes the current status of endemic nephropathy.

Whether silicon has a major, minor, or no role to play in endemic nephropathy, the potential direct nephrotoxicity of silicon compounds, especially simple silicates, is important to consider for those evaluating the role of silicon compounds in biology.[27] Although apparently not as clinically significant as the pneumoconioses induced by inhalation of similar compounds, a focus of interest continues to enlarge on long-term oral intake of silicon compounds in water, food, and drugs.

REFERENCES

1. **Allison, A. C.,** Silicon compounds in biological systems, *Proc. R. Soc. London Ser. B,* 171, 19–30, 1968.
2. **Allison, A. C., Clark, I. A., and Davies, P.,** Cellular interactions in fibrogenesis, *Ann. Rheum. Dis.,* 36 (Suppl.), 8–13, 1977.
3. **Austin, J. H., Rinehart, R., Williamson, T., Burcar, P., Russ, K., Nikaido, T., and Lafrance, M.,** Studies in aging of the brain. III. Silicon levels in postmortem tissues and body fluids, *Prog. Brain Res.,* 40, 485–495, 1973.
4. **Bailey, C. B.,** Silica excretion in cattle fed a ration predisposing to silica urolithiasis: Total excretion and diurnal variations, *Am. J. Vet. Res.,* 28, 1743–1749, 1967.
5. **Bailey, C. B.,** Renal function in cows with particular reference to clearance of silicic acid, *Res. Vet. Sci.,* 11, 533–539, 1970.
6. **Bailey, C. B.,** The precipitation of polymerized silicic acid by urine protein: A possible mechanism in the etiology of silica urolithiasis, *Can. J. Biochem.,* 50, 305–311, 1971.
7. **Barnes, J. M., Carter, R. L., Peristianis, G. C., Austwick, P. K. C., Flynn, F. V., and Aldridge, W. N.,** Balkan (endemic) nephropathy and a toxin-producing strain of *Penicillium verrucosum* var *cyclopium:* An experimental model in rats, *Lancet,* 1, 671–675, 1977.
8. **Baumann, H.,** Determination of silicic acid in biological material, *Z. Physiol. Chem.,* 319, 38–51, 1960.
9. **Baumann, H.,** Silic acid in human blood and urine, *Z. Physiol. Chem.,* 320, 11–20, 1960. 11–20, 1960.
10. **Bechtold, M. F., Vest, R. D., and Plambeck, L., Jr.,** Silicic acid from tetraethyl silicate hydrolysis. Polymerization and properties, *J. Am. Chem. Soc.,* 90, 4590–4598, 1968.
11. **Bennett, D. R. and Åberg, B., Eds.,** 2,6-*cis*-Diphenylhexamethylcyclotetrasiloxane chemistry, analytical chemistry, biological effects, and excretion, *Acta Pharmacol. Toxicol.,* 36(III), 1–147, 1975.
12. **Bischoff, F.,** Organic polymer biocompatibility and toxicology, *Clin. Chem.,* 18, 869–894, 1972.
13. **Bodganskaya, N. I.,** Poly(vinylpyridine-N-oxide) medication of experimental silicosis, *Gig. Tr. Prof. Zablo.,* 1973, 22–25, 1973.
14. **Bolton, R. E. and Davis, J. M. G.,** The short-term effect of chronic asbestos ingestion in rats, *Ann. Occup. Hyg.,* 19, 121–128, 1976.
15. **Bryson, G. and Bischoff, F.,** Silicate-induced neoplasms, *Prog. Exp. Tumor Res.,* 9, 77–164, 1967.
16. **Burger, H.,** Anomalies in the structural chemistry of silicon, *Angew. Chem. Int. Ed. Engl.,* 12, 475–486, 1973.
17. **Byczkowski, S. and Wrzesniowska, K.,** Studies on the physiological hair silicon content in men, *Toxicology,* 5, 123–124, 1975.
18. **Carlisle, E. M.,** Silicon as an essential element, *Fed. Proc. Fed. Am. Soc. Exp. Biol.,* 33, 1758–1766, 1974.
19. **Carlisle, E. M.,** Essentiality and function of silicon, in *Trace Element Metabolism in Animals,* Vol. 2, Hoekstra, W. G., Suttie, J. W., Ganther, H. E., and Mertz, W., Eds., University Park Press, Baltimore, Md., 1974, 407–423.
20. **Carlisle, E. M.,** Silicon, *Nutr. Rev.,* 33, 257–261, 1975.
21. **Carlisle, E. M.,** In vivo requirement for silicon in articular cartilage and connective tissue formation in the chick, *J. Nutr.,* 106, 478–484, 1976.
22. **Chedd, G.,** Half-way elements, *The Technology of Metalloids,* Doubleday, Garden City, N.Y., 1969, 9–31 and 85–102.
23. **Cook, P. M., Glass, G. E., and Tucker, J. H.,** Asbestiform amphibole minerals: Detection and measurement of high concentrations in municipal water supplies, *Science,* 185, 853–855, 1974.
24. **Cotton, F. A. and Wilkinson, G.,** *Advanced inorganic chemistry: A comprehensive text,* 3rd ed., Interscience, New York, 1972.
25. **Darley, W. M. and Volcani, B. E.,** Role of silicon in diatom metabolism, *Exp. Cell Res.,* 58, 334–342, 1969.
26. **Davis, J. M. G.,** Effects of polyvinylpyridine-N-oxide (P204) on the cytopathogenic action of chrysotile asbestos in vivo and in vitro, *Br. J. Exp. Pathol.,* 53, 652–658,.1972.
27. **Dobbie, S. J. W., Goray, M. J. B., and Kennedy, A. C.,** Silicon and the kidney, *Q. J. Med.,* 45, 691–692, 1976.
28. **Eaborn, C.,** *Organosilicon Compounds,* Academic Press, New York, 1960.
29. **Editorial,** Balkan nephropathy, *Lancet,* 1, 683–684, 1977.
30. **Ehrhart, L. A. and McCullaug, K. G.,** Silica urolithiasis in dogs fed an atherogenic diet, *Proc. Soc. Exp. Biol. Med.,* 143, 131–132, 1973.

31. Asbestos particles in food and drugs, *Fed. Regist.,* 38, 27076–27081, 1973.
32. **Fessenden, R. J. and Fessenden, J. S.,** The biological properties of silicon compounds, *Adv. Drug Res.,* 4, 95–132, 1967.
33. **Fregert, S.,** Studies on silicon in tissues with special reference to skin, *Acta Derm. Venereol. Suppl.,* 39, 1–91, 1959.
34. **Furia, T. E.,** in *CRC Handbook of Food Additives,* CRC Press, Cleveland, 1972, 791.
35. **Garson, L. R. and Kirchner, L. K.,** Organosilicon entities as prophylactic and therapeutic agents, *J. Pharm. Sci.,* 60, 1113–1127, 1971.
36. **Gendre, P.,** Ultrastructure of atheroma in the rabbit: Action of a silicon derivative. I. Effect on the lipidic plaque, *C.R. Soc. Biol.,* 163, 1373–1376, 1969.
37. **Gross, P.,** Is short-fibered asbestos dust a biological hazard?, *Arch. Environ. Health,* 29, 115–117, 1974.
38. **Gross, P., and Harley, R. A., Jr.,** The locus of pathogenicity of asbestos dust, *Arch. Environ. Health,* 27, 240–242, 1973.
39. **Haley, T. J.,** Asbestosis: A reassessment of the overall problem, *J. Pharm. Sci.,* 64, 1435–1449, 1975.
40. **Heffner, R. R., Page, R. C., and Frey, A.,** Urinary excretion of silica in humans following oral administration of magnesium trisilicate, *Am. J. Dig. Dis.,* 8, 219–221, 1941.
41. **Heinen, W.,** Time-dependent distribution of silicon in intact cells and cell-free extracts of *Proteus mirabilis* as a model of bacterial silicon transport, *Arch. Biochem. Biophys.,* 110, 137–149, 1965.
42. **Heinen, W.,** Ion accumulation in bacterial systems. III. Respiration-dependent accumulation of silicate by a particulate fraction from *Proteus mirabilis* cell-free extracts, *Arch. Biochem. Biophys.,* 120, 101–107, 1967.
43. **Heppleston, A. G.,** Observations on the mechanism of silicotic fibrogenesis. Inhaled particles, in *Proc. 3rd Int. 1970 Symp.,* Walton, W. H., Ed., Unwin Brothers, Old Woking, Surrey, England, 1971, 357–371.
44. **Holt, P. F., Yates, D. M., and Tomlin, D. M.,** Distribution of injected Si-31 in rats, *Biochem. J.,* 48, 14–15, 1951.
45. **Holt, P. F. and Lindsay, H.,** Isotactic and syndiotactic poly-(2-vinylpyridine 1-oxide) and 2-vinylpyridine oxide-2-*n*-propenylpyridine oxide copolymer. NMR and UV spectra, viscosities and reactions with silicic acids, *J. Chem. Soc. C,* 1012–1015, 1969.
46. **Hurtubise, R. J.,** Determination of silicon in streptomycin by atomic absorption, *J. Pharm. Sci.,* 63, 1128–1130, 1974.
47. **Iler, R. K.,** *Colloid chemistry of silica and silicates,* Cornell University Press, Ithaca, N.Y., 1955.
48. **Jankowiak, M. E. and LeVier, R. R.,** Elimination of phosphorus interference in the colorimetric determination of silicon in biological material, *Anal. Biochem.,* 44, 462–472, 1971.
49. **Joekes, A. M., Rose, G. A., and Sutor, J.,** Multiple renal silica calculi, *Br. Med. J.,* 1, 146–147, 1973.
50. **Jones, L. H. P., and Handreck, K. A.,** Silica in soils, plants and animals, in *Advances in Agronomy,* Vol. 19, Norman, A. G., Ed., Academic Press, New York, 1967, 107–148.
51. **Jorpes, J. E.,** *Jacob Berzelius, His Life and Work,* Almqvist and Wiksells Boktryckeri AB, Uppsala, Sweden, 1970, 156.
52. **Kasper, J. A., McCord, C. P., and Frederick, W. G.,** Toxicity of organic silicon compounds. I. Tetraethyl-ortho-silicate, *Ind. Med.,* 6, 660–664, 1937.
53. **Keeler, R. F. and Lovelace, S. A.,** The metabolism of silicon in the rat and its relation to the formation of artificial siliceous calculi, *J. Exp. Med.,* 109, 601–614, 1959.
54. **Keeler, R. F.,** Silicon metabolism and silicon-protein matrix interrelationship in bovine urolithiasis, *Ann. N.Y. Acad. Sci.,* 104, 592–611, 1963.
55. **Keeler, R. F. and Lovelace, S. A.,** The effect of urinary silicon concentration on the formation of siliceous deposits on bladder implants in steers, *Am. J. Vet. Res.,* 22, 617–619, 1961.
56. **King, E. J., Stacy, B. D., Holt, P. F., Yates, D. M., and Pickles, D.,** The colorimetric determination of silicon in the microanalysis of biological material and mineral dusts, *Analyst,* London, 80. 441–453, 1955.
57. **Kolev, K., Doitschinov, D., and Todorov, D.,** Morphological alterations in the kidneys by silicosis, *Med. Lavoro.,* 61, 205–212, 1970.
58. **Lentz, C. W.,** Silicate minerals as sources of trimethylsilyl silicates and silicate structure analysis of sodium silicate solutions, *Inorg. Chem.,* 3, 574–579, 1964.
59. **Leslie, J. G., Kao, K. T., and McGavack, T. H.,** Silicon in biological material. II. Variations in silicon contents in tissues of rats at different ages, *Proc. Soc. Exp. Biol. Med.,* 110, 218–220, 1962.
60. **LeVier, R. R.,** Distribution of silicon in the adult rat and rhesus monkey, *Bioinorg. Chem.,* 4, 109–115, 1975.

61. **Lewin, J. and Reimann, B. E. F.,** Silicon and plant growth, *Ann. Rev. Plant Physiol.,* 19, 289–304, 1968.
62. **Lichtenwalner, H. K. and Sprung, M. N.,** Silicones, *Encycl. Polym. Sci. Technol.,* 12, 464–569, 1970.
63. **Loeper, J.,** Studies of silicon in animal tissues, Ph.D. thesis, University of Paris, France, 1965.
64. **Loeper, J. and Loeper, J.,** Investigations on the role of silicon in the arterial wall, *C.R. Soc. Biol.,* 155, 468–470, 1961.
65. **Loeper, J., Loeper, J., and Beurlet, J.,** Studies on the early lesions of atherosclerosis, *La Presse Med.,* 71, 169–172, 1963.
66. **Macnab, G. and Harington, J. S.,** Haemolytic activity of asbestos and other mineral dusts, *Nature,* 214, 522–523, 1967.
67. **Markovic, B.,** Malignant diseases of urinary tract in Balkan countries, *Urol. Int.,* 27, 130–143, 1972.
68. **Markovic, B. L. and Arambasic, M. D.,** Experimental chronic interstitital nephritis compared with endemic human nephropathy, *J. Pathol.,* 103, 35–40, 1971.
69. **Marks, J.,** The neutralization of silica toxicity in vitro, *Br. J. Ind. Med.,* 14, 81–84, 1957.
70. **McGavack, T. H., Leslie, J. G., and Kao, K. T.,** Silicon in biological material. I. Determinations eliminating silicon as a contaminant, *Proc. Soc. Exp. Biol. Med.,* 110, 215–218, 1962.
71. **Mehard, C. W. and Volcani, B. E.,** Similarity in uptake and retention of trace amounts of [31]silicon and [68]germanium in rat tissues and cell organelles, *Bioinorg. Chem.,* 5, 107–124, 1975.
72. **Mehard, C. W. and Volcani, B. E.,** Evaluation of silicon and germanium retention in rat tissues and diatoms during cell and organelle preparation for electron probe microanalysis, *J. Histochem. Cytochem.,* 23, 348–358, 1975.
73. **Mehard, C. W. and Volcani, B. E.,** Silicon-containing granules of rat liver, kidney and spleen mitochrondria. Electron probe X-ray microanalysis, *Cell Tissue Res.,* 174, 315–328, 1976.
74. **Miskovits, G., Orban, E., Dubay, M., Appel, J., Ördögh, M., and Szabó, E.,** Comparative determination of silicon in lymph nodes by neutron activation and spectrophotometry, *Acta Morphol. Acad. Sci. Hung.,* 18, 73–78, 1970.
75. **Newberne, P. M. and Wilson, R. H.,** Renal damage associated with silicon compounds in dogs, *Proc. Natl. Acad. Sci. U.S.A.,* 65, 872–875, 1970.
76. **Nicholson, W. J., Maggiore, C. J., and Selikoff, I. J.,** Asbestos contamination of parenteral drugs, *Science,* 177, 171–173, 1972.
77. **Nielsen, F. H.,** New trace elements and possible applications in man, in *Trace Elements in Human Health and Disease,* Prasad, A. S., and Oberleas, D., Eds., Academic Press, New York, 1976, 390–393.
78. **Nikaido, T., Austin, J., Trueb, L., and Rinehart, R.,** Studies in aging of brain. II. Microchemical analysis of the nervous system in Alzheimer patients, *Arch. Neurol.,* 27, 549–554, 1972.
79. **Page, R. C., Heffner, R. R., and Frey, A.,** Urinary excretion of silica in humans following oral administration of magnesium trisilicate, *Am. J. Dig. Dis.,* 8, 13–15, 1941.
80. **Peace, B. W., Mayhan, K. G., and Montle, J. F.,** Polymers from the hydrolysis of tetraethoxysilane, *Polymer,* 14, 420–422, 1973.
81. **Policard, A. and Collet, A.,** Experimental studies of renal lesions induced by the excretion of silica, *J. Urol. Med. Chir.,* 60, 164–171, 1954.
82. **Policard, A., Collet, A., Moussard, D. H., and Pregermain, S.,** Early manifestations of alterations induced in kidney by silicic acid. Study with electron microscope, *J. Urol. Med. Chir.,* 66, 585–600, 1960.
83. **Policard, A., Collet, A., Daniel-Moussard, H., Pregermain, S., and Reuet, C.,** Deposition of silica in mitochondria: An electron microscopic study, *J. Biophys. Biochem. Cytol.,* 9, 236–238, 1961.
84. **Robock, K. and Klosterkotter, W.,** Investigations on the specific toxicity of different SiO_2 and silicate dusts, *Staub Reinhalt. Luft,* 33, 60–63, 1976.
85. **Rochow, E. G.,** The chemistry of silicon, in *Comprehensive Inorganic Chemistry,* Pergamon Press, Oxford, England, 1975, 1323–1467.
86. **Rohl, A. N., Langer, A. M., Selikoff, I. J., Tordini, A., and Klimentidis, R.,** Consumer talcums and powders: Mineral and chemical characterization, *J. Toxicol. Environ. Health,* 2, 255–284, 1976.
87. **Saldanha, L. F., Rosen, V. J., and Gonick, H. C.,** Silicon nephropathy, *Am. J. Med.,* 59, 95–103, 1975.
88. **Sattler, T. A., Dimitrov, T. S., and Hall, P. W.,** Relation between endemic (Balkan) nephropathy and urinary tract tumors, *Lancet,* 1, 278–280, 1977.
89. **Sauer, F., Laughland, D. H., and Davidson, W. M.,** Silica metabolism in guinea pigs, *Can. J. Biochem. Physiol.,* 37, 183–191, 1959.

90. **Sauer, F., Laughland, D. H., and Davidson, W. M.,** The silica content of guinea pig tissues as determined by chemical and isotopic techniques, *Can. J. Biochem. Physiol.,* 37, 1173–1181, 1959.

91. **Schafer, P. W. and Chandler, J. A.,** Electron probe x-ray microanalysis of a normal centriole, *Science,* 170, 1204–1205, 1970.

92. **Schlipköter, H. W., Dolgner, R., and Brockhaus, A.,** The treatment of experimental silicosis, *Ger. Med. Mon.,* 8, 509–514, 1963.

93. **Schwarz, K.,** in *New Trace Elements in Nutrition,* Mertz, W. and Cornatzer, W. E., Eds., Marcel Dekker, New York, 1971, 313.

94. **Schwarz, K.,** A bound form of silicon in glycosaminoglycans and polyuronides, *Proc. Natl. Acad. Sci. U.S.A.,* 70, 1608–1612, 1973.

95. **Schwarz, K.,** Recent dietary trace element research exemplified by tin, fluorine, and silicon, *Fed. Proc. Fed. Am. Soc. Exp. Biol.,* 33, 1748–1757, 1974.

96. **Schwarz, K.,** New essential trace elements (Sn, V, F, Si): Progress report and outlook, in *Trace Element Metabolism in Animals,* Vol. 2, Hoekstra, W. G., Suttie, J. W., Ganther, H. E., and Mertz, W., Eds., University Park Press, Baltimore, Md., 1974, 355–380.

97. **Schwarz, K.,** Silicon, fiber and atherosclerosis, *Lancet,* 1, 454–457, 1977.

98. **Schwarz, K., Ricci, B. A., Punsar, S., and Karvonen, M. J.,** Inverse relation of silicon in drinking water and atherosclerosis in Finland, *Lancet,* 1, 538–539, 1977.

99. **Sommer, L. H.,** *Sterochemistry, Mechanism and Silicon,* McGraw-Hill, New York, 1965.

100. **Speier, J. L.,** The Si–C bond and silicones, *Chemistry,* 37, 6–11, 1964.

101. **Voronkov, M. G.,** Biologically active compounds of silicon, *Russ. Chem. Rev.,* 38, 975–986, 1969.

102. **Voronkov, M. G., Zelchan, G. I., and Lukevics, E. J.,** *Silicon and Life,* Zinate, Riga, Latvia, 1971.

103. **Voronkov, M. G.,** Bio-organosilicon chemistry, *Chem. Br.,* 9, 411–415, 1973.

104. **Werner, D.,** The physiology and biochemistry of silicic acid, *Ber. Dtsch. Bot. Ges.,* 81, 425–429, 1968.

105. **Westman, A. E. R. and Murthy, M. K.,** Silica and silicates, *Encycl. Polym. Sci. Technol.,* 12, 441–464, 1970.

106. **WHO** Food Additive Series, 16th Report of the Joint FAO/WHO Expert Committee on Food Additives in Rome. Safety evaluation of some food additives, 4, Technical Report No. 557, WHO, Geneva, 1974, 37.

EFFECT OF NUTRIENT TOXICITIES IN ANIMALS AND MAN: TIN

R. A. Hiles, J. J. Franxman, C. O. Weiskittel, and D. W. Briggs

INTRODUCTION

Occurrence of Tin

Tin is found primarily in the ore cassiterite or tinstone (SnO_2). There are traces of tin in most soils and plants. Levels of 1.2 ppm in wheat flour, 1.6 ppm in bran, and 3.9 ppm in the outer pericarp of wheat have been reported.[1] Of the 30 trace metals of possible or probable biological significance, tin ranks 21st in abundance in the universe, 17th on the geosphere (3 ppm), 12th in the hydrosphere (3 ppb), and 8th in the body of man (0.4 ppm).[2] Sizable amounts of tin are present in the air of industrial cities. For example, amounts from 0.003 to 0.3 $\mu g/m^3$ were found in 60.6% of samples taken from 22 U.S. cities,[3] while 17% of samples from 6 major industrial centers had 0.03 to 0.3 μ/m^3.[4]

Until 1970, tin was generally regarded as a trace metal which is not necessary for life or health. This conclusion was probably based mainly on the prevailing, inadequate methods of tin analysis which showed "zero" levels of tin in newborn children and in natives of some foreign countries. In 1970, however, Schwarz and co-workers found that in rats maintained on purified amino acid diets in trace-element-controlled isolators, supplements of tin caused a significant increase in growth.[5] Tin (as stannic ion), at levels of 1 to 2 ppm of the diet, increased growth by 53 to 59%. These levels are similar to those normally present in foods, feeds, and tissues. The results suggest that tin may be a hitherto unrecognized essential trace element. Information concerning the evidence of a dietary need for tin in humans is not available; thus, its role as an essential element has not been proven.

Table 1 shows the analytical results of tests for tin in human tissues of subjects in the U.S.[2,6] Because of the wide individual variations, ranges are not shown. Reports of nondetectable levels in certain cases (for example, in the newborn) must be interpreted carefully, in view of the known analytical difficulties (such as the volatility of some tin salts during drying and ashing). Table 2 shows the occurrence of tin in several tissues in persons from various countries.[2,6]

Cases of Poisoning by Tin Compounds

Symptoms of acute tin poisoning (vomiting, diarrhea, nausea, abdominal cramps, and metallic taste)[7,8] have been known to occur in cases where as little as 100 to 150 mg of tin (as tin salts) has been ingested as part of a contaminated food product.[9-12] Still, the incidence of acute oral poisoning is low, and even as early as 1933 it was believed that "with less than two grains of tin per pound, there appears to be practically no danger of acute poisoning."[7] The level of 250 mg tin per kilogram (food) is generally considered to be the upper admissible limit of the metal in canned foods.[13]

A number of cases of pneumoconiosis (stannosis) due to inhalation of tin oxide dusts for long periods of time have been reported.[14-17] The pneumoconiosis was benign and symptomless, but was sharply recognizable by roentgenography. Tin deposited in the lungs in this form is considered nontoxic systemically.[14] Sn(II) oxidizes readily in air; therefore, doses or diets of Sn(II) salts reported in these studies may contain significant amounts of Sn(IV).

PHARMACODYNAMICS

Absorption, Distribution, and Excretion of Tin

Average daily human consumption of tin in food has been estimated to be between 1 and 45 mg tin.[2,18-21] Studies of tin in rats have shown a number of effects. The amount of tin absorbed from the gut is small; recent studies using radioisotopes show this absorption to be less than 3%.[22-24,24a,31] Stannous (II) tin is better absorbed than stannic (IV) tin, but the absorption of each species can be affected by both the anion complement of the metal ion and the food content of the gut.[23]

Excretion of systemic tin is primarily in the urine, but fecal elimination following biliary excretion plays a significant, though minor, role.[22-25] Liver, kidneys, and perhaps the adrenal glands are considered to be the major targets among the soft tissues for deposition of tin.[22-24] Bone appears to have the highest affinity for tin, and the turnover rate of tin in bone is slower than for other systems.[20,22-24,26,27] Moreover, bone appears to be the only tissue which accumulates tin during prolonged exposure.[23] After oral ingestion, tin also seems to persist to some extent in the intestines.[22,23,24a] Tin has been found in the lungs, but this occurrence of tin deposition is believed to result from inhalation of airborne tin metal or salts.[2,23] Accumulation of tin in the fetus does not occur to a significant extent.[2,23,28-30]

Table 3 shows the results of a number of studies investigating the pharmacodynamics of tin and its salts in animals and man. The analytical techniques used for the determination of tin may contain inherent difficulties (such as the volatilization of some tin salts). These analytical problems tend to lessen the certainty of the "quantitative" values listed. Interpretation and comparison of these results must be done with this thought in mind.

TOXICOLOGY OF TIN COMPOUNDS

Acute and Subacute Studies

Tin compounds have been administered to rats, mice, guinea pigs, rabbits, cats, dogs, and man to determine the toxic effects which may occur from such compounds. In this category of acute and subacute studies are investigations which were run for periods of 1 day to 13 weeks. Table 4 is taken from a study by deGroot et al.[34] in which rats were given various tin compounds at several dose levels as part of their diets. The rats were fed the tin compounds for periods from 4 weeks to 13 weeks. It should be kept in mind that Sn(II) diets may contain significant amounts of Sn(IV), because of the ease of oxidation of Sn(II).

Table 4 shows several results of interest in the 4-week feeding study of tin salts to rats.[34] There is quite a difference in the toxicities of the various tin compounds fed. Rats receiving stannous sulfide, stannous oleate, or stannic oxide, at dietary levels up to 1%, showed no noticeable effects. However, rats receiving stannous chloride, orthophosphate, sulfate, oxalate, or tartrate experienced growth retardation, anemia, and histological changes in their livers when the above salts constituted 0.3 or 1.0% of the diet. The variance in the toxicities of these compounds suggests that insoluble tin compounds have low toxicities but tin compounds which are soluble in water or dilute acid may produce toxic effects at dietary levels of 0.3% or higher. The variability in literature reports of the level of tin in foods necessary to produce toxic symptoms may be due to this greater toxicity of some tin compounds when compared to others.

The growth retardation caused by some tin salts in this study is not directly attributable to a lower food intake in some groups because food efficiency was lowered. This idea is also supported by a study in which Spanjers and deGroot noted lower body weights and lower hemoglobin levels in rats fed a 1% stannous chloride diet compared to rats on a control diet, even though food consumption was equal in both test groups.[34]

In the 13-week feeding study,[34] stannous oxide and stannous chloride were fed at levels up to 1% of the diet. The rats receiving the stannous oxide diet, even at the 1% level, showed no abnormalities or significant changes in any of the parameters measured.

Thus, stannous oxide seems to be a relatively harmless substance when added to the diet of rats.

The addition of stannous chloride to food, however, showed several toxic effects in the 13-week study, especially at the 1% dietary level. Rats at this level ate little food and abdominal distention was noted after only one week. Growth of animals was slow for the first several weeks and stopped completely after 4 to 6 weeks of the study. By week eight, several animals lost weight, and by week nine, four males had died. At the 0.3% diet level, poor appetite and abdominal distention were evident within 2 weeks, but all rats except one kept growing to some extent for the duration of the study. Rats fed the 0.1% or lower stannous chloride diets had normal body weight gains.

From the fourth week on, hemoglobin values were decreased in both the 0.3 and 1.0% groups. Hemoglobin values in rats receiving 0.1% stannous chloride or less were not significantly ($p < 0.05$) different from values of controls.

Other parameters (hematology, serum enzymes, and organ weights) showed few significant abnormalties in groups fed 0.3% or less stannous chloride by week 13. (The 1.0% group had been sacrificed at week nine, owing to their then morbid appearance.)

An autopsy of the rats at the 1% level showed distended intestines, slight ascites, small edematous pancreases, and grayish-brown livers. Histopathology showed moderate testicular degeneration, severe pancreatic atrophy, sponginess of part of the brain, acute bronchopneumonia, enteritis, and liver changes. The liver changes were characterized by a homogeneous appearance of the liver cell cytoplasm and a mild proliferation (oval cell type) of the bile duct epithelium.

Autopsy at 13 weeks showed only minor histopathological changes that were related to treatment. These were found in livers of rats from the 0.3% group. Other findings were considered unrelated to treatment.

The highest level of active tin compounds which did not produce toxic effects in these 4- and 13-weeks studies was 0.1%. This level corresponds to 450 to 650 ppm tin in the diet per kilogram body weight per day in a 90-day study. Results from another study suggest that the "no-effect level" of tin may be lower than 0.1% in diets which contain less iron.[34]

The safe level of tin in the diet seems low for a substance known to be poorly absorbed. This could be due to an effect of tin in the intestinal lumen, by which tin inhibits the absorption of dietary iron. Table 5 presents an overview of several other acute and subacute studies of tin toxicity.

Chronic Toxicity Studies

Several chronic studies have been conducted to determine the long-term toxic effects of tin compounds. Table 6 gives the results of these studies. Keep in mind once again that Sn(II) compounds oxidize easily to Sn(IV) in air. Thus doses of Sn(II) listed may contain some Sn(IV).

PHARMACOLOGY STUDIES INVOLVING TIN COMPOUNDS

Several studies have been done with tin compounds which show a pharmacologic effect in the species tested. Table 7 gives an overview of the results of these pharmacological studies.

SUMMARY

Tin deficiency has been produced in rats raised in an all-plastic isolator system and fed highly purified amino acid diets. Significant growth stimulation was observed in response to diets containing tin supplements at a level of 1 to 2 mg/kg;[5] therefore, tin may be considered an essential element for the rat. Information concerning the metabolic role of this element and evidence of a dietary need for tin in other species are lacking. The analytical determination of tin is complicated by the great volatility of some tin compounds. For this reason, the reported absence of tin from some soils, plants, and human tissues cannot be used as an argument against a possible biological function for this trace mineral.

Tin appears to be poorly absorbed in food. Rats fed diets containing 0.3% or more of stannous chloride, sulfate, oxalate, tartrate, or orthophosphate for 4 to 13 weeks exhibited growth retardation, anemia, and histological changes in the liver.[34] When tin was present in the diet at concentrations of 450 mg/kg or less, no toxic effects were apparent. The toxicity of tin may be greater if the diet is marginally deficient in iron. Lifetime studies with rats given diets containing tin as stannous chloride (5 mg/kg) suggest some toxicity at this dose level. Females had a decreased life span and both sexes showed increased incidence of fatty degeneration of the liver and of vascular changes in the renal tubules, as well as a moderate accumulation of tin in the tissues.[46] In mice, only tin accumulation was observed.[48,49] The subtle effects reported in rats exposed to long-term ingestion of tin deserve additional evaluation in relation to the levels which appear in human dietaries.

In studies on human volunteers, mild signs of toxicity were observed when tin was administered in fruit juices at concentrations up to 2000 ppm in amounts corresponding to 5 to 7 mg/kg of body weight.[26] A concentration of 0.25 g tin per kilogram food is generally considered the maximum permissible limit for tin in canned food. The question of permissible limits of tin is currently being investigated.

The daily intake of tin by an adult male has been estimated at 1 to 45 mg.[18-21] This estimate is higher than would be calculated from the tin content of individual unprocessed foods, suggesting that canning may have contributed a substantial amount of tin. Although in most countries cans are now coated with lacquer, many canned foods still contain higher concentrations of tin than uncanned foods. Quite high levels of tin in canned foods were often reported before coating the cans with lacquer was widely used. The use of tin foil in packaging foods can also be a source of increased tin content. Changes in packaging, the replacement of tin with other materials, and the use of coatings in canning have all contributed to the decrease in tin content of processed food. Except in unusual circumstances, ingested amounts are well below the level of 5 to 7 mg/kg body weight that has been shown to induce toxicity symptoms.

The recent demonstration that tin has a growth-stimulating effect in rats calls for a careful reexamination of our knowledge of this element. The published analytical data should be confirmed by modern methods, in view of the reported absence of tin in the tissues of newborns and in some foods. Tin deficiency is not known in man. Therefore, even the lower estimated daily intake may be assumed to meet the requirement, if any, of man for this substance. This intake is far less than the amounts producing toxicity.

Because organotin compounds do not normally occur in nature, they cannot be considered essential nutrients; therefore, they have not been included in this review. It should be noted, however, that many organotin compounds are highly toxic, often affecting the electron transport systems. Literature concerning the toxicity of organotin compounds is available.[59,60]

Table 1
TIN IN HUMAN TISSUES BY AGE, MEAN U.S. VALUES (ppm ASH)

ge group (years)	0	0–1	1–10	11–20	21–30	31–40	41–50	51–60	61–70	71–84
iney										
Concentration	0	57[a]	60[a]	33	20	34[a]	28[a]	22	34	32
Occurrence	0/16	9/10	19/20	10/15	17/21	36/43	22/26	26/27	13/14	8/9
ver										
Concentration	0	48[a]	61[a]	42	34	33[a]	25	35	38[a]	34[a]
Occurrence	0/21	11/12	22/23	14/14	20/21	39/43	27/28	25/25	13/13	9/9
ing										
Concentration	18	35	34	45[a]	27	31[a]	39[a]	53[a]	58	64[a]
Occurrence	1/10	5/6	5/6	10/13	18/21	34/35	27/27	28/29	13/13	9/9
um										
Concentration	0	101	98	80	174	116	97	53	172	140
Occurrence	0/8	1/2	1/1	9/10	12/13	26/28	16/17	12/12	6/6	3/4
tal occurrence (%)	1.8	86.7	94.0	82.7	88.2	79.8	93.9	97.8	97.7	93.5

ote: Concentrations of positive samples. Occurrence refers to the number positive per total cases.

Excluding values > 150.

aken from Schroeder, H. A., Balassa, J. J., and Tipton, I. H., *J. Chronic Dis.*, 17, 483, 1964. With permission.

Table 2
GEOGRAPHICAL DISTRIBUTION OF TIN IN
HUMAN KIDNEY, LIVER, AND LUNG (ppm ASH)

	No. of samples	% in which tin was found	Mean	Range	Median
Kidney					
U.S.	161	97	30	<5–480	20
Switzerland	9	33	5	<5–17	<5
Africa	53	19	5	<5–30	<5
Middle East	43	42	8	<5–55	<5
Far East	57	40	11	<5–110	<5
Liver					
U.S.	163	96	35	<5–300	24
Switzerland	9	78	6	<5–20	Tc[a]
Africa	49	25	4	<5–20	<5
Middle East	44	55	11	<5–140	Tc[a]
Far East	54	41	10	<5–74	<5
Lung					
U.S.	159	98	69	<5–920	39
Switzerland	7	100	37	23–62	28
Africa	50	60	10	<5–40	Tc[a]
Middle East	45	91	33	<5–200	17
Far East	57	93	65	<5–1200	29

Note: Mean and median values were obtained by assigning a value of one half the least detectable amount to those cases where tin was not detected.

[a] Trace, or barely detectable.

Taken from Schroeder, H. A., Balassa, J. J., and Tipton, I. H., *J. Chronic Dis.*, 17, 483, 1964. With permission.

Table 3
RESULTS OF VARIOUS PHARMACODYNAMIC STUDIES OF TIN AND ITS SALTS IN ANIMALS AND MAN

Species	Type of study[a]	Form of tin	Route of administration	Results and comments	Ref.
Rats	E	Canned foods	Oral	99% of tin intake was found in feces; none found in urine (2-day study)	26
Rats	E	Tin citrate in citric acid	Oral	96% of tin intake was found in feces; none found in urine (1-day study)	26
Rats	E	Tin (II) chloride	Oral	% Total tin found Time of experiment (days) — Feces — Urine 1 — 80.7 — 0.03 2 — 61.9 — 0.02 3 — 95.2 — 3.09 8 — 104.5 — 0.04 21 — 51.0 — 0.08	31
Rats	A, D, E	^{113}Sn(IV) citrate	i.m.	Complexing of ^{113}Sn with sodium citrate greatly enhanced absorption, with the amount remaining at injection site (i.m.) being 11% at 1 day and 3% at 60 days Urinary excretion of ^{113}Sn(IV) citrate only during the 24 hr after administration, accounting for 50% of the dose in the next 30 days, another 20% of the dose was eliminated by the digestive tract The skeleton was the most important deposition site; 30% of the dose had accumulated there by 1 day after dosing; two thirds of this skeletal tin was still in the skeleton 60 days later	22
Rats	A, D, E	^{113}Sn salts Tin (II) fluoride Tin (II) citrate Tin (II) pyrophosphate Tin (IV) fluoride Tin (IV) citrate	Oral (20 mg Sn/ kg body weight) i.v. (2 mg Sn/ kg body weight)	Absorption was found to be 2.8% for Sn(II) fluoride and citrate, and <1% for Sn(II) pyrophosphate; Sn(IV) fluoride and citrate absorption was only about 0.6%; within 48 hr, 50% of the absorbed tin was excreted; highest distribution of retained tin was in bone; soft-tissue retention was highest in liver and kidney (after repeated oral doses, only bone had an increase in accumulation of tin approximately proportional to systemic exposure) A single i.v. dose produced excretion of 30% of dose in urine; 11% of the Sn(II), but none of the Sn(IV) was eliminated in the bile	23

RESULTS OF VARIOUS PHARMACODYNAMIC STUDIES OF TIN AND ITS SALTS IN ANIMALS AND MAN

Species	Type of study[a]	Form of tin	Route of administration	Results and comments	Ref.
Rats	D, E	$^{113}Sn(II)Cl_2$ $^{113}Sn(IV)Cl_4$	Oral	Oral doses (single and repeated) produced only very small amounts of tin in urine; no characteristic distribution pattern was seen	24
			i.v.	Sn(II) content of bone was high, with significant amounts in liver also; the content of Sn(IV) in the bone was much lower, but the liver content of Sn(IV) was much higher than for Sn(II). Some ^{113}Sn was found in feces after the i.v. injection, but most ^{113}Sn was eliminated in the urine	
Rats	D	Tin (II) fluoride	Oral	SnF_2 was fed to pregnant rats, and fetuses were later analyzed for tin by emission spectroscopy; fetal tin level was higher in fetuses from those dams fed tin salts, but no consistent dose relationship was observed	28
Rabbits	A, D, E	Sodium stannous tartrate	s.c. (25 mg Sn/ kg body weight)	Subcutaneous injections of tin double salt were poorly absorbed; most of the injected tin was eliminated by means of the urine; a low level of tin was deposited in the liver	25
Man	E	Tin tartrate	Oral	After 5 days, the percentage of total tin intake was measured; 82—104% was found in the feces and 7.7—8.3% was excreted in the urine	32
Man	E	Canned foods	Oral	After 24 days, fecal excretion of tin equalled 96—119% of total tin intake; none was found in the urine	33

Table 4

MEAN VALUES FOR WEIGHT GAIN AND HEMATOLOGICAL PARAMETERS OF RATS FED A STOCK DIET SUPPLEMENTED WITH VARIOUS TIN COMPOUNDS AT LEVELS OF 0–1.0% FOR 4 WEEKS[a,b]

	Males				Females			
Dietary level (%)	Weight gain (g/rat)	Hb (g/100 ml)	HC (%)	RBC (10^6/ mm^3)	Weight gain (g/rat)	Hb (g/100 ml)	HC (%)	RBC (10^6/ mm^3)
			Stannic Oxide					
0.00	125	14.5	46.2	6.2	77	15.2	48.1	7.3
0.30	125	14.5	46.5	6.2	73	15.3	50.4	7.0
1.00	129	14.2	45.8	6.0	75	15.8	50.4	6.9
			Stannous Chloride					
0.00	132	14.2	–	6.0	78	15.5	–	7.0
0.30	93***	12.8*	–	5.7	68*	14.2**	–	6.6
1.00	47***	12.2**	–	5.9	43***	14.0*	–	6.6
			Stannous Orthophosphate					
0.00	125	14.4	46.7	6.9	63	15.1	50.6	7.3
0.30	87***	12.8**	42.4***	6.6	63	14.7	49.3	6.9
1.00	46***	12.3***	40.8***	6.4	37***	11.9***	42.4***	5.9***
			Stannous Sulfate					
0.00	129	14.3	45.5	6.3	82	14.4	46.8	7.9
0.30	108***	13.8	43.6*	6.5	79	13.6*	44.2*	7.1*
1.00	38***	12.1***	38.6***	6.1	46***	12.1***	42.3***	6.9***
			Stannous Sulfide					
0.00	126	14.3	45.8	6.2	75	14.8	48.2	6.6
0.30	125	14.3	46.1	6.1	83	14.1	48.8	6.6
1.00	131	14.3	47.4*	6.2	84	14.9	48.0	6.4

[a] Hb = hemoglobin; HC = hematocrit; RBC = red blood cells.

[b] Values are the means for groups of ten rats. Those marked with asterisks differ significantly (Wilcoxon's test) from those of controls: *$P < 0.05$; **$P < 0.01$; ***$P < 0.001$.

Taken from deGroot, A. P., Feron, V. J., and Til, H. P., *Food Cosmet. Toxicol.*, 11, 19, 1973. With permission.

Table 4 (continued)
MEAN VALUES FOR WEIGHT GAIN AND HEMATOLOGICAL PARAMETERS OF RATS FED A STOCK DIET SUPPLEMENTED WITH VARIOUS TIN COMPOUNDS AT LEVELS OF 0–1.0% FOR 4 WEEKS[a,b]

	Males				Females			
Dietary level (%)	Weight gain (g/rat)	Hb (g/100 ml)	HC (%)	RBC (10^6/ mm^3)	Weight gain (g/rat)	Hb (g/100 ml)	HC (%)	RBC (10^6/ mm^3)
Stannous Oleate								
0.00	127	14.2	45.7	6.8	84	15.5	48.0	7.0
0.30	129	14.3	46.1	6.1	83	14.1	48.8	6.6
1.00	129	14.7	47.4	6.7	81	15.2	48.0	7.1
Stannous Oxalate								
0.00	123	14.6	47.7	6.8	80	15.5	49.6	7.2
0.30	92***	12.4***	42.1***	5.8***	66***	13.0***	42.7***	6.3*
1.00	36***	11.1***	37.9***	5.5***	32***	11.7***	39.7***	6.2**
Stannous Tartrate								
0.00	128	13.4	44.6	6.5	74	14.1	46.9	7.3
0.30	114**	12.7*	42.5**	6.0	73	13.6	46.2	7.2
1.00	60***	12.0**	41.0***	6.4	50***	11.0**	38.7***	6.4

Note: Total food intake and food efficiency (weight gain [g] pet food consumed [g]) were measured, but showed no differences among groups. Dietary levels of 0.03 and 0.1% tin salts were administered, but results were not different in any parameters from central animals.

Table 5
ACUTE AND SUBACUTE STUDIES OF TIN TOXICOLOGY IN ANIMALS AND MAN

Species	Dose and time	Route of administration	Results and comments
Rats	150–5300 ppm $SnCl_2$ (as part of diet) for 4 to 6 weeks	Oral	Diets containing high levels of tin resulted in anemia unless large amounts of iron and copper were given; high tin levels also caused a growth depression in the rats, but this condition was not alleviated by enriching the diet with iron and copper; apparently, tin affects at least two different mechanisms: one related to hemoglobin synthesis and one related to growth
Rats	14–6000 mg $SnCl_2 \cdot 2H_2O$ kg body weight	i.v. i.p. p.o.	LD_{50} values (24-hr LD_{50} and 14-day LD_{50}) were obtained for stannous chloride (dihydrate) in rats (see below); signs of toxicity in rats included ataxia, muscle weakness, and CNS depression; the tin compound also produced pathologic changes in the rat kidney

Route of administration	24-hr LD_{50} (mg/kg)	14-day LD_{50} (mg/kg)
i.v.	29.3	15.8
i.p.	258.4	83.6
Oral (fasted rats)	2274.6	—
Oral (fed rats)	3190.1	—

Species	Dose and time	Route of administration	Results and comments
Rats	44.4 mg $SnCl_2 \cdot 2H_2O$/kg body weight	i.p.	Rat kidneys were studied with light and electron microscopy at 12 hr, 24 hr, 4 days, and 14 days after a single i.p. dose of stannous chloride Cytoplasmic swelling with fine vacuolization was the only abnormality noted in the proximal tubules at 12 and 24 hr; extensive necrosis of the proximal tubules and regeneration had occurred by 4 days; regeneration of the tubular cells was almost complete by day 7; occasional fibrous scars were noted in the cortex on day 14
Rats	7.6 mg $SnCl_2$/kg	s.c.	Testes of rats were examined histologically at 18 hr, 4 days, 8 days, or 12 days after single injection Results at 18 hr included shrunken tubules, edema, hyperemia, and markedly degenerated spermatozoa; at 4, 8, or 12 days, spermatogenesis had recovered except in the central tubules; necrosis persisted in the epididymis epithelial cells and in ductuli
Rats	7.6 mg $SnCl_2$/kg/ day for 30 days	s.c.	Histological examinations of rats testes were done at 2, 10, 21, and 30 days Severe damage to tubules occurred rapidly and spermatogenesis was almost suppressed in early stages; necrosis occurred in the head of the epididymis and in the ductuli; leukocyte infiltration occurred between tubules
Mice	2 mg/kg and 25 mg/kg $SnCl_4$	i.p.	A 2 mg/kg dose of stannic chloride administered i.p. to mice had a protective effect against a near lethal dose (25 mg/kg i.p.) of $SnCl_4$ given 24 hr later; the results of this study suggest that the tolerance acquired by pretreatment with the tin salt is a systemic phenomenon rather than a local one
Guinea pigs	40 mg $SnCl_2$/kg/ day for 6 days	i.m.	Produced nephritis, hepatitis, dyspnea, paralysis, death
Rabbits	40 mg $SnCl_2$/kg/ day for 6 days	i.m.	Produced nephritis, hepatitis, dyspnea, paralysis, death

Table 5 (continued)
ACUTE AND SUBACUTE STUDIES OF TIN TOXICOLOGY IN ANIMALS AND MAN

ecies	Dose and time	Route of administration	Results and comments	Ref.
e	60 mg SnCl$_2$/kg/ day for 8 days	i.m.	Produced hepatitis, nephritis, death	38
eons ts, cats, ogs, man	Fruit juices containing up to 2000 ppm tin	Oral	Orange juice and apple juice contaminated with up to 2000 ppm tin were administered to the species listed; the only toxic effects seen were ones involving gastrointestinal disturbances; the authors concluded that the gastrointestinal toxicity was caused by a local irritation of the mucous membrane of the alimentary tract rather than by a systemic absorption effect; some species seemed to acquire a tolerance to the juices	26
n	Canned foods	Oral	This is a review paper which lists some poisoning symptoms noticed after eating tin-contaminated canned foods; the symptoms include: lack of appetite, pressure feeling in the breast, diarrhea, vomiting, gastroenteritis, collapse, coma, colic, metallic taste, meteorism, tenesmus, throat dryness, nausea	13

Table 6
CHRONIC STUDIES OF TIN TOXICOLOGY IN ANIMALS AND MAN

Species	Dose and time	Route of administration	Results and comments
Rats	5 ppm $SnCl_2$ in drinking water for life of rats	Oral	Males showed lower body weight; females exhibited decreased survival and longevity and elevated glucose levels; both sexes had increased incidence of fatty changes in liver and renal tubular vacuolization; no criterion was altered in a favorable direction; these changes were found in the presence of only small amounts of tin in tissues
Rats	Sodium chloro-stannate as 2% of diet for 1 year	Oral	Three malignant tumors were found in 30 rats receiving the tin-containing diet; no malignancies were seen in 33 rats on a control diet; the difference in tumor incidence is regarded as being probably without significance
Mice	5 ppm $SnCl_2$ in drinking water for life of mice	Oral	Tin accumulated in the spleen and, to a lesser extent, in the heart, but no innate toxicity of tin was demonstrated in mice; additionally, stannous chloride was not carcinogenic
Mice	5000 ppm tin as sodium chloro-stannate or tin oleate in water or diet for 1 year	Oral	Mice receiving tin had a lower incidence of malignant tumors than did untreated controls; moreover, no other untoward effects were seen
Mice	5 ppm tin as $SnCl_2$ in drinking water for life of mice	Oral	No correlation was found between incidence of tumors and administration of tin
Rats Rabbits Dogs	250—1000 mg SnO_2/kg body weight	i.v.	Gross and histological examinations following long retention (4 months to 5 years) of SnO_2 particles have shown no fibrosis and no neoplasia and no other evidence of interference with life processes
Man	-	-	Concentrations of tin in cancerous and noncancerous tissues were measured by emission spectrography; no obvious correlation could be found between metal content of tissues and cancer

Table 7
PHARMACOLOGY STUDIES INVOLVING TIN COMPOUNDS

cies	Dose and time	Route of administration	Results and comments	Ref.
tro	—	—	Incubation of sheep pulmonary alveolar macrophages with 0.5 or 1.0 mM stannous ion resulted in a greater than 30% inhibition of the ATPase system of the cells	53
	200 mg SnF$_2$/kg diet or 100 mg NaF + 265 mg SnCl$_2$·2H$_2$O/kg diet or 265 mg SnCl$_2$·2H$_2$O/kg diet for 9 weeks	Oral	Rats receiving SnCl$_2$·2H$_2$O or SnCl$_2$·2H$_2$O plus NaF exhibited significant decreases in hepatic pyruvate and pyruvate dehydrogenase activity; these groups also had urinary cAMP levels significantly above those of control rats	52
	30—60 mg Sn/ kg body weight as SnCl$_2$ or SnCl$_4$ or Sn(OAc)$_2$	i.p.	Animals were sacrificed at 3 or 10 days after a single i.p. injection and various organs were analyzed for calcium concentration; kidney and spleen showed significant increases in calcium at both 3 and 10 days; these increases (kidney and spleen) were associated with decreased serum calcium levels	55
	30 mg Sn/kg body weight as SnCl$_2$	i.p.	Administration of stannous chloride significantly decreased the gastric secretion and decreased the serum calcium levels of the rats after 3 days; tetragastrin, administered subcutaneously 72 hr after SnCl$_2$ injection, inhibited the effects of SnCl$_2$ on the volume and total acidity of gastric secretion	56
	5—56 mg SnCl$_2$· 2H$_2$O/kg body weight	s.c.	Administration of stannous chloride significantly increased the activity of the enzyme heme oxygenase, both in the liver and in the kidneys	57
bits	Solutions of SnF$_2$ (≤0.5%) and SnCl$_2$ (≤2%)	Topical	Solutions of stannous fluoride or stannous chloride were applied to scratches made to the depth of the upper dermis on rabbit abdomens; both solutions produced destructive reactions; the authors postulate that the stannous salts may interfere with enzymes involved in the inflammatory process	54
nkeys	—	—	Pellets of tin (or various other metals) were implanted in the precentral motor cortex of monkeys in a series of chronic experiments; tin (and other metals) caused a mild and variable epileptogenic effect	58

REFERENCES

1. Kent, N. L., The occurrence of lead, tin, and silver in wheat and its milling products, *J. Soc. Chem. Ind.,* 61, 183, 1942.
2. Schroeder, H. A., Balassa, J. J., and Tipton, I. H., Abnormal trace metals in man: tin, *J. Chronic Dis.,* 17, 483—502, 1964.
3. Holden, H., Cadmium fume, *Ann. Occup. Hyg.,* 8, 1965.
4. Tabor, E. C. and Warren, W. V., Distribution of certain metals in the atmosphere of some American cities, *Arch. Ind. Health,* 17, 145, 1958.
5. Schwarz, K., Milne, D. B., and Vinyard, E., Growth effects of tin compounds in rats maintained in a trace element-controlled environment, *Biochem. Biophys. Res. Commun.,* 40, 22—29, 1970.
6. Tipton, I. H., Cook, D. B., Steiner, R. L., Boye, C. A., Perry, H. M., and Schroeder, H. A., Trace elements in human tissue. I. Methods, *Health Phys.,* 9, 89—101, 1963.

7.   Back, S., The presence of tin in food products, *Food Manuf.,* 8, 381–384, 1933.
8.   Warburton, S., Udler, W., Ewert, R. M., and Haynes, W. S., Outbreak of foodborne illness attributed to tin, *Public Health Rep.,* 77, 789–800, 1962.
9.   Kayser, *Beckurts Johresb.,* 1890.
10.   Luff, *Br. Med. J.,* 1, 833, 1890.
11.   Günther, *Z. Unters. Nahr. Genussm. Gebrauchsgegenstaende,* 2, 917, 1899.
12.   Mann, Dixon, *Forensic Med. Toxicol.,* 1922.
13.   Knorr, D., Tin-resorption, peroral toxicity and maximum admissable concentration in foods, *Lebensm. Wiss. Technol.,* 8, 51–56, 1975.
14.   Dundon, C. C. and Hughes, J. P., Stannic oxide pneumoconiosis, *Am. J. Roentgenol. Radium Ther.,* 63, 797–812, 1950.
15.   Cole, C. W. D., Davies, J. V. S. A., Kipling, M. D., and Ritchie, G. L., Stannosis in hearth tinners, *Br. J. Ind. Med.,* 21(3), 235–241, 1964.
16.   Bartak, F. and Tomecka, M., Stannosis (coniosis due to tin), *Proc. Int. Congr. Ind. Med., 9th, 1948,* 1949, 744–757.
17.   Dicken, D. Y. and Scott, M. J., An investigation into health conditions in a tin smelter with special reference to stannosis, *Med. J. Malaya,* 16(1), 1–13, 1961.
18.   Kent, N. L. and McCance, V. R. A., The absorption and excretion of 'minor' elements by man. Cobalt, nickel, tin, and manganese, *Biochem. J.,* 877–883, 1941.
19.   Tipton, I. H., Stewart, P. L., and Martin, P. G., Trace elements in diets and excreta, *Health Phys.,* 12, 1683–1689, 1966.
20.   Kehoe, R. A., Cholak, J., and Story, R. V., A spectrochemical study of the normal ranges of concentration of certain trace metals in biological materials, *J. Nutr.,* 19, 579–592, 1940.
21.   Bowen, H. J. M., *Trace Elements in Biochemistry,* Academic Press, New York, 1966, 101, 117, 165.
22.   Durbin, P. W., Scott, K. G., and Hamilton, J. G., The distribution of radioisotopes of some heavy metals in the rats, *Univ. Calif. Berkely Publ. Pharmacol.,* 3, 1–34, 1957.
23.   Hiles, R. A., Absorption, distribution, and excretion of inorganic tin in rats, *Toxicol. Appl. Pharmacol.,* 27, 366–379, 1974.
24.   Morimoto, S. and Yoneyama, Y., Fate of radioactive tin in rats, *J. Biochem.* (Tokyo), 49, 601–608, 1961.
24a. Fritsch, P., de Saint Blanquat, G., and Derache, R., Effect of various dietary components on absorption and tissue distribution of orally administered inorganic tin in rats, *Food Cosmet. Toxicol.,* 15, 147–149, 1977.
25.   Salant, W., Rieger, J. B., and Theuthardt, E. L. P., Absorption and fate of tin in the body, *J. Biol. Chem.,* 17, 265–273, 1914.
26.   Benoy, C. J., Hooper, P. A., and Schneider, R., The toxicity of tin in canned fruit juices and solid foods, *Food Cosmet. Toxicol.,* 9, 645–650, 1971.
27.   Arrhenius, G., Bramlette, M. N., and Picciotto, E., Localization of radioactive and stable heavy nucleides in ocean sediments, *Nature* (London), 180, 85, 1957.
28.   Theuer, R. C., Mahoney, A. W., and Sarett, H. P., Placental transfer of fluoride and tin in rats given various fluoride and tin salts, *J. Nutr.,* 101, 525–532, 1971.
29.   Flinn, F. B. and Inouye, J. M., Metals in our food, *JAMA,* 90, 1010–1013, 1928.
30.   Tipton, I. H., The distribution of trace metals in the human body, in *Metal Binding in Medicine,* Seven, M. J., Ed., J. B. Lippincott, Philadelphia, 1960, 27–42.
31.   Kutzner, J. and Brod, K. H., Investigation of the resorption and separation of tin after oral dosing with Sn^{113}, *Nucl. Med.,* 10, 286, 1971.
32.   Schryver, S. B., Some investigations on the toxicology of tin, with special reference to the metallic contamination of canned foods, *J. Hyg.,* 9, 253–263, 1909.
33.   Calloway, D. H. and McMullen, J. J., Fecal excretion of iron and tin by men fed stored canned foods, *Am. J. Clin. Nutr.,* 18, 1–6, 1966.
34.   deGroot, A. P., Feron, V. J., and Til, H. P., Short-term toxicity studies on some salts and oxides of tin in rats, *Food Cosmet. Toxicol.,* 11, 19–30, 1973.
35.   deGroot, A. P., Subacute toxicity of inorganic tin as influenced by dietary levels of iron and copper, *Food Cosmet. Toxicol.,* 11, 955–962, 1973.
36.   Conine, D. L., Yum, M., Martz, R. C., Stookey, G. K., Muhler, J. C., and Forney, R. B., Toxicity of sodium pentafluorostannite, a new anticariogenic agent. Comparison of the acute toxicity of sodium pentafluorostannite, sodium fluoride, and stannous chloride in mice and/or rats, *Toxicol. Appl. Pharmacol.,* 33, 21–26, 1975.
37.   Yoshikawa, H., Preventive effect of pretreatment with low dose of metals on the acute toxicity of metals in mice, *Ind. Health,* 8(4), 184, 1970.

38. Deschiens, R., Bertrand, D., and Romand, R., The toxicity and therapeutic dose of stannous chloride for the guinea pig, the rabbit, and the mouse, *Comptes Rendus*, 243, 2178–2180, 1956.

39. Roe, F. J. C., Boyland, E., and Millican, K., Effects of oral administration of two tin compounds to rats over prolonged periods, *Food Cosmet. Toxicol.*, 3, 277–280, 1965.

40. Herdson, P. B., Garvin, P. J., and Jennings, R. B., Time structural changes produced in rat liver by partial starvation, *Am. J. Pathol.*, 45, 157, 1964.

41. Komarkova, A., Zahor, Z., and Czabanova, C., Nephrocalcinosis in rats after forced weaning, *Experientia*, 25, 683, 1969.

42. MacKay, E. M. and Oliver, J., Renal damage following the ingestion of a diet containing an excess of inorganic phosphate, *J. Exp. Med.*, 61, 319, 1935.

43. Meyer, D. L. and Forbes, R. M., Effects of a thyroid hormone and phosphorus loading on renal calcification and mineral metabolism of the rat, *J. Nutr.*, 93, 361, 1967.

44. Yum, M. N., Conine, D. L., Martz, R. C., Forney, R. B., and Stookey, G. K., Renal tubular injury in rats induced by sodium pentafluorostannite, a new anticariogenic agent, *Toxicol. Appl. Pharmacol.*, 37, 363–370, 1976.

45. Hoey, M. J., The effects of metallic salts on the histology and functioning of the rat testis, *J. Reprod. Fertil.*, 12, 461–471, 1966.

46. Schroeder, H. A., Kanisawa, M., Frost, D. V., and Mitchener, M., Germanium, tin, and arsenic in rats: effects on growth, survival, pathological lesions and life span, *J. Nutr.*, 96, 37–45, 1968.

47. Mulay, I. L., Roy, R., Knox, B. E., Suhr, N. H., and Delaney, W. E., Trace-metal analysis of cancerous and non-cancerous human tissues, *J. Natl. Cancer Inst.*, 47, 1–13, 1971.

48. Schroeder, H. A. and Balassa, J. J., Arsenic, germanium, tin, and vanadium in mice: effects on growth, survival, and tissue levels, *J. Nutr.*, 92, 245–252, 1967.

49. Walters, M. and Roe, F. J. C., A study of the effects of zinc and tin administered orally to mice over a prolonged period, *Food Cosmet. Toxicol.*, 3, 271–276, 1965.

50. Kanisawa, M. and Schroeder, H. A., Life term studies on the effects of arsenic, germanium, tin, and vanadium on spontaneous tumors in mice, *Cancer Res.*, 27, 1192–1195, 1967.

51. Fischer, H. W. and Zimmerman, G. R., Long retention of stannic oxide, *Arch. Pathol.*, 88, 259–264, 1969.

52. Allman, D. W., Mapes, J. P., and Benac, M., Effect of nontoxic doses of F and Sn salts on rat liver metabolism in vivo, *J. Dent. Res.*, 54(1), 189, 1975.

53. Mustafa, M. G., Cross, C. E., Munn, R. J., and Hardie, J. A., Effects of divalent metal ions on alveolar macrophage membrane adenosine triphosphatase activity, *J. Lab. Clin. Med.*, 77(4), 563–571, 1971.

54. Stone, O. J. and Willis, C. J., The effect of stannous fluoride and stannous chloride on inflammation, *Toxicol. Appl. Pharmacol.*, 13, 322–338, 1968.

55. Yamamoto, T., Yamaguchi, M., and Sato, H., Accumulation of calcium in kidney and decrease of calcium in serum of rats treated with tin chloride, *J. Toxicol. Environ. Health*, 1(5), 749–756, 1976.

56. Yamaguchi, M., Naganawa, M., and Yamamoto, T., Decreased gastric secretion in rats treated with stannous chloride, *Toxicol. Appl. Pharmacol.*, 36, 199–200, 1976.

57. Kappas, A. and Maines, M. D., Tin: a potent inducer of heme oxygenase in kidney, *Science*, 192, 60–62, 1976.

58. Chusid, J. G. and Kopeloff, L. M., Epileptogenic effects of pure metals implanted in morot cortex of monkeys, *J. Appl. Physiol.*, 17, 697–700, 1962.

59. Piver, W. T., Organotin compounds: industrial applications and biological investigation, *Environ. Health Perspect.*, 4, 61–79, 1973.

60. Luijten, J. G. A., Applications and biological effects of organotin compounds, *Organotin Compounds*, 3, 931, 1972.

EFFECT OF NUTRIENT TOXICITIES IN ANIMALS AND MAN: VANADIUM

S. Leeson

Although vanadium (V) forms up to 300 ppm of the earth's crust, and its prevalence probably equals that of copper, lead, and zinc,[1] toxic effects in man and animals under natural conditions occur very infrequently. In man, vanadium toxicity is almost always associated with industrial processes, while in animals the only reported effects of a "natural" vanadium toxicity stems from the use of contaminated calcium phosphate in chick and laying hen diets.[2,3]

Fortunately, vanadium salts are poorly absorbed from the gastrointestinal tract and thus appear mostly in the feces.[4] Vanadium that is absorbed is excreted from the body in large part by the kidneys.[5,6] Although the threshold of intoxication appears variable in both humans[7] and animals,[8,9] relatively low concentrations in either the diet or drinking water are toxic. Table 1 shows the relative lethal doses of vanadium derived from various salts for some of the more common small laboratory mammals.

Once in the body, vanadium activity is closely related to phosphorus and/or lipid metabolism. Thus vanadium occurring in human tooth enamel and dentin is thought to be involved with an exchange of vanadium for phosphorus in apatite tooth structure,[10,11] while the uncoupling of oxidative phosphorylation by mitochondria in the presence of vanadium may be due to the formation of avanadylinstead of a phosphoryl intermediate, leading to an end product such as ADP-V rather than ATP.[12]

Vanadium has also been shown to inhibit cholesterol synthesis,[4,13,14] to effect the reactions of sulphur-containing compounds,[15,16] and to inhibit various mammalian enzyme systems.[17-19]

Table 2 gives a summary of some of the reported toxic effects of vanadium in man and animals in terms of growth, metabolism, and health. In addition to data provided in this table, the reader is referred to more extensive reviews of vanadium metabolism.[1,20,21]

Table 1
LETHAL DOSES OF VANADIUM (MG V_2O_5/KG BODY WEIGHT)

	Rabbit[a]	Guinea pig[b]	Rat[b]	Mouse[b]
Colloidal V_2O_5	1–2	20–28		87.5–117.5
Ammonium metavandate	1.5–2	1–2	20–30	25–30
Sodium orthovandate	2–3	1–2	50–60	50–100
Sodium pyrovandate	3–4	1–2	40–50	50–100
Sodium tetravandate	6–8	18–20	30–40	25–50
Sodium hexavandate	30–40	40–50	40–50	100–150
Vanadyl sulphate	18–20	35–45	158–190	125–150
Sodium vanadite		30–40	10–20	100–150

[a] Intravenous injection.
[b] Subcutaneous injection.

From Hudson, T. G. F., *Vanadium Toxicology and Biological Significance,* Elsevier, London, 1964, 69. With permission.

Table 2

TOXIC EFFECTS OF VANADIUM ON GROWTH, METABOLISM, AND HEALTH OF VARIOUS CLASSES OF ANIMAL AND MAN

Species	Age and/or weight of animal	Area of study	Source of V	Dose rate V	Observations	Ref.
Chick	0—4 weeks	Growth	Contaminated calcium phosphate (0.025—0.25%V)	10—20 ppm diet	Linear depression in growth rate with added levels of V	2
	0—2 weeks	Growth and mortality	Ammonium metavandate	0—8 ppm diet	Growth depression only observed with 8 ppm dietary V	2
		Growth and alleviation of toxicity	Ammonium metavandate	20 ppm diet	Growth depression due to 20 ppm V was alleviated by the replacement of 5% sucrose in a sucrose-fish meal diet with degosypolized cottonseed meal; no alleviation of toxicity recorded when other protein concentrates, such as soybean meal, corn-gluten meal, meat meal fish meal, casein, zein, or wheat gluten meal were substituted into the basal diet; addition of 0.25—0.5% ascorbic acid to the same basal diet also alleviated the toxicity due to 20 ppm V; vanadium from $VOSO_4$ and $VOCl_2$ with a valency of 4 was found to be as toxic as V from $NH_4 VO_3$ with a valency of 5	22
	0—3 weeks	Growth	Ammonium metavandate	10 ppm diet	Significant depression in growth rate	23
	0—30 days	Growth and mortality	Calcium vandate	40—600 ppm diet	Growth depression observed with >40 ppm V; below 200 ppm V feed utilization was also depressed while >200 ppm V caused 100% mortality by 30 days of age; 20 ppm V tolerated in terms of growth rate, feed utilization, and mortality	24
	0—2 weeks	Growth, mortality, and alleviation of toxicity	Ammonium metavandate and vanadyl chloride	10—25 ppm diet	10 ppm V had no effect on growth or mortality whereas 25 ppm depressed growth and caused significant mortality; V in 4⁺ and 5⁺ valency states found to be equally toxic; ethylenediamine-tetraacetate (EDTA) at 0.93 mmol/kg diet protected chicks against toxicity	25
	0—1 week	Association with phosphorylation	Ammonium metavandate	25 ppm diet	Chicks did not grow, became lethargic, and died, yet neither livers nor kidneys showed any gross or microscopic lesions; 25 ppm V uncoupled oxidative phosphorylation in mitochondria isolated from livers, and at a concentration of 0.1 m*M*	12

TOXIC EFFECTS OF VANADIUM ON GROWTH, METABOLISM, AND HEALTH OF VARIOUS CLASSES OF ANIMAL AND MAN

Species	Age and/or weight of animal	Area of study	Source of V	Dose rate V	Observations	Ref.
Chick (continued)	0—1 week (continued)				that vandate replaces the phosphate ion in reactions leading to the synthesis of ATP, resulting in the formation of a vanadyl instead of phosphoryl intermediate, and that the end product may be ATP-V rather than ATP	26
	0—4 weeks	Liver metabolism and cholesterol synthesis	Ammonium metavandate	100 ppm diet	Body and liver weight depressed while total liver lipid and cholesterol were elevated; results suggest increase in rate of cholesterol synthesis, as an explanation for the significant increase in plasma cholesterol found with such chicks; dietary V also enhanced hepatic lipogenesis and reduced the turnover of plasma total lipid and plasma fatty acids	
	0—4 weeks	Growth	Ammonium metavandate	0—113 ppm diet	In terms of growth rate, suggest chicks able to tolerate up to 20 ppm dietary V	27
Laying hen	1 year	Egg albumen quality	Contaminated calcium phosphate (0.025—0.25%V)	28 ppm diet	Albumen quality as measured by Haugh units, significantly decreased	3
	1 year	Egg albumen quality, egg production, and mortality	Ammonium metavandate	Graded levels 0—100 ppm diet	After 5 days treatment, 30 ppm V significantly depressed albumen quality in terms of Haugh units; 20 and 10 ppm V had similar effects after 13 and 29 days, respectively; egg production was immediately depressed by 100 ppm dietary V and after 3 weeks egg production was reduced to 20% and substantial feather loss was also recorded; hatchability of fertile eggs was decreased with 40 ppm V although embryonic examination revealed no specific abnormalities	3
	33 weeks	Egg quality	Ammonium metavandate	20—30 ppm diet	30 ppm V depressed egg production together with an indication of reduced feed intake; birds yielded eggs of an uncharacteristic smoothness, which broke clean and sharp like glass; body weight gain was depressed by the highest level of V; albumen quality in terms of height and Haugh unit measurement was significantly reduced	28

Table 2 (continued)

TOXIC EFFECTS OF VANADIUM ON GROWTH, METABOLISM, AND HEALTH OF VARIOUS CLASSES OF ANIMAL AND MAN

Species	Age and/or weight of animal	Area of study	Source of V	Dose rate V	Observations	Ref.
Mouse	Weaning—6 months	Growth and mortality	Vanadyl sulphate	1.4 μg/g diet	Growth rate and survival time was not significantly affected, observed a 5X increase in the V content of heart tissue to 25 μg/g wet weight	29
Rat	200 g	Growth and V distribution	Sodium metavandate	Daily injection of 1.25—2.5 mg sodium metavandat per kg body weight	Diarrhea and weight loss leading to eventual death with the highest dose rate; significant individual variation in tolerance of V dose observed; toxicity thought to be related to the accumulation of the metal in the liver and kidney, and possibly the ability of their tissues to reduce metavandate to vanadium	9
	200 g	Glucose metabolism	Ammonium metavandate	Intraperitoneal injection, 2 or 6 mg V per kg body weight	Disturbance of glucose metabolism as evidenced by reduced levels of $^{14}CO_2$ after 1 and 3 hr. of collection; $^{14}CO_2$ output not affected after this time; large biological variations observed	8
	Mature	Distribution V in body and alleviation of toxicity	^{48}V	10 μC orally	In comparison to young chicks, less V found in bone indicating an effect of maturity on distribution in this tissue; in the presence of EDTA, less V was absorbed from the intestinal tract, suggesting a possible reason for the alleviation of V toxicity by this compound	25
	0—4 years	Long term studies of toxicity	Vanadyl sulphate	5 ppm drinking water	No effect on growth, mature weight, or longevity; in females only, fasting serum glucose levels were significantly elevated; serum cholesterol was significantly increased in the mature male, and decreased in the mature female	14
Rabbit	Mature	Pulmonary defense mechanisms	Vanadium pentoxide, vanadium trioxide, vanadium dioxide	Particulate forms as specified in observations	Cell viability after 20 hr exposure reduced some 50% by 13 μg V/ml as the pentoxide, 21 μg V/ml as the trioxide, and 33 μg V/ml as the dioxide	30

TOXIC EFFECTS OF VANADIUM ON GROWTH, METABOLISM, AND HEALTH OF VARIOUS CLASSES OF ANIMAL AND MAN

Species	Age and/or weight of animal	Area of study	Source of V	Dose rate V	Observations	Ref.
Calf	6 months 127–185 kg	Toxicity	Ammonium metavandate	1–20 mg V/kg body weight daily, administered orally	Body weight not affected by 7.5 mg V/kg body weight, but higher doses produced a small depression in weight gain; no clinical symptoms evident in calves receiving 7.5 mg V/kg body weight; with higher doses, onset of symptoms were related to dose level; such animals showed persistent diarrhea with a gradual refusal to eat; the highest concentration of V was found in the kidney, followed by liver and spleen; animals receiving 20 mg V/kg body weight also showed signs of the element in the brain and heart muscle; V was not found in blood, lungs, skeletal tissue or testes	31
Man	Adult	Cholesterol synthesis	Diammonium oxytarttratovandate	100–125 mg/day for 6 weeks	No symptoms of toxicity or altered liver function; serum cholesterol level reduced by 20%	4
	Unspecified	Symptoms of toxicity	Sodium tetravandate	20 mg V_2O_5 intravenous	Symptoms shown at various times included excessive salivation and lacrimation, dissapearance of the pulse, cessation of respiration, diarrhea, and reduction in body temperature up to 3°F; suggest fatal dose of a salt such as sodium metavandate to be 0.43 mg V_2O_5/kg body weight	1
	Adult	Body retention	Sodium tetravandate	18–24 mg/week given as daily injection	81% of V excreted in the urine by the 7th day following last injection; 9% was excreted in the feces at the same time	6
	Unspecified	Toxicity	Ammonium vanadyl tartrate	50–125 mg V/d administered orally	All patients exhibited gastrointestinal difficulties with increased intestinal activity and occasional cramps; histological studies of white blood cells, platelets, and reticulocytes revealed no abnormalities; a purple–green tint developed on the tongue, occuring rapidly and dissapearing promptly; wide fluctuations in V urinary excretion rate suggested an unpredictable absorption rate	32

REFERENCES

1. **Hudson, T. G. F.,** *Vanadium Toxicology and Biological Significance,* Elsevier, London, 1964, 55–71.
2. **Berg, L. R.,** Evidence of vanadium toxicity resulting from the use of certain commercial phosphorus supplements in chick rations, *Poult. Sci.,* 42, 766–769, 1963.
3. **Berg, L. R., Bearse, G. E., and Merrill, L. H.,** Vanadium toxicity in laying hens, *Poult. Sci.,* 42, 1407–1411, 1963.
4. **Curran, G. L., Azarnoff, D. L., and Bolinger, R. E.,** Effect of cholesterol synthesis in normocholestermic young men, *J. Clin. Invest.,* 38, 1251, 1959.
5. **Talvitie, N. A. and Wagner, W. D.,** Studies in vanadium toxicology: distribution and excretion of vanadium in animals, *AMA Arch. Ind. Hyg. Occup. Med.,* 9, 414, 1954.
6. **Kent, N. L. and McCance, R. A.,** The absorption and excretion of 'minor' elements by man. I. Silver, gold, lithium, boron and vanadium, *Biochem. J.,* 35, 837–844, 1941.
7. **Bala, Y. M. and Kopylova, L. M.,** Value of vanadium in tuberculosis, *Probl. Tukerk.,* 49, 63–67, 1971.
8. **Meeks, M. J., Landolt, R. R., Kessler, W. V., and Born, G. S.,** Effect of vanadium on metabolism of glucose in the rat, *J. Pharm. Sci.,* 60, 482–483, 1971.
9. **Johnson, J. L., Cohen, H. J., and Rajagopalan, K. V.,** Studies of vanadium toxicity in the rat: lack of correlation with molybdenum utilization, *Biochem. Biophys. Res. Commun.,* 56, 940–946, 1974.
10. **Lowater, F. and Murray, M. M.,** Chemical composition of teeth, *Biochem. J.,* 31, 837, 1937.
11. **Soremark, R. and Anderson, N.,** Uptake and release of vanadium from intact human enamel following $V_2{}^{48}O_5$ application in vitro, *Acta Ondontol Scand.,* 20, 81, 1962.
12. **Hathcock, J. N.., Hill, C. H., and Tove, S. B.,** Uncoupling of oxidative phosphorylation by vandate, *Can. J. Biochem.,* 44, 983–988, 1965.
13. **Curran, G. L.,** Effect of certain transition group elements on hepatic synthesis of cholesterol in the rat, *J. Biol. Chem.,* 210, 765, 1954.
14. **Schroeder, H. A., Mitchener, M., and Nason, A. P.,** Zirconium, niobium, antimony, vanadium, and lead in rats: life term studies, *J. Nutr.,* 100, 59–68, 1970.
15. **Mountain, J. T., Delker, L. L., and Stokinger, H. E.,** Studies in vanadium toxicology: reduction in the cystine content of rat hair, *Arch. Ind. Hyg.,* 8, 406, 1953.
16. **Mountain, J. T., Stockwell, F. R., Jr., and Stokinger, H. E.,** Studies on vanadium toxicology; fingernail cystine as an early indicator of metabolic changes in vanadium workers, *Arch. Ind. Hyg.,* 12, 494, 1955.
17. **Mascitelli-Coriandoli, E. and Citterio, C.,** Effects of vanadium upon liver coenzyme A in rats, *Nature,* 183, 1527, 1959.
18. **Perry, H. M., Teitelbaum, S., and Schwartz, P. L.,** Effect of antihypertensive agents on amino acid decarboxylation and amino oxidation, *Fed. Proc., Fed. Am. Soc. Exp. Biol.,* 14, 113, 1955.
19. **Bergel, F., Bray, R. C., and Harrap, K. R.,** A model system for cysteine desulphydrase action: pyridoxal phosphate-vanadium, *Nature,* 181, 1654, 1958.
20. National Research Council Division of Medical Sciences Committee on Biological Effects of Atmospheric Pollutants, *Medical and Biological Effects of Environmental Pollutants – Vanadium,* N.A.S., Washington, D.C., 1974.
21. **Underwood, E. J.,** in *Trace Elements in Human and Animal Nutrition,* Academic Press, New York, 1971, 416–424.
22. **Berg, L. R. and Lawrence, W. W.,** Cottonseed meal, dehydrated grass and ascorbic acid as dietary factors preventing toxicity of vanadium for the chick, *Poult. Sci.,* 50, 1399–1404, 1971.
23. **Summers, J. D. and Moran, E. T., Jr.,** Interaction of dietary vanadium, calcium and phosphorus for the growing chicken, *Poult. Sci.,* 51, 1760–1761, 1972.
24. **Romoser, G. L., Dudley, W. A., Machlin, L. J., and Loveless, L.,** Toxicity of vanadium and chromium for the growing chick, *Poult. Sci.,* 40, 1171–1173, 1961.
25. **Hathcock, J. N., Hill, C. H., and Matrone, G.,** Vanadium toxicity and distribution in chicks and rats, *J. Nutr.,* 82, 106–110, 1964.
26. **Hafez, Y. and Kratzer, F. H.,** The effect of dietary vanadium on fatty acid and cholesterol synthesis and turnover in the chick, *J. Nutr.,* 106, 249–257, 1976.
27. **Nelson, T. S., Gillis, M. B., and Peeler, H. T.,** Studies on the effect of vanadium on chick growth, *Poult. Sci.,* 41, 519–522, 1962.
28. **Eyal, A.,** Vanadium Metabolism in the Laying Hen, M. Sci. thesis, University of Guelph, Guelph, Ontario, Canada, 1971.

29. **Schroeder, H. A. and Balassa, J.,** Arsenic, germanium, tin and vanadium in mice: effects on growth, survival and tissue levels, *J. Nutr.,* 92, 245–252, 1967.

30. **Waters, M. D., Gardner, D. E., and Coffin, D. L.,** Cytoxic effects of vanadium on rabbit alveolar macrophages in vitro, *Toxicol. Appl. Pharmacol.,* 28, 253–263, 1974.

31. **Platonow, N. and Abbey, H. K.,** Toxicity of vanadium in calves, *Vet. Rec.,* 82, 292–293, 1968.

32. **Dimond, E. G., Caravaca, J. C., and Benchimol, A.,** Vanadium excretion, toxicity, lipid effect on man, *Am. J. Clin. Nutr.,* 12, 49–53, 1963.

Other Nutrients

EFFECT OF NUTRIENT DEFICIENCIES AND TOXICITIES: WATER INTOXICATION

E. L. Overholt

Osmolality in normal man is remarkably constant despite large variations in water and solute, intake, and excretion. Every kilogram of body weight contains 285 to 290 mOsm solute, consisting primarily of salts of sodium in the extracellular fluids and potassium in the intracellular fluids. The identical osmolality of intracellular and extracellular fluids is produced by free movement of water across all cellular membranes. A gain or loss of free water will be shared by all major compartments in proportion to their relative sizes. One of the exceptions to this free movement is the control of water permeability at the distal portion of the nephron by antidiuretic hormone (ADH).

ADH is synthesized in the supraoptic nucleus of the hypothalamus. It is transported along with supraoptic-hypophyseal tract to the posterior pituitary where it is stored in secretory granules in association with a carrier protein, neurophysin. Elevation of blood osmolality by as little as 2% will result in stimulation of osmoreceptors located in the hypothalamus, which prompts release of ADH from its granules in the posterior pituitary.[1] Acute decrease in blood volume causes a decrease in stimulation of the stretch receptors located in the left atrium. This reduces vagal afferent activity to the hypothalamus with diminution of vago-inhibitory action on ADH release. In addition, acute lowering of arterial blood pressure activates carotid baroreceptors which, in turn, stimulates the release of large amounts of ADH. Pain, stress, and emotional influence are also capable of causing ADH release through neuropathways terminating in the hypothalamus. Hypoosmolality or increased left atrial pressure inhibits ADH release.

Normal adult posterior pituitary gland contains approximately 15 USP units of vasopressin (ADH). As little as 0.3 milliunits per kilogram of body weight per hour will elicit a normal antidiuretic effect. Larger amounts will only prolong this effect. ADH travels in the plasma in its free peptide form and has a halflife of 16 to 20 min. Because of the enormous reserve capacity of the kidney and the liver to inactivate ADH, syndromes associated with abnormal sustained levels in the circulation are not due to impaired inactivation of the hormone but to inappropriate release of ADH (IADH).

ADH acts on the renal medullary collecting ducts of the nephron. It is bound to the cell membrane where is activates adenyl cyclase which, in turn, acts within the cells to generate cyclic AMP from ATP. Cyclic AMP is believed to cause activation of a protein kinase with a subsequent phosphorylation occurring at the luminal .end of the cell membrane.[2] The renal tubule cell is rendered more permeable to water, thereby permitting diffusion of water along established concentration gradients into the hyperosmolar interstitial renal fluid. Administration of ADH to volunteers is followed by abrupt rise in urinary concentration, fall in urine volume, gain in body weight, and dilution of serum sodium and total solute concentration due to the retention of water. Following 3 days of vasopressin administration, there is a striking sodium diuresis in spite of severe hyponatremia. The cause of the urinary sodium loss has been attributed to the hypervolemic inhibition of both aldosterone secretion and proximal tubule sodium reabsorption (third factor effect) in association with an actual increase in the glomerular filtration rate.[3-8] Gross edema does not occur in spite of the total water retention because much of the water is distributed intracellularly. Moreover, retention of 3 to 4 l of fluid in the extracellular space is not likely to produce clinically evident edema. Continued administration of vasopressin eventually results in an "escape" from the maximal renal effect of the hormone, and a new steady state of water and sodium balance is established. The escape is dependent upon the sustained water retention eventually diluting the renal medulla and thereby markedly diminishing the concentrating ability of the kidney. Once the steady state has been reached, urine sodium and volume will reflect intake as in normal individuals. However, if there is further acute expansion of body fluid, sodium diuresis will reoccur. When vasopressin is discontinued, water diuresis occurs and sodium retention follows until the sodium deficit is corrected.

The symptomatology of excess ADH results from water retention proceeding to water intoxication. The symptoms correlate roughly with the degree of dilution of body fluids in which serum sodium concentration is an easily available index. The resulting hyponatremia reflects a hypo-osmolal state. Normal serum osmolality is maintained at 290± mOsm/kg serum water. Serum electrolytes, mainly sodium and its accompanying anions, contribute about 280 mOsm/kg. One can estimate the serum osmolality by simply doubling the serum sodium concentration in meq/l. Closer approximation is available by adding in the osmolal concentrations of glucose and urea nitrogen:

$$\text{Serum osmolality} = 2(\text{Na}+)\text{meq/l} + \frac{\text{glucose mg\%}}{18} + \frac{\text{BUN mg\%}}{2.8}$$

The nonelectrolytes glucose and urea together furnish about 10 mOsm/l except in azotemia and hyperglycemia. The osmolality of the serum, which is an even better index, may not be available in some hospital laboratories. The majority of patients have asymptomatic hyponatremia with serum sodiums above 120 meq/l. Symptoms of water intoxication usually occur when the sodium level is 110 to 115 meq/l or less.

When the serum sodium declines over a period of days, the signs of water intoxication are less dramatic, i.e., headaches, generalized muscular weakness, sleepiness, apathy, lethargy, and decreased tendon reflexes. If the sodium concentration declines more rapidly, one sees anorexia, nausea, vomiting, severe confusion, and eventually convulsions and coma. These symptoms are a regular occurrence in patients whose serum sodium concentrations fall below the 110 meq/l range. The hypo-osmolality and increase in total body water causes an increase of intracellular water with cellular swelling. This is primarily reflected in the interference of the function of the central nervous system. Diffuse electroencephalographic abnormalities may be seen. Pathologically, vacuolization of the cytoplasm, swelling of the nucleus, and liquefaction of the neurons have been observed.

As mentioned earlier, hypotonic (hyponatremic) plasma normally inhibits the secretion of ADH. Persistent ADH activity in the presence of coexisting hypotonic plasma is inappropriate (IADH). The clinical and laboratory features of the IADH syndrome are similar to the excess administration of ADH in volunteers:

1. Hyponatremia and hypotonicity of the body fluids in association with the urine hypertonic to plasma
2. Urinary excretion of significant quantities of sodium despite hyponatremia
3. Normal adrenal, renal, and pituitary function
4. Absence of severe stress, edema, dehydration, azotemia, or hypovolemia

Because of the increased total body water in the presence of normal renal and adrenal function, there is a low BUN (10 mg/100 ml or less) and low serum uric acid with a high uric acid clearance.

The clinical counterpart of chronic ADH administration was first recognized by Schwartz and associates in two patients with undifferentiated bronchogenic carcinoma.[9] These investigators emphasized that the continued antidiuretic state in the presence of serum hypotonicity, with normal renal and adrenal function, was inappropriate since serum hypotonicity normally inhibits secretion of ADH. Later, Bower and associates utilized a bioassay to detect increased ADH in the serum of the patient with undifferentiated bronchogenic carcinoma who manifested typical features of IADH syndrome to the point of water intoxication.[10] They demonstrated ADH in the circulating plasma as well as in the tumor.

In addition, aberrant production of ADH has been described in carcinoma of the duodenum, pancreas, ureter, thymoma and lymphoma.[2] It also may be associated with acute lobar pneumonia,[11,12] lung cavitation as with aspergillosis,[2] and advanced pulmonary tuberculosis.[13] How the pulmonary lesion induces secretion of ADH has not been elucidated.

A wide variety of conditions producing brain damage with the release of endogenous ADH have been reported. These include meningitis, head injuries, brain abscess, encephalitis, paroxysmal seizure disorders, subarachnoid hemorrhage, and brain tumor.[14-20] It also occurs with the central manifestations of Guillain-Barré syndrome[21] and acute intermittent porphyria.[22]

There is an increasing number of pharmaceutical agents which can induce changes similar to those seen in the IADH syndrome.[2] However, water intoxication will not result unless there is adequate water intake. In patients with diabetes mellitus, treated with either chlorpropamide or tolbutamide, there have been reports of water retention and hyponatremia.[2] This occurs more frequently when there is an underlying tendency to retain fluid as in congestive heart failure or cirrhosis of the liver. These drugs have no effect in patients with nephrogenic diabetes insipidus, a condition in which kidneys are totally refractory to ADH. Apparently, chlorpropamide potentiates the effect of ADH at the renal tubule. Less certain is its central mode of action with release of ADH from the posterior pituitary storage sites. To be included in the drug-induced causes are the antineoplastic agents of vincristine and cyclophosphamide. These drugs act directly on the supraoptico-hypophyseal system to release ADH. The tricyclic compounds, Carba-mazepine (Tegretal®), and Amitriptyline (Elavil®), as well as Thioridazine (Mellaril®) can induce hypotonic serum in patients with primary polydipsia. These drugs also stimulate the release of ADH.

Water intoxication can follow the i.v. infusion of oxytocin, especially when large volumes of fluids are given concomitently.[23-26] This drug is used to aid in delivery, to correct uterine atony, or to expel an incomplete abortion. Oxytocin like ADH is an octapeptide found in the human posterior pituitary gland. Both have very similar structures, differing only in the amino acids at the positions three and seven. They are considered to have distinctly separate actions when present in physiologic amounts. However, in pharmacological concentrations, the activity of oxytocin has been found to overlap that of vasopressin. The pure synthetic oxytocin has only 1/200th antidiuretic activity as compared to ADH. In physiological concentrations, oxytocin acts to increase myometrial contractility. There is a minimal antidiuretic activity when oxytocin is administered by a single injection.

Increased water intake in patients with hypopituitarism can induce water intoxica-tion.[28] These patients have a decrease in cortisol secretion as well as a limited production of aldosterone. In the absence of cortisol, the distal convoluted tubules are excessively permeable to water, thereby enhancing the effect of ADH on water reabsorption through the tubules. The recognition of hypopituitarism as the underlying problem in some patients with water intoxication is especially important. The usual treatment for most forms of water intoxication is water restriction. In hypopituitarism, more effective therapy is accomplished with the administration of cortisol. Another endocrine disorder, myxedema,[29] has on rare occasions been shown to produce water intoxication by mechanisms that have not been elucidated. This promptly responds to water restriction and thyroxine.

The treatment of water intoxication is water restriction of 200 to 400 cc/day.[27] However, with very low serum sodium concentrations, i.e., values below 115 meq/l, the cerebral complications require more rapid restoration of plasma tonicity to normal. This may be accomplished with the slow infusion of 3% sodium chloride. With correction of the symptom complex, the underlying causes previously discussed should be discovered and treated.

REFERENCES

1. **Maxwell, M. H. and Kleeman, C. R.,** *Clinical Disorders of Fluid and Electrolyte Metabolism,* 2nd ed., McGraw-Hill, New York, 1972, 215–295.
2. **Bartter, F. C.,** The syndrome of inappropriate secretion of antidiuretic hormone, *Dis. Mon.,* p. 1, November 1973.
3. **Jones, N. F. and Barraclough, M. A.,** The mechanism of increased sodium excretion during water loading with 2.5 per cent dextrose and vasopressin, *Clin. Sci.,* 25, 449, 1963.
4. **De Wardener, H. E. and Mills, I. H.,** Studies on the efferent mechanism of the sodium diuresis which follows the administration of intravenous saline in the dog, *Clin. Sci.,* 21, 249, 1961.
5. **Levinsky, N. G. and Lalone, R. C.,** The mechanism of sodium diuresis after saline infusion in the dog, *J. Clin. Invest.,* 42, 1261, 1963.
6. **Dirks, J. H. and Cirksena, W. J.,** The effect of saline infusion on sodium reabsorption by the proximal tubule of the dog, *J. Clin. Invest.,* 44, 1160, 1965.
7. **Cirksena, W. J. and Dirks, J. H.,** Effect of thoracic cava obstruction on response of proximal tubule sodium reabsorption to saline infusion, *J. Clin. Invest.,* 45, 179, 1966.
8. **Davis, J. O. and Howell, D. S.,** Effect of chronic pitressin administration on electrolyte excretion in normal dogs and in dogs with experimental ascites, *Endocrinology,* 55, 409, 1954.
9. **Schwartz, W. B. and Bennett, W.,** Syndrome of renal sodium loss and hyponatremia probably resulting from inappropriate secretion of antidiuretic hormone, *Am. J. Med.,* 23, 529, 1957.
10. **Bower, B. F. and Mason, D. M.,** Bronchogenic carcinoma with inappropriate antidiuretic activity in plasma and tumor, *N. Engl. J. Med.,* 271, 934, 1964.
11. **Roberts, H. J.,** Syndrome of hyponatremia and renal sodium loss probably resulting from inappropriate secretion of antidiuretic hormone, *Ann. Intern. Med.,* 51, 1420, 1959.
12. **Stormont, J. M. and Waterhouse, C.,** Severe hyponatremia associated with pneumonia, *Metabolism,* 11, 1181, 1962.
13. **Sims, E. A. H. and Welt, L. G.,** Asymptomatic hyponatremia in pulmonary tuberculosis, *J. Clin. Invest.,* 29, 1545, 1950.
14. **Peters, J. P. and Welt. L. G.,** Salt-wasting syndrome associated with cerebral disease, *Trans. Assoc. Am. Physicians,* 63, 57, 1950.
15. **Welt, L. G. and Seldin, D. W.,** Role of central nervous system in metabolism of electrolytes and water, *AMA Arch. Intern. Med.,* 90, 355, 1952.
16. **Goldberg, M. and Handler, J. S.,** Hyponatremia and renal wasting of sodium in patients with malfunction of central nervous system, *N. Engl. J. Med.,* 263, 1037, 1960.
17. **Epstein, F. H. and Levitin, H.,** Cerebral hyponatremia, *N. Engl. J. Med.,* 265, 513, 1961.
18. **Higgins, G. and Lewin, W.,** Metabolic disorders in head injury: survey of 76 consecutive cases, *Lancet,* 1, 61, 1954.
19. **Rapoport, S. and West, C. D.,** Salt losing conditions: renal defect in tuberculous meningitis, *J. Lab. Clin. Med.,* 37, 550, 1951.
20. **Harrison, H. F. and Fineberg, L.,** Disturbances of ionic equilibrium of intracellular and extracellular electrolytes in patients with tuberculosis meningitis, *J. Clin. Invest.,* 31, 300, 1952.
21. **Waldvogel, F., de Sousa, R. C., and Mach, R. S.,** Intoxication a l'eau due à une secretion inadequate d'hormone antidiuretique (syndrome de Schwartz-Bartter) d'origine idiopatique, *Schweiz. Med. Wochenschr.,* 97, 929, 1967.
22. **Ludwig. G. D.,** Hyponatremia in acute intermittent porphyria due to probable inappropriate secretion of antidiuretic hormone, *Clin. Res.,* 9, 340, 1961.
23. **Self, J.,** Water intoxication induced by oxytocin administration, *Am. J. Med. Sci.,* 252, 573, 1966.
24. **Pittman, J. G.,** Water intoxication due to oxytocin, *N. Engl. J. Med.,* 268, 481, 1963.
25. **Whalley, P. J. and Pritchard, J. A.,** Oxytocin and water intoxication, *JAMA,* 186, 601, 1963.
26. **Niswander, K. R. and Patterson, R. J.,** Hazards of elective induction of labor, *Obstet. Gynecol.,* 22, 228, 1963.
27. **Overholt, E. L.,** Water intoxication: its diagnosis and management, *Mil. Med.,* 133, 607, 1968.
28. **Bethune, J. E. and Nelson, D. H.,** Hyponatremia in hypopituitarism, *N. Engl. J. Med.,* 272, 771, 1965.
29. **Goldberg, M. and Reivich, M.,** Studies on the mechanism of hyponatremia and impaired water excretion in myxedema, *Ann. Intern. Med.,* 56, 120, 1962.

CALORIC EXCESS

S. E. Silvis

INTRODUCTION

Extensive studies have been conducted on nutritional requirements and the effects of general or specific nutritional deficiency. It is only in recent decades that the problems of dietary excess have been recognized. For the most part, this is a problem of prosperous countries. Caloric excess above the energy expenditure of the individual leads to obesity. It is obvious that the degree of this obesity will be in proportion to the time and magnitude of the difference between energy intake and energy expenditure. Different effects are present in sudden high caloric loads in the previously malnourished individual. They have been referred to as hyperalimentation syndrome.

Excesses of individual nutrients produce other adverse effects. In developed countries the form and composition of the caloric intake may be important in the etiology of arteriosclerosis and some malignancies. Excesses of refined sugars and fats or decreased fiber content especially have been indicted in these diseases. When imbalanced diets are given to the patients by either parenteral or tube feedings, harmful reactions may occur. Other specific excesses such as vitamins and amino acids will not be discussed in this contribution. These are ordinarily observed in diet fads or attempts to use normal dietary constituents in large amounts for therapeutic purposes.

TOTAL CALORIC EXCESS IN THE INDIVIDUAL WITH PREVIOUSLY NORMAL NUTRITION

When an individual with previously normal nutrition begins to consume calories beyond his energy expenditure, this caloric excess is deposited as fat. Undesirable accumulations of fat may be referred to as obesity. This initial concept of obesity is the ultimate in simplicity. When considered in depth it is apparent that some reserve energy stores are necessary and probably essential to the organism. The ideal quantity of these energy stores is very difficult to determine. It is clear that there must be an optimal amount of body fat and that amount is determined by the environmental situation confronting the individual. Therefore, "Ideal weight" is a conditional state. There are physiological advantages to ample fat stores for a man preparing for an emergency trek, a population entering a period of famine, a child entering a febrile illness or a growth spurt, or a woman during pregnancy (where the food supply is not abundant). Excessive thinness can be shown to be a detriment in all of these circumstances.[1-3] The development in the human of the ability to accumulate excess fat may relate to the periods of famine that our tribal ancestors endured. A linear relation between fat stores and survival during periods of starvation or fasting has clearly been shown in mammalian species studied under controlled conditions.[4-5] This is probably true in man as well.[6,8] We have descended from the survivors of famine.

Definition of Obesity

The majority of people, even in the food-abundant societies, adjust their energy balance with astounding sensitivity to remain within narrow limits of fatness. A few tend to be excessively thin and more than a few excessively fat. In the Framingham study[9] it was shown that the average individual gains approximately 4.2 kg between the ages of 20 to 50 years. This accuracy of food intake is illustrated by the following: If 2000 cal/day are eaten, and if this amount is 1% in excess of the energy expenditure, then a total of 20 cal/day would be available to form fat. This amounts to a weight gain of 0.8 kg/year. Projected over a 30-year period it would amount to 24 kg; hence, the average individual has an energy intake control of less than 1% error. Why do obese people escape this fine dietary control? This is the root of the problem of obesity.

At the present time no precise definition exists as to when the onset of fat storage becomes deleterious and can properly be called obesity. There are strong social, esthetic, folk, and medical differences as to what may be considered an ideal weight. If one looks at the art of western Europe (assuming that the artist was painting the female form attractive to his society), a marked difference is seen between the nudes of Rubens[10] and the models of the present-day fashion magazines. The women of the 16th century were two to three times heavier than what would be considered desirable weight today. Hence, the fat young lady of today may not be obese, but was only born in the wrong century! There are variations in social conditioning which are probably not related to climatic differences. It is obvious that for people living in extremely cold climates (such as the Eskimo), a greater degree of fatness would be desirable than in a tropical climate. The Abkhasians[11] who reside on the shores of the Black Sea and have been an identifiable people for probably 3 millennia, are a group of people who have extreme longevity. They have as part of their folklore the attitude that fatness is a sign of ill health and will inquire about the health of a young individual who begins to take on excess weight. At the opposite extreme are the Polynesian cultures of Samoa and Hawaii, where extreme obesity was considered esthetically desirable. One of the last queens of Hawaii apparently caused serious discussion as to whether she was an appropriate monarch for the kingdom because she weighed slightly under 300 pounds.[12] In neither of these extreme societies does the question of health seem to enter into their feelings. The Hawaiians apparently find the round, robust build desirable; the people in the Caucasus feel that somebody lean and trim has a desirable appearance. "A fat person would, of course, look ridiculous on a horse."[11] Notwithstanding the social differences, in the Western cultures, extreme obesity has been associated with ill health and early death for many centuries.[13]

In the latter part of the last century the American insurance companies recognized this problem and expended a massive effort in tabulating weights and longevities.[14-15] From these data arose the well-known life insurance ideal weight tables. Recently, extensive criticism has been given both to the concept of the ideal weight table and the manner in which these data were derived. First of all, the individuals studied are in no way a random sample, but represent the entire population applying for insurance coverage. There are probably major inaccuracies in the simple recording of height and weight which are done by insurance examiners. In addition, no attempt is made in the ideal weight tables of the insurance industry to separate overweight from obesity. The heavy-boned, heavy-muscled individual becomes heavy for his height and, therefore, obese. This individual may not actually have an increase in total or relative body fat. Clarification of the relationship of obesity to health and disease is an important matter.

Measurement of Obesity

It is apparent that insurance companies and the medical profession would like to define what is a desirable store of energy and when fat becomes a liability. Before this can be approached, one must measure body fat. One would like to study individuals with an increased content of body fat or with a rise in the ratio of fat to lean body weight. This can be precisely determined by measuring body density with the principle of Archimedes. This is impractical in examining large population groups because it requires total body immersion and measurement of the water displacement volume to calculate density. This technique will not gain wide application because, as Khosla and Lowe[16] have observed, total body immersion would "generally be regarded as too high a price to pay for a life insurance policy."

The measurement of skin-fold thickness with special calipers is extensively used in nutritional surveys.[17] This is not as simple or reproducible as it may seem; it only samples subcutaneous fat. It appears unlikely that physicians will obtain and use these calipers on a wide scale or with acceptable accuracy.[1]

What can be done with the widely recorded measurements of height and weight? A commonly used technique is relative weight, i.e., the ratio of the subject's weight to a standard weight for persons of that height and sex, corrected for age. The difficulty is in selecting an appropriate reference population. A popular reference is the weight from insurance examinations obtained from people who showed average or better than average survival, i.e., "ideal weight." A more noncommittal procedure is to use the average weight of similar individuals in the study population. Thus, the Framingham study refers to the "Framingham relative weight," which would be the patient's weight compared to individuals of the same height and sex included in that study.[18] Both of these procedures tend to distort the denominator of the relative weight to the extent that obesity is present in the reference population. This problem increases markedly with age during middle life.

Attempts have been made to avoid such reference standards by incorporating an expotential function of height intending to minimize the effect of height (H) on weight (W). Ideally, a power function of height should be found that would remove the contribution of height to weight over a wide range of height. The most simple of these is to use the ratio of W/H.[1] Two early improvements were W/H^2, called Quetelets index,[19] and the ponderal index, $\omega/\sqrt[3]{H}$.[29] Billewicz et al.[20] examined the relation of all three of these indices to obesity as measured by densometry and concluded that W/H underestimates obesity and the ponderal index overestimates it. Thus, W/H^2 performed the best, although these workers thought it tedious to calculate. This objection would be removed with the appearance of inexpensive electronic calculators. Khosla and Lowe[16] confirmed these conclusions in a population of men in Birmingham. The Quetelets index (W/H^2) showed a high correlation with obesity at all ranges of height (r = 0.85).

Reviewing all this evidence, Keyes et al.[14] concluded that W/H^2 (which they prefer to call body mass index) is the most useful indicator of obesity. Newens and Goldstein[21] have studied the problem of obesity in children. The assumption used for adults that obesity is independent of height or can be made so by an expotential function is not true in children.[22] Forbes[23] concluded that percentile standards of weight for height still offer the best means of classifying obesity in children.

Notwithstanding the previous discussion, gross obesity (which we will see later as the major health defect) is immediately and visually recognized by the clinician. The degree can be satisfactorily documented by the measurement of height and weight.

Genetic and Metabolic Causes of Obesity

The cause of obesity is the ultimate in simplicity, i.e., a situation in which energy intake exceeds energy expenditure. There is an old and treasured notion both in folk and professional medicine that obesity is caused by a metabolic defect, whether acquired or inherited. This makes obesity akin to an act of God. It removes all blame and postpones successful management until the biochemical lesion is understood. This fatalistic explanation of obesity was elevated to the role of a hypothesis by physiologists studying genetic obesity in laboratory animals.[24,27] The hypothesis that obesity is a result of an anatomical or biochemical lesion in the feeding center of the brain has been repeated so often that many came to believe it as a proven fact. Defects in lipolysis have been demonstrated in obese rodents. Although it is probably valid in some animal models, it is an unproven hypothesis in human beings.

It is difficult to show metabolic differences between obese and normal individuals.[27,28] The demonstration of increased heat production after eating was made many years ago and was called specific dynamic action.[29-33] Miller et al.[33] preferred to call this the "thermal effect" of food. They were able to show that thin and obese subjects differ in the production of heat after meals. In normal subjects, overeating leads to a 20 to 50% increase in heat production, which tends to dissipate the excessive intake. This energy dissipation is potentiated by exercise in persons with good energy equilibrium. In obese subjects, thermal dissipation of energy is less marked, so that more fat is laid down. Miller et al.[33] have demonstrated that the thermal effect of feeding amounts to 25% of the fasting metabolic rate. This amounts to the utilization of 7 to 10% of ingested energy and is proportional to meal size. When the meal is followed by exercise there is a doubling of the thermal effect. Hence, the lay belief of "running off" a large meal appears valid. These proposals deserve careful consideration because thermogenesis may explain part of the metabolic riddle of the difference between lean and fat people. It also suggests that programs for weight regulation should take advantage of the thermogenic action of exercise after eating. Perhaps more attention should be paid to thermogenesis in human obesity and less to appetite and regulation of intake. Drugs may be found that will safely influence thermogenesis and prove more useful than the ineffective appetite suppressors. Human obesity tends to be familial because of either genetic factors or early training. Withers[36] examined the correlation of parent-child weights, comparing natural children with adopted children. The correlation of body weight was significantly higher in natural children. The classic way to examine genetic vs. environmental factors is the study of twins reared together and apart. Identical twins are found to have closely correlated body weights. The mean adult weight difference of identical twins was 4.1 lb, as compared to 10 lb for fraternal twins and 10.4 lb for nontwin siblings. Identical twins reared apart had somewhat greater weight variation than those who were reared together, but they were more similar than either nonidentical twins or fraternal siblings. Clark[37] estimated that 69% of the total variability of body weight is accounted for by inheritance. Obesity in human beings has a genetic component, but strong environmental factors obscure the understanding of how genetic influences operate. Hypothalamic lesions have been shown to produce massive obesity in animals[38-41] (discussed in the section on feeding control). Genetic obesity has been demonstrated in rodents.[26,27] There are two common features of obesity in animal models that are of particular interest: (1) Obese mice show increased insulin levels and relative insulin resistance, and (2) they are unable to increase thermogenesis when exposed to cold and are apt to die. These clues suggest the need for more detailed examination of these metabolic functions in obese humans. Obesity in rodents is a homozygous recessive trait which would be expected to be an enzyme deficiency; however, none has been found. In spite of the enormous amount of work there is no biochemical explanation of these animal models. These features probably do not occur in humans.

Hormonal or "glandular" causes of obesity have been searched for both by the public and members of the medical profession. Weight gain is observed in both decreased thyroid function and increased adrenal function; however, these causes of obesity are rare and usually not severe.

Hypothalamic Regulation of Feeding

For many years it has been known that the hypothalamus is involved in the regulation of both food and fluid intake.[33,34] There is a region in the ventromedial hypothalamus that reduces feeding and an area in the lateral hypothalamus that elicits feeding and drinking.[38-44a] Only rarely have these defects been observed in humans. The first clues of the role of the hypothalamus in control of energy balance were obtained from lesion studies. When a bilateral lesion is made in the ventromedial hypothalamus, there is a marked increase in feeding during the first 2 to 3 weeks, with a characteristic rapid weight gain and increase in fatty tissue. Eventually, this weight gain plateaus and becomes static. This dynamic phase can be reestablished by restricting food. When food is again freely available the animal soon returns to its previous level of obesity, suggesting that a new set point for the regulation of energy stores has been established by the lesion in the ventromedial hypothalamus.[44 b-c]

Rats with lesions of the ventromedial hypothalamus are hypoactive as well as hyperphagic. This resulting reduction in energy expenditure also contributes to the obesity. This appears to be the reason that rats with hypothalamic lesions gain more weight with the same amount of food than the unoperative controls. Studies have been done which indicate that the mechanism for cessation of feeding is disrupted. These animals do not eat excessively when they must work to obtain the food, indicating there is probably not an increase in appetite.[44 d]

The hypothalamus is a very small structure with many functions. Lesions do not selectively destroy cells or fibers concerned with feeding and drinking. Many of the animals with these lesions often have defects of hypoactivity, apathy, and depression and exhibit motor deficits and disorders of locomotion. Lesions of the lateral hypothalamus in rats produce animals that do not eat or drink. The multiple effects of lesions and stimulation of the hypothalamus appear to be due to the relatively crude nature of these techniques in a very confined area.[41]

These centers may receive signals that initiate or stop feeding in two ways: by peripheral sensors in the viscera or by mediators carried in the blood to the central nervous system. (Thus, water, glucose, fat, protein, amino acids, the thermal effect of food, and the hormone cholecystokinin have been proposed as humeral regulators of feeding). It has been shown that cholecystokinin (but not secretin, gastrin, or glucagon) regulates satiety.[44 e-g] It remains unclear which of these many factors controls the overall appetite and energy expenditure balance.

The hypothalamus can no longer be considered the sole center for hunger and thirst. On the other hand, it is probably not appropriate to propose that the hypothalamus merely contains fibers of passage that are connecting structures of the midbrain and midforebrain, which subserve the ingestive responses. In animals with greater cerebrocortical function, ingestive patterns are less dependent on deficit signals. It appears that the forebrain structures from the cerebral cortex may play a relatively more important role in feeding and drinking.

In the section on the psychological aspects of obesity, it will be seen that defects in appetite, satiety, and cessation of feeding seem to have been demonstrated in some human subjects. A discrete lesion in the hypothalamus is a very rare cause of obesity in humans.

Cultural Factors in Obesity

The ability to become obese appears to be an almost universal trait in humans. Other mammalian species such as the pig or bear have a marked ability to accumulate fat, which is based on the feast-famine condition of their wild state. This trait has great survival value both to the individual and the species in the natural state. Many other animals (i.e., rodents) have very precise control of weight. With the development of modern agricultural and industrial societies in which food is abundant throughout the year, the tendency to consume more calories than is needed for the day becomes a liability. It is easy to understand why obesity is common where it is socially desirable. However, in the United States and western Europe, where obesity is regarded as a health hazard and cosmetically undesirable, the incidence remains extremely high. Some of the obese individuals come from family backgrounds where the trait has been perpetuated by familial endorsement. In one of the few experimental studies of obesity, Sims et al.[45] were able to persuade eight men to gain an average of 25% of their initial body weight. These individuals found that it was relatively easy to gain the weight; however, the weight reduction period was psychologically difficult. The adipose cells increased in size, but not number during the weight gain. As expected, there was an increase in serum lipids and a deterioration in the tolerance to oral or intravenous glucose. It is believed that the total number of fat cells in an individual is established at a very early age, and these cells persist all of that individual's life.[46] From this arises the theory of the predisposition of childhood obesity to obesity in later life; this relationship appears to be established.[47-49] The question of whether this is genetic or environmental is not settled.[50]

Although the obesity trait is almost universal in the human population, it is clear that cultural, social, and familial factors influence the manifestation. Of primary importance is the failure of the individual to control his excessive appetite. In addition, psychological factors may be important in some individuals.

Psychological Observations on Obesity

The extensive studies of the psychological aspects of obesity have been reviewed by Glucksman.[51] Bruch[52-60] extensively studied obesity in children. These children had more emotional disturbances. The obese child tends to be either an only child or the youngest member of a family. Frequently, the mothers in these families are the dominant figures and the fathers play subordinate roles. They characteristically offer their children food for nonnutritional reasons and commonly overemphasize the social tendency of using food to express affection. Bruch[57] observed that obese children have higher than average intelligence. They frequently suffer from social sequelae because of obesity and become shy, oversensitive, and withdrawn individuals.

A number of psychoanalytic investigators[61-67] have emphasized that obese persons encounter serious disturbances in the oral stage of psychosexual development. As a result of deficiencies of oral gratification in infancy, obese persons may become "fixed at the oral stage." One of the earliest pleasant contacts between parents and children is eating. Thus, individuals may continue to equate dining with being loved or a sense of security. Practically, this means that the individuals resort to eating to relieve emotional distress as adults.

Friedman[67] compared overweight, underweight, and normal women for 15 psychological variables and found that the overweight groups were significantly different from the normal weight in only two areas. They demonstrated (1) less objectivity and (2) more hypersensitivity. Weinberg et al.[68] compared obese and nonobese men using psychological testing and found no significant differences except that obese men have a higher verbal than performance I.Q. Stunkard[69-80] believes that obese persons as a group have significantly more psychopathology than the nonobese. There is not a unique or uniform personality pattern for the obese individuals as a group. Moreover, there are obese individuals without demonstrable psychopathology.[68] Mendelson's view[81] that there is a continuum characterized by increasing degrees of emotional instability in the obese population may best explain the data.

Bruch[56] suggests the following classification based on the relation between psychological patterns and obesity:

1. Absence of psychopathology. This category includes obese persons without emotional difficulties or persons whose neurotic conflicts are not fundamentally related to their obesity.

2. Developmental obesity. In this type, the obesity plays a central role in the individual's growth and development. These patients become obese in childhood, are preoccupied with size and weight, and may be unable to tolerate either frustration or delayed gratification. They frequently are unable to differentiate internal stimuli which indicate hunger, satiety, or other emotional states.

3. Reactive obesity. This type usually occurs secondary to a psychological trauma.

Overeating is utilized by these patients as a method of reducing anxiety, depression, or tension. Bruch[56] feels that some of these patients have an underlying fear of starvation which is restimulated with loss or disappointment. They eat both to alleviate the fear of starvation and to ward off the unpleasant symptomatology of depression.

4. Thin-fat people. These are individuals who have succeeded in weight reduction and are no longer obese. They are still preoccupied with body size, are chronically dissatisfied, and express neurotic conflicts by focusing on food.

Stunkard[67-71] has classified obese patients into the following categories based on their patterns of eating:

1. The night-eating syndrome. This is characterized by evening hyperphagia, insomnia, and morning anorexia. It occurs during periods of light stress and is intimately related to the development of obesity. He suggests that this syndrome may be related to functional impairment of the satiety mechanism.

2. Binge eating. This type of eating occurs during periods of light stress and is characterized by ingestion of large amounts of food over short periods of time. Psychodynamic meaning of this type of eating can usually be identified; it may be followed by feelings of self-condemnation.

3. Eating without satiety. Stunkard observed this type of eating in a patient after an attack of encephalitis. It bore no relation to appetite, hunger, or periods of stress. It occurred randomly and without feelings of self-condemnation. The patient had great trouble stopping.

Other classifications of psychological types of obesity have been presented. Swanson and Dinello[82,83] viewed obesity as a habituation syndrome. In this concept, food is compulsively used to produce sedation. There is a tendency for continually increasing consumption along with psychological dependency. Behavior changes occur when the customary food intake is restricted. A number of authors feel that there are abnormalities in the obese individual's perception of hunger. Some subjects experience an overwhelming appetite, but many overeat when not hungry.[81] Stunkard[73-75] points out that hyperphagia may result from disturbances in both the hunger and satiety mechanisms. For example, binge eaters reported an increased desire to eat, while those with nighttime eating reported an inability to stop. He showed experimentally that obese women fail to report hunger during the presence of gastric contractions. On the other hand, a group of nonobese women reported hunger during these same physiological phenomenon. Griggs and Stunkard[84] suggest that the bias in the perception of body functions can be improved by appropriate biofeedback and pattern reinforcements. Schachter[86] and Schachter et al.[86] attempted to test the hypothesis experimentally. They found that obese subjects ate the same amount of food regardless of whether their stomachs were full or empty. In contrast, nonobese subjects ate more when their stomachs were empty than when they were full. Subjects of normal weight ate more when they were at a low state of anxiety as opposed to a high state of tension. In an experiment in which the time was manipulated, obese subjects ate significantly more when they believed that more time had elapsed than actually had since the last meal.[87] Nisbett[88] demonstrated that the eating behavior of obese subjects, in contrast to nonobese individuals, is heavily influenced by external food stimuli. For example, obese subjects bought more food in the supermarket if they had recently eaten than if they were food deprived.[89,90] On the other hand, normal individuals bought more food as their state of deprivation increased.

From a clinical standpoint, hyperphagia appears to have symbolic importance as well as tension-reducing values for some individuals. Psychological factors involved in hyperphagia are still not well understood. The data suggest that overeating in some obese individuals is associated with biologically determined feeding patterns, altered perception of appetite and hunger, suppression of unpleasant psychological symptoms, and distortion of food gratification. Many obese individuals probably have a distortion of their "body image." Some patients use their body size in a neurotic, competitive fashion so that they can be the center of attention and more powerful than others. Obese individuals may use their imposing appearance to control and intimidate others, e.g., in primitive societies. In general, psychopathological use of body size in interpersonal relations is a function of being obese and neurotic. Some individuals do not have body-image disturbances. It is to be kept in mind that in a society in which obesity is undesirable, feelings of self-consciousness and embarrassment about obese body configuration can be expected.

The evidence to date suggests that psychological factors play an important role in some obese patients. It should be emphasized that the patients on whom psychological studies are conducted frequently have a recognizable problem associated with obesity that has brought them to the attention of a psychologist. The data indicate that in obesity there is an interrelationship among hereditary, physiological, and psychological factors, which may have varying importance. There are also many obese individuals in whom no psychological disturbances can be demonstrated.

Harmful Effects of Obesity

There are overwhelming data establishing that obese individuals are less healthy than individuals of normal weight.[1,47] Life insurance mortality tables[15] clearly show an increase over anticipated mortality when obesity is present. The following three points from this information should be emphasized:[91] (1) Little, if any, increased mortality is seen until the individual is more than 20% above average weight; (2) weight 30% above the average is associated with a doubling of mortality; and (3) the increased mortality shown in these tables may not be directly associated with excess fat, but may be related to coexisting disease. The various conditions that are believed to be affected by increased weight will be considered separately.

Cardiovascular Disease

Hypertension is more common in obese individuals.[19] A statistically significant percentage of hypertensive persons are obese.[92] Elevated blood pressure is a risk factor for coronary artery disease. Reduction of weight will often lower the blood pressure, but there are no good data showing that this improves the outlook for obese patients. The mechanism by which obesity produces high blood pressure is not known. It was felt that the increase in blood pressure in obese individuals, when measured by the conventional cuff, was an artifact of the increase in the size of that extremity. The present consensus is that the effect of arm size on the blood pressure is slight if the cuff bladder is large enough. If the cuff bladder completely encircles the obese arm the measurements are accurate.[19,47,92-95] Obesity has a real influence on blood pressure.

Alexander[96-98] studied blood pressure in morbidly obese patients (individuals more than twice normal weight). The degree of obesity and level of blood pressure were not well correlated. The hypertension in these huge patients was usually of moderate degree. Systemic vascular resistance was normal and the blood volume was increased in proportion to the extent of the obesity. The heart was enlarged due to left ventricular hypertrophy. He concluded that obesity places a hydraulic load on the heart by increasing the amount of tissue that must be perfused. The vascularity of fat is extensive and resembles that of muscle. Weight loss, of course, would reduce this workload.

The Framingham study showed that high blood pressure developed ten times more often in persons 20% or more overweight.[99-101] Men who maintained their weight over a 20-year period had one fifth the risk of hypertension during middle age. In this study obese patients had an increase in diabetes, cardiovascular accidents, and coronary artery disease during the follow-up period. It was demonstrated[99-101] that there remained a significant increment of increased coronary artery disease in obese patients even when the effects of the associated conditions such as diabetes and hypertension were removed. Whether this is associated with obesity or some other factor in the patient's diet is unclear. Obesity as a risk factor for coronary artery disease is controversial.[99-109] Some prospective studies of Chapman et al.,[109] Doyle et al.,[110] Kagan et al.,[111] and Paul et al.[102] indicated that with increased body weight there was an increase in coronary artery disease. This relationship almost disappears when allowance is made for hypertension. The increase was not observed in the Minnesota business and professional men who were followed for 15 years by Keyes et al.[19] It appears that the contribution of increased body weight to coronary disease is primarily manifested through other factors. It is of great importance to know whether weight reduction would either correct or prevent these health defects. Doyle et al.[110] suggest that the effect of obesity on blood pressure is mediated through the variable of salt intake. They found that obese hypertensive subjects were especially responsive to salt restriction. The lowered salt intake in the weight-reduction regimen lowers the blood pressure. The small hypotensive effect of weight loss makes weight reduction an improbable aid in managing coronary disease. It is difficult to demonstrate an effect on coronary artery disease when modest hypertension is reduced with drugs.

Alexander and Pettigrove[97] demonstrated that congestive heart failure occurs in the morbidly obese individual without hypertension or coronary artery disease. The majority of obese patients with congestive heart failure have associated hypertensive or arteriosclerotic heart disease. In addition, these individuals have fluid retention associated with their obesity, which may not be a manifestation of congestive failure.

In the Evans County, Georgia prospective study,[105] neither obesity nor excessive weight gain after age 20 was associated with coronary artery disease, but both were associated with cerebrovascular disease. Men overweight at age 20 and gaining 13 kg or more had three times more cerebrovascular disease than thin men. Correlation between blood pressure and either weight at age 20 or weight gain was of marginal significance. However, these authors believe that the action of weight gain was to increase the blood pressure, which led to cerebrovascular accidents. In a recent study[105a] measuring total body fat by tritium dilution, no significant correlation was found with coronary artery disease and obesity. Body fat was only a minor determinant of blood pressure and serum lipid levels.

Diabetes

The incidence of glucose intolerance and fasting hyperglycemia has been shown to be increased in obese populations.[92,112-113] Sims et al.[45] showed that glucose intolerance occured when normal individuals were experimentally encouraged to ingest huge quantities of food and to gain 25% of their normal body weights. This glucose intolerance disappeared as the individuals reduced their weights to more normal levels. It has been found in clinical medicine that mild diabetes will respond to weight reduction with the disappearance of both the clinical and biochemical findings. It is well known that cerebrovascular and renal complications of diabetes occur in mild diabetes associated with obesity.

Pulmonary Effects of Obesity

The classic pulmonary effect of obesity is the Pickwickian syndrome.[114-116] These patients have massive obesity, somnulence, and hypoventilation. It is now unclear whether they have a basic defect in the respiratory drive mechanism beyond the problem of massive weight.[114-117] A recent report[118] shows that the defect in the respiratory center persists after weight loss is accomplished. In the extremely obese patient lung capacity is less than normal, due to reduction in the vital capacity with the residual volume remaining constant. When the thoracic and abdominal girths are measured together with the lung volumes, a basic change in the mechanism of respiration is seen. There is a reduced ability of the obese patient to expand the chest. The amount of chest expansion per liter of inhaled air is only a fraction of that seen in the normal patient.[119] Therefore, the movement of the diaphragm accounts for more of the volume of air moved during the respiratory cycle than it does in normal individuals. When the movement of the lower ribs is studied fluoroscopically, their excursion is seen to be limited by as much as two thirds. The motion of the diaphragm is only slightly compromised. This change in respiration probably contributes to the patient's difficulty in handling pulmonary infections and anesthesia.[119] The induction of general anesthesia in the massively obese patient is frequently hazardous. Tracheal intubation is always necessary in these patients and may be difficult because of a short, thick neck. There is a constant shift of the anesthetic into the huge fat depots, which may cause a rapid reversal of anesthesia to lighter plains. Increased dosage can easily lead to toxic concentrations of the anesthetic agent. The recovery phase may be long because of slow release of anesthetic from fat stores and may be complicated by poor ventilation.[120-123] Despite the above problems with anesthesia,[124] a review of the complications of cholecystectomy in 400 obese patients showed no increased morbidity. The low mortality in the surgical therapies for obesity demonstrates that anesthesia problems can be satisfactorily handled in most patients.

Gastrointestinal Problems of Obesity

De La Vega and Ponce DeLeon[125] describe the association of gallstones, hiatus hernia, and diverticulosis of the colon. These problems probably have exogenous obesity as a common feature. The reduction of obesity in the patient with hiatus hernia and reflux esophagitis has been found to be effective in relieving this symptomatology. The increased incidence of gallstones in obese subjects is well documented. Further studies have shown an increase in the excretion of cholesterol in the bile of obese subjects with gallstones as compared to thin patients with cholelithiasis.[126,127] Liver disease of varying severity has been shown in patients with massive obesity. Approximately 90% of the morbidly obese patients who undergo jejunoileal bypass have fat in the liver. In addition, all degrees of liver disease, including cirrhosis, are seen.[128] As with all studies involving liver disease, it is very difficult to be certain that these patients are not ingesting a hepatotoxin, particularly alcohol. Nevertheless, there are some cases in which liver disease appears to be associated only with morbid obesity. On the basis of huge caloric excess, malabsorption has been shown to be present; however, this is probably very rare.

Obesity in Osteoarthritis

Large population studies[129-131] of patients with osteoarthritis have shown an increased incidence of obesity. A relationship has also been suggested between obesity and the development of the disease in specific weight-bearing joints.[132,133] An increase in osteoarthritis in nonweight-bearing joints has also been observed.[129,134] A recent study by Goldin et al.[135] investigated the incidence of osteoarthritis in morbidly obese individuals. An X-ray survey of multiple joints in 25 patients showed few signs of degenerative changes. Twenty percent of the patients had previously incurred traumatic injury requiring surgery to the cartilage of the knee. Goldin and co-workers felt that if obesity has a significant role in the development of the disease it should be seen in these massively obese individuals who average 45 years of age. There is logic in the recommendation that obese people with osteoarthritis lose weight to reduce the stress on weight-bearing joints. It appears unlikely that the obesity is actually etiological in this problem. Possibly, the arthritis limits activity and results in secondary obesity.

Therapy for Obesity

The therapy for obesity is the ultimate of simplicity: caloric intake must be reduced below caloric expenditure. Dietary treatment of the massively obese can be briefly summarized. It rarely works. We do better with the treatment of cancer than obesity.

As our natural appetites allow for great physical exertion (necessary in previous generations) obesity becomes almost inevitable in our sedentary society. It should be emphasized that a very large percent of the middle-aged population of this country is restricting calories to maintain a constant weight. In a very brief survey of the author's personal acquaintances, most individuals over 40 are restricting calories to control their weight. The tendency to ingest excess calories is more widespread than the incidence of obesity. If the individuals are at normal weight, the reduction in caloric intake is successful. If they are obese, they have failed in the past to balance energy requirements. In many individuals, dieting accomplishes the goal of weight stabilization by caloric restriction.

For patients with "morbid obesity," dietary therapies are almost uniformly unsuccessful. A succession of these dietary regimens appears in the lay press, each claiming to be the ultimate solution. The obese individual is told that he can eat all he wishes of fat, meat, citrus fruit, or bananas, as long as he avoids a sizeable list of common foods. He grows so tired of the special food that he makes the decision to eat less. Commercial considerations and financial gain also play a large role in the promotion of fad diets. It seems worth repeating that there is no consistent metabolic lesion that has been found to explain obesity. Only a small percentage of patients has an underlying disease.

In 1957 Bloom[136] proposed total fast as an initial therapy for obesity. He believed that total fast is more tolerable than diet restriction because of the ensuing ketosis which reduces the appetite. Metabolic studies have consistently shown that either total or partial starvation depletes the body of lean tissue and essential electrolytes as well as fat.[137,138] These changes cause profound alterations in mood and physiological function. Careful physicians recommend that if total fast is to be used as a treatment, it be done in a hospital, where mood changes and cardiac arrhythmias may be managed.[78,139,140] Fatalities have resulted from cardiac arrhythmias during total starvation.

Medications to suppress the appetite have been extensively used in the past; however, they have a brief effect of only a few weeks unless the dosage is increased. Numerous preparations of amphetamines, frequently with the addition of a sedative or tranquilizer were used in the 1950s. It is now realized that these preparations have only a mild anorectic effect and have a considerable risk of habituation. Over the past several years there has been renewed interest in medications that will suppress the appetite. One of these, femfluramine has been studied relatively extensively.[141] This is a trifluoromethyl substitution of amphetamine that has an appetite-suppressing activity. It has been in general use in the United Kingdom for several years. A report from South Africa showed it to be a common agent for drug abuse. It ranked fourth in frequency behind marijuana, amphetamines, and LSD and slightly above opiates and alcohol.[142] It appears that this will not be a breakthrough in the management of obesity and will have the same contraindications as the amphetamines.

Thyroid hormone is ineffective in weight reduction at less than thyrotoxic doses. Numerous combinations of medications have been used in "obesity" clinics. Most of these drugs have no physiological action in weight reduction, e.g., diuretics, which give an immediate fluid loss which is interpreted as weight loss. In a few "quack" clinics medications have been given for their toxic anorectic effect. There is no medication currently available that is of value in weight reduction.

Group therapy has been used to support attempts to maintain dietary control. The numerous lay organizations (Tops®, Weight Watchers®, etc.) are probably functioning through this mechanism.[143] Some appear to be using group reinforcement in a manner akin to Alcoholics Anonymous. Stunkard and McLaren-Hume[79] and Stunkard[80] reviewed a collection of psychotherapeutic approaches to obesity. This treatment was started on the assumption that there are basic behavior disorders in obese patients. Reports suggest that this therapy may be at least as effective as other therapies for obesity.

It is worth remembering that obesity may be wrongly accused as a cause of chronic disease. Probably the best evidence for the effectiveness of weight reduction comes from the insurance data.[91] Individuals who were issued risk insurance because of obesity may obtain a standard policy if the excess weight is lost. These individuals have been shown to have a normal survival. It is unclear that weight reduction has been shown to be a useful treatment for any chronic disease. The irony of this argument is that the treatments for obesity are so ineffective that the impact of weight reduction on any disease cannot be properly evaluated. Obesity is a relatively incurable disorder. This makes proper clinical trials impossible.

Surgical Therapy for Massive Obesity

The surgery for obesity that has had the widest use is intestinal bypass.[144] This operation has been used in some form for approximately 20 years. There has been considerable variation in the cases selected for its use. Scott et al.[145-147] together with Payne and DeWind,[148] have emphasized that these patients should be two to three times normal weight. They use the term "morbid obesity" to describe patients who have maintained this weight for 5 years or more, despite efforts at weight reduction. The treatment goal is to rehabilitate these patients socially, economically, psychologically, and physically for a prolonged period at a low risk. Intestinal bypass falls short of these goals in a number of areas. Many investigators feel that prospective candidates for intestinal bypass should have a clear understanding of the procedure and its consequences, including complications.[144] A thoughtful reflection on the operation by the patient is essential. The operation is contraindicated in the psychotic or alcoholic patient, in individuals with a compulsive desire to proceed immediately, and those having Cushing's syndrome or hypothyroidism.[149] It should not be used in patients with severe liver, renal, heart, or pulmonary disease which are not likely to be affected by weight loss. Patients are better-than-average candidates if certain complications related to obesity (such as diabetes mellitus, hypertension, hyperlipidemia, the Pickwickian syndrome, or osteoarthritis of the weight-bearing joints) are present.[150]

Operative Procedure for Jejunoileal Bypass

Kremen et al.[151] showed in 1954 that weight loss is associated with exclusion of the distal but not the proximal 50% of the small intestine in dogs. Payne et al.[152] began their clinical program in 1956. Their first 10 patients had jejunocolostomies. These patients were all more than 56 kg overweight and had partial or complete restoration of bowel continuity when the ideal weight was reached. Weight gain regularly followed reconstitution of normal bowel length. In 1969 Payne and DeWind[148] found it necessary to bypass all but 14 in. (35 cm) of jejunum and 4 in. (10 cm) of ileum in order to maintain weight reduction. It was not necessary to restore intestinal continuity in order to prevent excessive weight loss in this group of cases. Numerous modifications of the surgical procedure have been devised over the years. It is now generally agreed that jejunocolostomy is not well tolerated and some variations of an end-to-end jejunoileal anastomosis is preferable. The mean weight loss increases as the length of the bowel in continuity decreases from 80 to 40 cm; however, there is a wide variation for each bowel length. MacLean[144] has reviewed the results of intestinal bypass, collecting the data from a number of series. The mean weight loss varied between 30 and 40%. If, as stated previously, these patients start out at two to three times normal weight, they remain quite obese even after their weight has stabilized after the intestinal bypass.

Complications of Jejunoileal Bypass

Hospital mortality rates from this procedure varied between 0 and 11% and were primarily related to myocardial infarction, pulmonary embolism, wound and anastomotic complications, and liver failure. The complications in 94 patients were reported by Schwartz et al.[153] These included 28 complications in 94 patients, with one mortality. Ten complications consisted of serolipid wound drainage, which would be expected in these massively obese patients. They emphasized the importance of special care before, during, and after the operative procedure. Moss et al.[154] reported on the radiographic complications of bypass surgery in 89 patients and found that 10% had bowel obstruction, 5.6% cholecystitis, 3.4% peptic ulcer, and 6.7% megacolon.

Late complications of jejunoileal bypass are probably more serious. Most of these patients have attacks or continuous diarrhea, as would be expected. This is more severe with a jejunocolic bypass, which has been discontinued. Weight loss is very rapid during the early postoperative period and stabilizes in 1 to 2 years. In many units patients are instructed in the use of lomotil (diphenoxylate-hydrochloride), calcium, potassium, multivitamins, and vitamin B_{12}, which are given as supplements. Episodes of diarrhea not controlled by lomotil are usually managed by elemental diet.

The most serious complication of jejunoileal bypass is the aggravation of liver disease. Morbidly obese individuals frequently have liver biopsy abnormalities. The pathology may range from fatty infiltration to a well-developed cirrhosis. There is good evidence that the hepatic lesion may progress after the bypass. McGill et al.[155] reported 7 deaths from liver disease in 63 patients who had jejunocolostomy. This was the principal reason for abandoning this operation for obesity. The length of the jejunal circuit was 37 cm in three of these patients and 50 cm in three and the others were unknown. Although the frequency of progressive liver disease decreases with jejunoileostomy, it occurs after this procedure as well. There are reports of 13 deaths from hepatic failure or cirrhosis after jejunoileal shunts.[144] There is probably a much larger number of unreported cases. In some patients with liver failure, restoration of normal intestinal continuity returns the liver function and microscopic appearance to normal. The cause of liver failure and cirrhosis after intestinal bypass is unknown. A theory of hepatotoxins from the bowel was advanced by Brown et al.[156] They reasoned that the extremely long excluded bowel could permit the accumulation of bacteria with a production of toxic substances. Bondar and Pisesky[157] observed that dogs tolerated 80% resection of small bowel although they lost 25% of their weight. In contrast, an end-to-end jejunoileal bypass excluding 80% of the small intestine lead to death from liver failure in all of the dogs. Salmon[158] obtained different results in dogs and found that they could tolerate an end-to-end or end-to-side jejunoileal bypass for up to 4 years. Opposing this theory is a report by Buchwald et al.[159] that found no hepatotoxicity in extensive studies of patients undergoing partial ileal bypass for hyperlipidemia.

A more attractive theory to account for the liver failure would be nutritional deficiency, especially protein deficiency. Buchwald et al.[160] performed jejunoileal bypasses leaving 50 cm of functioning intestine in 650 patients without any deaths from liver failure. In contrast, Brown et al.[156] left only 40 cm in the circuit and noted severe liver failure in 6 of 36 patients. Heimburger et al.[161] successfully reversed severe fatty hepatic infiltration after intestinal bypass by using infusions of amino acids. The liver pathology reverted to normal and the serum albumin increased while the patients continued to lose weight. It is agreed by most workers that although liver enzyme abnormalities such as transaminase elevations frequently occur, liver failure can develop and progress without significant changes in liver-function studies. It is essential that the patients be followed with serial liver biopsies to detect the early development of liver disease.

There are other problems. Renal calculi have been reported in 32% of these patients.[162] Polyarthritis in 23% of the patients was reported by Shagrin et al.[163] The polyarthritis is usually an asymmetrical involvement, predominantly of the knees, ankles, and proximal intraphalyngeal joints and fingers, characterized by an acute synovitis. Treatment other than restoration of the bowel continuity has not been curative in any of these patients.

The primary problem with intestinal bypass is that the margin for error in this procedure is very small. A bypass shorter than 45 cm produces prohibitive complication rates of electrolyte problems, liver failure, and general malaise. On the other hand, a segment of longer than 50 cm is associated with a poor weight loss in almost all reports. It is difficult to measure the small bowel, and variations in tension on the bowel could make more than this magnitude (5 to 10 cm) of difference. Because of the numerous complications of the procedure and the narrow margin of error for a satisfactory result, jejunoileal bypass must be restricted to patients with extreme obesity who have not responded to other therapy.

Gastric Bypass

Because of the numerous complications of intestinal bypass, Mason and Ito,[164] Mason and Printen,[165,166] and Printen and Mason[167] developed a 90% gastric bypass designed to limit caloric intake while not interfering with digestion or absorption. Extensive experience with this bypass suggested that it is a suitable operation for the treatment of morbid obesity.[164-167] The stomach is divided so that the proximal 10% is in continuity with the esophagus, and this is anastomosed to the jejunum. The distal 90% of the gastric pouch is then closed and sutured to the undersurface of the diaphragm to prevent intussusception. This procedure is virtually free of the chronic complications observed frequently with small bowel bypass.

Hermreck et al.[168] reported on 75 patients with morbid obesity who underwent 90% distal gastric exclusions for weight reduction. The patients were selected with the exclusion of serious psychiatric disease, serious preexisting gastrointestinal disease such as duodenal ulcer, hiatus hernia with reflux esophagitis, or severe cardiopulmonary disease. The average weight of these patients was 121 kg (267 lb), the average height was 64.5 in., and the mean age was 31.4 years. The operation was postponed for at least 2 to 3 months after initial evaluation in order to discourage patients' requesting the surgery without careful consideration.

There were 20 surgical complications in the 74 patients. These included ten wound infections, three incisional hernias, one stomal obstruction, and one intestinal obstruction; four required splenectomy and one had a lacerated diaphragm. There were no fatalities. Fifteen patients developed late gastrointestinal complications including chronic vomiting (9), diarrhea (2), reflux esophagitis (2), and dysphagia (2). Complications of chronic vomiting and diarrhea appear to be related to overeating and are controlled by limiting the volume of meals. In addition, two patients developed vitamin deficiency and seven had inadequate weight loss. Two cases of severe dysphagia which required esophageal dilatation occurred after surgery. X-ray findings of compression of the distal esophagus were believed to be due to intramural hematoma. Both cases resolved with a single esophageal dilatation. Of the 54 patients who were followed for more than 1 year, 7 had inadequate weight losses. X-ray evaluation revealed dilatation of the jejunum in some of these patients; others appeared to eat continously to maintain their massive obesity. Of the 54 patients followed for 12 months there was a 35% decrease in weight as a result of gastric bypass. There was no statistically significant difference in weight reduction between small bowel and gastric bypass after 1 year. The authors[168] believe that this operation is probably superior to intestinal bypass. The disturbing complication of chronic vomiting was felt to be secondary to overeating and slow gastric emptying. When food intake is controlled carefully these symptoms rapidly abate. Despite counseling upon discharge, the patients' symptoms often recurred because of overeating. The most feared and lethal complication in Mason and Printen's series[165] was an anastomotic leak. The authors[164-168] agree that the operation is more difficult than intestinal bypass. It has a disadvantage over intestinal bypass in that the patients apparently have to cooperate in reducing their caloric intake or persistent vomiting may result. Gastric bypass has the advantage of not producing a malabsorption syndrome and the complications of liver disease. Because of the inherent risk of this procedure, it can only be recommended to patients with morbid obesity.

Hypothalamic Lesions for the Therapy of Obesity

Quaade[169] and Quaade et al.[170] from Denmark recently used exploring electrodes in an attempt to localize the hunger center in the hypothalamus of five obese patients. In three cases the convincing hunger response could be elicited by stimulation. In two of these patients, unilateral lesions were produced by electrocoagulation. A bilateral lesion was produced in the third patient. Body weight decreased slightly, but only temporarily. It was concluded that in humans as well as experimental animals, the lateromedial hypothalamus is the site of the hunger sensation. These authors were criticized for the experimental nature of the procedures.[171,172] They defended themselves[173] on the basis of the current unsatisfactory therapy for obesity and the fact that the patients were aware of the investigative nature of the undertaking. Although hypothalamic surgery (because it directly affects the appetite centers) might be the ultimate surgical answer to obesity, it seems unlikely that large numbers of people will or should resort to such drastic surgical therapy.

Summary of Surgical Therapies

In evaluating the surgical therapies that have been introduced for obesity, one is impressed that the immediate complication rate of the surgery is quite high. In addition, gastric bypass requires cooperation of the patients to reduce their caloric intake. The intestinal bypass allows the continued excessive food intake, but has major metabolic complications including severe and fatal liver disease. Severe diarrhea is frequently present. It is amazing that a sizeable number of the patients are desperate enough to accept this form of therapy. Surprisingly, patient satisfaction with the two operations appears to be approximately equal. With the intestinal bypass, patients are apparently willing to put up with the troublesome diarrhea; with the gastric bypass procedure, early satiety forces them to limit their caloric intake. The preliminary work on hypothalamic surgery does not show promise. It must be emphasized that surgical therapy for obesity must be limited to a very small number of individuals in whom defined health hazards are present. It certainly must not be done for cosmetic or social reasons.

CALORIC EXCESS IN THE PREVIOUSLY STARVED INDIVIDUAL

Specific and distinctive effects occur when a sudden high caloric load is imposed on a previously malnourished individual. Although famine has been a continuous part of human history, it was not studied until recent years. Adverse effects of sudden refeeding have been observed in the past, but have not been well documented. Starvation is rare in this country. In spite of this, large number of patients in the state of severe malnutrition are seen in our hospitals.[173a,173b] These are patients with a disabled gastrointestinal tract or prolonged anorexia for medical or psychological reasons. In the past, the underlying cause of this malnutrition would subside slowly and effective dietary intake would increase gradually. With the capability for total parenteral nutrition as described by Dudrick et al.[174] and Wilmore and Dudrick,[175] sudden caloric loads can now be delivered to these patients.

The Effects of Starvation in Mammals

The effects of starvation have been studied extensively and have been reviewed well.[137,138,176,177] With the onset of starvation the minimal glucose and glycogen stores of 150 to 300 g are rapidly depleted in normal humans. These minimum stores supply emergency sources of glucose for short periods of fasting. The glucose required during prolonged starvation is obtained from amino acids by glyconeogenesis.

The principal caloric source in starvation is fat. As the fat depots are depleted, an increasing amount of caloric needs are met by protein wasting, primarily from the skeletal muscle. At the same time, there is a profound decrease in caloric expenditure when the individual is allowed to regulate his activity. The reduction of caloric expenditure relates to both a decrease in metabolic rate and the activity of the individual. In prisoner-of-war camps, individuals on starvation diets usually lost weight very rapidly during the first 6 months.[6-8] Then multiple adaptive processes occurred and their weights tended to stabilize at 30 to 40% underweight. Hence, prisoners who were in the camp for from 6 months to 5 years had approximately the same weight at the time of their release. However, the group that had been there for 5 years had suffered a 60 to 80% mortality.[7]

In a number of species including man, death tends to occur at a 40 to 50% loss of body weight. It is usually preceded by apathy, loss of interest, and failure to eat the small amounts of food that are available. In animals the length of time that total fast can be tolerated is correlated with the initial obesity of the individual.[176,177] Massively obese humans have fasted for periods of many months without any apparent adverse effects.[138] They live off their fat and obtain their glucose requirements from protein reserves. In humans, the length of survival during starvation is not precisely related to obesity. Apparently, many other undefined factors significantly affect survival.[196] One of these factors is the adaptive mechanism that reduces caloric expenditure; this is better accomplished by some individuals than others.[6-8]

A large number of studies have been performed on enzyme levels and synthesis rates of various compounds during starvation and refeeding.[178-183] The enzymes necessary for urea synthesis, glyconeogenesis, and oxidating degradation of amino acids are increased. This would be anticipated because this pathway would be active during starvation. These studies have demonstrated increased lipolysis and decrease in activity of the glycolytic enzymes. Upon refeeding, normal activity of these enzymes is achieved rapidly, frequently increasing to above-normal levels. Winter[194] produced death by forced feedings of starved rats. Wilhemy et al.[195] showed that the stress of refeeding was greater when high-carbohydrate diets were fed.

Studies of cerebral metabolism during chronic starvation are not available. The report of Owen et al.,[196] on obese subjects during prolonged fasting showed that the major energy source for the brain in this situation was β-hydroxybutyrate and acetyl acetate. The glucose utilization by the brain dropped markedly from over 100 to approximately 24 g/day. Whether these data are applicable to chronic undernutrition as well as starvation remains to be determined. Agrawal et al.[197] showed a decrease in labeling of brain proteins and lipids from ^{14}C-labeled glucose, but not from ^{14}C-labeled L-lycine in starved rats. This suggests that there is a block in glucose metabolism. These observations illustrate that profound changes in the metabolism occur during starvation in the brain and liver. These changes appear to be reversible during voluntary refeeding. The majority of changes due to starvation are anticipated by an increase in the enzymes related to protein and fat breakdown and a decrease in the enzymes related to glucose metabolism. This is because the organisms shift from predominantly glucose metabolism to primarily fat metabolism.

Hypophosphatemia

This topic has recently been carefully reviewed.[197a] A decrease in serum phosphate is a uniform phenomenon when a sudden caloric load in the form of glucose is administered.[197a] It occurs whether or not amino acids are added to the intake.[198,199] It is not possible to determine if amino acids alone will produce a fall in phosphate because a large caloric load cannot be given due to amino acid toxicity. In addition, the available fat emulsion contains considerable phosphate, which masks the decrease in serum phosphate.[200] Hypophosphatemia has been reported with phosphate depletion secondary to nonabsorbable antacids. These patients have muscle weakness, anorexia, nausea, vomiting, malaise, bone pain, and diffuse numbness. Their symptoms are relieved by addition of phosphate. This hypophosphatemia is probably not related to that seen following a sudden caloric load in a starved individual.[201,202]

A decrease in serum phosphate is seen with a standard glucose-tolerance study. This decrease is more pronounced in patients with starvation, cirrhosis, and diabetes.[203-204a] The decrease in serum phosphate from insulin and glucose can be blocked by adrenergic blockers.[205] A decrease in serum phosphate is seen to a lesser degree in sepsis.[206] Common factors in these observed decreases in serum phosphate are unclear, but may be due to abrupt increases in glucose utilization. A profound fall in serum phosphate is seen in fructose intolerance when patients are given a load of fructose. This fall has been shown to be related to the accumulation of fructose-1-phosphate.[207,208]

Hypophosphatemia during hyperalimentation has been documented in both humans and animals by Lichtman et al.,[209] Travis et al.,[210] and this author's laboratory.[198,199] Abnormalities during hypophosphatemia have been demonstrated in the red cell,[209-213] white cell,[214] and platelets.[213] These abnormalities are corrected by the addition of phosphate to the infusion solutions and the maintenance of normal serum phosphate. In addition, the red cell abnormalities are correctable by the addition of adenosine and inorganic phosphate.[213] A shift in the hemoglobin oxygen association curve has been documented and the development of spherocytosis observed. The glucose metabolism in the red cell has been studied and a decrease in adenosine triphosphate, glucose-6-phosphate, and 2,3-diphosphoglyceric acid occurs during hypophosphatemia. Travis et al.[210] demonstrated an increase in the red cell "triose phosphate." He used this term to denote the combination of fructose diphosphate, dihydroglycerophosphate, and glycero-aldehyde-3-phosphate. The defects in white cell migration and phagocytosis may be significant in the infectious problems of hyperalimentation. Decrease in platelet adhesiveness has been demonstrated in dogs, but not in man.[213]

The cause of the decrease in serum phosphate is not clear. In all studies in which it was measured, urinary excretion of phosphate was very low. When nutrition is maintained parenterally, stool volume approaches zero; therefore, the low serum phosphate is not due to excessive excretion of phosphate from the body. In the studies of Travis et al.,[210] serum phosphate dropped from presumably normal levels of approximately 3.6 mg% to less that 1 mg% during hyperalimentation. The measured changes in phosphate intermediates in the red cell could not account for this marked change. Apparently, phosphate shifts from the serum into the tissues. Hemolytic anemia has been documented in animals and man during hyperalimentation. It appears that hypophosphatemia is symptomatic of a defect in glucose metabolism and is related to an intercellular shift of phosphate without external loss. In recent studies[215] we have demonstrated a similar hypophosphatemia in markedly starved dogs given intragastric high-calorie feedings. We[216] have also identified patients who developed a marked fall in the serum phosphate following oral ingestion of high-caloric diets. Thus, the hypophosphatemia of refeeding may be produced by both the oral and the intravenous route.

The marked hypophosphatemia is important in clinical medicine because of the profound effects on the cellular elements of the blood. In addition, it may alert the clinician to the other adverse effects of hyperalimentation syndrome.

Hyperalimentation Syndrome

Adverse effects of sudden refeeding were observed following release of prisoners from camps during the Second World War.[6-8] Some of these individuals seemed to progress well on refeeding, but would suddenly deteriorate and expire after 3 to 5 days. No detailed description of this phenomenon is available. This is understandable because when these camps were liberated, thousands of critically ill individuals were released. The medical facilities were overwhelmed with the treatment of multiple problems; for this reason, no detailed description of the refeeding syndrome was available.

When we treated patients at this institution[199] by utilizing intravenous hyper-alimentation, a fatal hyperalimentation syndrome was observed. This was reproduced in dog and rat models.[198] The basic characteristics of this syndrome consist of lassitude, numbness and tingling, and weakness of the extremities, which may proceed to coma and convulsions. Accompanying the syndrome is marked hypophosphatemia and fatty infiltration of the liver. These latter two phenomena are not etiologically related to the syndrome. Maintenance of normal serum phosphate in the animals does not prevent the fatal outcome. It occurs a number of days before the neurological symptoms and occurs to varying degrees in all starved individuals given high calorie loads. Fatty metamorphosis of the liver is clearly not etiologically related to the syndrome. It occurs in both normal and starved animals and humans when they are given large dextrose-caloric loads. It does not appear to cause symptomatology in normal animals or man. Coma is not related to an increase in serum osmolality, which remains normal. There is no appreciable increase in blood urea nitrogen or deterioration in renal function, and the syndrome can be produced in a protein-free infusion; therefore, amino acid imbalances do not play a significant role.

Observations have been made both in animals[218] and humans[219] that the dura tends to bulge in these patients. We[200] have been unable to document increased brain water content and in no other studies was the brain water content measured. Derr and Zieve[220] have measured abnormalities in the intermediates of glucose metabolism; however, their animal model consists of acutely starved rats receiving massive glucose infusion. The serum osmolality and blood sugars rose markedly in these animals. It is unclear whether this was the same phenomenon as described in patients and previous animal reports. It appears likely that they are dealing with either hyperosmolar coma or a combination of hyperalimentation syndrome and hyperosmolar coma.

The frequency of this problem in the hyperalimentation of starved individuals is uncertain. It occurs relatively infrequently following oral hyperalimentation and requires a very special set of circumstances found only in the alcoholic population in this country. In unpublished studies on dogs,[215,216] approximately 30% of the animals developed the syndrome at an intragastric infusion rate of 110 cal/kg/hr. The frequency in clinical medicine relates in part to the problem of recognition.

In reviewing the laboratory data of a 4-year period at the Minneapolis Veterans Administration Hospital, there were 110 patients with a serum phosphate of less than 1 mg%. Of the 20 patients in whom a confirmed serum phosphate of less than 1 mg% was present, only 7 had neurological examinations. These patients showed signs and symptoms compatible with the syndrome. Of these patients, four had no intravenous fluid therapy and three had small amounts of intravenous fluid. One patient has been observed prospectively to develop the hyperalimentation syndrome with oral intake.[221] There is now a total of 20 patients who have been described in the literature[199,217, 222,231] having all or parts of the hyperalimentation syndrome. They have been described under various titles and, possibly, some of the authors do not feel that they were associated with the high caloric intake.

At the present time no precise etiological mechanism has been determined for this condition. The concern that it is related to phosphate depletion cannot be supported by the experimental data in animals. In one patient the neurological symptomatology did not improve until the caloric content of the infusion was reduced.[235] In most of the reported patients it is unclear whether the caloric content was reduced (or it had definitely been reduced) before gradual recovery occurred. The current recommendation is to reduce caloric intake and replace the phosphate intravenously with potassium phosphate at the rate of 40 to 60 meq/day in an adult, or its equivalent in an infant.

Deficiency States Induced by Sudden Excessive Caloric Intake

A number of deficiency states have become manifest during intravenous hyper-alimentation. The requirements of certain metabolic cofactors varies with the caloric intake of the organism. The best example is the subclinical thiamine deficiency which becomes overt by forced feeding.[232] In addition, an interrelationship has been demonstrated between thiamine and magnesium deficiencies.[233] Thiamine deficiency has been postulated as a cause of the hyperalimentation syndrome. Although thiamine given with the infusion does not prevent the hyperalimentation syndrome, it is possible that the thiamine-dependent enzymes have a lag and are unable to replenish themselves rapidly enough to prevent clinical symptomatology. Animals have not been studied after preloading with thiamine, which would exclude this as a possibility. This experiment is currently being conducted in the author's laboratory.

Subclinical magnesium deficiency has been demonstrated by Caddell[234] in children with kwashiorkor. Magnesium deficiency has been observed during hyperalimentation. This appears to be related to the low quantities of magnesium in the infusion and the low stores of the element in the individual. Magnesium deficiency is not etiologically important in hyperalimentation syndrome because, as observed by this author, animals preloaded with magnesium and given magnesium during the infusion continue to develop the neurological symptomatology.

Essential fatty acid deficiency has been seen during hyperalimentation.[236,237] It requires from 2 to 3 weeks to develop and appears to be related to the deficiency of the infused mixture of essential fatty acids. Zinc[238,239] and copper[240,241] deficiencies during hyperalimentation have also been described. These deficiencies appear to be related to a previously depleted state and inadequate intake. They respond to replacement therapy.[236-242]

EXCESS OF FATS AND SUGARS OR DECREASED DIETARY FIBER

Refined sugar and saturated fats are the major caloric excesses producing obesity. These excesses produce a decrease in dietary fiber content. The question remains as to whether these substances may have specific adverse effects beyond their caloric excess.

There are good epidemiological data that a highly saturated fat diet is associated with a high incidence of coronary artery disease. The relationship of coronary artery disease to diet has been extensively studied for a number of decades.[96-111] These studies[19, 108-111] show a correlation between fat content of the diet and coronary artery disease. This is consistent in the European and American population groups. However, a number of African tribes consume a highly saturated fat diet without arteriosclerosis. Some tribes have cholesterol concentrations similar to populations prone to coronary artery disease; other tribes have low serum cholesterols.[243,244] All are free of cardiovascular disease,[245] this is suspected to be due to excellent physical condition.

Yudkin and Morland[246] and Yudkin[247,248] have implicated the ingestion of refined sugar in the incidence of coronary disease. The use of sugar is well correlated with the historical and epidemiological incidence of the disease. However, it is difficult to determine whether this is a cause-and-effect relationship, because population groups consuming diets high in saturated fats also tend to consume highly refined sugars.

Since the initial hypothesis that diets of high saturated fat produce arteriosclerosis was advanced decades ago, the use of polyunsaturated fats has increased almost threefold in the U.S. This has not decreased mortality from heart disease, which would be expected if polyunsaturated fat exerted a positive effect.[249] Dietary control of fat intake has failed to show a consistent decrease in the incidence of arteriosclerotic disease. This may be related to prior damage from a high-fat diet. It is possible that the diet must be changed in infancy or childhood.

An additional problem associated with a saturated fat diet has been carcinoma of the colon. The American incidence of colon carcinoma is very high, whereas the Japanese incidence is quite low. Japanese who migrate to this country and especially their children born in the U.S. have a much higher incidence of this disease than the Japanese who remain in their native country. The same trend is noted for Puerto Ricans upon their migration to the mainland.[251-253] Again, there are many other factors associated with the different social groups beside diets. Only cancer of the endometrium is associated with obesity itself.[250]

The question of the fiber in the diet has been raised by Burkitt and others[254-260] who have made observations of people in central Africa who eat high-fiber diets. They have almost no problem with appendicitis, diverticulosis of the colon, or carcinoma of the colon. As the rural Blacks of Africa become urbanized, the incidence of appendicitis, then diverticulosis, and 20 to 30 years later, carcinoma of the colon begins to increase. The researchers[254-256a] have postulated that the high-bulk diet produces frequent bowel movements and decreases the contact time of carcinogens with the mucosa of the colon. These conditions arose in the Western world beginning in the middle of the last century, when it became common practice to refine flour and markedly reduce the fiber content. There is strong evidence that excesses in either refined sugar or fat or decreases in fiber in the diet are related to the increased arteriosclerotic vascular disease and carcinoma of the colon in the Western world. These findings relate to the type of calories, but not necessarily to an excess in the overall caloric intake.

Protein excesses essentially occur only in artificial states of added amino acids. This subject has been reviewed at great length by Harper et al.[261] (discussing nonsodium glutamate) and will be discussed in the contribution on food additives. Dietary excesses of vitamins can occur with ingestion of some special foods, such as vitamin A toxicity due to polar bear liver.[262,263] Vitamin toxicities are almost exclusively related to ingestion of purified vitamins.[264,265] These are discussed elsewhere.

SUMMARY

Caloric excess is a problem of developed Western societies. The adverse effects of manifest obesity are well documented. Therapeutic approaches have been instituted and some of them are very dramatic; but, to date, no effective therapy has been devised with a low complication rate. The exact role of obesity in disease cannot be clarified until an effective therapy can be applied and controlled clinical trials conducted.

Caloric excess in the previously starved individual produces strikingly different phenomena. These include hypophosphatemia with hemolytic anemia as well as red and white cell defects. A fatal syndrome of paresthesias, confusion, coma, convulsions, and death (hyperalimentation syndrome) may be precipitated by these circumstances. This is important in treating the severely malnourished patients by intravenous therapy, and it may be of considerable importance in treating famine victims. Subclinical deficiency states may be manifested by a sudden increase in caloric intake in both clinical malnutrition and famine states.

ACKNOWLEDGMENT

The author wishes to thank the following: Dr. Rex Shafer and Dr. Roger Gebhard of the Minneapolis Veterans Administration Hospital for their careful review and constructive criticism of this chapter, Mrs. Harriet Aaker for her helpful assistance in the literature accumulation, Mrs. Karleen Luck and Miss Lynne LaBelle for their excellent stenographic assistance, and my wife Marilyn for her very careful proofreading of the final manuscript.

REFERENCES

1. Mann, G. V., The influence of obesity on health, *N. Engl. J. Med.,* 291, 178–185, 1974.
2. Fox, F. W., The enigma of obesity, *Lancet,* 2, 1487–1488, 1973.
3. Neel, J. F., Diabetes mellitus: a "thrifty" genotype rendered detrimental by "progress"?, *Am. J. Hum. Genet.,* 14, 353–362, 1962.
4. Ashworth, A., Malnutrition and metabolic rates, *Nutr. Rev.,* 28, 279–281, 1970.
5. Monemurro, D. G. and Stevenson, J. A. F., Survival and body composition of normal and hypothalamic obese rats in acute starvation, *Am. J. Physiol.,* 198, 757–761, 1960.
6. Zimmer, R., Weill, J., and Dubois, M., The nutritional situation in the camps of the unoccupied zone of France in 1941 and 1942 and its consequence, *N. Engl. J. Med.,* 230, 303–314, 1944.
7. Schnetker, M. A., Mattman, P. E., and Bliss, T. L., A clinical study of malnutrition in Japanese prisoners of war, *Ann. Intern. Med.,* 36, 69–96, 1951.
8. Burger, G. C. E., Drummond, J. C., and Sandstoad, H. R., Eds., *Malnutrition and Starvation in Western Netherlands, Sept. 1944–July 1945,* Parts 1 and 2, General State Printing Office, The Hague, Netherlands, 1948.
9. Gordon, T. and Kannel, W. B., Obesity and cardiovascular disease: the Framingham study, *Clin. Endocrinol. Metabl.,* 5, 367–375, 1976.
10. Wedgewood, C. V., *The World of Rubens,* Time-Life Books, New York, 1967.
11. Bennett, S., *Abkhasians, the Long-living People of the Caucasus,* Holt, Rinehart & Winston, New York, 1974.
12. Michener, J. A., *Hawaii,* Random House, New York, 1959.
13. Blackburn, H. and Parlin, R. W., Antecedents of disease, insurance mortality experience, *Ann. N. Y. Acad. Sci.,* 134, 965–1017, 1966.
14. Keyes, A., Fidanza, F., Karvonen, M. J., Kimura, N., and Taylor, H. L., Indices of relative weight and obesity, *J. Chronic Dis.,* 25, 329–343, 1972.
15. Anon., *Build and Blood Pressure Study,* Society of Actuaries, Chicago, 1959.
16. Khosla, T., and Lowe, C. R., Indices of obesity derived from body weight and height, *Br. J. Prev. Soc. Med.,* 21, 122–128, 1967.
17. Seltzer, C. C. and Mayer, J., A simple criterion of obesity, *Postgrad. Med.,* 38, A101–A107, 1965.
18. Dawber, T. C., Moore, F. E., and Mann, G. V., Coronary heart disease in the Framingham study, *Am. J. Public Health,* 47(Suppl.), 4–24, 1957.
19. Keyes, A., Aravanis, C., and Blackburn, H., et al., Coronary heart disease: overweight and obesity as risk factors, *Ann. Intern. Med.,* 77, 15–27, 1972.
20. Billewicz, W. Z., Kemsley, W. F. F., and Thomson, A. M., Indices of obesity, *Br. J. Prev. Soc. Med.,* 16, 183–188, 1962.
21. Newens, E. M. and Goldstein, H., Height, weight and the assessment of obesity in children, *Br. J. Prev. Soc. Med.,* 26, 33–39, 1972.
22. Garn, S. M. and Haskell, J. A., Fat thickness and developmental status in childhood and adolescence, *Am. J. Dis. Child.,* 99, 746–751, 1960.
23. Forbes, G. B., Lean body mass and fat in obese children, *Pediatrics,* 34, 308–314, 1964.
24. Mayer, J., Obesity, *Annu. Rev. Med.,* 14, 111–132, 1963.
25. Galton, D. J., An enzymatic defect in a group of obese patients, *Br. Med. J.,* 1, 1498–1500, 1966.
26. Bray, G. A. and York, D. A., Genetically transmitted obesity in rodents, *Physiol. Rev.,* 51, 598–646, 1971.
27. Mayer, J., The obese hyperglycemic syndrome of mice as an example of "metabolic" obesity, *Am. J. Clin. Nutr.,* 8, 712–718, 1960.

28. Drenick, E. J. and Dennin, H. F., Energy expenditure in fasting obese men, *J. Lab. Clin. Med.*, 81, 421–430, 1973.
29. Buskirk, E. R., Thompson, R. H., Lutwak, L., and Whedon, G. D., Energy balance of obese patients during weight reduction: influence of diet restriction and exercise, *Ann. N.Y. Acad. Sci.*, 110, 918–940, 1963.
30. Gulick, A., A study of weight regulation in the adult human body during over-nutrition, *Am. J. Physiol.*, 60, 371–395, 1922.
31. Neumann, R. O., Experimentelle Beitrage zur Lehre von dem taglichen Nahrungsbedarf des Menschen unter besonderer Berucksichtigung der notwendigen Eiweissmenge, *Arch. Hyg.*, 45, 1–87, 1902.
32. Gordon, E. S., Metabolic aspects of obesity, *Adv. Metab. Dis.*, 4, 229–296, 1970.
33. Miller, D. S., Mumford, P., and Stock, M. J., Gluttony: thermogenesis in overeating man, *Am. J. Clin. Nutr.*, 20, 1223–1229, 1967.
34. Nagamachi, Y., Effect of satiety center damage on food intake, blood glucose and gastric secretion in dogs, *Am. J. Dig. Dis.*, 17, 139–148, 1972.
35. Friedman, M. I., Effects of alloxan diabetes on hypothalamic hyperphagia and obesity, *Am. J. Physiol.*, 222, 174–177, 1972.
36. Withers, R. F. J., Problems in the genetics of human obesity, *Eugen. Rev.* (London), 56, 81–90, 1964.
37. Clark, P. J., The heritability of certain anthropometric characters as ascertained from measurement of twins, *Am. J. Hum. Genet.*, 8, 49–54, 1956.
38. Hetherington, A. W. and Ranson, S. W., Hypothalamic lesions and adiposity in the rat, *Anat. Rec.*, 78, 149–172, 1940.
39. Han, P. W., Energy metabolism of tube-fed hypophysectomized rats bearing hypothalamic lesions, *Am. J. Physiol.*, 215, 1343–1350, 1968.
40. Brobeck, J. P., Tepperman, J., and Long, C. N., Experimental hypothalamic hyperphagia in the albino rat, *Yale J. Biol. Med.*, 15, 831–853, 1943.
41. Han, P. W., Lin, C. H., Chu, K. C., Mu, J. Y., and Liu, A. C., Hypothalamic obesity in weanling rats, *Am. J. Physiol.*, 209, 627–631, 1965.
42. Lundbaek, K. and Stevenson, J. A. F., Reduced carbohydrate intake after fat feeding in normal rats and rats with hypothalamic hyperphagia, *Am. J. Physiol.*, 151, 530–537, 1947.
43. Anand, B. K., Nervous regulation of food intake, *Physiol. Rev.*, 41, 677–708, 1961.
44. Mayer, J., Regulation of energy intake and the body weight: glucostatic theory and the lipostatic hypothesis, *Ann. N.Y. Acad. Sci.*, 63, 15–43, 1955.
44a. Mogenson, G. J., Changing views of the role of the hypothalamus in the control of ingestive behaviors, in *Recent Studies of Hypothalamic Function*, Int. Symp. on Recent Studies of Hypothalamic Function, Calgary, 1973, Lederis, K. and Cooper, K. E., Eds., S. Karger, Basel, 1974, 268–293.
44b. Stevenson, J. A. F., Effects of hypothalamic lesions on water and energy metabolism in the rat, *Recent Prog. Horm. Res.*, 4, 363–394, 1949.
44c. Hoebel, B. G. and Teitelbaum, P., Weight regulation in normal and hypothalamic hyperphagic rats, *J. Comp. Physiol. Psychol.*, 61, 189–193, 1966.
44d. Miller, N. E., Bailey, C. J., and Stevenson, J. A. F., Decreased "hunger" but increased food intake resulting from hypothalamic lesions, *Science*, 112, 256–259, 1950.
44e. Gibbs, J., Falasco, J. D., and McHugh, P. R., Cholecystokinin-decreased food intake in rhesus monkeys, *Am. J. Physiol.*, 230, 15–18, 1976.
44f. Gibbs, J., Young, R. C., Smith, G. P., and Bourne, E. W., Cholecystokinin elicits satiety in rats with open gastric fistulas, *Nature*, 245, 323–325, 1973.
44g. Dafney, N. and Jacobson, E. D., Cholecystokinin and central nervous regulation of appetite, in *Symp. Gastrointestinal Hormones*, Thompson, J. C., Ed., University of Texas Press, Austin, 1975, 643–649.
45. Sims, E. A. H., Goldman, R. F., Gluck, C. M., Horton, E. S., Kelleher, P. C., and Rowe, D. W., Experimental obesity in man, *Trans. Assoc. Am. Physicians*, 81, 153–170, 1968.
46. Hirsch, J., Knittle, J. L., and Salans, L. B., Cell lipid content and cell number in obese and non-obese human adipose tissue, *J. Clin. Invest.*, 45, 1023, 1966.
47. Mann, G. V., The influence of obesity in health, *N. Engl. J. Med.*, 291, 226–232, 1974.
47a. Editorial, Infant and adult obesity, *Lancet*, 1, 17–18, 1974.
48. Hirsch, J. and Knittle, J., Cellularity of obese and non-obese human adipose tissue, *Fed. Proc. Fed. Am. Soc. Exp. Biol.*, 29, 1516–1521, 1970.
49. Widdowson, E. M. and Shaw, W. T., Full and empty fat cells, *Lancet*, 2, 905, 1973.
50. Ashwell, M. and Garrow, J. S., Full and empty fat cells, *Lancet*, 2, 1036–1037, 1973.
51. Glucksman, M. L., Psychiatric observations on obesity, *Adv. Psychosom. Med.*, 7, 194–216, 1972.

52. **Bruch, H.,** Obesity in childhood. III. Physiological and psychological aspects of the food intake of obese children, *Am. J. Dis. Child.,* 59, 739–781, 1940.
53. **Bruch, H. and Touraine, G.,** Obesity in children. V. The family frame of obese children, *Psychosom. Med.,* 2, 141–206, 1940.
54. **Bruch, H.,** Obesity in childhood and personality development, *Am. J. Orthopsychiatry,* 11, 467–474, 1941.
55. **Bruch, H.,** Psychiatric aspects of obesity in children, *Am. J. Psychiatry,* 99, 752–757, 1943.
56. **Bruch, H.,** *The Importance of Overweight,* Norton, New York, 1957.
57. **Bruch, H.,** Psychological aspects of obesity, *Psychiatry,* 10, 373–381, 1947.
58. **Bruch, H.,** Conceptual confusion in eating disorders, *J. Nerv. Ment. Dis.,* 133, 46–54, 1961.
59. **Bruch, H.,** Transformation of oral impulses in eating disorders, *Psychiatr. Q.,* 35, 458–481, 1961.
60. **Bruch, H.,** Psychological aspects of reducing, *Psychosom. Med.,* 14, 337–346, 1952.
61. **Alexander, F.,** The influence of psychological factors upon gastrointestinal disturbances, *Psychoanal. Q.,* 3, 501–539, 1934.
62. **Bychowski, G.,** On neurotic obesity, *Psychoanal. Rev.,* 37, 301–319, 1950.
63. **Hecht, M. B.,** Obesity in women, A psychiatric study, *Psychiatr. Q.,* 29, 203–231, 1955.
64. **Rascovsky, A., deRascovsky, M. W., and Schlossberg, T.,** Basic psychic structure of the obese, *Int. J. Psychoanal.,* 31, 144–149, 1950.
65. **Richardson, H. B.,** Obesity and Neurosis. A case report, *Psychiatr. Q.,* 20, 400–424, 1946.
66. **Schick, A.,** Psychosomatic aspects of obesity, *Psychoanal. Rev.,* 34, 173–183, 1947.
67. **Friedman, J.,** Weight problems and psychological factors, *J. Consult. Clin. Psychol.,* 23, 524–527, 1959.
68. **Weinberg, N., Mendelson, M., and Stunkard, A. J.,** A failure to find distinctive personality features in a group of obese men, *Am. J. Psychiatry,* 117, 1035–1037, 1961.
69. **Stunkard, A. J., Grace, W. J., and Wolff, H. G.,** The night-eating syndrome. A pattern of food intake among certain obese patients, *Am. J. Med.,* 19, 78–86, 1955.
70. **Stunkard, A. J.,** Eating patterns and obesity, *Psychiatr. Q.,* 33, 284–295, 1959.
71. **Stunkard, A. J.,** Hunger and satiety, *Am. J. Psychiatr.,* 118, 212–217, 1961.
72. **Stunkard, A. J., and Mendelson, M.,** Obesity and the body image. I. Characteristics of disturbances in the body image of some obese persons, *Am. J. Psychiatr.,* 123, 1296,–1300, 1967.
73. **Stunkard, A. J. and Burt, V.,** Obesity and the body image. II. Age at onset of disturbances in the body image, *Am. J. Psychiatr.,* 123, 1443–1447, 1967.
74. **Stunkard, A. J. and Wolff, H. G.,** Pathogenesis in human obesity, *Psychosom. Med.,* 20, 17–29, 1958.
75. **Stunkard, A. J.,** Research on a disease: strategies in the study of obesity, in *Physiological Correlates of Psychologic Disorders,* Roessler, R. L., and Greenfield, N. S., Eds., University of Wisconsin Press, Madison, 1962, 211–220.
76. **Stunkard, A. J.,** Obesity and the denial of hunger, *Psychosom. Med.,* 21, 281–289, 1959.
77. **Stunkard, A. J. and Koch, C.,** Interpretation of gastric motility. I. Apparent bias in the reports of hunger by obese persons, *Arch. Gen. Psychiatry,* 11, 74–82, 1964.
78. **Stunkard, A. J.,** The 'dieting depression'. Incidence and clinical characteristics of untoward responses to weight reduction regimens, *Am. J. Med.,* 23, 77–86, 1957.
79. **Stunkard, A. J. and McLaren-Hume, M.,** The results of treatment for obesity, *Arch. Inter. Med.,* 103, 79–85, 1959.
80. **Stunkard, A. J.,** Obesity, in *Comprehensive Textbook of Psychiatry,* Freedman, A. M., and Kaplan, H. I., Eds., Williams & Wilkins, Baltimore, 1967, 1059–1062.
81. **Mendelson, M.,** Psychological aspects of obesity, *Med. clin. North Am.,* 48, 1373, 1964.
82. **Swanson, D. W. and Dinello, F. A.,** Severe obesity as a habituation syndrome, *Arch. Gen. Psychiatry,* 22, 120–127, 1970.
83. **Swanson, D. W. and Dinello, F. A.,** Follow-up of patients starved for obesity, *Psychosom. Med.,* 32, 209–214, 1970.
84. **Griggs, R. C. and Stunkard, A. J.,** The interpretation of gastric motility. II. Sensitivity and bias in the perception of gastric motility, *Arch. Gen. Psychiatry,* 11, 82–89, 1964.
85. **Schachter, S.,** Cognitive effects on bodily functioning. Studies of obesity and eating, in *Neurophysiology and Emotion,* Glass, D. C., Ed., Rockefeller University Press, New York, 1967, 117–144.
86. **Schachter, S., Goldman, R., and Gordon, A.,** Effects of fear, food deprivation, and obesity on eating, *J. Pers. Soc. Psychol.,* 10, 91–97, 1968.
87. **Schachter, S. and Gross, L. P.,** Manipulated time and eating behavior, *J. Pers. Soc. Psychol.,* 10, 98–106, 1968.
88. **Nisbett, R. E.,** Determinants of food intake in obesity, *Science,* 159, 1254–1255, 1968.

89. **Nisbett, R. E.**, Taste, deprivation, and weight determinants of eating behavior, *J. Pers. Soc. Psychol.,* 10, 107–116, 1968.

90. **Nisbett, R. E. and Kanouse, D. E.**, Obesity, food deprivation, and supermarket shopping behavior, *J. Pers. Soc. Psychol.,* 12, 289–294, 1969.

91. **Hutchinson, J. J.**, Clinical implications of an extensive acturial study of build and blood pressure, *Ann. Intern. Med.,* 54, 90, 1961.

92. **Chiang, B. N., Perlman, L. V., and Epstein, F. H.**, Overweight and hypertension: a review, *Circulation,* 39, 403–421, 1969.

93. **Gordon, T. and Kannel, W. B.**, The effects of overweight on cardiovascular diseases, *Geriatrics,* 28 (8), 80–88, 1973.

94. **Amad, K. H., Brennan, J. C., and Alexander, J. K.**, The cardiac pathology of chronic exogenous obesity, *Circulation,* 32, 740–745, 1965.

95. **Mayer, J.**, Hypertension and obesity, *Postgrad. Med.,* 46, 253–254, 1969.

96. **Alexander, J. K.**, Chronic heart disease due to obesity, *J. Chronic Dis.,* 18, 895–898, 1965.

97. **Alexander, J. K. and Pettigrove, J. R.**, Obesity and congestive heart failure, *Geriatrics,* 22, 101–108, 1967.

98. **Alexander, J. K. and Peterson, K. L.**, Nutritional factors in cardiovascular disease, *Postgrad. Med.,* 44, 167–171, 1968.

99. **Kannel, W. B., Brand, N., Skinner, J. J., Jr., Dawber, T. R., and McNamara, P. M.**, The relation of adiposity to blood pressure and development of hypertension. The Framingham study, *Ann. Intern. Med.,* 67, 48–59, 1967.

100. **Gordon, T. and Kannel, W. B.**, Obesity and cardiovascular diseases: the Framingham study, *Clin. Endocrinol. Metab.,* 5, 367–375, 1976.

101. **Kannel, W. B., Labauer, E. J., Dawber, T. R., and McNamara, P. M.**, Relation of body weight to development of coronary heart disease: the Framingham study, *Circulation,* 35, 734–744, 1967.

102. **Paul, O., Lepper, M. H., Phelan, W. H., Dupertuis, G. W., MacMillan, H., McKean, H., and Park, H.**, Longitudinal study of coronary heart disease, *Circulation,* 28, 20–31, 1963.

103. **Pell, S. and D'Alonzo, C. A.**, Acute myocardial infarction in a large industrial population: report of a 6-year study of 1,356 cases, *JAMA,* 185, 831–838, 1963.

104. **Spain, D. N., Nathan, D. J., and Gellis, M.**, Weight, body type and the prevalance of coronary atherosclerotic heart disease in males, *Am. J. Med. Sci.,* 245, 63, 1963.

105. **Heyden, S., Hames, C. G., Bartel, A., Cassel, J. C., Tyroler, H. A., and Cornoni, J. C.**, Weight and weight history in relation to cerebrovascular and ischemic heart disease, *Arch. Intern. Med.,* 128, 956–960, 1971.

105a. **Weinsier, R. L., Fuchs, R. J., Kay, T. D., Triebwasser, J. H., and Lancaster, M. C.**, Body fat: its relationship to coronary heart disease, blood pressure, lipids and other risk factors measured in a large male population, *Am. J. Med.,* 61, 815–824, 1976.

106. Editorial, Obesity and coronary heart disease, *Br. Med. J.,* 1, 566–567, 1973.

107. **Montenegro, M. R. and Solberg, L. A.**, Obesity, body weight, body length, and atherosclerosis, *Lab. Invest.,* 18, 594–603, 1968.

108. **Fidanza, F.**, Dietary fat, obesity and coronary heart disease, *Nutr. Dieta,* 8, 200–209, 1966.

109. **Chapman, J. M., Goerke, L. S., Dixon, W., Loveland, D. B., and Phillips, E.**, The clinical status of a population group in Los Angeles under observation for two to three years, *Am. J. Public Health,* 47, 33–42, 1957.

110. **Doyle, J. T., Heslin, A. S., Hilleboe, H. E., Formel, P. F., and Korns, R. F.**, A prospective study of degenerative cardiovascular disease in Albany: report of three years' experience. I. Ischemic heart disease, *Am. J. Public Health,* 47, 25–32, 1957.

111. **Kagan, A., Dawber, T. R., Kannel, W. B., and Revotskie, N.**, The Framinghan study: A prospective study of coronary heart disease, *Fed. Proc. Fed. Am. Soc. Exp. Biol.,* 11(Suppl.), 52–57, 1962.

112. **Merimee, T. J.**, Obesity and hyperinsulinism, *N. Engl. J. Med.,* 285, 856–857, 1971.

113. **Grey, N. and Kipnis, D. M.**, Effect of diet composition on the hyperinsulinemia of obesity, *N. Engl. J. Med.,* 285, 827–831, 1971.

114. **Auchincloss, J. H., Jr., Cook, E., and Renzetti, A. D.**, Clinical and physiological aspects of a case of obesity, polycythemia, and alveolar hypoventilation, *J. Clin. Invest.,* 34, 1537–1545, 1956.

115. **Burwell, C. S., Robin, E. D., Whaley, R. D., and Bickelmann, A. G.**, Extreme obesity associated with alveolar hypoventilation – a Pickwickian syndrome, *Am. J. Med.,* 21, 811–818, 1956.

116. **Sharp, J. T., Henry, J. P., Sweaney, S. K., Meadows, W. R., and Pietras, R. J.**, Effects of mass loading the respiratory system in man, *J. Appl. Physiol.,* 19, 959–966, 1964.

117. **Miller, W. F. and Bashour, F. A.**, Cardiopulmonary changes in obesity, *Clin. Anesth.,* 3, 127–139, 1963.

118. **Kronenberg, R. S., Gabel, R. A., and Severinghaus, J. W.,** Normal chemoreceptor function in obesity before and after ileal bypass surgery to force weight reduction, *Am. J. Med.,* 59, 349–353, 1975.

119. **Lamberth, I. E.,** Obesity and anesthesia, *Clin. Anesth.,* 3, 56–66, 1968.

120. **Edelist, G.,** Extreme obesity, *Anesthesiology,* 29, 846–847, 1968.

121. **Gould, A. B., Jr.,** Effects of obesity on respiratory complications following general anesthesia, *Anesth. Analg.* (Cleveland), 41, 448–452, 1962.

122. **Schwartz, H.,** Problems of obesity in anesthesia, *N.Y. J. Med.,* 55, 3277–3281, 1955.

123. **Warner, W. A. and Garrett, L. P.,** The obese patient and anesthesia, *JAMA,* 205, 102–103, 1968.

124. **Pemberton, L. B. and Manax, W. G.,** Relationship of obesity to postoperative complications after cholecystectomy, *Am. J. Surg.,* 121, 87–90, 1971.

125. **De La Vega, J. N. and Ponce De Leon, A.,** Triade de Saint Medical cir. farm, in *Gastroenterology,* Bochus, H. L., Ed., Saunders, Philadelphia and London, 921.

126. **Mabee, T. M., Meyer, P., Denbesten, L., and Mason, E. E.,** The mechanism of increased gallstone formation in obese human subjects, *Surgery,* 79, 460–468, 1976.

127. **Freeman, J. B., Meyer, P. D., Printen, K. J., Mason, E. E., and Denbesten, L.,** Analysis of gallbladder bile in morbid obesity, *Am. J. Surg.,* 129, 163–166, 1975.

128. **Kern, W. H., Heger, A. H., Payne, J. H., and DeWind, L. T.,** Fatty metamorphosis of the liver in morbid obesity, *Arch. Pathol.,* 96, 342–346, 1973.

129. **Kellgren, J. H.,** Osteoarthrosis in patients and populations, *Br. Med. J.,* 2, 1–6, 1961.

130. **Kellgren, J. H. and Lawrence, J. S.,** Osteoarthrosis and disk degeneration in an urban population, *Ann. Rheum. Dis.,* 17, 388–397, 1958.

131. **Lawrence, J. S., Bremmer, J. M., and Bier, F.,** Osteoarthrosis. Prevalence in the population and relationship between symptoms and X-ray changes, *Ann. Rheum. Dis.,* 25, 1–24, 1966.

132. **Leach, R. E., Baumgard, S., and Broom, J.,** Obesity: its relationship to osteoarthritis of the knee, *Clin. Orthop.,* 93, 271–273, 1973.

133. **Law, W. A.,** Osteoarthritis of the hip, *Practitioner,* 193, 585–592, 1964.

134. **Silberberg, M., Jarrett, S. F., and Silberberg, R.,** Obesity and degenerative joint disease; experiments in 'yellow mice,' *Arch. Pathol.,* 61, 280–288, 1956.

135. **Goldin, R. H., McAdam, L., Louie, J. S., Gold, R., and Bluestone, R.,** Clinical and radiological survey of the incidence of osteoarthrosis among obese patients, *Ann. Rheum. Dis.,* 35, 349–353, 1976.

136. **Bloom, W. L.,** Fasting as an introduction to the treatment of obesity, *Metabolism,* 8, 214–220, 1959.

137. **Cahill, G. F., Herrera, M. G., Morgan, A. P., Soeldner, J. S., Steinke, J., Levy, P. L., Reichard, G. A., and Kipnis, D. M.,** Hormone-fuel interrelationship during fasting, *J. Clin. Invest.,* 45, 1751–1769, 1966.

138. **Duncan, G. G., Schless, G. L., and Cristofori, F. G.,** Contraindications and therapeutic results of fasting in obese patients, *Ann. N.Y. Acad. Sci.,* 131, 632–636, 1965.

139. **Bolinger, R. E., Lukert, B. P., Brown, R. W., Guevara, L., and Steinberg, R.,** Metabolic balance of obese subjects during fasting, *Arch. Intern. Med.,* 118, 3–8, 1966.

140. **Gilliland, I. C.,** Total fasting in the treatment of obesity, *Postgrad. Med. J.,* 44, 58–61, 1968.

141. **Symposium on femfluramine and derivates,** *Postgrad. Med. J.,* 51(Suppl. 1), 1–188, 1975.

142. **Levin, A.,** The non-medical misuse of femfluramine by drug dependent young South Africans, *Postgrad. Med. J.,* 51(Suppl. 1), 186–188, 1975.

143. **Stunkard, A., Levine, H., and Fox, S.,** Management of obesity: patient self-help and medical treatment, *Arch. Intern. Med.,* 125, 1067–1072, 1970.

144. **MacLean, L. D.,** Intestinal bypass operations for obesity: a review, *Can. J. Surg.,* 19, 387–398, 1976.

145. **Scott, H. W., Law, D. H., Sandstead, H. H., Lanier, V. C., Jr., and Younger, R. K.,** Jejunoileal shunt in surgical treatment of morbid obesity, *Ann. Surg.,* 171, 770–782, 1970.

146. **Scott, H. W., Sandstead, H. H., Brill, A. B., Burko, H., and Younger, R. K.,** Experience with new technic of intestinal bypass in treatment of morbid obesity, *Ann. Surg.,* 174, 560–572, 1971.

147. **Scott, H. W., Dean, R., Shull, H. J., Abram, H. S., Webb, W., Younger, R. K., and Brill, A. B.,** New consideration in use of jejunoileal bypass in patients with morbid obesity, *Ann. Surg.,* 177, 723–735, 1973.

148. **Payne, J. H. and DeWind, L. T.,** Surgical treatment of obesity, *Am. J. Surg.,* 118, 141–147, 1969.

149. **Burwell, C. S., Robin, E. D., Whaley, R. D., and Biekelman, A. G.,** Extreme obesity associated with alveolar hypoventilation – Pickwickian syndrome, *Am. J. Med.,* 21, 811–818, 1956.

150. **Walsh, R. E., Michaelson, E. D., Harkleroad, L. E., Zighelboim, A., and Sackner, M. A.,** Upper airway obstruction in obese patients with sleep disturbance and somnolence, *Ann. Intern. Med.,* 76, 185—192, 1972.

151. **Kremen, A. J., Linner, J. H., and Nelson, C. H.,** Experimental evaluation of nutritional importance of proximal and distal small intestine, *Ann. Surg.,* 140, 439—448, 1954.

152. **Payne, J. H., DeWind, L. T., Commons, R. R.,** Metabolic observations in patients with jejunocolonic shunts, *Am. J. Surg.,* 106, 273—289, 1963.

153. **Schwartz, M. Z., Varco, R. L., and Buchwald, H.,** Preoperative preparation, operative technique and postoperative care of patients undergoing jujunoileal bypass for massive exogenous obesity, *J. Surg. Res.,* 14, 147—150, 1973.

154. **Moss, A. A., Goldberg, H. I., Koehler, R. E.,** Radiographic evaluation of complications after jejunoileal bypass surgery, *Am. J. Roentgenol.,* 127, 737—741, 1976.

155. **McGill, D. B., Humphreys, S. R., Baggenstoss, A. H., and Dickson, E. R.,** Cirrhosis and death after jejunoileal shunt, *Gastroenterology,* 63, 872—877, 1972.

156. **Brown, R. G., O'Leary, J. P., and Woodward, E. R.,** Hepatic effects of jejunoileal bypass for morbid obesity, *Am. J. Surg.,* 127, 53—58, 1974.

157. **Bondar, G. F. and Pisesky, W.,** Complications of intestinal short-circuiting for obesity, *Arch. Surg.,* 94, 707—716, 1967.

158. **Salmon, P. A.,** Results of small intestine bypass operations for treatment of obesity, *Surg. Gynecol. Obstet.,* 132, 965—979, 1971.

159. **Buchwald, H., Lober, P. H., and Varco, R. L.,** Liver biopsy findings in seventy-seven consecutive patients undergoing jejunoileal bypass for morbid obesity, *Am. J. Surg.,* 127, 48—52, 1974.

160. **Buchwald, M., Schwartz, M. Z., and Varco, R. L.,** Surgical treatment of obesity, *Adv. Surg.,* 7, 235—243, 1973.

161. **Heimburger, S. L., Steiger, E., Lo Gerfo, P., Bieke, A. G., and Williams, M. J.,** Reversal of severe fatty hepatic infiltration after intestinal bypass for morbid obesity by calorie-free amino acid infusion, *Am. J. Surg.,* 129, 229—235, 1975.

162. **Dickstein, S. S. and Frame, B.,** Urinary tract calculi after intestinal shunt operations for treatment of obesity, *Surg. Gynecol. Obstet.,* 136, 257—260, 1973.

163. **Shagrin, J. W., Frame, B., and Duncan, H.,** Polyarthritis in obese patients with intestinal bypass, *Ann. Intern. Med.,* 75, 377—380, 1971.

164. **Mason, E. E. and Ito, C.,** Gastric bypass, *Ann. Surg.,* 170, 329—339, 1969.

165. **Mason, E. E. and Printen, K. J.,** Metabolic considerations in reconstitution of the small intestine after jejunoileal bypass, *Surg. Gynecol. Obstet.,* 142, 177—183, 1976.

166. **Mason, E. E., Printen, K. J., Hartford, C. E., and Boyd, W. C.,** Optimizing results of gastric bypass, *Ann. Surg.,* 182, 405—414, 1975.

167. **Printen, K. J. and Mason, E. E.,** Gastric surgery for relief of morbid obesity, *Arch. Surg.,* 106, 428—431, 1973.

168. **Hermreck, A. S., Jewell, W. R., and Hardin, C. A.,** Gastric bypass for morbid obesity: results and complications, *Surgery,* 80, 498—505, 1976.

169. **Quaade, F.,** Stereotaxy for obesity, *Lancet,* 1, 267, 1974.

170. **Quaade, F., Vaernet, K., and Larsson, S.,** Stereotaxic stimulation and electrocoagulation of the lateral hypothalamus in obese humans, *Acta Neurochir.,* 30, 111—117, 1974.

171. **Marshall, J. F.,** Stereotaxy for obesity, *Lancet,* 2, 106, 1974.

172. **Eth, S.,** Stereotaxy for obesity, *Lancet,* 1, 867, 1974.

173. **Quaade, F.,** Stereotaxy for obesity, *Lancet,* 1, 867, 1974.

173a. **Bistrian, B. R., Blackburn, G. L., Hallowell, E., and Heddle, R.,** Protein status of general surgical patients, *JAMA,* 230, 858—860, 1974.

173b. **Butterworth, C. E., Blackburn, G. L.,** Hospital malnutrition, *Nutr. Today,* 10, 8, 1975.

174. **Dudrick, S. J., Wilmore, D. W., Vars, H. M., and Rhoads, J. E.,** Long-term total parenteral nutrition with growth, development and positive nitrogen balance, *Surgery,* 64, 132—142, 1968.

175. **Wilmore, D. W. and Dudrick, S. J.,** Growth and development of an infant receiving all nutrients exclusively by vein, *JAMA,* 203, 860—864, 1968.

176. **Keys, A., Brozek, J., Henschel, A., Mickelsen, O., and Longstreet, T. H.,** The Biology of Human Starvation, University of Minnesota Press, Minneapolis, 1950.

177. **Young, V. R. and Scrunshaw, N. S.,** Physiology of starvation, *Sci. Am.,* 225, 14—21, 1971.

178. **Diller, E. R. and Harney, O. A.,** Interrelationship of sterol and fatty acid biosynthesis in rat liver slices as related to dietary lipid, *Biochemistry,* 3, 2004—2007, 1964.

179. **Coniglio, J. G. and Culp, F. B.,** The effect of fasting on fatty acid synthesis in cell-free preparation of rat mammary gland, *Biochm. Biophys. Acta,* 106, 419—421, 1965.

180. Mariana, A., Spadoni, M. A., and Tomassi, G., Effect of protein depletion on amino acid activating enzymes of rat liver, *Nature,* 199, 378–379, 1963.

181. Gaetani, S., Paolucci, A. M., Spondi, M. A., and Tomassi, G., Activity of amino acid-activating enzymes in tissues from protein depleted rats, *J. Nutr.,* 84, 173–178, 1964.

182. Munro, H. N. and Clark, C. M., The nutritional regulation of ribo-nucleic acid metabolism in the liver cell, *Proc. Nutr. Soc.,* 19, 55–60, 1960.

183. Campbell, R. M. and Kosterlitz, H. W., Effects of dietary protein, fat and choline on composition of liver cell and turnover of phospho-lipin and protein-bound phosphorus, *Biochim. Biophys. Acta,* (abstr.), 8, 644, 1952.

184. Weber, G., Study and evaluation of regulation of enzyme activity and synthesis in mammalian liver, *Adv. Enzyme Regul.,* 1, 1–35, 1963.

185. Weber, G., Singhal, R. L., Stamm, N. B., Lean, M. A., and Fisher, E. A., Synchronous behavior pattern of key glycolytic enzymes: glucokinase, phosphofructokinase, pyruvate kinase, *Adv. Enzyme Regul.,* 4, 59–81, 1966.

186. Johnson, B. C. and Sassoon, H. F., Studies of the induction of liver glucose-6-phosphate dehydrogenase in the rat, *Adv. Enzyme Regul.,* 5, 93–106, 1966.

187. Schimke, R. T., Differential effects of fasting and protein-free diets on levels of urea cycle enzymes in rat liver, *J. Biol. Chem.,* 237, 1921–1924, 1962.

188. McFarlane, I. G. and Von Holt, C., Metabolism and leucine in protein-calorie-deficient rats, *Biochem. J.,* 111, 565–571, 1969.

189. Smith, C. S. and Johnson, B. C., Glucose metabolism in the rat during starvation and re-feeding following starvation, *Proc. Soc. Exp. Biol. Med.,* 115, 438–444, 1964.

190. Young, J. W., Shrago, E., and Lardy, H. A., Metabolic control of enzymes involved in lipogenesis and glyconeogenesis, *Biochemistry,* 3, 1687–1692, 1964.

191. Tepperman, J. and Tepperman, H. M., Metabolism of glucose-1-C^{14} and glucose-6-C^{14} by liver slices of refed rats, *Am. J. Physiol.,* 200, 1069–1073, 1961.

192. Tepperman, H. M. and Tepperman, J., The hexosemonophosphate shunt and adaptive hyperlipogenesis, *Diabetes,* 7, 478–485, 1958.

193. Shrago, E., Lardy, H. A., Nordlie, R. C., and Foster, D. O., Metabolic and hormonal control of phosphoenolpyruvate carboxykinase and malic enzyme in rat liver, *J. Biol. Chem.,* 238, 3188–3192, 1962.

194. Winter, H. A., The effect of fasting upon the tolerance of normal rats to high carbohydrate diet administered by stomach tube, *Am. J. Physiol.,* 147, 228–236, 1946.

195. Wilhelmy, C. M., McGuire, T. F., and Waldmann, E. B., Stress of realimentation with protein or carbohydrate following prolonged fasting, *Am. J. Physiol.,* 168, 248–253, 1952.

196. Owen, O. E., Morgan, A. P., Kemp, H. G., Sullivan, J. M., and Herrera, M. G., Brain metabolism during fasting, *J. Clin. Invest.,* 46, 1589–1595, 1967.

197. Agrawal, H. C., Fishman, M. A., and Prensky, A. L., A possible block in the intermediary metabolism of glucose into proteins and lipids, *Lipids,* 6, 431–433, 1971.

197a. Knochel, J. P., The pathophysiology and clinical characteristics of severe hypophosphatemia, *Arch Intern. Med.,* 137, 203–220, 1977.

198. Silvis, S. E. and Paragas, P., Hyperalimentation syndrome, *J. Lab. Clin. Med.,* 78, 918–930, 1971.

199. Silvis, S. E. and Paragas, P., Paresthesias, weakness, seizures and hypophosphatemia in patients receiving hyperalimentation, *Gastroenterology,* 62, 513–520, 1972.

200. Silvis, S. E. and Voltin, R. F., unpublished data.

201. Lotz, M., Zisman, E., and Bartter, F. C., Evidence for a phosphorus-depletion syndrome in man, *N. Engl. J. Med.,* 278, 409–415, 1968.

202. Lichtman, M. A., Miller, D. R., and Freeman, R. B., Erythrocyte adenosine triphosphate depletion during hypophosphatemia in a uremic subject, *N. Engl. J. Med.,* 280, 240–244, 1969.

203. Corredor, D. G., Sabeh, G., Mendelsohn, L. V., Wasserman, R. E., Sunder, H. H., and Danowski, T. S., Enhanced postglucose hypophosphatemia during starvation therapy of obesity, *Metabolism,* 18, 754–763, 1969.

204. Danowski, T. S., Gillespie, H. K., Fergus, E. B., and Puntereri, A. J., Significance of blood sugar and serum electrolyte changes in cirrhosis following glucose, insulin, glucagon, or epinephrine, *Yale J. Biol. Med.,* 29, 361–375, 1956.

204a. Grzyb, S., Jelinek, C., and Sheldon, G. F., The phosphate depletion syndrome: relation to caloric intake and phosphate infusion, *Surg. Forum,* 24, 103–104, 1973.

205. Massara, F. and Camanni, F., Propranolol block of adrenaline-induced hypophosphatemia in man, *Clin. Sci.,* 38, 245–250, 1970.

206. Riedler, G. F. and Scheitlin, W. A., Hypophosphatemia in septicaemia: higher incidence in Gram-negative than in Gram-positive infections, *Br. Med. J.,* 1, 753–756, 1969.

207. **Froesch, E. R., Wolf, H. P., Bartsch, H., Prader, A., and Labhart, A.,** Hereditary fructose intolerance, *Am. J. Med.,* 34, 151–167, 1963.
208. **Dormandy, T. L. and Porter, R. J.,** Familial fructose and galactose intolerance, *Lancet,* 1, 1189–1194, 1961.
209. **Lichtman, M. A., Miller, D. R., Cohen, J., and Waterhouse, C.,** Reduced red cell glycolysis, 2,3-diphosphoglycerate and adenosine triphosphate concentration, and increased hemoglobin-oxygen affinity caused by hypophosphatemia, *Ann. Intern. Med.,* 74, 562–568, 1971.
210. **Travis, S. F., Sugarman, H. J., Ruberg, R. L., Dudrick, S. J., Delivora-Papadopoulos, M., Miller, L. D., and Oski, F. A.,** Alterations of red-cell glycolytic intermediates and oxygen transport as a consequence of hypophosphatemia in patients receiving intravenous hyperalimentation, *N. Engl. J. Med.,* 285, 763–768, 1971.
211. **Jacob, H. S. and Amsden, T.,** Acute hemolytic anemia with rigid cells in hypophosphatemia, *N. Engl. J. Med.,* 285, 1446–1450, 1971.
212. **Sugarman, H., Travis, S., Polleck, T., Ruberg, R., Dudrick, S., Delivoria-Papadopoulos, M., Miller, L., and Oski, F.,** Alterations in oxygen transport and red-cell metabolism as a consequence of hypophosphatemia in intravenous hyperalimentation, *Clin. Res.,* 19, 487, 1971.
213. **Yawata, Y., Hebbel, R. P., Silvis, S. E., Howe, R., and Jacob, H.,** Blood cell abnormalities complicating the hypophosphatemia of hyperalimentation: erythrocyte and platelet ATP deficiency associated with hemolytic anemia and bleeding in hyperalimented dogs, *J. Lab. Clin. Med.,* 84, 643–653, 1974.
214. **Craddock, P. R., Yawata, Y., VanSanten, L., Gilberstad, S., Silvis, S., and Jacob, H.,** Acquired phagocyte dysfunction: A complication of the hypophosphatemia of parenteral hyper-alimentation, *N. Engl. J. Med.,* 290, 1403–1407, 1974.
215. **Silvis, S. E.,** unpublished data, 1975.
216. **Silvis, S. E.,** unpublished data, 1977.
217. Editorial, Fatal hyperalimentation syndrome, *Nutr. Rev.,* 30, 121–126, 1972.
218. **Silvis, S. E., Gilberstadt, S., and Hauser, A.,** Cerebral edema occurs in hyperalimentation syndrome, *Gastroenterology,* 64, 803, 1973.
219. **O'Tauma, L. A.,** Raised intracranial pressure after hyperalimentation, *Lancet,* 2, 1101, 1973.
220. **Derr, R. F. and Zieve, L.,** Etiology of hyperalimentation coma, *N. Engl. J. Med.,* 288, 1080–1081, 1973.
221. **Silvis, S. E.,** unpublished data, 1977.
222. **Sand, D. W. and Pastore, R. A.,** Paresthesias and hypophosphatemia occurring with parenteral alimentation, *Am. J. Dig. Dis.,* 18, 709–713, 1973.
223. **Prins, J. G., Schrijver, H., and Staghouwer, J. H.,** Hyperalimentation, hypophosphatemia and coma, *Lancet,* 1, 1253–1254, 1973.
224. **Metzger, R., Burke, P., Thompson, A., Lordon, R., and Frimpter, W. G.,** Hypophosphatemia and hypouricemia during parenteral hyperalimentation with an amino acid and glucose preparation, *J. Clin Invest.,* 50, 65a–66a, 1971.
225. **Ruberg, R. L.,** Hypophosphatemia with hypophosphaturia in hyperalimentation, *Surg. Forum,* 22, 87–88, 1971.
226. **Allen, T. R., Ruberg, R. L., Dudrick, S. J., Long, J. M., and Steiger, E.,** Hypophosphatemia occurring in patients receiving total parenteral hyperalimentation, *Fed. Proc. Fed. Am. Soc. Exp. Biol.,* 30, 580, 1971.
227. **Ricour, C., Millot, M., and Balsan, S.,** Phosphorus depletion in children on long term total parenteral nutrition, *Acta Paediatr. Scand.,* 64, 385–392, 1975.
228. **Baughman, F. A., Jr. and Papp, J. P.,** Wernicke's encephalopathy with intravenous hyperalimentation: remarks on similarities between Wernicke's encephalopathy and the phosphate depletion syndrome, *Mt. Sinai J. Med. N.Y.,* 43, 48–52, 1976.
229. **Weintraub, M. I.,** Hypophosphatemia mimicking acute Guillain-Barre-Strohl syndrome: a complication of parenteral hyperalimentation, *JAMA,* 235, 1040–1042, 1976.
230. **Furlan, A. J., Hanson, M., Cooperman, A., and Farmer, R. G.,** Acute areflexic paralysis, *Arch. Neurol.,* 32, 706–707, 1975.
231. **Touloukian, R. J.,** Isomolar coma during parenteral alimentation with protein hydrolysate in excess of 4 gm/kg/day, *J. Pedatr.,* 86, 270–273, 1975.
232. **Bitter, R. A., Gubler, C. J., and Heninger, R. W.,** The effects of forced-feeding on blood levels of pyruvate, glucocorticoids and glucose and adrenal weight in thiamine-deprived and thiamine-antagonist treated rats, *J. Nutr.,* 98, 147–152, 1969.
233. **Zieve, L., Diozaki, W. M., and Stenros, L. E.,** The effect of magnesium deficiency on growth response to thiamine of thiamine-deficient rats, *J. Lab. Clin. Med.,* 72, 261–267, 1968.
234. **Caddell, J. L.,** Magnesium deficiency in protein-calorie malnutrition, *Ann. N.Y. Acad. Sci.,* 162, 874–890, 1969.
235. **Tombers, J.,** personal communication.

236. Caldwell, M. D., Jonsson, H. T., and Othersen, H. B., Jr., Essential fatty acid deficiency in an infant receiving prolonged parenteral alimentation, *J. Pediatr.*, 81, 894–898, 1972.

237. Richardson, T. J. and Sgoutas, D., Essential fatty acid deficiency in four adult patients during total parenteral nutrition, *Am. J. Clin. Nutr.*, 28, 258–263, 1975.

238. Kay, R. G., Tasman-Jones, C., Pybus, J., Whiting, R., and Black, H., A syndrome of acute zinc deficiency during total parenteral alimentation in man, *Ann. Surg.*, 183, 331–340, 1976.

239. Prasad, A. S., Halsted, J. A., and Madimi, M., Syndrome of iron deficiency anaemia. Hepatosplenomegaly, hypogonadism, dwarfism and geophagia, *Am. J. Med.*, 31, 532–546, 1961.

240. Karpel, J. I. and Peden, V. H., Copper deficiency in long term parenteral nutrition, *J. Pediatr.*, 80, 32–36, 1972.

241. Dunlap, W. M., James, G. W., and Hume, D. M., Anaemia and neutropenia caused by copper deficiency, *Ann. Intern. Med.*, 80, 470–476, 1974.

242. Vilter, R. W., Bozian, R. C., Hess, E. V., Zellner, D. C., and Petering, H. G., Manifestations of copper deficiency in a patient with systemic sclerosis on intravenous hyperalimentation, *N. Engl. J. Med.*, 291, 188–191, 1974.

243. Mann, G. V., Shaffer, R. D., Anderson, R. S., and Sandstead, H. H., Cardiovascular disease in the Masai, *J. Artheroscler. Res.*, 4, 289–312, 1964.

244. Shaper, A. G., Jones, K. W., Jones, M., Kyobe, J., Serum lipids in three nomadic tribes of northern Kenya, *Am. J. Clin. Nutr.*, 13, 135–146, 1963.

245. Mann, G. V., Shaffer, R. D., and Rich, A., Physical fitness and immunity to heart disease in Masai, *Lancet*, 2, 1308–1310, 1965.

246. Yudkin, J. and Morland, J., Sugar intake and myocardial infarction, *Am. J. Clin. Nutr.*, 20, 503–506, 1967.

247. Yudkin, J., Sucrose and cardiovascular disease, *Proc. Nutr. Soc.*, 31, 331–338, 1972.

248. Yudkin, J., Sugar and disease, *Nature*, 239, 197–199, 1972.

249. Pinckney, E. R., The potential toxicity of excessive polyunsaturates: do not let the patient treat himself, *Am. Heart J.*, 85, 723–726, 1973.

250. Blitzer, P. H., Blitzer, E. C., and Rimm, A. A., Association between teen-age obesity and cancer in 56,111 women: all cancers and endometrial carcinoma, *Prev. Med.*, 5, 20–31, 1976.

251. Wynder, E. L. and Shigematsu, T., Environmental factors of cancer of the colon and rectum, *Cancer* (Philadelphia), 20, 1520–1561, 1967.

252. Wynder, E. L., Kajitani, T., Ishikawa, S., Dodo, H., and Takano, A., Environmental factors of cancer of the colon and rectum, *Cancer* (Philadelphia), 23, 1210–1220, 1969.

253. Gregor, O., Toman, R., and Prusova, F., Gastrointestinal cancer and nutrition, *Gut*, 10, 1031–1034, 1969.

254. Burkitt, D. P., A deficiency of dietary fiber may be one cause of certain colonic and venous disorders, *Am. J. Dig. Dis.*, 21, 104–108, 1976.

255. Burkitt, D. P., Cancer of the colon and rectum, *Minn. Med.*, 55, 779–783, 1972.

256. Burkitt, D. P., Walker, A. R. P., and Painter, N. S., Effect of dietary fibre on stools and transit-times, and its role in the causation of disease, *Lancet*, 2, 1408–1411, 1972.

256a. Segel, I., Solomon, A., and Hunt, J. A., Emergence of diverticular disease in the urban South African black, *Gastroenterology*, 72, 215–219, 1977.

257. Walker, A. R. P., Walker, B. F., and Richardson, B. D., Bowel transit times in Bantu populations, *Br. Med. J.*, 3, 48–49, 1970.

258. Bremmer, C. G. and Ackerman, L. V., Polyps and carcinoma of the large bowel in the South African Bantu, *Cancer* (Philadelphia), 26, 991–999, 1970.

259. Hill, M. J., Drasar, B. S., Aries, V., Crowther, J. S., Hawksworth, G. and Williams, R. E. O., Bacteria and aetiology of cancer of large bowel, *Lancet*, 1, 95–100, 1971.

260. Walker, A. R. P., Diet and cancer of the colon, *Lancet*, 1, 593, 1971.

261. Harper, A. E., Benevenga, N. J., and Wohlhueter, R. M., Effects of ingestion of disproportionate amounts of amino acids, *Physiol. Rev.*, 50, 428–558, 1970.

262. Moore, T., Vitamin A deficiency and excess, *Proc. Nutr. Soc.*, 24, 129–135, 1965.

263. Yaffe, S. J. and Filer, L. J., Jr., American Academy of Pediatrics Joint Committee Statement Committee on Drugs and on Nutrition. The use and abuse of vitamin A, *Pediatrics*, 48, 655–656, 1971.

264. Mosher, L. R., Nicotinic acid side effects and toxicity: a review, *Am. J. Psychiatry*, 126, 1290–1296, 1970.

265. Schrauzer, G. N. and Rhead, W. J., Ascorbic acid abuse: effects of long term ingestion of excessive amounts of blood levels and urinary excretion, *Int. J. Vitam. Min. Res.*, 43, 201–211, 1973.

EFFECT OF NUTRIENT EXCESS IN ANIMALS AND MAN:
CARBOHYDRATES

Sheldon Reiser

GENERAL CONSIDERATIONS

This chapter will describe some of the metabolic effects observed after the intake of relatively large quantities of carbohydrates in the diets of animals and humans. When warranted, these metabolic changes will be discussed as they relate to human health and well-being. The material covered will deal primarily with studies in which the carbohydrates were fed orally. Some studies pertaining to nutritional excess will describe the effect of carbohydrate administered intragastrically, parenterally, or by tissue perfusion. Toxicological studies in which massive quantities of carbohydrate are administered are considered to be outside the scope of nutritional excess.

The manner in which an individual will react to a given amount of any nutrient is complex. It is recognized that other environmental factors such as the level of exercise, stress, and diet can influence the metabolic effects of dietary carbohydrates. Genetic factors can produce species differences and strain differences within a given species that modulate the effect of dietary carbohydrate. In animal studies, these genetic factors can be minimized. However, in humans the response to dietary carbohydrate may be under polygenic control and result in complex responses to a given dietary stress. Despite the difficulties inherent in generalizing on studies utilizing different experimental conditions, it appears that the excessive intake of various components of the carbohydrate diet produces metabolic changes that can profoundly affect human health.

INTRODUCTION

Carbohydrate supplies a large proportion of the caloric requirements of man. As much as 80% of the total calories consumed by people in underdeveloped countries is carbohydrate. In societies that are described as industrialized or westernized, carbohydrate supplies about 50% of the total caloric intake. It was estimated that carbohydrate provided 46% of the total caloric intake in the U.S. in 1974.[1]

A change in the dietary pattern of carbohydrate intake appears to be a characteristic of societies as they become more industrialized. In the U.S., the percent of the total dietary carbohydrate provided by sugars (carbohydrate in milk, fruit, and sweeteners) as compared to starch (carbohydrate in grain products and vegetables) has risen from 32% around 1910 until today sugars, predominantly sucrose, contribute more than 50% of the total carbohydrate.[2] This same time period has witnessed a drop in the average per capita consumption of flours and cereals from 300 to 141 lb/year and a rise in the average per capita consumption of refined sugar and other sweeteners from 87 to 126 lb/year. This pattern of carbohydrate consumption in industrialized societies indicates that there may be relative overconsumption of sugars at the expense of starch and fiber (the nondigestible portion of plants). In recent years several controversial hypotheses have implicated the increased ingestion of sucrose[3,4] and the decreased intake of fiber[5] as important factors in the etiology of various degenerative diseases characteristic of industrialized societies. Therefore, the main emphasis in this chapter will be directed toward a review of the metabolic effects observed after feeding sucrose, fructose, and components of fiber to animals and humans and to an evaluation of the relationship between these metabolic changes and disease states. Lactose and galactose will also be discussed in some depth since the excessive ingestion of these naturally occurring sugars has been shown to produce undesirable clinical manifestations. Less emphasis will be given to carbohydrates that are added to foods as stabilizers, emulsifiers, and thickeners.

SUCROSE

In industrialized societies, sucrose is estimated to provide an average of 15 to 20% of the total caloric intake.[2,6] Sucrose consumption in the U.S. is believed to be higher in specific age and sex groups (e.g., young males).[2] Sucrose intake in industrialized societies has remained relatively constant during the last 20 years. However, world production of sucrose has risen from 30 million tons in 1950 to 70 million tons in 1970 and is projected to reach 93 million tons by 1980.[7] These figures indicate a sharp increase in sugar consumption in the developing areas of the world. Whether these levels of sucrose intake represent a relative excess has been the subject of many animal and human studies. Most of these studies use levels of sucrose higher than the estimated average intake, fed over a short period of the life span of the experimental model. The interpretation that metabolic changes observed under these experimental conditions are applicable to humans consuming lower levels of sucrose over long periods of time is open to debate.

Sucrose is believed to be hydrolyzed to glucose and fructose by enzymes in the small intestine prior to the absorption of these monosaccharides. To attribute the metabolic effects observed after sucrose-feeding entirely to the metabolism of the component monosaccharides is not justified. The activities of various hepatic lipogenic enzymes were greater in rats starved for 2 days and then refed diets containing 31% or 40% sucrose as compared to invert sugar.[8,9] Rats meal-fed diets containing 45% sucrose as compared to an equivalent amount of invert sugar for 2 weeks showed significant increases in fasting serum insulin values and in food efficiency.[10] Rats fed diets containing 60% sucrose ad libitum for 2 weeks showed differences in the activities of hepatic aromatic hydroxylases and of cytochrome p-450 as compared to rats fed the same amount of invert sugar.[11] Blood fructose levels were higher in human volunteers after consuming 2 g/kg body weight of sucrose as compared to an equivalent amount of glucose and fructose.[12] These findings indicate that sucrose exerts an additional effect, presumably at the intestinal level, that produces metabolic changes not duplicated by the component monosaccharides. A similar situation appears to exist between maltose and glucose.[8-10,13]

Digestion and Absorption

Sucrose is hydrolyzed to glucose and fructose by sucrase (EC 3.2.1.26), an enzyme located in the intestinal brush border glycocalyx. The feeding of diets containing high levels of sucrose appears to produce an adaptive increase in sucrase activity. Rats starved for 3 days and then refed a diet containing 70% sucrose for 24 hours showed greater levels of sucrase activity than did rats refed a sucrose-free, high-casein diet.[14] This increase in sucrase activity was subsequently shown to be specific for sucrose. Rats starved for 3 days and then refed a diet containing 68% sucrose for 24 hours had significantly greater levels of sucrase activity than did rats refed a diet different only in that maltose replaced sucrose.[15] Sucrase activity in rats fed diets containing 70% carbohydrate was greater with sucrose than with either maltose or glucose.[16] Sucrase adaptation to sucrose also appears to occur in humans. Male volunteers consuming isocaloric diets differing in the nature and amount of dietary carbohydrate (40 to 80% of the calories) exhibited increased sucrase activity when fed sucrose as compared to glucose.[17,18]

Intestinal absorption processes also appear to respond to dietary sucrose. Rats accustomed to being fed a 65% sucrose diet showed increases in the intestinal absorption of sucrose and its constituent monosaccharides as compared with rats fed a stock diet.[19] The ad libitum or meal-feeding of diets containing 54% sucrose as compared to 54%

starch to rats for 8 to 12 weeks produced similar increases in sugar absorption.[20] Baboons showed an increased absorption capability for fructose and, to a lesser extent, glucose following a sucrose meal after being adapted to a 75% sucrose diet for 9 weeks.[21] On the basis of these intestinal events, it would be expected that animals adapted to sucrose-feeding would have the capacity to quickly absorb dietary sugars. The resultant steep increases in postprandial blood glucose with the associated strong stimulation of the insulin response is compatible with impaired glucose tolerance and increased lipogenesis.

Reproduction and Longevity

A diet containing 20% sucrose and 30% starch produced essentially complete reproductive failure when fed to female rats for 2 years.[22] Two other groups fed diets containing 20% lactose and 30% starch or 50% starch showed normal reproductive capacity. Two studies have suggested that feeding sucrose in place of starch can decrease the life expectancy of rats susceptible to kidney disease. Male Wistar rats fed a diet in which 15% of the starch was replaced by sucrose showed an average 80-day reduction in lifespan.[23] Under identical dietary conditions, female Wistar rats showed only a 22-day reduction in life expectancy. Male rats of the carbohydrate-sensitive BHE strain died an average of 151 days earlier when fed a diet containing 39% sucrose in place of 39% starch.[24] Renal disease was considered to be the primary cause of death.

Glucose Tolerance, Blood Insulin, and Tissue Sensitivity to Insulin

Numerous studies using rats have shown that the intake of high levels of sucrose produces an impairment of glucose tolerance, an increase in fasting blood insulin levels, and a decrease in the tissue sensitivity to insulin. These metabolic changes are characteristic of the symptomology found in maturity onset diabetes. Rats fed diets containing 67% sucrose as compared to 67% starch for 21 days showed increased levels of blood glucose 30, 60, and 90 min following an intragastric glucose load.[25] Increases in blood sugar were also observed after 40 days of feeding a 40% sucrose diet and after 50 to 100 days of feeding a 33% sucrose diet. Rats fed diets containing 70% sucrose as compared to 70% starch for 50 days exhibited a decreased adipose tissue sensitivity to insulin without increased levels of blood insulin.[26] Rats fed diets containing 65%[19] and 68%[27] sucrose for 4 to 11 weeks showed increased levels of fasting serum insulin when compared to rats fed a stock diet. In the latter study, a decreased insulin sensitivity of adipose tissue and diaphragm was also noted in the sucrose-fed rats. The ad libitum or meal-feeding of diets containing 54% sucrose as compared to 54% starch for 12 to 17 weeks resulted in increased levels of fasting serum insulin and decreased sensitivity of adipose tissue.[28] Meal-fed rats consuming a diet containing 56% sucrose as compared to 56% glucose for 24 days showed increased blood glucose levels 60, 90, and 120 min following a glucose tolerance test.[29] Feeding rats sucrose as compared to starch has been reported to have a distinct stimulatory action on the secretion of insulin by isolated pancreas perfused with glucose.[30] Rats stressed by immobilization were found to have a significantly greater level of blood glucose when consuming a diet containing 30% sucrose than when consuming a control diet.[31]

The hypothesis that sucrose is an etiological factor in diabetes has been tested in genetically selected rats.[32] Rats were bred on the basis of blood glucose levels following an intragastric glucose tolerance test. Animals with the greatest increase in blood glucose (Upward selection) were mated and those with the smallest increase (Downward selection) were mated. In succeeding generations of the Upward selection fed a diet containing 72% sucrose, the rats developed diabetes-like symptoms including high fasting blood sugars, hyperinsulinemia, peripheral insulin resistance, and retinal and renal vascular complications. This effect could be demonstrated with as little as 25% dietary sucrose fed over more extended time periods. In contrast, the sibling Upward selection fed a diet containing 72% starch did not show these symptoms. In the offspring of the Downward selection, the diabetes-like symptoms did not appear in rats fed either the 72% sucrose or starch diets. These results show the need for interaction between genetic and dietary factors for the expression of the metabolic defects characteristic of diabetes to become evident.

Human studies pertaining to the effect of dietary sucrose on indices of glucose tolerance have not been as conclusive. Human volunteers consuming a diet in which carbohydrate was mainly in the form of sucrose as compared to bread showed increased and extended glycemic response to an oral glucose tolerance test.[33] The intake of a diet in which carbohydrate was mainly in the form of sucrose as compared to starch increased serum insulin levels, especially during a glucose tolerance test, in 6 out of 19 human volunteers.[34] These results are consistent with the contention that the ingestion of sucrose produces adverse metabolic effects in a finite portion of the population that can be described as sucrose-sensitive. Human volunteers receiving 52 to 64% of the calories from sucrose as compared to starch had increased levels of insulin 2, 5, and 8 min after the intravenous administration of 25 g of glucose.[35] Insulin levels 60 min after a glucose load were significantly higher in young women who consumed a diet containing 42% as compared to 20% of the calories from sucrose.[36] If increased fasting blood insulin and insulin response to a carbohydrate load are considered to be indicators of impairment of glucose tolerance, then these studies suggest that sucrose feeding impairs glucose tolerance.

In contrast to these findings, no alteration in glucose tolerance was found in three subjects consuming carbohydrate either as sucrose or as uncooked maize starch.[37] Young male volunteers changed from a control diet to one containing 80% of the calories from sucrose showed significant improvement in the glycemic response to an oral glucose tolerance test.[38] The improvement in glucose tolerance was associated with slight reductions in plasma insulin values. The isocaloric exchange of 70% dietary starch with sucrose in diets containing 45% of the calories as carbohydrate fed to human volunteers significantly elevated fasting blood glucose and slightly increased the glycemic response to a glucose tolerance test.[39] However, plasma insulin levels were similar during these dietary periods. The isocaloric exchange of dietary sucrose and starch after 2-week feeding periods did not significantly change fasting serum insulin or the insulin response to a meal in five male volunteers.[40] Only small and inconclusive changes in glucose tolerance were reported after male volunteers consumed either a control or sucrose-free diet for 9 weeks.[41] Differences in the reported effects of sucrose intake on parameters of glucose tolerance in humans may be due to the age, sex, and body weight of the subjects, the nature of the other dietary components fed (e.g., saturated vs. unsaturated fat, fiber), the amount of sucrose fed, the length of time on the diet, and the genetic predisposition of the individuals.

Lipogenesis

In man and rat, the liver appears to be a primary site of the synthesis of lipid from carbohydrate. The activities of various hepatic enzymes involved in the conversion of carbohydrate to lipid have been reported to be increased in rats fed sucrose as compared to an equivalent amount of glucose or glucose polymers. The increased activity of the lipogenic enzymes produced by sucrose-feeding appears to be associated with an increased deposition of liver lipid.

Liver lipid was increased in rats fed diets containing approximately 80% sucrose as compared to diets containing an equivalent amount of either dextrin,[42] liquid glucose or dextrose,[43] or various starches.[44] The incorporation of radioactive acetate into total lipid and cholesterol was markedly greater in the liver after 20 weeks of the ad libitum or meal-feeding of diets containing 70% sucrose as compared to 70% starch.[45] The activity of hepatic fatty acid synthetase was reported to be quadrupled in rats changed from a starch diet to one containing 50% sucrose after only 3 days.[46] The activity of hepatic glucose-6-phosphate dehydrogenase (EC 1.1.1.49),[19,47-51] malic enzyme (EC 1.1.1.40),[19,48-50] pyruvate kinase (EC 2.7.1.40),[49-51] and total liver lipid[19,47,48] was increased in rats fed diets containing 50 to 72% sucrose as compared to diets containing an equivalent amount of glucose or starch. The increase in liver lipid due to sucrose-feeding appeared to be greater in the carbohydrate-sensitive BHE strain of rat than in the Wistar strain.[47] The transport of lipid from the liver to adipose tissue may contribute to the higher levels of blood triglyceride observed in animals fed sucrose.

Studies utilizing the removable fat tissue in the abdominal and retroperitoneal cavities indicate that the feeding of sucrose produces greater adiposity than does the feeding of starch. However, the greater deposition of body fat in animals fed sucrose as compared to starch has not been consistently observed.[52] Male baboons fed a diet containing 76% of the calories from sucrose for 26 weeks had 60% and 300% more omental, pericardial, and retroperitoneal fat than did baboons fed equivalent amounts of a partial starch hydrolysate and starch, respectively.[53] There was a greater amount of both liver fat and removable adipose tissue in the abdominal and retroperitoneal cavities in rats fed for 15 weeks a diet containing 68% sucrose as compared to 68% starch.[54] Replacement of 55% of dietary starch with sucrose resulted in a 16% to 26% increase in the perirenal fat content of rats after 2 weeks.[55] The ad libitum or meal-feeding of diets containing 54% sucrose as compared to 54% starch for 12 to 17 weeks resulted in a greater deposition of epididymal and perirenal fat.[28] These findings support the contention that the large intake of sucrose in industrialized or westernized societies contributes to the high incidence of human obesity. Since obesity has been generally regarded as an important risk factor in the etiology of both diabetes and heart disease, sucrose may be indirectly linked to these disease states.

Gluconeogenesis

Several studies indicate that the feeding of sucrose as compared to glucose or glucose polymers increases the activity of the important gluconeogenic enzyme glucose-6-phosphatase (G6Pase, EC 3.1.3.9) in the liver of rats. Male weanling rats exhibited greater hepatic G6Pase activity when fed a diet containing 60% sucrose as compared to 60% rice starch for 14 to 19 days.[56] Rats adapted to meal-feeding 1 hr/day showed greater hepatic G6Pase activity 4, 8, and 24 hr following a meal containing 65% sucrose as compared to 65% dextrin.[57] The feeding of diets containing 50%[47] and 72%[49] sucrose as compared to equivalent amounts of glucose or starch for 3 to 12 months increased the activity of hepatic G6Pase in three different strains of rat. The increased activity of G6Pase is not compatible with the efficient utilization of glucose via tissue phosphorylation and may contribute to a reduced glucose tolerance in sucrose-fed animals.

Blood Lipids

Increased levels of blood cholesterol and/or triglyceride are considered to be risk factors in the incidence of heart disease. The nature and amount of dietary carbohydrate have been shown to influence the levels of blood lipids both in experimental animals and humans. Recently there has been a growing interest in determining whether the excessive intake of sucrose relative to the equicaloric intake of other forms of carbohydrate contributes to the high incidence of heart disease in developed societies by raising blood lipid levels. The consensus of these studies is

1. Dietary sucrose increases serum cholesterol levels in experimental animals when cholesterol is added to the diet.
2. Dietary sucrose produces small increases in the serum cholesterol levels of

humans under specific experimental conditions, perhaps secondarily due to an increase in the prebeta lipoprotein fraction of blood which contains primarily triglyceride and a small amount of cholesterol.

3. Dietary sucrose significantly increases endogenous serum triglyceride levels in experimental animals and humans under a variety of conditions.

Cholesterol

As early as 1956, it was reported that feeding rats diets containing 56% sucrose as compared to 56% starch elevated serum cholesterol when 5% cholesterol was also included in the diet.[58] In the absence of dietary cholesterol, the increase in serum cholesterol due to sucrose was markedly reduced. Rats fed diets containing 55% carbohydrate as either sucrose, glucose, or pregelatinized potato starch for 4 weeks showed no differences in serum cholesterol.[59] When 1% cholesterol was added to the diets, the sucrose-fed rats showed large and significant increases in serum cholesterol as compared to the rats fed glucose or starch. Rats fed diets containing 1.5%[60] and 1%[61] cholesterol had higher levels of serum cholesterol when consuming diets containing 40 to 60% sucrose than when consuming diets containing equivalent amounts of starch over experimental periods as long as 180 days. In rats fed diets containing 2% cholesterol and 60% carbohydrate for 4 months, serum cholesterol was higher when the carbohydrate was sucrose than when it was glucose or starch from ragi, jowar, tapioca, and bajra.[62] However, in these studies rice and wheat starch were about as cholesteremic as was sucrose. In contrast to these results, lower serum cholesterol levels were found in rats fed diets containing 68% sucrose as compared to various forms of modified starch with 1% cholesterol added to the diets.[63] Increased levels of blood cholesterol in the absence of dietary cholesterol have been reported in rats fed diets containing 12%,[64] 39%,[65] 68%,[51] and 81%[43] sucrose as compared to diets containing an equivalent amount of starch. Chicks fed a diet containing 60% sucrose as compared to 60% glucose for 4 weeks showed no differences in plasma cholesterol.[66] When the diets also included 3% cholesterol, the sucrose-fed chicks showed 50 to 100% increases in plasma cholesterol. Similarly, diets containing 3% cholesterol produced much greater increases in serum cholesterol in both conventional and germ-free chickens when the 54% carbohydrate was supplied by sucrose than when it was supplied by either glucose or starch.[67] Serum cholesterol levels of rabbits fed a cholesterol diet containing 50% sucrose for 60 days were 50% higher than those from rabbits fed a diet containing 50% glucose.[66] No difference in serum cholesterol was observed in dogs fed a low cholesterol diet containing 45% carbohydrate either as sucrose or starch.[68] Three species of subhuman primates fed diets containing 0.5% cholesterol for 16 months showed no increase in serum cholesterol when the carbohydrate was 66% sucrose as compared to 66% dextrin.[69]

Sucrose-containing diets have been shown to produce higher serum cholesterol levels than diets containing isocaloric amounts of glucose and glucose polymers in a number, but not all, of human studies. Male subjects fed a chemically defined diet composed of amino acids, vitamins, minerals, ethyl linoleate, and 90% of the calories from glucose showed a decrease in serum cholesterol from 227 mg % on their usual diet to 160 mg % after 4 weeks.[70] When 25% of the glucose was replaced by sucrose, the serum cholesterol values rose an average of 48 mg % in 3 weeks. Since no difference in the effect of glucose and sucrose on serum cholesterol levels has been observed in other human studies utilizing natural dietary components,[71,72] it appears that the action of sucrose on serum cholesterol is profoundly influenced by the nature of the components comprising the rest of the diet. The supplementation of 50 g of sucrose to the diet of 30 atherosclerotic patients resulted in a significant increase in serum cholesterol of 19 mg % after 10 days.[73] Higher levels of serum cholesterol were observed during 4-week periods when humans were consuming a diet containing most of the 48% carbohydrate as simple sugars than when they were consuming a diet containing an equivalent amount of carbohydrate from vegetable and cereal starches.[74] Exchange of 450 to 500 g of raw cornstarch for sucrose in the diet of human volunteers resulted in an increase in serum cholesterol in male, but not young female, subjects.[37,75] Replacement of 200 g of dietary starch by sucrose produced a significant 14 mg % increase in the plasma cholesterol of young male volunteers after 2 weeks.[76] Humans fed a diet containing 35% of the calories from sucrose as compared to starch from bread showed increases in serum cholesterol which were not significant after 2 weeks (13 mg %) but were significant after 4 to 5 weeks (24 mg %).[77] The exchange of 16% of calories of starch from mixed vegetables, but not wheat flour, for an equivalent amount of sucrose significantly increased serum cholesterol levels in humans after 2 weeks.[78] Humans consuming diets containing carbohydrate as sucrose showed significant increases in serum cholesterol (8 to 18 mg %) over those observed when they were consuming diets containing equicaloric amounts of starch from vegetables and legumes[79,80] and from cereals and potatoes.[81] Serum cholesterol was significantly increased in subjects when the sucrose content of a typical western diet was doubled at the expense of complex carbohydrate to provide 34% of the total calories.[82] However, this increase was prevented by the replacement of 75% of the total fats by unsaturated fats. In contrast to these results, no significant differences were observed in the serum cholesterol of humans when starch from cereals and potatoes,[39] bread and potatoes,[80] rice,[83] or wheat[84] was replaced by equivalent amounts of sucrose. Young male volunteers showed an overall increase of 5 mg % in serum cholesterol after consuming a sucrose-free diet for 14 weeks.[85] However, in a subgroup of five subjects with predietary cholesterol values in excess of 220 mg %, there was a significant 9 mg % fall after the sucrose-free period.

In evaluating the significance of these modest increases in serum cholesterol found after sucrose feeding, a number of factors should be considered. Many of the human studies utilized replacement by sucrose of starch in the form of complex carbohydrates found in natural food. Under these conditions it is difficult to attribute changes in cholesterol to the increased intake of sucrose or to the decreased intake of dietary fiber or unsaturated fat. The major effect of dietary sucrose on blood lipids appears to be to increase endogenous triglyceride. The resultant increase of the prebeta lipoprotein fraction of blood which contains primarily triglyceride but also contains a small amount of cholesterol may indirectly produce small increases in blood cholesterol. The small and perhaps indirect increases of blood cholesterol reported after the intake of high levels of sucrose may nevertheless contribute to the numerous environmental factors producing high levels of blood cholesterol in developed societies.

Triglyceride

Studies with both experimental animals and humans have established that diets containing high levels of sucrose produce larger increases in endogenous blood triglyceride than do diets containing an equivalent amount of glucose or glucose polymers. The metabolism of the fructose moiety of sucrose is generally believed to be responsible for the increased lipogenesis that produces the higher triglyceride levels. The failure of some animal species to show increased levels of blood triglyceride after the intake of sucrose as compared to starch or glucose may be due to the metabolic fate of fructose.

In the rat, the feeding of sucrose as compared to either starch[51,61,64,65,86-88] or glucose[64,89,90] has been shown to result in large increases in endogenous blood triglyceride levels under a variety of experimental conditions. In the rat, as in man, fructose is poorly metabolized during its intestinal absorption and appears in the portal blood primarily as fructose with a small amount of lactate. In contrast, in the small intestine of the guinea pig fructose is mainly converted to glucose. The acute or prolonged feeding of fructose as compared to glucose did not increase serum triglyceride levels in the guinea pig.[91] It may therefore be possible to correlate the hypertriglycer-idemic effect of sucrose in various species with the postabsorptive appearance of nonmetabolized fructose in the circulation. The chicken[88,90] appears to be another species that does not respond to the intake of large amounts of sucrose with increased levels of blood triglyceride. Several studies utilizing nonhuman primates have shown that large amounts of sucrose administered either acutely[92] or fed in diets over a long period of time[69,93,94] produced greater increases in blood triglyceride than did equivalent amounts of glucose containing carbohydrates. The increase was more pronounced in male than in female primates.[93,94]

In human studies in which the intake of sucrose has been either eliminated[85] or reduced,[95,96] significant decreases in fasting serum triglyceride levels have occurred. The decreases were greater in subjects having the highest initial blood triglyceride levels.[85,96] In two of the studies,[95,96] the reduction in sucrose intake produced decreases in body weight which may have contributed to the decrease in triglyceride. However, it has been reported that weight gain induced in humans by the intake of excess calories produced increases in serum triglyceride when sucrose, but not starch, consumption was also increased.[76]

The increase in serum triglyceride in humans consuming diets in which sucrose isocalorically replaced starch is apparently dependent on a number of factors including the amount of sucrose fed, the age and sex of the subject, the nature of the dietary fat, and the genetic predisposition of the subject. Diets containing at least 30% of the calories as sucrose as compared to an equivalent amount of starch consumed for periods of time up to 4 weeks have consistently produced large and significant increases of endogenous blood triglycerides in predominantly male subjects.[34,35,37,74,76,82,84,97,98] Similar studies in which diets containing less than 30% of the calories as sucrose as compared to an equivalent amount of starch have not shown significant differences in fasting serum triglyceride levels.[39,40,81] In contrast to men and postmenopausal women,[99] fasting blood triglycerides of young women do not appear to increase after consuming diets containing as much as 70% of the total calories as sucrose.[75] The elevation of fasting blood triglycerides generally observed when the carbohydrate content of the diet is increased at the expense of fat similarly appears to be absent in young females.[100] The magnitude of the increase of blood triglycerides by sucrose is affected by the nature of the dietary fat. When 75% of the fat of a typical western diet was replaced by unsaturated fat, the increase in endogenous blood triglyceride produced by 34% sucrose calories in the diet was prevented.[82] Young men consuming a diet containing 60% sucrose, 30% fat, and 9% protein for 5 days showed significantly increased fasting blood triglycerides when the fat was saturated (cream) and not when it was unsaturated (sunflower oil).[101] These results may explain apparent contradictions found in the literature on the magnitude of the effect of dietary sucrose on blood triglycerides. These results also indicate that the usual diet consumed in developed societies with a polyunsaturated/saturated fatty acid ratio of about 0.25 is conducive to a large increase in serum triglycerides when high amounts of sucrose are also ingested.

The human studies described above have generally utilized normal subjects. There are segments of the population that appear to have a genetic predisposition that results in a large and permanent increase in blood triglycerides when consuming diets high in carbohydrate. This type of hyperlipemia has been described as carbohydrate-induced or Type IV[102] and has been shown to be associated with abnormal glucose tolerance, diabetes, and heart disease. It is estimated that about 9% of the population of the U.S. has this predisposition.[103] Numerous studies have demonstrated that in patients with carbohydrate-induced lipemia, diets containing sucrose as compared to an equivalent amount of starch will increase fasting blood triglyceride by as much as 300%.[104-108] Blood triglycerides were significantly increased in these patients with sucrose levels as low as 20 to 25% of the total caloric intake.[109] Although the levels of blood triglyceride were lower with diets containing unsaturated as compared to saturated fats, the magnitude of the sucrose increase of blood triglyceride was unaffected by the nature of the dietary fat in these individuals.[106,107] From these results, it is apparent that dietary sucrose is an important environmental factor in the expression of carbohydrate sensitivity in genetically susceptible subjects.

Blood Pressure

Rats ingesting a 5% sucrose + 1% NaCl solution were observed to have higher blood pressure than rats ingesting a 5% glucose + 1% NaCl solution.[110] Rats fed a diet high in sucrose for 10 weeks exhibited higher levels of systolic blood pressure than did rats fed a natural grain ration or diets high in fat or cornstarch.[111] The differences were greater when 1 to 2% NaCl was added to the drinking water. Rats fed a diet containing 5% lactose, 30% starch, and 10% sucrose for 15 weeks showed elevated systolic blood pressure as compared to rats fed 5% lactose and 40% starch.[112] Diastolic blood pressure was increased in 26 human volunteers over a 5-week period in which they consumed 200 g sucrose/day.[112] Only about 25% of the subjects responded to the high sucrose intake with elevated blood pressure. These results again suggest that individual response to sucrose depends on a genetic predisposition toward carbohydrate sensitivity. These results indicate that sucrose may be a hypertensive agent and thus links sucrose intake to diseases associated with hypertension such as heart disease.

Cariogenicity

In the U.S., approximately 95% of children have some form of tooth decay, and about 55% of the population have lost their teeth by the age of 55. It is clear from studies utilizing both experimental animals[113-118] and humans[119-122] that sucrose is among the most cariogenic of the sugars occurring in the diet. Since sucrose is the cariogenic substance which is most excessively ingested, it is considered to be the primary dietary culprit in the causation of caries. The magnitude of the sucrose-induced cariogenicity is complex and depends on genetic factors, frequency of consumption, duration of the exposure, the form in which the sucrose is fed (e.g., sticky vs. granular food), and the nature of other dietary constituents ingested. Individuals with hereditary fructose intolerance who must avoid all forms of dietary fructose and sucrose have substantially fewer dental caries than the general population,[123] but are not caries-free.[124]

FRUCTOSE

Fructose has now become a major dietary component due to the use of large amounts of sucrose and isomerized corn sweeteners. Fructose is also consumed in varying amounts in fruits, vegetables, and especially honey. On the basis of the consumption patterns of these dietary ingredients, the combined average fructose intake in the U.S. can presently be estimated at about 70 g/day.

Fructose-feeding has been found to produce adaptive changes in the digestive and absorptive capacities of the small intestine. Sucrase activity was increased in humans consuming diets containing 40% fructose as compared to 40% glucose after 7 days.[17] A twofold increase in fructose absorption was demonstrated after only 3 days in rats fed a 60%-fructose diet as compared to rats fed a stock diet.[125] Similarly, a significant twofold increase in the in vivo absorption of fructose was observed in rats following the feeding of a 10% solution of fructose for 3 days.[126] Several in vitro studies have shown that the presence of 10 to 25 mM fructose in the incubation media[127-129] or the preloading of the small intestine with fructose[130] will stimulate the intestinal absorption of amino acids such as valine,[127] glycine,[127] leucine,[128,129] and cycloleucine.[130] Fructose is absorbed in the small intestine at rates intermediate between those of the actively transported sugars and those of the sugars transported by passive diffusion. In contrast to glucose or sucrose, the intake of fructose as 40% of the dietary calories causes diarrhea in humans. Rats differ from humans in that they can tolerate fructose levels as high as 70% of the calories without significant diarrhea.[131]

As indicated in the section under sucrose, the intestinal metabolism of fructose is species-dependent and is minimal in humans and the rat. The efficient uptake and metabolism of fructose, primarily by the liver, prevents the concentration of fructose in the systemic blood from attaining levels greater than 30 to 40 mg % even after the ingestion of large amounts of sucrose or fructose. The cellular transport and metabolism of fructose in tissues such as liver, adipose, and kidney does not appear to require insulin nor does fructose alone potentiate the pancreatic release of insulin.[132] These metabolic characteristics have stimulated interest in the use of fructose in the dietary control of diseases that are associated with an improper insulin response to a glycemic stress, such as diabetes and hypoglycemia. Fructose has also been shown to accelerate the rate of ethanol metabolism in humans.[133-135] However, the intake of large amounts of fructose has been shown to produce undesirable metabolic and physiological effects in the following three areas of health-oriented concern: (1) hepatic lipogenesis and blood lipid levels, (2) adenine nucleotide utilization and uric acid production, and (3) lactic acidosis. In the very small percentage of individuals who suffer from fructose intolerance, due to the absence of either fructokinase (EC 2.7.1.4) or fructose-1-phosphate aldolase (EC 4.1.2.7), a small amount of fructose may represent a dangerous excess.

Hepatic Lipogenesis and Blood Lipids

Extensive studies using rats have shown that the feeding of diets containing at least 40% of the calories from fructose as compared to equivalent amounts of glucose or glucose polymers increases the levels of hepatic lipogenic enzymes,[9,51,90,136-138] liver lipid,[43,51,138-140] and the levels of blood lipids,[43,51,87,90,91,137,140,141] especially the triglyceride fraction. In contrast, fructose-feeding did not increase any of the above indices of lipogenesis in chicks.[90] Since these results are very similar to those obtained using sucrose, it appears that the fructose moiety of sucrose may be primarily responsible for sucrose-induced lipogenesis. The hepatic metabolism of fructose favors a lipogenic pathway. Fructose is converted to fructose-1-phosphate by a specific fructokinase which has greater activity than either hexokinase (EC 2.7.1.1) or glucokinase (EC 2.7.1.2)[142] and which shows an adaptive increase in activity due to fructose-feeding.[143-145] Fructose-1-phosphate is then cleaved by a specific aldolase to form dihydroxyacetone phosphate and glyceraldehyde. Further metabolism of glyceraldehyde requires reduction to glycerol followed by phosphorylation by glycerokinase (EC 2.7.1.30) to α-glycero-phosphate, a direct precursor of triglyceride. This metabolic sequence effectively produces substrates required for lipogenesis and circumvents two important control mechanisms in hepatic glycolysis: the reactions catalyzed by hexokinase and phospho-fructokinase (EC 2.7.1.11).

Several studies have shown that fructose is converted into lipogenic substrates and incorporated into lipid more rapidly than is glucose. Fructose intragastrically instilled into rats adapted to a diet containing 72% fructose was more readily converted to serum glycerides than was glucose instilled into glucose-adapted rats.[89] More fructose than glucose carbon was recovered in both the liver and serum triglycerides 1 hr after the intravenous injection of 1 ml of either a 20% glucose- or fructose-U ^{14}C solution.[91] Rats refed a diet containing 70% fructose for 2 days following a 2-day fast had greater concentrations of serum and liver triglycerides and higher levels of liver pyruvate and acyl CoA derivatives than did rats refed a diet containing an equivalent amount of glucose.[142] The incorporation of intravenously administered palmitic acid-1-^{14}C into plasma triglyceride was higher in rats adapted to fructose-feeding than in rats adapted to glucose-feeding.[146] Liver slices from rats fed a chow diet converted 3 to 19 times more uniformly labeled fructose than uniformly labeled glucose into lactic acid, pyruvic acid, carbon dioxide, fatty acids, and glyceride glycerol.[147] Rats maintained on a chow diet showed greater concentrations of plasma triglyceride and free fatty acids but lower amounts of plasma insulin and adipose clearing factor 3 hr after the gastric instillation of 3 ml of 100% fructose as compared to glucose.[148]

In humans the intake of large quantities of fructose is also associated with increased lipogenesis, especially in carbohydrate-sensitive individuals. Patients with carbohydrate-induced hypertriglyceridemia generally showed larger increases in serum triglycerides when they consumed 300 g/day of fructose than when an equivalent amount of glucose or starch was consumed.[149] Three out of five patients with hypertriglyceridemia showed increases in serum triglyceride levels when 75 to 90 g of fructose was added to the diet at the expense of starch.[109] Men and postmenopausal women, but not premenopausal women, showed significantly higher levels of fasting triglyceride after consuming a diet containing 3 g fructose/kg body weight for 5 days than when consuming a diet containing an equivalent amount of glucose.[99] Human liver biopsy samples showed a greater capacity for the glycolysis of fructose than glucose.[150] This was reflected by the more rapid hepatic conversion to fatty acids, carbon dioxide, and glyceride-glycerol of radioactive fructose as compared to radioactive glucose. This high hepatic lipogenic potential of fructose is consistent with the higher levels of blood triglyceride observed in humans after the consumption of diets containing large amounts of fructose or sucrose.

Adenine Nucleotides and Uric Acid

The metabolism of large quantities of fructose by the liver has been shown to produce a rapid decrease in adenine nucleotides, especially ATP, and inorganic phosphate under a variety of experimental conditions. The intravenous injection of 1 millimole of fructose into rats produced a 60% decrease in the levels of hepatic ATP and inorganic phosphate within 5 min.[151] In contrast, the intravenous injection of an equivalent amount of glucose, galactose, or ribose did not produce these hepatic changes. Rats injected intraperitoneally with 40 millimoles fructose/kg body weight showed a 60% reduction in liver ATP and a 75-fold increase in liver fructose-1-phosphate within 15 min.[152] The hepatic levels of ATP, total adenine nucleotides, and inorganic phosphate had decreased 67%, 58%, and 60%, respectively, 10 min after the perfusion of rat liver with a solution containing 10 mM fructose.[153] The phosphorylated compounds showing the greatest increase at this time were fructose-1-phosphate (40-fold) and α-glycerophosphate (8-fold). The large and rapid increase of hepatic α-glycerophosphate provides further evidence for the efficient conversion of fructose to lipogenic metabolites. The ATP level of rat liver slices incubated in the presence of 100 mM fructose for 1 hr was 50% lower than that observed with 100 mM glucose.[154] In humans, the intravenous administration of 10% fructose resulted in a decrease of 50% in total adenine nucleotides after 30 min; the largest decrease being in ATP.[155] Glucose infusion did not produce significant decreases in the nucleotide levels. The hepatic infusion of six times the physiological dose of fructose (6 g/kg/hr) produced structural changes in rat hepatocytes similar to those observed in humans with hereditary fructose intolerance after fructose loading.[156] The infusion of equivalent amounts of glucose, galactose, mannose, and mannitol did not alter hepatic structure. The structural changes can therefore be attributed to the decrease in hepatic ATP and/or the increase in hepatic fructose-1-phosphate. Fructose administered intravenously at the rate of at least 0.5 g/kg body weight/hr[157-161] or consumed at the level of 250 to 290 g/day[162] has been shown to produce hyperuricemia and increased urinary urate excretion in human subjects. Glucose[160,162] and galactose[160] administered at the same levels as fructose failed to produce comparable increases in blood uric acid.

The hyperuricemia observed after the administration of fructose can be explained on the basis of the properties of fructose metabolism in the liver. The phosphorylation of fructose has been shown to produce a rapid decrease of both hepatic ATP and inorganic phosphate. Since ATP inhibits 5' nucleotidase (EC 3.1.3.5) and inorganic phosphate inhibits AMP deaminase (EC 3.5.4.6), the AMP formed in the liver is readily converted to IMP and then to inosine. Inosine is then oxidized to uric acid which appears in the blood as the major product of hepatic adenine nucleotide degradation. Hyperuricemia occurs in hyperlipemia and has been associated with the incidence of myocardial infarction.[163]

Lactic Acidosis

The intravenous infusion of fructose at the rate of at least 0.5 g/kg body weight/hr has been shown to produce large increases in the blood lactate levels of humans.[159,160,164,165] The mortality rate from lactic acidosis has been estimated to be as high as 90%.[166] There appear to be individual differences in the ability of humans to convert fructose into lactic acid. In patients with no evidence of diabetes or liver disease, blood lactate concentrations after fructose infusion fall into two distinct groups.[165] The individuals in one group converted 25% of fructose to lactate while those in the other group converted 50%. There was a tendency in the individuals with the lower blood lactate levels to have higher levels of blood triglyceride, suggesting a relationship between the rate of fructose conversion to lactic acid and carbohydrate-induced hyperlipemia.

LACTOSE

In nursing mammals, lactose represents the primary source of dietary carbohydrate and provides up to 40% of the total dietary energy. The utilization of lactose is dependent upon digestion to glucose and galactose by lactase (EC 3.2.1.23), an enzyme found in the brush border membrane of the intestinal epithelial cell. After early childhood the intake of lactose rapidly decreases. In the United States, the average per capita intake of lactose is reported to be 25 g or about 3 to 4% of the caloric intake.[1]

Lactose Malabsorption

The major nutritional problem associated with the excess or even moderate intake of lactose-containing foods by humans is lactose malabsorption. This problem is precipitated by an acquired primary lactase deficiency that often develops after early childhood. It has been found that lactase deficiency is very high in many nonwhite ethnic groups including Asians, South and North American Indians, Africans, and Eskimos.[167,168] However, even in Caucasians lactase deficiency may reach 20% of the population.[167] A similar incidence of lactase deficiency was observed in blacks whether living in Africa or the United States[169,170] and in Jews whether living in Israel or North America.[171] These findings, together with the inability to show either adaptive increases after feeding humans diets high in lactose[172-175] or decreases in lactase activity after the prolonged feeding of lactose-free diets,[176,177] suggest that the lactase deficiency is due to genetic factors. In rats, however, it has been possible to produce an adaptive increase in lactase activity after feeding diets containing 10% lactose[178] or goat's milk.[179]

The primary clinical symptom of lactose malabsorption is diarrhea. The osmotic effect produced by the undigested lactose attracts water into the intestinal lumen. The bacterial metabolism of the nonabsorbed lactose in the large intestine contributes to the diarrhea by further increasing the osmolarity of the intraluminal contents. In addition, the lactic acid formed lowers the intraluminal pH, thus impairing water absorption.[180] Although lactose is not an essential component of the adult diet, the diarrhea associated with lactose intolerance can be serious enough to greatly decrease the transit time and/or to damage the small intestinal epithelium and thus prevent the proper utilization of the essential noncarbohydrate constituents found in dairy products (e.g., protein and calcium). Lactose intolerance is also a serious problem in diseases that damage the structure of the small intestinal epithelium (e.g., tropical sprue, celiac disease, and kwashiorkor). Although all of the disaccharidases are located in the brush border membranes, lactase appears to be located more superficially in the microvilli and is the first disaccharidase to be lost during digestive disturbances.[181]

GALACTOSE

Galactose formed by the digestion of lactose enters the circulation and is phosphorylated to galactose-1-phosphate by galactokinase (EC 2.7.1.6). Galactose-1-phosphate enters a major pathway of carbohydrate metabolism after conversion to glucose-1-phosphate by galactose-1-phosphate uridyl transferase (EC 2.7.7.10). The hereditary lack of the transferase in infants produces galactosemia, a condition characterized by the accumulation in the tissues of large excesses of galactose, galactose-1-phosphate, and galactitol. The latter compound is formed from galactose by aldose reductase (EC 1.1.1.21) and reduced pyridine nucleotide and cannot be further metabolized. Disease manifestations of galactosemia include cataract formation, liver and kidney dysfunction, and disturbed mental development due to organic brain damage. Since cataract formation is the only clinical symptom manifested in individuals with a hereditary lack of galactokinase, galactitol appears to be the specific causative agent. It has been estimated that galactosemia occurs as often as once in every 20,000 live births in the United States.[182] The main treatment for galactosemia involves removal from the diet of all foods that contain absorbable galactose, predominantly milk and milk products.

Cataract Formation

Some of the clinical disturbances found in galactosemic humans can be duplicated by feeding high levels of galactose to experimental animals. Early studies have shown that 68% of rats fed a diet containing 70% lactose developed cataracts after 22 weeks.[183] In contrast, rats fed an equivalent amount of either starch, maltose, dextrin, or sucrose developed no cataracts. Rats fed a diet containing 50% galactose showed cataract formation after only 2 weeks.[184] Subsequent studies have shown that a primary event in the formation of cataracts is the osmotic swelling of the lens fibers due to the accumulation of galactitol in the lens of rats fed diets containing 25 to 50% galactose.[185-188]

Brain Metabolism

Since mental retardation is the most significant finding in human galactosemia, a number of studies utilizing experimental animals have examined the effects on the brain of the feeding of large amounts of galactose. Chicks fed diets containing 15 to 20% galactose developed a syndrome characterized by shivering, generalized motor seizures, and death within 10 days.[189-192] Neurological degeneration in the brain of chicks fed galactose was also observed.[191] The greater sensitivity of female chicks to galactose-feeding is reflected in their higher mortality rate.[190,192,193] Galactose-1-phosphate, the postulated toxic agent in galactosemia, was found to be present in about twice the concentration in brain homogenates from female chicks as in male chicks.[192] Chicks fed 10% galactose in their drinking water for 2 days developed severe hyperosmolar dehydration.[194] Since hyperosmolar hyperglycemia has been shown to produce seizures in humans[195,196] and dogs,[197] the serum hyperosmolality produced by the high levels of blood galactose may contribute to the seizures observed in the chicks.

Chicks fed a diet containing 40% galactose as compared to glucose showed less rapid incorporation of intracranially injected phosphate into the brain and a decrease in the levels of ATP and glucose in the brain.[198] In another study,[199] galactose-toxic chicks demonstrated a decreased entry of intracardiac-administered glucose into the brain and a lower content of brain glucose. These results suggest that galactose toxicity is associated with an interference of glucose transport into the brain either through a competition with glucose for a common transport site or by reduction of the ATP content required for glucose transport. This contention is supported by the finding that in chicks fed a diet containing 40% galactose, the intraperitoneal injection of 1 ml of 1 M glucose prevented convulsions and tremors.[200] The remission of the symptoms by glucose administration was accompanied by increased levels of both ATP and glucose in the brain.

Changes in the metabolism of myoinositol have been associated with galactose toxicity. The concentrations of free and lipid-bound myoinositol in tissues of both galactosemic infants[201] and rats fed diets containing 35% galactose[202,203] were significantly lower than those of normal infants and rats fed galactose-free diets. Brain slices from rats fed a control diet converted five to seven times as much glucose into myoinositol as did brain slices from rats fed a diet containing 35% galactose.[203]

The brains of rats made galactosemic by the feeding of a diet containing 40% galactose from 19 to 80 days of age were found to contain less than 10% the serotonin receptor activity of controls.[204] The serotonin receptor was found to reside in the galactolipid fraction. A functional deficiency of serotonin due to the decreased amounts of receptor may be responsible for the mental retardation associated with galactosemia since lowered serotonin levels in infant mice have been shown to produce a learning defect.[205]

FIBER

Dietary fiber has been defined as the structural components of plants present in the cell walls of the leaf, stem, root, and seed that are resistant to digestion by the secretions of the human gastrointestinal tract.[206] Chemically, fiber is classified as a carbohydrate since it is primarily comprised of cellulose, hemicelluloses (i.e., xylans, arabinans, galactans, mannans, glucans, glycuronans), and pectic substances. In addition, lignin, a polymer based on phenylpropane units, is also classified as a component of dietary fiber.

Most of the values for the fiber content of food are given as crude fiber or the portion resisting extraction with sulfuric acid, sodium hydroxide, water, alcohol, and ether.[207] It is estimated that crude fiber values obtained by these extraction methods reflect only 16 to 20% of the nutritionally more relevant dietary fiber.[208] Dietary fiber has long been neglected as an important factor in human nutrition since it is a negligible source of energy and no specific deficiency symptoms develop in its absence. Recently, epidemiological observations coupled to an increasing knowledge of the physiological role of dietary fiber at the intestinal level have linked a variety of diseases prevalent in western civilizations and virtually absent in underdeveloped societies to a relative deficiency in dietary fiber.[5,206] The daily intake of crude fiber in the U.S. and Great Britain has been estimated to be in the range of 4 to 12 g[209-211] or 16 to 25% of the consumption in these countries 100 years ago.[210,212] In contrast, the fiber intake by rural South African Bantu tribesmen is approximately 25 g/day.[213] In this context, the intake of fiber in excess of that currently consumed in western civilizations may produce decreases in the prevalence of noninfective, degenerative diseases characteristic of these western societies.

Intestinal Function and Disease

It has been suggested that dietary fiber can decrease the incidence of diverticular disease, appendicitis, and constipation by increasing the bulk of the food and decreasing the transit time required for the food to pass through the gut. These functions are accomplished by the water-absorbing properties of fiber and by formation of volatile fatty acids from unabsorbed cellulose and hemicellulose by bacteria in the colon.[211]

Decreased transit times have been reported after inclusion of 15 to 30 g/day of unprocessed bran to the diet of normal human subjects[214-216] and diverticular patients.[217,218] Diverticular patients given either 18 g bran daily or a high-roughage diet with a bran supplement for 1 month also showed decreased transit times.[219] The particle size of the bran may be important, since 20 g of coarse bran daily for 4 weeks decreased transit time in diverticular patients while the same quantity of fine bran did not.[220] Brown bread was found to pass from the stomach and through the intestines of normal human subjects more rapidly than white bread.[221] Patients with diverticulosis treated with bran have generally shown improvement of the symptoms associated with this disease.[218,219,222] Sources of fiber other than bran also appear to be effective in decreasing transit time. The inclusion of 10.5 g bagasse (sugar cane fiber)[223] or five to seven oranges[224] in the daily self-selected diets of normal human subjects was found to decrease transit time.

It has been suggested that the decreased transit time produced by dietary fiber protects against colonic cancer by reducing the time that carcinogens, which may be produced by intestinal microflora, are in contact with colonic mucosa.[225] Indirect support for this hypothesis comes from the finding that some steroids may be metabolized by the colonic microflora of humans to carcinogenic substances.[226,227] Furthermore, the colonic flora of populations with low incidence of bowel cancer are different than those with higher incidence of the disease.[227] However, data for 37 countries indicate that cancer of the large intestine is correlated with dietary protein and fat and not with fiber.[228]

Cholesterol and Bile Acid Metabolism

There is evidence that dietary fiber increases the fecal excretion of bile salts in experimental animals and humans, presumably by its ability to absorb these compounds in the gastrointestinal tract. Since the conversion to bile salts is a major metabolic pathway for body cholesterol, the postulated role of dietary fiber in decreasing the incidence of heart disease is attributed to the hypocholesteremia that results from the inhibition of bile salt reabsorption. Pectin appears to be particularly effective in lowering cholesterol levels. The increased blood[55,229-232] and liver[55,229-231] cholesterol levels induced by feeding rats diets containing 0.5 to 1% cholesterol were prevented by the concurrent feeding of 2.5 to 5% pectin. When pectin was replaced by an equivalent amount of cellulose, plasma and liver cholesterol levels were not decreased.[229] The hypocholesteremic action of pectin appears to be associated with an inhibition of bile salt reabsorption. The feeding of a diet containing 5% pectin increased the fecal excretion of bile acids in rats by about 32%.[231] The presence of 0.18% pectin in the incubation medium significantly inhibited the in vitro absorption of taurocholic acid by sacs of inverted rat intestine.[231] Cockerels fed a diet supplemented with 5% pectin as compared to 5% cellulose for 18 months showed a significant retardation of spontaneous atherosclerosis.[233] The pectin-fed cockerels excreted about three times as much extractable lipid as did the cellulose-fed birds. The levels of plasma cholesterol of cockerels fed diets containing 3% cholesterol for 27 days were reduced 29 and 44% by the inclusion of 3% pectin and psyllium seed, respectively.[234] Of several cereals tested, rolled oats was found to be the most effective in decreasing the level of serum cholesterol in rats fed a 1%-cholesterol diet for 4 weeks.[235] Rats fed a diet containing 20% bagasse excreted more bile acid than did rats fed either a diet containing an equivalent amount of cellulose or rats fed a fiber-free diet.[236] No significant differences in serum cholesterol levels were observed in rats fed any of the three diets. The failure of dietary fiber to lower blood cholesterol levels in rats fed diets without added cholesterol has been consistently observed. Excretion of total steroids was significantly greater and plasma cholesterol concentrations significantly lower in rats fed a stock diet and a semisynthetic diet supplemented with 33% ground soybran than in rats fed an unsupplemented semisynthetic diet.[237]

Increased excretion of bile acids has also been noted in humans consuming fiber from a wide variety of foods. Male prisoners consuming 13.6 g fiber from maize, wheat, and vegetables had increased secretion of bile acids and sterols.[238] Serum cholesterol levels did not show any significant changes. Nuns consuming 10.5 g bagasse in addition to their normal diet for 12 weeks showed increased bile acid excretion, but no change in plasma cholesterol.[223] Males consuming a high-fat diet for 55 weeks showed increased excretion of bile acids and decreased levels of serum cholesterol when Bengal gram was also included in the diet.[239] Adding 100 g of cellulose to a cholesterol-containing diet consumed by young girls for 10 days increased fecal bile acid excretion and decreased serum cholesterol.[240] On the basis of these results and the studies with experimental animals, it appears that fiber lowers blood cholesterol levels only when the initial values are relatively high.

The ability of fiber to lower cholesterol levels in humans also appears to depend on the type and source of the fiber. Serum cholesterol levels were reduced in males consuming diets containing 6 to 36 g pectin daily for 2 to 4 weeks.[80,241-244] In contrast, an equivalent amount of cellulose[80,241] or wheat fiber[243] did not lower serum cholesterol. Cellulose consumed at levels as high as 60 g/day were reported to be ineffective in lowering serum cholesterol or triglyceride.[245] As a general rule, wheat fiber has not been shown to be effective in lowering serum cholesterol levels. Serum cholesterol levels were not decreased in human subjects consuming diets containing 3 to 100 g of wheat bran daily for 3 to 12 weeks.[243,246-252] Male subjects consuming diets containing additional fiber from legumes showed decreased levels of serum cholesterol.[79,80,253] Middle-aged men consuming self-selected diets supplemented with 140 g of rolled oats in bread daily for 3 weeks showed decreased levels of serum cholesterol.[235]

Obesity

It has also been proposed that the removal of fiber from foods leads to overconsumption and obesity.[3,254] Naturally sweet foods (except honey) are rich in fiber. The chewing and digestion of these foods produces salivary and gastric secretions that give a sensation of satiety due to the distension of the stomach. The water-retaining properties of fiber aid in producing this distension. In contrast, the ingestion of fiber-free carbohydrate generally requires less chewing and does not produce a comparable feeling of satiety. For example, to achieve the average daily per capita intake of refined sugars in the United States would require the consumption of about 3 lb of apples. Since obesity is generally considered to be an important contributary factor in the causation of diabetes, the relative deficiency of fiber has been indirectly linked to this disease.[3] The present evidence of a relationship between the intake of dietary fiber and the prevention of many serious diseases afflicting industrialized societies represents an intriguing aspect of carbohydrate nutrition that warrants extensive study.

Mineral and Nutrient Availability

A possible danger inherent in the excessive intake of fiber is a decreased intestinal availability of nutritionally marginal minerals. The decreased transit time produced by fiber could further decrease mineral absorption from the intestine. Calcium, magnesium, and phosphorus absorption were lower in human subjects consuming diets containing brown flour[255] or 69% extraction flour[256] than when consuming diets containing equivalent amounts of white flour or 92% extraction flour, respectively. Diverticular patients consuming self-selected diets supplemented with 24 g of wheat bran daily showed reduced urinary calcium excretion.[218] Serum calcium levels fell in one study in which subjects were receiving 18 to 100 g of unprocessed bran daily,[248] but not in another study in which the subjects were receiving 20 g bran daily.[257] The solubility and the uptake of zinc by strips of rat intestine were significantly greater from suspensions of zinc-labeled leavened bread than from zinc-labeled unleavened bread.[258] The increased physiological availability of zinc in whole-meal bread after fermentation is attributed in part to the action of yeast in destroying phytate. Increased fecal excretion of zinc, iron, and phosphorus, but not calcium, was found after subjects consumed 10 g of cellulose daily for 2 weeks.[259] Negative calcium, magnesium, zinc, and phosphorus balances were found after subjects consumed 50% of their calories from whole-meal bread for 20 days.[260] Young males consuming a self-selected diet supplemented with rolls baked with at least 3.3% added bran showed decreases in absorption of iron.[261] Males consuming diets supplemented with 36 g of wheat fiber daily for 3 weeks[252] or 20 g of bran daily for 6 weeks[257] showed reduced levels of serum iron.

The excessive intake of dietary fiber may also decrease the availability of other utilizable components of the diet. An increased fecal loss of energy, especially in the form of nitrogen and fat, was demonstrated in human subjects after consuming about 20 g of added fiber daily.[262]

THICKENERS, EMULSIFIERS, AND STABILIZERS

The food additives to be described in this section are all composed of complex polysaccharides obtained from natural sources. Many are composed of the same hexose or pentose units. They did not appear to produce adverse physiological effects even when fed at much higher levels than those presently consumed. Moreover, they appear to simulate the hypocholesteremic action of food pectin and therefore may be considered a beneficial component of the diet of industrialized societies.

Pectin

In addition to being a natural constituent of the cell walls and intracellular layers of certain plant tissues, pectin is also available in commercial form. Pectic substances consist of an associated group composed of arabinans, galactans, and galacturonic acid partially esterified with methanol and in part neutralized as salts of calcium and magnesium. The chief use of commercial pectin is as a thickener in the preparation of foods such as jellies, jams, and marmalades. The potential beneficial effects of pectin in lowering blood cholesterol levels have been previously discussed (see section on fiber).

Guar Gum

Guar gum is the ground seed endosperm of leguminous plants found in India, Pakistan, and the United States. The gum is a polysaccharide composed of mannose and galactose residues and has an average molecular weight in the range of 200,000 to 300,000. Because guar gum yields viscous dispersions in cold water, it has been extensively used in the food industry as an emulsion stabilizer, binder, or thickener and a texturizing agent.[263] Guar gum is used in amounts ranging from 0.02 to 0.78% in a wide variety of foods including breakfast cereals, processed vegetables, sweet sauces, cheese, imitation dairy products, milk products, fruit ices, fats and oils, snack foods, gravies, processed fruits, and gelatin puddings. The average daily intake of guar gum in the United States is estimated to be in the range of 10 to 110 mg/kg body weight and the maximum daily intake in the range of 23 to 290 mg/kg body weight. Guar gum was not toxic to rats when fed at levels as high as 30% of the diet[264] and produced no adverse effects on weight gain, liver, or kidneys when fed to rats at 5% of the diet for 6 months.[265] Guar gum appears to be extremely effective in lowering blood cholesterol levels when fed to experimental animals and consumed by humans. Plasma cholesterol was decreased 60% in cockerels fed a diet containing 3% cholesterol for 28 days when 3% guar gum was included in the diet.[234]

Rats fed a diet containing 1% cholesterol for 28 days showed reductions in both serum and liver cholesterol when the diet was supplemented with 5 to 10% guar gum.[266] Humans showed significant reductions in serum cholesterol levels after consuming a self-selected diet including 36 g of guar gum daily for 2 weeks.[243] In each of these studies,[234,243,266] guar gum was more effective in lowering cholesterol levels than was an equivalent amount of pectin. On the basis of these results, guar gum appears to be a nontoxic and potentially beneficial component of foods.

Gum Ghatti

Gum ghatti, also known as Indian gum since it is obtained as an exudate from a tree found in India and Sri Lanka (Ceylon), is basically the calcium salt of a complex polysaccharide containing galactose, mannose, rhamnose, arabinose, xylose, and galacturonic acid.[267] It is employed in foods primarily as a stabilizer for oil-in-water emulsions in frozen dairy products and nonalcoholic beverages in proportions of 0.045 to 0.2%. The average daily intake of gum ghatti in the United States is estimated to be in the range of <1 to 6 mg/kg body weight with maximum daily intake in the range of 1 to 16 mg/kg body weight. The oral toxicity for multiple doses of gum ghatti in the rat was reported to be greater than 5 g/kg body weight. Gum ghatti, added to the diet at a level of 3%, lowered the plasma cholesterol levels of cockerels fed diets containing 3% cholesterol for 28 days.[234] Gum ghatti therefore appears to have properties common to mucilaginous polysaccharides in being nontoxic and hypocholesteremic.

Gum Arabic

Gum arabic or gum acacia consists of the dried exudate from trees of various species which grow in arid and semiarid regions throughout the world. Gum arabic is a complex polysaccharide containing galactose, arabinose, rhamnose, glucuronic acid or 4-O-methyl-glucuronic acid, calcium, magnesium, and potassium.[263] Gum arabic is almost completely soluble in water, facilitating its use as a stabilizer, emulsifier, and thickening agent in foods. Gum arabic is found in soft and hard candy (28%), chewing gum (2 to 8%), snack foods (0.6 to 2.8%), imitation dairy products, fats and oils, frosting, and grain products (1%) and in sugar substitutes, fruit ices, nut products, gelatin puddings, baked goods, and alcoholic beverages (0.06 to 0.5%). The average daily intake of gum arabic in the United States is estimated to be in the range of 14 to 128 mg/kg body weight with maximum daily intake in the range of 113 to 329 mg/kg body weight. Rabbits fed diets containing 20% by weight of gum arabic had significant growth with no evidence of deleterious effects.[269]

Carrageenan

Carrageenan is a colloidal extractive of carrageen and other red algae. It is composed of a mixture of high molecular weight sulfated polysaccharides of galactose and anhydrogalactose. The acidic sulfate moiety is present in part as a mixture of sodium, potassium, calcium, and magnesium salts. Carrageenan is used chiefly as a suspending agent in foods, as a clarifying agent for beverages, and in controlling crystal growth in frozen confections. In all these applications, the concentration of carrageenan is less than 1% by weight of the product.[270] Rats and mice fed rations containing up to 15% by weight of carrageenan for 2 years showed no evidence of increased mortality or tumor formation.[270] Carrageenan was the most effective of 16 mucilaginous polysaccharides in decreasing plasma cholesterol levels in cockerels fed a diet containing 3% cholesterol for 28 days.[234] Rats fed diets containing 58% sucrose and 0.5% cholesterol for 42 days showed significant and equivalent decreases in serum cholesterol when the diet was supplemented with either 7% pectin or carrageenan.[271]

REFERENCES

1. Friend, B. and Marston, R., Nutritional review, *Natl. Food Situation,* 150, 26–32, 1974.
2. Page, L. and Friend, B., Level of use of sugars in the United States, in *Sugars in Nutrition,* Sipple, H. and McNutt, K., Eds., Academic Press, New York, 1974, 93–107.
3. Cleave, T. L., Campbell, G. C., and Painter, N. S., *Diabetes, Coronary Thrombosis and the Saccharine Disease,* 2nd ed., John Wright and Sons, Bristol, England, 1969.
4. Yudkin, J., Diet and coronary thrombosis: Hypothesis and fact, *Lancet,* 2, 155–162, 1957.
5. Burkitt, D. P., Walker, A. R. P., and Painter, N. S., Dietary fiber and disease, *J.A.M.A.,* 229, 1068–1074, 1974.
6. Wretlind, A., World sugar production and usage in Europe, in *Sugars in Nutrition,* Sipple, H. and McNutt, K., Eds., Academic Press, New York, 1974, 81–92.
7. Mayer, J., The bitter truth about sugar, in Congressional Record United States Senate, 1976, pp. S10492–S10494.
8. Michaelis, O. E., IV and Szepesi, B., The mechanism of a specific metabolic effect of sucrose in the rat, *J. Nutr.,* 104, 1597–1609, 1974.
9. Michaelis, O. E., IV, Nace, C. S., and Szepesi, B., Demonstration of a specific metabolic effect of dietary disaccharides in the rat, *J. Nutr.,* 105, 1186–1191, 1975.
10. Michaelis, O. E., IV, Scholfield, D. J., and Nace, C. S., Demonstration of the disaccharide effect in starved-refed and meal-fed rats, *Fed. Proc.,* 35, 520, 1976.
11. Basu, T. K., Dickenson, J. W. T., and Parke, D. V., Effect of dietary substitution of sucrose and its constituent monosaccharides on the activity of aromatic hydroxylase and the level of cytochrome P-450 in hepatic microsomes of growing rats, *Nutr. Metab.,* 18, 302–309, 1975.
12. Macdonald, I. and Turner, J. L., Serum-fructose levels after sucrose or its constituent monosaccharides, *Lancet,* 1, 841–843, 1968.
13. Naismith, P. J., Differences in the metabolism of dietary carbohydrates studied in the rat, *Proc. Nutr. Soc.,* 30, 259–265, 1971.
14. Blair, D. G., Yakimets, W., and Tuba, J., Rat intestinal sucrase. II. The effects of rat age and sex and of diet on sucrase activity, *Can. J. Biochem. Physiol.,* 41, 917–929, 1963.
15. Deren, J. J., Broitman, S. A., and Zamcheck, N., Effect of diet upon intestinal disaccharidases and disaccharide absorption, *J. Clin. Invest.,* 46, 186–195, 1967.
16. Reddy, B. S., Pleasants, J. R., and Wostmann, B. S., Effect of dietary carbohydrates on intestinal disaccharidases in germ free and conventional rats, *J. Nutr.,* 95, 413–419, 1968.
17. Rosensweig, N. S. and Herman, R. H., Control of jejunal sucrase and maltase activity by dietary sucrose or fructose in man: A model for the study of enzyme regulation in man, *J. Clin. Invest.,* 47, 2253–2262, 1968.
18. Rosensweig, N. S. and Herman, R. H., Dose response of jejunal sucrase and maltase activities to isocaloric high and low carbohydrate diets in man, *Am. J. Clin. Nutr.,* 23, 1373–1377, 1970.
19. Reiser, S., Michaelis, O. E., IV, Putney, J., and Hallfrisch, J., Effect of sucrose feeding on the intestinal transport of sugars in two strains of rats, *J. Nutr.,* 105, 894–905, 1975.
20. Reiser, S., Hallfrisch, J., Putney, J., and Lev, F., Enhancement of intestinal sugar transport by rats fed sucrose as compared to starch, *Nutr. Metab.,* 20, 461–470, 1977.
21. Crossley, J. N. and Macdonald, I., The influence in male baboons, of a high sucrose diet on the portal and arterial levels of glucose and fructose following a sucrose meal, *Nutr. Metab.,* 12, 171–178, 1970.
22. Witnah, C. H. and Bogart, R., Reproductive capacity of female rats as affected by kinds of carbohydrate in the ration, *J. Agric. Res.* (Washington, D.C.), 53, 527–532, 1936.
23. Dalderup, L. M. and Visser, W., Influence of extra sucrose in the daily food on the life-span of Wistar albino rats, *Nature,* 222, 1050–1052, 1969.
24. Durand, A. M. A., Fisher, M., and Adams, M., The influence of type of dietary carbohydrate: Effect on histological findings in two strains of rats, *Arch. Pathol.,* 85, 318–324, 1968.
25. Cohen, A. M. and Teitelbaum, A., Effect of dietary sucrose and starch on oral glucose tolerance and insulin-like activity, *Am. J. Physiol.,* 206, 105–108, 1964.
26. Vrána, A., Slabochová, Z., Kazdová, L., and Fábry, P., Insulin sensitivity of adipose tissue and serum insulin concentration in rats fed sucrose or starch diets, *Nutr. Rep. Int.,* 3, 31–37, 1971.
27. Blazquez, E. and Quijada, C. L., The effect of a high-carbohydrate diet on glucose, insulin sensitivity and plasma insulin in rats, *J. Endocrinol.,* 44, 107–113, 1969.
28. Reiser, S. and Hallfrisch, J., Insulin sensitivity and adipose tissue weight of rats fed starch or sucrose diets ad libitum or in meals, *J. Nutr.,* 107, 147–155, 1977.
29. Romsos, D. R. and Leveille, G. A., Effect of meal frequency and diet composition on glucose tolerance in the rat, *J. Nutr.,* 104, 1503–1512, 1974.

30. Pfeiffer, E. F., Obesity, islet function and diabetes mellitus, in *Lipid Metabolism, Obesity, and Diabetes Mellitus: Impact Upon Atherosclerosis,* Greten, H., Levine, R., Pfeiffer, E. F., and Renold, A. E., Eds., George Thieme, Stuttgart, 1974, 143–151.

31. Sneer, A., Dinu, M., Herscovici, B., Papp, E., and Grigore, R., L'influence d'un regime alimentaire riche en sucre associee an "stress" sur le niveau de la glycemie et de l'equilibre lipidique du sang, *Rev. Med. Chir. Soc. Med. Natlas.,* 79, 235–240, 1975.

32. Cohen, A. M., Teitelbaum, A., Briller, S., Yanko, L., Rosenmann, E., and Shafrir, E., Experimental models of diabetes, in *Sugars in Nutrition,* Sipple, H. and McNutt, K., Eds., Academic Press, New York, 1974, 483–511.

33. Cohen, A. M., Teitelbaum, A., Balogh, M., and Groen, J. J., Effect of interchanging bread and sucrose as a main source of carbohydrate in low fat diet on the glucose tolerance curve of healthy volunteer subjects, *Am. J. Clin. Nutr.,* 19, 59–62, 1966.

34. Szanto, S. and Yudkin, J., The effect of dietary sucrose on blood lipids, serum insulin, platelet adhesiveness and body weight in human volunteers, *Postgrad. Med. J.,* 45, 602–607, 1969.

35. Nestel, P. J., Carroll, K. F., and Havenstein, N., Plasma triglyceride response to carbohydrates, fats and calorie intake, *Metab. Clin. Exp.,* 19, 1–18, 1970.

36. Kelsay, J. L., Behall, K. M., Holden, J. M., and Prather, E. S., Diets high in glucose or sucrose and young women, *Am. J. Clin. Nutr.,* 27, 926–936, 1974.

37. Macdonald, I. and Braithwaite, D. M., The influence of dietary carbohydrates on the lipid pattern in serum and in adipose tissue, *Clin. Sci.,* 27, 23–30, 1964.

38. Anderson, J. W., Herman, R. H., and Zakim, D., Effect of high glucose and high sucrose diets on glucose tolerance of normal men, *Am. J. Clin. Nutr.,* 26, 600–607, 1973.

39. Dunnigan, M. G., Fyfe, T., McKiddie, M. T., and Crosbie, S. M., The effects of isocaloric exchange of dietary starch and sucrose on glucose tolerance, plasma insulin and serum lipids in man, *Clin. Sci.,* 38, 1–9, 1970.

40. Mann, J. I. and Truswell, A. S., Effects of isocaloric exchange of dietary sucrose and starch on fasting serum lipids, postprandial insulin secretion and alimentary lipaemia in human subjects, *Br. J. Nutr.,* 27, 395–405, 1972.

41. Fry, A. J., The effect of a 'sucrose-free' diet on oral glucose tolerance in man, *Nutr. Metab.,* 14, 313–323, 1972.

42. Litwack, G., Hankes, L. V., and Elvehjem, C. A., Effect of factors other than choline on liver fat deposition, *Proc. Soc. Exp. Biol. Med.,* 81, 441–445, 1952.

43. Allen, R. J. L. and Leahy, J. S., Some effects of dietary dextrose, fructose, liquid glucose and sucrose in the adult male rat, *Br. J. Nutr.,* 20, 339–347, 1966.

44. Marshall, M. W. and Womack, M., Starches, sugars and related factors affecting liver fat and nitrogen balances in adult rats fed low levels of amino acids, *J. Nutr.,* 57, 193–202, 1955.

45. Fábry, P., Poledne, R., Kazdová, L., and Braun, T., The effect of feeding frequency and type of dietary carbohydrates on hepatic lipogenesis in the albino rat, *Nutr. Dieta,* 10, 81–90, 1968.

46. Veech, R. L., in *Sweeteners Issues and Uncertainties,* National Academy of Sciences, Washington, D.C., 1975, 113–115.

47. Chang, M. L. W., Lee, J. A., Schuster, E. M., and Trout, D. L., Metabolic adaptation to dietary carbohydrate in two strains of rats at three ages, *J. Nutr.,* 101, 323–330, 1971.

48. Michaelis, O. E., IV and Szepesi, B., Effect of various sugars on hepatic glucose-6-phosphate dehydrogenase, malic enzyme and total liver lipid of the rat, *J. Nutr.,* 103, 697–705, 1973.

49. Cohen, A. M., Briller, S., and Shafrir, E., Effect of long term sucrose feeding on the activity of some enzymes regulating glycolysis, lipogenesis and gluconeogenesis in rat liver and adipose tissue, *Biochim. Biophys. Acta,* 279, 129–138, 1972.

50. Roggeveen, A. E., Geisler, R. W., Peavy, D. E., and Hansen, R. J., Effects of diet on the activities related to lipogenesis in rat liver and adipose tissue, *Proc. Soc. Exp. Biol. Med.,* 147, 467–470, 1974.

51. Naismith, D. J. and Rana, I. A., Sucrose and hyperlipidaemia, *Nutr. Metab.,* 16, 285–294, 1974.

52. Bender, A. E. and Damji, K. B., Some effects of dietary sucrose, *World Rev. Nutr. Diet.,* 15, 104–155, 1972.

53. Brook, M. and Noel, P., Influence of dietary liquid glucose, sucrose and fructose on body fat formation, *Nature,* 222, 562–563, 1969.

54. Laube, H., Klör, H. U., Fussgänger, R., and Pfeiffer, E. F., The effect of starch, sucrose, glucose and fructose on lipid metabolism in rats, *Nutr. Metab.,* 15, 273–280, 1973.

55. Chang, M. L. W. and Johnson, M. A., Influence of fat level and type of carbohydrate on the capacity of pectin in lowering serum and liver lipids of young rats, *J. Nutr.,* 106, 1562–1568, 1976.

56. Carroll, C., Influences of dietary carbohydrate-fat combinations on various functions associated with glycolysis and lipogenesis in rats, *J. Nutr.*, 79, 93–100, 1963.

57. Freedland, R. A. and Harper, A. E., Initiation of glucose 6-phosphatase adaptation in the rat, *J. Nutr.*, 89, 429–434, 1966.

58. Portman, O. W., Lawry, E. Y., and Bruno, D., Effect of dietary carbohydrate on experimentally induced hypercholesteremia and hyperbetalipoproteinemia in rats, *Proc. Soc. Exp. Biol. Med.*, 91, 321–323, 1956.

59. Staub, H. W. and Thiessen, R., Jr., Dietary carbohydrate and serum cholesterol in rats, *J. Nutr.*, 95, 633–638, 1968.

60. Fillios, L. C., Naito, C., Andrus, S. B., Portman, O. W., and Martin, R. S., Variations in cardiovascular sudanophilia with changes in the dietary levels of protein, *Am. J. Physiol.*, 194, 275–279, 1958.

61. Qureshi, P., Akinyanju, P. A., and Yudkin, J., The effect of an 'atherogenic' diet containing starch or sucrose upon carcass composition and plasma lipids in the rat, *Nutr. Metab.*, 12, 347–357, 1970.

62. Vijayagopalan, P. and Kurup, P. A., Effect of dietary starches on the serum aorta and hepatic lipid levels in high fat high cholesterol-fed rats. II. Nature of the starch and hypolipidaemic activity, *Atherosclerosis*, 16, 247–256, 1972.

63. Anderson, T. A., Effect of carbohydrate source on serum and hepatic cholesterol levels in the cholesterol-fed rat, *Proc. Soc. Exp. Biol. Med.*, 130, 884–887, 1969.

64. Mukherjee, S., Basu, M., and Trivedi, K., Effect of low dietary levels of glucose, fructose and sucrose on rat lipid metabolism, *J. Atheroscler. Res.*, 10, 261–272, 1969.

65. Taylor, D. D., Conway, E. S., Schuster, E. M., and Adams, M., Influence of dietary carbohydrates on liver content and on serum lipids in relation to age and strain of rat, *J. Nutr.*, 91, 275–282, 1967.

66. Grant, W. C. and Fahrenbach, M. J., Effect of dietary sucrose and glucose on plasma cholesterol in chicks and rabbits, *Proc. Soc. Exp. Biol. Med.*, 100, 250–252, 1959.

67. Kritchevsky, D., Kolman, R. R., Guttmacher, R. M., and Forbes, M., Influence of dietary carbohydrate and protein on serum and liver cholesterol in germ-free chicken, *Arch. Biochem. Biophys.*, 85, 444–451, 1959.

68. Schultz, A. L. and Grande, F., Effects of starch and sucrose on the serum lipids of dogs before and after thyroidectomy, *J. Nutr.*, 94, 71–73, 1968.

69. Lang, C. M. and Barthel, C. H., Effects of simple and complex carbohydrates on serum lipids and atherosclerosis in non-human primates, *Am. J. Clin. Nutr.*, 25, 470–475, 1972.

70. Winitz, M., Graff, J., and Seedman, D. A., Effect of dietary carbohydrate on serum cholesterol levels, *Arch. Biochem. Biophys.*, 108, 576–579, 1964.

71. Anderson, J. T., Grande, F., Matsumoto, Y., and Keys, A., Glucose, sucrose and lactose in the diet and blood lipids in man, *J. Nutr.*, 79, 349–359, 1963.

72. Shammaá, M. and Al-Khalidi, U., Dietary carbohydrates and serum cholesterol in man, *Am. J. Clin. Nutr.*, 13, 194–196, 1963.

73. Pleshkov, A. M., Effect of long-term intake of easily absorbed carbohydrates (sugars) on blood lipid level in patients with atherosclerosis, *Ter. Arkh.*, 35, 66, 1963; *Fed. Proc.*, 23(2), T334–T336, 1964.

74. Hodges, R. E. and Krehl, W. A., The role of carbohydrates in lipid metabolism, *Am. J. Clin. Nutr.*, 17, 334–336, 1965.

75. Macdonald, I., The lipid response of young women to dietary carbohydrates, *Am. J. Clin. Nutr.*, 16, 458–463, 1965.

76. Naismith, D. J., Stock, A. L., and Yudkin, J., Effect of changes in the proportions of the dietary carbohydrates and in energy intake on the plasma lipid concentrations in healthy young men, *Nutr. Metab.*, 16, 295–304, 1974.

77. Groen, J. J., Balogh, M., Yaron, E., and Cohen, A. M., Effect of interchanging bread and sucrose as main source of carbohydrate in a low fat diet on the serum cholesterol levels of healthy volunteer subjects, *Am. J. Clin. Nutr.*, 19, 46–58, 1966.

78. Anderson, J. T., Grande, F., Foster, N., and Keys, A., Different dietary carbohydrates and blood lipids in man, Proc. 9th Int. Congr. Nutrition, 64, 1972; cited by Grande, F., Sugars in cardiovascular disease, in *Sugars in Nutrition*, Sipple, H. and McNutt, K., Eds., Academic Press, New York, 1974, 401–437.

79. Keys, A., Anderson, J. T., and Grande, F., Diet-type (fat constant) and blood lipids in man, *J. Nutr.*, 70, 257–266, 1960.

80. Grande, F., Anderson, J. T., and Keys, A., Effects of carbohydrates of leguminous seeds, wheat and potatoes on serum cholesterol concentration in man, *J. Nutr.*, 86, 313–317, 1965.

81. McGandy, R. B., Hegsted, D. M., Myers, M. L., and Stare, F. J., Dietary carbohydrates and serum cholesterol levels in man, *Am. J. Clin. Nutr.*, 18, 237–242, 1966.

82. **Mann, J. I., Watermeyer, G. S., Manning, E. B., Randles, J., and Truswell, A. S.,** Effects on serum lipids of different dietary fats associated with a high sucrose diet, *Clin. Sci.,* 44, 601–604, 1973.

83. **Irwin, M. I., Taylor, D. D., and Feeley, R. M.,** Serum lipid levels, fat, nitrogen, and mineral metabolism of young men associated with kind of dietary carbohydrate, *J. Nutr.,* 82, 338–342, 1964.

84. **Akinyanju, P. A., Qureshi, R. U., Salter, A. J., and Yudkin, J.,** Effect of an "atherogenic" diet containing starch or sucrose on the blood lipids of young men, *Nature,* 218, 975–977, 1968.

85. **Roberts, A. M.,** Effects of a sucrose-free diet on the serum-lipid levels of men in Antarctica, *Lancet,* 1, 1201–1204, 1973.

86. **Bruckdorfer, K. R., Kang, S. S., and Yudkin, J.,** The hyperlipaemic effect of sucrose in male and female rats, *Proc. Nutr. Soc.,* 31, 11A, 1971.

87. **Chevalier, M. M., Wiley, J. H., and Leveille, G. A.,** The age-dependent response of serum triglycerides to dietary fructose, *Proc. Soc. Exp. Biol. Med.,* 139, 220–222, 1972.

88. **Bruckdorfer, K. R., Kari-Kari, B. P. B., Khan, I. H., and Yudkin, J.,** Activity of lipogenic enzymes and plasma triglyceride levels in the rat and the chicken as determined by the nature of the dietary fat and dietary carbohydrate, *Nutr. Metab.,* 14, 228–237, 1972.

89. **Macdonald, I. and Roberts, J. B.,** The incorporation of various C^{14} dietary carbohydrates into serum and liver lipids, *Metab. Clin. Exp.,* 14, 991–999, 1965.

90. **Waterman, R. A., Romsos, D. R., Tsai, A. C., Miller, E. R., and Leveille, G. A.,** Effects of dietary carbohydrate source on growth, plasma metabolites and lipogenesis in rats, pigs and chicks, *Proc. Soc. Exp. Biol. Med.,* 150, 220–225, 1975.

91. **Bar-On, H. and Stein, Y.,** Effect of glucose and fructose administration on lipid metabolism in the rat, *J. Nutr.,* 94, 95–105, 1968.

92. **Macdonald, I. and Roberts, J. B.,** The serum lipid response of baboons to various carbohydrate meals, *Metab. Clin. Exp.,* 16, 572–579, 1967.

93. **Coltart, T. M. and Crossley, J. N.,** Influence of dietary sucrose on glucose and fructose tolerance and triglyceride synthesis in the baboon, *Clin. Sci.,* 38, 427–437, 1970.

94. **Coltart, T. M. and Macdonald, I.,** Effect of sex hormones on fasting serum triglycerides in baboons given high sucrose diets, *Br. J. Nutr.,* 25, 323–331, 1971.

95. **Rifkind, B. M., Lawson, D. H., and Gale, M.,** Effect of short term sucrose restriction on serum-lipid levels, *Lancet,* 2, 1379–1381, 1966.

96. **Mann, J. I., Hendricks, D. A., Truswell, A. S., and Manning, E.,** Effects on serum-lipids in normal men of reducing dietary sucrose or starch for five months, *Lancet,* 1, 870–872, 1970.

97. **Antar, M. A. and Ohlson, M. A.,** Effect of simple and complex carbohydrates upon total lipids, nonphospholipids and different fractions of phospholipids of serum in young men and women, *J. Nutr.,* 85, 329–337, 1965.

98. **Hodges, R. E., Krehl, W. A., Stone, D. B., and Lopez, A.,** Dietary carbohydrates and low cholesterol diet: Effects on serum lipids in man, *Am. J. Clin. Nutr.,* 20, 198–208, 1967.

99. **Macdonald, I.,** Influences of fructose and glucose on serum lipid levels in men and pre- and post-menopausal women, *Am. J. Clin. Nutr.,* 18, 369–372, 1966.

100. **Beveridge, J. M. R., Jagannathan, S. N., and Connell, W. F,** The effect of the type and amount of dietary fat on the level of plasma triglycerides in human subjects in the postabsorptive state, *Can. J. Biochem.,* 42, 999–1003, 1964.

101. **Macdonald, I.,** Inter-relationship between the influences of dietary carbohydrates and fats on fasting serum lipids, *Am. J. Clin. Nutr.,* 20, 345–351, 1967.

102. **Fredrickson, D. S., Levy, R. I., and Lees, R. S.,** Fat transport in lipoproteins – an integrated approach to mechanisms and disorders, *N. Engl. J. Med.,* 276, 273–281, 1967.

103. **Wood, P. D. S., Stern, M. P., Silver, A., Reaven, G. M., and von der Groeben, J.,** Prevalence of plasma lipoprotein abnormalities in a free-living population of the Central Valley, California, *Circulation,* 45, 114–126, 1972.

104. **Kaufmann, N. A., Poznanski, R., Blondheim, S. H., and Stein, Y.,** Changes in serum lipid levels of hyperlipemic patients following the feeding of starch, sucrose and glucose, *Am. J. Clin. Nutr.,* 18, 261–269, 1966.

105. **Kuo, P. T., Feng, L., Cohen, N. N., Fitts, W. T., Jr., and Miller, L. D.,** Dietary carbohydrates in hyperlipemia (hyperglyceridemia); hepatic and adipose tissue lipogenic function, *Am. J. Clin. Nutr.,* 20, 116–125, 1967.

106. **Little, J. A., Birchwood, B. L., Simmons, D. A., Antar, M. A., Kallos, A., Buckley, G. C., and Csima, A.,** Interrelationship between the kinds of dietary carbohydrate and fat in hyperlipoproteinemic patients. I. Sucrose and starch with polyunsaturated fat, *Atherosclerosis,* 11, 173–181, 1970.

107.   Antar, M. A., Little, J. A., Lucas, C., Buckley, G. C., and Csima, A., Interrelationship between the kinds of dietary carbohydrate and fat in hyperlipoproteinemic patients. III. Synergistic effect of sucrose and animal fat on serum lipids, *Atherosclerosis,* 11, 191–201, 1970.

108.   Smith, U., Cahlin, E., and Scherstén, T., Sucrose feeding in man. Effects on lipolysis and antilipolytic action of insulin in the adipose tissue, *Acta Med. Scand.,* 194, 147–150, 1973.

109.   Nikkilä, E. A., Influence of dietary fructose and sucrose on serum triglycerides in hypertriglyceridemia and diabetes, in *Sugars in Nutrition,* Sipple, H. and McNutt, K., Eds., Academic Press, New York, 1974, 439–448.

110.   Hall, C. E. and Hall, O., Comparative effectiveness of glucose and sucrose in enhancement of hypersalimentation and salt hypertension, *Proc. Soc. Exp. Biol. Med.,* 123, 370–374, 1966.

111.   Beebe, C. G., Schemmel, R., and Mickelsen, O., Blood pressure of rats as affected by diet and concentration of NaCl in drinking water, *Proc. Soc. Exp. Biol. Med.,* 151, 395–399, 1976.

112.   Ahrens, R. A., in *Sweeteners Issues and Uncertainties,* National Academy of Sciences, Washington, D.C., 1975, 96–99.

113.   Schafer, W. G., The caries-producing capacity of starch, glucose, and sucrose diets in the Syrian hamster, *Science,* 110, 143–144, 1949.

114.   Haldi, J., Wynn, W., Shaw, J. H., and Sognnaes, R. F., The relative cariogenicity of sucrose when ingested in the solid form and in solution by the albino rat, *J. Nutr.,* 49, 295–305, 1953.

115.   Stephan, R. M., Effects of different types of human foods on dental health in experimental animals, *J. Dent. Res.,* 45, 1551–1561, 1966.

116.   Grenby, T. H., Investigations in experimental animals on the carogenicity of diets containing sucrose and/or starch, *Caries Res.,* 1, 208–221, 1967.

117.   Frostell, G., Keyes, P. H., and Larson, R. H., Effect of various sugars and sugar substitutes on dental caries in hamsters and rats, *J. Nutr.,* 93, 65–76, 1967.

118.   Grenby, T. H. and Hutchison, J. B., The effects of diets containing sucrose, glucose or fructose on experimental dental caries in two strains of rats, *Arch. Oral Biol.,* 14, 373–380, 1969.

119.   Gustafsson, B. E., Quensel, C.-E., Lanke, L. S., Lundqvist, C., Grahnen, H., Bonow, B. E., and Krasse, B., The Vipeholm dental caries study, *Acta Odontol. Scand.,* 11, 232–364, 1954.

120.   Winter, G. B., Hamilton, M. C., and James, P. M. C., Role of the comforter as an aetiological factor in rampant caries of the deciduous dentition, *Arch. Dis. Child.,* 41, 207–212, 1966.

121.   Kroll, R. G. and Stone, J. H., Nocturnal bottle-feeding as a contributory cause of rampant dental caries in the infant and young child, *J. Dent. Child.,* 30, 454–459, 1967.

122.   Bibby, B. G., The cariogenicity of snack foods and confections, *J. Am. Dent. Assoc.,* 90, 121–132, 1975.

123.   Newbrun, E., Sucrose, the arch criminal of dental caries, *J. Dent. Child.,* 36, 239–248, 1969.

124.   Marthaler. T. M. and Froesch, E. R., Hereditary fructose intolerance. Dental status of eight patients, *Br. Dent. J.,* 123, 597–599, 1967.

125.   Mavrias, D. A. and Mayer, R. J., Metabolism of fructose in the small intestine. I. The effect of fructose feeding on fructose transport and metabolism in rat small intestine, *Biochim. Biophys. Acta,* 291, 531–537, 1973.

126.   Crouzoulon-Bourcart, C., Crouzoulon, G., and Pérès, G., Recherches sur l'absorption intestinale des hexoses. II. Effets de quelques regimes simples sur l'absorption intestinale du fructose in vivo, *C.R. Soc. Biol.,* (Paris), 165, 1071–1074, 1971.

127.   Reiser, S. and Christiansen, P. A., Intestinal transport of amino acids as affected by sugars, *Am. J. Physiol.,* 216, 915–924, 1969.

128.   Reiser, S. and Christiansen, P. A., The properties of the preferential uptake of L-leucine by isolated intestinal epithelial cells, *Biochim. Biophys. Acta,* 225, 123–139, 1971.

129.   Reiser, S., Michaelis, O. E., IV, and Hallfrisch, J., Effects of sugars on leucine and lysine uptake by intestinal cells from rats fed sucrose and stock diets, *Proc. Soc. Exp. Biol. Med.,* 150, 110–114, 1975.

130.   Alvarado, F., Amino acid transport in hamster small intestine: Site of inhibition by galactose, *Nature,* 219, 276–277, 1968.

131.   Hill, R., Baker, N., and Chaikoff, I. L., Altered metabolic patterns induced in the normal rat by feeding an adequate diet containing fructose as sole carbohydrate, *J. Biol. Chem.,* 209, 705–716, 1954.

132.   Curry, D. L., Curry, K. P., and Gomez, M., Fructose potentiation of insulin secretion, *Endocrinology,* 91, 1493–1498, 1972.

133.   Merry, J. and Marks, V., Effect on performance of reducing blood-alcohol with oral fructose, *Lancet,* 2, 1328–1330, 1967.

134.   Patel, A. R., Paton, A. M., Rowan, T., Lawson, D. H., and Linton, A. L., Clinical studies on the effect of laevulose on the rate of metabolism of ethyl alcohol, *Scott. Med. J.,* 14, 268–271, 1969.

135. **Brown, S. S., Forrest, J. A. H., and Roscoe, P.,** A controlled trial of fructose in the treatment of acute alcoholic intoxication, *Lancet,* 2, 898–900, 1972.
136. **Chevalier, M. M., Wiley, J. H., and Leveille, G. A.,** Effect of dietary fructose on fatty acid synthesis in adipose tissue and liver of the rat, *J. Nutr.,* 102, 337–342, 1972.
137. **Bruckdorfer, K. R., Khan, I. H., and Yudkin, J.,** Fatty acid synthetase activity in the liver and adipose tissue of rats fed with various carbohydrates, *Biochem. J.,* 129, 439–446, 1972.
138. **Sugawa-Katayama, Y. and Morita, N.,** Effects of a high fructose diet on lipogenic enzyme activities in some organs of rats fed ad libitum, *J. Nutr.,* 105, 1377–1383, 1975.
139. **Maruhama, Y. and Macdonald, I.,** Some changes in the triglyceride metabolism of rats on high fructose or glucose diets, *Metab. Clin. Exp.,* 21, 835–842, 1972.
140. **Waddell, M. and Fallon, H. J.,** The effect of high-carbohydrate diets on liver triglyceride formation in the rat, *J. Clin. Invest.,* 52, 2725–2731, 1973.
141. **Baron, P., Griffaton, G., and Lowy, R.,** Metabolic inductions in the rat after an intraperitoneal injection of fructose and glucose, according to the nature of the dietary carbohydrate. II. Modification after seven months of diet, *Enzyme,* 12, 481–498, 1971.
142. **Zakim, D., Pardini, R. S., Herman, R. H., and Sauberlich, H. E.,** Mechanism for the differential effects of high carbohydrate diet on lipogenesis in the rat liver, *Biochim. Biophys. Acta,* 144, 242–251, 1967.
143. **Stifel, F. B., Rosensweig, N. S., Zakim, D., and Herman, R. H.,** Dietary regulation of glycolytic enzymes. I. Adaptive changes in rat jejunum, *Biochim. Biophys. Acta,* 170, 221–227 1968.
144. **Rosensweig, N. S., Stifel, F. B., Herman, R. H., and Zakim, D.,** Dietary regulation of the glycolytic enzymes. II. Adaptive changes in human jejunum, *Biochim. Biophys. Acta,* 170, 228–234, 1968.
145. **Shakespeare, P., Srivastava, L. M., and Hübscher, G.,** Glucose metabolism in the mucosa of the small intestine, *Biochim. J.,* 111, 63–67, 1969.
146. **Nikkilä, E. A. and Ojala, K.,** Induction of hypertriglyceridemia by fructose in the rat, *Life Sci.,* 4, 937–943, 1965.
147. **Pereira, J. N. and Jangaard, N. O.,** Different rates of glucose and fructose metabolism in rat liver tissue *in vitro, Metab. Clin. Exp.,* 20, 392–400, 1971.
148. **Cryer, A., Riley, S. E., Williams, E. R., and Robinson, D. S.,** Effects of fructose, sucrose and glucose feeding on plasma insulin concentrations and on adipose-tissue clearing-factor lipase activity in the rat, *Biochem. J.,* 140, 561–563, 1974.
149. **Kaufmann, N. A., Poznanski, R., Blondheim, S. H., and Stein, Y.,** Effect of fructose, glucose, sucrose and starch on serum lipids in carbohydrate induced hypertriglyceridemia and in normal subjects, *Isr. J. Med. Sci.,* 2, 715–726, 1966.
150. **Zakim, D., Herman, R. H., and Gordon, W. C., Jr.,** The conversion of glucose and fructose to fatty acids in the human liver, *Biochem. Med.,* 2, 427–437, 1969.
151. **Mäenpää, P. H., Raivio, K. O., and Kekomäki, M. P.,** Liver adenine nucleotides: Fructose-induced depletion and its effect on protein synthesis, *Science,* 161, 1253–1254, 1968.
152. **Burch, H. B., Max, P., Jr., Chyu, K., and Lowry, O. H.,** Metabolic intermediates in liver of rats given large amounts of fructose or dihydroxyacetone, *Biochem. Biophys. Res. Commun.,* 34, 619–626, 1969.
153. **Woods, H. F., Eggleston, L. V., and Krebs, H. A.,** The cause of hepatic accumulation of fructose 1-phosphate on fructose loading, *Biochem. J.,* 119, 501–510, 1970.
154. **Romsos, D. R. and Leveille, G. A.,** Effect of dietary fructose on in vitro and in vivo fatty acid synthesis in the rat, *Biochim. Biophys. Acta,* 360, 1–11, 1974.
155. **Bode, C., Schumacher, H., Goebell, H., Zelder, O., and Pelzel, H.,** Fructose induced depletion of liver adenine nucleotides in man, *Horm. Metab. Res.,* 3, 289–290, 1971.
156. **Adachi, F., Yu, D. T., and Phillips, M. J.,** An ultrastructural study of fructose-induced hepatic cell injury, *Virchows Arch. B:,* 10, 200–209, 1972.
157. **Perheentupa, J. and Raivio, K.,** Fructose-induced hyperuricaemia, *Lancet,* 2, 528–531, 1967.
158. **Heuckenkamp, P.-U. and Zöllner, N.,** Fructose-induced hyperuricaemia, *Lancet,* 1, 808–809, 1971.
159. **Fox, I. H. and Kelley, W. N.,** Studies on the mechanism of fructose-induced hyperuricemia in man, *Metab. Clin. Exp.,* 21, 713–721, 1972.
160. **Narins, R. G., Weisberg, J. S., and Myers, A. R.,** Effects of carbohydrates on uric acid metabolism, *Metab. Clin. Exp.,* 23, 455–465, 1974.
161. **Raivio, K. O., Becker, M. A., Meyer, L. J., Greene, M. L., Nuki, G., and Seegmiller, J. E.,** Stimulation of human purine synthesis de novo by fructose infusion, *Metab. Clin. Exp.,* 24, 861–869, 1975.
162. **Emmerson, B. T.,** Effect of oral fructose on urate production, *Ann. Rheum. Dis.,* 33, 276–280, 1974.

163. Jacobs, D., Hyperuricaemia and myocardial infarction, *S. Afr. Med. J.*, 46, 367–369, 1972.
164. Sahebjami, H. and Scalettar, R., Effects of fructose infusion on lactate and uric acid metabolism, *Lancet,* 1, 366–369, 1971.
165. Cook, G. C. and Jacobson, J., Individual variation in fructose metabolism in man, *Br. J. Nutr.,* 26, 187–195, 1971.
166. Tranquada, R., Lactic acidosis, *Calif. Med.,* 101, 450–461, 1964.
167. Gray, G. M., Intestinal digestion and maldigestion of dietary carbohydrates, *Annu. Rev. Med.,* 22, 391–404, 1971.
168. Rosensweig, N. S., Adult lactase deficiency: Genetic control or adaptive response, *Gastroenterology,* 60, 464–467, 1971.
169. Bayless, T. M. and Rosensweig, N. S., Topics in clinical medicine: Incidence and implications of lactase deficiency and milk intolerance in white and negro populations, *J.A.M.A.,* 197, 968–972, 1966.
170. Cook, G. C. and Kajubi, S. K., Tribal incidence of lactase deficiency in Uganda, *Lancet,* 1, 725–730, 1966.
171. Leichter, J., Lactose tolerance in a Jewish population, *Digestive Diseases,* 16, 1123–1126, 1971.
172. Cuatrecasas, P., Lockwood, D. H., and Caldwell, J. R., Lactase deficiency in the adult, *Lancet,* 1, 14–18, 1965.
173. Newcomer, A. D. and McGill, D. B., Disaccharidase activity in the small intestine: Prevalence of lactase deficiency in 100 healthy subjects, *Gastroenterology,* 53, 881–889, 1967.
174. Keusch, G. T., Troncale, F. J., Thavaramara, B., Prinyanout, P., Anderson, P. R., and Bhamarapravathi, N., Lactase deficiency in Thailand: Effect of prolonged lactose feeding, *Am. J. Clin. Nutr.,* 22, 638–641, 1969.
175. Gilat, T., Russo, S., Gelman-Malachi, E., and Aldor, T. A. M., Lactase in man: A nonadaptable enzyme, *Gastroenterology,* 62, 1125–1127, 1972.
176. Kogut, M. D., Donnell, G. N., and Shaw, K. N. F., Studies of lactose absorption in patients with galactosemia, *J. Pediatr.,* 71, 75–81, 1967.
177. Rosensweig, N. S. and Herman, R. H., Diet and disaccharidases, *Am. J. Clin. Nutr.,* 22, 99–102, 1969.
178. Bolin, T. D., McKern, A., and Davis, A. E., The effect of diet on lactase activity in the rat, *Gastroenterology,* 60, 432–437, 1971.
179. Goldstein, R., Klein, T., Freier, S., and Menczel, J., Alkaline phosphatase and disaccharidase activities in the rat intestine from birth to weaning. I. Effect of diet on enzyme development, *Am. J. Clin. Nutr.,* 24, 1224–1231, 1971.
180. Christopher, N. L. and Bayless, J. M., Role of the small bowel and colon in lactose-induced diarrhea, *Gastroenterology,* 60, 845–852, 1971.
181. Littman, A. and Hammond, J. B., Diarrhea in adults caused by deficiency in intestinal disaccharidases, *Gastroenterology,* 48, 237–249, 1965.
182. Hansen, R. G., Bretthauer, R. K., Mayes, J., and Nordin, J. H., Estimation of frequency of occurrence of galactosemia in the population, *Proc. Soc. Exp. Biol. Med.,* 115, 560–563, 1964.
183. Mitchell, H. S. and Dodge, W. M., Cataract in rats fed on high lactose rations, *J. Nutr.,* 9, 37–49, 1935.
184. Yudkin, A. M. and Arnold, C. H., Cataracts produced in albino rats on ration containing a high proportion of lactose or galactose, *Arch. Ophthalmol.,* 14, 960–966, 1935.
185. Friedenwald, J. S. and Rytel, D., Contributions to the histopathology of cataract, *AMA Arch. Ophthalmol.,* 53, 825–831, 1955.
186. van Heyningen, R., Formation of polyols by the lens of the rat with sugar cataract, *Nature,* 184, 194–195, 1959.
187. Kinoshita, J. H., Merola, L. O., Satolv, K., and Dikmak, E., Osmotic changes caused by the accumulation of dulcitol in the lenses of rats fed with galactose, *Nature,* 194, 1085–1087, 1962.
188. Kinoshita, J. H., Cataracts in galactosemia, *Invest. Ophthalmol.,* 4, 786–799, 1965.
189. Rutter, W. J., Krichevsky, P., Scott, H. M., and Hansen, R. G., The metabolism of lactose and galactose in the chick, *Poult. Sci.,* 32, 706–715, 1953.
190. Nordin, J. H., Wilken, D. R., Bretthauer, R. K., Hansen, R. G., and Scott, H. M., A consideration of galactose toxicity in male and female chicks, *Poult. Sci.,* 39, 802–812, 1960.
191. Rigdon, R. H., Couch, J. R., Creger, C. R., and Ferguson, T. M., Galactose intoxication pathologic study in the chick, *Experientia,* 19, 349–352, 1963.
192. Mayes, J. S., Miller, L. R., and Myers, F. K., The relationship of galactose-1-phosphate accumulation and uridyl transferase activity to the differential galactose toxicity in male and female chicks, *Biochem. Biophys. Res. Commun.,* 39, 661–665, 1970.

193. **Parkhurst, G. W. and Mayes, J. S.,** Galactose toxicity and activities of the galactose-metabolizing enzymes during development of the chick, *Arch. Biochem. Biophys.,* 150, 742–745, 1972.
194. **Malone, J. I., Wells, H. J., and Segal, S.,** Galactose toxicity in the chick: Hyperosmolality, *Science,* 174, 952–954, 1971.
195. **Maccario, M., Messio, C. P., and Vastola, E. F.,** Focal seizures as a manifestation of hyperglycemia without Ketoacidosis. A report of seven cases with review of the literature, *Neurology,* 15, 195–206, 1965.
196. **Maccario, M.,** Neurological dysfunction associated with nonketotic hyperglycemia, *Arch. Neurol.,* 19, 525–534, 1968.
197. **Wierzuchowski, M.,** Overflow diabetes and toxic phenomena due to the infusion of glucose in normal dogs, *J. Physiol.,* 87, 85P–86P, 1936.
198. **Kozak, L. P. and Wells, W. W.,** Effect of galactose on energy and phospholipid metabolism in the chick brain, *Arch. Biochem. Biophys.,* 135, 371–377, 1969.
199. **Malone, J. I., Wells, H. J., and Segal, S.,** Galactose toxicity in the chick: Hyperosmolality or depressed energy reserves?, *Science,* 176, 816–817, 1972.
200. **Knull, H. R., Wells, W. W., and Kozak, L. P.,** Galactose toxicity in the chick: Hyperosmolality or depressed brain energy reserves?, *Science,* 176, 815–816, 1972.
201. **Wells, W. W., Pittman, T. A., Wells, H. J., and Egan, T. J.,** The isolation and identification of galactitol from the brains of galactosemia patients, *J. Biol. Chem.,* 240, 1002–1004, 1965.
202. **Quan-Ma, R. and Wells, W. W.,** The distribution of galactitol in tissues of rats fed galactose, *Biochem. Biophys. Res. Commun.,* 20, 486–490, 1965.
203. **Wells, H. J. and Wells, W. W.,** Galactose toxicity and myoinositol metabolism in the developing rat brain, *Biochemistry,* 6, 1168–1173, 1967.
204. **Wild, G., Woolley, D. W., and Gommi, B. W.,** Effects of experimental galactosemia on the measured serotonin receptor activity of rat brain, *Biochemistry,* 6, 1671–1675, 1967.
205. **Woolley, D. W. and van der Hoeven, T.,** Serotonin deficiency in infancy as a cause of a mental defect in experimental phenylketonuria, *Int. J. Neuropsychiatry,* 1, 529–544, 1965.
206. **Trowell, H. C.,** Crude fibre, dietary fibre and atherosclerosis, *Atherosclerosis,* 16, 138–140, 1972.
207. **Kent-Jones, D. W. and Amos, A. J.,** General analytical procedure for cereals, in *Modern Cereal Chemistry,* 6th ed., Food Trade Press, London, 1967, 564.
208. **Southgate, D. A. T.,** Determination of carbohydrates in foods. II. Unavailable carbohydrates, *J. Sci. Food Agric.,* 20, 331–335, 1969.
209. **Hardinge, M. G., Chambers, A. C., Crooks, H., and Stare, F. J.,** Nutritional studies of vegetarians. III. Dietary levels of fiber, *Am. J. Clin. Nutr.,* 6, 523–525, 1958.
210. **Trowell, H.,** Ischemic heart disease and dietary fiber, *Am. J. Clin. Nutr.,* 25, 926–932, 1972.
211. **Cummings, J. H.,** Progress report dietary fibre, *Gut,* 14, 69–81, 1973.
212. **Scala, J.,** Fiber the forgotten nutrient, *Food Technol.* (Chicago), 28, 34–36, 1974.
213. **Lubbe, A. M. A.,** A comparative study of rural and urban Venda males: Dietary evaluation, *S. Afr. Med. J.,* 45, 1289–1297, 1971.
214. **Harvey, R. F., Pomare, E. W., and Heaton, K. W.,** Effects of increased dietary fibre on intestinal transit, *Lancet,* 1, 1278–1280, 1973.
215. **Payler, D. K.,** Food fibre and bowel behavior, *Lancet,* 1, 1394, 1973.
216. **Payler, D. K., Pomare, E. W., Heaton, K. W., and Harvey, R. F.,** The effect of wheat bran on intestinal transit, *Gut,* 16, 209–213, 1975.
217. **Findlay, J. M., Smith, A. N., Mitchell, W. D., Anderson, A. J. B., and Eastwood, M. A.,** Effects of unprocessed bran on colon function in normal subjects and in diverticular disease, *Lancet,* 1, 146–149, 1974.
218. **Brodribb, A. J. M. and Humphreys, D. M.,** Diverticular disease: Three studies, *Br. Med. J.,* 1, 424–430, 1976.
219. **Taylor, I. and Duthie, H. L.,** Bran tablets and diverticular disease, *Br. Med. J.,* 1, 988–990, 1976.
220. **Kirwan, W. O., Smith, A. N., McConnell, A. A., Mitchell, W. D., and Eastwood, M. A.,** Action of different bran preparations on colonic function, *Br. Med. J.,* 4, 187–189, 1974.
221. **McCance, R. A., Prior, K. M., and Widdowson, E. M.,** A radiological study of the rate of passage of brown and white bread through the digestive tract of man, *Br. J. Nutr.,* 7, 98–106, 1953.
222. **Painter, N. S., Almeida, A. Z., and Colebourne, K. W.,** Unprocessed bran in treatment of diverticular disease of the colon, *Br. Med. J.,* 2, 137–140, 1972.
223. **Walters, R. L., Baird, I. M., Davies, P. S., Hill, M. J., Drasar, B. S., Southgate, D. A. T., Green, J., and Morgan, B.,** Effects of two types of dietary fibre on faecal steroid and lipid excretion, *Br. Med. J.,* 2, 536–538, 1975.

224. **Walker, A. R. P.,** Effect of high crude fiber intake on transit time and the absorption of nutrients in South African Negro schoolchildren, *Am. J. Clin. Nutr.,* 28, 1161–1169, 1975.

225. **Burkitt, D. P.,** Epidemiology of cancer of the colon and rectum, *Cancer,* 28, 3–13, 1971.

226. **Hill, M. J., Crowther, J. S., Drasar, B. S., Hawksworth, G., Aries, V., and Williams, R. E. O.,** Bacteria and aetiology of cancer of large bowel, *Lancet,* 1, 95–100, 1971.

227. **Hill, M. J.,** Steroid nuclear dehydrogenation and colon cancer, *Am. J. Clin. Nutr.,* 27, 1475–1480, 1974.

228. **Drasar, B. S. and Irving, D.,** Environmental factors and cancer of the colon and breast, *Br. J. Cancer,* 27, 167–172, 1973.

229. **Wells, A. F. and Ershoff, B. H.,** Beneficial effects of pectin in prevention of hyper-cholesterolemia and increase in liver cholesterol in cholesterol-fed rats, *J. Nutr.,* 74, 87–92, 1961.

230. **Wells, A. F. and Ershoff, B. H.,** Comparative effects of pectin N. F. administration on the cholesterol-fed rabbit, guinea pig, hamster and rat, *Proc. Soc. Exp. Biol. Med.,* 111, 147–149, 1962.

231. **Leveille, G. A. and Sauberlich, H. E.,** Mechanism of the cholesterol-depressing effect of pectin in the cholesterol-fed rat, *J. Nutr.,* 88, 209–214, 1966.

232. **Phillips, W. E. J. and Brien, R. L.,** Effect of pectin, a hypocholesterolemic polysaccharide, on vitamin A utilization in the rat, *J. Nutr.,* 100, 289–292, 1970.

233. **Fisher, H., Griminger, P., and Weiss, H. S.,** Avian atherosclerosis: Retardation by pectin, *Science,* 146, 1063–1064, 1964.

234. **Fahrenbach, M. J., Riccardi, B. A., and Grant, W. C.,** Hypocholesterolemic activity in mucilaginous polysaccharides in White Leghorn cockerels, *Proc. Soc. Exp. Biol. Med.,* 123, 321–326, 1966.

235. **de Groot, A. P., Luyken, R., and Pikaar, N. A.,** Cholesterol-lowering effect of rolled oats, *Lancet,* 2, 303–304, 1963.

236. **Morgan, B., Heald, M., Atkin, S. D., and Green, J.,** Dietary fibre and sterol metabolism in the rat, *Br. J. Nutr.,* 32, 447–455, 1974.

237. **Balmer, J. and Zilversmit, D. M.,** Effect of dietary roughage on cholesterol absorption, cholesterol turnover and steroid excretion in the rat, *J. Nutr.,* 104, 1319–1328, 1974.

238. **Antonis, A. and Bersohn, I.,** The influence of diet on fecal lipids in South African white and Bantu prisoners, *Am. J. Clin. Nutr.,* 11, 142–155, 1962.

239. **Mathur, K. S., Kahn, M. A., and Sharma, R. D.,** Hypocholesterolaemic effect of Bengal gram: A long-term study in man, *Br. Med. J.,* 1, 30–31, 1968.

240. **Shurpalekar, K. S., Doraiswamy, T. R., Sundaravalli, O. E., and Narayana, R.,** Effect of inclusion of cellulose in an "atherogenic" diet on the blood lipids of children, *Nature,* 232, 554–555, 1971.

241. **Keys, A., Grande, F., and Anderson, J. T.,** Fiber and pectin in the diet and serum cholesterol concentration in man, *Proc. Soc. Exp. Biol. Med.,* 106, 555–558, 1961.

242. **Palmer, G. H. and Dixon, D. G.,** Effect of pectin dose on serum cholesterol levels, *Am. J. Clin. Nutr.,* 18, 437–442, 1966.

243. **Jenkins, D. J. A., Leeds, A. R., Newton, C., and Cummings, J. H.,** Effect of pectin, guar gum, and wheat fibre on serum-cholesterol, *Lancet,* 1, 1116–1117, 1975.

244. **Durrington, P. N., Manning, A. P., Bolton, C. H., and Hartog, M.,** Effect of pectin on serum lipids and lipoproteins, whole-gut transit-time, and stool weight, *Lancet,* 2, 394–396, 1976.

245. **Huth, K. and Fettel, M.,** Bran and blood-lipids, *Lancet,* 2, 456, 1975.

246. **Eastwood, M. A.,** Dietary fibre and serum-lipids, *Lancet,* 2, 1222–1224, 1969.

247. **Eastwood, M. A., Kirkpatrick, J. R., Mitchell, W. D., Bone, A., and Hamilton, T.,** Effects of dietary supplements of wheat bran and cellulose on faeces and bowel function, *Br. Med. J.,* 4, 392–394, 1973.

248. **Heaton, K. W. and Pomare, E. W.,** Effect of bran on blood lipids and calcium, *Lancet,* 1, 49–50, 1974.

249. **Bremner, W. F., Brooks, P. M., Third, J. L. H. C., and Lawrie, T. D. V.,** Bran in triglyceridemia: A failure of response, *Br. Med. J.,* 3, 574, 1975.

250. **Connel, A. M., Smith, C. L., and Somsel, M.,** Absence of effect of bran on blood-lipids, *Lancet,* 1, 496–497, 1975.

251. **Durrington, P., Wicks, A. C. B., and Heaton, K. W.,** Effect of bran on blood-lipids, *Lancet,* 2, 133, 1975.

252. **Jenkins, D. J. A., Hill, M. S., and Cummings, J. H.,** Effect of wheat fiber on blood lipids, fecal steroid excretion and serum iron, *Am. J. Clin. Nutr.,* 28, 1408–1411, 1975.

253. **Luyken, R., Pikaar, N. A., Polman, H., and Schippers, F. A.,** The influence of legumes on the serum cholesterol level, *Voeding,* 23, 447–453, 1962.

254. **Heaton, K. W.,** Food fiber as an obstacle to energy intake, *Lancet,* 2, 1418–1421, 1973.

255. McCance, R. A. and Widdowson, E. M., Mineral metabolism on dephytinized bread, *J. Physiol.,* 101, 304–313, 1942.
256. McCance, R. A. and Widdowson, E. M., Mineral metabolism of healthy adults on white and brown bread dietaries, *J. Physiol.,* 101, 44–85, 1942.
257. Persson, I., Raby, K., Fonns-Bech, P., and Jensen, E., Bran and blood-lipids, *Lancet,* 2, 1208, 1975.
258. Reinhold, J. G., Parsa, A., Karimian, N., Hammick, J. W., and Ismail-Beigi, F., Availability of zinc in leavened and unleavened whole meal wheaten breads as measured by solubility and uptake by rat intestine in vitro, *J. Nutr.,* 104, 976–982, 1974.
259. Reinhold, J. G., Ismail-Beigi, F., and Faradji, B., Fibre vs phytate as determinant of the availability of calcium, zinc and iron of breadstuffs, *Nutr. Rep. Int.,* 12, 75–85, 1975.
260. Reinhold, J. G., Faradji, B., Abadi, P., and Ismail-Beigi, F., Decreased absorption of calcium, magnesium, zinc and phosphorus by humans due to increased fiber and phosphorus consumption as wheat bread, *J. Nutr.,* 106, 493–503, 1976.
261. Bjorn-Rasmussen, E., Iron absorption from wheat bread, influence of various amounts of bran, *Nutr. Metab.,* 16, 101–110, 1974.
262. Southgate, D. A. T. and Durnin, J. V. G. A., Caloric conversion factors. An experimental reassessment of the factors used in the calculation of the energy value of human diets, *Br. J. Nutr.,* 24, 517–535, 1970.
263. Klose, R. E. and Glicksman, M., Gums, in *Handbook of Food Additives,* 2nd ed., Furia, T. E., Ed., Chemical Rubber Co., Cleveland, 1972, 295–359.
264. Krantz, J. C., Jr., Carr, C. J., and de Farson, C. B., Guar polysaccharide as a precursor of glycogen, *J. Am. Diet. Assoc.,* 24, 212, 1948.
265. Booth, A. N., Hendrickson, A. P., and DeEds, F., Physiologic effects of three microbial polysaccharides on rats, *Toxicol. Appl. Pharmacol.,* 5, 478–484, 1963.
266. Riccardi, B. A. and Fahrenbach, M. J., Effect of guar gum and pectin N.F. on serum and liver lipids of cholesterol-fed rats, *Proc. Soc. Exp. Biol. Med.,* 124, 749–752, 1967.
267. Glicksman, M., Natural plant exudates, in *Gum Technology in the Food Industry,* Academic Press, New York, 1969, 94–129.
268. Stanford Research Institute, Study of Mutagenic Effects of Gum Ghatti (FDA 71-13), Compound Report No. 6 prepared under DHEW Contract No. FDA 71-267, Menlo Park, Calif., 1972, p. 1–29.
269. Hove, E. L. and Herndon, J. E., Growth of rabbits on purified diets, *J. Nutr.,* 63, 193–199, 1957.
270. Nilson, H. W. and Wagner, J. A., Feeding test with carrageenan, *Food Res.,* 24, 235–239, 1959.
271. Tsai, A. C., Elias, J., Kelley, J. J., Lin, R-S.C., and Robinson, J. R. K., Influence of certain dietary fibers on serum and tissue cholesterol levels in rats, *J. Nutr.,* 106, 118–123, 1976.

EFFECT OF PROTEIN TOXICITY OR
EXCESS IN ANIMALS AND MAN

J. C. Dougherty

Proteins in the diet supply the amino acids required for the growth and maintenance of animal tissues and for numerous other metabolic functions. In many parts of the world, the availability of dietary protein is restricted because of its high cost. For this reason, most nutritional studies have emphasized the determination of the minimal daily requirement or computation of the recommended daily allowance of protein. As the income level rises and protein is more available, the question of optimum intake becomes important. High-protein diets are recommended in some weight reduction programs, in training diets for athletes, and for patients recovering from burns and trauma. These are among the situations where the potential harmful or toxic effects of high-protein diets must be considered. Adverse effects could be acute or chronic. They could be systemic or could cause damage to specific organs. The effects could be modified by disease or by physiological changes in the animal or man.

High is a relative term. It can be defined either in relation to the recommended allowance, in relation to the average intake of the population under consideration, or in terms of the absolute proportion of total caloric intake. The basic dietary allowance for the adult man is 0.5 g of protein per kilogram of body weight per day. To this is added allowances for variations in quality, as well as variations in individual metabolism. This brings the recommended allowance to 0.9 g/kg/day.[1] This amounts to 8 to 10% of the caloric intake. The allowance for infants and children is three to four times higher. The quality of a protein is determined by its essential amino acid content. As the proportions of amino acids in the protein approach the composition of tissue, the biological value increases. Eggs, milk, meat, and fish have high biological value.

The pool of labile proteins and amino acids is small. Intake in excess of the amount required for growth or maintenance is metabolized. The amino group is removed and the residue metabolized in the same manner as carbohydrates and fats. The removed ammonia groups are converted into urea (by the liver) for excretion by the kidney. Urea is the main contributor to solute excretion. The consumption of protein above 15% of the total caloric intake leads to increased water requirements. Metabolism of 100 cal of carbohydrate or fat required 50 g of water, while 100 cal of protein requires 350 g of water.[2] The presence of high solute loads in the lumen of the renal tubules interferes with the concentrating mechanisms and results in an osmotic diuresis.

SYSTEMIC TOXICITY

In general, it is quite difficult to demonstrate any toxicity of dietary protein. In most instances, the host must be compromised in some way. In acute feeding experiments, adult rats can consume a diet of 80% casein and survive. When the protein is solubilized by enzyme hydrolysis, a single dose can be lethal. The LD_{50} is 26 ± 2 g/kg.[3] Death is caused by acute dehydration. Administration of undenatured, reconstituted spray-dried egg white produces a similar lethal dehydration syndrome.[4] The excessive amounts of amino acids are metabolized to urea. Its excretion results in an osmotic diuresis. If the experimental animals were given free access to water, it is probable that there would have been no ill effects. Acute feeding experiments have been performed in normal young men by Rafoth and Onstad in an attempt to assess hepatic functional capacity. The test meal consisted of 40% milk protein and 60% soy protein. Test meals of 1, 2, or 4 g of protein per kilogram of lean body mass (60, 120, and 240 g of protein) were fed to hydrated individuals. In addition, one subject ingested 20 g of protein every 20 min for consecutive hours (420 g total). The only ill effects noted were nausea and bloating for 1 to 2 hr.[5]

The concept that high-protein diets are harmful developed in part from the observations of investigators such as Chittenden and Leaf.[6,7] The American physiologist Chittenden kept healthy men on diets containing as little as 40 g of protein per day for years. He asserted that the low-protein diet decreased the strain on the kidneys, reduced vascular disease, and increased physical and mental vigor. The observations of Leaf lend support to this view. In a fascinating account of his visits to communities of long-lived individuals in Ecuador, Pakistan, and the Soviet Caucasus, he described the generally low protein and low energy content of the local diets. Total protein intake is in the range of 35 to 50 g/day. In addition, and perhaps significantly, the contribution of animal fat is also very low. His conclusion was that low-protein diets are beneficial. This view is also held by Eckholm and Record.[8] They have spoken out against the affluent diet and the disease it apparently produces and have called for a return to a more prudent diet with a lower protein content.

The evidence that man can survive and be fit on a low-protein diet does not automatically prove that diets containing large amounts of protein are harmful. Multiple observations support the view that ample dietary protein is not injurious and may be beneficial. The studies of Voit in 1881 and Rubner in 1914 have been cited.[7] Each of these men suggested that 145 g of protein was a suitable daily allowance. They believed that high-protein diets promoted vigor and physical efficiency. Davidson and Passmore also point out that the Australian range rider, the gaucho of the South American plains, and the Massai warrior of Central Africa maintain their health on diets which contain up to 300 g of protein per day. Glatzel states that there are no health problems to be expected in a healthy adult eating a predominantly or exclusively protein diet.[9] He based his opinion upon the extensive observations by scientists at the Norsk Polar Institute in Oslo. They studied Eskimos who subsisted on their natural diet. The Eskimo diet was essentially meat and fish. Negligibly small amounts of greens were consumed. The group was fit and had neither more nor less arteriosclerosis than white people of the same age. The Arctic explorer Stefansson lived and subsisted ten winters and seven summers with the Canadian Eskimos. During this time, his food was strictly of animal origin with the exception of some berries preserved in fish oil and some lichen from the stomachs of slain animals. He reported feeling better and more fit than ever before. Subsequently, he lived on a strictly meat diet under the supervision of the physiologist Dubois at Bellevue Hospital in New York City. There were no objective or subjective problems in regard to his physical fitness.

The level of dietary protein does influence several blood chemistry determinations. Elevations of blood urea nitrogen (BUN), uric acid, and albumin were observed in over 50% of the members of a professional ice hockey team who were on a high-protein diet.[10] The direct correlation between the level of dietary protein and BUN has been shown in rats, dogs, sheep, cows, and man.[11-15] As protein intake increases, the BUN rises. In birds where uric acid is the end product of nitrogen metabolism, a similar response is seen.[16] Rising levels of urea nitrogen have been considered evidence of increased work for the kidney or evidence of renal damage.[17] These opinions can result from a misunderstanding of the mechanisms of urea excretion. In a steady state, urea excretion is equal to urea production. The mechanism of renal excretion is filtration with a variable amount of passive back diffusion dependent upon water reabsorption in the proximal tubule. If urea production increases, the serum level will rise until the filtered load (plasma concentration times glomerular filtration rate) again approximates the amount excreted and a new steady state is established.[18] The glomerular filtration rate (GFR) in mammals is relatively constant. Demands for increased excretion are not met by a rise in filtration rate, but are met, as described, by a rise in serum concentration of the product excreted by the filtration process. In the presence of progressive renal disease, as the GFR falls, the BUN will rise in a person on a constant protein intake. In normal individuals, there is no evidence that high-protein diets impair renal function nor do they increase the work of the kidney. Urea is excreted by filtration. The hydrostatic pressure required for this is independent of the blood urea nitrogen level. There is no requirement for more work or energy consumption in people on a high-protein diet.

Animal feeding studies have been performed to examine some of the relationships between dietary protein and atherosclerosis. Lofland et al. reported on studies in white Carneau pigeons fed synthetic diets containing varying levels of protein, fat, and cholesterol.[19] High-protein diets produced elevated serum cholesterol levels when cholesterol was present in the diet. The experimental evidence suggested that protein and cholesterol interact in determining the level of aorta cholesterol and atherosclerosis index in these birds. Subsequently, Little and Angell performed similar experiments varying only the protein level from 10 to 40%.[20] They found no significant changes in either the serum cholesterol concentration or the atherosclerosis indices. Their study suggests that dietary fat is more important than high protein intake in producing atherosclerotic lesions in these pigeons.

The protein requirement of infants is greater than that of adults because amino acids must be available for tissue growth. Gordon and associates reported that protein feedings of 6 g/kg/day were optimum for premature infants.[21] This was the formula used in the feeding of low birthweight infants in the nursery of The New York Hospital for many years. Subsequently, Davidson and co-workers demonstrated that a diet of 4 g/kg/day was as effective.[22] They also showed that blood urea nitrogen concentrations directly reflected the levels of protein intake. The major hazard of high-protein or overconcentrated diets in infancy is the presentation of an excessive solute load to the functionally immature kidney.[23] The ability to form a concentrated urine is not well developed in the first months of life. The potential hazards have been stressed by Davies.[24] The high anabolic rate of the infants offers some protection, but the renal solute load becomes an important factor when protein intake comprises more than 20% of the caloric intake. Infants on such a diet are at increased risk if they develop illnesses which lead to increased extrarenal fluid losses. Severe dehydration, hypernatremia, and azotemia develop. Mortality rates of 20% can occur and permanent brain damage may result.[24] An example of the tragic consequences has been reported by Abrams et al.[25] A 7-week-old infant was fed undiluted evaporated milk for 5 days. Severe hypertonic dehydration and hyperglycemia developed. Renal failure, disseminated intravascular coagulation, gangrene of the legs, and coma were added complications. These authors recommended that the dangers of overconcentrated formulas in infant feedings should be widely publicized. Infants are at additional risk because they cannot communicate their thirst. The cry of thirst may be misinterpreted as the cry of hunger and more high-solute formula may be offered. This aggravates rather than relieves the problem.

Dehydration syndromes resulting from excessive protein intake and insufficient water intake can also occur in adults and can be life threatening. The tube feeding syndrome is an example.[26,27] The major manifestations are the same as seen in infants: dehydration, hypernatremia, azotemia, and falling level of consciousness. It is called the tube feeding syndrome because it occurs predominantly in unconscious patients given high-protein feedings via nasogastric tube. It can occur also in conscious patients who have difficulty communicating their thirst. Patients who have a tracheostomy coincidental to extensive head and neck surgery for carcinoma of the pharynx or larynx are a high risk group.[27] In their review of the literature, Galut and co-workers found that the protein content of the tube feedings in affected patients varied from 72 to 216 g/day.[26] The average intake was 150 g. Fluid intake averaged 2800 ml/day with a range of 1800 to 5100 ml.

Recognition of severe dehydration may be delayed because the osmotic diuresis produces "adequate" volumes of urine. Oliguria is usually the first indicator of volume depletion. With severe water deficits, the serum sodium concentration may reach toxic levels.[26] Because water is lost from all fluid compartments, the signs of plasma volume depletion, tachycardia and hypotension appear late. If extrarenal fluid losses are also present, severe dehydration will occur more quickly. The mechanism of dehydration is the same as in the acute feeding experiments in rats. The ammonia groups removed from the amino acids are converted to urea. The high urea loads presented to the kidney produce excessive water loss by their osmotic diuretic effect.

The morbidity and mortality which result from high protein intake in these unique situations are not the result of kidney failure. The injury results from an excessive solute load being presented to the kidney in situations where thirst, the major compensatory mechanism, cannot be effectively expressed. Prevention is clearly preferable to treatment. When the syndrome develops, therapy is directed towards decreasing the protein intake and replacing the water deficit safely.[6,7] Special precautions must be taken when tube feedings are given to patients who have preexisting renal disease, are comatose, or who for other reasons cannot communicate their thirst or take effective measures to relieve it.

The acute systemic toxic effects of high protein intake are seen only when water intake is also inadequate. There is little evidence to support the view that higher than adequate levels ingested over long periods increase morbidity or mortality. In some situations, excessive intake can be shown to lead to specific organ damage. More important in clinical medicine are those conditions which so reduce protein tolerance that average or normal levels of intake produce toxic or adverse effects.

SPECIFIC ORGAN DAMAGE

The liver and the kidney are two organs which may be damaged by high protein intake. The fatty liver and kidney syndrome (FLKS) is a disease of unknown etiology which affects growing chickens. It occurs in epidemics. Early studies suggested that it was related to the quality and quantity of dietary protein. Blair et al. fed young chicks diets containing 16.8 to 26.6% protein.[28] There was a higher death rate on the lower protein diets. When the experiment was repeated using only the highest and lowest protein content diets, there was no mortality. From this observation the authors concluded that environmental factors played a significant role. Subsequently, Hemsley and co-workers fed chicks diets containing 16.5% fishmeal from various sources.[29] The mortality rate was near 5% on all rations. They concluded that the protein content of all diets was unsatisfactory.

Whitehead et al. studied the interrelationships between the levels of dietary biotin, fat, and protein.[30] The clinical severity of biotin deficiency was least when dietary protein was low. An increase in protein intake would unmask the biotin deficiency. In this instance, an adequate protein intake brought out the signs of a nutritional deficiency. In this sense then, a normal intake was toxic. The highest levels of dietary protein were fed by Evans and co-workers.[16] They were concerned with the time-course response of several liver and kidney gluconeogenic enzymes in chicks fed high-protein diets. On a fresh weight basis, the protein content was 22.9 and 77.9%, respectively, in the control and experimental groups. The high-protein diet was also carbohydrate free. Chickens fed the experimental diet exhibited an initial increase in relative liver size, total protein, and soluble protein. The liver size and soluble protein levels returned to control levels by the beginning of the second week. Total protein in the liver (mg/100 g body weight) was increased considerably.

Liver hypertrophy has been reported in rats fed high levels of protein.[31] The intentional production of enlarged fatty livers is the result of force feeding geese. Cooked corn is the main foodstuff used for cramming. It may enhance the imbalance of essential amino acids and the protein/energy ratio. In their studies, Nitsan et al. fed diets of 16 and 28% protein during the prefattening phase; in this range there was no effect on body and liver weights.[32] Fattening caused an increase in body, liver, and kidney weights. Both the increase in energy and, to a lesser extent, the protein excess are responsible for the accumulation of triglycerides in the liver.

Adequate or high-protein diets have beneficial effects on renal structure and function. Kidney size is increased in animals and birds. Renal function, as measured by renal blood flow, glomerular filtration rate, and renal concentrating ability, is improved.[11,13,15] The major adverse effect has been described in rats. Protein overload nephropathy may develop. Moise and Smith performed a unilateral nephrectomy on rats and then fed groups a diet containing from 20 to 90% casein.[35] There was practically a direct proportion between the degree of renal hypertrophy and the level of dietary protein. Significant glomerular and tubular changes developed after 90 days on the high-protein diets. Lalich produced glomerular and tubular degeneration in 150 days by feeding a 40% protein diet to animals with a single kidney.[36] Unilateral nephrectomy accelerates the development of glomerular sclerosis but is not a prerequisite.[36]

Commercial diets for rats contain 25% protein. Bras and Ross fed diets containing 30 to 50% casein for 600 days before significant glomerulosclerosis could be observed.[37] Alterations recognized by light microscopy include hyaline droplet formation in tubular cells, hyperplasia of the epithelium, and cast formation. Subsequent studies have defined the ultrastructural details of the renal injury.[38] While the visceral epithelium is undergoing hypertrophy, hyperplasia, cytoplasmic edema formation, and disruption, there is continuous production of basement membrane material by parietal epithelium. Mesangial cells produce more matrix which results in thickening of the basement membrane. Each of these alterations contributes to the eventual obliteration of the urinary spaces in the glomeruli. The pathological changes are not unique to protein overfeeding. Similar changes may be produced by alloxan-induced diabetes, chronic adrenocorticoid administration, and X-irradiation.[36]

The mechanism by which protein overload induces renal pathology has not been clarified. Heredity plays an important role in the susceptibility of various rat strains. Sprague-Dawley and Long-Evans rats develop glomerulosclerosis while Wistar-Furth rats do not.[38,39] Potential mechanisms in susceptible animals include extraneous contaminants in the protein or amino acid imbalances.[35] Some amino acids are nephrotoxic. When injected intraperitoneally, degeneration of proximal tubules and proteinuria are induced.[40] Glycine dietary supplements enhance the nephrotoxicity of commercial and prepared diets. In the uninephrectomized rat model, multiple variables exert an appreciable influence upon glomerular function and morphological change.[41]

Nephrocalcinosis occurs in several strains of rats maintained on semipurified or stock pellet diet.[42] Subsequent studies have demonstrated that increasing the casein content from 25 to 40% reduced the incidence of renal calcification. Decreasing the casein content to 16% increased the severity of the nephrocalcinosis.[43] Acidification of the urine by feeding ammonium chloride produced similar results. The authors suggested that proteins exerted a protective effect by the increased excretion of acid derived from the S-amino acids. Chicks were fed a diet of 77.9% protein by Evans and co-workers.[16] Uric acid clearance increased, an apparent adaptation at the tubular level for increased uric acid secretion. The birds were sacrificed after 12 weeks. Hypertrophy of both the liver and the kidneys was present. On histological examination, there was renal tubular swelling and occasional areas of inflammation and necrosis in some birds. The observed lesions were confined to the juxtamedullary area of the kidney. This damage might be related to the elevated serum uric acid levels.

PROTEIN TOXICITY IN THE PRESENCE OF ORGAN DAMAGE

Normal levels of dietary protein may produce symptoms in patients with acute or chronic liver disease. The metabolic encephalopathy produced is characterized by a variable depression of the level of consciousness, personality changes, a flapping tremor, and increased respiration. A common name of the syndrome, portal-systemic encephalopathy, describes the underlying pathogenesis. Blood is shunted around the liver or past nonfunctioning liver cells. Materials which are ordinarily removed from the portal blood by the liver reach the brain. Of the several chemicals which may be responsible for the encephalopathy, ammonia appears to be the most important and has been the most extensively studied. Blood ammonia is derived from bacterial deamination of amino acids or the hydrolysis of urea. In patients with cirrhosis and chronic liver disease, the level of blood ammonia parallels the severity of the changes in mental function. Ammonia intoxication or hepatic encephalopathy is seen more frequently after operations which divert the portal blood from the bowel into the systemic circulation. These operations are generally performed for the relief of portal hypertension and life-threatening bleeding from esophageal and gastric varices. Because of the limited ability of these cirrhotic patients to metabolize a nitrogen load or detoxify ammonia, therapy is directed towards limiting the availability of these substances.

Restriction of dietary protein is the cornerstone of therapy. In comatose patients, protein can be eliminated completely from the diet or intravenous fluids. Less severe restriction is sufficient for less severe intoxications. Ambulatory patients can generally tolerate 20 g of protein per day. Nonabsorbable antibiotics (neomycin) may be used to suppress the urease-producing bacterial flora in the gut, thus reducing ammonia production from this source. Another line of treatment is the oral administration of lactulose which probably traps ammonia in the colon.[44] As this is accomplished, dietary protein is increased to about 40 g/day. The goal of therapy is to balance protein and ammonia availability with the metabolic tolerance of the patient.

In the presence of renal insufficiency, restriction of dietary protein is indicated. Protein in adequate amounts has been demonstrated to be necessary for optimal renal function; high-protein diets increase the morbidity and mortality of renal insufficiency. Addis recognized the need for dietary restriction in uremic patients.[1] Other clinicians have demonstrated the benefits of selected and restricted protein diets in this same group of patients.[46,47] Burton reviewed the general therapeutic principles.[48] The goal of dietary therapy is to bring into balance the excretory demands placed upon the kidney and its functional capacity. The major excretory products of protein metabolism are urea and hydrogen ion (nonvolatile acids). Urea is excreted by filtration, hydrogen ion by complex tubular secretory mechanisms. As renal mass is lost, the excretory capacity decreases. The level of protein intake must be matched to the excretory capacity so that intolerably high blood levels of urea and severe acidosis are not produced.

Giordano's report in 1963 placed the diet therapy of chronic renal insufficiency on a sound scientific basis.[47] He demonstrated that if the essential amino acids are supplied, then urea can be recycled into protein synthesis. The result is a decrease in BUN, less acidosis, and a disappearance of many uremic symptoms.[49] These modified and selectively restricted protein diets do not improve renal function nor change the course of the renal disease. Symptomatic relief is obtained. Symptoms can be expected to recur when renal function deteriorates further.[6] Since these diets do not change renal function, they are begun when the patient is symptomatic and must be tailored to the individual patient.[50] Of interest is the fact that after the usefulness of dietary therapy was demonstrated in patients with uremia, the same principles were applied to the dietary treatment of dogs with nephritis.[51]

A certain amount of amino acids normally appears in the urine; the total amount is usually less than 2 to 3% of the total urinary nitrogen. Abnormal aminoaciduria results from an acquired or hereditary disturbance of cellular metabolism or transport of amino acids.[52] The severity of the resultant disease is often related to the level of dietary protein. Dietary manipulation can be the cornerstone of medical therapy. It is convenient to classify the inborn errors of amino acid metabolism according to the major metabolic pathways involved. Disorders of phenylalanine and tyrosine, transsulfuration, the branched-chain amino acids, the urea cycle, and the imino acids are the five major groups.

Phenylkeytonuria (PKU) is the prototype of the aminoacidurias associated with mental retardation. This disorder, due to the failure of conversion of phenylalanine to tyrosine, is inherited as an autosomal recessive trait. The major features are mental retardation, microcephaly, convulsions, and failure to thrive. Early detection and treatment with a semisynthetic diet low in phenylalanine are effective in preventing the development of the clinical features. The special diet is required to about age five. Infants born to mothers with the disorder can be assumed to be heterozygotes and would be expected to escape the ravages of classic genetically determined PKU. However, the mother will have elevated serum levels which adversely affect fetal development. The seriousness of this problem is indicated by the studies of Mabry et al.[53] These investigators found that each of the 25 children born to six mothers with the disorder was retarded. Since dietary management of PKU is not necessary beyond early childhood, the reidentification and treatment of these pregnant women are medical challenges. The task is made more difficult because prepared low phenylalanine foods are generally unpalatable. Other aminoacidurias which are treated by reduction of the dietary intake of specific amino acids are hereditary tyrosinemia, homocystinuria from the disorders of transsulfuration, and maple syrup urine disease, a metabolic derangement of the branched-chained amino acids.

The disorders of the urea cycle have a different clinical picture. The manifestations are easy to understand if the urea cycle is thought of as a mechanism for the removal of ammonia produced by the metabolism of protein. Enzyme deficiencies result in the accumulation of ammonia. The result is episodic vomiting, lethargy, coma, and convulsions. In argininosuccinic aciduria, which is the most common disorder in this class, mental retardation is an invariable finding. A low-protein diet will ameliorate the chemical abnormalities in this group of diseases.[54] Histidinemia is a metabolic disease of an essential amino acid. Over 30 patients with this disease have been reported. Most are mentally retarded and have a speech defect that exceeds the degree of retardation. In theory, early identification and treatment with a diet low in histidine should be beneficial. As yet, however, there are no reports of success. An aspect to be considered is the therapeutic dilemma of the need to restrict an amino acid which is required for growth and development.

Disproportionate amounts of amino acids can produce adverse effects in animals and man. These may be due to toxicities, antagonisms, or imbalances. This subject has been extensively reviewed by Harper.[55] Most of the adverse effects in animals have been demonstrated by feeding an excess of the specific amino acid. The toxicity is due to the amino acid and is not a protein toxicity. There is little evidence to suggest that individual amino acids are present in foodstuffs in quantities large enough to cause adverse effects in adult man. The situation in infants is different. Levy et al. reported 18 infants with hypermethioninemia.[56] Protein intake varied between 7 and 16 g/kg/day. Three infants had sleeplessness, and one of these was also irritable. None had hemolytic anemia or growth retardation, which have been seen in animals with hypermethioninemia. In each, symptoms resolved when dietary protein was decreased to 3 to 4 g/kg/day. Very high protein diets may bring out the immaturity of the infants' metabolic functions.

Protein hydrolysates are used for intravenous nourishment. The infused amino acids are the source of nitrogen to support protein anabolism. The goal of total parenteral nutrition had been sought since the effective use of intravenous glucose solutions in World War I. The first investigator to achieve the goal was Dudrick in 1967.[57] In a skillfully designed experiment, he showed that the simultaneous administration of adequate glucose along with a nutritionally balanced mixture of protein hydrolysate, vitamins, minerals, and salts would support the normal growth of puppies. Skeletal and muscular growth was normal. The technique has since achieved wide clinical use. The most convenient nitrogen source for such infusates is a hydrolysate of complete proteins such as casein or fibrin. More recently, mixtures of synthetic L-amino acids have been utilized. The hydrolysates have the disadvantages of containing variable amounts of amino acids and large amounts of di- and tripeptides.

The synthetic amino acids allow specific tailoring of the solution but have the disadvantage of being present as chloride or hydrochloride salts. Complications or toxicities of the hydrolysate or amino acid solutions include: hyperchloremic metabolic acidosis, serum amino acid imbalances, hyperammonemia, and prerenal azotemia.[58,59] Metabolic acidosis is infrequently seen with the use of protein hydrolysates; it was frequent after the introduction of crystalline amino acid solutions.[58] The metabolism of 1 mol of the chloride or hydrochloride salt of an amino acid results in the liberation of 1 mol of chloride or hydrochloric acid. The problem was compounded by the addition of the chloride salts of other cations. The complication can be avoided in patients with normal renal function by the use of phosphate or acetate salts of potassium and the bicarbonate or lactate salts of sodium.

Hyperammonemia is the second potential complication. It has been reported after the use of both protein hydrolysates and crystalline amino acid infusions. In protein hydrolysates, 8 to 10% of nitrogen is present as ammonia nitrogen. In patients with normal liver function, ammonia intoxication has not been a significant clinical problem. Severe ammonia intoxication has been reported in neonates.[59] Chan recommends that half-strength solutions be used in premature infants who are more prone to develop acidosis.[60] Blood ammonia levels as high as 823 mg/100 ml were recorded; in some infants, grand mal seizures developed. Dudrick reported that in adults, the blood ammonia concentration rose with the administration of increasing amounts of crystalline amino acid solutions. He attributed this disorder to the relative deficiency of arginine in the infusions solutions. This deficiency impairs the functioning of the urea cycle. The addition of 0.5 to 1.0 mmol/kg/day of arginine prevented hyperammonemia.

Isosmolar coma has also been reported as a complication of total parenteral nutrition in infants.[61] The more common cause is nonketotic hyperglycemic coma. Touloukian reported the development of frank coma in eight of ten infants receiving more than 4 g/kg/day of protein hydrolysate. There was no evidence of dehydration, serum osmolality, and electrolyte concentrations were normal, as was blood glucose. Coma developed within 24 hr of the time the infusion rate was increased to more than 4 g/kg/day. Discontinuation of the infusion resulted in recovery within several hours. From these studies it is apparent that the potential toxicity of proteins is increased when they are infused intravenously either as protein hydrolysates or crystalline amino acid solutions.

REFERENCES

1. Food and Nutrition Board National Research Council, Recommended Dietary Allowances, 8th ed., National Academy of Sciences, Washington, D.C., 1974, 76–103.
2. Goodhart, R. S. and Shils, M., *Modern Nutrition in Health and Disease,* Lea and Febiger, Philadelphia, 1973.
3. Boyd, E. M., Food and drug toxicity, *J. Clin. Pharmacol.,* 8, 281, 1968.
4. Boyd, E. M., Predictive drug toxicity, assessment of drug safety before human use, *Can. Med. Assoc. J.,* 98, 278, 1968.
5. Rafoth, R. D. and Onstad, G. R., Urea synthesis after oral protein ingestion in man, *J. Clin. Invest.,* 56, 1170, 1975.
6. Leaf, A., Getting old, *Sci. Am.,* 229, 44, 1973.
7. Davidson, S. and Passmore, R., *Human Nutrition and Dietics,* Williams & Wilkins, Baltimore, 1969, 74–85.
8. Eckholm, E. and Record, F., The affluent diet: Worldwide health hazard, *Futurist,* 11, 32, 1977.
9. Glatzel, H., Is a predominantly protein diet hazardous to health?, *Med. Welt.,* 27, 1114, 1976.
10. Oppenheim, I. A. and Shields, D. F., Influence of exercise and high protein diet on blood chemistry, *J. Med. Soc. N.J.,* 70, 640, 1973.
11. Dicker, S. E., Effect of protein content of the diet on the glomerular filtration rate of young and adult rats, *J. Physiol.,* 108, 197, 1949.
12. Anderson, R. S. and Edney, A. T. B., Protein intake and blood urea in the dog, *Vet. Rec.,* 84, 348, 1969.
13. Rabinowitz, L., Gunther, R. A., Shoji, E. S., Freedland, R. A., and Avery, E. H., Effects of high and low protein diets on sheep renal function and metabolism, *Kidney Int.,* 4, 188, 1973.
14. Manston, R., Russell, A. M., Dew, S. M., and Payne, J. M., Influence of dietary protein upon blood composition in dairy cows, *Vet. Rec.,* 96, 497, 1975.
15. Pullman, T. N., Alving, A. S., Dern, R. J., and Lansdowne, M., The influence of dietary protein intake on specific renal functions in man, *J. Lab. Clin. Med.,* 44, 320, 1954.
16. Evans, R. M., Scholz, R. W., and Mongin, P., Effects of a high protein "carbohydrate free" diet on liver and kidney constituents and kidney function in chicks, *Comp. Biochem. Physiol.,* 40A, 1029, 1971.
17. Buskirk, E. R., Diet and athletic performance, *Postgrad. Med.,* 61, 229, 1977.
18. Dougherty, J. C., Influence of high protein diets on renal function, *J. Am. Diet. Assoc.,* 63, 392, 1973.
19. Lofland, H. B., Clarkson, T. B., and Goodman, H. O., Interactions among dietary fat, protein and cholesterol in atherosclerotic susceptible pigeons. Effects on serum cholesterol and aortic atherosclerosis, *Circ. Res.,* 9, 919, 1961.
20. Little, J. M. and Angell, E., Dietary protein level and experimental aortic atherosclerosis, *Atherosclerosis,* 26, 173, 1977.
21. Gordon, H. H., Levine, S. Z., Bauer, C., and Dann, M., Feeding premature infants: Comparison of human and cow's milk, *J. Dis. Child,* 73, 442, 1947.
22. Davidson, M., Levine, S. Z., Bauer, G., and Dann, M., Feeding studies in low-birthweight infants, *J. Pediatr.,* 70, 695, 1967.
23. Edelman, C. M., Jr. and Spitzer, A., The maturing kidney: A modern view of well-balanced infants with imbalanced nephrons, *J. Pediatr.,* 75, 509, 1969.
24. Davies, D. P., Protein intake, osmolality homeostasis and renal function in infancy, *Postgrad. Med. J.,* 51 (Suppl. 3), 25, 1975.
25. Abrams, C. A. L., Phillips, L. L., Berkowitz, C., Blackett, P. R., and Priebe, C. J., Jr., Hazards of overconcentrated milk formula, *JAMA,* 232, 1136, 1975.
26. Gault, M. H., Dixon, M. E., Doyle, M., and Cohen, W. M., Hypernatremia, azotemia and dehydration due to high protein tube feeding, *Ann. Intern. Med.,* 68, 778, 1968.
27. Walike, J. W., Tube feeding syndrome in head and neck surgery, *Arch. Otolaryngl.,* 89, 533, 1969.
28. Blair, R., Bolton, W., Duff, R. H., Fatty liver and kidney disease in broiler chickens receiving diets with varying contents of protein, *Vet. Rec.,* 84, 41, 1969.
29. Hemsley, L. A., Robertson, G., and Smetana, P., A trial to produce the fatty liver and kidney syndrome of young chickens, *Vet. Rec.,* 86, 722, 1970.
30. Whitehead, C. C., Bannister, D. W., Evans, A. J., Siller, W. G., and Wight, P. A. L., Biotin deficiency and fatty liver syndrome in chicks given purified diets containing different fat and protein levels, *Br. J. Nutr.,* 35, 115, 1976.

31.   Harper, A. E., Effect of variations in protein intake on enzymes of amino acid metabolism, *Can. J. Biochem.,* 43, 1589, 1965.

32.   Nitsan, Z., Nir, I., Dror, Y., and Bruckental, L., The effect of forced feeding and of dietary protein level on enzymes associated with digestion, protein and carbohydrate metabolism in geese, *Poult. Sci.,* 52, 474, 1973.

33.   Fraser, H. S. and Alleyne, G. A. O., The effect of malnutrition on the pattern of growth in the rat kidney and the renal response to acidosis, *Br. J. Nutr.,* 31, 113, 1974.

34.   Klain, G. J. and Hannon, J. P., Kidney response to cold stress and high protein intake, *Proc. Soc. Exp. Biol. Med.,* 152, 393, 1976.

35.   Moise, T. S. and Smith, A. H., Effect of high protein diets on kidneys, *Arch. Pathol.,* 4, 530, 1927.

36.   Lalich, J. J., Faith, G. C., and Harding, G. E., Protein overload nephropathy, *Arch. Pathol.,* 89, 548, 1970.

37.   Bras, G. and Ross, M. H., Kidney disease and nutrition in the rat, *Toxicol. Appl. Pharmacol.,* 6, 247, 1964.

38.   Lalich, J. J. and Allen, J. R., Protein overload nephropathy in rats with unilateral nephrectomy. II. Ultrastructural study, *Arch. Pathol.,* 91, 372, 1971.

39.   Marshall, M. W. and Lehman, R. P., Influence of heredity on response of inbred rats to diet, *Metabolism,* 16, 763, 1967.

40.   Newburgh, L. H. and Marsh, P. L., Renal injuries by amino acids, *Arch. Intern. Med.,* 36, 682, 1925.

41.   Lalich, J. J., Burkholder, P. M., and Paik, W. C. W., Protein overload nephropathy in rats with unilateral nephrectomy. A correlative light, immunofluorescence and electron microscopical analysis, *Arch. Pathol.,* 99, 72, 1975.

42.   Cousins, F. B. and Geary, C. P. M., Sex determined renal calcification in rats, *Nature,* 211, 980, 1966.

43.   Goulding, A. and Malthus, R. S., Effect of the protein content of the diet on the development of nephrocalcinosis in rats, *Aust. J. Exp. Biol. Med. Sci.,* 48, 313, 1970.

44.   McLaren, D. S., Dietary protein in clinical practice, *Practitioner,* 212, 441, 1974.

45.   Addis, T., *Glomerular Nephritis,* Macmillan, New York, 1949, 258–300.

46.   Kolff, W. J., Forced high caloric low protein diet with proteins of high biological value and the treatment of uremia, *Am. J. Med.,* 12, 667, 1952.

47.   Giordano, C., Use of exogenous and endogenous urea for protein synthesis in normal and uremic subjects, *J. Lab. Clin. Med.,* 62, 231, 1963.

48.   Burton, B. T., Current concepts of nutrition and diet in diseases of the kidney, *J. Am. Diet. Assoc.,* 65, 623, 1974.

49.   Anderson, C. F., Nelson, R. A., Margie, J. D., Johnson, W. J., and Hunt, J. C., Nutritional therapy in adults with renal disease, *JAMA,* 223, 68, 1973.

50.   Franklin, S. S., Gordon, A., Kleeman, C. R., and Maxwell, M. H., Use of a balanced low protein diet in chronic renal failure, *JAMA,* 202, 141, 1967.

51.   Edney, A. T. B., Observations on the effects of feeding a low protein diet to dogs with nephritis, *J. Small Anim. Pract.,* 11, 281, 1970.

52.   Frimpter, G. W., Aminoacidosis due to inherited disorders of metabolism, *N. Eng. J. Med.,* 289, 835, and 895, 1973.

53.   Mabry, C. C., Denniston, J. C., and Coldwell, J. G., Mental retardation in children of phenylketonuric mothers, *N. Eng. J. Med.,* 275, 1331, 1966.

54.   Hambraeus, L., Hardell, L. I., Westphal, O., Lorentsson, R., and Hjorth, G., Argininosuccinic aciduria, *Acta Paediatr. Scand.,* 63, 525, 1974.

55.   Harper, A. E., Effects of disproportionate amounts of amino acids, in *Toxicants Occurring Naturally in Foods,* Publ. 1354, National Academy of Sciences, Washington, D.C., 1966, 138.

56.   Levy, H. L., Shih, V. E., Madigan, P. M., Karolkewwicz, V., Carr, J. R., Lum, A., Richards, A. A., Crawford, J. D., and MacCready, R. A., Hypermethioninemia with other hyperaminoacidemias, *Am. J. Dis. Child.,* 117, 96, 1969.

57.   Dudrick, S. J., Wilmore, D. W., and Vars, H. M., Long-term total parenteral nutrition with growth in puppies and positive nitrogen balance in patients, *Surg. Forum,* 18, 356, 1967.

58.   Heird, W. C., Dell, R. B., Driscoll, J. M., Grebin, B., and Winters, R. W., Metabolic acidosis resulting from intravenous alimentation mixtures containing synthetic amino acids, *N. Engl. J. Med.,* 287, 943, 1972.

59.   Dudrick, S. J., MacFadyen, B. N., VanBuren, C. T., Ruberg, R. L., and Maynard, A. T., Parenteral hyperalimentation: Metabolic problems and solutions, *Ann. Surg.,* 176, 259, 1972.

60.   Chan, J. C. M., Asch, M. J., Lin, S., and Hays, D. M., Hyperalimentation with amino acid and casein hydrolysate solutions, *JAMA,* 220, 1700, 1972.

61.   Touloukian, R. J., Isosmolar coma during parenteral alimentation with protein hydrolysate in excess of 4 gm/kg/day, *J. Pediatr.,* 86, 270, 1975.

EFFECT OF NUTRIENT TOXICITIES IN
ANIMALS AND MAN: AMINO ACIDS

R. W. Longton

INTRODUCTION: DIETARY AMINO ACIDS

The impact of amino acid excess on animals and man remains a challenging area for research. Animals and men randomly select protein-containing foods for their diet without adequate knowledge of the specific amino acid content of food proteins. The potential biologic impact of this random selection on the body or its functions is not fully known. Previous research has shown that animals or man, given optional selection from a low- or high-quality protein diet, do not always select the high-quality protein diet. Occasionally, a potentially pathogenic diet is selected.

Food is rarely analyzed for amino acid content. Although the amino acid composition of many proteins is known, levels contained in various types of food are not readily available or published for the consumer. However, free amino acid levels in normal human blood or plasma are known, and these norms are available for reference.[136,257] Blood and urine samples are rarely examined for amino acid levels except when there is strong evidence of metabolic disturbance.

Perhaps there are safeguards, such as individual accommodation or tolerance to the randomly selected diets, provided by a variety of foods with contrasting protein content. In addition, there may be short-term tolerance to temporary elevations of certain amino acids found in protein selection. There is also a strong possibility that these food proteins balance one another so that amino acid levels in the body remain in a steady state or physiological balance. Hopefully, these are levels consistent with good health.

On the other hand, there are diet-restricted groups, fad eaters, weight reducers, or isolated groups who have selectively narrowed their range of protein sources. This reduced range may cause amino acid excess, with subsequent functional impairment. For example, a group of villagers in Thailand are prone to bladder stones (crystalluria). These villagers were shown to produce crystalluria when fed diets supplemented with hydroxyproline, a precursor of oxalate.[285]

The research of many investigators has revealed numerous metabolic changes and pathologies, indicating the complexity of interplay of metabolic factors arising from excessive intake of a single amino acid. There are sufficient facts available to conclude that an excess of certain single amino acids produces pathology in animals and man.

AMINO ACID EXCESS

Amino acid excess is a pathologic nutritional state which includes toxicity, structural changes, or metabolic alterations. This is caused by the increased consumption of a single amino acid in addition to the normal percentage of that acid in the basic diet. A 1 to 3% increase of this amino acid in the standard diet, consumed daily for about 2 weeks, produces these alterations or pathologies. These pathologies result from both the essential and nonessential amino acid groups as well as from the L and D isomers of amino acids. The L amino acids are more pathologic than D amino acid enantiomorphs; however, there are also D amino acids which are more pathologic than the L amino acid enantiomorph. Individual tissues or particular organs of man and certain animals are vulnerable targets for an amino acid excess. Additionally, there appears to be vulnerability to specific L or D forms of amino acid.

Amino acid excess can also result from an intraperitoneal, intravenous, intramuscular, or intradermal injection of a single amino acid in solution or in suspension at a fixed concentration. The injected excess amino acid can be essential or nonessential and of either the L or D form. After several injections, 1 ml of a 1-M solution will be pathologic to small animals.

AMINO ACID AND DISEASE: BLOOD AND URINE

Specific amino acids or related amino acid metabolites are found in high concentration in the blood and urine of patients with certain systemic diseases. The levels of amino acids or related metabolites in these patients are generally three to five times higher than those in the blood or urine of healthy people. Patients with elevated amino acid levels have anatomical structural changes in one or more of the following body components: skin, bone, joints, heart, lung, liver, pancreas, brain, and kidney. Mental retardation is also a common problem associated with high levels of a single amino acid. Autopsies performed on animals or persons afflicted with an amino acid excess may show lesions or pathology in either the brain, vital organs, skeleton, or skin. These pathologies include extensive demyelinization, lipid and phospholipid changes in the brain or spinal cord, liver damage, atrophy of pancreatic acid, disorganization of the epiphyseal (growth) cartilage, thickening of skin, loss of hair, inhibited spermatogenesis, and decreased growth.

A vascular lesion of an arteriole located in a joint of a Swiss white mouse injected with 0.5 mmol of tyrosine daily for 10 days is shown in Figure 1. Extensive coagulation necrosis, including pyknosis of mitochondria (subcellular energy generation and storage components) of the proximal convoluted tubule of the rat kidney produced by DL-serine, is shown in Figures 2 and 3. Rabbit skin response to daily topical exposure of aspartic acid, glutamic acid, and ornithine, using carbamate as a negative control, is shown in Figure 4.

AMINO ACID TRANSPORT AND EXCHANGE

The transport and exchange of amino acids are altered by amino acid excess; furthermore, the transport of one amino acid has been shown to affect the transport of others.[43,61,207,210,237,259,299] One amino acid produces conditions which change the levels of others.[44,241] This newly derived amino acid profile becomes a further superimposed influence on amino acid transport and exchange.

With regular nutrition, there is generally a normal profile of plasma amino acids. Specific amino acids are preferentially put out or taken up by the brain, kidney, gut, liver, and muscles to provide a net balance or profile.[76] Alanine and glutamine appear to play a key role in the overall exchange of amino acids between various tissues.[79,182]

Muscle provides the greatest release of amino acids, mainly in the form of alanine and glutamine (see Figure 5). The cardiac muscle also contributes to the net negative balance of amino acids and release of α-amino nitrogen consistent with muscle tissues. Muscles regularly take up smaller quantities of amino acids, usually cystine, glutamic acid, and serine in the uneven cyclical function.[79,80,182]

Conversely, alanine is taken up in large quantities by the liver, while glutamine is absorbed by the gut, providing a positive balance of amino acids.[38,39,78,80,81,82] Serine, threonine, and glycine complete the group of amino acids taken up preferentially, but to a lesser extent, by the liver. Alanine and serine are potent gluconeogenic precursors in the liver.[77,78,240] Serine is taken up moderately by the gut, while alanine, an important source of α-nitrogen transport, is released.[80,82,190]

The serine taken up by the liver, gut, and muscle is put out by the kidney, as is alanine.[222] Conversely, glutamine, proline, and glycine are taken up by the kidney.

The branched amino acids valine, leucine, and isoleucine are taken up by the brain. Valine, unlike the other two amino acids, is taken up rapidly and in vast quantities.[18] Leucine and isoleucine are taken up in moderately large amounts and are oxidized easily.[217] Smaller amounts of most other amino acids are also taken up by the brain.

This normal transport and exchange would be altered by an amino acid excess. For example, valine, leucine, and isoleucine were reduced 38 to 64% in immature and adult rat brains when phenylalanine was increased.[193]

MEMBRANE PERMEABILITY

The toxicity of aromatic chemicals was defined in terms of molecular partitioning across polar to nonpolar interfaces. This reaction of the aromatic compounds was considered to be related to membrane permeability of the aryl compounds.[31,38,41,209,239] The ratio of concentrations of the aryl compounds in n-octanol and in aqueous phosphate buffer provided partition coefficients and chemical reactivity indices, such as the Hammett Sigma values for these compounds.[119,120]

The partition coefficients, developed from the compatibility of an aryl compound with the polar or the nonpolar phases, were placed into the following relationship as a measurement of arthritic index.[239]

A.I. = 1.3 (lnP − lnPo)^2 + 3.2 or (regression efficient) = 0.76

where

A.I. = Arthritic Index
P = Partition coefficient
Regression coefficient = the fit of the data to the equation

Skatole (3-methylindole), a metabolite of tryptophan metabolism, was shown to have effective membrane permeability and to be involved with arthritic injury,[90,209,238] thus providing an example of an aryl biologic metabolite which can be toxic. The partition coefficient or optimal value of lnPo is 5.984, selected from the value obtained with skatole.[90] Molecules with partition coefficients (lnP) larger than the optimal (lnPo) would be slowed or stopped for a polar phase, and molecules with partition coefficients smaller than the optimal would be slowed or stopped by the nonpolar phase.[238] Amino acids or their metabolites having a partition coefficient in the range of 4.5 to 7.4 would be expected to provide potential joint injury, as measured by the arthritic index.[238]

AMINO ACID POOLS

There are individual pools of amino acids in each of the tissues and organs of the body. In a healthy state, each of these pools has a characteristic balance. The level of each amino acid in these tissues is measured individually and does not necessarily parallel the plasma level of that amino acid.[43] Therefore, the impact of an amino acid excess would vary from tissue to tissue or organ to organ, depending upon the particular endogenous amino acid pool.[177]

The net effect of amino acids in the brain, blood, and liver was different after the intraperitoneal injection of a phenylalanine load into male rats.[41] Additionally, the net transport of several amino acids from blood to brain was inhibited by phenylalanine.[31,32,209] Valine, methionine, leucine, and isoleucine were reduced 50% in the intracerebral pools after the intake of phenylalanine.[31,32,192]

ENZYME AMOUNT AND ACTIVITY

Amino acid excess does influence the quantity and activity of certain enzymes. These enzymes may be decreased by the increase of a specific amino acid, related metabolites, or other amino acids and metabolites affected by the amino acid excess. For example, increased methionine fed to rats caused a decrease in cystathionine synthetase activity to 30% of that measured in the control rat liver.[60] The metabolic relationships are shown in Figure 6.[84] The cystathionase activity, on the other hand, increased twofold with a methionine excess. The metabolic result was accumulation of homocysteine and depressed conversion of the carbonyl carbon of methionine to CO_2. Excess methionine and accumulation of homocysteine appear to be in part responsible for arteriosclerosis, cardiovascular disease, psychiatric disorders, liver disease, and oncogenesis.[52,62,143,191] Further clinical findings in the area of cystathionine-involved synthetase deficiency, methionine metabolism, and disease are shown in Table 1.

Changes in enzymatic activity or quantity can be genetically transmitted to offspring. Inborn errors are genetically controlled defects in enzymes that control levels of specific amino acids. Distribution of cystathionine-synthetase activity among homozygotes, heterozygotes, and controls are shown in Figure 7.

HORMONAL ACTIVITY

Amino acid excess influences the release of various hormones which regulate or modify body functions. The elevated amino acid, or agonist, reacts with other nutrients in this hormonal regulation. The glandular elements may be modified by nonamino acid components such as glucose before reaction with amino acid agonists. Glucose drastically decreases the affinity of the agonists to α-cell receptor sites, thus inhibiting the release of glucagon, as shown in Figure 8. On the other hand, glucose increases the affinity of the agonists to the β-cell receptor sites, thus enhancing the release of insulin.

The release mechanism of a hormone such as insulin may differ from the specificity of the amino acid itself. Leucine and arginine affect insulin release differently, as shown in Table 2. The relative effectiveness of glucagon release through the stimulation of activity is shown in Figure 9, and the relative effectiveness of insulin release through the stimulation by amino acid excess in decreasing order of activity is shown in Figure 10. Thus, hormones such as glucagon and insulin are modified in activity by amino acid excess but modified at individual levels.

The released glucagon and insulin affect vascular levels of glucose and protein, as shown in Figure 11. The amino acid in excess, as well as other amino acid levels, is altered by the amino acid in excess as shown in Table 3. There is a metabolic interplay or feedback mechanism between levels of amino acid, insulin, glucagon, and glucose. This is illustrated by a study of amino acid content in infants born with low blood glucose as compared with those who have normal blood glucose. There are apparent differences in alanine, glycine, proline, lysine, taurine, and valine levels in the blood of infants with the various levels of glucose content as shown in Figure 12.

VITAMIN AND MINERAL CHANGES

Amino acids, metabolites, and related enzymes react with and are dependent upon vitamins and minerals for utilization and function. Vitamins and minerals are effective as cofactors or catalysts in biological reactions. The interplay of vitamins and minerals with these components generally provide physiologic functions which, under adversity of amino acid excess, become pathologic to the host.

Amino acid excess, for example, increases the severity of vitamin deficiencies.[160,]

[183,213] The superimposed amino acid excess on vitamin deficiency can cause severely retarded growth and death. Examples of amino acid excess superimposed on vitamin deficiency are shown in Table 4.

Amino acid excess can also alter the binding of essential minerals needed for enzymatic activity in the maintenance of health, thus predisposing the host to pathologic conditions. Copper, for instance, is critical to the oxidation deamination of lysine, which is necessary for collagen formation and function.[42,200,214,215] Copper deficiency also seems to lead to increased vascular fragility, spontaneous vascular rupture, and aortic aneurisms.[215,216,264] Lysine, histidine, or dipeptides such as glycyl-histidine or histidyl-glycine were shown to bind with copper.[137,287] The Cu^{2+} binds with α-amino and the carboxyl group.[287] Two types of Cu^{2+} $(His)^2$ complexes were considered under equilibrium, with the Cu^{2+} exclusively bound with four nitrogens.[287] In the dipeptide, an oxygen of the glycyl-carbonyl group is involved in the Cu^{2+} binding.[287]

GROWTH MODIFICATION

Growth retardation in a variety of animals generally results from an amino acid excess. An increased single amino acid in a range of 1 to 3% of the diet generally will cause retardation of growth, which occurs when L-methionine is as low as 1.25% of the diet or L-glutamic acid is as high as 15%.[153,188] Depressed food intake may be a contributing factor to growth retardation.[22,23,124]

The quality of protein in the diet plays an important role in growth response to an amino acid excess.[160] Generally, the lower the amount of protein in the diet or the lower the quality of the amino acid content in the protein (low quality protein), the greater the retardation from the amino acid excess. Vitamin reserves in animals are important in the resistance of growth retardation due to amino acid excess.

Supplementation of the amino acid excess by other amino acids, or in some cases vitamins, has been known to alleviate growth retardation.[23,153,160,242,252,268] Excessive methionine appears to produce a deficiency of threonine and glycine,[153,242] which alleviates the growth depression produced by excess methionine.[153] Serine or glycine aids the catabolism of methionine, thus lowering the blood level of methionine and leading to increased food consumption and subsequent growth.[23] More information on growth modification caused by amino acid excess is shown in Table 5.

FETAL MODIFICATION

There are relationships in maternal amino acid levels which result in a derived amino acid excess caused by dietary deficiencies of other amino acids. The amino acids which are increased in relation to those remaining appear to correlate with a decreased fetal growth, cranial volume, motor development, and mental development. Valine and threonine have a negative correlation with birth weight, as shown in Table 6. Elevated valine or threonine in maternal blood predisposes the fetus to decreased birth weight. Decreased fetal cranial volume would be expected when threonine, histidine, or glutamic acid was elevated in maternal blood. Motor development would likely be diminished from elevated glutamic acid and proline. Finally, the rise of maternal blood glutamine or histidine could possibly signal retarded mental development of the fetus. Histidinemics genetically transmit an apparently related event which begins to affect the fetus in utero, and later, the histidine elevations are found in the blood of the mentally retarded offspring.[291]

STRUCTURAL AND METABOLIC CHANGES

Various tissues and organs, both human and animal, degenerate or malfunction from an amino acid excess obtained through the diet or from an external source. The degeneration, malfunction, or toxicity is targeted to specific tissues or organs by small amino acid increases of 2 to 5%. The increased amino acid can be essential or nonessential and either in the L or D form. The specific amino acid alters metabolism in defined ways that contribute to pathological development.

Nutritional pathosis results from an amino acid excess that specifically alters the metabolism of the host, particularly in individual tissues or organs. Disease was experimentally induced in rats from an excess of phenylalanine, resulting in depleted cerebral lipids of varying severity.[4,47,89] The phenylalanine excess caused anatomical changes resembling a metabolic disease (enzyme deficiency) found in man.[9,45,46,88,100,181,197] Elevated phenylalanine is accompanied by faulty metabolism of lipids, as well as poor formation and maintenance of myelin. The incorporation of carbohydrates into structure and their transport (in carbohydrates such as galactose) for incorporation into cerebral lipids is inhibited by phenylalanine levels.[258] Other nutritional pathoses are summarized in Table 7.

MUTANT PROTEINS AND FALSE HYBRID PROTEINS

Numerous mutant proteins have been found in nature or produced in the laboratory. Most of these involve the change of a single amino acid in the backbone of the protein molecule. The substitution is attributable to the change of a single nucleotide base in the messenger RNA codon transcribed from a corresponding substitution of a nucleotide base in DNA.[148,266] The codon change signals incorporation of the replacement amino acid into the modified protein, most of which are globulins. Further investigations and discoveries have been made since 1957.[87,141,149,169,170]

Nucleotide base changes of purine to purine ($A \rightleftharpoons G$) or pyrimidine to pyrimidine ($U \rightleftharpoons C$) are called transitions. Nucleotide base changes of purine to pyrimidine or vice versa ($A \rightleftharpoons U, A \rightleftharpoons C, G \rightleftharpoons U, G \rightleftharpoons C$) are called transversions.[94] In the laboratory, mutations were produced by a transition of adenine to hypoxanthine (which pairs with cytosine) or cytosine to uracil by deamination, using nitrous acid. Mutations were expressed after tobacco mosaic virus coat was exposed to nitrous acid.

Substitute amino acids may be incorporated into protein in the presence of any amino acid excess. Ambiguous amino acid incorporation of leucine, serine, or valine was obtained from rabbit reticulocytes in the presence of excess phenylalanine.[34] These proteins are perhaps false hybrids rather than mutants since genetic reference DNA is not altered. These proteins variated through translation error have an altered amino acid composition which provides for potential structural and functional changes. Lesions were produced on both soft and hard tissue with a similar collagen modification when mice were given excess tyrosine or histidine.[176]

PROTEIN SYNTHESIS OR DEGRADATION

Protein synthesis is regulated by a variety of factors, such as nutrient uptake, availability of amino acids, RNA synthesis, number of ribosomes, mRNA binding capacity to ribosomes, ribosomal functional controls, and availability of energy compounds.[17,53,72,131,144,245,275] These factors vary according to the status of the cell, such as dividing, stationary, untransformed, or transformed.[72,126,131,272,275,282]

Certain amino acids stimulate the attachment of mRNA to the ribosome and contribute to the functional control of protein synthesis.[144,250] The incorporation of amino acids in protein synthesis is dependent upon their availability and concentration.[117,144,265] The concentration of the individual amino acid has an important effect on ribosomal activity in protein synthesis.[118]

An excess of L-aspartic or D-aspartic amino acid did not significantly decrease protein synthesis in fetal rabbit skin cells, although there was considerable cell death with additions of D-aspartic acid.[53] This toxicity did not correlate with protein synthesis which, together with degradation from an amino acid excess, depends upon the response of specific tissues or organs.[42] For example, protein synthesis and degradation in the liver, muscle, and brain of a rat fed a high tyrosine and low protein diet differed.[142] Liver polysomes shifted to heavier ribosomal aggregates with increased protein synthesis. Under these tyrosine conditions, the polysomes disaggregated, and protein synthesis was decreased in the muscles and brain.[142] With a phenylalanine excess, similar polysomal disaggregation and decreased synthesis was produced in a rat brain.[12,166] With a tyrosine load,[142] increased protein degradation was produced in rat liver. Perhaps the proteolysis in the liver was signaled by and responded to insulin, glucagon, and the amino acid excess.[204,304]

LEUCOCYTE RESPONSE

Leucocytes generally function in a defensive or phagocytic role in man or animals. Some lymphocytes contribute to immunological defense by cellular involvement as T-cells or by release of molecular products from these T-cells or β-cells. Some β-cell products are immunoglobulins, which further react with foreign substances. T-Cells can be activated by peptides and synthetic combinations of amino acids. β-Cells also can be turned on by ampholytes, as observed in our laboratory by Knudsen et al. Polymorpho-nuclear leucocytes and neutrophiles generally engulf or release digestive enzymes which degrade foreign substances through the release of lysosomal products. These products may be stimulated by an amino acid excess.[238,239]

Acute inflammation of rabbit skin and joints was produced from lysates harvested after intraperitoneal injection of glycogen, recovery of lysates, and reintroduction of the lysates into skin and joints.[300] The synovium became infiltrated with round cells; the synovial lining cells became hypertrophied and hyperplastic; and this was followed by pannus formation with cartilage degradation.[300] In separate studies, aryl metabolites of amino acids such as indole or skatole were used to induce lysis of neutrophiles, which further released autolytic or degradative substances contained in the lysosomes.[50,238,239]

IMMUNE RESPONSE

An amino acid excess can alter immune response in several ways. Most of the factors that form the elements of the immune system — namely cells or molecules — are responsive to the internal metabolism of the host. Immune response is composed of a complex system of numerous cellular populations in various maturated stages of categories of T- and β-cells and the immunoglobulin protein types of the humoral phase. The cells respond to mitogens or foreign molecules called antigens by providing products such as lymphokines or antibodies.

Debilitating systemic diseases such as lupus erythematosus, myasthenia gravis, scleroderma, rheumatoid arthritis, rheumatic fever, Hodgkin's disease, or Lesch-Nyhan syndrome (deficiency of hypoxanthine-guanine phosphoribosyl transferase activity) involve the immune system.[8,105,199,209,263,277] The autoimmune responses in these diseases represent an abnormal injurious reaction between self-focused T-cells or antibodies and self-antigens or host-generated substances.[57] In studying Hodgkin's disease, several investigators have implicated quantitative and qualitative disturbances of T-cell function, lipoid nephrosis with proteinuria from reagenic response involving IgE, or circulating abnormal lymphokine toxic to the glomerular basement membrane.[3,98,115,260]

In Lesch-Nyhan syndrome, the T-cell function used to test antigens and contact sensitisers was normal.[8] The functional response of β-cells was abnormal with low peripheral β-cell counts and low serum Ig and isohemagglutinin levels.[8] Adenosine deaminase deficiency appears to contribute to depressed pyrimidine metabolism, which affects β-cell function.[230] Lymphocyte function, in diseases such as homocystinuria, was consistent with the seriousness of the metabolic deficiency, since mitogen stimulated low-level release of cystathionine synthetase, the enzyme depressed in that disease.[104]

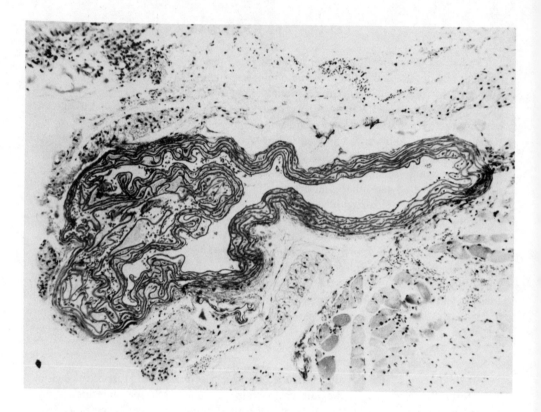

FIGURE 1.   Arteriole located in a joint of a Swiss White mouse after intraperitoneal injection of 0.5 mmol L-tyrosine for 10 days. A portion of the vessel wall has a focal-thickened mural invagination. The laminar distortion contained a substance which appeared to be mucoid. (From Longton, R. W., Ph.D. thesis, Northwestern University, Evanston, Ill., 1971, 76.)

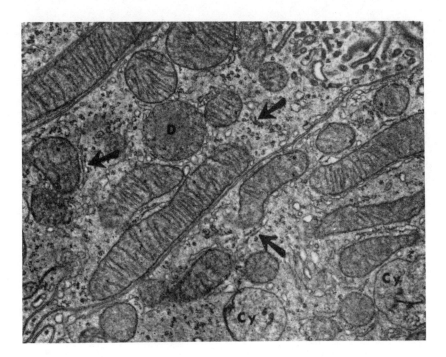

FIGURE 2.   Portion of a normal cell in the proximal tubular convolution of a rat kidney. In lower left corner, basement membranes and infolded plasma cellular membranes appear. In upper right corner, cross sections of microvilli of brush border are evident, beneath which, there are small vacuolar and tubular structures. Apart from elongated mitochondria, two cytosomes (Cy) and one dense body (D) are seen. RNP particles are visible throughout the cytoplasm. Arrows point to endoplasmic reticulum. (Magnification × 16,400.) (From Wachstein, M. and Besen, M., *Am. J. Pathol.*, 44, 395, 1964. With permission.)

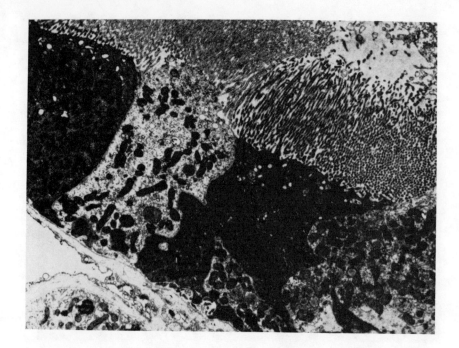

FIGURE 3.   Three hours after the intraperitoneal injection of DL-serine. Extensive coagulation necrosis has occurred in several cells, and vacuolar changes appear in a cell even in the vacuolated area. Tips of microvilli in the central upper part of the micrograph exhibit vacuolation, and tips overlying the diffusely pyknotic area to the right show condensation. (Magnification × 5,780.) (From Wachstein, M. and Besen, M., *Am. J. Pathol.*, 44, 395, 1964. With permission.)

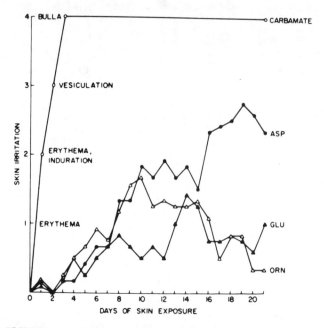

FIGURE 4.   Rabbit skin response to daily exposure to 0.5 ml of a 1 mM concentration of amino acid. Rabbit skin was exposed to carbamate (6 test animals), ASP (12 test animals), GLU (12 test animals), and ORN (6 test animals). Scoring was according to modified Draize: 0, negative reading; 1, definite erythema of the entire patch area; 2, erythema and induration; 3, vesiculation; and 4, bullous lesion. The mean values from scoring of the skin were placed in the chart. (From Longton, R. W., *Experientia*, 29, 1013, 1973. With permission.)

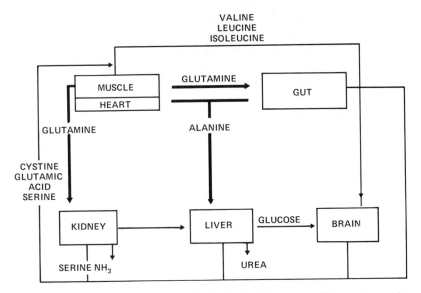

FIGURE 5. Amino acid exchange in the brain, kidney, gut, liver, and muscle provides a net balance or profile of plasma amino acids. Specific amino acids are preferentially put out or taken up by these essential organs, as illustrated. (From Felig, P., *Annu. Rev. Biochem.*, 44, 933, 1975. With permission. Copyright © 1975 by Annual Reviews, Inc. All rights reserved).

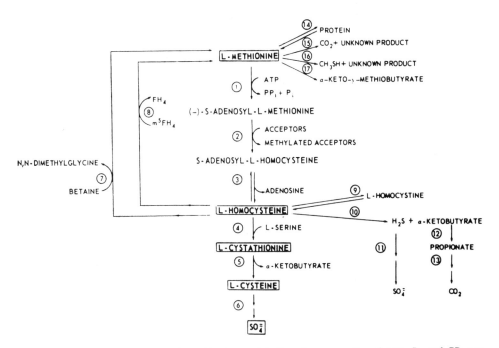

FIGURE 6. Current concepts of methionine metabolism in mammalian tissues. P_i and PP_i are inorganic phosphate and pyrophosphate, respectively. FH_4 is tetrahydrofolic acid and m^5FH_4 is N^5-methyltetrahydrofolic acid. (From Finkelstein, J. O., Methionine metabolism in animals, *Metabolism*, 23, 387–398, 1974. With permission.)

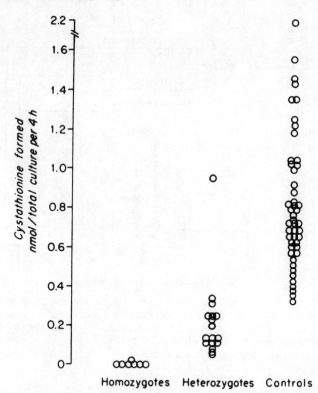

FIGURE 7.   Distribution of the levels of cystathionine-synthase activity among homozygotes, heterozygotes, and controls. Each reaction was incubated for 4 hr at 37°C and contained in 0.1 ml of lymphocyte extract (protein concentration varied between 0.2 and 0.7 mg) and additional components. Enzyme activity is expressed as nanomole cystathionine formed per total culture per 4 hr (1 nmol = 50,000 cpm). (From Goldstein, J. L., Campbell, B. K, and Gartler, S. M., *J. Clin. Invest.,* 52, 220, 1973. With permission.)

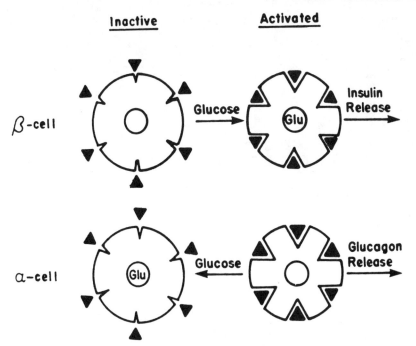

FIGURE 8.   Activation and inhibition of amino acid receptors by glucose in β-cells and α-cells, respectively. Amino acid sites are depicted arbitrarily by a set of notches, amino acids by triagonal shapes, and the glucose site by a central circle. The glucose molecule is represented by the abbreviation GLU. In the β-cells, addition of glucose makes the amino acid sites accessible to the agonists. However, in the α-cells, addition of glucose drastically decreases affinity of the agonists for their receptor sites. The "activated" receptor leads to hormone release. (From Pagliara, A. S., Stillings, S. N., Hover, B., Martin, D. M., and Matschinsky, F. M., *J. Clin. Invest.*, 54, 830, 1974. With permission.)

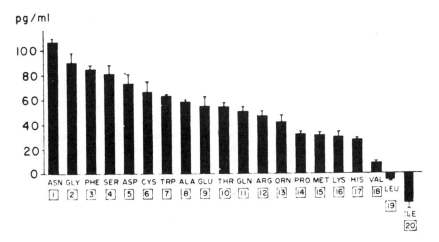

FIGURE 9.   Glucagon-stimulating activity (GSA) of 1 mmol/kg of 20 L-amino acids in the dog, presented in decreasing order of activity. Numbers refer to rank of GSA. (From Rocha, D. M., Faloona, G. F., and Unger, R. H., *J. Clin. Invest.*, 51, 2347, 1972. With permission.)

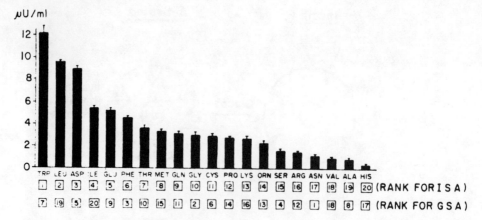

FIGURE 10.   Insulin-stimulating activity (ISA) of 1 mmol/kg of 20 L-amino acids in the dog, presented in decreasing order of activity. (From Rocha, D. M., Faloona, D. F., and Unger, R. H., *J. Clin. Invest.*, 51, 2349, 1972. With permission.)

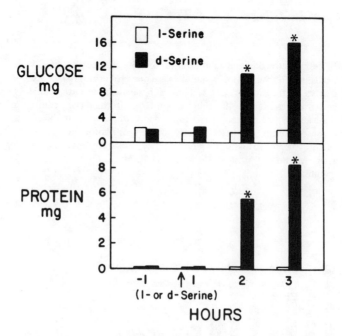

FIGURE 11.   Hourly urine protein and glucose before and after treatment with L- or D-serine in rats. Asterisks represent significant differences (P < 0.01). (From Carone, F. A. and Ganote, C. E., *Arch. Pathol.*, 99, 661, 1975. Copyright 1975, American Medical Association. With permission.)

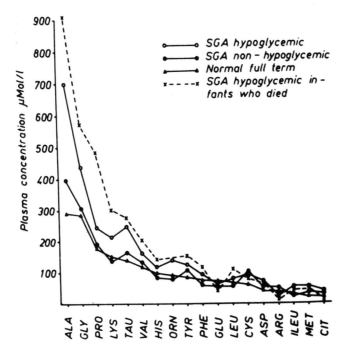

FIGURE 12. Mean plasma concentrations of 17 amino acids obtained 3 to 12 hr after birth in normal full-term, nonhypoglycemic SGA, and hypoglycemic SGA infants. Mean concentrations of four hypoglycemic infants who died are also indicated (X). (From Mestyan, J., Soltesz, G., Schultz, K., and Horvath, M., Hyperaminoacidemia due to the accumulation of gluconeogenic amino acid precursors in hypoglycemic small-for-gestational age infants, *J. Pediatr.*, 87, 409, 1975. With permission.)

Table 1
CLINICAL FINDINGS
IN CYSTATHIONINE
SYNTHASE DEFICIENCY

Ectopia lentis
Skeletal abnormalities
    Osteoporosis
    Dolichostenomelia
    Scoliosis
    Pectus excavatum or carinatum
Arterial and venous thromboses
Mental retardation
Abnormal hair pigmentation

From Finkelstein, *Metabolism*, 23, 391, 1974. With permission.

Table 2
COMPARISON OF INCREASES IN PLASMA INSULIN UPON
INTRAVENOUS ADMINISTRATION OF LEUCINE AND ARGININE TO HEALTHY SUBJECTS
AND TO PATIENTS WITH FUNCTIONING ISLET CELL TUMORS

| Amino acid infused (mM/kg) | Subjects | No. | Plasma insulin (microunits/ml) mean ± SEM | | | |
			Mean of maximal increases	P	Mean sum of increases during first hour	P
Leucine (1.52)	Healthy subjects	20	10.1 ± 1.6	<0.001	37.1 ± 6.3	<0.001
	Patients with islet cell tumor	9	61 ± 15.0	<0.001	232.3 ± 63.1	<0.001
Arginine (2.39)	Healthy subjects	20	82 ± 14.4	Not significant	267 ± 42.8	Not significant
	Patients with islet cell tumor	9	76 ± 18.9	Not significant	275 ± 78	Not significant

From Fajans, Floyd, Knopf, Guntsche, Rull, Thiffault, and Conn, *J. Clin. Endocrinol.*, 27, 1602, 1967. With permission.

Table 3
HOURLY URINARY EXCRETION OF
AMINO ACIDS

Amino acids	Hours (μmol/24 hr)			
	−1	1	2	3
Taurine	0.08	0.13	0.88	0.39
Hydroxyproline	−	0.41	1.6	0.67
Aspartic acid	0.04	0.08	−	0.20
Threonine	0.06	0.69	1.2	1.1
Serine	0.03	361.0	101.0	9.1
Proline	−	0.23	1.2	0.86
Glutamic acid	−	1.22	4.4	3.8
Citrulline	−	0.78	0.63	0.58
Glycine	1.05	4.24	3.86	2.19
Alanine	0.05	2.17	2.5	1.62
α-Amino-n-butyric acid	−	10.2	3.05	0.26
Valine	−	2.15	0.63	0.36
Methionine	0.02	0.09	0.26	0.28
Isoleucine	0.03	0.19	0.42	0.39
Leucine	0.06	0.42	0.74	0.67
Tyrosine	−	0.19	0.31	0.26
Phenylalanine	−	0.09	0.32	0.32

From Carone and Ganote, *Arch. Pathol.*, 99, 661, 1975. With permission.

Table 4
CHANGES CAUSED BY AMINO ACID EXCESS
IN ANIMALS OR MAN

Amino acid increased in diet (or intake)	Subject	Vitamin relationship	Toxicity	Ref.
Cystine	Rat	Riboflavin deficiency	Increased toxicity	183
Methionine[a]	Rat	Pyridoxine deficiency	Increased cystathionine excreted (3 times more cystathionine excreted with L than DL); growth depression	63, 251, 261
Phenylalanine[a]	Chick	Nicotinic acid deficiency	Severe growth depression	10
Tryptophan[a]	Rat	Riboflavin deficiency; pyridoxine deficiency	Increased deathrate Increased excretion of xanthurenic acid	171, 183
Tyrosine	Rat	Riboflavin deficiency; nicotinic acid deficiency	Increased toxicity	183, 213

[a] Essential for man.

Table 5
GROWTH MODIFICATION CAUSED BY AMINO ACID EXCESS IN ANIMALS OR MAN

Amino acid increase in diet (or intake)	Stereo-isomer	Subject	Amount of increase (%)	Diet protein level	Growth modification	Counteractivity factor	Ref.
Alanine	L	Chick	2.5		None		2
	DL	Chick	2.5		Depression		2
			4.0		Little or none		11, 139
	D	Chick			Depression		189
Arginine	L	Rat	5.0	Low or adequate protein	None		252
	L	Rat	4.0	Low protein	Depression		123
			4.0	Adequate	None		147
			7.5	15% Casein	Severe depression		255
Asparagine		Chick	5.0	12% Casein	None		267
Aspartic acid	L	Rat	5.0	Low protein	50% Depression		252
	L	Rat	5.0	Low protein	50% Depression		252
	DL	Rat	5.0	Low protein	Severe depression		252
			5.0	Adequate protein	Depression		108
	DL	Chick	4.0	Adequate protein	Depression		139
	L	Chick	4.0		None		145
	D	Chick	2.0		Depression more toxic than L		145
					Depression more toxic than L		189
	D	Rat			Depression more toxic than L		24
Cysteine	L	Rat	2.42	10% Casein	Depression; 50% mortality within 2 weeks		24
Cystine	L	Rat	1.0	18% Casein	None		24
			1.2	10% Caseine	None		49
			1.45	12% Casein	None		56
			1.5	8% Casein	None		251
			2.0	8% Casein	Depression		54, 55
			0.6–0.9	15% Casein	Depression stopped	Yeast extract	56
			2.50–3.75	8% Casein	Growth		
			4–20	Adequate—high protein	Depression and death		
	L	Chick	4.0	Adequate	Depression; some death		139
			5.0	20% Casein	100% Death		108
			5.0	Low protein	Severe depression		252
	L	Rat	5.0	8% Casein	100% Death		56
			6.4	Adequate	Severe depression; no death		248
			10	High protein	Many deaths		

Table 3 (continued)

GROWTH MODIFICATION CAUSED BY AMINO ACID EXCESS IN ANIMALS OR MAN

Amino acid increase in diet (or intake)	Stereo-isomer	Subject	Amount of increase (%)	Diet protein level	Growth modification	Counteractivity factor	Ref.
Glutamic acid	L	Rat	5.0	Low to moderate	None		108, 130, 183, 252, 280
			7.0	Adequate	Slight depression		279
			10	Low to moderate	Slight depression		128, 278
	L	Chick	10–20	Adequate	Slight depression		128, 278
			4.5–10	Low	Moderate depression		85, 267
	DL	Chick	15	Adequate	None to slight depression		161, 188
	D	Chick	7.5	Adequate	Significant depression		188
			3.75	Adequate	Moderate depression		188
			5.0	Adequate	Depression		188
Glutamine	L	Rat	6.0	Adequate	None		129
Glycine	L	Rat	3.0	Adequate	None		224
			5.0	Adequate	None		22
			5.0	6% Casein	50% Depression		252
			6.0	15–20% Casein	Depression		133
			10	4% Casein	Depression		278
	L	Chick	2.0	Adequate low nicotine	Depression	Nicotinic acid	114
			4.0	Adequate low nicotine	Depression	Nicotinic acid	144
			5.0	Adequate	None		10, 196, 207, 208
			8.0	Adequate low nicotinic acid	Depression	None	114
			8.0	18% Casein	Depression		113
			9.0	23% Soybean	Depression		180
Histidine	L	Rat	10	Adequate	Slight depression		110
			2.0	Adequate	Slight depression		123
			2.0	Low protein	Severe depression		123
			4.0	Adequate	Slight depression	Quality of protein	22, 123
	DL	Rat	10	Adequate	Severe depression; death	None	183, 184
			1.0	Adequate	Slight depression		305
			4.0	Adequate	Severe depression		305
	L	Chick	1.6	Low protein	Depression		269
Hydroxyproline	L	Rat	3.0	Adequate	Depression	Arginine	268
			2.4	Low protein	None	Glycine	253
Isoleucine[a]	L	Rat	3.4	Inadequate	Depression		28, 302
			5.0	Adequate	Slight depression		252
			6.0	9% Casein	Slight depression		25, 26, 122

[a] Essential for man.

Table 5 (continued)

GROWTH MODIFICATION CAUSED BY AMINO ACID EXCESS IN ANIMALS OR MAN

Amino acid increase in diet (or intake)	Stereo-isomer	Subject	Amount of increase (%)	Diet protein level	Growth modification	Counteractivity factor	Ref.
Leucine[a]	L	Rat	3–5 Excess	Low protein	Depression	Isoleucine	125
			3.0	9% Casein	Depression	Isoleucine, valine	25, 26, 122
			5.0	6% Casein	Severe depression		252
			5.0	40% Casein	None	40% Casein	252
	DL	Rat	4.0	Adequate low vitamin B_{12}	Depression	Vitamin B_{12}	139
			5.0	Adequate	Depressed more than L isomer		122, 252
	L	Rat	5.0	Low protein	Depressed less than L isomer		122, 252
			5.0	20% Casein	None		108
Lysine[a]	L	Rat	5.0	Low protein	Depression		245
		Chick	2.0	Casein-gelatin	Depression	Arginine	147
	L		5.0	Low protein	Depression		125
	DL	Chick	5.0	20% Casein	None		108
Methionine[a]	L	Rat	2.0	Low protein	Severe depression		22, 252
			2.0	10% Soybean	Depression		252
			2.0	24% Casein	Slight depression		49
			2.5	18% Casein	Severe depression		49
			3.0	10% Casein	Very severe depression		22
	DL	Rat	1.5	8% Casein	None		150
			2.0	8% Casein	Depression		150
			3.0	8% Casein	Severe depression		150
Phenylalanine[a]	DL	Chick	0.84	13% Protein	Depression	Choline	160
	L	Rat	4.0	Adequate	Depression		133
			4.0	Low protein	Depression		213, 306
			5.0	Adequate	Depression		59, 123, 157, 235, 294
			7.0	Adequate	Severe depression; death		67, 294
	L	Hamster	5.0	Adequate	Depression		138
	L	Chick	1.0	Adequate	Slight depression		109
			4.0	Adequate	Depression		125
			10	Adequate	Severe depression		110
	D	Rat	4.0	Low protein	Depression less than L		306

GROWTH MODIFICATION CAUSED BY AMINO ACID EXCESS IN ANIMALS OR MAN

Amino acid increase in diet (or intake)	Stereo-isomer	Subject	Amount of increase (%)	Diet protein level	Growth modification	Counteractivity factor	Ref.
Proline	L	Rat	3.0	14% Casein	None		1
			3.0	Adequate	None		302
			5.0	Adequate	Slight depression		133
			5.0	Low protein	Slight depression:		252
	L	Chick	4.0	Adequate	None		125
Serine	L	Rat	2.1	10% Casein	Very slight depression		22
			4.2	10% Casein	Slight depression		22
			6.3	10% Casein	Depression		22
	DL	Rat	2.0	Adequate	None		108
			5.0	Adequate	None		108
			5.0	Low protein	Slight depression		252
	DL	Chick	5.0	Adequate	None		207, 208
	D	Chick			None		262
Threonine[a]	L	Rat	1.25	6% Casein	Depression	Tryptophan	5
			2.4	Adequate	None		22
	DL	Rat	3.2	Low protein	Depression		248
			5.0	Low protein	Depression		252
			6.0	Low protein	Depression		123
			4.0	Adequate	None		305
			6.0	Adequate	Depression		123
			8.0	Adequate	Depression		305
Tryptophan[a]	L	Man	1–2 g	Adequate	None	Low pyridoxine	116
	L	Rat	1.0	Adequate	None		108, 123, 248, 252, 308
			1.5	Low protein	Slight depression		123, 248, 305
			3–5	Adequate	Depression		22, 59, 212, 252
	L	Chick	High	Adequate	Depression		146
	D	Rat	1.0	Adequate	None		108
	DL	Man	10 g/week	Adequate	None		16

Table 5 (continued)

GROWTH MODIFICATION CAUSED BY AMINO ACID EXCESS IN ANIMALS OR MAN

Amino acid increase in diet (or intake)	Stereo-isomer	Subject	Amount of increase (%)	Diet protein level	Growth modification	Counteractivity factor	Ref.
Tyrosine	DL	Rat	20	Low protein	Severe depression; death		278
	L	Rat	1–2	Low protein	Depression		185, 256
			3.0	Low protein	Slight depression		5, 27
			5.0	Low protein	Severe depression; death		29, 142
			10	Adequate	Severe depression		185, 187
	L	Chick	4–5	Adequate	Depression		10, 134, 139, 207
			10	High protein	Severe depression		185, 187
Valine[a]	D	Rat	5	Adequate	Depression less than L		186
	L	Rat	5.0	Low protein	Slight depression		252
				High levels	Depression		273
	DL	Rat	2.0	Low protein	Slight depression		121, 252
	D	Rat	2.0	Low protein	Slight depression		121, 252

Table 6
FETAL MODIFICATION CAUSED BY ELEVATIONS OF AMINO ACIDS IN ANIMALS AND MAN

Amino acid increased	Subject	Dietary increase	Plasma increase	Fetal modification	Counteracting factor	Ref.
Glutamic acid	Man		x	Decreased motor development; decreased cranial volume		201
Glutamine	Man		x	Decreased mental development		201
Histidine	Man		x	Decreased cranial volume; decreased mental development		201
Leucine[a]	Rat	Injected		Death; abnormalities		228
Lysine[a]	Rat	x		Many deaths		51
Methionine[a]	Rat	x		Decreased weight		168, 286
Phenylalanine[a]	Chick	x		Decreased reproductive activity		281
Proline	Man		x	Decreased motor development		201
Threonine[a]	Man		x	Decreased birth weight; decreased cranial volume		201
Valine[a]	Man		x	Decreased birth weight		201

[a] Essential for man.

Table 7

NUTRITIONAL PATHOSIS FROM AMINO ACID EXCESS IN ANIMALS AND MAN

Amino acid increased in diet (or intake)	Stereo-isomer	Subject	Anatomical structural change	Metabolic interaction	Counteracting factor	Ref.
Alanine	L	Man		Glucagon secretion stimulated (increased in starvation); glucagon secretion stimulated in diabetic, but not obese diabetic; insulin release stimulated		96, 297, 298
				Beta-hydroxybutyrate decreased; fewer free fatty acids; antiketotic; delayed release of growth hormone; increased glutamine, glycine, proline, serine, and threonine; increased α-amino-n-butyric acid; decreased isoleucine, leucine, and valine; gluconeogenic		96, 97, 198, 224, 241
		Dog		Glucagon secretion stimulated; very slight insulin release stimulated		205, 237
Arginine		Rat		Glucagon release stimulated		226
	L	Man		Insulin release greater than with leucine stimulation; glucagon release stimulated		14, 73, 74, 284
		Dog		Glucagon release stimulated; very slight insulin release stimulated		151, 237
		Rat		Glucagon released stimulated; insulin release stimulated; pancreatic α-cells release glucagon, especially in absence of glucose; pancreatic β-cells release insulin, especially in presence of glucose		99, 255
Asparagine	L	Dog		Highest glucagon release stimulated; very slight insulin release stimulated		237
Aspartic acid	L	Monkey	Erythema, induration, and vesiculation of skin			177
		Dog		High insulin release stimulated; glucagon release stimulated		237
		Mouse	Brain damage; retinal and hypothalamic lesions			219
		Rabbit	Erythema, induration, and vesiculation of skin			177
	D	Rabbit	Erythema and some induration of skin			177
	L	Guinea pig	Erythema, induration, and vesiculation of skin			177
Cysteine	L	Dog		Glucagon release stimulated; insulin release stimulated		237
Cystine	L	Mouse	Brain damage; retinal and hypothalamic lesions			219
		Rat	Kidney damage and degeneration (nephrosis)		Yeast extract	54, 55, 112, 172
			Kidney necrosis; tubular and glomerular injury;			56, 162
			Kidney hemorrhage			173
			Liver cirrhosis, hemorrhage, and necrosis			68—70
		Dog		Glucagon release stimulated; insulin release stimulated		220, 221 237
Glutamic acid	L	Monkey	Brain damage			218
		Dog		Glucagon release stimulated; insulin release stimulated		179, 218, 219
		Rat	Brain damage			
		Mouse	Brain damage; degeneration of retina; hypothalamic neurons destroyed			

NUTRITIONAL PATHOSIS FROM AMINO ACID EXCESS IN ANIMALS AND MAN

Amino acid increased in diet (or intake)	Stereo-isomer	Subject	Anatomical structural change	Metabolic interaction	Counteracting factor	Ref.
Glutamine	L	Dog		Glucagon release stimulated; insulin release stimulated		237
Glycine	L	Man	Kidney stones	Gluconeogenic		198, 307
			Mental retardation			195
		Dog		High glucagon release stimulated; moderate insulin release stimulated		237
Histidine	L	Rat	Giant cells in liver; soft friable bones	Histidinemia	Methionine	177, 292
		Mouse	Liver and spleen damage; soft friable bones			176
		Dog	Kidney damage			211
		Rabbit	Kidney damage			211
		Monkey	Hyperlipemia			158, 159
		Dog		Glucagon release stimulated; very slight to no insulin release stimulated		151, 237
Hydroxyproline	L	Rat	Systemic toxicity	Oxalcrystalluria, increased oxalate and hydroxy-proline excretion	Aspartic acid and serine	1
		Man	Urinary bladder		Orthophosphate	64, 185
Isoleucine[a]	L	Man	stone	Ketogenic, gluconeogenic; yields CO_2, proprionyl CoA, and acetyl-CoA		111
			Convulsions, respiratory disturbances, coma. Mental retardation, neurochemical disturbances of neural lipids (i.e., cerebrosides, glycoproteins, and proteolipids)			270
		Dog		Glucagon release stimulated; no insulin stimulated		237
		Rat		Mutual antagonism of branch-chain amino acids; branch-chain ketoaciduria	Leucine valine	58, 274
Leucine[a]	L	Man	Pellagra	Depresses tryptophan and nicotinic acid metabolism		106, 107
				Altered excretion of tryptophan products (i.e., increased quinolinic acid; decreased N-methyl6-pyridone-3-carboxamide and 5-hydroxyindole acetic acid)		21
				Stimulates insulin; produces hypoglycemia	Diazoxide and trichlormethiazide	48, 66, 73, 75
		Monkey	Pellagra	Niacin deficiency; ketogenic (i.e., produces aceto-acetate and acetyl-CoA)		19, 111
		Dog	Black tongue	Niacin deficiency; high insulin release stimulated (glucose independent); no glucagon release stimulated	Nicotinic acid	20, 83, 237
		Rat		Depresses 5-hydroxytryptophan (serotonin) in brain; blood brain barrier transport interference		95, 194, 310
				Mutual antagonism of branch-chain amino acids; branched-chain ketoaciduria; increased branch-chain α-ketoacid dehydrogenase	Isoleucine and valine	58, 274, 301
Lysine[a]	L	Dog		Low moderate insulin release stimulated; low moderate glucagon release stimulated		237
		Rat		Lysine accumulation in plasma and muscle		147, 252
		Chick	Weakness, tremor, and poor feathers	Lysine accumulation in plasma		11, 155, 312
		Man	Neurological disturbance			229

[a] Essential for man.

Table 7 (continued)
NUTRITIONAL PATHOSIS FROM AMINO ACID EXCESS IN ANIMALS AND MAN

Amino acid increased in diet (or intake)	Stereo-isomer	Subject	Anatomical structural change	Metabolic interaction	Counteracting factor	Ref.
Methionine[a]	L	Dog		Moderate insulin release stimulated; low glucagon release stimulated		237
	DL	Rat	Kidney enlargement; dilatation of kidney tubules		Arginine	35, 164
	L	Rat	Kidney enlargement; dilatation of kidney tubules		Glycine	242, 243, 244
	L	Rat	Damage to pancreatic acinar cells		Arginine or glycine; arginine and glycine	154, 165, 276
			Liver atrophy and necrosis			70, 71, 276
						102, 249
Phenylalanine[a]	DL	Chick	Perosis			153
	L	Chick	Changes in behavior and mental functioning		Glycine and threonine	160
		Monkey	Changes in behavior and mental functioning			152, 293
		Rat		Increased serine-threonine dehydratase activity; reduction in plasma threonine		152, 231, 232, 253, 254
		Rat		Reduction of plasma threonine		237
		Dog		Increased choline requirement		
		Dog		High moderate insulin release stimulated; high glucagon release stimulated		15, 132
	DL	Rat		Increased plasma phenylalanine (not so in rats)		30, 103, 140
	L	Chick		Increased plasma and urinary phenylpyruvate		193
		Rat		Increased brain phenylalanine		65, 92, 140, 303
		Hamster		Phenylalanine-hydroxylase activity decreased		138
		Rat		Phenylalanine-hydroxylase activity decreased		93, 311
		Rat		Tryptophan hydroxylation severely depressed in liver		
		Rabbit		Antibody synthesis depressed		247
Proline	L	Man		Gluconeogenic		198
		Dog		Moderate glucagon release stimulated; moderate insulin release stimulated		237
Serine	L	Dog		High glucagon release stimulated; very slight insulin release stimulated		237
	D	Rat	Kidney necrosis			283, 296
	DL	Rat	Kidney damage — corticomedullary junction (not L-serine)	Amino aciduria (not L-serine)		202, 203, 288, 289, 290
		Rat	Kidney damage to proximal convoluted tubule			163
Threonine[a]	D	Rat	Kidney damage; acute necrosis of proximal tubules		Glutathione glycine, and methionine	40, 86
	L	Dog		Glucosuria, amino aciduria, and proteinuria; Moderate insulin release stimulated; moderate glucagon release stimulated	Homocystine	237
		Rat	Cataracts (severity increased by tyrosine)		Tryptophan	5, 283

NUTRITIONAL PATHOSIS FROM AMINO ACID EXCESS IN ANIMALS AND MAN

Amino acid increased in diet (or intake)	Stereo-isomer	Subject	Anatomical structural change	Metabolic interaction	Counteracting factor	Ref.
Tryptophan[a]	L	Man	CNS disturbances	Increased serotonin	Monamine oxidase or decarboxylase inhibitor	135, 234, 295
		Rat		Splash enzyme effect (i.e., elevation of ornithine transaminase, phenylalanine, hydroxylase, phosphoenol-pyruvate carboxykinase, and tyrosine transaminase)		91, 92, 156, 227
		Rat	Cataracts	Increased plasma tryptophan and tissue tryptophan		132, 252, 271, 283
	DL	Cattle	Pulmonary emphysema	Highest insulin release stimulated; moderate glucagon release stimulated		37
	L	Dog				237
Tyrosine	L	Rat	Hip- and knee-joint cartilage discoloration; increased blood pressure		Increased plasma tyrosine	7, 33, 185
		Rat	Lesions in skin, bone, pancreas, liver, kidney, and testes		Cystine, glycine, methionine, or tryptophan, cortisol, thiouracil	6, 85, 176, 177, 186, 256, 278
		Rat	Protein synthesis in liver; protein degradation in muscle and brain polysomal aggregation in liver; polysomal disaggregation in muscle and brain	Increased excretion of homogentasis acid; tyrosine transaminase high and homogentasic oxidase low		36, 142, 167 174, 175, 256
		Mouse	Lesions, including collagen changes in skin and bone, mural invagination of arteriole in knee joint, perivascular aggregation of eosinophiles, adhesions of internal organs			176, 177
		Rabbit	Kidney and liver damage			256
		Dog	Kidney damage			256
		Man		Gluconeogenic		198
		Dog		Neither insulin nor glucagon release stimulated produced succinyl-CoA; glyconeogenic		237
Valine[a]	L	Rat		Mutual antagonism of branched-chain amino acids; branched-chain ketoaciduria	Isoleucine and leucine	58, 274

REFERENCES

1. **Abernathy, R. P. and Miller, J.,** Effects of imbalances or antagonisms among nonessential amino acids on growth and nitrogen utilization by rats, *J. Nutr.,* 86, 231–235, 1965.
2. **Adkins, J. S., Sunde, M. L., and Harper, A. E.,** The development of a free amino acid diet for the growing chick, *Poult. Sci.,* 41, 1382, 1962.
3. **Aisenberg, A. C.,** Studies on delayed hypersensitivity in Hodgkin's disease, *J. Clin. Invest.,* 41, 1964–1970, 1962.
4. **Akopyan, Z. I.,** Phenylalanine hydroxylase from the brain of white rats, *Vopr. Med. Khim.,* 14, 23, 1968.
5. **Alam, S. Q., Becker, R. V., Stucky, W. P., Rogers, Q. R., and Harper, A. E.,** Effect of threonine on the toxicity of excess tyrosine and cataract formation in the rat, *J. Nutr.,* 89, 91–96, 1966.
6. **Alam, S. Q., Boctor, A. M., Rogers, Q. R., and Harper, A. E.,** Some effects of amino acids and cortisol on tyrosine toxicity in the rat, *J. Nutr.,* 93, 317–323, 1967.
7. **Alam, S. Q., Rogers, Q. R., and Harper, A. E.,** Effect of tyrosine and threonine on free amino acids in plasma, liver, muscle, and eye of the rat, *J. Nutr.,* 89, 95–105, 1966.
8. **Allison, A. C., Hovi, T., Watts, R. W. E., and Webster, A. D. B.,** Immunological observations on patients with Lesch-Nyhan Sydrome, and on the role of de-novo purine synthesis in lymphocyte transformation, *Lancet,* 2, 1179–1183, 1975.
9. **Alvord, E. C., Stevenson, L. D., Vogel, F. S., and Engle, R. L., Jr.,** Neuropathological findings in phenylpyruvic oligophrenia (Phenylketonuia), *J. Neuropathol. Exp. Neurol.,* 9, 298, 1950.
10. **Anderson, J. O. and Combs, G. F.,** Effect of single amino acid excesses on glucose metabolism and chick growth, as influenced by the dietary amino acid balance, *J. Nutr.,* 46, 161–170, 1952.
11. **Anderson, J. O., Combs, G. F., Groschke, A. C., and Briggs, G. M.,** Effect on chick growth of amino acid inbalances in diets containing low and adequate levels of niacin and pyridoxine, *J. Nutr.,* 45, 345–360, 1951.
12. **Aoki, K. and Siegal, F. L.,** Hyperphenylalanemia disaggregation of brain polyribosomes in young rats, *Science,* 168, 129–130, 1970.
13. **Artom, C., Fishmann, W. H., and Morehead, R. P.,** The relative toxicity of L- and DL-serine in rats, *Proc. Soc. Exp. Biol. Med.,* 60, 284–287, 1945.
14. **Assan, R., Rosselin, G., and Dolais, J.,** Effets sur la glucagonémie des perfusions et ingestions d'acides aminés, *J. Annu. Diabetol. Hotel Dieu,* 7, 25–41, 1967.
15. **Auerback, V. H., Waisman, H. A., and Wyckeff, L. B., Jr.,** Phenylketonuria in the rat associated with decreased temporal discrimination learning, *Nature* (London) 182, 871–872, 1958.
16. **Baker, E. M., Canham, J. E., Nunes, W. T., Sauberlich, H. E., and McDowell, M. E.,** Vitamin B_6 requirement for adult men, *Am. J. Clin. Nutr.,* 15, 59–66, 1964.
17. **Bandman, E. and Gurney, T.,** Differences in the cytoplasmic distribution of newly synthesized poly(A) in serum-stimulated and resting cultures of BALB/c T3 cells, *Exp. Cell. Res.,* 90, 159–168, 1975.
18. **Battistin, L., Grynbaum, A., and Lajtha, A.,** The uptake of various amino acids by the mouse brain in vivo, *Brain Res.,* 29, 85–99, 1971.
19. **Belavady, B., Madhavan, T. V., and Gopalan, C.,** Experimental production of niacin deficiency in adult monkeys by feeding diets, *Lab. Invest.,* 18, 94–99, 1968.
20. **Belavady, B., Madhavan, T. V., and Gopalan, C.,** Production of nicotinic acid deficiency (black tongue) in pups fed diets supplemented with leucine, *Gastroenterology,* 53, 749–753, 1967.
21. **Belavady, B., Spikantia, S. G., and Gopalan, C.,** The effect of the oral administration of leucine on the metabolism of tryptophan, *Biochem. J.,* 87, 652–655, 1963.
22. **Benevenga, N. J. and Harper, A. E.,** Alleviation of methionine and homocystine toxicity in the rat, *J. Nutr.,* 93, 44–52, 1967.
23. **Benevenga, N. J. and Harper, A. E.,** Effect of glycine and serine on methionine metabolism in rats fed diets high in methionine, *J. Nutr.,* 100, 1205–1214, 1970.
24. **Benevenga, N. J., Harper, A. E., and Rogers, Q. R.,** Effect of an amino acid imbalance on the metabolism of the most limiting amino acid in the rat, *J. Nutr.,* 95, 434–444, 1968.
25. **Benton, D. A., Harper, A. E., and Elvehjem, C. A.,** Effect of isoleucine supplementation on the growth of rats fed zein or corn diets, *Arch. Biochem. Biophys.,* 57, 13–19, 1955.
26. **Benton, D. A., Harper, A. E., Spivey, H. E., and Elvehjem, C. A.,** Leucine, isoleucine, and valine relationships in the rat, *Arch. Biochem. Biophys.,* 60, 147–155, 1956.
27. **Bernhart, F. W. and Zilliken, A.,** Effect of dietary carbohydrate and chlortetracycline on growth and excretion of phenols in rats fed extra tyrosine, *Arch. Biochem. Biophys.,* 82, 462–471, 1959.

28. Birnbaum, S. M., Winitz, M., and Greenstein, J. P., Quantitative nutritional studies with water soluble chemically defined diets. III. Individual amino acids as sources of "nonessential" nitrogen, *Arch. Biochem. Biophys.*, 72, 428–436, 1957.

29. Boctor, A. M. and Harper, A. E., Tyrosine toxicity in the rat. Effect of high intake of *p*-hydroxyphenylpyruvic acid and of force feeding high tyrosine diet, *J. Nutr.*, 95, 535–540, 1968.

30. Boggs D. E. and Waisman, H. A., Biochemical correlates in rats with phenylketonuria, *Arch. Biochem. Biophys.*, 106, 307–311, 1964.

31. Boggs, D. E., DeRopp, R. S., and McKean, C. M., The effects of high circulatory phenylalanine on brain amino acids, *Fed. Proc. Fed. Am. Soc. Exp. Biol.*, 23, 146, 1964.

32. Boggs, D. E. and McKean, C. M., The effects of a high phenylalanine diet on blood and brain amino acids, *Fed. Proc. Fed. Am. Soc. Exp. Biol.*, 24, 316, 1965.

33. Bondurant, R. E. and Henry, J. B., Pathogenesis of ochronosis in experimental alkaptonuria of the white rat, *Lab. Invest.*, 14, 62, 1965.

34. Bose, K. K., Woodley, C. L., Chatterjee, N. K., and Gupta, N. K., Poly r-U directed ambiguous amino acid, incorporation in rabbit reticulocyte system, *Biochem. Biophys. Res. Commun.*, 37, 179, 1969.

35. Brown, J. H. and Allison, J. B., Effects of excess dietary DL-methionine and/or L-arginine on rats, *Proc. Soc. Exp. Biol. Med.*, 69, 196-198, 1948.

36. Butts, J. S., Sinnhuber, R. O., and Dunn, M. S., The metabolism of L-tryosine, *J. Biol. Chem.*, 140, xxii-xxiii, 1941.

37. Carlson, J. R., Dyer, I. A., and Johnson, R. J., Tryptophan-induced interstitial pulmonary emphysema in cattle, *J. Vet. Res.*, 92, 1983–1989, 1968.

38. Carlsten, A., Hallgren, B., Jagenburg, R., Svanborg, A., and Werkö, L., Myocardial metabolism of glucose, lactic acid, amino acids, and fatty acids in healthy human individuals at rest and at different work loads, *Scand. J. Clin. Lab. Invest.*, 13, 418–428, 1961.

39. Carlsten, A., Hallgren, B., Jagenburg, R., Svanborg, A., and Werkö, L., Arteriohepatic venous differences of free fatty acids and amino acids, *Acta Med. Scand.*, 181, 199–207, 1967.

40. Carone, F. A. and Ganote, C. E., D-Serine nephrotoxicity, *Arch. Pathol.*, 99, 658–682, 1975.

41. Carver, M. J., Influence of phenylalanine administration on the free amino acids of brain and liver in the rat, *J. Neurochem.*, 12, 45, 1965.

42. Chou, W. S., Savage, J. E., and O'Dell, B. L., Role of copper in biosynthesis of intramolecular cross-links of chick tendon collagen, *J. Biol. Chem.*, 244, 5785–5789, 1969.

43. Christensen, H. N., Amino acid transport and nutrition, *Fed. Proc. Fed. Am. Soc. Exp. Biol.*, 22, 1110, 1963.

44. Christensen, H. N., Stricher, J. A., and Elbinger, R. L., Effects of feeding individual amino acids upon the distribution of other amino acids between cells and extracellular fluid, *J. Biol. Chem.*, 172, 515–524, 1948.

45. Crome, L., The association of phenylketonuria with leucodystrophy, *J. Neurol. Neurosurg. Psychiatry*, 25, 149, 1962.

46. Crome, L. and Pare, C. M. B., Phenylketonuria a review and a report of the biological findings in four cases, *J. Ment. Sci.*, 106, 862, 1960.

47. Clark, J. T. R. and Lowden, J. A., Hyperphenylalanine: Effect on the developing rat brain, *Can. J. Biochem.*, 47, 291–295, 1969.

48. Cochrane, W. W., Payne, W. W., Simpkins, N. J., and Woolf, L. I., Familial hypoglycemia precipitated by amino acids, *J. Clin. Invest.*, 35, 411–422, 1956.

49. Cohen, H. P., Choitz, H. C., and Berg, C. P., Response of rats to diets high in methionine and related compounds, *J. Nutr.*, 64, 555–569, 1968.

50. Cohen, Z. A. and Hirsch, J. G., Isolation and properties of specific cytoplasmic granules of rabbit polymorphonuclear leucocytes, *J. Exp. Med.*, 112, 983–1004, 1960.

51. Cohlan, S. Q. and Stone, S. M., Effects of dietary and intraperitoneal excess of L-lysine and L-leucine on rat pregnancy and offspring, *J. Nutr.*, 74, 93–95, 1961.

52. Cohn, C. K., Dunner, D. L., and Axelrod, J., Reduced catechol-θ-methyltranferase activity in red blood cells of women with primary affective disorder, *Science*, 170–1324, 1970.

53. Cole, J. S., III and Longton, R. W., L-Aspartic acid, protein synthesis, and cell viability in cultured mammalian cells, *J. Dent. Res.*, 55, 712, 1976.

54. Cox, C. J. and Hudson, L., The nephropathogenic action of cystine. II. The dietary control of cystine nephrosis, *J. Nutr.*, 2, 271–276, 1930.

55. Cox, G. J., Smythe, C. V., and Fishback, C. F., The nephropathogenic action of cystine, *J. Biol. Chem.*, 82, 95–103, 1929.

56. Curtis A. C., Newburgh, L. H., and Thomas, F. H., The toxic action of cystine on the kidney, *Arch. Intern. Med.*, 39, 817–827, 1927.

57. **Dameshek, W.,** The significance of autoimmune disease, in *The Thymus,* Wolenstenholm, G. E. W., and Porter, R., Eds., Little, Brown, Boston, 1966, 476.

58. **Dancis, J., Levitz, M., and Westall, R. G.,** Maple syrup urine disease: Branched-chain ketoaciduria, *Pediatrics,* 25, 72–79, 1960.

59. **Daniel, R. G. and Waisman, H. A.,** The effects of excess amino acids on the growth of the young rat, *Growth,* 32, 255–265, 1968.

60. **Daniel, R. G. and Waisman, H. A.,** Adaptation of the weanling rat to diets containing excess methionine, *J. Nutr.,* 99, 299–306, 1969.

61. **Darmenton, P. and Woebe, J. F.,** Influence of L-lysine on glucose transport across the isolated small intestine of the rat, *C. R. Seances Soc. Biol.,* 160, 768, 1966.

62. **Datta, R. K. and Datta, B.,** Role of methylated nucleic acids in carcinogenesis, *Exp. Mol. Pathol.,* 10, 129, 1969.

63. **Debey, H. J., Snell, E. E., and Baumann, C. A.,** Studies on the interrelationship between methionine and vitamin B_6, *J. Nutr.,* 46, 203–214, 1952.

64. **Dhanamitta, S. A., Valyasevi, A., and Van Reen, R.,** Studies on bladder stone disease in Thailand, IX. Effect of orthophosphate and fat-free powdered milk supplementations on the occurrence of crystalluria, *Am. J. Clin. Nutr.,* 20, 1387, 1967.

65. **Dierks-Ventling, C., Wapnir, R. A., and Braude, M. C.,** Biochemical and behaviorial effects of a commercial "low phenylalanine" diet in growing rats, *Proc. Soc. Exp. Biol. Med.,* 127, 121–127, 1968.

66. **DiGeorge, A. M. and Auerback, V. H.,** Leucine-induced hypoglycemia: A review and speculations, *Am. J. Med. Sci.,* 240, 792–801, 1960.

67. **Dolan, G. and Godin, C.,** Phenylalanine toxicity in rats, *Can. J. Biochem.,* 44, 143–145, 1966.

68. **Earle, D. P., Jr. and Victor, J.,** Cirrhosis of the liver caused by excess dietary cystine, *J. Exp. Med.,* 73, 161–172, 1941.

69. **Earle, D. P., Jr. and Victor, J.,** The effect of various diets on the liver damage caused by excess cystine, *J. Exp. Med.,* 75, 179–189, 1942.

70. **Earle, D. P., Jr. and Kendall, F. E.** Liver damage and urinary excretion of sulfate in rats fed 1-cystine, d1-methionine, and cysteic acid, *J. Exp. Med.,* 75, 191–195, 1942.

71. **Earle, D. P., Jr., Smull, K., and Victor, J.,** Effects of excess dietary cysteic acid, d1-methionine, and taurine on the rat liver, *J. Exp. Med.,* 76, 317–323, 1942.

72. **Engelhardt, D. L.,** An inhibitor of protein synthesis in cytoplasmic extracts of density-inhibited cells, *J. Cell. Physiol.,* 78, 333–343, 1971.

73. **Fajans, S. S., Floyd, J. C., Jr., Knopf, R. F., Guntsche, E. M., Rull, J. A., Thiffault, C. A., and Conn, J. W.,** A difference in mechanism by which leucine and other amino acids induce insulin release, *J. Clin. Endocrinol.,* 27, 1600–1606, 1967.

74. **Fajans, S. S., Floyd, J. C., Jr., Knopf, R. F., and Conn, J. W.,** Effect of amino acids and proteins on insulin secretion in man, *Recent Prog. Horm. Res.,* 23, 617–662, 1967.

75. **Fajans, S. S., Floyd, J. C., Jr., Knopf, R. F., Rull, J. A., Guntsche, E. M., and Conn, J. W.,** Benzothiadiazine suppression of insulin release from normal and abnormal islet tissue in man, *J. Clin. Invest.,* 45, 481–492, 1966.

76. **Felig, P.,** Amino acid metabolism in man, *Annu. Rev. Biochem.,* 44, 933–955, 1975.

77. **Felig, P., Marliss, E., Owen, O. E., and Cahill, G. F., Jr.,** Blood glucose and gluconeogenesis in fasting man, *Arch. Intern. Med.,* 123, 293–298, 1969.

78. **Felig, P., Owen, O. E., Wahren, J., and Cahill, G. F., Jr.,** Amino acid metabolism during prolonged starvation, *J. Clin. Invest.,* 48, 584–594, 1969.

79. **Felig, P., Pozefsky, T., Marliss, E., Cahill, G. F., Jr.,** Alanine: Key role in gluconeogenesis, *Science,* 167, 1003–1004, 1970.

80. **Felig, P. and Wahren, J.,** Amino acid metabolism in exercising man, *J. Clin. Invest.,* 50, 2703–2714, 1971.

81. **Felig, P. and Wahren, J.,** Influence of endogenous insulin secretion on splanchnic glucose and amino acid metabolism in man, *J. Clin. Invest.,* 50, 1702–1711, 1971.

82. **Felig, P., Wahren, J., and Raf, L.,** Evidence of interorgan amino acid transport by blood cells in humans, *Proc. Natl. Acad. Sci. U.S.A.,* 70, 1775–1779, 1973.

83. **Findlay, J. A. and Ashcroft, S. J. H.,** Cells of the islets of Langerhans, in *The Cell in Medical Science,* Vol. 3, Beck, F. and Lloyd, J. B., Eds., Academic Press, London, 1975, 293–306.

84. **Finkelstein, J. D.,** Methionine metabolism in mammals. The biochemical basis for homocystinuria, *Metabolism,* 23, 387–398, 1974.

85. **Fisher, H., Grininger, P., and Lutz, H.,** The amino acid requirement of laying hens. Amino acid balance in low protein diets, *Poult. Sci.,* 39, 173–175, 1960.

86. **Fishman, W. H. and Artom, C.,** Serine injury, *J. Biol. Chem.,* 145, 345–346, 1942.

87. Fitch, W. M., Evidence suggesting a nonrandom character to nucleotide replacements in naturally occuring mutations, *J. Mol. Biol.*, 26, 499–507, 1967.

88. Foote, J. L., Allen, R. J., and Agranoff, B. W., Fatty acids in esters and cerebrosides human brain in phenylketonuria, *J. Lipid Res.*, 6, 518–524, 1965.

89. Foote, J. L. and Tao, R. V. P., The effects of P-chlorophenylalanine and phenylalanine in brain ester-bound fatty acids of developing rats, *Life Sci.*, 7 (II), 1187–1192, 1968.

90. Forbes, J. C. and Neal, R. C., The production of chronic arthritis by indole and other products of tryptophan putrefaction, *J. Lab. Clin. Med.*, 22, 921–924, 1937.

91. Foster, D. O., Lardy, H. A., Ray, P. D., and Johnson, J. B., Alteration of rat liver phosphoenolpyruvate carboxykinase activity by L-tryptophan in-vivo and metals in-vitro, *Biochemistry*, 6, 2120–2128, 1967.

92. Freedland, R. A., Krakowski, M. C., and Waisman, H. A., Influence of amino acids on rat liver phenylalanine hydroxylase activity, *Am. J. Physiol.*, 206, 341–344, 1964.

93. Freedland, R. A., Wadzinski, I. M., and Waisman, H. A., The effect of aromatic amino acids on the hydroxylation of tryptophan, *Biochem. Biophys. Res. Commun.*, 6, 227–231, 1961.

94. Freese, E., The difference between spontaneous and base analogue induced mutations of phage T_4, *Proc. Natl. Acad. Sci. U.S.A.*, 45, 622–633, 1959.

95. Geller, E. and Yuwiler, A., Brain amine decrease in leucine-fed rats, *J. Neurochem.*, 14, 725–731, 1967.

96. Genuth, S. M., The effects of oral alanine administration in fasting obese subjects, *Metabolism*, 22, 927, 1973.

97. Genuth, S. M. and Castro, J., Effect of oral alanine on blood deltahydroxybutyrate, and plasma glucose, insulin, free fatty acids, and growth hormone in normal and diabetic subjects, *Metabolism*, 23, 375–386, 1974.

98. Gerber, M. A. and Paronetto, F., I E in glomeruli of patients with nephrotic syndrome, *Lancet*, 1, 1097–1099, 1971.

99. Gerich, J. E., Charles, M. A., and Grodsky, G. M., Characterization of the effects of arginine and glucose on glucagon and insulin release from the perfused rat pancreas, *J. Clin. Invest.*, 54, 833–841, 1974.

100. Gerstl, B., Malamud, M. D., Eng., L. F., and Hyman, R. B., Lipid alteration in human brains in phenylketonuria, *Neurology*, 17, 51–59, 1967.

101. Ghadimi, H., Partington, M. W., and Hunter, A., A familial disturbance of histidine, *N. Engl. J. Med.*, 265, 221–224, 1961.

102. Girard-Globa, A., Robin, P., and Forestier, M., Long-term adaptation of weanling rats to high dietary levels of methionine and serine, *J. Nutr.*, 102, 209–218, 1972.

103. Goldstein, F. B., Biochemical studies on phenylketonuria. I. Experimental hyperphenylalanemia in the rat, *J. Biol. Chem.*, 236, 2656–2661, 1961.

104. Goldstein, J. L., Campbell, B. K., and Gartler, S. M., Homocystinuria: Heterozygote detection using phytohemagglutinin-stimulated lymphocytes, *J. Clin. Invest.*, 52, 218–221, 1973.

105. Good, R. A., Peterson, R. D. A., Martinez, C., Sutherland, D. E. R., Kellum, M. J., and Finstad, J., The thymus in immunology: with special reference to autoimmune disease, *Ann. N.Y. Acad. Sci.*, 124, 73–94, 1965.

106. Gopalan, C., Leucine and pellagra, *Nutr. Rev.*, 26, 323–326, 1968.

107. Gopalan, C. and Srikantia, S. G., Leucine and pellagra, *Lancet*, 1, 954–957, 1960.

108. Graham, C. E., Hier, S. W., Waitkoff, H. K., Saper, S. M., Bibler, W. G., and Pentz, E. I., Studies on natural and racemic amino acids with rats, *J. Biol. Chem.*, 185, 97–102, 1950.

109. Grau, G. R., Interrelations of phenylalanine and tyrosine in the chick, *J. Biol. Med.*, 170, 661–669, 1947.

110. Grau, C. R. and Kamel, M., Amino acid imbalance and growth requirements for lysine and methionine, *J. Nutr.*, 41, 89–101, 1950.

111. Greenberg, D. M., Ed., Carbon catabolism of amino acids, in *Metabolic Pathways*, Vol. 2, 2nd ed., Academic Press, New York, 1961, 79–162.

112. Griffith, W. H. and Wade, N. J., Choline metabolism. II. The interrelationship of choline, cystine, and methionine in the occurrence and prevention of hemorrhagia degeneration in young rats, *J. Biol. Chem.*, 132, 627–637, 1940.

113. Groschke, A. C., Anderson, J. O., and Briggs, G. M., Peculiar enlargement of eyeballs in chicks caused by feeding a high level of glycine, *Proc. Soc. Exp. Biol. Med.*, 69, 488–491, 1948.

114. Groschke, A. C. and Briggs, G. M., Inhibitory action of certain amino acids on chicks receiving nicotinic acid-low diets, *J. Biol. Chem.*, 165, 739–740, 1946.

115. Han, T. and Sokal, J. E., Lymphocyte response to phytohemaglutinin in Hodgkin's disease, *Am. J. Med.*, 48, 728–734, 1970.

116. **Hankes, L. V., Brown, R. R., Lippincott, S., and Schmaller, M.,** Effects of L-tryptophan load on the metabolism of tryptophan 2C^{14} in man, *J. Lab. Clin. Med.,* 69, 313–324, 1967.
117. **Hanking, B. M. and Roberts, S.,** Stimulation of protein synthesis *in vitro* by elevated levels of amino acids, *Biochim. Biophys. Acta,* 104, 427, 1965.
118. **Hanking, B. M. and Roberts, S.,** Influence of alterations in intracellular levels of amino acids on protein synthesizing activity of isolated ribosomes, *Nature* (London), 207, 862, 1965.
119. **Hansch, C. and Fujita, T.,** p-o-π Analysis, A method for the correlation of biological activity and chemical structure, *J. Am. Chem. Soc.,* 86, 1616–1626, 1964.
120. **Hansch, C., Muir, R. M., Fujita, T., Maloney, P. O., Geiger, F., and Striech, M.,** Correlations of biological activity of plant growth regulators and chloromycetin derivations, with Hammett constants and partition coefficients, *J. Am. Chem. Soc.,* 85, 2817–2824, 1963.
121. **Harper, A. E.,** Amino acid balance and imbalance. I. Dietary level of protein and amino acid imbalance, *J. Nutr.,* 68, 405–418, 1959.
122. **Harper, A. E., Benton, D. A., and Elvehjem, C. A.,** L-leucine and isoleucine antagonist in the rat, *Arch. Biochem. Biophys.,* 57, 1–12, 1955.
123. **Harper, A. E., Becker, R. B., and Stucki, W. P.,** Some effects of excessive intake of indispensable amino acids, *Proc. Soc. Exp. Biol. Med.,* 121, 695–699, 1966.
124. **Harper, A. E., Benevenga, N. J., and Wohlhueter, R. M.,** Effects of ingestion of disproportionate amounts of amino acids, *Physiol. Rev.,* 50, 428–558, 1970.
125. **Harper, A. E., Benton, D. A., Winje, M. E., and Elvehjem, C. A.,** Leucine-isoleucine antagonism in the rat, *Arch. Biochem. Biophys.,* 51, 523–524, 1954.
126. **Hassel, J. A. and Englehardt, D. L.,** Translational inhibition in extracts from serum-deprived animal cells, *Biochim. Biophys. Acta,* 324, 545, 1973.
127. **Hassel, J. A. and Engelhardt, D. L.,** The regulation of protein synthesis in animal cells by serum factors, *Biochemistry,* 15, 1375–1381, 1976.
128. **Hepburn, F. N. and Bradley, W. B.,** The glutamic acid and arginine requirement for high growth rate of rats fed amino acid diets, *J. Nutr.,* 84, 305–312, 1964.
129. **Hepburn, F. N. and Bradley, W. B.,** Effect of glutamine or inhibition of rat growth by glycine and serine, *J. Nutr.,* 94, 504–510, 1968.
130. **Hepburn, F. N., Calhoun, W. K., and Bradley, W. B.,** A growth response of rats to glutamic acid when fed an amino acid diet, *J. Nutr.,* 72, 163–168, 1960.
131. **Hershko, A., Mamont, P., Shields, R., and Tomkins, G. M.,** Pleiotypic response, *Nature New Biol.* (London), 232, 206–211, 1971.
132. **Hier, S. W.,** Influence of ingestion of single amino acids on the blood level of free amino acids, *J. Biol. Chem.,* 171, 813–820, 1947.
133. **Hier, S. W., Graham, G. E., and Klein, D.,** Inhibitory effect of certain amino acids on growth of young male rats, *Proc. Soc. Exp. Biol. Med.,* 56, 187–190, 1944.
134. **Hill, D. C., Slinger, S. J., and Marcellus, F. N.,** Tolerance of chicks for diets high in tyrosine, *Poult. Sci.,* 24, 234–236, 1945.
135. **Hodge, J. V., Oates, J. A., and Sjoerdsma, A.,** Reduction of the central effect of tryptophan by a decarboxylase inhibitor, *Clin. Pharmacol. Ther.,* 5, 149–155, 1964.
136. **Holden, J. T., Ed.,** *Amino Acid Pools,* Elsevier, Amsterdam, 1961, 850.
137. **Horsley, W., Sternlicht, H., and Cohen, J. S.,** Carbon-13 magnetic resonance studies of amino acids and peptides, *J. Am. Soc.,* 92, 680–686, 1970.
138. **Horwitz, I. and Waisman, H. A.,** Some biochemical changes in hamsters fed excess phenylalanine diets, *Proc. Soc. Exp. Biol. Med.,* 122, 750–754, 1966.
139. **Hsu, P. T. and Combs, G. F.,** Effect of vitamin B$_{12}$ and amino acid imbalances on growth and levels of certain blood constituents in the chick, *J. Nutr.,* 47, 73–91, 1952.
140. **Huang, I., Tannenbaum, S., Blume, L., and Hsia, D. Y.,** Metabolism of 5-hydroxyindole compounds in experimentally produced phenylketonuric rats, *Proc. Soc. Exp. Biol. Med.,* 106, 533–536, 1961.
141. **Ingram, V.,** Gene mutations in human haemoglobin: The chemical difference between normal and sickle cell haemoglobin, *Nature* (London), 180, 326–328, 1957.
142. **Ip, C. and Harper, A. E.,** Protein synthesis in liver, muscle, and brain of rats fed a high tyrosine-low protein diet, *J. Nutr.,* 105, 885–893, 1975.
143. **Israelstam, D. M., Sargent, T., Finley, N. N., Winchell, H. S., Fish, N. B., Motto, J., Pollycove, M., and Johnson, A.,** Abnormal methionine metabolism in schizophrenic and depressive states. A preliminary report, *J. Psychiatr. Res.,* 7, 185–190, 1970.
144. **Jefferson, L. S. and Karner, A.,** Influence of amino acid supply on ribosomes and protein synthesis on perfused rat liver, *Biochem. J.,* 111, 703–712, 1969.
145. **Jones, J. D.,** Lysine-arginine antagonism in the chick, *J. Nutr.,* 84, 313–321, 1964.

146. Jones, J. D., Petersburg, S. J., and Burnett, D. C., The mechanism of lysine-arginine antagonism in the chick: Effect of lysine on digestion, kidney arginase and liver transamidinase, *J. Nutr.,* 93, 103–116, 1967.

147. Jones, J. D., Walters, R., and Burnett, P. C., Lysine-argine-electrolyte relationships in the rat, *J. Nutr.,* 89, 171–188, 1966.

148. Jukes, T. H., Relations between mutations and base sequences in amino acid code, *Proc. Natl. Acad. Sci. U.S.A.,* 48, 1809–1815, 1966.

149. Jukes, T. H. and Gatlin, L., Recent studies concerning the coding mechanism, *Prog. Nucleic Acid Res. Mol. Biol.,* 11, 303–350, 1971.

150. Kade, C. F., Jr. and Shepherd, J., The inhibitory effect of excess methionine on protein utilization, *Fed. Proc. Fed. Am. Soc. Exp. Biol.,* 7 (abstr.), 291, 1948.

151. Kaneto, A. and Kosaka, K., Stimulation of glucagon secretion by arginine and histidine infused intrapancreatically, *Endocrinology,* 88, 1239–1245, 1971.

152. Karrer, R. and Cahilly, G., Experimental attempts to produce phenylketonuria in animals: a critical review, *Psychol. Bull.,* 64, 52–64, 1965.

153. Katz, R. S. and Baker, D. H., Methionine toxicity in the chick: Nutritional and metabolic implications, *J. Nutr.,* 105, 1168–1175, 1975.

154. Kaufman, N., Klavins, J. V., and Kinney, T. D., Pancreatic damage induced by excess methionine, *Arch. Pathol.,* 70, 331–337, 1960.

155. Kelly, M. and Scott, H. M., Plasma lysine titers in the chick in relation to source of lysine and mode of administration, *J. Nutr.,* 94, 326–330, 1968.

156. Kenney, F. T. and Flora, R. M., Induction of tyrosine-α-ketoglutarate transaminase in rat liver. I. Hormonal nature, *J. Biol. Chem.,* 236, 2699–2702, 1961.

157. Kerr, G. R. and Waisman, H. A., Dietary induction of hyperphenylalanemia in the rat, *J. Nutr.,* 92, 10–18, 1967.

158. Kerr, G. R., Wolf, R. C., and Waisman, H. A., Hyperlipemia in infant monkeys fed excess L-histidine, *Proc. Soc. Exp. Biol. Med.,* 119, 561–562, 1965.

159. Kerr, G. R., Wolf, R. C., and Waisman, H. A., A disorder of lipid metabolism associated with experimental hyperhistidinemia in Macaca mulata, in *Some Recent Developments in Comparative Medicine,* Academic Press, New York, 1966, 371–392.

160. Ketola, H. C. and Nesheim, M. C., Influence of dietary protein and methionine levels on the requirement for choline by chickens, *J. Nutr.,* 104, 1484–1489, 1974.

161. Klain, G. J., Scott, H. M., and Johnson, B. C., The amino acid requirement of the growing chick fed crystalline amino acids, *Poult. Sci.,* 37, 976–977, 1958.

162. Klavins, J. V., Pathology of amino acid excess. II. Effects of administration of excessive amounts of sulfur-containing amino acids: L-cystine, *Br. J. Exp. Pathol.,* 44, 516–519, 1963.

163. Klavins, J. V. and Johansen, P. V., Pathology of amino acid excess. IV. Effects and interactions of excessive amounts of dietary methionine, homocystine, and serine, *Arch. Pathol.,* 79, 600–614, 1965.

164. Klavins, J. V., Kinney, T. D., and Kaufman, N., Histopathologic changes in methionine excess, *Arch. Pathol.,* 75, 661–673, 1963.

165. Klavins, J. V. and Peacocke, I. L., Pathology of amino acid excess. III. Effects of administration of excessive amounts of sulfur-containing amino acids: methionine with equimolar amounts of glycine and arginine, *Br. J. Exp. Pathol.,* 45, 533–547, 1964.

166. Lamar, C. and Imm, B., Effect of specific amino acids on in vitro protein synthesizing system from rats, *J. Nutr.,* 101, 1589–1594, 1971.

167. Lanyer, F., Experimental alkaptonuria in the albino rat, *Z. Phys. Chem.,* 278, 155–164, 1943.

168. Leathem, J. H., Oslapas, R., and Fisher, C. J., Protein nutrition and the biochemical composition of the uterus, *Fertil. Steril.,* 19, 206–212, 1968.

169. Lehmann, H. and Carrell, R. N., Variations in the structure of human haemoglobin with particular reference to the unstable haemoglobins, *Br. Med. Bull.,* 25, 14–23, 1969.

170. Lehmann, H. and Huntsman, R. G., *Man's Haemoglobin,* p. 188–204, North-Holland (Publishing Company), Amsterdam, 1974.

171. Lepkovsky, S., Roboz, E., and Haagensmit, A. J., Xanthurenic acid and its role in tryptophan metabolism of pyridoxine-deficient rats, *J. Biol. Chem.,* 149, 195–201, 1943.

172. Lewis, H. B., The metabolism of sulfur. IX. The effect of repeated administration of small amounts of cystine, *J. Biol. Chem.,* 65, 187–195, 1925.

173. Lillie, R. D., Histopathologic changes produced in rats by the addition to the diet of various amino acids, *U.S. Public Health Rep.,* 47, 83–93, 1932.

174. Lin, E. C. C. and Knox, W. E., Role of enzymatic adaptation in production of experimental alkaptonuria, *Proc. Soc. Exp. Biol. Med.,* 96, 501–505, 1957.

175. Lin, E. C. C. and Knox, W. E., Adaptation of the rat liver tyrosine-α-ketoglutarate transaminase, *Biochim. Biophys. Acta,* 26, 85–88, 1957.

176. **Longton, R. W.,** Connective Tissue Response to Stress Associated with Intraperitoneally Injected Amino Acid During Biosynthesis of Collagen in Swiss White Mice, Ph.D. thesis, Northwestern University, Evanston, Ill., 1971.

177. **Longton, R. W.,** Production of lesions in various tissues by amino acid, *Experientia,* 29, 1013–1015, 1973.

178. **Longton, R. W., Cole, J. S., III, and Grays, R.,** Reculture of electrofocused foetal rabbit skin cells, *Differentiation,* 2, 43–46, 1974.

179. **Lucas, D. R. and Newhouse, J. P.,** The toxic effect of sodium L-glutamate on the inner layers of the retina, *AMA Arch. Opthalmol.,* 58, 193–201, 1957.

180. **Machlin, L. J., Lankenau, A. H., Denton, C. A., and Bird, H. R.,** Effect of vitamin B_{12} and folic acid on growth and uricemia of chickens fed high levels of glycine, *J. Nutr.,* 46, 389–398, 1952.

181. **Malamud, N.,** Neuropathology of phenylketonuria, *J. Neuropathol. Exp. Neurol.,* 25, 254–268, 1966.

182. **Marliss, E. B., Aoki, T. T., Pozefsky, T., Most, A., and Cahil, G. F., Jr.,** Muscle and splanchnic glutamine and glutamate metabolism in postabsorptive and starved man, *J. Clin. Invest.,* 50, 814–817, 1971.

183. **Martin, G. J.,** Toxicity of amino acids as influenced by riboflavin deficiency, *Proc. Soc. Exp. Biol. Med.,* 63, 528–529, 1946.

184. **Martin, G. J.,** The toxicity of histidine in the normal and avitaminotic rat, *Exp. Med. Surg.,* 5, 191–195, 1947.

185. **Martin, G. J.,** The hypertensive effect of diets high in tyrosine, *Arch. Biochem.,* 1, 397–401, 1943.

186. **Martin, G. J.,** Toxicity of tyrosine in normal and vitamin deficient animals, *J. Am. Pharma. Assoc.,* 36, 187–190, 1947.

187. **Martin, G. J. and Hueper, W. C.,** Biochemical lesions produced by diets high in tyrosine, *Arch. Biochem.,* 1, 435–438, 1943.

188. **Maruyama, K., Harper, A. E., and Sunde, M. L.,** Effects of D-, DL-, and L-glutamic acid on chicks, *J. Nutr.,* 105, 1012–1019, 1975.

189. **Maruyama, K., Sunde, M. L., and Harper, A. E.,** Effect of D-alanine, and D-aspartic acid on the chick, *J. Nutr.,* 102, 1441–1452, 1972.

190. **Matsutaka, H., Aikawa, T., Yamanoto, H., and Ishikawa, E.,** Gluconeogenesis and amino acid metabolism. III. Uptake of glutamine and output of alanine and ammonia by non-hepatic splanchnic organs of fasted rats and their metabolic significance, *J. Biochem.* (Tokyo), 74, 1019–1029, 1973.

191. **McCully, K. S.,** Macromolecular basis for homocysteine-induced changes in proteoglycan structure in growth and arteriosclerosis, *Am. J. Pathol.,* 66, 83–96, 1972.

192. **McKean, C. M. and Boggs, D. E.,** Influence of high concentrations of phenylalanine on the amino acids of cerebrospinal fluid and blood, *Proc. Soc. Exp. Biol. Med.,* 122, 987–991, 1966.

193. **McKean, C. M., Boggs, D. E., and Peterson, N. A.,** The influence of high phenylalanine and tyrosine on the concentration of essential amino acids in brain, *J. Neurochem.,* 15, 235–241, 1968.

194. **McKean, G. M., Schanberg, S. M., and Giarman, N. J.,** Amino acidemias: Effects on maze performance and cerebral serotoni, *Science,* 157, 213–215, 1967.

195. **Meister, A.,** Some disorders of amino acid metabolism in man, in *Biochemistry of the Amino Acids,* Vol. 2, 2nd ed., Academic Press, New York, 1965, 1021–1084.

196. **Menge, H. and Combs, G. F.,** Action of vitamin B_{12} in counteracting glycine toxicity in the chick, *Proc. Soc. Exp. Biol. Med.,* 75, 139–142, 1950.

197. **Menkes, G. A.,** Cerebral lipids in phenylketonuria, *Pediatrics,* 37, 967–978, 1966.

198. **Mestyan, J., Soltész, G., Schultz, K., and Horváth, M.,** Hyperaminoacidemia due to the accumulation of gluconeogenic amino acid precursors in hypoglycemic small-for-gestational age infants, *J. Pediatr.,* 87, 409–414, 1975.

199. **Miller, J. F. A. P.,** Role of the thymus in immunity, *Br. Med. J.,* 2, 459–464, 1963.

200. **Miller, E. J., Martin, G. R., Mecca, C. E., and Piez, K. A.,** The biosynthesis of elastin cross-links. The effect of copper deficiency and a lathrogen, *J. Biol. Chem.,* 240, 3623–3627, 1965.

201. **Moghissi, K. S., Churchill, J. A., and Kurrie, D.,** Relationship of maternal amino acids and proteins to fetal growth and mental development, *Am. J. Obstet. Gynecol.,* 123, 398–410, 1975.

202. **Morehead, R. P., Fishman, W. H., and Artom, C.,** Renal injury in the rat following the administration of serine by stomach tube, *Am. J. Pathol.,* 21, 803–812, 1945.

203. **Morehead, R. P., Fishman, W. H., and Antom, C.,** The nephrotoxic action of DL-serine as related to certain dietary factors, *Am. J. Pathol.,* 22, 385–393, 1946.

204. Mortimore, G. E., Neely, A. N., Cox, J. R., and Guinivan, R. A., Proteolysis in homogenates of perfused rat liver. Response to insulin glucagon, and amino acids, *Biochem. Biophys. Res. Commun.*, 54, 89–95, 1973.

205. Muller, W. A., Faloona, G. R., and Unger, R. H., The effect of alanine on glucagon secretion, *J. Clin. Invest.*, 50, 2215–2218, 1971.

206. Munch, B. G., Amino acid transport by the small intestine of the rat. The transintestinal transport of tryptophan in relation to the transport of neutral and basic amino acids, *Biochim. Biophys. Acta,* 126, 299–307, 1966.

207. Naber, E. C., Cravens, W. W., Baumann, C. A., and Bird, H. R., The relation of dietary supplements and tissue metabolites to glycine toxicity in the chick, *J. Nutr.,* 60, 75–85, 1956.

208. Naber, E., Snell, C., and Cravens, W. W., The effect of folic acid on glycine toxicity in the chick, *Arch. Biochem. Biophys.,* 37, 158–163, 1952.

209. Nakoneczna, I., Forbes, J. C., and Rogers, K. S., The arthrogenic effect of indole, skatole, and other tryptophan metabolites in rabbits, *Am. J. Pathol.,* 57, 523–532, 1969.

210. Neame, K. D., Phenylalanine as inhibitor of transport of amino acids in brain, *Nature,* (London), 192, 173–174, 1961.

211. Newburgh, L. H. and Marsh, P. L., Renal injuries by amino acids, *Arch. Intern. Med.,* 36, 682–711, 1925.

212. Nimini, M. E., Hom, D., and Bavetta, L. A., Dietary composition and tissue protein synthesis. II. Tryptophan toxicity and amino acid deficiency in collagen synthesis, *J. Nutr.,* 78, 133–138, 1962.

213. Niven, C. F., Jr., Washburn, M. R., and Sperling, G. A., Growth retardation and corneal vascularization with tyrosine and phenylalanine in a purified diet, *Proc. Soc. Exp. Biol. Med.,* 63, 106–108, 1946.

214. O'Dell, B. L., Bird, D. W., Ruggles, D. L., and Savage, J. E., Composition of aorta tissue from copper-deficient chicks, *J. Nutr.,* 88, 9–14, 1966.

215. O'Dell, B. L., Elsden, D. F., Thomas, J., Partridge, S. M., Smith, R. H., and Palmer, R., Inhibition of demosine biosynthesis by a lathrogen, *Biochem. J.,* 96, 35, 1965.

216. O'Dell, B. L., Hardwick, B. C., Reynolds, G., and Savage, J. E., Connective tissue defect in the chick resulting from copper deficiency, *Proc. Soc. Exp. Biol. Med.,* 108, 402–405, 1961.

217. Odessey, R. and Goldberg, A., Oxidation of leucine by rat skeletal muscle, *Am. J. Physiol.,* 223, 1376–1383, 1972.

218. Olney, J. W., Brain lesions, obesity and other disturbances in mice treated with monosodium glutamate, *Science,* 164, 719–721, 1969.

219. Olney, J. W. and Ho, O. L., Brain damage in infant mice following oral intake of glutamate, aspartame, or cysteine, *Nature,* (London), 227, 609–610, 1970.

220. Olney, J. W. and Sharpe, L. G., Brain lesions in an infant rhesus monkey treated with monosodium glutamate, *Science,* 166, 386–388, 1969.

221. Olney, J. W., Sharpe, L. G., and Feigin, R. D., Glutamate-induced brain damage in infant primates, *J. Neuropathol. Exp. Neurol.,* 31, 464–488, 1972.

222. Owen, E. E. and Robinson, R. R., Amino acid extraction and ammonia metabolism by human kidney during prolonged administration of ammonium chloride, *J. Clin. Invest.,* 42, 263–276, 1963.

223. Page, E. and Gingras, R., Glycine toxicity and pyridoxine requirements in the white rat, *Trans. R. Soc. Can. Sect. V,* 40, 119–122, 1946.

224. Pagliara, A. S., Karl, I. E., DeVivo, D. C., and Kipnia, D. M., Hypoalaninemia: a concomitant of ketotic hypoglycemia, *J. Clin. Invest.,* 51, 1440–1449, 1972.

225. Pagliara, A. S., Stillings, S. N., Hover, B., Martin, D. M., and Matschinsky, F. M., Glucose modulation of amino acid-induced glucagon and insulin release in the isolated perfused rat pancreas, *J. Clin. Invest.,* 54, 819–832, 1974.

226. Pek, S., Fajans, S. S., Floyd, J. C., Knopf, R. K., and Conn, J. W., Effects upon plasma glucagon of infused and ingested amino acids and of protein meals in man, *Diabetes,* 18, 328, 1969.

227. Peraino, C., Blake, R. L., and Pitot, H. C., Studies on the induction and repression of enzymes in rat liver. III. Induction of ornithine s-transaminase and threonine dehydrase by oral intubation of free amino acids, *J. Biol. Chem.,* 240, 3039–3043, 1965.

228. Persaud, T. V. N., Development abnormalities induced by the amino acid leucine, *Naturwissenschaften,* 56, 37–38, 1969.

229. Phear, E. A., Ruebner, B., Sherlock, S., and Summerskill, W. H., Methionine toxicity in liver disease and its prevention by chlortetracycline, *Clin. Sci.,* 15, 93–117, 1956.

230. Pickering, R. J., Pollara, B., and Meuwissen, H., Meeting report: Workshop on severe combined immunological deficiency disease and adenosine deaminase deficiency, *Clin. Immunol. Immunopathol.,* 3, 301–303, 1974.

231. **Polidora, V. J.,** Behavioral effects of "phenylketonuria" in rats, *Proc. Natl. Acad. Sci. U.S.A.,* 57, 102–106, 1967.

232. **Polidora, V. J., Boggs, D. E. and Waisman, H. A.,** A behavioral deficit associated with phenylketonuria in rats, *Proc. Soc. Exp. Biol. Med.,* 113, 817–820, 1963.

233. **Prensky, A. L. and Moser, H. W.,** Brain lipids, proteolipids, and free amino acids in maple syrup urine disease, *J. Neurochem.,* 13, 863–874, 1966.

234. **Quay, W. B.,** Effect of dietary phenylalanine and tryptophan on pineal and hypothalamic serotonin levels, *Proc. Soc. Exp. Biol. Med.,* 114, 718–721, 1963.

235. **Rendina, G., Ryan, M. F., Oslong, J., Tuttle, J. M., and Giles, C. E.,** Some biochemical consequences of feeding excesses of phenylalanine to rats, *J. Ment. Defic. Res.,* 11, 153–168, 1967.

236. **Robinson, J. W. L.,** Interactions between neutral and dibasic amino acids for uptake by the rat intestine, *Eur. J. Biochem.,* 7, 78–89, 1968.

237. **Rocha, D. M., Faloona, G. F., and Unger, R. H.,** Glucagon stimulating activity of 20 amino acids in dogs, *J. Clin. Invest.,* 51, 2346–2351, 1972.

238. **Rogers, K. S.,** Rabbit erythrocyte hemolysis of lipophilic aryl molecules (33737), *Proc. Soc. Exp. Biol. Med.,* 130, 1140–1142, 1969.

239. **Rogers, K. S., Forbes, J. C., and Nakoceczna, J.,** Arithrogenic properties of lipophilic, aryl molecules (33949), *Proc. Soc. Exp. Biol. Med.,* 131, 670–672, 1969.

240. **Ross, B. D., Hems, R., and Krebs, H. A.,** The rate of gluconeogenesis from various precursors in the perfused rat lever, *Biochem. J.,* 102, 942–951, 1967.

241. **Rossini, A. A., Aoki, T. T., Ganda, O. P., Soeldner, J. S., and Cahill, G. F., Jr.,** Alanine-induced amino acid interrelationships, *Metabolism,* 24, 1185–1192, 1975.

242. **Roth, J. S. and Allison, J. B.,** The effect of feeding excess glycine L-arginine, and DL-methionine to rats on a casein diet, *Proc. Soc. Exp. Biol. Med.,* 70, 327–330, 1949.

243. **Roth, J. S. and Allison, J. B.,** The effects of feeding excess DL-methionine and choline to rats on a casein diet, *J. Biol. Chem.,* 183, 173–178, 1950.

244. **Roth, J. S., Allison, J. B., and Milch, L. J.,** The sulfur balance of rats fed excess DL-methionine plus glycine or DL-alanine, *J. Biol. Chem.,* 186, 113–118, 1950.

245. **Rudland, P. S.,** Control of translation in cultured cells. Continued synthesis and accumulation of messenger RNA in nondividing cultures, *Proc. Natl. Acad. Sci. U.S.A.,* 71, 750–754, 1974.

246. **Russell, W. C., Taylor, M. W., and Hogan, J. M.,** Effect of excess essential amino acids on growth of the white rat, *Arch. Biochem. Biophys.,* 39, 249–253, 1952.

247. **Ryan, W. L. and Carver, M. J.,** Inhibition of antibody synthesis by L-phenylalanine, *Science,* 143, 479–480, 1964.

248. **Salmon, W. D.,** The significance of amino acid imbalance in nutrition, *Am. J. Clin. Nutr.,* 6, 487–494, 1958.

249. **Sanchez, A. and Swenseid, M. E.,** Amino acid levels and enzyme activity in tissues of rats force-fed diets differing in methionine content, *J. Nutr.,* 99, 145–151, 1969.

250. **Sarma, D. S. R., Bongiorno, M., and Sidransky, H.,** Influence of L-tryptophan on mouse liver polyribosomes, *Fed. Proc. Fed. Am. Soc. Exp. Biol.,* 27, 1136, 1968.

251. **Sarma, P. S., Snell, E. E., and Elvehjem, C. A.,** The bioassay of vitamin B4 in natural materials. *J. Nutr.,* 33, 121–128, 1947.

252. **Sauberlich, H. E.,** Studies on the toxicity and antagonism of amino acids for weanling rats, *J. Nutr.,* 75, 61–72, 1961.

253. **Savage, J. R. and Harper, A. E.,** Influence of gelatin on growth and liver pyridine nucleotide concentration of the rat, *J. Nutr.,* 83, 158–164, 1964.

254. **Schalock, R. L. and Klopter, F. D.,** Phenylketonuria enduring behaviorial deficits in phenylketonuric rats, *Science,* 155, 1033–1035, 1967.

255. **Schimke, R. T.,** Studies on factors affecting the levels of urea cycle enzymes in rat liver, *J. Biol. Chem.,* 238, 1012–1018, 1963.

256. **Schweizer, W.,** Studies on the effect of L-tyrosine on the white rat, *J. Physiol.* (London), 106, 167–176, 1947.

257. **Scriver, S. R., Clow, C. L., and Lamm, P.,** *Am. J. Clin. Nutr.,* 24, 876–890, 1971.

258. **Shah, S. N. and McKean, C. M.,** Effect of chronic and acute phenylalanine injections on the biosynthesis of glycolipids in the developing rat brain, *Fed. Proc. Fed. Am. Soc. Exp. Biol.,* 27, 488, 1968.

259. **Shah, S. N., Peterson, N. A., and McKean, C. M.,** Effects on phenylalanine and its metabolites on cerebral lipid metabolism in developing rats, *Fed. Proc. Fed. Am. Soc. Exp. Biol.,* 28, 558, 1969.

260. **Shalhoub, R. J.,** Pathogenesis of lipoid nephrosis: A disorder of T-cell function, *Lancet,* 2, 556–560, 1974.

261. Shannon, B. M., Smiciklas-Wright, H., Itzcovitz, J., and Shearer, H. L., Jr., Cystathionine excretion in relation to vitamin B_6 nutriture of rats: Strain differences and effects of DL- or L-methionine supplementation, *J. Nutr.*, 105, 1334–1340, 1975.

262. Shen, T. F., Bird, H. R., and Sunde, M. L., Effect of excess dietary L-, DL- and D-serine on the chicks, *Poult. Sci.*, 52, 1168–1174, 1973.

263. Sherman, M., Bacteriological and immunological basis of rheumatic fever and rheumatic heart disease, *Public Health Serv. Publ.*, U.S. Department of Health Education, and Welfare, #1516, 24, 1966.

264. Shields, G. S., Carnes, W. H., Cartwright, G. E., and Wintrobe, M. M., The dietary induction of cardiovascular lesions in swine, *Clin. Res.*, 9, 62, 1961.

265. Simpson, M. V. and Tarver, H., Studies of protein synthesis *in vitro*. III. Further observations on the incorporation of methionine into liver protein, *Arch. Biochem.*, 25, 384–395, 1950.

266. Smith, E. L., Nucleotide base coding and amino acid replacements in proteins, *Proc. Natl. Acad. Sci. U.S.A.*, 48, 859–864, 1962.

267. Smith, R. E., Effect of arginine upon the toxicity of excesses of single amino acids in chicks, *J. Nutr.*, 95, 547–553, 1968.

268. Snetsinger, D. C. and Scott, H. M., Efficacy of glycine and arginine in alleviating the stress induced in dietary excesses of single amino acids, *Poult. Sci.*, 40, 1675–1681, 1961.

269. Snetsinger, D. C. and Scott, H. M., The relative toxicity of intraperitoneally injected amino acids and the effect of glycine and arginine thereon, *Poult. Sci.*, 40, 1681–1687, 1961.

270. Snyderman, S. E., Maple syrup urine disease, in *Amino Acid Metabolism and Genetic Variation*, Nyhan W. L., Ed., McGraw-Hill, New York, 1967, 171–183.

271. Snyderman, S. E., Holt, L. E., Jr., Norton, R. M., and Roitman, E., Effect of high and low intake of individual amino acids on the plasma aminogram, in *Protein Nutrition and Amino Acid Patterns*, Leathem, J. H., Ed., Rutgers University Press, New Brunswick, N.J. 1968, 19–31.

272. Soeiro, R. and Amos, H., Arrested protein synthesis in polysomes of cultured chick embryo cells, *Science*, 154, 662–665, 1966.

273. Spolter, P. D. and Harper, A. E., Leucine-isoleucine antagonism in the rat, *Am. J. Physiol.*, 209, 513–518, 1961.

274. Spolter, P. D. and Harper, A. E., Effects of leucine-isoleucine and valine antagonism and comparison with the effect of methionine on rat liver regeneration, *Arch. Biochem. Biophys.*, 100, 369–377, 1963.

275. Stanners, C. P. and Beckerl, H., Control of macromolecular synthesis in proliferating and resting Syrian hamster cells in monolayer culture. I. Ribosome function, *J. Cell. Physiol.*, 77, 31–42, 1971.

276. Stekol, J. A. and Szaran, J., Pathological effects of excessive methionine in the diet of growing rats, *J. Nutr.*, 77, 81–90, 1962.

277. Strauss, A. J. L. and Van Der Geld, H. W. R., The thymus and human disease with autoimmune concomitants with special reference to myasthenia gravis, in *The Thymus*, Wolstenholme, G. E. W. and Porter, R. Eds., Little, Brown, Boston, 1966, 416.

278. Sullivan, M. X., Hess, W. C., and Sebrell, W. H., Studies on the biochemistry of sulfur. XII. Preliminary studies on amino acid toxicity and amino acid balance, *U.S. Public Health Rep.*, 47, 75–83, 1932.

279. Swendseid, M. E., Hickson, J. B., and Friedrich, B. W., Effect of non-essential nitrogen supplements on growth and on the amino acid content in plasma and muscle of weanling rats fed a low-protein diet, *J. Nutr.*, 78, 115–119, 1962.

280. Swendseid, M. E., Villalobos, J., and Friedrich, B., Ratios of essential-to-nonessential amino acids in plasma from rats fed different kinds and amounts of proteins and amino acids, *J. Nutr.*, 80, 90–102, 1963.

281. Tamimie, H. S. and Pscheidt, G. R., Influence of feeding 5% L-phenylalanine on reproductive performance of pullets, *Am. J. Physiol.*, 211, 955–958, 1966.

282. Todaro, G. J., Lazar, G. K., and Green, H., The initiation of cell divison in a contact-inhibited mammalian cell line, *J. Cell. Comp. Physiol.*, 66, 325–334, 1965.

283. Totter, J. R. and Day, P. L., Cataract and other ocular changes resulting from tryptophan deficiency, *J. Nutr.*, 24, 159–166, 1942.

284. Unger, R. H., Aguilar-Parada, E., Mueller, W. A., and Eisentraut, A. M., Studies of pancreatic alpha-cell function in normal and diabetic subjects, *J. Clin. Invest.*, 49, 837–848, 1970.

285. Valyasevi, A., Dhanamitta, S., and Van Reen, R., Studies of bladder stone disease in Thailand. XVI. Effect of 4-hydroxy-L-proline and orthophosphate supplementations on urinary composition and crystalluria, *Am. J. Clin. Nutr.*, 26, 1207–1211, 1973.

286.   **Viau, A. T., Leathem, J. H., and Wannemacher, R. W., Jr.,** Pregnancy and excess dietary methionine in rats, *Fed. Proc. Fed. Am. Soc. Exp. Biol.,* 28, 638, 1969.

287.   **Voelter, W., Sokolowski, G., Weber, U., and Weser, U.,** The initial binding of Cu (II) to some amino acids and dipeptides: A ^{13}C nuclear-magnetic-resonance study, *Eur. J. Biochem.,* 58, 159–166, 1975.

288.   **Wachstein, M.,** Nephrotoxic action of DL-serine in the rat. I. The localization of the renal damage, the phosphatase activity and the influence of age, sex, time and dose, *Arch. Pathol.,* 43, 503–514, 1947.

289.   **Wachstein, M.,** Nephrotoxic action of DL-serine in the rat. II. The protective action of various amino acids and some other compounds, *Arch. Pathol.,* 43, 515–526, 1947.

290.   **Wachstein, M.,** Influence of DL-methionine and other substances on the nephrotoxic action of DL-serine, *Nature,* (London), 159, 236, 1947.

291.   **Wachstein, M., and Besen, M.,** Electron microscopy of renal coagulative necrosis due to DL-serine, with special reference to mitochondrial pyknosis, *Am. J. Pathol.,* 44, 383–400, 1964.

292.   **Waisman, H. A.,** Experimental models of inborn errors of metabolism and some enzyme relationships, in *Enzymes in Mental Health,* Martin G. J., and Kisch, B., Lippincott, Philadelphia, 1966, 69–83.

293.   **Waisman, H. A. and Harlow, H. F.,** Experimental phenylketonuria in infant monkeys, *Science,* 147, 685–695, 1965.

294.   **Wang, H. L. and Waisman, H. A.,** Experimental phenylketonuria in rats, *Proc. Soc. Exp. Biol. Med.,* 108, 332–335, 1961.

295.   **Wang, H. L. and Waisman, H. A.,** Effect of dietary phenylalanine and tryptophan on brain serotonin, *Arch. Biochem. Biophys.,* 97, 181–184, 1962.

296.   **Wise, E. M., Jr. and Elwyn, D.,** Hyperaminoaciduria in rats following D-serine administration, *Proc. Soc. Exp. Biol. Med.,* 121, 982–986, 1966.

297.   **Wise, J. K., Handler, R., and Felig, P.,** The glycemic response to alanine: index of glucagon secretion in man, *Clin. Res.,* 20, 561, 1972.

298.   **Wise, J. K., Handler, R., and Felig, P.,** Evaluation of alpha-cell function by infusion of alanine in normal, diabetic, and obese subjects, *N. Engl. J. Med.,* 288, 487–490, 1973.

299.   **Wiseman, G. and Ghadially, F. N.,** Studies on amino acid uptake by RD$_3$ sarcoma cell suspensions *in vivo, Br. J. Cancer,* 9, 480–485, 1955.

300.   **Wiseman, G., Spilberg, I., and Krakauer, T.,** Arthritis induced in rabbits by lysates of granulocyte lysosomes, *Arthritis Rheum.,* 11, 103–115, 1969.

301.   **Wohlhueter, R. M. and Harper, A. E.,** Coinduction of rat liver branched-chain x-keto acid cehydrogenase activities, *J. Biol. Chem.,* 245, 2391–2401, 1970.

302.   **Womack, M. and Rose, W. C.,** The role of proline, hydroxyproline, and glutamic acid in growth, *J. Biol. Chem.,* 171, 37–50, 1947.

303.   **Woods, M. N. and McCormick, D. B.,** Effects of dietary phenylalanine on activity of phenylalanine hydroxylase from rat liver, *Proc. Soc. Exp. Biol. Med.,* 116, 427–430, 1964.

304.   **Woodside, K. H., Ward, W. F., and Mortimore, G. E.,** Effects of glucagon on general protein degradation and synthesis in perfused rat liver, *J. Biol. Chem.,* 249, 5458–5463, 1974.

305.   **Wretlind, K. A. J.,** The effect of synthetic amino acids essential for growth on the body-weight of growing rats, and the synthesis of the amino acids used, *Acta Physiol. Scand.,* (17 Suppl.), 59, 1–101, 1949.

306.   **Wretlind, K. A. J.,** The availability for growth and the toxicity of L- and D-phenylalanine, *Acta Physiol. Scand.,* 25, 276–285, 1952.

307.   **Wyngaarden, J. B.,** Glycinuria, in *The Metabolic Basis of Inherited Disease,* Stanburg, J. B., Wyngaarden, J. B., and Fredrickson, D. S., Eds., McGraw-Hill, New York, 1960, 1295–1301.

308.   **Yang S. P., Tilton, K. S., and Ryland, L. L.,** Utilization of a delayed lysine or tryptophan supplement for protein repletion of rats, *J. Nutr.,* 94, 178–184, 1968.

309.   **Yum, M. N., Edwards, J. L., and Kleit, S.,** Glomerular lesions in Hodgkin's disease, *Arch. Pathol.,* 99, 645–649, 1975.

310.   **Yuwiler, A. and Geller, E.,** Serotonin depletion by dietary leucine, *Nature* (London), 208, 83–84, 1965.

311.   **Yuwiler, A., Geller, E., and Slater, G. G.,** On the mechanism of brain serotonin depletion in experimental phenylketonuria, *J. Biol. Chem.,* 240, 1170–1174, 1965.

312.   **Zimmerman, R. A. and Scott, H. M.,** Interrelationship of plasma amino acid levels and weight gain in the chick as influenced by suboptimal and superoptimal dietary concentration of single amino acids, *J. Nutr.,* 87, 13–18, 1965.

Index

INDEX

A

Abdomen, 351, 413
Abomasum, 164, 235—236, 257, 259, 332
Abortion, see Pregnancy
Acetate, 35, 79, 91, 92, 159, 209, 297
   radioactive, 412
Acetic acid analogues, iodothyronines, 235
Acetrizoate, 226
Acetyl acetate, 394
Acetylcholine, 3, 10—11, 17, 59, 114
Acetyl CoA, 212
Acid hydroxylase, 200
Acid phosphatase, 199, 287
Acinar tissues, 296
ACTH, 37
Actinolite, 341
Acyl CoA derivatives, 418
Additives, food, see Food industry, use of fortifiers and
      preservatives by
Adenine, 452
Adenine nucleotide, 418—420
Adenosine, 395
Adenosine deaminase, 287, 454
Adenosine triphosphatase, 238, 287, 344, 419—420, 422
   sodium-potassium, 344
Adenosine triphosphate, 109, 162, 199, 236—237, 258,
      332, 365—367, 375, 395
   ATP-V, 363—367
Adenylcyclase, 37, 375
ADH, see Antidiuretic hormone
Adipose cells, number of, related to obesity, 384
Adipose clearing factor, 419
Adipose tissue, effect on, see also Obesity, 29, 35—38
   carbohydrates, 411, 413, 418, 419
   fat-soluble vitamins, 88
   fats, see Caloric excess; Obesity
   fatty acids, 35—37
   triglyceride lipolysis, 35—38
Adrenal glands, effect on
   caloric intake irregularities, 382
   fat-soluble vitamins, 79
   minerals, 147, 187
   trace elements, 284, 323, 329, 350
   water intoxication, 376
   water-soluble vitamins, 55—56
Adrenalin, 204
Adrenergic blockers, 395
Adrenocortical steroids, 89
Aeces aegypti, biotin toxicity, 48
Aerosols, see also Sprays, 153
Air, see also Pollution, 219, 227, 286, 288, 300. 349—351
Alanine, 448, 450, 457, 463, 464, 470
Albumin, 34, 98, 204, 237, 314, 331, 337, 367, 391, 438
Alcohol, 14—15, 182, 188, 299, 388, 422
Alcohol dehydrogenase, 208
Alcoholic beverages, 195, 426
Aldehydes, 200
Aldolase, 418
Aldose reductase, 421

Aldosterone, 128—130, 375, 377
Alfalfa, 156, 163, 208, 229, 230
Algae, 200, 208
   brown, 225—226
   red, 426
Alginates, 225
Alimentary tract, see also Gastrointestinal tract, 226, 236,
      239, 250—252, 260, 284, 359, 395—397
Alkali, 150
Alkali disease, see Selenium, alkali disease
Alkali metal sulfides, 158
Alkaline phosphatase, 32—34, 208, 293, 295, 330
Alkaloids, 198, 210
N-Alkylesters, 99
Allergeas, 17, 239—241
Allergic response
   fat-soluble vitamins, 93—95
   tests, 239—241
   trace elements, 211—212, 239—242, 252, 300
   water-soluble vitamins, 14, 16—17, 25, 61, 67—69
Allicin, 165
Allium, 165
Alloxan, 66, 441
Allylthiosulfonate, 165
Aluminosilicate, 342
Aluminum silicate, 344
Alveoli, 16, 251, 334
Alzheimer's disease, see Silican, Alzheimer's disease
Amines, 200
Amino acid metabolites, see also specific metabolites by
      name, 448—450, 453, 455—458, 461
   aryl, 453
D-Amino acid oxidase, 12
Amino acids
   branched-chain, see also specific branched-chain types
      by name, 443, 449, 457
   caloric intake irregularities and, 383, 394—396, 398
   carbohydrates and, 414, 417
Catabolism, 159
   crystalline solutions, 434
   crystalluria and, 437
   death caused by, 461, 464, 466, 468, 469
   dietary considerations, 447
   diseases, blood and urine content in, 448, 455, 456
   enzyme amount and activity influenced by, 450, 457,
      458, 461
   excess, effects of, 447—473
   fetal modification by, 451, 469
   growth modification by, 161—164, 451, 464—468,
      470—473
   Hodgkin's disease and, 453
   hormonal activity and, 450, 459, 460, 463
   immune response and, 453—454
   iodated, 233
   Lesch-Nyhan syndrome and, 453—454
   leukocyte response and, 453
   membrane permeability and, 449
   metabolism, 121
   metabolites, see Amino acid metabolites; specific
      metabolites by name

B

ficiency in animals, 208—210
ilepsy, experimental, produced by, 205—206
matopoietic effect of, 207—208
a, 203, 206—207
n and, 186, 210—212
etabolism, 210—211
ncreas and thyroid, effect on, 203—205
xicity in man, 211—212
mor production by, 206—207
amin B₁₂ and, 203—212
tamin D and, 211
alt blue, 203
alt chloride, 203—204, 206, 207
arboxylase, 7, 11, 16, 18
kerel, toxicity studies, see also Chicken
arbohydrates, 423, 425, 426
a butter, 105
, 221
eine sulfate, 16
liver, vitamin A in, 73
liver oil, vitamin D in, 87
nzyme A, 13, 203, 209, 212
nzyme B₁₂, 203
nzymes, 13, 25—26, 34, 203
site, 341
oitrogens, 247
agens, 65—66, 74, 81, 92, 200, 451, 452, 473
oidal silicon dioxide, 342
loids, 99, 153, 250, 256, 260, 261
on, see also Intestines, 254, 260, 388, 398, 422, 423, 442
oring substances, see Food colors
nmercial diets for rats, protein content, 441
nplement, 240, 252
MT, see Catechol-O-methyltransferase
aception, see Reproduction
adensed milk, see Milk, condensed
ndiments, 325
afections, 223—224, 231, 262
agenital malformations, see Fetus; Pregnancy; Teratogens
atrast media, see Radiographic contrast media; Radiological contrast media
oley's anemia, see Iron, Cooley's anemia
oper
lloys, 196
mino acids and, 451
nemia, 195, 199
alcium and, 200
aloric intake irregularities and, 397
death caused by, 195—197, 199
eficiency, 156—157, 451
lusts, toxicity, 196—197
umes, toxicity, 196—197
on, 196
ron and, 178, 188
D₅₀, 198
oading, 200
nedicinal properties, 195
nolybdenum and, 198—199
ulfate and, 156—157
in and, 358
oxicity
in animals, 198—200

in man, 205—207
vineyard sprayer's lung, 207
vitamin C, effect on excretion, 68
Wilson's disease, 195
zinc and, 292—298, 301
⁶⁴Copper, 294
Copper carbonate, 197
Copper oxide, 196
Copper salts, see also specific salts by name, 196, 197, 200
Copper sulfate, 195—198
Copper, zinc-thionein, 297
Cordials (beverages), 158
Corn, see also Maize, 163, 440—441
Cornea, see also Eye, 197
Cornflakes, 124
Corn gluten meal, 366
Corn-soybean rations, 111
Cornstarch, 415, 417
Corn sweeteners, 417
Corpora lutea, 47
Corticoids, 141
Cortisol, 89, 377, 473
Cortisone, 89
Cosmetics, 300
Cottonseed meal, 229, 366
Cough mixtures, 241
Coupling of iodinated tyrosyls, 296
Cow, toxicity studies
amino acids, 473
fat-soluble vitamins, 73, 78—79, 89
liver, vitamin A in, 79, 80
minerals, 103, 153—157, 159, 161, 162, 164, 166—167
protein, 438
trace elements, 198—199, 208—210, 220—224, 229—230, 233—234, 236, 237, 255, 257—258, 295—297, 301, 310—312, 331, 332, 339, 340, 342, 343, 369
water-soluble vitamins, 31
Crab, 124, 221
Cranberry, 221
Cream, 222, 223, 416
Creatine, 73—94
Creatinine, 111, 162, 232
Cristobalite, 341
Crocidolite, 341
Cross-linking agents, 341
Cruciferae, sulfur in, 160—161
Crystalline amino acid solutions, see Amino acids, crystalline solutions
Crystalloid fluids, 98
Crystalluria, see Amino acids, crystalluria and
Cucumber, 221
Cupric thiomolybdate, 156
Cuprous iodide, 225, 229
Curare-like action of nutrients, 3—10, 114
Curry powder, 130
Cyanide, 160
Cyclic AMP, see AMP
Cyclic sulfur, 154
Cycloleucine, 417
Cyclophosphamide, 377
Cynodon dactylon, 230
Cystathionase, 450
Cystathione, 161—162, 165

H

I

M

O

, 251, 229, 312, 338
olled, 423, 424
sity, see also Adipose tissue; Caloric excess
 ltural factors involved in, 383—384
 rbohydrates and, 413, 424
 finition, 379—380
 t, sugar, and fiber intake related to, 379, 397—398
 netic and metabolic causes of, 381—382
 armful effects of, 386—388
 ypothalamic regulation of feeding, 382—383
 easurement of, 380—381
 sychological considerations, 384—386
 erapy for, 389—390
  surgical, 390—393
 ans, 154, 219
 ctanol, 449
  see also specific oils by name, 158, 219, 341, 425, 426
 eed, 161
 eal, 229
 actory nerve, 157
 e, 124
 ve oil, 37—38
 nivorous animals, nicotinc acid excretion by, 29
 PM, see 2-Methyl-4-amino-5-hydroxy-
    methylpyrimidine
 ion, 165, 221
 al, see Opaline silica
 aline silica, 341, 343
 ange, 423
 ange juice, 182, 359
 ganic calcium salts, see Calcium salts, organic
 ganic sulfur compounds, see Sulfur, organic
    compounds
 ganophosphates, 111
 gano-selenium compounds, 301—315
 ganosilicon compounds, see Silicones
 ganosiloxanes, see Silicones
 ganotin compounds, see also Tin, 352
 gans, see specific organs by name
 nithine, 448, 456
 thophosphates, toxicity of, 109—111
 thosilicic acid, 342
 molality, 385—386, 396, 420, 421, 438—440, 444
 teoarthritis, caloric intake irregularities related to, 388
 teoblasts, see also Cancer, 338
 abin, 134
 vary, effect on, see Genital organs; Reproduction
 verweight condition, see Obesity
 xalate, 97, 447
 xalic acid, 68
 xidants, 220
 xidative phosphorylation, 236—238, 258
 xides, see specific oxides by name
 xidizers, 160, 254, 321
 xycholesterol, 34
 xygen, 34, 177, 207—208, 461
 xyhemaglobin, 141
 xythiamine, 12
 xythiamine chloride, 5
 xytocin, 377
 yster, 221, 291, 299

P

Packaging, food, see Food industry, packaging materials
    used by
Paint, 196
Palate, 74, 80, 81
Pallidum, 285
Palmitate, 80, 81
Palmitic acid-1-^{14}C, 418—419
Pancreas, effect on
    amino acids, 478, 502, 503
    carbohydrates, 411, 418
    minerals, 164, 165, 189
    trace elements, 204, 291, 293—297, 300, 329, 330, 351
    water intoxication, 377
    water-soluble vitamins, 66
Pancreatic acid, 448
Pancreatic islet, 151, 203—204
Pantothenate, 12
Pantothenic acid, 13
Papillocarcinoma, see also Carcinoma, 344
Paprika, 130
Paraaminosalicyclic acid, 99
Paranasal sinuses, 242
Parasiticides, 153
Parasympathetic nervous system, 11
Parathormone, 110
Parathyroid, 88, 110
Parenchyma, 141, 199, 240, 347
Parietal cells, 254
Parotids, 242, 252
Partition coefficient, amino acids, 449
Passive transfer test, 239
PBI levels, 255
Pea, 123—124, 221
Peanut butter, 300
Pear, 221
Pectic substances, 422, 425
Pectin, 423—426
Penicillin, 147, 226, 332
Pentacalcium orthoperiodate, 229
Pentose, 26, 425
Pentose phosphate cycle, 26—27
Pentasulfides, 158
Pepper, 130
Peptides, 314, 453
Perchlorate, 236, 247
Percomorph liver oil, vitamin D in, 87
Pericardial fluid, 199
Periosteum, 78
Peripheral nervous system, effect on
    fat-soluble vitamins, 76
    minerals, 116
    trace elements, 197
Perisinusoidal cells, 74
Perivascular tissues, 240
Peroxidase, 177, 235, 236, 245, 246
Persulfates, 160
    tolerance to, 166
Pesticides, 198
Petroleum, 159
PGE$_1$, 37
pH, effect on

W

CRC PUBLICATIONS OF RELATED INTEREST

CRC HANDBOOKS:

CRC FENAROLI'S HANDBOOK OF FLAVOR INGREDIENTS 2nd Edition
Edited, translated, and revised by: **Thomas E. Furia** and **Nicolo Bellanca,** Dynapol, Palo Alto, California.
This two-volume Handbook is an update of the 1st Edition and includes comprehensive review chapters by recognized experts in the field. It is extensively indexed for easy use.

CRC HANDBOOK OF FOOD ADDITIVES, 2nd Edition
Edited by **Thomas E. Furia,** Dynapol, Palo Alto, California.
Nearly 1000 pages of pertinent food additive information is offered, reflecting the important changes that have occurred in recent years in the area of food additives.

CRC UNISCIENCE PUBLICATIONS:

FOOD ANALYSIS: Analytical Quality Control Methods for the Manufacturer and Buyer, 3rd. Edition
By **R. Lees M.R.S.H., A.I.F.S.T.,** food industry consultant. This book brings together methods of analysis which are of most value to the factor control chemist.

LOW CALORIE AND DIETETIC FOODS
Edited by **Basant K. Dwivedi, Ph.D.,** Estee Candy Company, Inc.
This book includes all aspects of low calorie and dietetic foods including discussions on fructose, applications and commercial potential of sweetening agents, and the present status of food products for people with special dietary requirements.

MAN, FOOD AND NUTRITION
Edited by **Miloslav Rechcigl, Jr., M.N.S., Ph.D., F.A.A.A.S., F.A.I.C.,F.W.A.S.,** Agency for International Development, U.S. Department of State.
This interdisciplinary treatise offers a comprehensive and integrated critical review of the nature and the scope of the world food problem. It presents strategies and discusses various technological approaches to overcoming world hunger and malnutrition.

TOXICITY OF PURE FOODS
By **Eldon M. Boyd, Ph.D.** (deceased), Queen's University, Kingston, Ontario. Edited by **Carl E. Boyd, M. D.,** Health and Welfare, Canada.
A systematic study of the toxicity of pure foods is described in this volume.

WORLD FOOD PROBLEM
Edited by.,**Miloslav Rechcigl, Jr., M.N.S., Ph.D., F.A.A.A.S., F.A.I.C.,F.W.A.S.,**Agency for International Development, U.S. Department of State.
This is a comprehensive and up-to-date bibliography on all important facets of the world food problem, encompassing such areas as the availability of natural resources and the present future sources of energy.

CRC MONOTOPIC REPRINTS:

FLEXIBLE PACKAGING OF FOODS
By **Aaron L. Brody, S. B., M.B.A., Ph.D.,** Arthur D. Little Inc.
The aim of this book is to describe food products employing flexible packaging, the requirements dictating the flexible packaging being used and the current state of flexible packaging in the food industry.

FREEZE-DRYING FOODS

By **C. Judson King, B.E., E.M., Sc.D.**, University of California.

This review concentrates on several papers published in recent years, relates them to the rest of the field, gives an evaluation of their findings, and the conclusions they draw.

SOYBEANS AS A FOOD SOURCE, 2nd Edition

By **W. J. Wolf, B.S., Ph.D.**, and **J. C. Cowan, A.B., Ph.D.**, Northern Marketing and Nutrition Research Division, Peoria, Illinois.

The conversion of soybeans to food is summarized in this book. Emphasis is given to the protein content of soybeans because of the current high level of interest.

STORAGE, PROCESSING, AND NUTRITIONAL QUALITY OF FRUITS AND VEGETABLES

By **D.K. Salunkhe, B.S., M.S., Ph.D.**, Utah State University.

This book reviews the nutritional value and quality of fruits and vegetables as influenced by chemical treatments, storage and processing conditions.

THE USE OF FUNGI AS FOOD AND IN FOOD PROCESSING, Parts 1 and 2

By **William D. Gray, A.B., Ph.D.**, Northern Illinois University. Edited by Thomas E. Furia, Dynapol, Palo Alto, California.

This two-volume work describes how filamentous fungi have been used in the past and present as a food source in various areas of the world.

CRC CRITICAL REVIEW JOURNALS:

CRC CRITICAL REVIEWS IN FOOD SCIENCE AND NUTRITION

Edited by **Thomas E. Furia**, Dynapol, Palo Alto, California.

RI
A